ENERGY, ENVIRONMENT AND GREEN BUILDING MATERIALS

PROCEEDINGS OF THE 2014 INTERNATIONAL CONFERENCE ON ENERGY, ENVIRONMENT AND GREEN BUILDING MATERIALS, NOVEMBER 28–30, 2014, GUILIN-GUANGXI, P.R. CHINA

Energy, Environment and Green Building Materials

Editor

Ai Sheng
Information Science and Engineering Technology Research Association (ISET), Hong Kong, China

CRC Press
Taylor & Francis Group
Boca Raton   London   New York   Leiden

CRC Press is an imprint of the
Taylor & Francis Group, an **informa** business

A BALKEMA BOOK

CRC Press/Balkema is an imprint of the Taylor & Francis Group, an informa business

© 2015 Taylor & Francis Group, London, UK

Typeset by diacriTech, Chennai, India

Published by: CRC Press/Balkema
           P.O. Box 11320, 2301 EH Leiden, The Netherlands
           e-mail: Pub.NL@taylorandfrancis.com
           www.crcpress.com – www.taylorandfrancis.com

ISBN: 978-1-138-02718-3 (Hardback)
ISBN: 978-1-315-73540-5 (eBook PDF)

Energy, Environment and Green Building Materials – Sheng (ed.)
© 2015 Taylor & Francis Group, London, ISBN 978-1-138-02718-3

Table of contents

Preface xi

Organizing committee xiii

Analysis of heavy metal pollution sources and their water environmental
behavior in Guangxi Province, China 1
J. Cheng, W.B. Qian & C.C. Cao

Current situation of rural ecological planning in North China—the
Baoding Shijiatong village 5
S.Y. Zhao, S.J. Fu & Y.H. Hao

Research on domestic sewage treatment by "Activated Sludge" combined "Biofilm" 9
Y.P. Li, W. Hong, Y. Li, Z.B. Wang & Z. Han

DSM-based modeling framework of emergency management 13
P. Zhang, L.B. Zhang, B. Chen, L. Ma & X.G. Qiu

Formation mechanism and controlling factors of secondary pore in clastic rock 19
Y.R. Li, X.Y. Fan, R.G. Jiang, M.T. Li & X. Xiao

Thermal and mechanics research on a 15μm umbrella-like structure microbolometer 25
L. Zhang, C. Chen & T. Wang

Optimal control of anoxic selector based on activated sludge with different phosphorus removal ability 29
Z.X. Peng & Z.G. He

Fuzzy evaluation for optimal selective disassembly in recycling of end of life product 35
Z.Q. Zhou, P.L. Hu, Y.F. Gong & G.H. Dai

Plate pillars formed by drawing 39
X. Zeng, Y. Hu, Q. Xia, Y.K. Cheng & Y.B. Duan

Research on relationship between compression damage and load-carrying capacity for RC columns 43
X.M. Chen, J. Duan, H. Qi & Y.G. Li

Practical method for fire-resistant design of steel members based on critical temperature 49
J. Xing

Seismic characteristics study of six-pylon cable-stayed bridge with corrugated steel webs 53
F.M. Wang, G.L. Deng, X.Z. Dang & W.C. Yuan

High-performance lightweight over-strength UHMWPE UD cloth preparation and its application 59
X.H. Meng, G. Lu & W.F. Da

Study on the rendering algorithm of spatial planar circle through coordinate transformation 63
J.Y. Li, G.X. Yuan, Y.M. Zhang & Z.Q. Huang

Investigation on Urban color and city image in Wuhan 69
L.Y. Li

Research on a novel deflection surface concentrator 73
C.F. Ye, Q.L. Qiu, Y. Wu, & J. Zuo

Wear resistance characterization and analysis of hard particle
reinforcement in composite Fe-based alloy +WC+ Cr_3C_2 laser cladding 79
J. Xu, G.B. Li, X.Z. Du, J. Hao, W.H. Tang, H.B. Wang & H.Y. Zhao

Research on whole industry chain of housing industry modernization in big data era 85
X.M. Yang

Key-technology research of the image retrieval based on the feature points and invariant moments 89
Z.M. Wang & B.Q. He

Analysis of slope ecological protection substrate moisture evaporation characteristics 93
X.J. Zhang, Z.Y. Xia, W.J. Hu & L.L. Zhang

Effecting analysis of the temporary traffic regulation on the traffic flow in urban expressway 97
M. Wang, J.F. Wang, R.X. Xiong & X.D. Yan

Study on rural domestic sewage treatment modes of Ningbo 101
J. Wei, J. Yang, Y.L. Wang & D.H. Hu

Phenolic wastewater treatment by pulsed electrolysis with alternative current 105
X.M. Zhu, N.L. Nguyen, B. Sun, Z.R. Li & P.H. Zhan

Fundamental theory and impact factors of art city 109
Jie Zhao

Study of spatial color reconciling in urban color planning—based on the
thinking of urban color planning of Harbin 115
Z. Bian & Y.W. Zhang

Influence factors of fluorine concentration in groundwater in Gaomi area, Shandong province, China 119
J.G. Feng, Z.J. Gao, J.F. Duan & M.J. Shi

Study on the training of safety signs in coal mines 123
W. Jiang, Y.N. Liu & S.H. Shi

Constitution analysis of pollution sources in urban rivers of Hangzhou 129
J. Yang, J. Wei, M. Yuan, Y. Lu & F.F. Chen

Thermal effects on the geometric nonlinearity of beam structures 135
J. Duan & Y.G. Li

Material balance calculation and application of modified step-feed process 139
Q. Chen, W. Wang, C.X. Wang, Y.Z. Peng, X. Zhang & C.H. Shao

Study of phase change material based on solar building 145
W.H. Li, Ch.W. Zhang, M.Y. Zhang, H.T. Feng & Y. Liu

2D Numerical simulation of flow field around bridge piers in Bend River based on FLUENT 149
W.C. Pan & M.J. Liu

Research on control system of multidimensional dynamic indoor environment 155
F.Z. Nian & D.W. Shao

Research on the FLUENT simulation of the single aluminum pore growth 161
M. Zhang, C.J. Chen & X.N. Wang

Exploration for geographic information of an ancient village based on GIS technique—
Da Liangjiang Village in Jingxing County of Hebei Province as an example 167
C.Y. Wang, G.Q. Li & Y. Yang

Research on a sewage treatment system with low operation cost applied in rural areas 173
C.Y. Wang, G.Q. Li & N.Z. Wang

Numerical studies of the oxidation process of Hg by ozone injection 179
B. Li, J.Y. Zhao & J.F. Lv

Studies on image processing technology of chalkiness detection in early indica rice 183
P. Cheng, Q. Liu, Y. Su & D.S. Cheng

Properties and coke deactivation of Pt-Sn-Na/ZSM-5 catalysts for propane dehydrogenation to propylene 189
C.X. Wang & Z.Q. Wang

Turbidity variation in a detention pond after storing runoff 193
C. Tang & T.Y. Guo

Experimental study on kitchen garbage hydrolysis conditions 199
G.S. Cheng, Y. Zhao & C.Y. Luo

Study of uniaxial constitutive model of concrete in code and development in ABAQUS 203
H. Qi, Y.G. Li, X.M. Chen, J. Duan & J.Y. Sun

An ecosystem-level ecological risk assessment framework for lake watershed:
A case study of upstream Taihu lake watershed 207
N. Li, J.C. Qi, R.R. Han, B.H. Zhou & Y.Y. Huang

Study of preconditioning methods for flows at all speeds 217
L. Li, G.P. Chen, G.Q. Zhu & W. Zhang

Finite element simulation and optimization design of biodegradable seedling pot 223
X.D. Liu, S.J. Li, L.J. Han, Q.R. Jing, D.Y. Li & Z.Y. Qiu

Design of robust sliding mode observer for a class of linear system with
uncertain time delay: A linear matrix inequality method 227
F.X. Xie, L. Chen & A. Zhou

Groundwater environmental impact assessment of Red Bull beverage due to
mining water in Guangdong Province 233
L.S. Tang & H.T. Sang, Y.L. Sun & Z.G. Luo

Experimental study of Cl and F releasing of straw combustion 239
Y. Zhao, Y. Xiang, G.S. Cheng, Y. Wang, Y. Zhang, C.Q. Dong, Z.M. Zheng & H.L. Zhang

Properties of gelatin and poly (3-hydroxybutyrate-co-3-hydroxy-valerate) blends 243
Y. Li, J.K. Duan, Y.J. Wang, L. Jiang, K.Q. Shi & S.X. Shao

Treatment of simulated acid mine drainage by air cathode microbial fuel cell 247
C.F. Cai, F.Z. Qi, J.P. Xu, H.C. Zhu & L. Jiang

Biochemicals of acid mine drainage using sulfate-reducing bacteria activity on a column reactor 251
C.F. Cai, L. Jiang, C.F. Wu & F.Z. Qi

Intelligent monitoring system design of coal mine ventilation based on internet of things technology 255
W.C. Li, G.F. Hao, G.Q. Li & H.J. Wang

Public opinion propagation model based on deffaunt model 259
X. Zhou, B. Chen, Z.C. Song, L. Ma & X.G. Qiu

Effects of learners' motivation in second language acquisition 263
J.B. He & Z. Wang

Large section roadway excavation rapid technology 269
Z.F. Yao, W.B. Wei, B. Shi & Z. Wei

Structural design of cyclonic microbubble flotation column system 273
B.Q. Dai, J.J. Yuan, X.Y. Liu & J.X. Ge

Effects of inorganic ions on phosphorus removal with Fe/Mn oxide formed *in situ* by
$KMnO_4$-Fe^{2+} process 277
K. Liu, J.H. Sun & T. Zhou

Scale fingerprint clustering-based gait tracking algorithm under
high-frequency gait mode switching environment 281
Y. Liu, S.L. Wang, L. Wang & L.L. Li

Positive scheme numerical study for gas jet in the direction pipe 285
J.L. Zhong & G.G. Le

Risk identification and evaluation for decision-making in integrated
urban-rural water supply system constructions 289
Y.H. Mao, B. Yang & H.Y. Li

Research and simulation on the integration of marine SCR and exhaust muffler 293
Y. Chen & L. Lv

Pedestrian seamless positioning algorithm based on low-cost GPS / IMU 301
Z.S. Tian, H.M. Huang, Z.W. Yuan & M. Zhou

Nitrogen oxide emission comparison of biodiesel and diesel in diesel engine 307
Q.L. Zhang

Research on electrolysis factors of galvanizing slag's soluble anode effect on current efficiency 311
X.F. He, Y.G. Li, J. Chen & Z.Y. Cai

Treatment of phenol wastewater by titanium catalyst 315
R. Wang, Y.X. Xiao & J.C. Gu

Study on the beneficiation process of a fine ilmenite in YunNan province 319
Q.R. Yang, G.F. Zhang & P. Yan

Utilization status and expectation of phosphogypsum 323
Q.R. Yang, P. Yan & G.F. Zhang

Life prediction of the subsea tunnel based on chloride migration model 327
X.H. Lv

Optimization of control parameters based on an improved detailed
urea-SCR model 331
T. Feng & L. Lu

Research for the influence of light transmissivity of air humidity in fixed distance 335
Y.N. Wu, Z.J. You, J.N. Ma & L. Chen

Evaluation of rural territorial functions: A case study of Henan Province, China 339
C. Fu

Thermal environment simulation analysis of a new regenerated glass pumice external wall
insulation building in hot summer and cold winter zone 349
S. Shu, Q. Gu, X. Zhou & B. Li

Analysis on the space–time difference characteristics of inbound tourism in Gansu province   355
J.P. Yan & X.Z. Wang

Research on multi-level water bloom decision-making method based on information
entropy of vague set   361
Q.W. Zhu, X.Y. Wang, L. Wang, J.P. Xu & Y.T. Bai

Development and review of application of low-noise asphalt concrete in urban roads   367
T.F. Nian, P. Li, L. Yang, Y.L. Zhang & Z.C. Liu

Research of spatiotemporal schedule model of agents' behavior in artificial society   373
L. Ma, B. Chen, L.B. Zhang, P. Zhang, X. Zhou & X.G. Qiu

Research status of magnetic plasma mass separator   379
H.L. Zhao, B.H. Jiang, C.S. Wang, H.C. Liu & Z.L. Zhang

Antimicrobial effect mechanism of polyphenols against campylobacter jejuni   385
Q. Tian, X.J. Du, R. Xue, J.J. Gen, X.F. Xie, B. Liang & J.P. Wang

Improving the mechanical properties of two aluminum alloys by thermal cycling treatment   389
C.Y. Chen

Exploring the relationship between buildings' temperature and
eco-environment components around them in Futian District, Shenzhen, China   393
X.Q. Sun & M.M. Xie

Research on damage identification of grid structures based on frequency response function   399
M.H. Wang, S.Y. Dong, C.M. Ji & X.T. Yang

Damage detection of cable force relaxation based on HHT   403
M.H. Wang, Y.S. Liu, S.S. Shi & X.T. Yang

State-feedback control for stochastic high-order nonlinear systems with time-varying delays   407
Z.G. Zhong

Numerical simulation on structure and anti-explosion performance of new ASA building plates   411
H.J. Wang, W.Z. Liao, M. Li & S.J. Shen

Analyses of the exergetic efficiency of solar energy heat pump system   417
Y.Q. Di, W. Zhao & H.Y. Di

Heat recovery ventilator technology development and application in the public buildings   423
L. Yuan

Active distribution network application practice for grid   427
H.G. Zhao, X.L. Yang, M. Li, M. Chen & Z. Tang

Commercial complex advantages and development in Nanning City   433
D.Y. Li, Y.Q. Li, Q.J. Zeng & J.C. Yu

HPLC method for determination of concentration of Oxiracetam in human plasma   439
X. Xie

A domain-expert oriented modeling framework for unconventional emergency   443
Z.C. Fan, X.G. Qiu, L. Liu, P. Zhang & X.Y. Zhao

The influence of David Pepper's ecological theory of socialism on China's era   447
H.R. Chen

The theory of David Pepper about the blueprint of the construction of the ecological socialism theory   451
H.R. Chen, G.F. Liu & S.F. Chen

Application of a 3D modeling software—Cityengine in urban planning 455
Y.W. Luo, J. He & H.J. Liu

On the relationship between walk space form of college campus and summer thermal environment 459
L. Qin, J. He, Y.W. Luo, Y.Y. Yao, B. Chen, Y.J. Meng & Y.H. Wu

Analysis of outdoor thermal environments of universities in summer in hot and humid areas 465
Y.Y. Yao, J. He, Y.W. Luo, L. Qin, Y.J. Meng, B. Chen & Y.H. Wu

Study on biogas production of co-digestion of vegetable waste with other materials 471
R. Feng, J.P. Li, J.Y. Yang & D.D. Zhou

Microstructures and photocatalytic properties of metal ions doped nanocrystalline TiO_2 films 475
H.Y. Wang & J.M. Yu

Numerical simulation of flow-field coupling with the six degree of freedom
topology-changeable motion 479
F.B. Yang, D.W. Ma & Q.Q. Xia

Natural conditions suitability analysis of renewable energy building application:
Case study of hot-humid climate nanning 483
D.Y. Li, J. He, X. Xu & Y.Q. Li

Leisure space of hospital building design in hot and humid areas—based on the
analysis of environment-behavior 487
D.Y. Li, J. He, X. Xu & Y.Q. Li

Applicability of natural ventilation technology to public buildings in South China 493
H.J. Liu, J. He & Y.W. Luo

Architectural design strategies for bio-climatic design in hot-summer and cold-winter region 499
H.J. Liu, J. He & Y.W. Luo

On strategies of developing eco-tourism around Poyang Lake eco-tourism circle 503
J.B. He & Z. Wang

The analysis of properties of oxide scale on the surface of typical superheater material 507
M.C. Zheng, J.F. Xiao, S. Liu & Z.P. Zhu

Adsorption of Cr(VI) from aqueous solution by attapulgite 511
J. Ren, F.J. Zhang, Y.N. Li, L. Tao & Y.Z. Zhang

Numerical simulation of airflow distribution in air-conditioned room of office building
based on ANSYS 515
R.H. Ma

A dual interfaces and high-speed solid state recorder design 519
S. Li, Q. Song, Y. Zhu & J.S. An

Research on urban residential pattern oriented to the aging society 523
Q. Liu & L.C. Fang

Author index 527

Energy, Environment and Green Building Materials – Sheng (ed.)
© 2015 Taylor & Francis Group, London, ISBN 978-1-138-02718-3

Preface

The 2014 International Conference on Energy, Environment and Green Building Materials (EEGBM2014) was held November 28-30, 2014, in Guilin, Guangxi. EEGBM2014 provided a valuable opportunity for researchers, scholars and scientists to exchange their new ideas and application experiences face to face together, to establish business or research relations and to find global partners for future collaboration. The papers in this book are selected from more than 500 papers submitted to the 2014 International Conference on Energy, Environment and Green Building Materials (EEGBM2014). The contributions cover the topics of New Energy, Energy Efficiency and Energy Management, Building Materials and Environmental Engineering and Management, among others. The conference will promote the development of the Energy, Environment and Green Building Materials, strengthening international academic cooperation and communications.

We would like to thank the conference chairs, organization staff, and the members of International Technological Committees for their hard work. Thanks are also given to CRC Press / Balkema (Taylor & Francis Group).

We are looking forward to seeing all of you next year at EEGBM 2015.

Yizhong Wang
Tianjin University of Science and Technology, China

Energy, Environment and Green Building Materials – Sheng (ed.)
© 2015 Taylor & Francis Group, London, ISBN 978-1-138-02718-3

Organizing committee

ORGANIZER

Information Science and Engineering Technology Research Association (ISET), Hong Kong, China

CONFERENCE CO-CHAIRS

Aisen, *Information Science and Engineering Technology Research Association (ISET)*
Yizhong Wang, Professor, *Tianjin University of Science and Technology, China*

COMMITTEE

Jihe Zhou, *Professor, Chengdu Sport University, China*
Hongmin Gao, *Professor, Beijing Institute of Technology, China*
Chunguang Xu, *Professor, Beijing Institute of Technology, China*
Haitao Li, *Professor, Southwest Petroleum University, China*
Zhiming Liu, *Professor, Liao Ning Institute of Science and Technology, China*
Shanglin Hou, *Professor, Lanzhou University of Technology, China*
N. K. Sharma, *Professor, The Glocal University, India*
Kanglin Wei, *Professor, Chongqing University, China*
Je-Ee Ho, *Professor, I-Lan University, Taiwan*
Chunpeng Li, *Professor, Quanzhou Normal University, China*

Energy, Environment and Green Building Materials – Sheng (ed.)
© *2015 Taylor & Francis Group, London, ISBN 978-1-138-02718-3*

Analysis of heavy metal pollution sources and their water environmental behavior in Guangxi Province, China

Jie Cheng & Wei-Bin Qian
College of Environmental Science and Engineering, Guilin University of Technology, Guilin, China

Chang-Chun Cao
College of Environmental Science and Engineering, Guilin University of Technology, Guilin, China
The Guangxi Talent Highland for Hazardous Waste Disposal Industrialization, Guilin, China

ABSTRACT: Non-ferrous metals (density more than $4.5 g/cm^3$) are also called heavy metals. Non-ferrous metal mine tailing ponds and slag of hydrometallurgy were the main pollution source of heavy metals. Heavy metal migration, such as lead and cadmium, was closely related to Natural Organic Matter (NOM) and the mineral processing reagents (Synthetic Organic Compounds (SOC)). After entering the surface water system, a complex of heavy metals and humic acid is the main pattern of their migration. Finally, they are transported to the sea by sediment or dissolution. However, heavy metals of deposition may return to the upper water, or finally enter into the human body by the food chain. So it is very important to understand the behavior of heavy metals of drinking water sources in water environmental safety and food safety.

KEYWORDS: Heavy mental, humic acid, processing reagent, water environmental behavior

1 HEAVY METALS, ORGANICS, AND ORE TAILING

Each year, metal and non-metallic ore mining amount to 10 billion tons, the majority of which is in the form of waste rock and tailings left in mines. Non-ferrous metal resources in China are of polymetallic ore and have a low overall recovery ratio. Nowadays, the tailings of accumulation reach 8.04 billion tons, growing at 140 million tons per year. These tailings and yard are both huge artificial deposits and heavy metal pollution sources. Ore processing wastewater and open-air rainwater of mines leach these rock mineral particles. While under the action of water and oxygen, sulfur mineral particles can form acid mine water, further soaking and leaching a particle size of less than 200 mesh. Heavy metal sulfide under the action of acid mine water, rain water, and the mineral processing reagents constantly seeped from the tailings and entered into the ground or surface water system[1,2,3]. Heavy metals in the supergene environment were generally insoluble in water and usually in-situ deposit, unless increasing their solubility and mobility by binding to organic matter. Therefore, researching of effects among heavy metal, mineral processing reagents (SOC), and humic acid (NOM) is of great importance to understand its activation migration behavior in the mine-water system.

2 MINERAL PROCESSING REAGENTS— SYNTHETIC ORGANIC COMPOUNDS AND HEAVY METALS

Metal sulfide minerals were hydrophilic minerals, which must be separated from gangue mineral by flotation agent. According to chemical structures, those with heteropolarity were acid, alkali, or salt; those with nonpolarity were hydrocarbon: aliphatic hydrocarbon, naphthenic hydrocarbon, and aromatic hydrocarbon; bipolarity of straight chain aliphatic alcohol, pyridine, quinoline, pyrrole, phenol, and carbonyl compounds. Considering the decline of heavy metals in ore grade, complex composition, as well as the original comprehensive utilization of waste slag, hydrometallurgy is the first choice for many mining smelting nowadays, which uses extraction agents to soak powder that has been smashed, levigated, and roasted[4]. Most of these agents contains organic phosphorus, organic amines, hydroxamic acids, and other organic compounds, some of which were typical of persistent toxic organics and a large dosage of mine mineral processing reagent. The whole world consumed more than 4 million tons of the mineral processing reagents, in the flotation process of 2 billion tons of ore per year, whereas our country, as experts estimated, consumed more than 1 million tons. After underground mine ores have been crushed, smashed, and levigated, they enter into mineral smelting processing; processing conditions will change from reductive into the surface oxidation environment. Under the effect of water and oxygen, the sulfide ore is easily oxidized, forming acid mine water. Then, raw ore particles react again, leading to the faster leaching of heavy metal. There were a lot of remaining reagents in mineral processing wastewater .Though under the condition of the surface oxidation, they will decompose, and basic functional groups[5,6] still exist. Further, surrounding rock conditions of ore deposits were different, so it still needs further working as to whether surrounding rocks with different proportions of carbonate rock type and silicon aluminum promote the migration of heavy metal elements in tailings or restrain their activities.

3 NATURAL ORGANIC MATTER— HUMIC ACID AND HEAVY METALS

Humus is a kind of natural organic matter, the total surface amount of which is $6*10^{12}$ tons. It is the carrier and medium of earth supergene geochemical effects and stable polymer. Space forms of fulvic acid from acidic to alkaline, in order, were fibrous, reticular, sponge and flake, and fine granular. Humic acid in acidic solution did not dissolve in water, and in alkaline environment it had a similar space form as fulvic acid, which was vulnerable to environmental conditions, [7,8,9] such as acidity–alkalinity, salt concentration, cation valence, and so on. Through its space structure with a variety of models, it still has some common features: The aromatic ring of humic acid was a center skeleton structure, and many groups were connected to the center skeleton through the branched chain such as aliphatic chain, carboxyl, hydroxyl, and so on, and the center of aromatic ring was in the shape of a spherical mesh, which could adsorb various metal ions and organic matter. The effects of humic acid and metal ions depends on the way of functional groups binding to the metal ions. Currently, carboxyl and hydroxyl functional groups were of the main function, when binding to the alkali metal and alkaline earth metal ion; the binding mode is ion exchange, whereas with transition, metals and heavy metals mainly form chelates. The complexing capacity of the fulvic acid with stronger acid is larger than that of the humic acid, and humic acid often forms insoluble chelates with heavy metals and transition metals. If the binding constants of fulvic acid and humic acid forming complexes or chelates with various metal ions are obtained at normal temperature and pressure, then their behaviors in aqueous solution would be completely controllable and predictable. So there are many models about hot metal ions binding to humic acid theoretically, and the study of humic acid and heavy metal ions has been the focus all these

years. But studies showed that, under the same conditions of temperature and metal ions, humic acid complexation constants differ by several orders of magnitude; results were mutual conflict and indecisive, due to no definite structure and composition of humic acid. According to the behavior of humic acid in soil, there was attenuation of heavy metal ions in soil, reducing plant roots on the metal ion procurability, by complexing to slow down liberation of metal ions. The role of humic acid in river system for heavy metal ions should also have a similar function. Heavy metal ions and their compounds, which flowed from mine tailings fields and hydrometallurgy, then entered into the river. In the water system, under conditions of existing humic acid with suspended solids, colloidal, and dissolved state, how to allocate and transform, when there are changes in seasonal water temperature, acidity–alkalinity, redox conditions, rock composition, and ion composition in water; how they affect their distribution between the three-phase state; and if, through research, to understand its laws of distribution and migration activation of heavy metals—organic matter among water and suspended particles—effective measures could be taken to deal with pollution behaviors of heavy metals in the environment [10]; these could accurately estimate the environmental capacity of heavy mental in the river system.

Development of tin polymetallic deposit tailings of Dachang town began in the Tang and Song dynasties. Pollutions of heavy metals in Longjiang river and Diaojiang river, Hechi were of long standing; sudden accidents happened from time to time. Studying migration types of the heavy metal cadmium and lead emerged from the tailings and torage yards of wet metallurgy, as well as their variation paths and destination after entering into the water system. Guangxi is located in the south of China with a warm humid climate, flourishing vegetation, sufficient rainfall, high yield of the surface humic acid, and great loss of soil organic matter; as a result, a large quantity of humic acid enters into the water system and interacts with those clay particles and heavy metal complex, which enter into the water system simultaneously. There are acidity–alkalinity, oxidation reduction potential, rock composition, conditions of water dynamic change, studies on the characteristics of distribution, migration, precipitation, and activation among heavy metal cadmium, lead, tin, and zinc in humic acid, colloidal particles, clay particles, and stream sediment.

4 INTERACTION

Interaction between natural organic matter in water and metal is very complicated. According to the chemical properties of humic acid and existence forms, distribution characteristics of metal elements, it mainly can be divided into the following functions:

4.1 *Chelation*

From the view of geochemistry, chelation is one of the most important functions of humic acid. Almost all the metal cations in natural water can chelate with humic acid (HA), and their chelating capability is generally consistent with the Irving–Williams sequence, namely $Mo^{2+}<Mn^{2+}<Fe^{2+}<Cd^{2+}<Co^{2+}<Ni^{2+}<Cu^{2+}<Zn^{2+}$, [9,10]chelating capability between divalent transitional metal and humic acid is very strong; 1g weathered coal humic acid can complex with 188mg Cu /g HA, 487 mg Pb/g HA, 165 mg Zn/g HA, 247 mg Cd/g HA, and 586 mg Fe/g HA. Only when COOH group and phenolic OH group of humic acid are in dissociation or —NH₂ is exposed can humic acid provide coordination atoms to chelate with metal cation, so the chelation of humic acid is controlled by pH. Thus, alkaline environment contributes to the formation of complexes; 97% of carboxyl is dissociated and can chelate metal cations when pH is at 8. In the partial alkaline marine environmental, heavy metals can easily concentrate to humic acid.

4.2 *Adsorption*

Adsorption is an important property of colloidal humic acids; fulvic acid in aqueous solution can not only strongly chelate with Cu but also adsorb a lot of Cd^{2+}. Suspended organic particles in natural water also showed strong adsorption properties, such as the partition coefficient of Hg in suspended solid particles (which cover organic matter) and water is 1.34-1.88*10⁵. [12]Studies had also shown that 40–90% of Hg were absorbed by suspended solids in Hg contaminated Ji canal.

4.3 *Surface complexation*

Surface complexation of sediment–water interface is one cause of accumulation of heavy metal to the sediment. Surfaces of sediments contain rich humic acid, and this humic acid was combined by organic–clay complex, which contains multipolarity continuously catching free metal ions from the bottom water and then chelating them.

4.4 *Coagulation*

The gel property of humic acid is the precondition of coagulation. Humic acid and its metallo-chelate with many negative charges, as well as repulsive force in water solution hinders the mutual cohesion between particles, after adding electrolyte. Colloid precipitates quickly because of electricity neutralization. Humic acid precipitates whereas humic acid coprecipitates

with metal elements. For example, more than 95% of Fe and Cd would coprecipitate with humic acid in solution with 0.001 mol $CaCl_2$; one or more of the earlier functions would be suitable for the interaction between humic acid and metal in a natural water system. Studies had shown that in the solution containing fulvic acid, Pb^{2+}, and Zn^{2+}, 32.5% of Pb and fulvic acid would be in adsorption state [13];whereas 67.4% of it would be in chelating state. Interactions between organic and metal were also related to the water system; chelation between humic acid and metal in fresh water could lead to increasing of metals solubility in the water, playing a role in delaying metal precipitation. Under the conditions of humic acid, the concentration of the metal elements precipitate must be 8–43 times higher than that without humic acid. This delayed action was confronted with the adsorption of Fe and Mn oxides, hydroxyl, and clay mineral in sediment, whereas they played the same role in seawater and led metal to the enrichment of the sediments, because of the flocculation.

5 ANALYSIS OF THE BEHAVIOR AND FATE OF HEAVY METAL-ORGANIC MATTER

5.1 *Solution*

Behaviors of heavy metal were restricted by many factors in water; take Cd for example, where dissolution is the first step of metal migration, so the concentration of Cd2 + is controlled by solubility product in aqueous solution: [14,15]

$$CdS \rightarrow Cd^{2+} + S^{2-} \tag{1}$$

$$Cd(OH)_2 \rightarrow Cd^{2+} + 2OH^- \tag{2}$$

$$Cd^{2+} + Cl^- \rightarrow CdCl^+ \tag{3}$$

$$Cd^{2+} + HA + 2e- \rightarrow Cd—HA \tag{4}$$

$$Cd^{2+} + \text{organic colloid} \rightarrow Cd^{2+}\text{-organic colloid} \tag{5}$$

$$\Sigma Cd = Cd^{2+} + CdCl^+ + Cd—HA + Cd^{2+}—\text{organic colloid} \tag{6}$$

$$X = S + I + R + J \tag{7}$$

where value S is controlled by the solubility product; therefore, the change of value S is comparatively small. Value I is associated with natural inorganic ligand concentration in natural water; in the same system of water (fresh water or seawater), due to the buffer action of natural water, total inorganic ligand concentration of polluted and unpolluted water changes little. But when different water is mixed with each other, such as the river water flowing into the sea, the inorganic ligand concentration, especially the concentration of Cl⁻, increases many times, so value I is mainly controlled by the water system. Value R is affected by organic ligands of water system, especially the concentration of humic acid; similarly, value J is related to the amount of colloidal particles and suspended particles in water. Here, value I and value R were not only related to the ligand concentration but also related to their complex stability constant.

$$\beta_1 == (CdCl^+) / (Cd^{2+}) (Cl^-) \tag{8}$$

$$\beta_2 == (Cd—HA) / (Cd^{2+}) (HA) \tag{9}$$

5.2 *Distribution*

For the same kind of metal, generally with β_2, $\beta_1 > 10^5$, so only needs (HA) / (Cl⁻) > 10^{-5}, organic complex can be dominant; that is why humic acid—Cd complex is the main form of Cd in freshwater and estuarine water. Water self-purification tends to have a longer distance from the pollution source to the downstream for rebalance of natural water. Three main changes of heavy metals flowed from land freshwater to the sea: The concentration change of ligand, the concentration of suspended colloidal particles and polymer ligands such as humid acid was sharply reduced from freshwater to seawater, but inorganic ligands such as Cl were obviously increased. Metal elements in combination with the change of state, such as Cd of humic acid Cd complexus, accounted for the total Cd above 90 % in freshwater; whereas in the estuary with higher salinity, the binding state of humic acid Cd complexus fell sharply, which accounts for only 10% of the total Cd in normal salinity (35 ‰), and Cd (OH)$_2$ and CdCO$_3$ complexus were the main forms at this time. Laws of element accumulation dispersion from fresh water to sea water obviously presented Br, Sr, Li, Rb, and so on; alkali and alkaline earth metals tended to accumulate in sea water and disperse in river water; toxicity elements Cu, Pb, Zn, and Cd tended to be dispersed in sea water and then accumulated in river water. [16,17] When river water with a low salinity contained rich organic matter and suspended particles mixing with seawater with a high salinity in estuary, humic acid with colloidal properties would flocculate suspended particles immediately. Those metals, which were easily bound to humic acid or absorbed by suspended particles, such as Cu,Pb,Zn,Cd, and so on, would chelate with humic acid and suspended particles, forming sediments with rich heavy metal in estuary. However, those humic acid alkali metal and alkaline earth metal with poor chelating capability will be residues in sea water. In the equation, value R and value J tend toward the minimum; the value X

depends on the value S and value I, so Cd and other heavy metals tend to be scattered in the sea.

6 CONCLUSIONS

Heavy metals absorb on the suspended particles in freshwater or are condensed into water sedimentary layers after agglutinating themselves; finally, they enter into the sea in the form of bed and suspended load. But in this process, heavy metals, which have been deposited in the river sediment, may be reactivated in the water. Some heavy metals binding to fulvic acid may be dissolved into intakes of drinking water. Under the existing condition of water treatment technology, heavy metals binding to fulvic acid at microgram level were hard to be flocculated and deposited or intercepted by quartz and filters.

REFERENCES

[1] Andrew Miller, Linda Figueroa, Thomas Wildeman (J) Applied Geochemistry 2011, 26:125–132.
[2] Augh Davies, Paul Weber, Phil Lindsay, etal (J) Applied Geochemistry 2011, 26:2121–2133.
[3] F. M. Romero • L. Nu´n˜ez • M. E. Gutie´rrez etal (J) Arch Environ Contam Toxicol 2011, 60:191–203.
[4] Silvana Santomartino, John A. Webb (J) Applied Geochemistry 2007, 22:2344–2361.
[5] W.H. Strosnider, R.W. Nairn (J) Journal of Geochemical Exploration 2010, 105 34–42.
[6] Roger St. C. Smart a, Stuart D. Miller b, Warwick S. Stewart etal. (J) Science of the Total Environment 2010, 408:3392–3402.
[7] Liping Weng a, Willem H. Van Riemsdijk etal (J) Journal of Colloid and Interface Science 2006, 302:442–457.
[8] Gilles Guibaud a,*, Eric van Hullebusch b, Francois Bordas etal (J) Bioresource Technology 2009, 100:2959–2968.
[9] Ping Zhou, Hui Yan, Baohua Gu (J) Chemosphere 2005, 58:1327–1337.
[10] Ping Zhou, Hui Yan, Baohua Gu (J) Chemosphere 2005, 58:1327–1337.
[11] SiripatSuteerapataranon, MurielBouby, Horst eckeis, etal. (J) Water Research 2006, 40:2044–2054.
[12] Pablo Lodeiro, A Roberto Herrero, A and Manuel E.eta (J) Environ. Chem. 2006, 3:400–418.
[13] Vladimir Ivezić, Åsgeir Rossebø Almås, Bal Ram Singh (J) Geoderma 2012, 170:89–95.
[14] P. VILLAVERDE, D. GONDAR, J. ANTELO etal (J) European Journal of Soil Science, June 2009, 60:377–385.
[15] Calin David, Sandrine Mongin, Carlos Rey-Castro etal Competition effects in cation binding to humic acid: Conditional affinity spectra for fixed total metal concentration conditions (J) Geochimica et Cosmochimica Acta 2010, 74:5216–5227.
[16] Takumi Saito,*Luuk K. Koopal, Shinya Nagasaki etal (J) J. Phys. Chem. B 2008, 112:1339–1349.
[17] Waiter J. Weber, Jr, Paul M. McGinley, and Lynn E. Katd (J) Envion. Sci. Technol. 1992, 26:1955–1962.

4

Energy, Environment and Green Building Materials – Sheng (ed.)
© 2015 Taylor & Francis Group, London, ISBN 978-1-138-02718-3

Current situation of rural ecological planning in North China—the Baoding Shijiatong village

Shi-Yong Zhao & Su-Juan Fu
Hebei Academy of Building Research, Shijiazhuang, China

Yu-Hang Hao
Hebei Building Research Engineering Co. Ltd., Shijiazhuang, China

ABSTRACT: Taking Baoding Shijiatong village as an example, the current rural planning situation in North China was studied. It was proposed to establish a sustainable eco-friendly new countryside based on ecological planning principles, the overall layout, public service facilities, road works, housing, and environmental planning measures.

KEYWORDS: North China; Planning Situation; Ecological Planning

1 INTRODUCTION

Actively and steadily promoting urbanization of towns and villages has become an important national strategy. It is an important part of balancing urban and rural development, improving rural livelihood, building a harmonious society, and building a new socialist countryside. It is one of the effective ways to expand domestic demand, promote economic restructuring, maintain stable and rapid economic development, and solve the "three rural issues" [1]. There are large rural populations in north China; the task of new rural construction is more urgent and arduous. As the head of the new rural construction, the new rural planning has been paid more attention to [2]. Taking Baoding Shijiatong village planning and construction status as an example, this paper will explore the research status of new rural ecological planning in North China.

2 GENERAL SITUATION OF SHIJIATONG VILLAGE

Shijiatong village subordinates Xi shanbei village Yi county, located at the foot of Langya mountain. It is about 4 km east of Xi shanbei village government and 2.5 km west of Langya mountain. There is a road from east to west through the village, and it is the main channel into the Langya mountain.

3 PLANNING SITUATION OF SHIJIATONG VILLAGE

3.1 *Using dispersion of village space*

Shijiatong village consists of east and west groups: One group is the main component of village in the west, and the east group is a smaller scale. The east group has only about 30 families. The village belongs to the typical mountainous countryside. The northwest terrain of Shijiatong village is higher than southeast, more than 80% of the total area is mountainous territory, climate is variability, and farmland is nervous, so the village is not suitable for large-scale settlements. The undulating terrain and scattered land determine the layout of Shijiatong village. It combines the characteristics of strip layout and scattered layout. It has a "spatial dispersion using" feature. In order to facilitate farming, Shijiatong village farmers' homestead is adjacent to their agricultural land. With the rapid development of urbanization in rural areas, the traditional "small and complete" mode of production and life have changed. A large number of buildings' function has lost, and the land failed in timely replacement, resulting in the space utilization and over dispersal in Shijiatong village.

3.2 *Uneven construction quality*

Shijiatong village experiences some common problems in rural housing construction in North China through research, such as the uneven quality of village houses and the arbitrary quality of residential buildings. The problems are summarized as follows:

1 The buildings in the village are built by villagers, so they are entirely constructed in accordance with their own wishes. Due to the lack of planning and design of housing construction, it results in architectural color, pattern, and style tending to be similar, but not in harmony with the surrounding environment. The houses right from the ancient house to the villa forms the appearance according to which we can see the new house but cannot see the new village.

2 Residential scattered, layout disorder. Because Shijiatong village is a mountain village, the local villagers choose building places near the land location for farming needs. This situation makes residential scattered layouts and causes inconvenience, poor accessibility information with other villagers, and external traffic. At the same time as the migrant workers increased, there was "more than one room" phenomenon in part of the family. There are many hollow houses phenomenon in a village, and the hollow room layout is scattered. This is not conducive to unified planning and layout.

3 Rural residential function is not complete. Residential design is unable to meet the increasing requirements of residents with regard to the material and cultural life.

Figure 1.   Local villagers self-built housing.

3.3   Road traffic unpaired system

The road is from east to west through the village leading to Langya mountain. The road is from north to south through the village leading to the forest park. These two roads form "a horizontal, one vertical" two trunk roads, and other roads consist of streets in

Figure 2.   Uneven architectural quality.

the village. In addition to the two trunk roads, most of the roads are an unpaired system. The roads are narrow. There are no ramp and parking lots or other transportation facilities in the village. It will not be able to meet the access requirements. Due to mountainous countryside vertical height and some of the roads being too steep, the roads in Shijiatong village make it difficult for local villagers to move around. Some roads are muddy, which results in it becoming inconvenient for villagers to get around on rainy days.

3.4   Lack of public service facilities and infrastructure

The public service facilities and infrastructure in Shijiatong village are not completed. The quality of water in local villages is poor; most of the water does not strictly deal with small water supply or well water. The construction of water plant is very backward, and it cannot meet the supply of clean water in rural areas. In the field of sewage, the vast majority of sewage are natural discharge, and there are no corresponding measures for the domestic sewage and production wastewater. Those result in serious water pollution and the deterioration of environment. Although local schools and clinics exist, poor facilities cannot meet the actual needs of the local countryside.

3.5   Poor ecological environment

There are no centralized garbage piling up points in Shijiatong village. Although a few dustbins are placed around the village, the utilization rate is not high. Each discarded garbage caused the village to become dirty and messy.

Figure 3. Garbage piled up points reunification scene of Shijiatong village.

4 ECOLOGICAL PLANNING PRINCIPLES

1 Adjust measures to local conditions and give play to advantage

Insist on the reality of Shijiatong village. Clear dominant industry. Accelerate scientific and technological progress. Give full play to location advantage and natural ecological advantages.

2 Adhere to the principle of unity ecological and economic benefits

Follow the law of economic development and ecological laws. Highlight two industries of "features fruit" and "leisure travel" in Shijiatong village. Focus on the construction simultaneously of village appearance, rural civilization, and democratic management. Achieve environmental benefits as well as economic and social harmonization[3].

3 Overall planning, step-by-step implementation

In order to protect the cultivated land, optimize the environment, conduct comprehensive development, and support the construction. These should clear development orientation, determine the development goals rationally, and achieve success in stages.

5 ECOLOGICAL PLANNING PROGRAM MEASURES

5.1 Overall layout planning of Shijiatong village

Maintain the original east and west as two groups of the structure model in Shijiatong village: The eastern group is on a smaller scale; it should be transformed and upgraded in the current situation combined with tourism road widening; Western group is the main component of Shijiatong village, and it should be comprehensive reform combined with village center and the construction of new dwellings. Shijiatong village will be built into a new eco-system in rural villages of "one, two axes, six area."

5.2 Plan of public service facilities

The public service facilities center and villagers culture square will be built combined with the transformation and construction of the village in the center of Shijiatong village. That will form Shijiatong center landscape area combined with river training("one"). There will be villagers' fitness square, plaza, basketball court, and parking lots in the center landscape area; it forms a place that is suitable for villagers' living ecological service facilities.

5.3 Plan of road project

Shijiatong village is a mountainous countryside in North China. It has the rural mountain road traffic characteristics in common: The road is narrow and curved, the density of the road is low, and the road condition is poor. Combined with the characteristics of mountainous countryside, as much as possible to retain the original road pattern in Shijiatong village, the two trunk roads of "one horizontal, one vertical" in the village will be broadened. Simultaneously, improve the road layout, add line density, and improve the road conditions. That will increase the accessibility and comfort.

5.4 Plan of road project

On the one hand, the cluster scale of the village should be increased in rural residential planning. It reflects the principle of intensive use of land, convenient production, and living. Second, it should suit local conditions in rational distribution, improving living conditions, the protection of historical culture, and rural landscape. On the other hand, it should guide farmers to build houses reasonably and to continue to control their own local characteristics and per capita land area. So the residential planning in Shijiatong village can be divided into the region of the old houses transformation and new residential demonstration zone. It should adopt the following three measures for old houses in the region of the old houses transformation: One is for internal residential streets to sort out the status quo, second is for the security risk status for protection engineering process, and the third is a part of the construction whose quality is poor for transformation and renovation.

5.5 *Plan of ecological environment*

1 Implement the hill afforestation project. There are 3600 acres of barren hills in Shijiatong village. We should adhere to the principle of combination of ecological forest and economic forest in barren hill greening projects. There are mainly pinus tabulaeformis and platycladus orientalis on the top of the mountain. The center of the mountains is suitable for planting prunus armeniaca. Colorful Koelreuteria paniculata, five maple, Huang Lu, and other species also can be planted in the central part of the mountains. We should strive toward the four seasons change in color. Persimmon, Walnut Hill, and other higher economic benefit trees will be built in the area beneath the mountains[4].

2 It can be better to plant trees at the edge of the village, roadside, and riverside. The rainwater should be accumulated and water should be conserved. Road greening is taking the road as the main line. Joe, irrigation, and grass flowers should be in a reasonable layout. The trees along the road can be into the net, the air can be fresh, and the environment will be beautiful.

3 Garbage collection means using centralized waste disposal system processing. The garbage can be collected by garbage bags according to classification and transported regularly. Set closed bins and garbage bags in the group entrances, streets, squares, and other places. The radius of service garbage collection points should be not more than 70 meters. The garbage can be transported by garbage trucks unified by township government. Blocks must be kept clean, and sanitation workers should clean throughout the day.

Waste landfill site will be constructed in the southeast of the village. Waste storage facilities will be built among the village, which can achieve domestic waste collection treatment centralization. Rural living garbage will be transported to landfills. This serves as a perfect village cleaning mechanism.

6 CONCLUSION

The development of rural urbanization in China is rapid during the "Twelfth five-year" period, but resource and environmental problems also exist in the development. Rural ecological planning and construction can realize sustainable development and environment improvement of social economy. The new rural landscape planning process should be combined with the rural characteristics, and ecological principles should be applied to protect the ecological environment and natural resources. That will establish a sustainable development and a new ecological countryside.

ACKNOWLEDGMENT

The authors are grateful for the task (2013BAJ10B09) for providing funds.

REFERENCES

[1] Chun Yi. Exploration and practice of urban and rural areas based on the new rural planning[D]. central south university, 2008.
[2] Kai Wang. Introducing the concept of ecological village planning[D]. Suzhou University of Science and Technology, 2010.
[3] Huili Wang. New socialist countryside construction planning Hebei Research[J]. Hebei Agricultural University, 2007, 6:31.
[4] Chenguang Zhang. Construction of new rural planning Hebei Province under the background of rapid urbanization[D]. Hebei Agricultural University, 2010.
[5] Zhenzhen Yu. Mountainous Rural Landscape Planning Research[D]. Shandong Agricultural University, 2008.

Energy, Environment and Green Building Materials – Sheng (ed.)
© 2015 Taylor & Francis Group, London, ISBN 978-1-138-02718-3

Research on domestic sewage treatment by "Activated Sludge" combined "Biofilm"

Yan-Ping Li
Stage Grid of China Technology College, Jinan, Shandong, China

Wei Hong
Shandong Academy of Environmental Science, Jinan, Shandong, China

Yong Li
Stage Grid Shandong Electric Power Company, Jinan, Shandong, China

Zhen-Bing Wang & Zhen Han
Stage Grid of China Technology College, Jinan, Shandong, China

ABSTRACT: The engineering with a processing capability of 2000 m^3/d via combination process of activated sludge and biological membrane shows that this combined process is of stable operation and has great treatment effect for the domestic sewage. In the system inlet water, the conditions are depicted as given next: COD$_{Cr}$ 195~296mg/L, SS 68~176mg/L, ammonia nitrogen 19.4~43.9mg/L, and total phosphorus 2.21~3.59mg/L; whereas conditions of outlet water after treatment are as follows: COD$_{Cr}$ 18~33mg/L, SS 5~8mg/L, ammonia nitrogen 0.06~0.37mg/L, and total phosphorus 0.01~0.06mg/L. Main pollutant indexes of outlet water quality are better than those in Level I of Discharge Standard of Pollutants for Municipal Waste Water Treatment Plant (GB18918-2002). All treated water should be used for dust suppression on the road of the park, greening, and landscape water supplement.

KEYWORDS: Activated Sludge; Biological Membrane; Domestic Sewage; Operation Effect

1 INTRODUCTION

Municipal domestic sewage means sewage generated during daily life by citizens with characteristics of large discharge and strong volatility. With development of urbanization in China, the processing rate of municipal sewage is constantly being improved. In current treatment process of municipal domestic sewage, the activated sludge method is widely adopted, such as A^2/O process, inverted A^2/O process, SBR process, and CASS process.

This engineering uses combined process of activated sludge and biological membrane to treat domestic sewage. The activated sludge adopts traditional inverted A^2/O process, whereas the biological membrane adopts modified polyurethane porous carrier bio-reactor to construct a domestic sewage treatment engineering with a processing capability of 2000m^3/d. All outlet water of the engineering should be used for dust suppression on the road of the park, greening, and landscape water supplement so as to realize water reclamation. This paper studies the effects of domestic sewage after treatment with

combined process so as to instruct the engineering practice and to promote the development of reclamation of domestic sewage.

2 SEWAGE TREATMENT TECHNICAL FLOW PROCESS

The designed processing capacity of this engineering is 2000m^3/d with adoption of combined process of activated sludge and biological membrane for the main process; the sewage treatment technical flow process is shown in Fig. 1:

Sewage water runs into the screen ditch after collection via pipe network. Suspended matter will be screened by the rough and fine grids, and water will run into the adjusting tank. The adjusting tank is used to balance the water quality fluctuation of the discharge in the Hengyuan Park. Submersible sewage pump is installed in the adjusting tank so as to lift the sewage to anaerobic/anoxia/aerobiotic activated sludge pond (inverted A^2/O process). Highly effective removal of pollutant indexes such as COD,

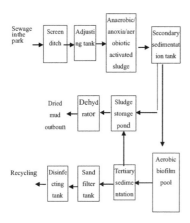

Figure 1. Sewage treatment flow chart.

ammonia nitrogen, total nitrogen, and total phosphorus should be realized via metabolism of activated sludge microorganisms in the anaerobic/anoxia/aerobiotic environment. The outlet water of activated sludge tank will enter into the secondary sedimentation tank to realize separation of water and sludge. A part of precipitated sludge will flow back to the activated sludge tank, and the rest will be discharged into sludge storage pond. The outlet water from the secondary sedimentation tank will flow into the aerobic bio-film pool, which is installed with a modified polyurethane porous carrier bio-reactor. Dissolved pollutants in the sewage should be further removed via metabolism of bio-film on the carries. The outlet water from the aerobic bio-film pool will enter into the tertiary sedimentation tank to realize separation of fell bio-film and sewage sediment. The sedimentary fallen bio-film will be discharged into the sludge storage tank. The outlet water from tertiary sedimentation tank will be lifted into the sand filter tank so as to remove suspended matter in the sewage. The outlet water from the sand filter tank will enter into the disinfecting tank. Malignant bacteria such as colon bacillus in the sewage should be disinfected with chlorine dioxide. Reclaimed water after disinfection should be used for dust suppression on the road of the park, greening, and landscape water supplement via recycling system. Sludge collected by sludge storage tank should be conveyed for outside treatment after dehydration by stacked spiral dehydrator.

3 REMOVAL EFFECT OF MAIN POLLUTANTS IN SEWAGE BY THE COMBINED PROCESS

This engineering started production from March 2013 with a stable operation. The average inlet water amount is $1400m^3/d \sim 1800m^3/d$. According to

analysis of operation data in typical seasons such as summer and winter, the daily routine testing data in two months from June 1st 2013 to June 31st 2013 and from December 1st 2013 to December 31st 2013 are analyzed so as to discuss the removal effect of main pollutants in domestic sewage by this process, such as COD_{Cr}, ammonia nitrogen, and total phosphorus.

3.1 Removal rate of COD_{Cr} for sewage by the combined process

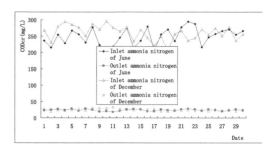

Figure 2. Content change of COD_{Cr} in inlet and outlet water of combined process.

Content change of COD_C in summer (June) and winter (December) (typical two months) is shown in Fig. 2. We can see from Fig. 2 that when the highest content of process system in inlet water in June is $COD_{Cr}294mg/L$ the lowest content is $COD_{Cr}195mg/L$ and average level is $COD_{Cr}24mg/L$; the highest content in the outlet is $COD_{Cr}28mg/L$; the lowest content is $COD_{Cr}18mg/L$; and the average level is $COD_{Cr}24mg/L$. The removal rate by combined process is 90.3%; the highest content of process system in inlet water in December is $COD_{Cr}296mg/L$ whereas the lowest content is $COD_{Cr}206mg/L$ and the average level is $COD_{Cr}260mg/L$; the highest content in outlet is $COD_{Cr}33mg/L$; and the lowest content is $COD_{Cr}21mg/L$ and the average level is $COD_{Cr}26mg/L$. The removal rate by combined process is 90.0%. When the content of OD_{Cr} in the combined process fluctuates heavily, the COD_{Cr} in the outlet water is stable, demonstrating that the combined process has great impact resistance against COD_{Cr} in the sewage.

3.2 Removal rate of ammonia nitrogen for sewage by the combined process

Content change of ammonia nitrogen in summer (June) and winter (December) (typical two months) is shown in Fig. 2. We can see from Fig. 2 that when the highest content of process system in inlet water in June is ammonia nitrogen 41.7mg/L the lowest content

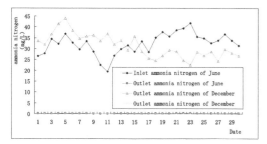

Figure 3.　Content change of ammonia nitrogen in inlet and outlet water of combined process.

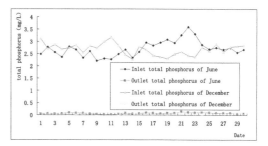

Figure 4.　Content change of total phosphorus in inlet and outlet water of combined process.

is ammonia nitrogen 19.4mg/L and average level is ammonia nitrogen 32.3mg/L; the highest content in outlet is ammonia nitrogen 0.28mg/L; the lowest content is ammonia nitrogen 0.06mg/L; and the average level is ammonia nitrogen 0.16mg/L. The removal rate by combined process is 99.5%; the highest content of process system in inlet water in December is ammonia nitrogen 43.9mg/L whereas the lowest content is ammonia nitrogen 24.2mg/L and the average level is ammonia nitrogen 31.2mg/L; the highest content in outlet is ammonia nitrogen 0.37mg/L; and the lowest content is ammonia nitrogen 0.13mg/L and the average level is ammonia nitrogen 0.26mg/L. The removal rate by combined process is 99.2%. When the content of ammonia nitrogen in the combined process fluctuates heavily, the ammonia nitrogen in the outlet water is stable, demonstrating that the combined process has great impact resistance against ammonia nitrogen in the sewage. In the traditional treatment process of sewage, when the North area experiences a season of low temperature, nitrification performance of biochemical treatment system will be slightly weakened. The removal rate of ammonia nitrogen of this system in cold December still remains at 99.2%, demonstrating that this process has better low-temperature nitrification performance than traditional process.

3.3　Removal rate of total phosphorus for sewage by the combined process

Domestic sewage should be used as reclaimed water after treatment, especially for landscape supplement. The total phosphorus has a significant effect on the re-use of reclaimed water [4]. Excessive multiplication of alga will happen due to relatively high content of total phosphorus in the water, leading to turbidity of water quality, stink or death of fish due to oxygen deficit, which will seriously affect landscape effect.

　Content change of total phosphorus in summer (June) and winter (December) (typical two months) is shown in Fig. 4. We can see from Fig. 4 that when the highest content of process system in inlet water

in June is total phosphorus 3.59mg/L the lowest content is total phosphorus 2.21mg/L and average level is total phosphorus 2.71mg/L; the highest content in outlet is total phosphorus 0.13mg/L; the lowest content is total phosphorus 0.01mg/L; and the average level is total phosphorus 0.07mg/L. The removal rate by combined process is 97.4%; the highest content of process system in inlet water in December is total phosphorus 3.16mg/L whereas the lowest content is total phosphorus 2.27mg/L and the average level is total phosphorus 2.68mg/L; the highest content in outlet is total phosphorus 0.09mg/L; the lowest content is total phosphorus 0.02mg/L; and the average level is total phosphorus 0.05mg/L. The removal rate by combined process is 98.1%. When the content of total phosphorus in the combined process fluctuates heavily, the total phosphorus in the outlet water is stable, demonstrating that the combined process has great impact resistance against total phosphorus in the sewage.

4　CONCLUSION

1　The combined process with integration of activated sludge and biological membrane has great treatment effect for domestic sewage. The system inlet is under conditions as given next: COD_{Cr} 195~296mg/L, ammonia nitrogen 19.4~43.9mg/L, and total phosphors 2.21~3.59mg/L; the outlet water after treatment is in conditions where COD_{Cr} 18~33mg/L, ammonia nitrogen 0.06~0.37mg/L, and total phosphors 0.01~0.06mg/L. Main pollutant indexes of outlet water quality are better than those in Level IA of Discharge Standard of Pollutants for Municipal Waste Water Treatment Plant (GB18918-2002).

2　The combined process with integration of activated sludge and biological membrane has significant impact resistance against indexes such as COD, ammonia nitrogen, and total phosphors in the domestic sewage. When the inlet water in the system fluctuates heavily, the quality of outlet water is still relatively stable.

3 For the combined process with integration of activated sludge and biological membrane for domestic sewage, the outlet water will be used for dust suppression on the road of the park, greening, and landscape water supplement. No adverse effect is detected.

REFERENCES

[1] C.P. WANG, T. WEI, M.X. ZHENG, etc. *Up-flow Anoxic/Oxic Process for Nitrogen Removal in Domestic Sewage Treatment [J]*, China Water Supply and Drainage 2014, 30 (5):88–91.

[2] F. MA, P. LI, X.Q. ZHANG etc. *Impact Factors and Properties of SBR Rector Simultaneous Nitrification and Denitrification[J]*. Journal of Harbin Institute of Technology (New Series, 2011, 43(8):55–60.

[3] Rosenberger S, Labbs C, Lesjean B, etal. Impact of Colloidal and Soluble Organic Material on Membrance in Membrane Bioreactors for municipal Wastewater Treatment *[J]*.Water Research, 2006, 40(4):710–720.

[4] R. YAN, W.B. ZHANG, Q. WANG etc. *Treating the Domestic Sewage from Scenic Spots Using Integration Contact Oxidation Reactor [J]*.Environmental Engineering, 2011(S1):26–28.

Energy, Environment and Green Building Materials – Sheng (ed.)
© 2015 Taylor & Francis Group, London, ISBN 978-1-138-02718-3

DSM-based modeling framework of emergency management

Peng Zhang, Lao-Bing Zhang, Bin Chen, Liang Ma & Xiao-Gang Qiu

College of Information and Management, National University of Defense Technology, China.

ABSTRACT: This paper focuses on the Modeling and Simulation (M&S) process of Public Health Emergency Management (PHEM), which is based on ACP approach and public security triangle theory. Emergency management always involves multi-disciplines, and models are always characterized by diversity, hierarchy on a large scale. In this paper, Domain-Specific Modeling (DSM) is introduced to integrate the complex domain knowledge to support the M&S in PHEM. This paper establishes the modeling framework, and meta-models are designed using GME (Generic Modeling Environment) tools. In the case study, a public health emergency scenario is constructed based on these meta-models, including disease model and intervention model.

KEYWORDS: Emergency Management; DSM; KD-ACP; Epidemic

1 INTRODUCTION

Unconventional emergency events such as H1N1 influenza always break out on a large scale, and the boundary is always uncertain. The modeling issue of epidemic has attracted high attention since the first SIR model proposed in 1970. Moreover, transportation (Cheng et al. 2012) and social networks (Ge et al. 2011) also play critical roles in influenza. PHEM is a typical emergency management, with a view to explore the epidemic transmission mechanisms and intervention measures.

According to public security triangle theory, disaster-bearing body, emergency event, and intervention measures are contained in the emergency management. It is always a complex system combined with many social issues, which cannot be simulated in the traditional way. Professor Wang put forward the ACP approach (2004), including artificial society, computational experiment, and parallel execution, which can explore the micro-to-macro transition mechanism in the social systems. In artificial society, actors' features can depict the heterogeneity of individuals. It provides a new way to study the complex social problems, such as emergency management. However, models are the foundation of artificial society, and modeling problems are the key to emergency management.

Most agent-based modeling platforms always just consider the geography or space factors of individuals (Fatima 2004), such as Repast and Swam. However, they always cannot integrate the physiology, psychology, and demography characters together to support the M&S in PHEM. Therefore, our team develops the computational experiment platform (KD-ACP),

which can support the modeling, analysis, and simulation process of emergency management (Chen 2014). Modeling environment is the most important part of the platform, and DSM approach is introduced to construct the modeling framework.

2 THE M&S PLATFORM IN EMERGENCY MANAGEMENT

2.1 *KD-ACP platform*

As shown in Fig. 1, artificial society is composed of a series of basic models and statistics data. Models are driven by data, so models and data should match each other. Simulation data can have an impact on the real social system, and dynamic data from the real system can revise the operational state of artificial society. The process ensures the parallel evolution of the artificial society and real system.

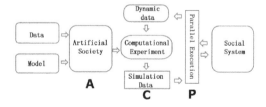

Figure 1.   Methodology and workflow of ACP approach.

As shown in Fig. 2, the integrated emergency management M&S platform (KD-ACP) combines the DSM technology, ACP approach, and public security triangle theory, which can support the construction of artificial society and computational experiment. ACP provides a good way to study emergency management, in which DSM solves the problems in the construction of emergency scenario. Model development environment is the most important part of the KD-ACP platform, and DSM-based modeling framework provides the ability to establish the domain meta-models.

The platform contains parallel engine, M&S tool sets, data collector, and a series of databases (Zhang 2014). It provides the ability to support a large-scale computational experiment that is used to study complex social problems, such as environment pollution and influenza spreading.

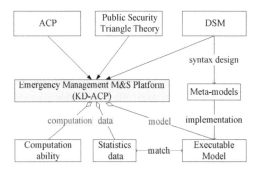

Figure 2. Emergency management of M&S platform.

2.2 *Domain-specific modeling*

Modeling is one of the most universal scientific description techniques that describes the world in terms of pre-established syntax (Li 2013). Models describe systems in a precise way, using diagrams, rules, symbols, signs, and so on. Domain-specific modeling (DSM) integrates the domain knowledge to support the modeling process.

According to model-driven architecture (MDA), there are four modeling layers, including M0, M1, M2, and M3. The main task of DSM is to use elements at M2 layer to build the classes at M1 layer, which are considered domain models. Our research just focuses on the M0, M1, and M2 layers, and it tries to build the domain classes at M1 layer with the modeling elements at M2. Then, instant models can be established to support the M&S in PHEM.

Meta-modeling is introduced to design the domain specific language (DSL), and the syntax contains the basic elements to describe artificial society and emergency management process, including population, environment, resource, and main action sequences. Meta-models contain all possible modeling elements about the domain and the relevant service components. All models are developed under the unified

modeling framework, so that they can interoperate with each other.

3 MODELING FRAMEWORK

3.1 *Model characters and modeling architecture*

Public health emergency directly affects the public health and results in loss to the society. Influenza is always affected by social contact among individuals. Intervention measures change the spreading process of disease and health state of individuals. Artificial society provides a visual society to study these problems, and the emergency management models have the following characteristics:

- Multidiscipline: Models always refer to multidisciplines, such as psychology, behavior, social culture, and social system.
- Large scale: Artificial society is always composed of massive entities, including millions of individuals or objects.
- Hierarchy: Artificial society always has multi-hierarchy, such as individual, family, community, and city.
- Diversity: Individuals may belong to different organizations and play different roles. They interact with each other or the environment, and they develop the diversity of the artificial society.
- Dynamic evolution: All individuals keep active in the artificial society, which is dynamically developing and evolution.

The model architecture of emergency management mainly contains three types, including environment, population, and event. Population is the core component of artificial society, and the behavior characters are always expressed by activities. Environment provides the basic services to the population, such as transportation, building, and climate. Event model can have an impact on population, and it change the states of individuals or groups.

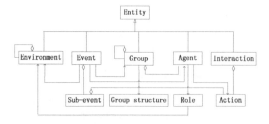

Figure 3. Artificial society modeling elements.

The main modeling elements include environment, event, agent, group, and interaction. As shown in

Fig. 3, the relationships among entities are described. Agents may dynamically play different roles in various types of groups, which have their own structure and constrain the role of agents. An event is composed of sub-events, which may act on the environment, agent, or group. Interactions among entities are composed of a series of actions, which can develop the diversity of artificial society.

3.2 *Meta-models design*

With the help of GME tool, artificial society meta-models are established, mainly including population and environment. Population meta-models contain agent, group, social relationship networks, and activities schedule, whereas environment meta-models contain buildings, climate, and transportation.

Under different scenarios or situations, individuals may dynamically play different roles to adapt to the environmental, social, or their own changes. Based on individual models, social structures are designed to describe the group characteristics. In order to study the emergency management, we should add the domain relevant attributes into meta-models. For example, disease and intervention relevant attributes are contained to support the study of PHEM, as shown in Table 1.

Table 1. Syntax constraint of population meta-models property.

Attribute	Property	Syntax Constrain
Basic	gender	{male, female}
	age	(0,100)
	popuType	{student, worker,...}
Geographic	latitude	(0,180)
	longititude	(0,180)
	belongEnv	{house, hospital,...}
Disease	healthStatus	{health, infected,...}
	infectTime	(0,86400)
	diseasePhase	{incubation,...}
Intervention	isolation	{isolated, no}
	immune	{immune, no}
	treatment	{treatment, no}

Schedule is designed to simulate the daily activities of individuals. In the schedule, activity is banded with time and location, and all behaviors are specified with time and geography information. Schedule is always characterized by agent role and situation. Agent role depends on population type and its current

group, whereas situation includes workday, holiday, and emergent period. Agents with different role or situation may have different schedules, and the emergency or intervention process just changes the agent schedule from the normal pattern to emergency pattern or intervention pattern.

Social network plays an indispensable role in disease spreading, and the topologies of social networks can affect the transmission process. In general, social networks have multi-layer structures. Each individual may have families, friendships, colleagues, consortiums, classmates, and other relationships simultaneously. In the influenza spreading, individuals are infected with disease through contact with each other in special social networks. Social network can be represented by the concept of group, which has its type, size, and group structure.

Physical environment provides individuals with the activity space, such as school, government, and hospital. Usually, environment entities must contain the geography information (longitude and latitude), function, and capacity. Each environment has population density limitation and contact frequency among individuals, both of which the key factors to impact the spreading process.

Transportation system has a significant impact on society for its tight connection with traveling behaviors. People may maintain contact with each other in the moving process, along with the disease spreading among crowds. People may choose to walk, or to use the car, bus, bike, and subway to their destination. In our studies, the choice is mainly determined by the distance between the two points.

Emergency can have an impact on the artificial society dynamically. Actually, disease model is the emergency model in PHEM. The basic attributes of emergency meta-models include event time, place, intensity, and scope. Disease model usually includes the symptom, infectious rate, spreading characters, and so on. Disease phases and their switchover conditions are also described in the meta-models, which contain susceptible state, incubation state, symptom state, recovering state, recovered state, and death state.

Intervention model can have an impact on the artificial society and revise the abnormal state or situation to the normal. The trigger condition, affect object, effective scale, and effective delay time are the basic elements. The affect objects include population and environment. In the PHEM, interventions such as closing public place, immunization, isolation, hospital treatment, and wearing marks are designed. Each intervention has the trigger condition, based on time or situations. Of course, measures will be cancelled if the epidemic is under control.

4   PUBLIC HEALTH EMERGENCY SCENARIO

4.1   *Emergency management process*

In artificial society, all models are independent and operate according to the internal mechanism. As dynamic parts of the computational experiment, event model and intervention model can be   included   in the artificial society and the operation state can be changed, as shown in Fig. 4.

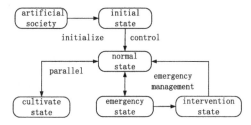

Figure 4.   Emergency management base on artificial society.

In general, the disaster-bearing body is the component of artificial society, which may be the population or environment. Loading emergency event and intervention to artificial society can help us study the process of emergency management.

4.2   *Emergency model*

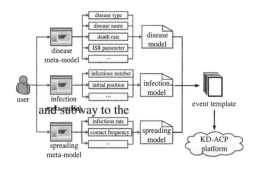

Figure 5.   Emergency models in PHEM.

In the PHEM, the emergency model is designed based on meta-models. It contains three parts: disease model, infection source model, and spreading model, as shown in Fig. 5. Then, we can fill parameters and get instance models, which can be injected into the artificial society on the KD-ACP platform, and support the simulation execution.

The SEIR model of H1N1 epidemic is established, which contains the susceptible state, incubation state, symptomatic state, recovering state, and recovered state in the disease period. In the experiment, the minority is initialized to incubation state, whereas the majority is susceptible state. The contact between the susceptible ones and infected ones may lead to the influenza outbreak. Once infected by influenza, agents usually may go through the whole period from the susceptible state to recovered state. Intervention may change the duration, and infected agent will recover in a shorter time. Of course, minority agent also may die if they are seriously ill and cannot get immediate treatment.

4.3   *Intervention model*

In the PHEM, there are several types of intervention measures to control the epidemic. Intervention measures include drug intervention and non-drug intervention. Drug intervention always includes vaccine injection, hospital therapy, and so on. Non-drug intervention includes wearing masks, closing public spaces, and patient isolation.

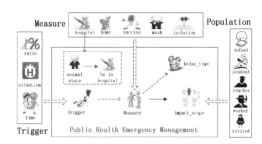

Figure 6.   Intervention measure in PHEM.

As shown in Fig. 6, hospital treatment, isolation, self-resistance, closing public place, and reducing public activities are designed in the interventions. When infection rate or epidemic situation reaches a threshold, the corresponding intervention will be trigged. Taking the hospital treatment as an example, the individual may be sent to hospital if the body's temperature reaches 38C. Intervention scope contains the impact on population types and quantity. Of course, the combination of measures is supported in the experiment, which may achieve a better effect in the emergency management. For example, if a person takes a vaccine and simultaneously wears a mask, they may be infected with influenza at a low probability.

5 CONCLUSION AND DISCUSSION

This work provides a meta-modeling method to construct the domain special language (DSL), which can support artificial society modeling in PHEM. The model architecture of artificial society mainly contains population, environment, and event. Agent is the basic entity in artificial society, and schedule meta-models are designed to depict the individual behaviors and daily activities.

Though our method works well to some extent, there are still some limitations. First, artificial society is a complex system, and many factors affect the operation of the visual system. In our study, we just consider the main characteristics of the individual and environment, which cannot contain all the possible items. Second, the meta-models have disadvantages in describing the behavior, methods, and functions. Although schedule can describe the daily activities, it is also difficult to depict the action atom. In order to promote the modeling work, we should integrate enough domain knowledge to enrich the meta-models.

ACKNOWLEDGMENTS

This work was supported by the National Nature and Science Foundation of China under Grant (Nos. 91024030 and 91324013).

REFERENCES

Cheng, Z., Qiu, X., Zhang, P., & Meng, R. (2012). An Agent-Based Artificial Transportation System Framework for H1N1 Transmission Simulation. *In System Simulation and Scientific Computing*, pp. 313–321.

Ge, Y., Duan, W., Qiu, X., & Huang, K. (2011). Agent based modeling for H1N1 influenza in artificial campus. In Emergency Management and Management Sciences, *2011 2nd IEEE International Conference*, pp. 464–467.

Fei-Yue, W. (2004). Artificial Societies, Computational Experiments, and Parallel Systems: A Discussion on Computational Theory of Complex Social-Economic Systems. *Complex Systems and Complexity Science*, 4, 25–35.

Fatima, S. S., Wooldridge, M., & Jennings, N. R. (2004). An agenda-based framework for multi-issue negotiation. Artificial Intelligence, 152(1):1–45.

Chen, B., Ge, Y., Zhang, L., Zhang, Y., Zhong, Z., & Liu, X. (2014). A Modeling and Experiment Framework for the Emergency Management in AHC Transmission. *Computational and mathematical methods in medicine*, 2014.

Zhang P. (2014). A Framework of Computational Experiment within Special Artificial Society Scene. *International journal of modeling and optimization*, 2(4):104–110.

Li, X., Lei, Y., Wang, W., Wang, W., & Zhu, Y. (2013). A DSM based multi-paradigm simulation modeling approach for complex systems. *In 2013 Winter Simulation Conference*, pp. 1179–1190.

Energy, Environment and Green Building Materials – Sheng (ed.)
© 2015 Taylor & Francis Group, London, ISBN 978-1-138-02718-3

Formation mechanism and controlling factors of secondary pore in clastic rock

Y.R. Li
Geophysical Institute of Henan Oil Field, Zhengzhou, China

X.Y. Fan
China University of Geoscience, Wuhan, China

R.G. Jiang
Daqing Oil Field, Daqing, China

M.T. Li
China Oil Field Services Limited, Sanhe, China

X. Xiao
Geophysical Institute of Henan Oil Field, Zhengzhou, China

ABSTRACT: This paper discussed the recent research results from the aspects of formation mechanism and controlling factors of secondary pore. The formation mechanism includes a combination of fluid and mineral, including feldspar and acidic solution, carbonate minerals and acidic solution, quartz and alkaline solution, and laumontite and acidic solution. Controlling factors of secondary pore include temperature, PH, pressure, sedimentary facies, tectonic factor, transformation of clay minerals, and biological action.

KEYWORDS: Secondary pore; Development mechanism; Pressure; Sedimentary facies

1 INTRODUCTION

Secondary porosity refers to the pores or cracks generated from consolidated sediments experiencing a variety of secondary actions, such as dissolution, recrystallization, and tectonic stress. The existence of secondary porosity in favor of improving reservoir properties, especially deep reservoir property, its development, and preservation, is vulnerable to a variety of factors, such as abnormal pressure, temperature, PH, sedimentary environment, and tectonism, which result in the complexity of the development of secondary porosity. This paper, on the basis of summarizing the results of previous studies, further discusses development mechanism of dissolution-typed secondary pore and influencing factors.

2 FORMATION MECHANISM

Secondary porosity is essentially the result of a series of physical, chemical, and biological effects, under the influence of a variety of factors, such as temperature, PH, pressure, and so on, and the macro

control of sedimentary environment and tectonism occurs between soluble mineral and dissolving fluid. From the perspective of oil and gas field development, the soluble minerals forming secondary porosity are carbonate minerals, feldspar, quartz, laumontite, sulfate minerals, and so on. Dissolving fluid is mainly carbonate, organic acids, water (meteoric water and precipitated mineral interlayer water), alkaline solution, and so on .

Determining the selective matching relationship between fluids and minerals based on chemical reaction principle, minerals and fluids can be divided into the following four matching relationships: ①. feldspar and acidic solution;②. carbonate minerals and acidic solution; ③. quartz and alkaline solution; and ④. laumontite and acidic solution.

2.1 Feldspar and acidic solution

Feldspar is the most important erodible mineral formation of secondary porosity (Fig 1). The solutes of acidic solution include carbonic acids and organic acids, which have an important contribution to the formation of dissolution-type secondary

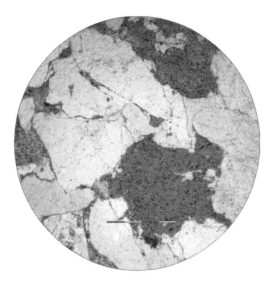

Figure 1. Feldspar dissolved partially by acid Well Shan 372 -3543.49m-orthogonal polarization.

Figure 2. Calcite cement dissolved by inorganic acid Well Shu 208-3944.5m-orthogonal polarization.

pore. Experimental studies of chemical dissolution conducted by many scholars in China and outside of China confirmed that the influence of organic acids on aluminum silicate and carbonate minerals is greater than on the carbonate (Crossey, L.C et al, 1984 &Huang, F.T.et al, 1998).

2.1.1 *Feldspar and organic acids*

Two mechanisms exist for organic acids dissolving feldspar (Xie, J.R. et al, 2000): First, carboxylic acids and Al^{3+} binding, leading to the dissolution of aluminum silicate minerals, forming dissolution pores. Carboxylic acids has a stronger complexing ability for Al^{3+}, so Al^{3+}will migrate with pore fluid in the form of complexes.

The second mechanism involves organic acids reacting with feldspar, leading to a decrease in the mineral volume and an increase in the porosity (taking K-feldspar as example).

2.1.2 *Feldspar and carbonic acids*

From the perspective of cause, carbonic acids can be divided into two kinds: organic origin and inorganic origin. In the process of thermal maturation, deoxidation of kerogen and other oxygen-containing organics generates CO_2, which dissolves in water and forms organic origin carbonic acids (Lundegard, P. D et al,1984).

2.2 *Carbonate minerals and acidic solution*

Carbonate minerals can also be dissolved by organic and inorganic acids (Fig 2), which have developed in

many regions. For example, there is a good negative correlation between secondary porosity zone and carbonate minerals development zone in the Paleogene sandstone in Dongying Sag. The sum of secondary porosity and carbonate mineral content change in a little interval is studied (Zhang, Q et al, 2003). There are two mechanisms of organic acids dissolving carbonate minerals(Lundegard, P.D et al,1984):①Carbonic acids is generated in CO_2, which comes from decarboxylation of organic acids, dissolving in water, so that the carbonate minerals are dissolved to form dissolution pores;②The decomposition of organic acids generates H^+ dissolving carbonate minerals and forming dissolution pores.

2.3 *Quartz and alkaline solution*

In the early secondary porosity research, quartz is rarely taken into consideration. Many scholars had discussed the possibility of dissolution of quartz grains and other related issues, and they think that in certain conditions quartz can be dissolved and point quartz dissolution rate in different environments (K. Pye et al, 1985). In recent years, studies of quartz in an alkaline environment that can form a certain scale of secondary porosity is a breakthrough in the secondary porosity research field. Under alkaline conditions, dissolved silica can form secondary porosity (about 10% to 35% of the total porosity) (Qiu, L et al, 2002). Zhong, D.K (2006) and other researchers believe that not only in alkaline conditions but also in neutral and even acidic solution, SiO_2 dissolution can still occur, but the silica is dissolved in neutral and acidic medium lacking quantitative research.

2.4 *Laumontite and acidic solution*

Secondary porosity formed from Laumontite dissolution is relatively rare, so there are relatively few previous studies of its dissolution. But that does not mean that Laumontite cannot form a lot of secondary porosity. The proportion of secondary porosity formed from Laumontite dissolution of secondary porosity changes in a wide range, from small till about 90%. In China, secondary porosity in the formation of laumontite dissolution is more typical in Songliao Basin, Yanchang Formation in northern Shanxi (Zhu, G.H, 1985), which may be related to the higher contents of organic acids.

In general, there are several causes of secondary porosity in the same clastic rocks and feldspar dissolution is the most common, followed by carbonate dissolution, laumontite acidic dissolution, and quartz alkaline dissolution, which are rare, but under certain geological conditions, they are prone to occur more locally and can be the main causes of secondary porosity.

3 CONTROLLING FACTORS

Most reservoirs dominated by secondary porosity are not influenced by one factor, but several factors influenced each other. The factors involved are as follows: temperature, PH value, pressure, sedimentary facies, structural factors, and clay mineral transformation.

3.1 *Temperature and PH*

Effect of temperature on the development of secondary porosity is significant. In most cases, the increases of temperature promote dissolution in the acidic dissolution mechanism. Silica dissolution only takes place at a higher temperature in alkaline fluid mechanism condition. Organic acids and carbon dioxide released during the organic maturation dissolve in the water to form the inorganic acids, which are the main acid types.

Temperature and PH are often matched to control dissolution. Let us take Weixinan depression Liushagang group as a case: Quartz cement will not precipitate due to lower temperatures on the section when the depth <1900m; the peak of quartz cement formed under suitable temperature on the section of 2200–2400m. However, quartz cement is dissolved when buried deep > 2500m due to high temperature. Carbonate cements exhibit the same distribution (Figure 3).

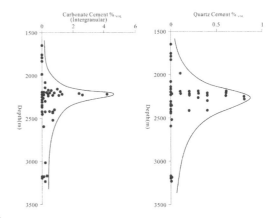

Figure 3. Vertical distribution characteristic of the Liushagang formation cement in southern Weizhou depression.

3.2 *Pressure*

Most research focuses on the abnormal high pressure, and it has a close relationship with the development of secondary porosity belt (You, J et al, 1997& Cui, Y et al, 2002). The role of abnormal high pressure played in the formation of secondary porosity is as follows:

①Protective effect of abnormal pressure on the secondary porosity; fluids share some of the overburden pressure that should be supported by the particles, reducing the reservoir compaction and cementation(Miao, J.Y et al,2000).②Abnormal pressure restrains the organic evolution, meanwhile delaying the discharge of organic acids and CO_2 effectively(McTavish, R.A,1978 & Hao, F.et al, 1995& Wang,Y. et al,2006), which leads to a full dissolution reaction to the formation of more secondary porosity.③Abnormal high pressure can produce a lot of micro-cracks in the course of episodic expulsion.

3.3 *Sedimentary facies*

Sedimentary facies affect the development of secondary porosity as the result of many factors as follows: different mineral assemblages formed under different sedimentary environment, reservoir properties, and spatial distribution of sands and source rocks. For example, sand bodies developed in the delta plain formed under strong hydrodynamic conditions during deposition, which has less miscellaneous group content; intersection of delta front sand and former Delta shale, resulting in extensive contacts of reservoir sands and source rocks in favor of various acids, CO_2, and hydrocarbons directly into the sandstone, which is conducive to secondary porosity development. Let us take Liushagang formation of Weixinan Depression as a case: Different sedimentary microfacies have different degrees of secondary porosity, which has a good positive correlation (Figure 4) and permeability.

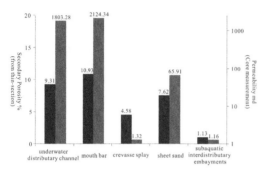

Figure 4. Relationship between sedimentary facies and secondary pore of the Liushagang formation in southern Weizhou depression.

3.4 *Tectonic activity*

Control of tectonic activity on secondary porosity is mainly as follows:

1 Sandbodies besides fault channel under long-term effects of vicinity of a large number of acidic aqueous solution were in favor of dissolution. Besides, subsurface fluid can flush dissolved substance away through the fault, thus contributing to the ongoing dissolution.
2 Leaching of atmospheric fresh water usually developed in the surface and in a shallow burial site or at the unconformity. Denudation caused by tectonic uplift can enhance the leaching of atmospheric fresh water, which leads to the formation of secondary porosity (Zhang, F.S et al, 2008).
3 Tectonic fluid system would destroy the original matching dissolve pattern, which would lead to the direction in favor of developing secondary porosity reservoir.

3.5 *Transformation of clay minerals*

As the depth increases, transformation of different clay minerals into each other is conducive to the formation of secondary porosity during late diagenesis:

1 Clay minerals are helpful for organic matter and kerogen transferring into hydrocarbon; both the process amounts of CO_2 are produced. Besides, a reaction between clay minerals and carbonates may also form a large number of CO_2, which cause acidic water formation, thus contributing to

dissolving the soluble component and thus the formation of secondary porosity.
2 Dehydration of clay minerals promotes the formation of an abnormal high pressure zone; low water salinity between its prolapse layers enhances the ability of a chemical to dissolve the pore fluid, prompting the dissolution of carbonate and feldspar.
3 Clay mineral diagenesis contracts formation of micro-cracks, the dissolution of scale, increasing the porosity of sandstone cracks (Zhang, F.S et al, 2004).

4 CONCLUSION

Three main conclusions are made through the formation mechanism and control factors:

1 Formation mechanisms of secondary porosity are summarized into feldspar and acidic solution, the acidic solution of carbonate minerals, quartz and alkaline solutions, laumontite, and acidic solution;
2 The development of secondary porosity is controlled by temperature, PH value, pressure, sedimentary facies, structural factors, the clay mineral transformation, and biological effects of different research areas. Temperature, pressure, and PH directly affect the development of secondary porosity; pressure, sedimentary facies, structural factors; the clay mineral transformation indirectly controlled the formation of secondary porosity, and they can strengthen the degree of development of secondary porosity.
3 Despite decades of research on the formation mechanism of secondary porosity gaining some achievements, it does not reach a mature quantitative stage. There are still many problems and challenges in the study of secondary porosity, such as the formation mechanism of alkaline fluid, how abnormal low pressure can affect the development of secondary porosity, and how deep thermal fluid affects secondary porosity development and preservation.

REFERENCES

Crossey L.C. & Frost B. R. & Surdam R. C. 1984. Secondary Porosity in Laumontite-Bearing Sandstones, Clastic Diagenesis. *AAPG Memoir 37*, 225–238.
Huang, F.T. & Zou, X.F. & Zhang, Z.X. 1998. Effect of major acids in formaton water on physical properties of reservoir. *P.G.O.D.D*, 17(3):7–9.
Xie, J.R & Kong, J.X. 2000. Mechanism of secondary porosity in sandstone. *Natural Gas Exploration and Development*, 23(1):52–55.

Lundegard, P. D. & Land L. S. & Galloway W. E. 984. Problem of secondary porosity: Frio Formation (Oligocence), *Texas Gulf Coast.Geology*, 12(1):399–402.

Zhang, Q. & Zhong, D.K. & Zhu, X.M.2003. Porosity evolution and genesis of secondary in Paleogene clastic reservoirs in Dongying Sag. *Oil & Gas Geology*, 24(3): 281–285.

K. Pye, D. H. 1985.Krinsleyt. Formation of secondary porosity in sandstones by quartz framework grain dissolution. *Nature*, 317(5):54–55.

Zhong, D.K. & Zhu, X.M. & Zhou, X.X. 2006.Phase of secondary pore generation and dissolution mechanism-Taking Silurian Asphaltic Sandstone in Central Tarim Basin as an Example. *Natural Gas Industry*, 26(9):21–24.

Qiu, L. & W. Jiang, Z.X. &Chen, W.X. 2002 A new type of secondary porosity quartz dissolution porosity. *Acta sedimentological Sinica*, 24(4):621–627.

Zhu, G.H.1985.Formation of laumontite sandbodies with secondary porosity and their relationship with hydrocarbons. *Acta Petrolei Sinica*, 6(1):1–8.

You, J. & Zheng J.M & Zhou J.S. 1997.Anormal pressure and abnormal prosity and hydrocarbon reservoir in the deep strata. *China Off shore Oil and Gas: Geology*, 11(4)249–253.

Cui, Y& Zhao, C.L. 2002.Porosity generation and the relationship associated with overpressure leakage. Journal of Chengdu University of Technology: *Natural Science Edition,* 29 (1):49–52.

Miao, J.Y. & Zhu, Z.Q & Liu, W.R.2000. Relationship between temperature-pressure and secondary pores of deep reservoirs in Eogenne at Jiyang Depression. *Acta Petrolei sinica*, 21(3):36–40.

McTavish, R.A.1978. Pressure retardation of vitrinite digenesis, off shore north-west Europe. Nature, 271(16):648–650.

Hao, F.& Sun Y.C. & Li S.T.1995.Overpressure retardation of organic matter maturation and petroleum generation.A case study from the Yinggehai and Qiong dong nan Basins, South China sea. *AAPG Bulletin*, 79(4):551–562.

Wang, Y. & Zhong, J.H. & Chen, H. 2006.Vertical distribution and genesis of the secondary pore in deep formation of Paleogene at Dongpu Sag. *Petroleum Exploration and Development*, 33(5):576–580.

Wikinson, M.D. & Haszeldine, R.S.Couples, G.D.1997. Secondary porosity generation during deep burial associated with overpressure leak-off:Fulmar Formation,United KingdomCentral Graben. *AAPG Bulletin*, 81(5):803–813.

Zhang, F.S. & Zhu, Y.H. & Wang. F.R. 2008. Forming mechanism of secondary pores in deep buried reservoirs of Junggar Basin. *Acta Sedimentologica Sinica*, 26(3): 469–478.

Chen, Y.Q. & Yu.X.H & Zhou, X.G.2004. Research on diagenetic evolution succession and occurrence of secondary porosity of lower Tertiary in different structural belt of Dongying Depression. *Natural Gas Geoscience*, 15(1):68–74.

Energy, Environment and Green Building Materials – Sheng (ed.)
© 2015 Taylor & Francis Group, London, ISBN 978-1-138-02718-3

Thermal and mechanics research on a 15μm umbrella-like structure microbolometer

L. Zhang, C. Chen & T. Wang

State Key Laboratory of Electronic Thin Films and Integrated Devices, University of Electronic Science and Technology, China

ABSTRACT: Thermal parameters of a small-sized microbolometer(≤15μm) has a direct impact on large-scale uncooled Infrared Focal Plane Arrays (IRFPAs). To obtain the thermal parameters of the microbolometer, finite element analysis is used in this paper. An accurately three-dimensional structure of a 15μm VO_X microbolometer with an umbrella-like structure is modeled. Dynamic thermal, static mechanics, finite element analysis, and theoretical calculations are studied. Thermal capacity, thermal conductance, and time constant of the model are obtained, which presented good thermal properties. The deformation and residual stress are very small, which means good mechanical properties of the double-layer structure. Analysis results in this paper offered a constructive reference for designers.

KEYWORDS: Thermal conductance; Uncooled microbolometer; Thermal time constant; FEA; Mechanics properties

1 INTRODUCTION

Uncooled microbolometer is widely used in civil and military applications due to its room temperature operation capability, small size, low power dissipation, light weight, and superior reliability [1,2,3]. The detection principle of microbolometer pixel is based on the absorption of infrared radiation, which changes the temperature of the IR-sensitive membrane. The change of electrical resistance due to this temperature change can be read out by an integrated circuit of microbolometer [2].

The thermal parameters of microbolometer play an important role in its performance. The thermal conductance (G) represents how well the IR-sensitive membrane is thermally isolated from the substrate. Smaller thermal conductance means better thermal isolation results in higher sensing speed. However, smaller thermal conductance reduces frame rate of microbolometer. Therefore, careful consideration of the design and optimization of thermal conductance is required without compromising speed and frame rate.

With the increasing of the number of pixels in IRFPAs, the pixel size of microbolometer is decreasing. To obtain a high-fill factor without reducing thermal isolation, an umbrella-like structure microbolometer that has a double layer is introduced. So good mechanics properties of the complex umbrella-like structure can prevent serious deformation.

In this paper, an accurate three-dimensional model of a 15μm umbrella-like microbolometer is found.

Dynamic thermal analysis and mechanics of the model are simulated by using finite element model (FEM), and then thermal parameters and mechanics properties can be obtained.

2 THERMAL CONDUCTANCE DESIGN

The responsivity is an important parameter of IR detectors; it is defined as the output voltage divided by the incident radiant power and it can be expressed as Eq(1):

$$S_v = \frac{\partial V_o}{\partial P} = \frac{\alpha \cdot \eta \cdot V_{fid} \cdot T_{int}}{\sqrt{G^2 + \omega^2 C^2} \cdot R_b \cdot C_{int}} \quad (1)$$

where α, η, I_b, R_b, G_{eff}, ω, T_{int}, and C_{int} are reference temperature coefficient of resistance, absorption coefficient, bias current, resistance of the microbolometer plate, thermal conductance, modulation angular frequency, integrate time, and integrate capacitor, respectively.

The thermal time constant, τ_{eff}, is an important parameter that is a measurement of the time required for the thermistor to respond to a change in the ambient temperature. The technical definition of thermal time constant is the time required for a thermistor to change 63.2% of the total difference between its initial and final body temperature when subjected to a step-function change in temperature, under zero power conditions [1]. It is also given by [1]:

$$\tau = \frac{C}{G} \qquad (2)$$

The thermal conductance is usually estimated by solving Fourier's first law of conduction and is approximately given by Eq. (4) [6]:

$$G \approx 2\sum_{i=1}^{n} \frac{k_i \cdot A_i}{l_i} \qquad (3)$$

where n is number of parallel layers forming arms; k_i, A_i, and l_i are the thermal conductivity, cross-sectional area, and length of thermal conductance of the layer of region at the arms, respectively.

The thermal capacitance of microbolometer can be expressed as [1]:

$$C = \sum_{i=1}^{n} v_i c_i \rho_i \qquad (4)$$

where n is number of parallel layers forming microbolometer; v_i, c_i, and ρ_i are the volume, heat capacity, and density of layer of the region of the microbolometer, respectively.

3 MODELING AND SIMULATION

3.1 Modeling

As we can see, thermal conductance and thermal capacitance are related to the geometry of microbolometer. By using Eq.3 to calculate thermal conductance, only the arms' contribution is taken into consideration; whereas in Eq.4, the calculation of thermal capacitance can be more precise than that of thermal conductance considering the volume of microbolometer is easier to get. So the calculated result of thermal capacitance is more accurate than thermal conductance of microbolometer.

By using dynamic thermal simulation based on the accurate three-dimensional modeling of microbolometer, the thermal time constant closely related to the geometry can be evaluated. And more precisely, the result of the thermal conductance can be achieved.

In this paper, a finite element model (FEM) is used for solving the dynamic thermal simulation as well as for building accurate three-dimensional modeling of microbolometer by using the MEMS analysis software Intellisuite. Mask and three-dimensional modeling of microbolometer is schematically shown in

Figure. 1(a), and three-dimensional model is meshed as shown in Figure. 1(b).

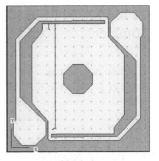

(a) Mask

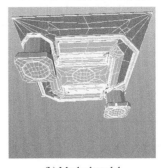

(b) Meshed model

Figure 1. Model of 15μm umbrella-like structure microbolometer.

The size of the model that is used for thermal capacitance value calculation can be obtained in the software. The material properties of every layer are defined according to the actual process parameters of materials and are listed in Table 1.

Table 1. Material properties of microbolometer.

materials	Thermal conductivity W/m·K	heat capacitivity J/cc·K	Density g/cm3
Si3N4	2	3.33	2.3
NiCr	30	3.92	8.5
VOx	5	5.00	4.34

Two thermal impulses are shown in Figure. 2, in which infrared radiation flux is defined as $2.4\ nw/\mu m^2$, and it is load on the surface of three-dimensional modeling of microbolomete for modeling a two-frame image. The initial temperature is defined as 30°C. According to the technical definition of thermal time constant, time constant can be easily and accurately read in temperature response curve. Thermal capacitance can be also accurately calculated from Eq.4.

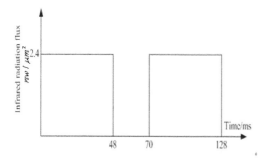

Figure 2.   Thermal impulse.

The bottom of microbolometer is fixed during static mechanics analysis. The stress of each thin film is defined, which is shown in Table 2.

Table 2.   Mechanics properties of microbolometer.

film	Young's Modulus/ Gpa	Poisson's ratio	Stress Type	Stress Value/ Mpa
Si3N4	300	0.26	Compressive	200
NiCr	200	0.312	Tensile	250
VOx	80	0.25	Tensile	70

3.2   Simulation results

The rule of thumb for the adjustment of the time constant of the detector is [5],

$$\tau \leq \frac{1}{2 \times frame\ rate} \qquad (5)$$

where the frame rate is commonly given as 60 fps for high-speed applications [6]. Therefore, the time constant should not exceed 8.3msec. According to the technical definition of thermal time constant, it can be easily and exactly read out; corresponding time constant is 7.81msec, which is shown in Fig. 3(b).

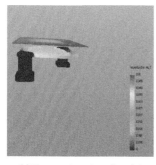

(a) Temperature rise of model

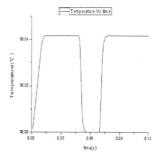

(b) Dynamic thermal curve of the model

Figure 3.   Thermal response and dynamic thermal curve of the model.

Thermal capacitance value is calculated from Eq.4 as 1.638e-9J/K. And then, thermal conductance is 2.10e-7W/K, which is comprehensive and provides an accurate description of thermal isolation performance of this model.

Max residual stress of the leg, their layer and top absorption layer are about 20–80 Mpa shown in Fig. 3 (a) which results in max deformation being 0.0966μm, as shown in Figure. 3 (b).Max residual stress regions of the total umbrella-like structure are two electrodes, which are shown in Fig. 3 (a) as 14300.7Mpa. Two electrodes are connected with

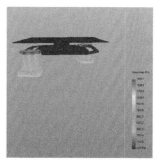

(a) residual stress

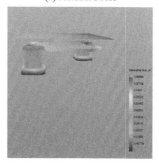

(b) deformation

Figure 4.   Static mechanics analysis.

substrate and deformation could be ignored. With small residual stress and deformation on the leg, their layer and top absorption layer indicate good mechanical properties of the model.

4 CONCLUSION

In this paper, dynamic thermal, static mechanics and finite element analysis of microbolometer is presented. Through the establishment of precise three-dimensional modeling of microbolometer, as well as defining the actual material parameters, dynamic thermal simulation is analyzed. Static mechanics analysis shows that the deformation and residual stress are very small, which indicates good mechanical properties of the double-layer structure. The result of dynamic thermal, static mechanics analysis is closer to the actual device, which is constructive for the design and fabrication of microbolometer.

REFERENCES

[1] R.T.R. Kumar, B. Karunagaran, D. Mangalaraj, et al, Determination of thermal parameters of vanadium oxide uncooled microbolometer infrared detector, Int. J. Infrared Millimeter. 24 (3), 2003, pp:327–334.
[2] Hong, C. W., Xin, J. Y. & Guang, H., IR microbolometer with self-supporting structure operating at room temperature. Infrared Physics& Technology, 45, pp. 53–57, 2004.
[3] Jerominek, H., Francis, P. & Nicholas., Micro machined uncooled VOx based, IR bolometer array, SPIE, Vol. 2746. pp. 60–71.
[4] Effa Dawit: Electro-Thermal Mechanical Modeling of Microbolometer for Reliability Analysis (Ms.,Waterloo,Canda 2010). p.83.
[5] C. Li, G. Skidmore, C. Howard, E. Clarke, and C. J. Han, "Advancement in 17 Micron Pixel Pitch Uncooled Focal Plane Arrays," Proc. of SPIE, vol 7298 (2009).
[6] W. Radford, D. Murphy, M. Ray, S. Propst, A. ennedy, and K. Soch,"320x240 microbolometer focal plane array for uncooled applications".

Energy, Environment and Green Building Materials – Sheng (ed.)
© *2015 Taylor & Francis Group, London, ISBN 978-1-138-02718-3*

Optimal control of anoxic selector based on activated sludge with different phosphorus removal ability

Z.X. Peng & Z.G. He
College of Water Conservancy & Environmental Engineering, Zhengzhou University, Zhengzhou, China

ABSTRACT: The reasons that anoxic selector fails in control filamentous bulking were investigated based on different inoculated sludge. Sequencing Batch Reactors (SBR) were operated as anoxic/aerobic pattern under different COD: NO_x^--N ratios. Sludge settleability, phosphorus, and COD removal performance were analyzed. The results show that phosphorus removal ability and carbon substrate utilization pathway are decisive for anoxic selector. When phosphorus removal deteriorates, anoxic selector will fail in improving settleability. Even when phosphorus removal is excellent, whether sludge bulking can be suppressed depends on the COD: NO_x^--N ratio. When it is smaller than 33.00, residual $NO12^-$-N by incomplete denitrification is harmful for anoxic selector operation. When it is in the range of 33.00~66.00, filamentous bulking can be suppressed. When it is bigger than 66.00, the excess COD will induce filamentous bacteria to proliferate in the next aerobic period.

KEYWORDS: anoxic selector; activated sludge; SBR; filamentous bacteria; sludge bulking

1 INTRODUCTION

The activated sludge process is widely applied in wastewater treatment plant, but it is usually troubled by sludge bulking. Most sludge bulking is induced by filamentous bacteria over-proliferation compared with floc-forming bacteria (Paolo et al, 2000). Till date, many methods such as adding oxidants or coagulants have been established to control filamentous bulking, but it is prone to recur when agents are finished. Otherwise, the cost of agents is expensive (Alejandro et al, 2003. Agridiotis et al, 2007). Recently, selector has become the promoted strategy to enhance sludge settleability (aerobic selector, anoxic selector and anaerobic selector) (Wu et al, 2003). As the discharge standard for nutrients, which are the major cause of eutrophication, is stricter than ever before (Auvray et al, 2006), the anoxic selector own nutrient removal ability is becoming more popular. Although many efforts have been made to optimize anoxic selector operation, there are still reports showing failure events in controlling sludge bulking (Eikelboom, 1998.).

For anoxic selector, the readily biodegradable COD (RBCOD) to NO_x^--N ratio is a main design parameter (Martins et al, 2004). When it is too big, the excess COD will promote filamentous bacteria proliferate in the aerobic stage (Kruit et al, 2002). Otherwise, denitrification intermediates such as nitric oxide (NO) will inhibit floc-forming bacteria (Casey et al, 1999). Although the second condition cannot be accepted

as a general law (Andreasen & Nielsen, 2000), it, indeed, existed in some low F/M biological nutrition removal (BNR) systems. In addition, low dissolved oxygen (DO), soluble little molecular organics, low temperature, low pH, long SRT, and so on also can induce filamentous sludge bulking (Vaiopoulou et al, 2007; Tsang et al, 2006), As the earlier introductions, this study investigated the operation effects of anoxic selector under different COD: NOx--N ratios. It is hoped to supply some experimental guidance for practical application.

2 MATERIALS AND METHODS

2.1 Operation patterns

The study was performed using 5 SBRs. Each SBR consists of lucite with a 12-L working volume, a 200-mm diameter, and a 700-mm working height. The 8 h cycle consisted of 0.5h feeding (anoxic mixing), 2.0h aerobic reaction, and 5.5h settling/decanting/idling. The influent NOx--N concentrations are shown in Table.1 and Table. 2. During feeding period, 3L wastewater was added by peristaltic pump. At the end of aerobic period, 100 mL waste sludge was withdrawn to maintain the SRT at 12.5 d (assuming no solids left the system through the effluent). MLSS was controlled in the range of 1500–3000 mg·L^{-1}. Temperature was maintained at 22~24°C by thermostatic heaters. DO, pH, and ORP probes were placed to monitor the parameters.

2.2 Wastewater and sludge

Synthetic wastewater with the following composition was used as the feeding solution: 663.8 mg of $CH_3COONa \cdot 3H_2O$ (326.9 mg·L^{-1} as COD), 166.9 mg NH4Cl (43.7 mg·L^{-1} as NH$_4^+$-N), 37.6 mg of KH_2PO_4 (8.6 mg·L^{-1} as PO$_4^{3-}$–P), (0~150mg) NaNO$_2$ (0~30mg/L as NO$_2^-$–N), (0~870mg) KNO$_3$ (0~120 mg·L-1 as NO3–N), 375 mg of NaHCO$_3$ (400 mg·L-1 as CaCO3 alkalinity), 40 mg of $CaCl_2 \cdot 2H_2O$, 80 mg of $MgSO_4$, and 0.3 mL of nutrient solution per liter (Smolders et al, 1994). The inoculation sludge came from two sources. One part was from an AO reactor (weak phosphorus removal). It was assigned into SBR1, SBR2, SBR3, and SBR4 for two experimental phases (shown in Table.1). The other part was from an AAO reactor (excellent phosphorus removal). It was assigned into the SBR5 (shown in Table 2). Both inoculation sludge had excessive filamentous bacteria. The SVI were around 360 mL·g-1 (MLSS 2600 mg·L^{-1}) and 500 mL·g^{-1} (MLSS 1800 mg·L^{-1}), respectively.

Table 1. Experimental stages of inoculated sludge with weak phosphorus removal ability.

Reactor	NO$_2^-$-N mg·L^{-1}	NO$_3^-$-N mg·L^{-1}	COD/ NO$_x$-N	cycle number
SBR1	0	0	-	0~87
SBR2	0	10	33.00	0~87
SBR3	0	20	16.50	0~87
SBR4	0	40	8.25	0~87
SBR1	0	60	5.50	0~87
SBR2	0	80	4.13	0~87
SBR3	0	100	3.30	0~87
SBR4	0	120	2.75	0~87

Table 2. Experimental stages of inoculated sludge with excellent phosphorus removal ability.

Reactor	NO$_2^-$-N mg·L^{-1}	NO$_3^-$-N mg·L^{-1}	COD/ NO$_x$-N	cycle number
SBR5	0	0	-	0~78
	0	20	16.50	79~99
	0	10	33.00	100~135
	0	5	66.00	136~168
	5	0	110.00	169~192
	10	0	55.00	193~222
	20	0	27.50	223~264
	30	0	18.33	265~279

2.3 Analytical methods

Samples were taken 3–4 times per week. The samples for analysis were immediately filtered using a 0.45-mm filter membrane to separate the bacterial cells from the liquid. COD, MLSS, MLVSS, SV, SVI, NH$_4^+$-N, NO$_2^-$-N, NO$_3^-$-N, and PO$_4^{3-}$–P were measured according to standard methods (APHA, 1992). Microbial morphology was observed by using an optical microscope (OLYMPUSBX51). DO, pH, and ORP were monitored by a WTW Multi 340i DO meter.

3 RESULTS AND DISCUSSION

3.1 Effects of influent COD: NO$_x$-N ratio on sludge settleability under weak phosphorus removal

The COD: NO$_x$-N ratio is important for anoxic selector operation effect (Martins et al, 2004). Sodium acetate, a small molecular substrate, was the only carbon source. According to Activated Sludge Model (ASM) NO1, to reduce 1g NO$_3^-$–N to N$_2$ 2.86 g COD is needed. Considering the heterotrophic bacteria assimilation effect, 2.86/(1-Y$_H$) g COD is used, where Y$_H$ is the yielding coefficient. According to ASM1, it can be taken as 0.67 (Henze et al, 1987). Then, 8.67g COD will be consumed to denitrify 1g NO3–N to N$_2$. Otherwise, some COD can be converted to internal storage products such as poly-β-hydroxybutyrate (PHB) (Majone et al, 2007), so it is difficult to determine the precise COD: NO3–N ratio for complete denitrification. The anoxic selector operation effects on excessive filamentous bulking are shown in Fig. 1.

Each day, an SBR operated 3 cycles; about 37 cycles were run in an SRT (SRT=12.5d). This can be regarded as a standard to judge whether the selector can take effect. When COD: NO$_x$-N ratios were infinity, 33.00, 16.50, and 8.25, sludge settleability in all selectors showed only a slight improvement in 87 cycles. During this phase, carbon source was excessive for denitrification, and there was no NO$_3^-$-N left at the end of anoxic period. The inhibition effects on filamentous bulking were limited.

During the second phase, the carbon source was insufficient; sludge settleability not only showed any improvement trend but also deteriorated all along. Its inhibition effects were even worse than the first phase. When COD was limited, denitrification could not proceed completely. As a result, obvious NO$_x$-N was left, especially the NO$_2^-$-N. According to the Nitric oxide (NO) hypothesis (Casey et al, 1999), the denitrification intermediates accumulated in the floc-forming bacteria such as NO will inhibit its proliferation under aerobic period. But the inhibition effect did not exist for filamentous bacteria. The experiment was consistent with this hypothesis.

The phosphorus removal was unfavorable (shown in Fig.2). Theoretically speaking, to release 1g PO_4^{3-}—P, only 0.67~0.84 g COD is needed (Peng et al, 2010). In the first phase, although carbon source was in excess, because the SRT was too big, phosphorus removal performance just slightly improved in SBR1, SBR2, and SBR3. In the second phase, no carbon source was left for PO_4^{3-}—P release process, and phosphorus removal ability almost disappeared.

3.2 Effects of influent COD: NOx–N ratio on sludge settleability with excellent phosphorus removal

Based on the earlier analysis, bigger COD: NO_x^--N ratio can contribute to inhibit filamentous bulking. In SBR5, all the ratios were bigger than 8.67.

It was run for 279 cycles under 8 different COD: NO_x^--N ratios, and operation effects were shown in Fig. 3. Settleability obviously varied, and two improvement processes and two deterioration processes appeared. In the first 78 cycles, SVI fluctuated around 550 mL·g[-1]. When C:N was 16.50, because denitrification competed with phosphorus release for carbon source, phosphorus removal was seriously suppressed (shown in Fig.4). As a result, sludge settleability showed almost no change. From the 100th cycle to the 135th cycle, with an improvement in phosphorus removal, SVI showed a slight decrease. When C:N was 66.00, sludge settleability obviously improved. SVI dropped from 510 mL·g[-1] to 333 mL·g[-1] sharply. When C:N increased to 110.00, SVI gradually stopped decreasing. Instead, sludge settleability deteriorated again after the 180th cycle. When C:N increased to 55.00, SVI increase trend gradually stopped. After the 210th cycle, it decreased again. When C:N was 27.50, sludge settleability obviously improved.

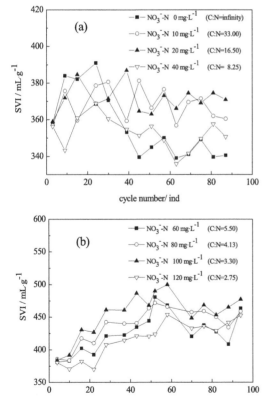

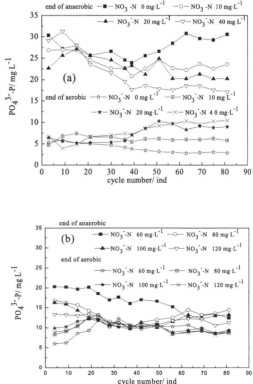

Figure 1.   Sludge settleability variations of the inoculated sludge with weak phosphorus removal ability (a) COD: NO_3^-–N = infinity, 33.00,16.50, 8.25; (b) COD: NO3$^-$–N = 5.50, 4.13,3.30,2.7.

Figure 2.   Phosphorus removal performance of the inoculated sludge with weak phosphorus removal ability (a) COD: NO_x^-–N = infinity, 33.00,16.50,8.25; (b) COD: NO_x^-–N = 5.50,4.13,3.30,2.75.

But with phosphorus removal deteriorated, SVI again increased. When C:N was 18.33, both phosphorus removal and sludge settleability deteriorated.

Phosphorus removal had a very close relationship with sludge settleability. SVI decreased only when the sludge had excellent phosphorus removal ability. The reason is that polyphosphate-accumulating organisms (PAOs) are rod and cocci, not filamentous (Seviour et al, 2003). In addition, the activated sludge buoyancy density is an important factor to determine settling rate. Generally speaking, the average biomass density varied in the range of 1.022~1.056 g·mL^{-1}. But the density of storage products such as polyphosphate, glycogen, and PHB is estimated to be as high as 1.23, 1.25, and 1.15, respectively (Schuler & Jang, 2007). Polyphosphate in PAO can reach till 15% of its cell dry weight. So activated sludge is more prone to settle when PAO are in excess. However, only excellent phosphorus removal cannot ensure a well sludge settleability. When C:N is 110.00, although phosphorus removal is excellent, the residual COD still restrains anoxic selector.

Although all C:N ratios were bigger than 8.67, denitrification rate was restricted by many factors such as biomass, temperature, pH, and so on. So NO_x^-–N cannot be denitrified completely under each C:N condition. When NO_2^-–N is accumulated, sludge settleability will show deterioration trend. For example, 27.50 and 33.00 were close; the NO_2^-–N accumulation deteriorated both phosphorus removal and sludge settleability when C:N was 27.50. However, when C:N was 33.00, NO_2^-–N was negligible. This experiment investigated the impacts of COD: NO_x^-–N ratio on anoxic selector operation effect. However, microscopic parameters such as species population and internal carbon source were deficient. In addition, only sodium acetate was used as carbon source. Whether the conclusions can be applied in practical use needs further investigation.

4 CONCLUSIONS

1 The phosphorus removal ability and carbon source utilization pathway are two determining factors for anoxic selector operation effect.
2 Along with the activated sludge with poor phosphorus removal, anoxic selector cannot take effective inhibition on filamentous bulking. Even if the phosphorus removal were excellent, whether anoxic selector can take effect is determined by the RBCOD: NO_x-N ratio. The recommended ratio is in the range of 33.00~66.00.
3 NO_2^--N left by incomplete denitrification is fatal for anoxic selector. It usually occurs with phosphorus removal deterioration. As a result, filamentous bacteria cannot be suppressed.
4 The carbon source utilization pathway depends on the RBCOD: NO_3^--N ratio. To avoid excess carbon substrate induce filamentous bacteria proliferate, the ratio should not be bigger than 66.00. Otherwise, to avoid NO_2^--N accumulation, it should not be smaller than 33.00.

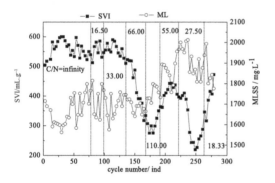

Figure 3. Sludge settleability variations of the inoculated sludge with excellent phosphorus removal ability.

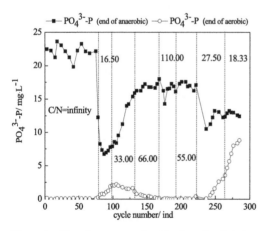

Figure 4. Phosphorus removal of the inoculated sludge with excellent phosphorus removal ability.

ACKNOWLEDGMENT

This research was supported by national key science and technology special projects of water pollution control and management during the twelfth five-year plan period (2012ZX07204-001-02).

REFERENCES

Agridiotis V, Forster C F, Carliell-Marquet C. 2007. Addition of Al and Fe salts during treatment of paper mill effluents to improve activated sludge settlement characteristics. Bioresource Technology, 98(15): 2926–2934.

Andreasen K, Nielsen P. 2000. Growth of Microthrix parvicella in nutrient removal activated sludge plants: studies of in situ physiology. Water Research, 34 (5): 1559–1569.

Alejandro C, Edgardo M C, Leda G, et al. 2003. Modeling of chlorine effect on floc forming and filamentous micro-organisms of activated sludges. Water Research, 37(9): 2097–2105.

APHA (American Public Health Association). 1992. Standard Methods for the Examination of Water and Wastewater (18th ed). Washington DC, USA.

Auvray F, van Hullebusch E D, Deluchat V, et al. 2006. Laboratory investigation of the phosphorus removal (SRP and TP) from eutrophic lake water treated with aluminium. Water Research, 40(14): 2713–2719.

Casey T G, Wentzel M C, Ekama G A. 1999. Filamentous organism bulking in nutrient removal activated sludge systems. Paper 11: a biochemical/microbiological model for proliferation of anoxic-aerobic (AA) filamentous organisms. Water S A, 25 (4): 443–451.

Eikelboom D H, Andreadakis A, Andreasen K. 1998. Survey of filamentous populations in nutrient removal plants in four european countries. Water Science Technology, 37, (4–5): 281–289.

Henze M, Grady C P L Jr, Gujer W, et al. 1987. "Activated sludge model No.1". (IAWPRC scientific and tech report No.1). London: IAWPRC.

Kruit J, HulsbeekJ, Visser A. 2002. Bulking sludge solved?!. Water Science Technology, 46 (1–2): 457–464.

Majone M, Beccari M, Dionisi D, et al. 2007. Effect of periodic feeding on substrate uptake and storage rates by a pure culture of Thiothrix (CT3 strain). Water research, 41(1): 177–187.

Martins A M P, Pagilla K, Heijnen J J, et al. 2004. Filamentous bulking sludge—a critical review. Water Research, 38 (4): 793–817.

Paolo M, Donatella D, Giorgia G. 2000. Survey of filamentous microorganisms from bulking and foaming activated-sludge plants in Italy. Water Research, 34 (6): 1767–1772.

Peng Z X, Peng Y Z, Gui L J, et al. 2010. Competition for single carbon source between denitrification and phosphorus release in sludge under anoxic condition, Biotechnology and Bioengineering, 18 (3): 472–477.

Schuler A J, Jang H. 2007. Causes of variable biomass density and its effects on settleability in full-scale biological wastewater treatment systems. Environmental Science and Technology, 41(5): 1675–1681.

Seviour R T, Mino T, Onuki M. 2003. The microbiology of biological phosphorus removal in activated sludge systems. FEMS Microbiology Reviews, 27(1): 99–127.

Smolders G J F, van de M J, van Loosdrecht M C M, et al. 1994. Stoichiometric model of the aerobic metabolism of the biological phosphorus removal process. Biotechnology Bioengineering, 44(7): 837–848.

Tsang Y F, Chua H, Sin S N, et al. 2006. A novel technology for bulking control in biological wastewater treatment plant for pulp and paper making industry. Biochemical Engineering Journal, 32(3): 127–134.

Vaiopoulou E, Melidis P, Aivasidis A. 2007. Growth of filamentous bacteria in an enhanced biological phosphorus removal system. Desalination, 213(1–3): 288–296.

Wu F S, Peng Y Z, Wang W B. 2003. P and N removals and Bio-selector. Water Wastewater engineering, 29 (12): 32–34.

Energy, Environment and Green Building Materials – Sheng (ed.)
© *2015 Taylor & Francis Group, London, ISBN 978-1-138-02718-3*

Fuzzy evaluation for optimal selective disassembly in recycling of end of life product

Z.Q. Zhou
Postdoctoral Workstation of Jiangsu Huahong Technology Stock Co. Ltd., Wuxi, China
School of Mechanical Engineering, Changshu Institute of Technology, Changshu, China

P.L. Hu & Y.F. Gong
Jiangsu Huahong Technology Stock Co. Ltd., Wuxi, China

G.H. Dai
School of Mechanical Engineering, Changshu Institute of Technology, Changshu, China

ABSTRACT: In order to protect the natural environment, the end-of-life product should be treated rationally, because the complete disassembly will lack practical feasibility when confronted with complex products such as from the disassembly of an end-of-life motor vehicle. Hence, selective disassembly should be applied in this type of end-of-life product. In this paper, a liaison diagram and a fuzzy evaluation-based approach for selective disassembly is presented. After the liaison diagram is determined, the graphic of possible disassembly sequence can be generated. Then, a fuzzy evaluation approach is presented instead of the traditional quantitative calculation in order to find the optimal disassembly path among the several possible disassembly sequences.

KEYWORDS: Production cycle; Disassembly; Reuse; Re-manufacturing

1 INTRODUCTION

Selective disassembly is an effective approach for the purpose of maintaining and recycling. For most of the end-of-life products, complete disassembly is not necessary, because with deepening of the degree of disassembly, the costs will be sharply increased and the economic benefit of the disassembly is almost none.

In the current research of selective disassembly, on the one hand, the strength of constraint between parts is neglected; on the other hand, the cost of disassembly is evaluated by the time cost in operation or numbers of direction changing of tools occurring within the operation. However, in the recycling of end-of-life products, it is very difficult to acquire the cost of disassembly by this means. And the strength of constraint as well should be taken into account in order to find an optimal disassembly sequence.

This paper will focus on the recycling oriented disassembly, including the target selection and disassembly sequence generation method. The assembly structure of end-of-life product is described by liaison diagram, which can be acquired from the CAD model of the product directly. The recovery information, such as material, mass, and value, is attached to each part and subassembly. Note that most of the information is given by the fuzzy quantities, including price, quality, and costs of disassembly. During the stage of disassembly sequence generation, the optimized disassembly sequence should be determined by choosing from the several possible disassembly paths. In this paper, a fuzzy evaluation approach is used to distinguish the optimal sequence and the non-optimal sequence. The advantage of this approach is that the works of inputting information can be greatly simplified, and they are more feasible for practical application.

2 FUZZY EVALUATION FOR OPTIMAL DISASSEMBLY SEQUENCE

2.1 *Fuzzy evaluation for disassembly operation*

For the sake of evaluating the cost of disassembly of end-of-life products, quantitative performance is necessary. However, precise or crisp data of performance cannot be acquired easily. Usually, the precise value of the performance is unavailable according to some uncertain factor. For example, the time cost of releasing a screw will be increased if it is rusted. So, fuzzy number is appropriate for representing the grade of connection between the parts. In the theory of fuzzy set, the triangular fuzzy number is most commonly used because of its intuitive and high efficiency for computation. A triangular fuzzy number can be defined as a triplet (a,b,c), as shown in Fig. 1, in which membership function is defined in Eq.(1).

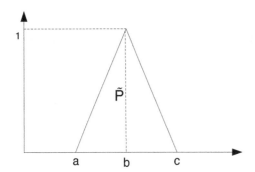

Figure 1.   Triangle fuzzy numbers P.

$$P(x) = \begin{cases} 0, & x \leq a \\ \dfrac{x-a}{b-a}, & a \leq x \leq b \\ \dfrac{c-x}{c-b}, & b \leq x \leq c \\ 0, & x \geq b \end{cases} \quad (1)$$

In the process of fuzzy evaluation, the operation of fuzzy number is needed, which obeys the following rules:

Fuzzy number addition

$$\tilde{P}_1 \oplus \tilde{P}_2 = \left(a_1 + a_2,\ b_1 + b_2,\ c_1 + c_2 \right) \quad (2)$$

Fuzzy number is multiplied with a coefficient of real number

$$\lambda \otimes \tilde{P} = (\lambda a_1,\ \lambda b_1,\ \lambda c_1) \quad (3)$$

In this study, the disassembly operations are mainly evaluated according to the disassembly time and the tools required; that is, the difficulty of operation, which is used to reflect the quantities of disassembly cost. Based on the fuzzy theory, fuzzy number should be given with linguistic variables. We propose five levels of scale that are defined with linguistic variables such as very low, low, medium, high, and very high. Linguistic variable that are very low are used to describe the connection that two parts that are connected with each other have only by two surfaces fit together. That is, one of the parts can be removed directly and easily. Linguistic variables are used to describe that some of the parts are connected by the snap-in method. This type of connection is most commonly used in electronic products. The linguistic variable medium is used to describe the connection by using fasteners. That is, if the two parts are connected by screws or bolts, the special tools are needed to loose the connection and remove

the parts. The linguistic variable is highly used to describe the strong connection such as bearing and shaft. A hydraulic tool will be needed to disassembly the part. The linguistic variable is very highly used to describe the connection with high strength, which is more than the linguistic variable of "high" and that is a rare instance.

The distribution of possible values of a linguistic variables is usually given with membership function. Several methods are used to compose membership function for linguistic variables, such as statistic, exemplification, and experience of experts [17]. In this study, we use the method of exemplification and experience of experts to compose the membership function. For the assembly illustrated in Fig. 1, the membership function of the earlier mentioned linguistic variable is shown in Fig. 2. The disassembly cost is represented with a normalized real data in [0,1] on the horizontal axis, which should be adjusted according to the characteristic of end-of-life products. According to the fuzzy set on the horizontal axis, the function expression of linguistic variable can be acquired with Eq.(1).

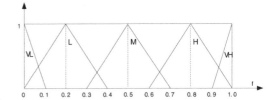

Figure 2.   Membership function of grade of connection.

2.2   Copying old text into new file

The objective of fuzzy evaluation for liaison mainly is to choose the optimal disassembly sequence by comparing the cost of the operation. The cost of disassembly is represented by Eq. (4):

$$C = \sum \lambda_i P_i \quad (4)$$

where C is the cost of disassembly for processing current liaison, and Pi represents the cost of disassembly with liaison of index of i in fuzzy number. Coefficient λi represents the number of the same connection. For example, if there are six screws between part A and part B, λi is set with 6. Besides, the operation of Eq.(3) is fuzzy number addition and multiplication introduced in Eqs. (2) and (3).

Afterward, costs of several possible disassembly sequences are calculated with Eq.(4). The ranking of this fuzzy numbers is needed. Several methods have been proposed for ranking the fuzzy numbers; even though the method center of gravity is the most commonly used, it is only suitable for symmetric fuzzy numbers. However, after operation, the fuzzy number usually becomes unsymmetrical, so we use another method to rank the fuzzy number. First, all the fuzzy numbers are defuzzified into a real number i by Eq.(5). In this equation, coefficient η is the subjective state of decision-makers. Here, we set η with 0.4, which represents the optimistic state where l, m, and n are the elements of triplet defined in Eq.(1).

$$i = \eta * \frac{l+m}{2} + (1-\eta) * \frac{m+u}{2} \qquad (5)$$

With the earlier mentioned method, the minimum number associated with the liaison in the same hierarchy can be found. According to the possible disassembly path and fuzzy evaluation, the optimal disassembly sequence can be determined.

3 CASE STUDY FOR SIDE MIRROR

3.1 *Structure of side mirror*

To test the earlier mentioned method, a case study is carried out using end-of-life products. Generally, end-of-life vehicle is a typical end-of-life product that needs to be recycled. The parts and components are made from many different types of materials, and all the parts and components have several different recycling usage, remanufacturing, reuse, material, and so on. Here, we just use a side mirror as an instance to illustrate the fuzzy evaluation approach-based selective disassembly sequence generation.

The components relationship of the side mirror is shown in Fig. 3. It is driven by two small DC motors

inside the body. There is a worm gear on the top of each motor. And the worm connected with the internal base can be driven by the worm gear. Then, two joint pins are installed in the hole of each worm with thread. So, by rotating the worm, the joint pin can protrude or withdraw from the internal base. Since the mirror holder is connected with the internal base by a ball joint, two joint pins are also connected with the mirror by a ball joint. Hence, by controlling the rotational angle of the motor, the direction of the mirror holder can be changed. Thus, the perspective of mirror is changed.

3.2 *Disassembly of side mirror*

In this study, a rule is defined to simplify the structure of end-of-life products. A group of parts that have the same assembly relationship can be simplified as a single part. For example, four screws between the mirror and the internal base are transformed as single vertices G in Fig. 4. Similarly, two motors are simplified as single vertices C, two worms are simplified as single vertices D, and two joint pins are simplified as single vertices E. Since the mirror is made of the glass that has little value for recycling, and it can be separated from the mirror base by shredding and sorting, so mirrors are simplified as a single vertice I. Based on the earlier mentioned method, the liaison diagram of the side mirror is generated as shown in Fig. 10. The relationship of the obstacle of the side mirror is given in Table 1.

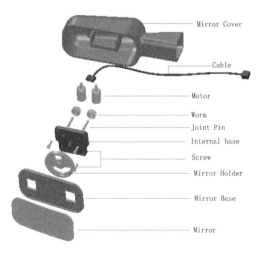

Figure 3. Structure of side mirror.

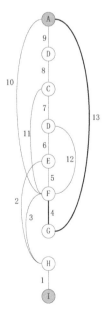

Figure 4.   Liaison diagram of side mirror.

Table 1. Parts relationship with obstacle of side mirror.

No.	Symbol	Name of part	Number	Obstacle
1	A	Mirror cover	1	
2	B	Cable	1	F or A
3	C	Motor	2	F or A
4	D	Worm gear	2	F or A
5	E	Joint pin	2	F or A
6	F	Internal base	1	I or A
7	G	Screw	4	I
8	H	Mirror holder	1	I
9	I	Mirror base and mirror	1	

The grade of connection of side mirror is given by fuzzy number in Table 2.

Table 2. Linguistic variable of side mirror.

Liaison	Grade	Fuzzy number
1	Low	0, 0.2, 0.4
2	Low	0, 0.2, 0.4
3	Low	0, 0.2, 0.4
4	Medium	0.3, 0.5, 0.7
5	Very low	0, 0, 0.1
6	Medium	0.3, 0.5, 0.7
7	Very low	0, 0, 0.1
8	Low	0, 0.2, 0.4
9	Very low	0, 0, 0.1
10	Very low	0, 0, 0.1
11	Low	0, 0.2, 0.4
12	Low	0, 0.2, 0.4
13	Medium	0.3, 0.5, 0.7

According to the earlier mentioned searching algorithm and fuzzy evaluation, a prototype is implemented with Delphi 7.0 and database with Access 2003. Then, the disassembly sequence is automatically generated by the prototype. There are four possible disassembly sequences, and all are listed in Table 3. It is obvious that the sequence of 1-(2-3)-4-13-10-11-8-7 is the optimal one.

4 CONCLUSIONS

The grades of constraint between components are described by the fuzzy number-based linguistic variable, which is more practical and more convenient than the method of time cost in performance. Then, a fuzzy evaluation approach is presented for choosing an optimal disassembly sequence among several possible sequences. The advantage of this approach is that the work of collecting the accurate information about the disassembly of end-of-life products can be neglected. Thus, it is convenient for practical application.

Table 3. Fuzzy evaluation of side mirror disassembly.

No	Disassembly sequence	Fuzzy number of evaluation	i
1	1-(2-3)-4-13-10-9-8-11-7	2.4, 5.4, 8.8	5.82
2	1-(2-3)-4-13-10-11-8-7	2.4, 5.0, 7.9	5.35
3	1-(2-3)-4-13-10-5-6-7-(11-8)	3.0, 6.4, 10.3	6.89
4	1-(2-3)-4-13-10-12-7-(11-8)	2.4, 5.8, 9.5	6.23

ACKNOWLEDGMENT

This work was financially supported by the 333 High-Level Talent Engineering of Jiangsu Province (BRA2012158).

REFERENCES

[1] Yuan mao Huang, YU-CHUNG LIAO. Disassembly processes with disassembly matrices and effects of operations[J]. *Assembly Automation*, 2009, 129(4): 348–357.
[2] Chulho chung, QINGJIN PENG. An integrated approach to selective-disassembly sequence planning[J]. *Robotics and Computer-Integrated Manufacturing*, 2005, 121(4):475–485.
[3] SHANA S.SMITH, WEI-HSIANG CHEN. Rule-based recursive selective disassembly sequence planning for green design[J]. *Advanced Engineering Informatics*, 2011, 125(1):77–87.
[4] Zhang xiufen, ZHANG SHUYOU. Product disassembly sequence planning based on particle swarm optimization algorithm[J], *Computer Integrated Manufacturing Systems*, 2009, 115(3):508–514. (in Chinese).
[5] WU HAO, ZUO HONGFU. Selective disassembly sequence planning based on improved genetic algorithm[J]. *ACTA Aeronautica et Astronautica Sinica*, 2009, 130(5):952–958. (in Chinese).
[6] J.F. WANG, J.H.LIU, S.Q.LI, et al. Intelligent selective disassembly using the ant colony algorithm[J], *Artificial Intelligence for Engineering Design, Analysis and Manufacturing*, 2003, 17(4):325–333.
[7] H.SRINIVASAN, R.GADH, A geometric algorithm for single selective disassembly using the wave propagation abstraction[J], *Computer Aided design*, 1998, 130(8):603–613.
[8] JIANJUN YI, BIN YU, LEI DU, et al. Research on the selectable disassembly strategy of mechanical parts based on the generalized CAD model[J], *International Journal of Advanced Manufacturing Technology*, 2008, 37(5):599–604.
[9] LU ZHONG, SUN YOUCHAO, et al. Wu Haiqiao,Disassembly sequence planning for maintenance based on metaheuristic method[J]. *Aircraft engineering and aerospace technology:an International Journal*, 2011, 83(3):138–145.

Energy, Environment and Green Building Materials – Sheng (ed.)
© *2015 Taylor & Francis Group, London, ISBN 978-1-138-02718-3*

Plate pillars formed by drawing

X. Zeng, Y. Hu, Q. Xia & Y.K. Cheng
Mechatronic Department, Yibin Vocational and Technical College, Yibin, Sichuan, China

Y.B. Duan
Push Die & Mold Co. Ltd, Yibin, Sichuan, China

ABSTRACT: In this paper, the blank shape of the plate pillar was determined by the result of finite element simulation and the enterprise actual production. Then according to the structural characteristic of the part to design the binder and addendum, the draw beads were designed basis on binder and addendum. According to the CAE software, dynaform is used to predict the defect. By means of practice and simulation, the solution was been proposed to correct the forming defect. Finally, the actual forming test proved that the forming project that was shown in this paper was feasible.

KEYWORDS: plate pillar, addendum, draw bead, finite element simulation

1 INTRODUCTION

The plate pillars were compounded with the C column into a whole structure covering parts; their main function is to protect passengers when accidents happen. So it is strictly required for it to possess stiffness and strength. Due to this product belonging to the non-standard pieces, no test standards were developed, but their strength and stiffness were measured. Higher stiffness and strength plate pillar means that when accidents happen the plastic deformation and damage would hardly appear. So more reliable safety protection is provided for passengers. By means of using high-strength material and large plastic deformation, an effective method is utilized to improve the performance of its security.

2 PRODUCT MORPHOLOGY ANALYSIS

Figure 1 is the plate pillar in application; it is a symmetrical arrangement, and it can be found that the morphology of the part exhibits a large difference. Based on the product structure analysis, the drawing process consists of the following compounds: drawing, punching, inclining, and punching. The forming parts with a large drawing height difference and the punching area were quite different, and the interference was likely to occur due to the hole being closer to the edge. The difficulty of punching was increased

by the hole on the side so that the parts belong to the complexity class and it is hard to form.

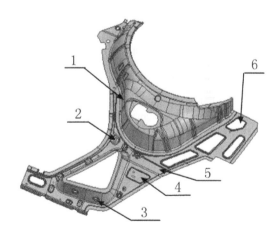

Figure 1. A plate pillar parts.

3 BINDER AND ADDENDUM DESIGN

The plate pillar, which is shown in Fig. 1, was difficult to form; in response to the forming needs and material flow, the binder and addendum were set. Figure 2 was the case of addendum and binder's distribution: The area that was marked A was addendum, and the area that was labeled B was binder.

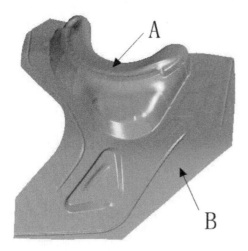

Figure 2.   Supplement of plate pillar parts.

4   DRAW BEAD DESIGN

During the drawing, the different areas needing material supplement were evidently distinct, due to the complex outline. Wrinkle must have been avoided in this drawing, so the draw bead was used. According to the outline of the product and the material flow control, the draw bead was designed in Fig. 3.

There were 4 draw beads that were marked in Fig. 3 but that did not completely package the product outline; they were designed by actual needs. The area close to the draw bead that was marked B needed more material supply, so this draw bead was designed relatively minimal. In the material supplement, the relatively small area in the draw beads was designed relatively large.

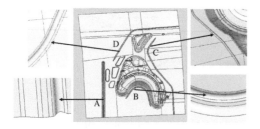

Figure 3.   Draw bead layout.

5   FORMING SIMULATION ANALYSIS

Figure 4 depicts the drawing process of the CAE model that was established in dynaform software.

Figure 4e depicts the blanking blank shape of the plate pillar in drawing. The blank shape had been optimized by the actual product in enterprises and the CAE simulation.

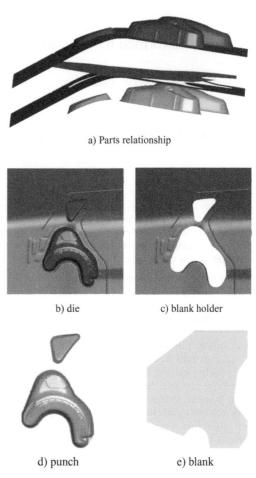

a) Parts relationship

b) die   c) blank holder

d) punch   e) blank

Figure 4.   CAE model of drawing process.

Figure 5 represented forming a limit diagram of plate pillar drawing simulation across dynaform software. The forming limit diagram was divided by different colors. Among them, pale pink and pink represented wrinkle and severe wrinkle; it could be divided into two different regions according to forming limit diagram. Zone 1 was located at drawing bead edge; fortunately, wrinkling in this area did not have any bad effect on the quality of the product. Zone 2 was inside the product similar to the area A that was singed. It should be controlled, since this would result in waste occurring. It is worth noting that sometimes the light wrinkling was allowed. The red area represents crack, and it exists in the highest area of fillet drawing that was marked B. Once cracks appear in the work area, they indicate waste; therefore, they need to be controlled. The yellow area represents existence fracture risk; similar to the red area, it also exists in the highest area of fillet drawing. All the same, it belongs

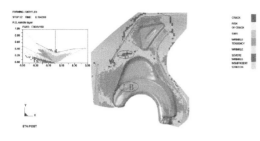

Figure 5. FLD of drawing.

to the risk area and should be strictly controlled as well. The blue region represents wrinkle risk; it belongs to a critical state and also needs attention to be paid to it. The green region represents the security area; as shown in Fig. 5, it mostly appeared in the drawing area. The area was expected more; the more blue area, the less-quality problems would occur.

According to forming limit diagram in Fig. 5, the forming defect forcer on the area was marked A and B, so that the measures could be taken to solve the problems. The area that was marked A was the wrinkle region; solution one was to adjust the binder feeding resistance, and solution two was to take away a part of the plot material by increasing the shallow boss height or shape. A was surrounded by underutilized drawing area, so the wrinkle was caused by resistance being too small. The regional B was the risk to occur in cracks, and the remedy was to increase the radius on the area of forming edge, leaving the edge area forming in the next process, so that the shaping process was set after drawing. In addition to the deficiencies besides the gray areas, it means defect sufficient deformation; based on actual production experience, it could be corrected by the mold.

According to the actual business process test results, the forming program and its amendments were feasible; a qualified product could be got, as shown in Fig. 6.

6 CONCLUSIONS

The binder that was designed could be better adapted to the mold closing bending, and it could also smoothly supply the material for the drawing forming. The layout of draw bead and the selection of the cross-section of draw bead height size and the geometry of the slab design could result in reasonably adjusting material flow. The quality problem wrinkle and crack appearing in numerical experiments need further discussion, and the solution measures had been raised by plastic forming a theoretical and empirical analysis. The practice in Fig. 6 proved that the measure could effectively solve the forming problems.

REFERENCES

[1] Luo Zhengzhi. *Analysis of Flanging Forming Springback of Sheet Metal Based on Numerical Simulation*[D]. Master's thesis of Chongqing University. 2005.5.
[2] Xie Houxun, Zhu Maotao. *Influences of Virtual Male Die Speed on the Stamping Simulation Result*[J]. *Machinery Design and Manufacture* 2002 (6): 77–79.
[3] Liu Kejin. *Research on Experiments of Sheet Stamping Springback and Contrastive Analysis of Numerical Simulation* [D] Master's thesis of Hunan University. 2004.4.
[4] Liang Yan, Xie Zhizhou. *Analysis of Factors Influencing the Numerical Simulation Precision of Sheet Forming Springback*[J]. Forging Equipment and Manufacturing Techniques 2006(5): 55–58.

Figure 6. Wheel cover plate products inspection.

Energy, Environment and Green Building Materials – Sheng (ed.)
© 2015 Taylor & Francis Group, London, ISBN 978-1-138-02718-3

Research on relationship between compression damage and load-carrying capacity for RC columns

X.M. Chen, J. Duan, H. Qi & Y.G. Li
China State Construction Technical Center, Beijing, China

ABSTRACT: The compression damage was considered the basic criterion for the damage assessment of concrete members in high building. In order to consider the relationship between micro damage of material and macro damage of member, the damage of columns in a framework was researched by using a concrete damaged plastic constitutive model. The numerical results show that the damage of members will not exhibit a direct relationship with the micro-damage criterion, and a criterion about the variation of member's load-carrying capacity is much more necessary and visualized.

KEYWORDS: damaged plastic, load-carrying capacity, elastic-plastic, time-history analysis

1 INTRODUCTION

The nonlinear property of concrete is the focus of analysis for concrete structures. The constitutive model should be used according to the different failure modes. In ABAQUS, three main models can be chosen for concrete: They are concrete brittle cracking, concrete smeared cracking, and concrete damaged plasticity.

Concrete brittle cracking model is designed for applications in which the behavior is dominated by tensile cracking and assumes that the compressive behavior is always linear elastic. It must be used with the linear elastic material model, which also defines the material behavior completely before cracking. Thus, it is obvious that this theory is not suitable for those members under compression in high buildings.

For concrete smeared cracking model, although it takes an isotropically hardening yield surface for the dominantly compressive stress and uses the concept of oriented damaged elasticity concepts (smeared cracking) to describe the reversible part of the material's response after cracking failure for the analysis of reinforcement concrete, it is designed for applications only in which the concrete is subjected to essentially monotonic straining at low confining pressures; so it may be more suitable for nonlinear static analysis, such as PUSH.

Compared with the models mentioned earlier, the concrete damaged plasticity is the most suitable model for nonlinear time-history analysis of high buildings. It uses concepts of isotropic damaged elasticity in combination with isotropic tensile and compressive plasticity to represent the inelastic behavior of concrete, and it is designed for applications in which concrete is subjected to monotonic, cyclic, and/or dynamic loading under low confining pressures.

The biaxial plastic-damage model in ABAQUS is based on the models proposed by Lubliner & Oliver [1989] and by Lee & Fenves [1998], in which skeleton curves and damage curves should be defined by users. For beam elements, the uniaxial concrete constitutive model of Mander [1988] is widely used; except for this, Guo-Zhang's model [1985] and the model proposed by national code [2010] are also taken as the preferred scheme by Chinese engineers.

By taking beam elements as an example, these theories based on the micro damage of concrete may be more accurate for revealing the failure mechanism of material, but compared with the plastic hinge model, this micro damage will not always show a direct correlation with the macro damage. This is because an integral point in the section being damaged seriously may not mean the whole section will be obviously degraded. Therefore, it is valuable to research the relationship between micro damage and the macro load-carrying capacity.

In this paper, a framework under seismic wave is analyzed by elastic-plastic time-history method by using ABAQUS. Based on the concrete damaged plastic model proposed by ABAQUS [2006], columns under different amplitude of seismic wave are researched. And then some relationships between compression damage and the load-carrying capacity can be shown by numerical results.

2 FIBER-BUNDLE ELEMENT AND CONCRETE DAMAGED PLASTIC MODEL

The columns and beams in buildings are modeled by fiber-bundle elements as shown in Fig. 1(a).

The section is divided into many fibers, and the mechanical property of each fiber can be presented by an integration point. The integration scheme and the output request of the section can be defined according to needs. For a rectangular section, the default integration scheme is shown in Fig. 1(b) and the default output section points are the four corners.

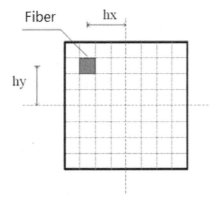

Figure 1.　(a) Section of fiber-bundle element.

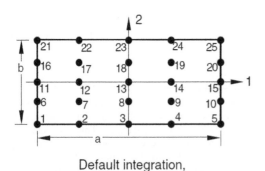

Default integration, beam in space

Figure 1.　(b) Default integration for beam section.

Actually, the damaged plastic model is only used for the axial deformation of each integration point, the deformation induced by shear force has nothing to do with the axial deformation, and the shear stiffness will always keep linear elastic; the shear stiffness of section can be written as follows:

$$K = kGA \qquad (1)$$

where G is the elastic shear modulus or moduli, A is the cross-sectional area of the beam section, and k is the shear factor.

The uniaxial constitutive model at each integration point is defined as follows (Figure 2):

$$\sigma_t = (1-d_t)E_0(\varepsilon_t - \tilde{\varepsilon}_t^{pl})$$
$$\sigma_c = (1-d_c)E_0(\varepsilon_c - \tilde{\varepsilon}_c^{pl}) \qquad (2)$$

where σ_t and σ_c are the stresses of tension and compression;

d_t and d_c are the damages variables of tension and compression;

E_0 is the initial modulus of the material;

ε_t and ε_c are the strains of tension and compression;

$\tilde{\varepsilon}_t^{pl}, \tilde{\varepsilon}_c^{pl}$ are the equivalent plastic strains.

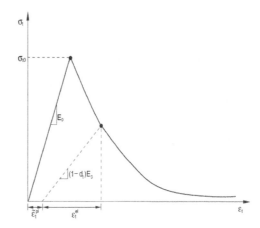

Figure 2.　(a) Skeleton of tension.

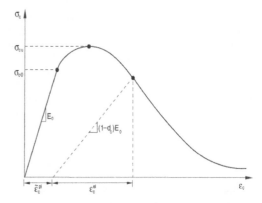

Figure 2.　(b) Skeleton of compression.

The section forces can be formulated by section integration as follows:

$$M_x = \sum_{i=1}^{n} A_i h_y \sigma_i$$

$$M_y = \sum_{i=1}^{n} A_i h_x \sigma_i$$

$$N = \sum_{i=1}^{n} A_i \sigma_i \qquad (3)$$

3 NUMERICAL EXAMPLE OF FRAMEWORK

As shown in Fig. 3, a reinforced concrete framework is analyzed based on fiber-bundle model and concrete damaged plastic model; the columns denoted in Figure 3(b) are taken as examples of members for numerical analysis. The section of columns is 500mm*500mm, and the grade of concrete is C35.

Figure 3.   Structural model.

The seismic waves loaded on the fixed nodes are shown in Fig. 4, in which the maximum amplitude is 220gal; the other two amplitudes of 100gal and 310 gal are also considered.

For the damping system, stiffness proportional damping is neglected and mass proportional damping is taken as follows:

$$\alpha = \frac{4\pi\xi}{T} \qquad (4)$$

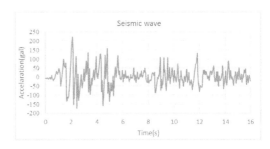

Figure 4.   Seismic wave.

where ξ is the damping ratio, and T is the period of the first mode.

The compression damage of Col5 and Col14 denoted in Fig. 3 is shown in Fig. 5. Being the damage of the whole section, it is the average of 25 integration points.

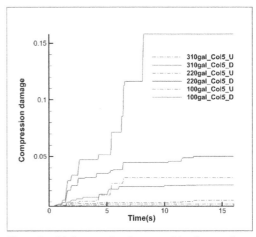

Figure 5.   (a) Compression damage at 5th column ends.

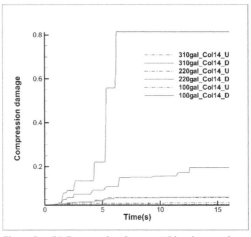

Figure 5.   (b) Compression damage at 5th column ends.

45

Usually, the concrete member may be classified as being destroyed seriously as the compression damage reaches 0.3 and the member may collapse as the compression damage reaches 0.8. To prove this conclusion, the load-carrying capacity should be researched from macro-scale. The deflections at column top and the time-histories of axial force are shown in Fig. 6.

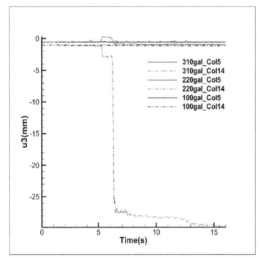

Figure 6.   (a) Time histories of u3.

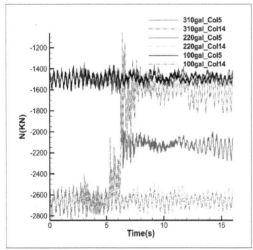

Figure 6.   (b) Time histories of axial force.

When the maximum amplitude of seismic wave is 310gal, it can be seen that the axial load capacity of the 14th column degraded rapidly as the compression damage exceeded about 0.55 and then the axial force of the 5th column increased but still without serious damage. In the other cases, this did not happen.

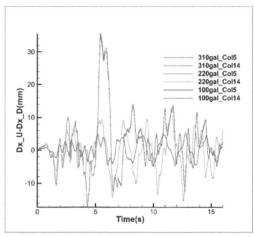

Figure 7.   (a) Relative displacement of column ends.

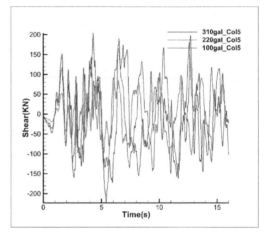

Figure 7.   (b) Shear force of 5th column.

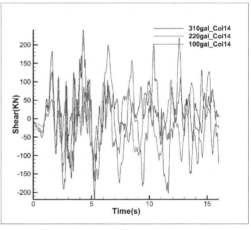

Figure 7.   (c) Shear force of 14th column.

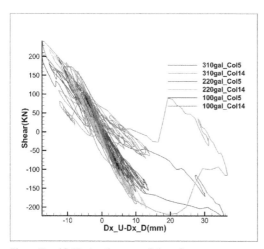

Figure 7. (d) Hysteretic curve of shear force.

The responses of shear force are also presented in Fig. 7. It is obvious that the shear force of the 14th column also dropped simultaneously and the hysteretic curve shows that the shear stiffness was significantly degraded.

4 SUMMARY

The concrete damaged plastic model is the main constitutive form of elastic-plastic time-history analysis for high buildings, so the micro damage of concrete is taken as the most important criterion for the damage of members, but the damage of members may be influenced by many other factors. The numerical result in this paper shows that load-carrying capacity of members may not drop as the compression damage exceeds the critical limit widely used by many researchers, and by taking the load-carrying capacity into account from the macro perspective, the damage assessment of members may be more visualized.

REFERENCES

[1] ABAQUS Inc. 2006. *ABAQUS User Manual*, V6.5.5.
[2] China architecture & building press. 2010. *Code for design of concrete structures.*, Beijing, China. (In Chinese).
[3] Guo Z.H, Zhang X.Q., Wang C.Z. 1985. Stress-strain curve of concrete confined with hoops under cycle loading.industrial structure, 12:16–20 (in Chinese).
[4] Lee J, Fenves G.L. 1998. Plastic-damage model for cyclic loading of concrete structure. ASCE Journal of Engineering Mechanics, 124(3): 892–900.
[5] Lubliner J., Oliver S. 1989. A Plastic-damage model for concrete. International Journal of Solids Structures, 25:299–326.
[6] Mander, J.B., Priestly, M.J.N., Park, R. 1988. Theoretical Stress—Strain Model for Confined Concrete Journal of Structural Engineering, ASCE, 114(8): 1804–1826.

Energy, Environment and Green Building Materials – Sheng (ed.)
© *2015 Taylor & Francis Group, London, ISBN 978-1-138-02718-3*

Practical method for fire-resistant design of steel members based on critical temperature

J. Xing
The Chinese People's Armed Police Academy, Langfang, Hebei, China

ABSTRACT: Based on the high temperature mechanical model of common structure steel in China, the relationship between critical temperature and critical stress was derived, and its corresponding relationship was displayed in a data table form. A simplified equation of critical stress at a high temperature was proposed. Taking an axial compression steel member, for example, the error of critical stress induced by the simplified method was less than 10%. Practical methods and procedures of fire-protection layer thickness based on critical temperature were introduced, which will provide reference for the convenience of engineering application.

KEYWORDS: critical temperature; critical stress; steel member; fire-protection layer thickness

Steel materials used in construction projects are non-combustible, but the material properties such as strength and elastic modulus will decrease as the temperature increases. Bareness steel structures have extremely weak fire resistance plus high thermal conductivity and therefore effective fire protection is almost always needed. Currently, fire-resistant design of structures in China is based on the fire resistance test of steel members, that is, determining the fire resistance of structural members by the test time. It will be safe as long as the tested fireproof time is equal or greater than the required fire resistance rating. In fact, the loading conditions are different among steel members in structures and therefore the fireproof time obtained from a separated test on a member cannot fully represent the fire resistance of the member in the practical structure. Accordingly, it is scientific and economic to adopt a design approach considering actual loading conditions of steel members and thermal properties of fire protection materials.

High-temperature mechanical performance of steel is a basic parameter for fire resistance design. Various mechanistic models are nowadays adopted in China for fire resistance design as different testing methods, and theories are used in other countries. The high-temperature mechanic model was adopted in this paper while considering actual engineering conditions in China, and it was also systematically tested by research fellows. The method based on fire-protection layer thickness of steel members at a critical temperature was introduced apart from the corresponding simplified equations and procedures for practical engineering design.

1 HIGH-TEMPERATURE STRENGTH OF COMMON STRUCTURE STEEL

High-temperature performance of steel has been systematically tested in Tongji University. Based on the testing results along with a comparison with relevant standards in England and other countries in Europe, the yield strength of steel at a high temperature can be predicted by the following equation:

$$f_{yT} = \eta_T f_y \tag{1}$$

$$\eta_T = \begin{cases} 1.0 & 20°C \leq T_s \leq 300°C \\ 1.24 \times 10^{-8} T_s^3 - 2.096 \times 10^{-5} T_s^2 \\ +9.228 \times 10^{-3} T_s - 0.2168 & 300°C < T_s < 800°C \\ 0.5 - T_s / 2000 & 800°C \leq T_s \leq 1000°C \end{cases} \tag{2}$$

where f_y is the yield strength of steel at room temperature; f_{yT} is the yield strength of steel at a temperature of T_s; and η_T is the strength reduction factor of steel at a high temperature. The strength of normal steel varies little at a temperature less than 300°C, whereas it is depleted at a temperature higher than 800°C. Therefore, the strength-reduction factor is going to be studied at a temperature between 300°C and 800°C, which can be expressed by Equation 2.

2 COMPUTATION OF FIRE-PROTECTION LAYER THICKNESS BASED ON CRITICAL TEMPERATURE

2.1 Critical stress and critical temperature

The critical temperature refers to the temperature on the section of a steel member in the limit state of fire resistance, denoted by T_c. From the definition of critical temperature, it can be inferred that a steel member will not fail in fire when the temperature is less than the critical temperature. The maximum normal stress acting on the section of a steel member in the limit state of fire resistance is defined as critical stress of the steel member in fire, denoted by σ_{crT}.

When a steel member reaches its limit state, that is, when the load effect is equivalent to structural resistance, it can be known from the definition of critical stress that the load effect on the section is σ_{crT} and the structural resistance is the design strength of steel member at a high temperature. The following equation can be derived:

$$\sigma_{crT} = \frac{f_{yT}}{\gamma_R} = \frac{\eta_T f_y}{\gamma_R} = \eta_T f \tag{3}$$

where γ_R is the resistance coefficient of steel; f is the design strength of steel at room temperature.

The definition of critical temperature indicates that T_c is equal to T_s in Equation 2 at a critical state. Combining Equations 2 and 3, the relationship between critical stress and critical temperature can be rewritten as follows:

$$\frac{\sigma_{crT}}{f} = 1.24 \times 10^{-8} T_c^3 - 2.096 \times 10^{-5} T_c^2 + 9.228 \times 10^{-3} T_c - 0.2168 \tag{4}$$

The calculation of the critical temperature by Equation 4 requires solving a cubic equation, which is inconvenient for engineering applications. The

Table 1. Critical temperature T_c(°C) corresponding to different critical stress.

σ_{crT}/f	0.3	0.4	0.5	0.6	0.7	0.8	0.9	1.0
T_c	664	621	581	543	503	460	409	300

author uses numerical method to avoid the difficulty in solution and lists the relationship between critical stress and critical temperature in Table 1, which can be referred to in engineering design.

2.2 Simplified method of critical stress

It can be seen from Table 1 that the primary parameter used to determine the critical temperature of a steel member is the critical stress at a high temperature. The critical stress of an axial compression steel member at a high temperature can be calculated by Equation 5.

$$\sigma_{crT} = \frac{N}{\phi_T A} \tag{5}$$

where ϕ_T is the stability coefficient of axial compression steel member at a high temperature. An iterative method has to be chosen to solve Equation 5 since ϕ_T is a function of temperature T, which results in very complicated calculation and thus is unsuitable for engineering practice. Reference 2 reports that there is no significant difference between ϕ_T and the stability coefficient ϕ of axial compression steel member at room temperature, whereas ϕ has been tabulated in related standards. Therefore, the author simplified Equation 5 by substituting ϕ_T with ϕ. The error induced by the simplification was summarized in Table 2.

Table 2. Error in critical stress calculated by ϕ instead of ϕ_T (%).

slenderness ratio	Temperature (°C)							
	300	350	400	450	500	550	600	650
50	-0.6	-0.6	-0.5	-0.2	0.2	0.4	0.5	-0.2
100	-4.3	-4.8	-3.7	-1.6	1.1	3.6	3.9	-1.7
150	-6.9	-7.6	-6.0	-2.7	1.9	6.4	6.9	-2.8
200	-7.9	-8.6	-6.9	-3.1	2.2	7.5	8.0	-3.2

In Table 2, a negative value indicates that critical stress calculated by the simplified method is less than the actual value (i.e. in a dangerous state) and a positive value denotes that the resulting critical stress obtained from the simplified method is larger than the actual value (i.e. in a safe state). The results in Table 2 show that the errors induced by the simplified method mentioned earlier are always less than 10% and the smaller the slenderness ratio, the smaller the error. Therefore, the critical stress of a steel member at high temperature can be calculated by the same equations at room temperature as listed in Table 3. Likewise, the critical stresses for members subjected to other load types can also be determined by the simplified equation as shown in Table 3.

Table 3. Simplified equations of critical stress for steel members at high temperature.

types of steel members	equations of critical stress
axial tension member	$\sigma_{crT} = N/A$
axial compression member	$\sigma_{crT} = N/\varphi A$
flexural member (statically determinate beam)	$\sigma_{crT} = M_x/\gamma_x W_x$

2.3 Elevated temperature of steel members in fire

Fire-resistant design of steel structures based on critical temperature has to ensure the temperature of steel members does not exceed the critical tempera-ture at the specified fire resistant time. Therefore, it is necessary to calculate the elevated temperature of steel members in fire. Temperature of steel members with fire protection under standard elevated tempera-ture can be calculated by Equation 6:

$$T_s(t+\Delta t) = T_s(t) + \frac{\alpha \Delta t}{C_s \rho_s}\left[T_g(t) - T_s(t)\right] \qquad (6)$$

$$\alpha = \frac{\lambda}{D}\cdot\frac{S}{V}\cdot\frac{1}{1+\xi} \qquad (7)$$

where α is section-material coefficient; T_g is a aver-age room temperature given by ISO834 standard fire temperature-time curve; C_s, ρ_s are specific heat and density of steel, respectively; λ is thermal conductiv-ity of fire protection materials; D is the thickness of fire-protection layer; S/V is section coefficient; and ξ is a parameter of energy absorption for fire protection cover.

If fire-resistant time t and critical temperature T_c in Equation 6 are given, for a specified time incre-ment Δt, α can be obtained with the iterative method, as listed in Table 4.

Table 4. Section-material coefficient α [W/(m³·°C)].

T_c	t (min)						
(°C)	60	75	90	105	120	135	150
350	700	532	426	353	300	261	229
400	842	638	510	422	359	311	274
450	1000	756	603	498	423	366	322
500	1177	886	705	582	493	427	375
550	1377	1033	819	674	571	493	433

2.4 Calculation of fire-protection thickness

First, σ_{crT} is calculated according to the corresponding equations in Table 3 based on its load type and then T_c is determined from Table 1; followed by choice of α with T_c and required fire-resistant time t from Table 4. Finally, fire-protection thickness D can be solved by Equation 7. If fire protection cover can meet Equation 8, it can be regarded as light fire-protection and the heat absorption ability can be negligible. Thus, it can be assumed that $\xi = 0$.

$$\xi = \frac{c\rho SD}{2c_s \rho_s V} \leq \frac{1}{4} \qquad (8)$$

3 PRACTICAL FIRE DESIGN OF A STEEL MEMBER

An axial compression column was created in a work-shop with a fire endurance rating of grade two, made of Q235, with a section of plain rolled I-steel, type of I50a($A=119cm$, $i_y=3.07cm$, $f=205N/mm^2$), section coefficient $S/V=135m^{-1}$, effective length in weak axis is $3m$, and design axial strength of the steel column is N=1100kN. Thick-layered fire coating is adopted for fire protection, thermal conduction coefficient $\lambda=0.09W/(m\cdot°C)$, specific heat $C=900J/(kg\cdot°C)$, den-sity $\rho=450kg/m^3$. The required thickness of fire-pro-tection layer is calculated.

(1) Slenderness ratio of the steel column, $\lambda=300/3.07=97.7$, stability coefficient, $\varphi=0.57$, referred to Appendix C in Reference 4.

(2) Critical stress, $\sigma_{crT} = N/\varphi A = 162N/mm^2$; critical temperature can be obtained by linear interpolation from Table 1, $T_c=464°C$.

(3) Steel columns in a building with fire endur-ance rating grade 2 are required to have a fire limit of 2.5 hours based on the Code. Referring Table 4, with $T_c=464°C$, $t=150min$, it can be found that $\alpha=337$ $W/(m^3\cdot°C)$.

(4) Calculation of required thickness of fire-protection layer. Assuming a light fire-protection layer, $D=(\lambda/\alpha)\cdot(S/V)=0.036m$; ξ can be calculated by Equation 8. The result shows that with $\xi<0.25$, it can be inferred that the fire-protection layer is light and thus it is not necessary to perform any adjustments, that is, the required fire-protection thickness is 36mm.

The earlier computation indicates that the steel column in this workshop will not fail in a fire when the temperature is less than 464°C and a fire-resistant coating of 36mm can ensure it will not fail within 2.5 hours under standard elevated temperature.

4 CONCLUSIONS

1 High-temperature mechanic model of normal structural steel, systematically tested in Tongji University, was adopted in this paper, which is more suitable for actual engineering conditions in China. Since strength of a normal steel member varies little at a temperature less than 300°C while depleting rapidly at a temperature higher than 800°C, the strength reduction factor is only studied for a temperature range between 300°C and 800°C.

2 Relationship between critical temperature and critical stress is derived from the implication of a steel member's fire resistance capacity in the limit state and is also tabulated for the convenience in application. Critical temperature can be directly found in this table during computation.

3 A simplified method of solving critical stress at a high temperature is proposed based on the fact that the calculation of critical stress at a high temperature is very complicated; there is no significant difference between the results of critical stress at a high temperature and those at room temperature. The errors induced by the simplified method for an axial compression member are always less than 10%. And the smaller the slenderness ratio is, the smaller the error induces. The simplified approach is simple in expression and understanding and thus used in engineering application.

4 Procedure of computation of fire-protection layer thickness based on critical temperature is presented. Engineers can determine the thickness of the fire-protection layer by looking up the table and some simple calculation according to the actual load types and thermal physical parameters of protection materials. It is not necessary to perform fire resistance tests on steel members. Once the thermal parameters of fire protection materials are measured, the design requirements of steel members under various load types can be met, which results in a more scientific and economic design approach.

REFERENCES

China Association for Engineering Construction Standardization. 2006. *Technical Code for Fire Safety of Steel Structure in Buildings*. Beijing: China Planning Press.

Li, GuoQiang. & Han, LinHai. & Lou, GuoBiao. 2006. *Fire-resistant Design of Steel and Steel-concrete Composite Structure*. Beijing: China Architecture & Building Press.

Li, GuoQiang. & Jiang, ShouChao. & Lin, GuiXiang. 1999. *Fire-resistant Calculation and Design of Steel Structure*. Beijing: China Architecture & Building Press.

Ministry of Construction of the People's Republic of China. 2003. *Code for Design of Steel Structures*. Beijing: China Planning Press.

Qu, LiJun. 2003. *Performance-based Design on Fire Safety of Building Structures*. Langfang: The Chinese People's Armed Police Academy, Langfang, Hebei, China.

Energy, Environment and Green Building Materials – Sheng (ed.)
© *2015 Taylor & Francis Group, London, ISBN 978-1-138-02718-3*

Seismic characteristics study of six-pylon cable-stayed bridge with corrugated steel webs

F.M. Wang
State Key Laboratory of Disaster Reduction in Civil Engineering, Tongji University, China

G.L. Deng
Nanchang Municipal Public Group, Jiangxi Province, China

X.Z. Dang & W.C. Yuan
State Key Laboratory of Disaster Reduction in Civil Engineering, Tongji University, China

ABSTRACT: The seismic response characteristics of a six-pylon cable-stayed bridge with corrugated steel webs are studied by the response spectrum method. Through a comparison of the dynamic characteristics and seismic response characteristics of a six-pylon cable-stayed bridge with corrugated steel webs with the one with concrete webs, the conclusion that the resistant earthquake properties of the former are superior to those of the latter is drawn. In the meantime, the seismic isolation project using Cable-Sliding Friction Aseismic Bearing (CSFAB) is presented. The seismic response analysis of the six-pylon cable-stayed bridge with corrugated steel webs is done using the nonlinear time history analysis method, and the study shows that the design project using the CSFAB is suitable to meet the earthquake resistance needs of a multi-pylon cable-stayed bridge with corrugated steel webs.

KEYWORDS: Corrugated Steel Web, Six-Pylon Cable-stayed Bridge, Dynamic Characteristics, Cable-sliding Friction Aseismic Bearing

1 INTRODUCTION

The composite box girder with corrugated steel webs is a new structure with a distinct advantage in structure stress and structure construction, so it has been developed rapidly at both home and abroad. Since the first box girder bridge with corrugated steel webs was built in France in 1987, this new bridge structure has experienced extensive development worldwide (Chen et al. 2005). However, the study on the composite box girder with corrugated steel webs is currently focused on static analysis, and the dynamic study relatively lags behind.

The concern about multi-pylon cable-stayed bridge has been rising since the Jiashao Bridge, which is the first multi-pylon cable-stayed bridge that was built in our country. Compared with conventional cable-stayed bridge, multi-pylon cable-stayed bridge not only is better in terms of safety and cost effectiveness but also reduces safely risk and difficult degree during construction. The static and dynamic characteristic of multi-pylon cable-stayed bridge has been analyzed by many domestic scholars, and some research results have been done (Xiao et al. 2011), but the seismic response study of multi-pylon cable-stayed bridge with corrugated steel webs was little.

In this paper, the comparative analysis of dynamic characteristics and seismic response of a six-pylon cable-stayed bridge with corrugated steel webs and a general multi-pylon cable-stayed bridge with concrete webs has been done using the finite element software. For the bridge with corrugated steel webs, the seismic isolation program using the CSFAB system has been designed. CSFAB is excellent in durability and the cable is easy to replace, which provides a new way for the multi-pylon cable-stayed bridge to withstand earthquakes and provides a reference for similar bridges for seismic research.

2 ENGINEERING GENERAL SITUATION

Figure 1 shows the effect picture of the one six-pylon cable-stayed bridge. The span arrangement is 79 m + 5 × 150 m + 79 m, and deck width is 40.5 m. The beam structure is a composite box girder with corrugated steel webs. The bridge approach is a double separate deck bridge, whose span arrangement is 5 × 50 m and the bridge deck width is 2 × 18.25 m. The structural system of this six-pylon cable-stayed bridge takes the form of the tower and beam consolidated and the pier and beam separated. The main tower is a reinforced concrete structure. The cable layout adopts the single cable plane and fan arrangement. Each tower has a total of 2 × 18 cables, and the full bridge has a total 2 × 108 cables. The foundation of this bridge is a borehole cast-in-place concrete pile; the pile diameter of the main tower is 2 m, and that of the side pier is 1.5 m.

Figure 1. Effect picture of the six-pylon cable-stayed bridge.

Figure 2 shows two cross-sections of box girder with corrugated steel webs and concrete webs, respectively. The beam height changes as quadratic parabola, and a thickened segment is set up at the end of beams and at the top of piers. The thickness of corrugated steel webs and concrete webs is 12 mm and 600 mm, respectively.

(a) Girder cross-section with corrugated steel webs

(b) Girder cross-section with concrete webs

Figure 2. Cross-sections of two girders.

The seismic isolation design project of the six-pylon cable-stayed bridge with corrugated steel webs is the cable-sliding friction aseismic bearing design (Yuan et al. 2010). In order to get the basic performance parameters, the quasi-static test of a CSFAB whose design-carrying capacity is 8MN has been done, and the hysteretic curve has been obtained as shown in Fig. 3. Figure 4 shows the photo of the test bearing.

In this seismic isolation project, sliding CSFAB was set on pier T15 and T19. The free displacement of cables was selected as 0.12 m, the elastic stiffness value was 1×106 kN/m, and the friction coefficient was 0.02. The layout of bearings on the other piers was similar to that of the conventional project.

In this way, the displacement of bearings could be controlled and the lateral torsion of beam could be limited. Under minor and moderate earthquakes, the shear bolt in a fixed-type bearing should not break, and normal operations of the bridge should be carried out. Under a severe earthquake that causes the shear bolt to break, the fixed-type bearing functions as a sliding-type bearing mitigate the transmission of earthquake forces and dissipate seismic energy, whereas the excessive relative displacement between the superstructure and the pier can be restrained by the cable components.

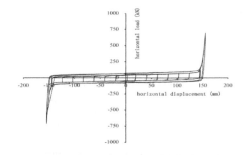

Figure 3. Hysteretic curve of CSFAB.

Figure 4. Photo of the test CSFAB.

3 NUMERICAL MODELING

In this paper, the spatial finite element model of the full bridge was established and implemented by using the SAP2000 software, as shown in Fig. 5. Beam elements were chosen to simulate the beams, towers, and piers; in a similar manner, spatial truss elements simulated cables, and spring elements simulated group piles to consider the interaction between soil and pile. All bearings were simulated by three-dimensional bearing elements, and the CSFAB was simulated by combination of the Plastic (wen) and polyline elastic element (Yuan et al. 2012).

According to the dead load result of SAP2000 calculation, the proportion of superstructure dead load

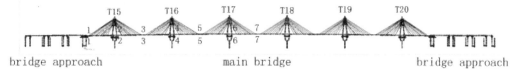

bridge approach   main bridge   bridge approach

Figure 5.   Finite element model of six-pylon cable-stayed bridge.

and the whole structure dead load of the bridge with corrugated steel webs are 77% and 82%, respectively, in those of the bridge with concrete webs. Compared with box girder with concrete webs, the ratio of the bending stiffness in X-axis and Y-axis of the box girder with corrugated steel webs is 70% and 60%, respectively, and the ratio of torsional stiffness is about 73%.

3.1   Ground motion inputs

Response spectrum was generated according to seismic safety evaluation report of the bridge site and the ground motion parameters report, and the nearby seismic field was not considered. Table 1 shows two time history waves. In the next calculation, the peak acceleration of ground motion was adjusted as 0.4 g, and damping ratio was taken as 5%.

Table 1.   Parameters of ground motions.

Seismic wave	1992 Capemend	1994 Northridge
Station	Petrolia	Rinaldi
Magnitude (Mw)	7.1	6.7
PGA (g)	0.662	0.472
PGV (cm/s)	89.7	73.0
PGD (cm)	29.5	19.7

3.2   Response spectrum results analysis

In this section, the primary mission is the comparison analysis of both bridges with different webs, so the results of response spectrum were analyzed only under E1. The input combinations of ground motion are longitudinal/lateral direction + vertical direction, and vertical ground motion took the horizontal direction of 0.65 times. Tables 2 and 3 show the force comparison at the bottom of piers of the bridge with both webs under the response spectrum effect. Tables 4 and 5 show the internal force comparison of key sections of both bridges, and the key sections have been marked, as shown in Fig. 5.

From Table 2, the force ratio of T17 pier is less, because the bearing on the top of T17 pier is fixed bearing and the horizontal seismic forces are subjected by this pier. In the meantime, the friction

Table 2.   Force ratio of piers in the case of longitudinal and vertical direction.

Piers	Axial force	Shear force	Bending moment
T15	0.78	0.99	0.98
T16	0.75	1.00	0.98
T17	0.74	0.83	0.85

Table 3.   Force ratio of piers in the case of lateral and vertical direction.

Piers	Axial force	Shear force	Bending moment
T15	0.66	0.89	0.89
T16	0.60	0.90	0.89
T17	0.65	0.82	0.80

effect of bearings at the top of other piers was not considered; therefore, the result difference between two structures is not great. In the case of lateral direction + vertical direction ground motion as shown in Table 3, the fixed bearing is set at the top of each pier, so the influence of superstructure on the seismic force of the tower end is basically the same. Overall, the internal forces proportion of piers of the bridge with corrugated steel webs is 80%–90% in those of the bridge with concrete webs.

From Tables 4 and 5, we can see, in both cases, that the result difference between two structures of the key

Table 4.   Force ratio of key sections in the case of longitudinal and vertical direction.

Key sections	Horizontal bending moment	Vertical bending moment	Torque
1-1	0.55	0.69	0.70
2-2	0.89	0.39	0.71
3-3	0.33	0.73	0.72
4-4	0.87	0.78	0.67
5-5	0.65	0.94	0.68
6-6	0.75	0.85	0.81
7-7	0.61	0.94	0.68

Table 5. Force ratio of key sections in the case of lateral and vertical direction.

Key sections	Horizontal bending moment	Vertical bending moment	Torque
1-1	0.69	0.72	0.71
2-2	0.88	0.89	0.47
3-3	0.65	0.78	0.45
4-4	0.64	0.74	0.57
5-5	0.57	0.73	0.44
6-6	0.70	0.87	0.86
7-7	0.48	0.66	0.56

section 2–2, 4–4, and 6–6 is not great, because these key sections set thickened segment, which causes the weight of these sections on both structures to be basically the same. The average difference between longitudinal and horizontal bending moment is 76% and 66%, respectively, and 65% of torque. Thus, it can be seen that the cross-section internal forces of the multi-pylon cable-stayed bridge with corrugated steel webs are less than those of the bridge with concrete webs.

3.3 Nonlinear dynamic time-history results analysis

Nonlinear time history analysis has been conducted by using the input ground motions, in which longitudinal ground motion was only considered. In order to make a comparative study, the case of conventional bearing layout was simultaneously analyzed. Tables 6 and 7 show the comparative results of the forces at the bottom of pier T15 and pier T17 in two input motion cases. When the bearing layout was considered the conventional system, the shear force and bending moment at the bottom of the fixed pier T17 were great, which far exceed their capacity. However, in the case of CSFAB design, CSFAB was set on pier

Table 6. Comparative results of piers under Capemend wave.

Condition	Piers	Shear force (MN)	Bending moment (MN·m)
Model ①	T15	34.0	421.3
	T17	125.6	2669.1
Model ②	T15	59.4	987.7
	T17	42.2	514.7
(②−①)/①	T15	74.7	134.4
	T17	−66.4	−80.7

Table 7. Comparative results of piers under Northridge wave.

Condition	Piers	Shear force (MN)	Bending moment (MN·m)
Model ①	T15	43.1	634.8
	T17	170.7	3620.6
Model ②	T15	58.7	985.1
	T17	33.4	561.5
(②−①)/①	T15	36.2	55.2
	T17	−80.4	−84.5

T15 and T19, which could share the seismic forces effectively; thus, the bending moment of pier T17 felled an average of more than 50%. The seismic forces of pier T15 significantly increased, but the forces were still less than its capacity.

The maximum beam displacement and maximum relative displacement between pier and beam are shown in Table 8. The maximum relative displacement between pier and beam of two waves was 19.2 cm and 16.7 cm, respectively, and it was controlled within 20cm, which shows the performance advantage of CSFAB in spacing limit.

Table 8. Maximum displacements (cm).

Ground motion	Capemend	Northridge
Beam displacement	22.2	17.6
Relative displacement between pier and beam	19.2	16.7

4 CONCLUSIONS

1 For the internal forces and displacements of piers and beams under earthquake, the multi-pylon cable-stayed bridge with corrugated steel webs is less than the bridge with concrete webs, which shows that the structural performance under earthquake of multi-pylon cable-stayed bridge with corrugated steel webs is superior to that with concrete webs.

2 Under moderate or strong earthquakes, the fixed piers of the bridge that towers rigidly connect with the girders and the piers that are separated with the girders are vulnerable components. By adopting the seismic isolation design, the seismic force can be significantly reduced. When the CSFAB was set on the piers distributed symmetrically by taking the fixed pier, the seismic force of the fixed pier was reduced, and the force of the piers that set

CSFAB increased at an acceptable range, whereas the relative displacement between pier and beam could be limited in a manageable level.

ACKNOWLEDGMENTS

This research is supported by the National Natural Science Foundation of China under Grants No. 51278376 and 51478339; this support is gratefully acknowledged.

REFERENCES

Chen, B.C. & Huang, Q.W.. 2005. A summary of application of prestressed concrete box-girder bridges with corrugated steel webs, *Highway*. 7: 45–53.

Xiao, M.K., Wang, X.W. & Liu, G.. 2011. Static load test and calculation on multi-tower cable-stayed bridge with tie-down cables, *Journal of Civil, Architectural & Environmental Engineering*. 33: 43–49.

Yuan, W.C., Cao, X. J. & Rong, Z. J.. 2010. Development and experimental study on cable-sliding friction aseismic bearing, *Journal of Harbin Engineering University*. 31: 1593–1600.

Yuan, W.C., Wang, B.B. & Cheung, P.. 2012. Seismic performance of cable-sliding friction bearing system for isolated bridges, Earthquake Engineering and Engineering Vibration. 11: 173–183.

Energy, Environment and Green Building Materials – Sheng (ed.)
© 2015 Taylor & Francis Group, London, ISBN 978-1-138-02718-3

High-performance lightweight over-strength UHMWPE UD cloth preparation and its application

X.H. Meng
Unit 91872 of PLA, Beijing, China

G. Lu
Department of Navy Equipment, Beijing, China

W.F. Da
Ningbo Da-cheng Advanced Material Co. LTD, Zhejiang, China

ABSTRACT: A self-developed differentiated UHMWPE fiber preparation technology and surface modification technology are applied in developing UD cloth preparation process in order to produce uniform and stabilized fiber. By screening thermoplastic elastics, a hybrid elastic matrix resin system is developed to improve the inter-facial bonding properties and anti-aging properties between the fiber and matrix resin. An organic/inorganic hybrid approach is opted to develop nano-enhanced hybrid elastic matrix resin in order to form a physical network that can pass shock load. Ultimately, the performance of UD fabric can reach an internationally advanced level.

KEYWORDS: UHMWPE fiber, new elastic matrix resin system, shock resistance performance, UD fabric preparation technology

1   INTRODUCTION

UHMWPE fiber has many outstanding characteristics such as low density, high weather resistance, chemical resistance, and so on. UHMWPE fiber has been widely used in areas such as logistic equipment, aerospace, marine engineering, security, and bio-medical materials, and bulletproof is an important area where UHMWPE fiber is applied [1]. The fiber is mostly used in body armor, and it has occupied the major of the international body armor fiber market[2][3]. Based on a self-developed UHMWPE fiber preparation technology and surface modification technology, a high-performance lightweight over-strength UHMWPE UD cloth preparation technology is researched in the production of high-performance and lightweight UHMWPE UD fiber and body armor.

2   IMPACT ON UHMWPE UD CLOTH'S BULLETPROOF PERFORMANCE OF ELASTIC MATRIX RESIN SYSTEM

As pre-production and research have shown, the performance of elastic matrix resin system plays a key role in the production of high-performance, lightweight UHMWPE fiber UD cloth. Thus, according to production experience and technology accumulated in UD fabric preparation, the structure and performance of a variety of commercial-grade thermoplastic elastomeric are systematically studied. A mixed elastic matrix resin system is developed that is mainly composed of poly (styrene-isoprene-styrene), which is helpful in improving the inter-facial bonding properties and anti-aging properties between the system and UHMWPE fibers. Thus, the shock resistance performance of UHMWPE fiber UD cloth and bulletproof armors is effectively improved.

2.1   *Impact on UHMWPE fiber UD cloth's bulletproof performance of resin content*

According to molding process, the preparation process for UD cloth belongs to the type of dipped composite molding, so resin content (gel content) is one of the key process parameters that requires strict control. Arrangement equipment is used to dip the fiber and control the resin content after unwinding, then the fiber is wounded into unidirectional sheets, and finally the sheets are formed to 0° / 90° stack at size of 1m × 1m by cutting. Resin content is calculated as follows:

$$\text{Resin content} = \frac{m_1 - m_0}{m_1} \times 100\%$$

where m_0 is weight of the fibers measured before and after each piece of sheet is laminated; m_1 is weight of sheet after drying.

Then, UD fabric sheet is cut into targets at a size of 200mm × 200mm; the surface density ≤ 6.5kg / m^2. Next, the fiber goes through thermo-molding process and ballistic tests; the experimental results are shown in Fig. 1:

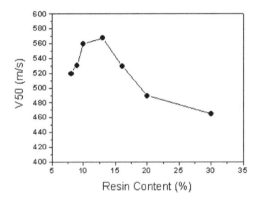

Figure 1.   Impact on UHMWPE UD cloth's bulletproof performance of resin content.

It can be seen from Fig.1 that resin content has a significant impact on UHMWPE UD cloth's bulletproof performance; the best content for resin is 13%. When the indicator is too low, it is difficult to uniformly disperse the resin in continuous UD cloth. There will be no good bond between the resin and fiber, and the fiber thus stays in a relatively loose state. Thus, the slip process is likely to happen during the bomb attack process even if the fiber strength is high; therefore, the effective number of fibers participating in the bulletproof process will be reduced, the gap is enlarged, resulting in degradation of bulletproof performance. When the resin content is too high, the number of fibers per unit volume will be reduced, the improvement of the interface will be difficult, and the bulletproof performance of UD fabric will be also weakened. After extensive experimental studies, the best resin content range is set to 12–14%.

2.2   Impact on UHMWPE UD cloth's bulletproof performance of areal density

UD fabric has significant characteristics of the layered structure when it is used to constitute bulletproof composite material; the aggregation and distribution of fibers in every layer of the base material can also affect the performance of UD fabric cloth. The impact on UHMWPE fiber UD cloth's bulletproof performance of areal density is researched given the resin content; the experimental results are shown in Fig. 2:

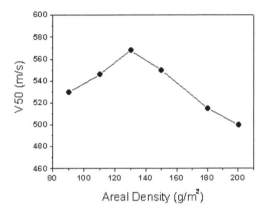

Figure 2.   Impact on UHMWPE UD cloth's bulletproof performance of areal density.

Similar to surface density, it can be seen from Fig. 2 that there is also an optimal density range for areal density; the bulletproof performance of the target reaches a maximum when the areal density is 130g/m^2. This is because when the density is too low, the fibers are not uniformly arranged, the gaps among yarns are relatively large, which leads to weakened bulletproof performance; whereas if the surface density goes too high, it will not be helpful to improve the bulletproof performance of the cloth. This is probably because the fibers have a tendency to aggregate when the density is too high, resulting in uneven structure and stress concentration in the UD cloth, and the situation does no good to improve bulletproof performance. The areal density for UD Fabric is finally set to 130g / m^2 or so.

3   IMPROVEMENT OF BULLETPROOF AND RESISTANT PERFORMANCE PROPERTY OF UD FABRIC AND ITS PRODUCTION

3.1   Modification analysis of bulletproof and resistant performance for UD fabric based on hybrid elastic matrix resin

To further improve the bulletproof performance of UD fabric, nano-organic/inorganic hybrid composite technology is used to modify new elastic matrix resin in order to improve the resistant property of composite materials and to produce lightweight composite materials.

The dispersion of three kinds of nanophase particles with different shapes in the SIS thermoplastic elastomers is analyzed. The particles are fibrous multi-walled carbon nano-tubes with large aspect ratio, layered organic-modified montmorillonite, and spherical surface-modified silica particles.

Performance of nanocomposites is mainly determined by the dispersion state of nanoparticles in the matrix resin; the dispersion state is vital to enhance inter-facial adhesion between nanoparticles and the matrix resin and to avoid large-scale defects resulting from reunited particles. Based on this principle, according to the surface characteristics of various nanophase materials, corresponding surface modification methods are brought forward. A surface modification technology is used to graft the alkyl group to the surface of the nanoparticles, so that the nanophase particles can be stably dispersed in the solution of new elastomer, and the inter-facial adhesion between nanoparticles and the elastic resin can also be improved.

Specific modification method is as follows: After the nitric acid oxidation, carboxyl group is grafted

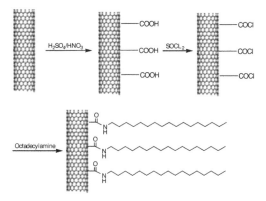

Figure 3. Octadecylamine-modified carbon nano-tubes reaction route.

to the surface of multi-wall carbon nano-tubes (MWCNT); the carboxylic acid group is converted to acid chloride via the acid chloride. Octadecylamine is then reacted with the surface of the MWCNT; in the end, the surface of the MWCNT is covered with long-chain alkyl groups. The reaction process is shown in Fig. 3. The hydrophilic amino group of organic cationic modifier molecules and sodium ions in montmorillonite are exchanged; hydrophobic long-chain alkyl groups are then introduced into the inter-layer, which is able to increase the compatibility between the elastomer and montmorillonite and achieve dispersion effect at nanometer stage.

3.2 Production process of high-performance UHMWPE fiber UD cloth

Numerous studies show that stable uniformity and tension control of UHMWPE fiber arrangement is a key factor in improving the performance of UD fabric's bulletproof performance. The uniformity of the fiber and tension can be enhanced by utilization of improved fiber arrangement equipment and technology, so that the microstructure of UD cloth will be kept uniform.

Based on experimental studies and production experience, the existing UD fabric preparation process is improved and optimized, and the new process is as follows:

Through the process described, the prepared UD cloth will have a uniform and dense microstructure as shown in Fig. 5 and Fig. 6. On this basis, the areal density of newly developed UD cloth is decreased to 130 g/m2 while ensuring its bulletproof performance by utilization

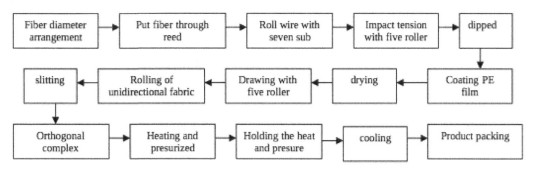

Figure 4. Production line of high-performance UHMPE fiber UD cloth.

of optimized preparing technology such as coating, drying, and pressing stereotypes. The performance can reach the same performance indicators of latest series SB-3A UD cloth produced by DSM company, so the newly developed UHMWPE fiber UD cloth can lay a solid foundation for producing high-performance and lightweight composite bulletproof material.

4 PREPARATION OF SOFT UHMWPE FIBER BULLETPROOF LAYER

Based on production experience, the existing soft body armor preparation process is improved and optimized; the following is the optimized preparation process:

Figure 5. High-performance lightweight UD fabric appearance.

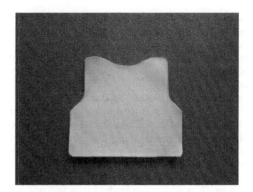

Figure 7. Trimmed bulletproof layer of UHMWPE fiber UD cloth.

1 to draw up bulletproof process card according to the type of body armor and protection class (controlled);
2 to design model following controlled process card of this batch (controlled);
3 to draw lines according to the process model;
4 to orthogonally stack bulletproof UD cloth according to the required number of layers;
5 to lay designated model above the multi-orthogonal overlapping UD cloth and trim the cloth according to lines;
6 to sew the cropped bulletproof according to requirement determined by the alternative process card (controlled);
7 double-sided adhesion or three-sided adhesion of sewed bulletproof layers.
8 to amend and cut the UD layers according to the standard dashes;
9 to sew jacket with rectified bulletproof layer plus bush.

Meanwhile, ergonomic principles are taken into consideration during the design and manufacture of body armor. The basic design principle is to balance the protection degree and human comfort, in order to achieve the best layout of body armor and the lowest possible weight when confronted with varied missionary risk, so that the body can move without restrict and better perform combat mission.

In addition, by optimization, the fabric of body armor jacket (600 × 600D high-strength polyester fiber fabric through treatment of waterproof, fire retardant, and abrasion) and foam materials are selected; versatile body armor is developed, which has excellent properties such as bulletproof, waterproof, fire retardant, wearable, and buoyancy aid. Figure 7 showed the trimmed bulletproof layers of UHMWPE fiber UD cloth.

5 CONCLUSIONS

UHMWPE fibers are widely used in production of bulletproof materials because of their good ballistic properties. Based on impact analysis of the resin content and areal density on UHMWPE fiber UD cloth's bulletproof performance, the impact on UHMWPE UD cloth's bulletproof and resistance performance of three kinds of nanoparticles with different shapes are studied. Finally, a new preparation process of UHMWPE fiber UD cloth is based on elastic matrix resin system. The bulletproof and resistant performance of UHMWPE fiber UD cloth is promoted, and a new technology fabric is prepared to improve the ballistic impact resistance without sheeting to achieve polyethylene Preparation UD fabric products. The massive production of high-performance and super-lightweight UHMWPE fiber UD cloth is achieved.

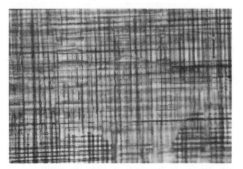

Figure 6. Optical micrograph of high-performance lightweight fabric UD cloth.

REFERENCES

[1] Shen Guangyi. Performance superiority of carbon nanotube fibers in new bulletproof body armor[J] Chinese personal protective equipment, 2010 (2): 27–30.
[2] Guan Xinjie. The application of high performance fiber in manufacture of body armor[J] nonwovens, 2010: 18 (6): 21–23.
[3] Chen Ning, Fanfeng Bin, Huang Jiqing Development of New Dual-purpose Body Armor[J]. Medical Equipment, 2011: 32 (11): 43–45.

Energy, Environment and Green Building Materials – Sheng (ed.)
© *2015 Taylor & Francis Group, London, ISBN 978-1-138-02718-3*

Study on the rendering algorithm of spatial planar circle through coordinate transformation

J.Y. Li, G.X. Yuan, Y.M. Zhang & Z.Q. Huang

School of Resources and Environment, North China University of Water Resources and Electric Power, Zhengzhou, Henan, China

ABSTRACT: A circle can be easily drawn on a two-dimensional plane, but it is not easy to render it in three-dimensional space. In this paper, from the equation of circle, sphere surface, and spatial plane, the concept of spatial planar circle is proposed first. And then, another concept, that of the spatial circular plane is derived, which has properties such as shape, size, location, and direction. On this basis, the circle in the two-dimensional planar local coordinate system can be transformed to the three-dimensional spatial global coordinate system through transformation of rotation and translation. Finally, the rendering algorithm of spatial planar circle is implemented. And the correctness and effectiveness of the algorithm is validated by simulating rock fractures that exist widely in nature. This algorithm will lay a solid foundation for the application of spatial circular plane in many fields.

KEYWORDS: Coordinate transformation; Spatial planar circle; Spatial circular plane; Rendering algorithm

1 INTRODUCTION

Modeling and rendering of a curve and a surface has been one of the fundamental problems in computer graphics. Among many APIs of graphics, line segments and polygons are usually adopted as the basic elements. And for the drawing of curves and surfaces, the integral method is usually applied. That is, the curves and the surfaces can be simulated by the integral of a series of line segments and triangles.

For the drawing algorithms of a circle, a planar circle in the two-dimensional space is studied much more (TANG Di et al. 2004). But for a circle in the three-dimensional space, especially when the coordinates of the discrete points cannot be obtained directly, how to draw a spatial planar circle quickly and efficiently is the key issue to be currently resolved.

In many practical applications, the location, the orientation, and the size of a spatial planar circle tend to be obtained (Jian-yong Li et al. 2012). The problem of a spatial planar circle is that it has to be studied in depth, and it has a wide range of applications in many areas. The method of modeling and quick and efficient rendering of a spatial planar circle is proposed and implemented through the transformation of coordinates.

2 SPATIAL PLANAR CIRCLE AND SPATIAL CIRCULAR PLANE

For a general spatial polygon, its location, shape, and size can be denoted by the coordinates of a series of vertices, and its direction can be represented by its normal vector n through the cross-product of two adjacent edge vector u and v, i.e., $n = u \times v$. In practical applications, the location and the direction of a spatial plane can be measured. However, the shape and the size is difficult to be determined because of lack of adequate information of vertices coordinates. Circle, as a closed curve, can express a finite spatial plane only by two parameters, an origin and a radius. Therefore, circle is widely applied in many practical areas.

2.1 *Curve equation of a circle and surface equation of a sphere*

For a circle whose center is at (a,b) and radius is R, it can be represented by

$$(x-a)^2 + (y-b)^2 = R^2 \tag{1}$$

For a sphere whose center is (a,b,c) and radius is R, its surface equation can be expressed as

$$(x-a)^2 + (y-b)^2 + (z-c)^2 = R^2 \tag{2}$$

2.2 *Equation of a spatial plane*

As shown in Fig. 1, the general equation of a spatial plane can be expressed by

$$Ax + By + Cz + D = 0 \qquad (3)$$

Or in another form

$$(A,B,C)\cdot(x,y,z) = -D \qquad (4)$$

In the above, (x, y, z) is the coordinate of any point on the plane, and (A, B, C) are the components of the normal vector of the plane.

The angle between the normal vector of the spatial plane and the one of the horizontal plane is denoted by α. The azimuth of the projection of the normal vector on the horizontal plane is denoted by β, as shown in Fig. 2. α and β are also known as dip angle and dip direction angle in geology. Therefore, the normal vector can be represented by

$$\boldsymbol{n} = (A, B, C) = (\sin\alpha\sin\beta, \sin\alpha\cos\beta, \cos\alpha) \qquad (5)$$

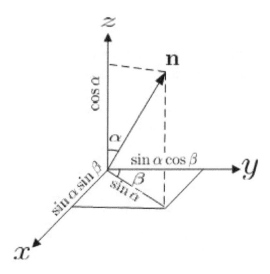

Figure 2. Normal vector of a spatial plane.

Simultaneously, a special type of a plane with boundaries is defined by the spatial planar circle, that is, spatial circular plane. The boundary of the plane is a circular curve in the spatial plane. Both spatial planar circle and spatial circular plane have a wide range of applications in many areas.

Obviously, the boundary contour of a circular plane can be described with a circular curve. It is very easy to be implemented for a graphics system.

3 TRANSFORMATION FROM 2-D PLANAR CIRCLE TO 3-D SPATIAL CIRCLE

If a spatial planar circle is to be drawn in a graphics system, the center (a, b, c), the radius R, the dip angle α, and the dip direction angle β tend to be the known input information. It is very easy to realize the rendering algorithm of a circle in the two-dimensional space. However, if a circle is to be drawn in the three-dimensional space, a straightforward idea is that a two-dimensional circle is implemented first and then a three-dimensional circle is realized through the transformation of coordinates. Generally speaking, this requires two rotations and one translation to be completed.

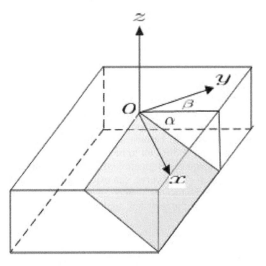

Figure 1. Spatial plane.

2.3 Spatial planar circle and spatial circular plane

A spatial planar circle can be obtained by cutting through a surface of a sphere with a spatial plane. Therefore, the equation of the spatial planar circle can be expressed with

$$\left. \begin{array}{l} (x-a)^2 + (y-b)^2 + (z-c)^2 = R^2 \\ Ax + By + Cz + D = 0 \end{array} \right\} \qquad (6)$$

A spatial planar circle has the location, the direction of the spatial plane and the shape, and the size of the plane circular curve.

3.1 Two kinds of coordinate systems

Two kinds of coordinate systems are first provided: two-dimensional local coordinate system and three-dimensional global coordinate system. The origin of the local coordinate system is located on the center of the plane circle. The x-axis is parallel to the dip direction of the plane. The y-axis is parallel to the trend direction of the plane, as shown in Fig. 3. For the 3-D global coordinate system, its x-axis points to the east, y-axis to the north, and z-axis upward, as shown in Fig. 4.

The general idea is that the points in the 2-d local coordinate system are represented by the 3-D global coordinate system through coordinate transformations.

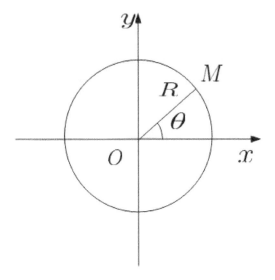

y

M

R

θ

O

x

Figure 3.   2-D local coordinate system.

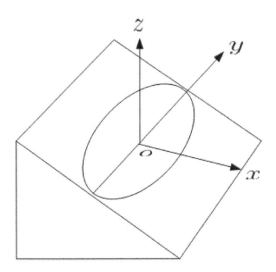

z

y

O

x

Figure 4.   3-D global coordinate system.

3.2   Generation of a planar circle

First, a plane circle can be determined that is located on the origin $O(0,0)$. Then, any point

$M(x_0,y_0)$ on the circular curve can be expressed in the local coordinate system as

$$\left.\begin{array}{l} x_0 = R\cos\theta \\ y_0 = R\sin\theta \end{array}\right\} \tag{7}$$

It is specified that the z-axis of the local Cartesian system is perpendicular to the plane xOy upward. Therefore, the value of z component of any point M is 0, that is, $z_0 = 0$.

3.3   Transformation of rotation around the y-axis

The local coordinate system is rotated counterclockwise around the y-axis. The angle of rotation is α. The y component of coordinate remains unchanged. Any point (x_0,y_0,z_0) in the original system can be transformed to a new coordinate (x_1,y_1,z_1) in the new system.

$$\left.\begin{array}{l} x_1 = x_0\cos\alpha + z_0\sin\alpha \\ y_1 = y_0 \\ z_1 = z_0\cos\alpha - x_0\sin\alpha \end{array}\right\} \tag{8}$$

3.4   Transformation of rotation around the z-axis

After rotation around the y-axis, the new coordinate system rotates continuously around z-axis counterclockwise. The angle of rotation is $90^\circ - \beta$. The new x-axis points to the east, and the z component of the coordinate remains unchanged. Any point (x_1,y_1,z_1) in the old coordinate system is changed to (x_2,y_2,z_2) in the new system after rotation.

$$\left.\begin{array}{l} x_2 = x_1\sin\beta - y_1\cos\beta \\ y_2 = y_1\sin\beta + x_1\cos\beta \\ z_2 = z_1 \end{array}\right\} \tag{9}$$

3.5   Transformation of translation

After the rotation around the z-axis, the origin of the current coordinate system can be translated to (a,b,c). Then, any point (x_2,y_2,z_2) on the planar circle can be transformed to (x_3,y_3,z_3) in the new coordinate system after translation.

$$\left.\begin{array}{l} x_3 = x_2 + a \\ y_3 = y_2 + b \\ z_3 = z_2 + c \end{array}\right\} \tag{10}$$

The current (x_3, y_3, z_3) is the point on the spatial planar circle. A series of such points can be generated according to this method. Thereafter, all points are connected in turn and closed. The spatial planar circle can be drawn. The shape of the circle depends on the number of the sampling points. If the points are less, the shape drawn is that of a spatial planar polygon. The more the points, the smoother the arc line, and the circle is more realistic.

4 ALGORITHM DESIGN AND CASE STUDY

4.1 *Processes of algorithm design*

According to the equation of the circle and the coordinate transformation of spatial points, any point on the planar circular curve is generated first, whose center is located on the origin $(0,0,0)$ and radius is R. Thereafter, the new coordinate of the point in the three-dimensional spatial Cartesian system can be obtained through rotation twice and translation once. With the increase of the sampling angle, a series of points in three-dimensional space will be generated. These points are connected line by line successively, and the spatial planar circle will be obtained. The center of the circle is (a,b,c), and the radius is R. Its dip angle is α, and dip direction angle is β. The algorithm steps are described as follows.

Step 1: Input the center, radius, dip angle, and dip direction angle of the circle.

Step 2: Set the initial values of variables of sampling angle: $\theta=0$.

Step 3: Generate a point (x_0, y_0) on the circle whose center is at $(0,0)$ and radius is R.

Step 4: Rotate around the y-axis to α angle, from (x_0, y_0, z_0) to (x_1, y_1, z_1).

Step 5: Rotate around the z-axis to $(90^\circ - \beta)$ angle, from (x_1, y_1, z_1) to (x_2, y_2, z_2).

Step 6: Translate to (a,b,c), from (x_2, y_2, z_2) to (x_3, y_3, z_3).

Step 7: $\theta_i = \theta_{i-1} + \Delta$, Δ is the increment of sampling angle, and it controls the precision.

Step 8: If $\theta_i < 360^\circ$, return to step 3; otherwise, proceed to step 9.

Step 9: Draw out the spatial planar circle by connecting the points line by line.

4.2 *Case study*

In geology, the joints can be simulated by a spatial circular plane. In a project of geotechnical investigation, the location, the size and the direction of 100 joints were measured, in which the data of the first 5 joints are shown as in Table 1.

Table 1. Location, size, and direction of the joints in the rock mass.

	X(m)	Y(m)	Z(m)	R(m)	$\alpha(^\circ)$	$\beta(^\circ)$
1	0.42	87.86	64.43	34	67	195
2	69.58	95	59.96	16	67	210
3	16.76	98.65	74.28	14	67	210
4	48.87	58.04	86.31	0	67	243
5	36.83	71.18	1.33	17	67	210

According to the proposed algorithm, the spatial planar circle can be drawn both quickly and efficiently (Fig. 5). Then, the joints dispersed in the fractured rock mass can be simulated by the corresponding spatial circular plane (Fig. 6).

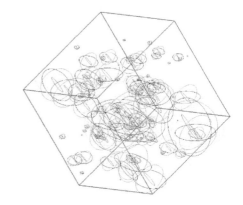

Figure 5. Spatial planar circle.

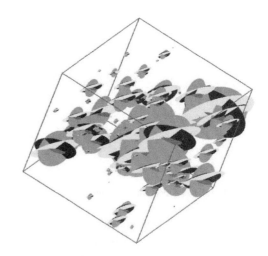

Figure 6. Spatial circular plane.

5 CONCLUSION

Based on the equation of circle, equation of sphere surface, and equation of spatial plane, the concept spatial planar circle is proposed. A circle can be obtained by cutting sphere surface with a spatial plane. Furthermore, the concept of spatial circular plane is derived, which has properties such as shape, size, location, and direction and it has a wide range of applications in many areas. Thereafter, a circle in the two-dimensional local coordinate system can be transformed into the three-dimensional global coordinate system. Finally, the algorithm of drawing the spatial planar circle is realized both quickly and efficiently. The spatial circular plane can be used to simulate the joints dispersed in the rock mass. The practical example proved the algorithm correct and effective. The method will lay a solid foundation for the wide application of the spatial circular plane in many areas.

ACKNOWLEDGMENTS

The study is supported by the Distinguished Research Project of North China University of Water Resources and Electric Power (201208) and the project of the National Natural Science Foundation of China (41402269).

REFERENCES

Jian-yong Li, Jian Xue, Jun Xiao, et al. 2012. Block Theory on the Complex Combinations of Free Planes. *Computers and Geotechnics*. 40:127–134.

Ke Xinli. 2008. Coordination Conversion in Drawing 3D Pipe Using OpenGL. *GEOSPATIAL INFORMATION* 6(3): 53–55.

LI Jian-yong, XIAO Jun, WANG Ying. 2010. Simulation method of rock stability analysis based on block theory. *Computer Engineering and Applications* 46(21): 4–8.

LI Ying-shuo, YANG Fan, YUAN Zhao-kui. 2013. A detection method for 3D circle fitting. Science of Surveying and Mapping. Science of Surveying and Mapping 38(6): 31–32.

Ma Xiaping. 2012. FITTING METHOD FOR SPACE PLANE CIRCLE PARAMETERS WITH CONSTRAINT CONDITIONS. *JOURNAL OF GEODESY AND GEODYNAMICS* 32(6): 86–89.

TANG Di, CHAI Jun-hong. 2004. A Rapid and Efficient Integer Algorithm for the Generations of Circles. *Journal of Liaoning Normal University (Natural Science Edition)* 27(1): 31–33.

WANG Zhixi, WANG Runyun. 2004. Improvement of Bresenham's Circle Generation Algorithm. *Computer Engineering* 30(12): 178–180.

YANG Wei, CHEN Jia-xin, LI Ji-shun. 2009. Two-step spatial circle fitting method based on projection. *Journal of Engineering Design* 16(2): 117–121.

Zhang Xiaopeng, Chen Zezhi. 1996. AN ALGORITHM OF GENERATING A CIRCLE VITH PIECEWISE BEZIER CURVES. JOURNAL OF XI'AN *MINING INSTITUE* 16(1): 85–87.

ZHU Xiao-lin, GAO Cheng-hui, HE Bing-wei, et al. 2010. An Improved Method of Hough Transform Circle Detection Based on the Midpoint Circle-Producing Algorithm. *JOURNAL OF ENGINEERING GRAPHICS* 6: 29–33.

Zou Jingui, Chen Jian. 2012. Research and Application of the Arithmetic Based on Space Vector in 3D Circle Fitting. *Journal of Geomatics* 37(6): 3–5.

Energy, Environment and Green Building Materials – Sheng (ed.)
© 2015 Taylor & Francis Group, London, ISBN 978-1-138-02718-3

Investigation on Urban color and city image in Wuhan

L.Y. Li

School of Art and Design, Wuhan University of Science and Technology, Wuhan, P.R. China

ABSTRACT: Color planning plays a very important role in displaying urbanscape and promoting city image. The paper lists the existing color problems in Wuhan and points out the necessity of changing current situation if Wuhan government wants to promote city image and be more competitive in the future. In this paper, several suggestions are proposed in the hope of supporting instructive ideas for urban color construction in Wuhan. First, basic research should be enhanced for urban color planning; second, urban color theme should be established; third, color pollution in outdoor advertisements should be controlled and reduced; last, universal education of urban color planning should be enforced.

KEYWORDS: Urban color; city image; Wuhan

1 INTRODUCTION

In big cities of China, McDonald's always catches people's eyes with its big yellow "M." However, in Sedona of Arizona, the United States, there is a McDonald's with a blue "M," which is due to the strict rules enacted by the municipal government to keep the whole city landscape harmonious. The government believes that the big shining yellow "M" is alien to the native culture and will damage the beautiful and quiet desert scenery of the city. As a result, McDonald's has no choice or option but to paint its logo "M" blue. In addition, McDonald's is forced to put its logo "M" on the wall of the restaurant, which is usually on a big advertising board highly suspended at the gate of McDonald's, since the local government stipulates specific rules that buildings should not be built over a certain height, so does an advertising board. Through the case of McDonald's in Sedona, it can be easily seen that close attention has been paid by the government to keeping urban colors harmonious. Compared with the government of Sedona, local governments in China pay little attention to the choice of urban colors, and some are not even aware of the relationship between urban colors and city image.

2 URBAN COLORS AS AN ESSENTIAL COMPONENT OF CITY IMAGE

2.1 Concept of city image

City image is the general impression that people have of the city's striking and unique characteristics, and it is also the public opinion on natural conditions, spatial structures, streets, buildings, green belt in the city and folklore, and values and civilization inside the culture. The public naturally connect with the picture in their minds that they had earlier with the city and thus they cherish a sense of acceptance and yearning. As a significant factor that works on the city's subsistence, competition, and development, city image is a strategic resource, a core value, and an intangible asset for its harmonious development.

The establishment and maintenance of city image is a comprehensive system, including five subsystems, namely ethos identity, behavior identity, visual identity, space identity, and, finally, management and promotion, all of which are closely related and interact with each other.

2.2 Importance of urban colors in promoting city image

City is an important carrier for civilization. City image is a direct reflection of certain regional, cultural, and national characteristics. Urban colors, as one of the most important components of city image, pose as one of the most vivid forms representing city image. First, proper choice of urban colors will do good to the establishment of the city's unique and spectacular scenery since colors are explicit to the public and are one of the most striking factors that contribute to the establishment of city image. Orderly, harmonious, and readily identifiable colors will be helpful in establishing distinctive features for a city. Second, urban colors are the main carrier of regional culture, which reveals the nature of a regional culture from certain aspects. Last but not the least, well-planned urban colors can improve environmental quality and help enhance the city's competitive strength from multiple levels, thus leaving visitors with a striking impression of the city.

3 EXISTING COLOR PROBLEMS IN WUHAN

In 2003, Wuhan Urban Planning Administration promulgated two important documents, "Technical Guidelines for Construction Colors in Wuhan" and "Regulations on Construction Colors in Wuhan." The two documents put forward specific methods and provisions on the control of building colors in Wuhan and have made contribution to developing integrated, harmonious, and orderly urban colors. However, from the current situation, color problems in Wuhan still exist and mainly focus on several aspects.

3.1 Lack of color "theme"

When we mention Santorini in Greek, simple white geometrical buildings against a sky-blue sea will immediately emerge in our mind. Speaking of the capital of Iceland, Reykjavik, we will be reminded of the limitless vitality of colorful buildings against blue sky and azure sea. However, if visitors are asked for their impression of the color theme of the city Wuhan, they may find it a difficult task because urban colors in Wuhan are quite disordered and lack a main tune. In order to make their products or services more prominent, some commercial companies use highly saturated or metallic colors as the main color of their buildings, ignoring their discoordination against surrounding buildings and the environment.

3.2 Color abuse in advertisements

A major source that causes urban colors in Wuhan to be disordered is that shopping centers misuse colors in their commercial advertisements. Whether in central business districts or shopping streets in residential communities, advertisers overuse strong colors to make a sharp contrast. Taking the Walking Street in Optics Valley, for example, we will find the building facades are covered with huge-scale advertisements whose colors are striking but fragmented without considering keeping concert with each other.

3.3 Improper color design on public vehicles

Public vehicles, mainly taxi and bus, are mobile platforms that are intended to display a city's spirit. However, the taxis in Wuhan are chaotically painted with many colors instead of the same color. Therefore, when you stand on the street in Wuhan, you may feel dazzled at the sight of taxis with different colors in the city. Buses in Wuhan also share similar problems, and the advertisements on the body of buses are huge and messy with striking colors, moving around the city and making a nuisance everywhere.

The problems just mentioned pertaining to Wuhan have a negative impact and prevent Wuhan from developing a good image. A question that we must answer is how to enhance the city's color identity and clear up the incongruousness of mobile colors in the city.

4 PROPOSALS FOR WUHAN'S URBAN COLOR PLANNING

4.1 Enhancing basic research of urban color planning

Urban color planning should be combined with urban master planning. Bureau for Municipal Design should start the relevant investigations, including the studies of the current situation of the city's urban colors, the historical evolution, structural layout, and functional division of the city. Simultaneously, the color inclination that the public hold toward Wuhan should not be neglected. The purpose of investigating the current situation of the city's urban colors is to have a good understanding of the characteristics and existing problems of Wuhan's urban color so as to provide pertinent programs for reconstruction and planning. Investigation of the historical evolution of the city will help in providing a better understanding of color marks in different periods of the city's evolution and making typical local culture stand out in the color planning of urban development. Clarifying the city's structural layout and functional divisions is especially necessary for Big Wuhan since different divisions play different roles in the city operations. We should employ the characteristics of different colors to enhance the functional features of different divisions when making urban color planning. Having a good understanding of the color inclination of the public hold means that we should figure out the likes, needs, and desires of the public hold for the urban color. It is an essential reference for the urban color planning.

4.2 Establishing color theme for the city

Many cities abroad have their specific landmark colors, such as beige in Paris and Khaki in London. Many domestic cities also start planning their urban colors. Nanjing government agrees on the proposal to set "cyan, grey and white" as the theme of the urban colors. Harbin government is determined to choose beige and white as the theme of the urban colors.

Wuhan is a typical subtropical inland city with an abundant surface water system and four distinct seasons. Compared with coastal cities, Wuhan is rich and diverse in its environmental background, which helps provide basic hues for the choice of the urban color. Since Wuhan is a city with cold winter and sultry summer, we should not select bright colors or cool tones so as not to exert a negative impact on the public's mentality. Therefore, the main tone of Wuhan should be lively and lightly shallow warm colors

4.3 Eliminating color pollution through the regulation of outdoor commercial advertising

Reasonably employing colors is the key to regulating outdoor advertisements in the central business districts and shopping streets in residential communities. When planning, we should seriously study the scaling, positional, and color relation between the billboards and surrounding buildings, and simultaneously, the official departments concerned should pull down the advertisements whose color or size poses damage to the surrounding environment. Moreover, specific rules should be enacted for controlling outdoor commercial advertising and business companies should also have a deep understanding and acknowledgment of the importance of urban color planning.

4.4 Enforcing the popularity of color education

In many developed countries, the public have achieved a social consensus for the protection of urban colors, and the governments also attach great importance to improving their people's awareness of protecting urban colors. However, not enough attention is paid to urban color in China. The protection and planning of urban colors proposed by a few of scholars and experts are not universally supported by the public and most people are not aware of the concept about urban color. Even some college students majoring in Architecture Design or Landscape Design, along with concerned practitioners, pay little attention to

the importance of urban colors. There are still many gaps to be filled in color research, consultation, and practical applications.

5 SUMMARY

The city Wuhan is promoting its image to enhance competitive power through designing diverse architecture forms and enforcing strict but reasonable urban planning. However, if there is no appropriate planning for urban color construction, it is impossible for Wuhan to exhibit its inner beauty. Wuhan is undergoing a fast economic development with a new construction upsurge, expanding city area and rising status, and it is a significant issue for the local government to well plan urban color construction so as to show the charm of the city. If Wuhan government can employ harmonious urban colors to depict the world as an energetic modern city with abundant natural resources, a rich historical heritage, and convenient urban facilities, it will promote the city's popularity and attraction for the outside world to a great extent.

REFERENCES

[1] Wei Xiang, Xiong Xiangning, Huang Shenghui, investigation on Urban Color Planning of Wuhan, Planner. [J]. (2003).
[2] Yang Fen, Searching for Lost Colors: Investigation on Urban Color. [D]. Wuhan: Hubei University of Technology, (2006).
[3] An Ping, Research on the Planning of City's Colorscape: A Case Study of Tianjin Central District. [D]. Tianjin: Tianjin University, (2010).
[4] Jiang Fang, Huang Weixiu, Analysis of landscape design of Urban Colors: A Case Study of Urban Color in Wuhan, Journal of Hubei University of Education. [J]. (2009).

Energy, Environment and Green Building Materials – Sheng (ed.)
© *2015 Taylor & Francis Group, London, ISBN 978-1-138-02718-3*

Research on a novel deflection surface concentrator

C.F. Ye, Q.L. Qiu, Y. Wu, & J. Zuo
Key Laboratory of Energy Thermal Conversion and Control of Ministry of Education, School of Energy and Environment, Southeast University, Nanjing, Jiangsu, China

ABSTRACT: Based on the disadvantages of complexity and high cost in the traditional parabolic trough mirror manufacturing process, a novel deflection surface concentrator was studied. First, the hot-bending process was introduced, and then a dynamically theoretical model of the whole hot-bending process was created based on the strong viscoelasticity when the glass was under high temperature. Second, the surface contour and roughness experiments were carried out after the hot-bending process, which showed that the deflection surface had a precise surface contour and superior surface characteristics. Finally, the concentrating characteristics caused by the tailor ratio, initial deflection, and incidence angle were analyzed, respectively, which can guide the design of deflection surface concentrator and further mirror groups in the project.

KEYWORDS: solar energy; deflection surface concentrator; surface roughness; concentrating characteristic; mirror group

1 INTRODUCTION

Reflectors are an essential part of CSP (concentrating solar power) systems, and they account for about 20% of the total cost [1]. The commonly materials used in trough mirror include glass, polymer composite material, and aluminum mirror, among which thick glass mirrors can maintain their reflectance very well in CSP environments, leading to a wide application in the current commercial trough projects (SEGS/NS1/APS/AndaSol-1) [2].

According to the difference in glass hot bending, two main methods are employed in the manufacturing of traditional parabolic trough mirrors, namely gravity method and compression molding. Gravity method refers to the replication of the profile of the high-precision mold surface under gravity at the softening temperature, which is relatively convenient but the error is big. The mirror manufactured through compression molding is accurate, but the process is complex. Moreover, the glass plates are in contact with the mold in both the mirror manufacturing methods under softening temperature, which destroyed the smooth surface formed in the float process, leading to a high cost of polishing process [3, 4]. Therefore, in order to simplify the production process and reduce production cost further, much in-depth research in concentrator structure and hot-forming process has been conducted by experts and scholars. Among these, a zigzag surface of parabolic trough concentrator was studied in [5,6], and a mechanical bending method was put forward to make cambered surface solar mirror in [7]; in addition, a novel deflection surface concentrator was studied in [8–10].

For the purpose of reducing the cost of mirrors and for further promoting the development of solar energy heat utilization system, the concentrating characteristics of deflection surface concentrator were studied in this paper. And the concentrating characteristics caused by tailor proportion, initial deflection, and incidence angle were analyzed, respectively, hoping the results can guide the design of deflection surface concentrator and mirror group in the project.

2 PRINCIPLE AND MODEL

Solar deflection surface concentrator is a novel non-imaging concentrator that possesses the advantage of a simple and economic production process. As shown in Fig. 1, the glass plate supported by hollow mold bends under its own weight by heating [11]. Within the scope of transformation temperature, the viscosity of the glass falls sharply, and behaves similar to viscoelasticity between elasticity

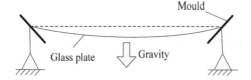

Figure 1. Diagram of glass plate hot bending.

and viscosity. Therefore, it is possible to express the dynamic change of glass surface contour in the whole hot-bending process based on the viscoelastic model.

The glass property can be described by Burgers model when the bending temperature is 420–560°C; whereas it should be described by Maxwell model at 560–630°C [12]. Based on our experience in manufacturing, reasonable hot-bending temperature generally ranges between 500°C and 550°C. Therefore, the Burgers model, which contains all the properties of the viscoelastic material [13], is suitable to describe the thermoforming process of the glass plate. This model is shown in Fig. 2.

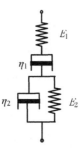

Figure 2. Burgers model.

Under constant stress σ, the creep equation $\varepsilon(t)$ and creep compliance $J(t)$ with increasing t obey the following equations:

$$\varepsilon(t) = \frac{\sigma}{E_1} + \frac{\sigma}{E_2}[1 - \exp(-\frac{t}{\tau_s})] + \frac{\sigma}{\eta_1} \quad (1)$$

$$J(t) = \frac{\varepsilon(t)}{\sigma} = \frac{1}{E_1} + \frac{1}{E_2}[1 - \exp(-\frac{t}{\tau_s})] + \frac{1}{\eta_1} \quad (2)$$

where E_1, E_2, η_1, and η_2 are elastic modulus and viscosities of the viscoelastic material, and τ_s is defined as the term "retardation time" to describe creep rate in the viscoelastic deformation.

In the process of hot bending, ignoring the effect of glass width, the glass plate can be simplified as a viscoelastic beam. At a uniform temperature field, according to the corresponding principle of viscoelastic mechanics,

the glass plate creep deformation equation can be obtained:

$$y(t,x) = \frac{qx}{24I}(x^3 - 2lx^2 + l^3)\{\frac{1}{E_1}$$
$$+ \frac{1}{E_2}[1 - \exp(-\frac{t}{\tau_s})] + \frac{1}{\eta_1}\} \quad (3)$$

where $y(t, x)$ is the deflection at horizontal coordinate x at time t; q is uniform load, $\rho g w h$, where ρ, g, w, and h are density, acceleration of gravity, glass width, and glass thickness, respectively; l is the hot-bending span or glass length; and I is sectional moment of inertia of the plate glass, $wh3/12$.

3 EXPERIMENTS

3.1 Deflection surface contour experiment

In the experiment, the glass plate with geometries of 1500 mm × 1000 mm × 4 mm (length × width × height) was regarded as the experiment object. First, the glass plate was placed in the hot-bending furnace and heated; second, after the hot bending was finished, the glass was removed from the furnace placing vertically to reduce the internal stress when the glass temperature was reduced to about 100°C; at last, the deflection surface contour was measured by the caliper after the temperature was near air temperature.

During the measure test, every 0.15 m was taken as a test point, and a set of theoretical and measured deflection was shown in Table 1.

Table 1. Deflection error between theoretical and measured deflection.

Span length (m)	Theoretical result (mm)	Measured result (mm)
0	0	0
0.15	12.76	13.2
0.3	24.11	24.0
0.45	32.99	33.0
0.6	38.61	38.5
0.75	40.5	40.5
0.9	38.51	38.2
1.05	32.79	32.1
1.2	23.83	23.0
1.35	12.42	13.0
1.496	0	0

The curvature error between the measured and theoretical result is evaluated by the deflection variation in Equation 4:

$$\Delta Zi = Zi_m - Zi_t \tag{4}$$

where Zi_m and Zi_t are measured deflection and theoretical deflection, respectively. The root mean square (*RMS*) of the deflection error is given by Equation 5:

$$RMS = \sqrt{\frac{1}{n}\sum_{i=1}^{n}\Delta Z_i^2} \tag{5}$$

where n is the total number of points. As shown in Fig. 3, *RMS* of deflection errors ranges between 0.4 and 0.8 mm. Compared with hot-bending maximum deflection, the surface error is small, so the deflection surface possesses a precise surface contour.

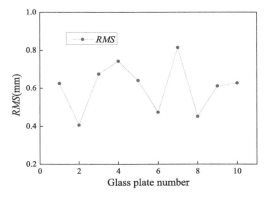

Figure 3.   Surface contour *RMS* errors of the hot-bending deflection surface mirrors.

3.2   Surface roughness experiment

For concentrating reflector, a high mirror reflectivity is needed to avoid the generation of diffuse reflection. According to the modern optical film theory [14], if the reflective surface wants to avoid producing diffuse reflection of a certain wavelength of light waves, the roughness of reflecting surface *Ra* must meet the following equation:

$$Ra \leq \frac{\lambda}{16} \tag{6}$$

About 98% of the solar radiation energy is concentrated in the wavelength band between 0.3 μm and 3 μm [15]. So if the specular reflection is required to produce, the surface roughness of the mirror must be less than 18.75 nm according to Equation (6).

In the experiment, 10 pieces of 1500 mm × 1000 mm × 4 mm glass plates before and after the hot-bending were tested and statistically analyzed. And 20 measure points were randomly drawn from the front and back of each piece of the glass plate. The roughness measuring instrument selected for the present experiment was MahrPerthometerM1 (Fig. 4).

Figure 4.   Roughness measuring instrument—Mahr Perthometer M1.

The roughness values *Ra* are obtained by two sides of the 10 pieces of the glass plate for statistical analysis, and the average values of 20 points for each side of the mirror are shown in Table 2.

Table 2.   Surface roughness before and after hot bending.
(Unit : mm)

Glass number	Concavity before bending	Convexity before bending	Concavity after bending	Convexity after bending
1#	10.3	9.25	8.9	10.4
2#	10.65	10.2	8.6	11.15
3#	7.9	8.05	8.55	12.45
4#	7.25	7.65	9.55	12.75
5#	10.6	8.75	14.4	11.45
6#	11.5	10.35	15.1	10
7#	7.55	9.2	11.35	10.65
8#	8.05	10.2	10.9	9.25
9#	8.45	8.9	10	12.4
10#	9.25	12.45	10.6	14.65
Average *Ra*	10.7		10	

From Table 2, it can be found that the glass surface roughness before and after hot bending meets the requirement less than 18.75 nm, indicating that the mirror has an excellent surface property. The average value of *Ra* changes from 10.7 nm to10.0 nm after the hot bending, which may be caused by viscoelasticity of the glass at a high temperature.

4 ANALYSIS OF CONCENTRATING CHARACTERISTICS

4.1 Influences of tailor ratio on concentrating characteristics

Through the analysis of reflected beams of the reflector, we found that the reflected beams gathered most closely from most of the mirror in the middle, but the rest of the edge of the mirror loosely focused on the beam, resulting in a wide focal spot. Therefore, it is possible to significantly improve concentrating effect by intercepting the edge of the mirror. Based on a mirror with an initial glass span of 1500 mm and a deflection of 60 mm, the influences of tailor ratio on concentrating characteristics were analyzed. As shown in Fig. 5, with the decreasing span after cutting the focal spot width and focal length decrease, the concentrating ratio changes reversely. The focal spot width ranges from 197 mm at the span of 1500 mm to 46 mm at the span of 860 mm, showing that the concentrating effect is significantly improved. But taking a big tailor ratio that causes wasting of the mirror is uneconomical. Taking into account the size of common absorbed pipe diameter of parabolic trough, about a 1/3 tailor ratio is accepted in project design.

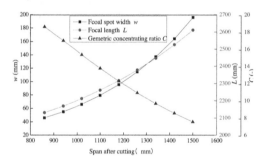

Figure 5. Influences of span after cutting out concentrating characteristics.

4.2 Influences of deflection on concentrating characteristics

Deflection surface concentrator is a novel non-imaging concentrator, and its concentrating characteristics have corresponding changes with the change of curve parameters. So, it can be developed with a variety of usage, such as the short focal length concentrating mirror, mirror group, and the long focal length linear Fresnel reflector. The influences of the initial deflection on the concentrating characteristics within the scope of 40 mm-70 mm were analyzed next, based on the mirror at a span of 1500 mm with a 255 mm tailoring quantity on both sides.

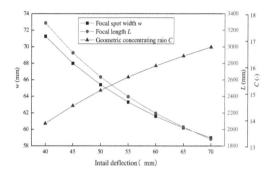

Figure 6. Influences of initial deflection on concentrating characteristics.

As shown in Fig. 6, with the increase in initial deflection, the focal spot width and the focal length decrease, whereas geometric concentrating ratio increases from 13 to 17. Within the scope of analysis, the focal spot width varies between 59 mm and 72 mm, and the focal length changes within the range of 1900 mm and 3300 mm. For a single deflection surface concentrator, the analysis results are useful to guide the design of a single deflection surface mirror in the project.

4.3 Influences of incident angle on concentrating characteristics

The single deflection surface concentrator has good concentrating characteristics, but it is limited to achieving a high concentrating ratio. In order to achieve the high concentrating ratio, mirror groups with combination of mirrors are put forward. In the process of mirror group design, the incidence angle and position of the mirror need to be suitably selected based on the design requirements. So the analysis of the influences of incident angle under different deflection on the focusing effect can provide a guide for the design of the deflection surface mirror group. The analysis conditions are as follows: the glass span of 1500 mm under the optimal tailor scheme with a constant 990 mm span; the initial deflection y ranges within 40 mm and 70 mm; and the incidence angle ranging from 0° to 25°.

As shown in Fig. 7, the influences of incident angle on focal width and focal length are display in the following manner: With the increase of incident angle, the optimal focal spot changes a little and has a slightly decreasing trend, but the focal length decreases sharply. In the engineering design of the deflection surface mirror group, according to the requirement of design, incident angle is calculated, and then appropriate initial deflection is chosen.

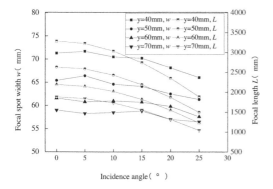

Figure 7. Influences of incident angle on focal width and focal length.

This process is repeated, and the whole design of the mirror group will be completed.

5 MIRROR GROUP DESIGN

Design demonstration: Take the solar thermal used absorber pipe with a diameter of 70 mm and a symmetrical distribution of 8-column deflection surface mirror group as an example. The design parameters are as follows: focal length is 2000 mm; focal spot width is 63 mm; the middle symmetry mirrors spacing (operation channel width) is 300 mm; and the rest of the adjacent mirror column spacing is 10 mm.

According to the position relationships between mirrors, first of all, roughly calculate the rotation angle of the first mirror and choose the appropriate initial deflection in Fig. 7. Then, fine tune the parameters of the mirror to meet the design requirements and the other mirrors can be completed with the same method one after another. Figure 8 was a completed mirror group, which had a focal spot width of 63 mm and a focal length of 2000 mm. The total aperture of the mirror group was 7247.59, and its geometric concentrating ratio reached 115.

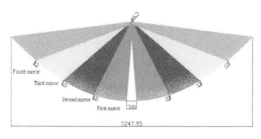

Figure 8. Mirror group of deflection surface.

6 CONCLUSIONS

In this paper, the main conclusions are as follows:

1 The deflection surface mirror has a precise surface contour and superior surface characteristics.
2 Considering economy and concentrating effect, the tailor ratio should be controlled in about 1/3.
3 With the increasing initial deflection, the focal spot width and the focal length decrease, whereas geometric concentrating ratio increases from 13 to 17.
4 Influences of incident angle on focal width and focal length under different deflection can be used to guide the design of deflection surface mirror group.

ACKNOWLEDGMENT

The authors would like to thank the National Natural Science Foundation of China (No. 51278190) for the financial support rendered.

REFERENCES

[1] H. Price, Assessment of parabolic trough and power tower solar technology cost and performance forecasts. Sargent & Lundy LLC Consulting Group, National Renewable Energy Laboratory, Golden, Colorado. 2003.
[2] D.W. Kearney, Parabolic trough collector overview. Parabolic trough workshop. 2007, 200.
[3] Y. Chen, A.Y. Yi, Design and fabrication of freeform glass concentrating mirrors using a high volume thermal slumping process. Solar Energy Materials and Solar Cells, 2011, 95(7): 1654–1664.
[4] L. Su, Experimental and Numerical Analysis of Thermal Forming Processes for Precision Optics. Ohio State University, 2010.
[5] D. Ning, Developments of parabolic trough for solar energy system. Acta Energiae Solaris Sinica, 2003, 24(5): 616–619.
[6] J.X. Ma, D. Ning, Y.Y. Hong, et al. Developments in zigzag parabolic like for solar energy system. Acta Energiae Solaris Sinica, 2004, 25(5): 643–646.
[7] S.S. Xu, Q.M. Li, J.T. Zheng, Faming Zhuanli Shenqing Gongkai Shuomingshu, CN102495440A (2012) (in Chinese).
[8] X.Y. Li, Y.M. Zhang, W. Gu, et al. Experiment research of the deflection trough concentrators array. Acta Energiae Solaris Sinica, 2012, 33(003): 391–396.
[9] R. Kuang, Y. Wu, J. Zheng, et al. Analysis of optical properties and geometrical factors of linear deflection surface concentrators. Solar Energy Materials and Solar Cells, 2014, 121: 53–60.
[10] Y. Zhang, W. Gu, X.Y. Li, et al. Experiment Research On Solar Cooker with Deflection Trough Concentrators and Heat Pipe. 13th International Conference on

Non-Conventional Materials and Technologies: Novel Construction Materials and Technologies for Sustainability, 13NOCMAT 2011.

[11] A. Aronen, R. Karvinen, Explanation for edge bending of glass in tempering furnace. Proceedings of Glass Performance Days. 2009: 575–579.

[12] H.J. Shan, Y.L. Zhang, M.K. Zhou, The Differences of Rheological Behaviors of Glass under Compression and Tension. Glass& Enamel. 2004, 32(4).

[13] D. Roylance, Engineering viscoelasticity. Department of Materials Science and Engineering–Massachusetts Institute of Technology, Cambridge MA, 2001, 2139: 1–37.

[14] J.F. Tang, P.F. Gu, X. Liu, Modern optical thin film technology. Zhejiang University press, 2006:386–388.

[15] J.A. Due, W. A. Beckman. Solar engineering of thermal processes. 3rd ed.USA,Hoboken: New Jersey:John Wiley & Sons,Inc.2006:3–9,325.

Energy, Environment and Green Building Materials – Sheng (ed.)
© *2015 Taylor & Francis Group, London, ISBN 978-1-138-02718-3*

Wear resistance characterization and analysis of hard particle reinforcement in composite Fe-based alloy +WC+ Cr_3C_2 laser cladding

J. Xu, G.B. Li, X.Z. Du, J. Hao, W.H. Tang & H.B. Wang
China North Material Science and Engineering Technology Group Corporation, NingBo, China

H.Y. Zhao
Institute of Electrical and Information Engineering, Department of Materials Science and Engineering, China University of Mining and Technology, Beijing, China

ABSTRACT: In this work, WC+Cr_3C_2-reinforced iron-based alloy wearing resistant coatings was fabricated on the surface of 45 steel by high-power CO_2 laser. Comparative abrasive tests between laser cladding before and after adding WC+Cr_3C_2 have been performed by following M200 sliding wear-test method. The results showed that dark gray fine dendrites and gray particle WC were distributed in the cladding layer. The main mechanism during the wear process was abrasive wear. The reinforcing phase played a leading role in withstanding loads and pinning, which strengthened the matrix material during the wear process, making the wear volume of the coating layer substantially lower.

KEYWORDS: laser cladding; wear resistance; coating; Fe-base alloy; WC+ Cr_3C_2

1   INTRODUCTION

During the past two decades, the power laser in industrial applications has been rapidly developed, and it is now used in a variety of applications, such as in the wear, corrosion, and manufacturing working environments [1,2]. Compared with traditional hardfacing, laser cladding can not only obtain a hard, wear-resistant metallurgical surface combined with substrate but also be observed in a typical dilution of 1–2%, whereas the usual dilution of hardfacing is 5–10%[3]. Because of the few volumes of melt and the extremely fast cold vector, the size of the grain of laser cladding layer is much smaller, and the grain refinement process is significantly fulfilled. Therefore, the surface performance of the softer components could be substantially improved with low dilution and dimensional stability; laser cladding appears as a very exciting method of surface hardening[4].

Nowadays, there are mainly three common available cladding materials: Co matrix, Ni matrix, and Fe matrix[5]. Although the Co matrix and Ni matrix alloy cladding layers have better oxidation and abrasion resistances, they are too expensive to be widely used. Since Fe matrix alloy is similar to steel materials in composition, the firmly combined interface could be formed, and it could reduce the usage of precious metals such as Co and Ni. More and more attention is being paid to Fe-based alloy laser cladding[6]. To increase the life of the components submitted to serious working conditions, the addition of various kinds of hard particles into self-fluxing alloy forming composite coatings with good hardness, abrasion resistance, impact bearing, and erosion or corrosion resistance has been investigated and developed[7,8].

In this paper, a comparative study is conducted between Fe-based alloy laser cladding coatings to which appropriate WC + Cr_3C_2 are added or not. The wear resistances of the two coatings were evaluated, and the principle of the wear resistance of hard particles has been preliminarily presented.

2   EXPERIMENTAL APPROACH

2.1   *Samples preparation*

The claddings composed of Fe-base matrix reinforced with WC + Cr_3C_2 particles were deposited on a carbon steel substrate (Fe + 0.45Cwt.%). The square samples were 12 × 16 × 10 (mm), with a trough 12 × 10 × 2.5 (mm) on the surface 12 × 16 (mm). The samples were washed and sanded. To produce the composite laser claddings, the premixed powders described in Table 1 were dried and coated into the trough earlier. The binder was an aqueous solution of sodium silicate.

The TJ-5000CO_2 laser was employed to provide a continuous wave till 5.0 KW. The process parameters were tried many times to obtain claddings with good bonding and minimal dilution. The final parameters are shown in Table 2.

Table 1. Elements of cladding material and nomenclature of samples.

Elements of cladding material (wt%)	Nomenclature	
	1	2
B	0.4	0.3
C	1.15	2.64
Si	0.36–0.54	0.48–0.57
W	0.01	8.45
Cr	16.36	18.65
Mo	0.2	0.15
Ni	9.55–10.00	9.23–9.78
Fe	bal	bal

Table 2. Laser cladding processing parameters.

Parameters	Quantitative value
Laser power (kW)	2.0
Scanning speed (mm/min)	200
Spot diameter (mm)	2.5
Preset thickness (mm)	1.5
Overlape rate (%)	20

In order to be characterized by optical microscopy and phase analysis, the cladded samples were polished smoothly by abrasive paper and then to a mirror by using diamond paste with a grain size of 1 μm. The reagent was used to etch the laser cladding layers to reveal the microstructure (20%volHNO$_3$ + 30%volHF + 50%volH$_2$O).

2.2 Wear resistance experiment

Before the abrasion test, the surface of each clad layer was ground to a smooth finish using a 600-grit diamond wheel. The hardness of the specimens was then measured using an indenter for the HV test. The microhardness was measured by HBRVU-187.5 Hardness Tester. To obtain a characteristic hardness for the laser cladding coatings, a mean value was obtained based on five measurements. The slide wear test was finished by M-200 wear testing machine, load 50 kg, speed 240 r/min, and time 60 min. The grinding ring was made of hard metal. The ring was φ 40 mm × 10 mm, (HRA)84. The lubricant was ChangCheng 4502 compressor oil. The optical microscope was used to measure the length of each of the wear samples every 3 μm. The abrasion test results are reported as volume loss in cubic microns. From these results, it is observed that the materials with a higher abrasion resistance have a lower volume loss.

According to the formula[9]:

$$V = b \left[r^2 \sin^{-1}(\frac{d}{2r}) - \frac{d}{2} \sqrt{r^2 - (\frac{d}{2})^2} \right]$$

The volume of wear could be calculated, and the evaluation of abrasion resistance could be achieved. In the formula, V stands for volume loss, r stands for the radius of the ring, d stands for the width of the simples, and b stands for the length of the wear mark.

3 RESULT AND DISCUSSION

3.1 Microstructural observations

The microstructure of a typical laser clad material reinforced with hard particles is presented in Fig. 1. The cladded region is formed by a fine microstructure. As illustrated by the dendrites oriented perpendicularly to the free surface of the coating, the solidification is initiated at the clad/substrate interface and oriented toward the surface of the clad region (following the direction of the heat flux).

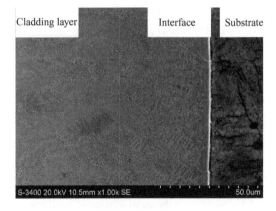

Figure 1. Laser cladding sample, clad/substrate interface.

Around the WC particles, modifications are observed in the clad material. The dendrites are oriented in the radial direction around the particles as they appear to be similar to chrysanthemum (Fig. 2). Because of the larger granularity and high melting point, partial melting of WC particles could be observed after laser cladding whereas all Cr$_3$C$_2$ particles were melted and disappeared. A good dispersion of the WC particles is obtained in the cladding, through the thickness and perpendicular to it (see Fig. 3). The particles are not melted, and their original shape was conserved. No micro-cracks were observed in the cladding, although the samples are applied on cold substrates and on rigid components. This is because the laser processing parameters have been repeatedly experienced and suitably chosen.

Figure 2. Typical laser cladding material with WC and Cr$_3$C$_2$ particles.

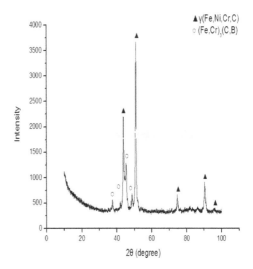

Figure 4. XRD pattern of coating without hard particles.

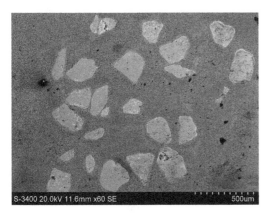

Figure 3. Laser cladding material—distribution of the spherical fused tungsten carbides.

3.2 Phase composition of the cladding layers

The phase composition of the cladding layers is shown in Fig. 4 and Fig. 5 by XRD. As illustrated in the pictures, there are γ (Fe, Ni,Cr,C), and (Fe,Cr)$_3$(B,C) made up by Fe-Cr-B-C in the layer before the addition of the particles. (Fe,Cr)$_3$(B,C) is similar to Fe$_3$C in atom structure. Cr and B took the places of Fe and C in Fe$_3$C partly since there are adjacent positions of the related elements in the periodic table of chemical elements. Compared with the coatings without reinforcing phase, the new phases of the cladding layers, including WC and Cr$_3$C$_2$, are (Fe,Cr)$_{23}$(B,C)$_6$ and unmelted WC. With the increase of Cr and C by adding Cr$_3$C$_2$, new carbide is combined to be formed. Likewise, (Fe,Cr)$_{23}$(B,C)$_6$ is similar to Cr$_{23}$C$_6$ in atom structure and Fe and B partly took the places of Cr and C in Fe$_3$C.

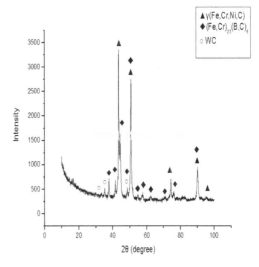

Figure 5. XRD pattern of coating with hard particles.

3.3 Wear resistance and analysis

To evaluate the wear resistance of the reinforced laser cladding samples produced and the samples with no hard particles, comparative abrasion tests were performed using oil lubrication and hard metal wheel abrasion test. The microhardness was measured, and the abrasion tests were performed. The microhardness of the layer with hard particles achieved HV750, whereas the layer without reinforcing phase was only HV550. The abrasion tests results obtained are presented in Fig. 6 and Fig. 7.

81

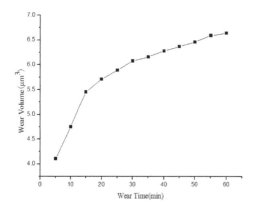

Figure 6. Curve of the worn volume with time of coatings with hard particles.

As presented, the ultimate volume of removed material for the Fe alloy coating is higher than the one including WC and Cr_3C_2. It could also be found that 15 mins after the starting of the experiment, the coatings with WC and Cr_3C_2 reached their stable wear stage. However, the coatings without WC and Cr_3C_2 had almost reached their stable wear stage at the beginning of the experiment. Wear curve fitting shows that the wear rate in stable wear stage of the composite coatings is much lower than that of the Fe-alloy coatings. We can infer that there would be a long period before the severe wear stage coming, so the coating with WC+Cr_3C_2 represented a better wear resistance. The Fe-WC+ Cr_3C_2 laser cladded samples showed much higher hardness and much better wear resistance when compared with coated samples. This phenomenon verified that there is a general correspondence between hardness and wear resistance.

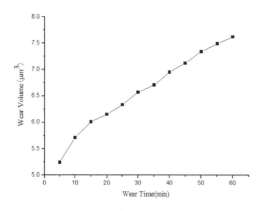

Figure 7. Curve of the worn volume with time of coatings without hard particles.

Pictures of the specimen surfaces, after the abrasion test, are presented in Fig. 8 for comparison.

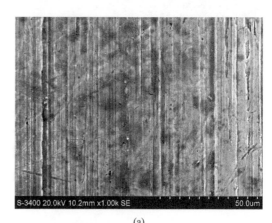

Figure 8. Pictures of the sample surfaces after the abrasion test. (a) Coating of Fe-alloy, (b) Coating with WC+ Cr_3C_2.

As illustrated in Fig. 8a and b, a higher density of the wear marks is obtained for the laser cladding without hard particles. The finer dendritic microstructure of the coating that has formed is the reason for hard particles. Cr_3C_2 was melted during the laser cladding; however, re-crystallizing in the grain boundary as reinforcing phase $(Fe,Cr)_{23}(B,C)_6$ existing in the cladding layer, which prevented the grains from growing larger[10]. And because of its grain boundary strength, the plow of the material was reduced significantly. The partly melted WC particles exist as little projections, resisting the wear from the ring. This reduces the chance for the coating to touch the hard metal and, consequently, reduces the final volume loss. Both the reinforcing phases also play a role in withstanding the loads and strengthening the matrix material in the wear process. This prevented the plastic deformation of the material and simultaneously obstructed the crack propagation, making the wear volume of the coating layer be much lower than the one with no reinforcing phase.

From the pictures of the worn surface of the clad layers, the wear mechanism during the process appears to be a removal of the binder phase. It should be abrasive wear from the furrows and tiny pieces of the debris[11]. The WC particle appears to be intact, and no evidence of fragmentation of the WC particles is observed.

4 CONCLUSIONS

1 The rapid solidification rate combined with the low dilution obtained with the laser cladding has a significant impact on the wear resistance of the coating produced.
2 The laser cladding coatings' abrasive wear resistance is obviously higher by adding WC+ Cr_3C_2 than that of the coatings with no hard particles.
3 The main wear mechanism during the process was abrasive wear. Unmelted WC bear the loads and are recrystallizad as ($Fe,Cr)_{23}(B,C)_6$, pinning down the grain boundaries. They improved the wear resistance of the laser cladding coatings in two different ways.

ACKNOWLEDGMENTS

The author especially acknowledges the materials analyzing laboratory, Tsinghua University for the use of their XRD facilities. MaFeng and FanLei are highly appreciated for their useful contribution in the microstructure analysis.

REFERENCES

[1] Riabkina-Fishman M, Babkin E, LeVin P, etal. Laser Produced functionally graded tungsten carbide coatings on M2 high-speed tool steel[J]. Materials Science and Engineering, 2001: A302: 106–114.
[2] Michael Factor, Itzhak Roman. Vickers microindentation of WC-12%Cothermal spry coating, Part 1: Stantistical analysis of microhardness data[J], surface and coating technology, 2001, 132: 181–193.
[3] So H, Chen C T, Chen Y A. Wear behavior of laser clad satellite alloy6[J]. Wear, 1996, 192(l-2): 78–84.
[4] Modonald G, Hpendricks R C. Effect of thermal cycling on ZrO_2-Y_2O_3 thermal barrier coating[J]. Thin Solid Films, 1980, 73: 491–496.
[5] Riabkina-Fishman M, Babkin E, LeVin P, etal. Laser Produced functionally graded tungsten carbide coatings on M2 high-speed tool steel[J]. Materials Science and Engineering, 2001, A302:106–114.
[6] Steen W. Laser Surface Cladding[C]. In: Clifton W Drapered. Laser Surface Treatment of Metals, London: Pergamon Press, 1988: 369–387.
[7] Ricciardi G, CantelloM. Surface Treatment of Automobile Parts by RTM[J]. Laser Beam Surface Treating and Coating, 1988, 957: 66–74.
[8] W.Liu, J.N.DuPont. Effects of substrate crystallographic orientations on crystal growth and microstructure development in laser surface-melted super alloy single crystals: Mathematical modeling of single-crystal growth in a melt pool (Part11)[J]. Acta Materialia, 2005, 53(5): 1545–1558.
[9] MANNA I, MAJuMDAR J D, CHANDRA B R, et al. Laser surface cladding of Fe-B-C, Fe-B-Si and Fe-BC—SiAl-C on plain carbon steel[J]. Surface and Coatings Technology, 2006, 201(1–2): 434–440.
[10] Komvopoulos K. Effect of process parameters on the microstructure geometry and microhardness of laser clad coating materials[J]. Material Science Forum.1994, 163: 417–421.
[11] Das D K. Surface roughness created by laser alloying of aluminum with nickel[J]. Surface and Coating Technology, 1994, 64: 11–15.

Energy, Environment and Green Building Materials – Sheng (ed.)
© 2015 Taylor & Francis Group, London, ISBN 978-1-138-02718-3

Research on whole industry chain of housing industry modernization in big data era

X.M. Yang

Rongzhi College of Chongqing Business and Technology School, Chongqing, China

ABSTRACT: Through the main problem restricting China's housing industry modernization of research, we find a series of problems of China Housing Industry Modernization: insufficient number of affiliates to drive, failure to reach corporate group structure, and existing inability to connect to a complete industrial chain. This paper presents big data for innovative contributions to the housing industry that will be reflected in the formation of the true meaning of the industrial chain, thus enhancing the whole industry chain development mode and countermeasures.

KEYWORDS: Big data; modernization of housing industry; whole industry chain

1 INTRODUCTION

The housing industry was popular in the 1960s due to the Japanese concept of Europe. Its basic model is built depicting industrialization and housing component parts standardization. Due to the traditional way of production house "Three high" problem (high energy consumption, material consumption, high pollution), the situation is even worse in China. Depending on the relevant data, civil energy is about 49.5% of the total energy consumption in our country (building materials production accounts for about 20%, about 1.5% of the construction of energy consumption, and energy use accounts for about 28%). Through housing industry modernization, it will largely reduce labor inputs, resource consumption, and energy consumption. This goal will be reached in the green, energy-saving, environmental protection and sustainable development. Therefore, the housing industry has been the focus of concern in China's real-estate construction and building industry.

2 DEVELOPMENT STATUS QUO OF CHINA HOUSING INDUSTRY MODERNIZATION, AND THE MAIN PROBLEM RESTRICTING ITS DEVELOPMENT

2.1 *Modern developments in the housing industry in China*

Housing industry modernization: It is based on the market-oriented demand; it relies on building materials and light industry, for factory production of numerous residential parts and with site assembly as a basis. It joins the real-estate design, components and parts production, construction, sales and after-sales service, and other aspects of a complete industrial system. It is a form of organization that is used to achieve the production and operation of production, supply, and integration.

China launched the construction of industrialization in 1968–1978, but the development is relatively slow. "Housing Industrialization Base pilot scheme" was promulgated in 2006. Housing industrialization process started only after another advance in the provinces. According to the data: As of April 2014, China has set up six industrial pilot cities, two industrial base Park, and 41 residential development components and parts manufacturers for the industrial base; it has assessed more than 320 national demonstration projects. Currently, the pilot results are reflected, but there is a huge gap compared with developed countries. The proportion of the housing industry modernization was more than 60%, accounting for new buildings in developed countries, whereas China is less than 1%. Housing industrialization process has lagged far behind the growth rate of the construction area, compared with nearly a decade of rapid growth in housing construction area.

2.2 *Main issues constraining the development of the housing industry modernization*

2.2.1 *Extensive development model*
Currently, the housing industry is mainly dominated by the local government to promote, taking the road of their extensive development, the lack of co-ordination at national-level planning. Countries, mainly through the establishment of pilot cities and industrial base

city for the layout, encourage enterprises as a part of the production base for the industrialization. Regardless of the pilot cities, the industrial base city, the number of producers is a limited number of terms, and residential industry modernization effect is uneven around the country. In addition, the industrialization of residential architecture and construction of the system is most directly applied to architecture and a technology roadmap to Europe, Japan, and other countries; there are also adaptive problems. National and provincial standards for the housing industry have standardized construction deficiencies, lack of technical standard housing industry design, manufacture, inspection, and other aspects of the whole process; the housing industry still cannot form the technical specifications system.

2.2.2 *Insufficient involvement of social subjects*

According to foreign experience, the housing industry production reached 100,000 square meters later. The cost is 1.25 times that of traditional construction costs; but before that, the construction cost will be more than 40% of traditional construction costs. Thus, large-scale development is the way out of the housing industry modernization. Due to the rapid development of the real-estate market in recent years, a real-estate company's business focus is primarily on expanding business scale and expanding market share, and thus there is little pressure to control costs. Real-estate companies are unwilling to invest effort and money into housing industry modernization. Currently, the number of firms involved in the implementation process is very limited. Data show the following: Housing industry base is mainly divided into developing business alliances type (group type), parts manufacturers, and integrated pilot city type. Currently, the country only has approved the establishment of 41 production companies for the industrial base, and companies involved in the housing industry are still less than one hundred implementations. It is not conducive to the rapid development of the housing industry modernization.

2.2.3 *Not forming a complete industrial chain in the true sense*

The industry chain is a four-dimensional concept that includes the value chain, business chain, supply chain, and space chain. Exchange on the upstream and downstream relationships and mutual value exists in a large number of industrial chains; the nature of the industry chain is used to describe an intrinsic link with a certain enterprise group structure. From the perspective of a complete industrial chain, participants include not only developers, research and design enterprises, processing enterprises, construction companies, and other assembly upstream and downstream businesses, including logistics, energy, home, electrical, decoration, and other industries. Housing industry places more emphasis on residential construction industry chain integration, including design, production, sales, and services (such as residential buildings standardized sets of component parts, operating service to the community, etc.). Due to the current housing industry being dominated and mainly implemented by the government, individual large companies (such as Vanke, ambitious, etc.) are involved in the development and production, construction, and sales. Although all aspects of the housing industry modernization have involved, but because the participants are more single, the whole process is mainly composed of full implementation of these large-scale enterprises. So, we need to bring in all aspects of associated companies (such as market research, product research, design, production, sales, service and other companies), which are insufficient. Thus, they fail to reach corporate group structure and still cannot connect to a complete industrial chain.

In summary, this situation makes modern housing industry lag behind the pace of development in China; it cannot form a complete industrial system that has a broad social base and is market based, and it is not conducive to accelerating the implementation of the follow-up process of the housing industry.

3 INNOVATION AND COUNTERMEASURES OF BIG DATA TO BRING MODERNIZATION TO HOUSING INDUSTRY

3.1 *Innovative contributions to the modernization of housing industry by way of large data*

With the advent of the era of cloud, big data has attracted more and more attention. Subversion and innovation of Big data are reflected in almost every role in every industry. Big data has become not only the third addition to the productivity of labor and capital but also an important manifestation of the national strategic resources and competitiveness. Big data plays a huge role in the housing industry modernization, mainly in market research, product development and standardization, processing and distribution, service, and others. It can result in economies of scale, greatly reducing costs while improving management capacity. According to statistics, big data can reduce product development and assembly costs by more than 50 percent; working capital is reduced by more than 7% in the manufacturing sector. By analogy, the construction industry can be closer to finesse manufacturing by big data. In addition to these contributions, the big data for the housing industry innovative contribution will be reflected in the formation of the industrial chain in the true sense: It can attract more social enterprises to participate in the housing industry modernization and to achieve economies of scale modernization of housing industry to achieve the true purpose of reducing construction costs. By reducing construction costs, it will further attract more participants to join them to form a truly virtuous circle, to promote the rapid development of the housing industry modernization(Figure).

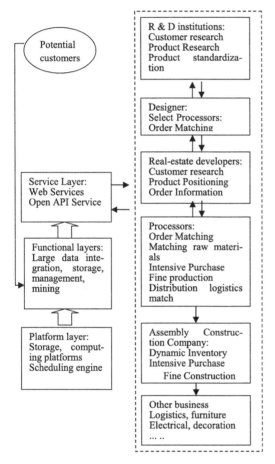

Figure 1. Housing industry modernization whole industry chain development model.

First, product development and standardization: McKinsey's 2012 reports shows big data for the manufacturing sector to reduce 50% of product development costs. Building parts data platform is built; housing user behavior, preferences, as well as related parts of residential consumption data are collected. By analyzing vast amounts of data, it is beneficial to find the key points of difference and commonality, so that standardization of products designed could be carried out.

Second, customer research and product positioning: being collected by potential customer behavior data and preference data, using big data for populations very specific segments, so that products and services are precisely tailored to meet the needs of users.

Third, the supply chain: By analyzing the potential customer with a large data, you can determine the need for different types of products trends, whether it is conducive to the real-estate development process of product positioning, forecasting market needs scale. It can predict the market needs and determine the size of the production and supply capacity based on large data, so processing enterprises will be able to get rid of the original simple order production model, the formation of market operation in the true sense. Manufacturers do not need to directly face the construction business, they simply provide products to vendors. Logistics, warehousing, and distribution involved in the housing industry may also decide the amount of warehouse construction and distribution services, logistics, and distribution organization based on analysis of large data.

Fourth, assembly areas: Since accurate forecasting big data, logistics and production, and assembly convergence are more accurate, the situation will be close to zero inventory; assembly companies will no longer demand more money to stock a lot of parts. You can use big data to improve the allocation and coordination of human and material resources in the implementation process; gaps in the management of the construction industry are also much smaller than in other industries.

Fifth, the related downstream business areas: Big data can allow enterprises to have full understanding of its customers and the up-downstream business; the market value of the product or service has been timely introduced. Companies can obtain additional data needed by its big data and promote the formation of cooperative-based ecosystems. It enables all stakeholders to benefit from being conducive to the whole industry chain of house production modernization with a market base to be built. The chain can achieve a real sense of social enterprises to participate in the whole industry chain, instead of the current development model in which government-led, individual enterprises promote the pilot model.

3.2 Housing industry modernization whole industry chains development countermeasures in the big data era

First, to further strengthen the modernization of overall planning in the housing industry, as well as architecture big data platform: Local governments should abandon the existing fragmented mode of implementation of the development path, and they should co-ordinate the planning and overall progress at the national level. Due to various government departments accumulating a wealth of data resources, housing information, land information, the user's personal information, market information, and other aspects of running constitute the underlying data of real estate. These data have been partially analyzed with large data characteristics. So the government can use this existing platform architecture big data platform, and the introduction of a market-oriented operation mechanism provides information.

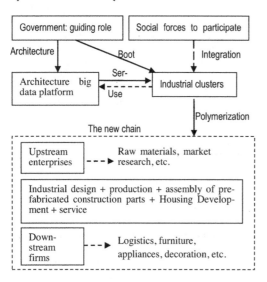

Figure 2. Development strategies of housing industry modernization whole industry chain.

Second, we need to attract social forces to participate in the integration formation of industrial clusters: The central government and local governments should constitute the division of labor in guiding the formation of the house industry modern industrial clusters. The central government should thoroughly research the industry support policies (such as direct financial compensation or prefabricated VAT exemption and other tax incentives, the relaxation rate control volume and other non-financial policy, etc.) and study how to design the framework of industrial clusters polymerization mechanism. Local governments should strengthen policy enforcement. Simultaneously, the government should adjust residential construction-related technical specifications and material requirements,

the new architecture of the house to increase rigid implementation efforts. By foregoing policies to guide the market, companies have to understand that there are more opportunities in the modernization of the housing industry. It attracts more design, processing, construction, and other companies involved in supporting industries and, ultimately, achieves the effect of the integration of the formation of industrial clusters.

Third, the use of big data platform to build new housing industry modernization of the industrial chain: The manner in which policies guide the market can only be resolved by early rapid advance modernization of housing industry problems. To make a long-term healthy development of the follow-up, you need to market your own role. Through the use of big data platform, companies can promptly seize the pulse of the market, to see more market prospects. As more and more enterprises participate, the industry will reach a certain size. It will greatly improve its economic competitiveness. In addition to design, manufacture, construction, and other basic industries support, it will further attract its upstream and downstream enterprises to actively participate and will eventually form a new industry chain polymeric of housing industry modernization.

ACKNOWLEDGMENTS

The authors thank the following foundation item: the Chongqing City Board of Education Science and technology research project (KJ120729); Research projects of the humanities and Social Sciences in Rongzhi college of Chongqing business and technology school (20128001).

REFERENCES

Li, P. & Lv, X. 2014. Bige data-aided real estate macro decision. *Macroeconomic management*(8):34–36.
Jin, H.. 2014. Research on Business Model Innovation for the Era of Big Data. *Science & Technology Progress and Policy*31(7):15–19.
Chen, F. F. 2013. The research of housing industrialization promoting mechanism. *Journal of Anhui Institute of Architecture & Industry*21(3):61–66.
Ding, L & Cui, J. 2013. Analytical thinking housing industry in China. *Business Times*(50):130–131.
Li, R.S. & Gong, J. 2014. Develop History and Experience of Residential Industrialization of Developed Countries. *Chinese & Overseas Architecture*(2):58–60.

Energy, Environment and Green Building Materials – Sheng (ed.)
© 2015 Taylor & Francis Group, London, ISBN 978-1-138-02718-3

Key-technology research of the image retrieval based on the feature points and invariant moments

Z.M. Wang & B.Q. He
Nanchang Institute of Technology, Nanchang, China

ABSTRACT: Image retrieval based on content (CBIR) technologies is one of the hotspots in current research. Through extracting a variety of image features, one kind of similarity measure methods is used to realize the image retrieval. This image retrieval algorithm is based on feature layer, mainly related to the feature extraction and matching. As the image database increases day by day, how to extract one effective feature to describe image information is one research focus. Therefore, finding one feature that can perfectly describe image information while simultaneously having a low dimension is the key to improving the accuracy of image retrieval and reducing the retrieval time.

KEYWORDS: image retrieval, characteristics, match, feature layer

1 KEY TECHNOLOGY OF IMAGE RETRIEVAL

Image retrieval algorithm in this paper is based on image feature layer, mainly related to the image feature extraction and matching. The aim is to extract one feature that can perfectly describe the image information, which has a low dimension characteristic, enhance the precision of image retrieval, and reduce the retrieval time. The key technology of content-based image retrieval is the feature extraction [1], feature matching [2], and algorithm evaluation [3].

2 MAIN FEATURES FOR CBIR

2.1 Color features [4]

Color is the most common feature of all things in nature. The wavelength and intensity of the electromagnetic wave can have very big differences. When people can feel the wavelength range from about 312.30 nm to 745.40 nm, it is called visible light. If the electromagnetic waves are arranged together by wavelength (or frequency) size, we can get the spectrum of light source. An object's spectrum decides its optical properties, including its color.

2.2 Texture features [5]

Texture is a reflection of the same phenomenon in the image visual features, reflects the shared intrinsic attributes in the surface of objects, and is usually manifested as local irregular and macroscopic rule. Texture features include the following: 1) through the gray distribution and its surrounding neighborhood pixels space to express; 2) local texture with different levels repeat of information.

2.3 Shape features [6]

Shape feature is a kind of important image features that can well reflect targets' characteristics. Typically, shape feature representation methods have two kinds: One kind is the contour feature, and the other is regional characteristics. The shape feature description methods commonly include the following: 1) boundary characteristic method. The method obtains the shape parameters based on boundary feature description. 2) Fourier shape descriptors method. Fourier shape descriptors use the Fourier transform of the object boundary as the shape description method, and place the two-dimensional problem in the one-dimensional problem. 3) Invariant moments. Invariant moments are widely used as a shape description method in the field of image retrieval. Commonly used moments include Hu moment, Zernike moment, and wavelet moment.

2.4 Geometric features [7]

Geometric features are extended from the concept in geometry. The point, line, surface, and their relationship to each other belong to the category of geometric characteristics. Common geometric features include the following contents:

1 centroid
An object's center location can be expressed by the center of its area. The two-value image quality

distribution is uniform; its center of shape and mass is the same. Proposing a pixel's position is (x_i, y_i), its centroid can use the formula 2-1 to calculate:

$$\bar{x} = \frac{1}{mn} \sum_{i=0}^{n-1} \sum_{j=0}^{m-1} x_i; \qquad \bar{y} = \frac{1}{mn} \sum_{i=0}^{n-1} \sum_{j=0}^{m-1} y_j \qquad (2\text{-}1)$$

2 direction

If the object is elongated, it can take the longer axis as the direction. Usually, the two minimum moments (equivalent axle shaft minimum principal axis of inertia in the two-dimensional plane) of the object are defined as the direction; that is to say, to find a straight line to make E have the minimum value:

$$E = \int \int r^2 f(x, y) \, dx \, dy \qquad (2\text{-}2)$$

3 COMMONLY USED METHODS OF FEATURE EXTRACTION

According to the different characteristics, feature extraction can be divided into point features, moments feature, and edge features extraction method.

3.1 *Point features*

Point features are the most basic; they refer to the gray signals that are significantly changed in the two-dimensional direction of the point, such as corner, dot, and so on. Point features can be used in applications such as image registration and matching, object description and recognition, beam calculation, moving object tracking, and many other fields. The point feature extraction operator is called interest operator or favorable operator, namely using some kind of algorithm to extract people interested parts from the image for a certain purpose. In image analysis and computer vision field, an effective extracting method is selected according to different applications.

1 Moravec feature points

Moravee operator[8] has four main directions, and the maximum or minimum variance points are chosen as feature points.

First, calculate interest value (IV) of each pixel.

Second, given an experience threshold, choose the interest value greater than the threshold point as a candidate. The selection of threshold should include the candidate feature points needed, and it should not contain too many non-feature points.

Third, select the extreme point of the candidate as feature points.

In addition to the earlier methods, the edge contour can be also used to extract feature points.

2 Forstner feature points

The Forstner operator [9] is an effective method to extract point features from the image. By Robert gradient calculation of each pixel and gray covariance matrix of a window as the center, look for the smallest circular points as feature points in the image.

First: calculate Robert gradient of each pixel.

Second: calculate the covariance matrix of gray in the 1 * 1 window.

Third: calculate the interest value Q and W.

Fourth: determine the candidates.

Fifth: select the extreme point.

3 Harris feature points

Harris point [10] is currently one of the most widely used feature points. Its advantages include simple calculation; the point is uniform and reasonable, and it is not sensitive to image rotation and noise interference. Harris corner extraction steps are as follows:

First: use horizontal, vertical differential operators for filtering, then get I_x, I_y.

$$m = \begin{bmatrix} I_x^2 & I_x I_y \\ I_x I_y & I_y^2 \end{bmatrix}, \ I_x^2 = I_x * I_x, I_y^2 = I_y * I_y \qquad (3\text{-}3)$$

Second: use Gauss filter for m, then get the new matrix *M*.

$$M = G * \begin{bmatrix} I_x^2 & I_x I_y \\ I_x I_y & I_y^2 \end{bmatrix}, \ G = \exp(-\frac{x^2 + y^2}{2\sigma^2}) \qquad (3\text{-}4)$$

Third: use the formula 3-5 to calculate the response value R. for each pixel.

$$R = \det(M) - k * trace(M)^2 \qquad (3\text{-}5)$$

Fourth: if R is greater than the threshold T, the point is recognized as a corner; otherwise, discard it.

3.2 *Invariant moment feature*

Moment can express many important features. In the field of image processing, because of good performance, it has been widely used in image registration, segmentation, and target recognition

1 Hu moments

In 1962, M.K. Hu proposed 7 orders of moments. These have the advantage of simple calculation, good rotation, and scale invariance, but their anti-noise performance is poor, and there is no affine invariance.

In fact, only the first two moments M1 and M2 are maintained well, whereas other moments will bring big errors. Hu moments' calculation speed is very fast, but the recognition accuracy is low.

2 Zernike moments

Zernike moment is an image projection function in orthogonal polynomials, and its expression is:

$$V_{nm}(x,y)=V_{nm}(\rho,\theta)=R_{nm}(\rho)\exp(jm\theta) \qquad (3-6)$$

where 'n' is a positive integer or zero, $R_{nm}(\rho)$ is radial polynomial, and M is a positive or negative integer and $n-|m|=even, |m|\le n$.

N order moments is defined as the formula 3-7:

$$A_{nm}=\frac{n+1}{\pi}\int\int_{x^2+y^2\le 1}f(x,y)V_{nm}^{*}(\rho,\theta)dxdy \qquad (3-7)$$

Zernike moment is a group of orthogonal matrix, with rotation invariance, and it can be constructed arbitrarily in a high order, so it has a better recognition effect than Hu moment.

3.3 Edge feature

Edge feature is the basic characteristic of the image. Since 1960, many algorithms of image edge detection have been proposed, and they can be roughly divided into the following categories: (1) using a derivative of image features; (2) using two-order derivative of image features; (3) the optimal edge detector (filter); (4) other algorithms, such as Robet, Canny, and so on.

4   EVALUATION METHODS OF RETRIEVAL

After the completion of the image retrieval, the method used should be checked to ensure whether it is reasonable or effective. In view of the fact that the retrieval evaluation methods play an important role in the system of image retrieval, many experts and scholars in this field have carried out deep research.

4.1   Recall rate and precision [13]

Recall rate is the success rate of the retrieval system, which can be expressed as follows:

$$P=\frac{a}{a+c} \qquad (4-1)$$

where 'a' is the related numbers, and 'c' is the total numbers in the system.

The use of strong specificity language can improve the retrieval precision, but the recall rate will fall.

Precision is a measure used to test signal-noise ratio, and it can be expressed as follows:

$$P=\frac{a}{a+b} \qquad (4-2)$$

where 'a' is the related numbers, and 'c' is the total retrieval numbers.

4.2   Ranking evaluation method [14]

Suppose N is the return number of query images, N_R is the relevant number of retrieved images, P_r the is sorting serial number, and $\hat{N}_R$ is the real number of relevant images, the evaluation parameters are as follows:

The average ranking of relevant retrieved images:

$$K_1=\frac{1}{N_R}\sum_{r=1}^{N_R}P_r \qquad (4-3)$$

The ideal average ranking of the related image:

$$K_2=\frac{\hat{N}_R}{2} \qquad (4-4)$$

Related image loss rate of R is as follows:

$$r=\frac{N_R}{\hat{N}_R} \qquad (4-5)$$

4.3   Tau coefficient

Tau coefficient can be defined as the formula 4-6:

$$H=\frac{V_1-V_2}{V} \qquad (4-6)$$

where V represents the total possible number of sequence, V_1 represents the sequence number, and V_2 represents the sequence number ranked not by order. This method considers the number of ordered pairs, but for missing detection, H and tau may be high.

4.4   Consumption time

Time consumption is the total time between the input and retrieval completion process. If an algorithm's efficiency is high, its time consumption tends to be less. On the contrary, if the retrieval time is long, the search efficiency of this algorithm is low.

5 SUMMARY

Image retrieval is an open research topic. This paper introduces the current image retrieval techniques, including feature extraction method, different kinds of similarity measure method, and evaluation algorithm. The retrieval features commonly include color, shape, texture, and other characteristics. Through using different characteristics, it can achieve complementary advantages and improve the retrieval effect and performance. The similarity measure method is one of the key techniques used. Different methods of measurement will produce different search results. Because there is no method for all features, the choice of a suitable measurement method is one of the key research topics for image retrieval algorithm. In addition, the evaluation method is one important factor for the performance of the algorithm, so it is also an important research focus.

REFERENCES

[1] Liu Li, Kuang Gangyao. Image texture feature extraction methods[J]. Chinese Journal of Image and Graphics, 2009, 14 (4): 622–635.

[2] Yang Heng, Wang Qing. An efficient matching algorithm for local image features[J]. Journal of Northwestern Polytechnical University, 2010, 28 (2): 291–297.

[3] Shi Lukui, Zhang Jun, Gong Xiaoteng. The manifold learning algorithm evaluation model based on the field to maintain [J]. Computer Application, 2012, 32 (9):2516–2519.

[4] Wang Juan, Kong Bing, Jia Qiaoli. Image Retrieval Technology based on Color Feature[J]. Computer System Application, 2011, 20 (7): 160–164.

[5] Zhang Gang, Ma Zongmin. A texture feature extraction method using Gabor wavelet [J]. Chinese Journal of Image and Graphics, 2010, 15 (2): 247–254.

[6] Wei Dongxing, Chen Xiaoyun, Xu Rongcong. The Image Shape Feature Extraction Method based on Corner Detection [J]. Computer Engineering, 2010, 36 (4): 220–222.

[7] Yang Cheng, Lu Rong, Fan Yong, Chen Niannian. Rapid Extraction Algorithm of Image Geometric Feature Parameters[J]. Computer Engineering and Science, 2012,34 (7): 124–129.

[8] Kumari M.S, Shekar B H. The use of Moravec-operator for text detection in document images and videoframes [C]. International Conference on Recent Trends inInformation Technology, 2011: 910–914.

[9] Wang Li-Qiang, Hao Ying. Radon Transform and Forstner Operator Applying in Buildings Contour Extraction[C]. Sixth International Conference on Fuzzy Systems and Knowledge Discovery, 2009: 415–419.

[10] W Zhao, S Gong, C Liu, X Shen. Adaptive Harris Corner Detection Algorithm[J]. Computer Engineering, 2008, 34 (10): 212–215.

[11] He Qiang, Yan Li. Edge detection algorithm based on LOG and Canny operator [J]. Computer Engineering, 2011, 37 (3): 210–212.

[12] Zhao Na. An fast edge extraction design and realization of KIRSCH operator[J]. Software Guide, 2009, 8(5): 188–190.

[13] Chen Guangying, Zhang Qianli, Li Xing. Recall and precision control in the anomaly detection [J]. Control and Decision, 2004, 19 (4): 478–480.

[14] Wang Lei, Kang Zhi, Lou Xinyuan. An evaluation method of priority ordering for testing[J]. Journal of Chongqing Institute of Technology, 2007, 21(2): 61–64.

Energy, Environment and Green Building Materials – Sheng (ed.)
© *2015 Taylor & Francis Group, London, ISBN 978-1-138-02718-3*

Analysis of slope ecological protection substrate moisture evaporation characteristics

X.J. Zhang
College of Hydraulic & Environmental Engineering, China Three Gorges University, Yichang, China

Z.Y. Xia
Collaborative Innovation Center for Geo-Hazards and Eco-Environment in Three Gorges Area, Yichang, China

W.J. Hu & L.L. Zhang
College of Civil Engineering & Architecture, China Three Gorges University, Yichang, China

ABSTRACT: Concrete vegetation for ecological protection technology is a developing technology for rock slope protection and greening slope ecological protection, and it has been widely applied and developed in ecological restoration. It will reinforce the exposed part of the mountain and combined with the slope vegetation restoration, achieving the combination of ecological restoration and the projection of the slope, it will solve the major problems of infrastructure and ecological environment combined.

KEYWORDS: Slope ecological protection substrate, Water evaporation, Water Cycle

1 INTRODUCTION

Ecological protection technique of vegetation concrete is a type of technology that is used for slope protection. It aims at reproducing the habitat for vegetation growth by spraying a layer of base course material similar to natural soil on the slope and mixing some proper plant seeds or planting material in it, which finally helps the plants to grow by turns on the slope. The water in the base course material not only is the crucial factor for the vegetation growth and the construction of ecological systems but also plays a key role in the Eco-Protection technology and its system. In other words, whether the protection works or not depends on the water retention property of the base-course material. On the other hand, the growth of vegetation may be inhibited because the base course material for the slope protection is usually made ten-centimeter thick, in which the water may be not enough for the growth of vegetation. Therefore, studying the moisture evaporation property of the base course material is of great significance.

Vegetation concrete is a special engineering soil; the research of its moisture evaporation characteristics is still in its infancy. Scholars at home and abroad not only do a lot of work for the soil water

cycle but also provide a theoretical basis for the study of its moisture evaporation characteristics.

2 PHYSICAL PROCESSES OF SOIL MOISTURE EVAPORATION

Soil evaporation is a special form during the soil water movement. In this stage, owing to the soil being in contact with the atmosphere, the soil water movement is closely related to the atmospheric conditions. Soil evaporation is affected by outside weather conditions and inherent soil factors. Outside weather conditions contain solar radiation, temperature, ground temperature, humidity, wind speed, precipitation, and infiltration methods. Internal factors contain soil moisture, groundwater level, soil texture and structure, and soil color; surface characteristics of evaporation depend on the water vapor pressure difference between soil surface and atmospheric surface. When the vapor pressure difference is positive, the soil moisture evaporation occurs. If the difference is zero, the evaporation is zero; if the difference is negative, the water vapor in the atmosphere is transformed into the soil characteristics, and the soil transport capacity is determined by capillaries.

3 FACTORS THAT AFFECT SOIL EVAPORATION

3.1 Soil color and surface characteristics

The capacity of the soil to absorb solar radiation is related to the color, and soil color also affects soil evaporation. The darker the soil, the more heat the soil absorbs, and the much higher are the temperature and evaporation that are obtained. According to the relevant reference information, loess soil evaporation is larger than that of the white soil. Brown soil and black soil evaporation is also higher than that of the white soil. Slope and roughness of the surface also affect the evaporation. If the slope is toward the sun and rough surface, the evaporation will increase.

3.2 Radiation, air temperature, and ground temperature

Continuous evaporation must have a continuous supply of energy for latent heat of vaporization, and the solar radiation is the energy source of the latent heat of vaporization. Therefore, when other conditions are the same, the larger the amount of solar radiation, the greater the amount of evaporation. Simultaneously, air temperature and ground temperature also has a significant impact on evaporation. Temperature determines the saturated water content in air and the speed of the water vapor diffusion. Ground temperature determines the activity level of soil moisture.

3.3 Humidity

Relative humidity is an extremely important factor affecting evaporation. If the relative humidity in the atmosphere is high, the water vapor is close to saturation and this keeps the soil moist for a long time. The greater the humidity gradient above the ground, the more intense the soil evaporation.

3.4 Precipitation and infiltration

The moisture that can evaporate in soil is related to the rainfall intensity and infiltration. In reality, after a rainfall, the water starts to evaporate and exhibits infiltration. In this condition, simultaneously in different parts of the soil, these two processes are separately going on: First, the surface water evaporation results in upward movement of soil moisture. Second, there is water infiltration or gravity drainage. At this point, in the deeper parts of the gravity gradient and water suction gradient, a middle plane of zero flux is formed in the middle of the soil. This is known as zero-flux surface, and the surface is gradually moving toward the lower profile. Therefore, both evaporation and rainfall infiltration processes are considered in this study. On the surface, the two processes are in different positions, but in fact, they are interconnected.

3.5 Soil moisture

Soil moisture is a source of soil moisture evaporation. When the soil moisture is high, soil evaporation will be greater than the surface water evaporation. The moisture is fully recharged when soil evaporation is consumed.

4 SOIL MOISTURE EVAPORATION CALCULATION AND DETERMINATION

Evapotranspiration is the process of moisture being released into the atmosphere, including soil evaporation and plant transpiration, which is difficult to distinguish under normal circumstances and often putting the two together, this is called "evapotranspiration." Evaporation process deals with the components of the Earth's surface water balance and heat balance, and the evaporation links the Earth's surface (water) with heat balance[1]. Plant transpiration is an important part of the process of life, and it has a close relationship with plant physiological processes and biological production. With the advance of technology levels, laboratory equipment, and means of observation, people make a series of theoretical methods and empirical formula to estimate evapotranspiration. The main methods of estimating land surface evapotranspiration are hydrology, micro meteorology method, plant physiology, infrared remote sensing methods, and SPAC integrated simulation method At home and abroad, the hydrology method is based on water equilibrium; at a certain area of soil depth within a period, it is used to measure the total evapotranspiration indirectly. This is also called "water balance method," as it can measure the spatial scales as small as a few square meters to a few square kilometers large. The biggest advantage of the method is that it is not limited by weather conditions; the disadvantages are that the measurement time is relatively long and it is difficult to reflect the daily dynamic changes of evapotranspiration. With continuous development of computer technology and meteorological instruments, micro-meteorology has become a more common evapotranspiration calculation method; it mainly contains energy balance method, aerodynamics law, energy balance and aerodynamic joint formula, eddy covariance, and infrared remote sensing method. Energy balance method, which is also Bowen ratio-energy balance method (BRER), has a clear physical concept and simple calculation, and it has no special requirements and restrictions on the atmosphere. Typically, it is often considered a test criterion to other evapotranspiration calculation methods with its high precision under the circumstances of an open and uniform underlying surface. In 1948, Penman combined the energy balance principle and aerodynamics, and presented the famous Penman formula to calculate potential evapotranspiration for the first time. Monteith[2] presented canopy evapotranspiration computing model, which is the famous Penman-Monteith model, on the basis of Penman and Covey work. This model has a good physical basis as its full consideration of the physical characteristics of the atmospheric patterns and physiological characteristics of vegetation evapotranspiration; opens up a new way for studying non-saturated land surface evapotranspiration; and is widely studied and applied for its more clear understanding of change process and its impact mechanism of evapotranspiration. Eddy covariance method can directly measure sensible heat and latent heat turbulent fluctuation values on the underlying surface and calculate evapotranspiration. Compared with other micrometeorological methods, it has the most complete and reliable physical theory[3], and high accuracy; it not only requires expensive probes, data acquisition, and computer systems but also causes serious observation error for the disturbance of the pulsating ultrasound probes and the air flow. Since the 1970s, with the continuous development of remote sensing technology, it appears that remote sensing technology methods are used to calculate evapotranspiration of vegetation, which is called "infrared remote sensing method." It calculates evapotranspiration by spectral characteristics of vegetation and micro-meteorological parameters infrared information[4~5]. Plant physiology method is mainly used for measuring the whole or part of the amount of water consumption of plants, but it is difficult to use it in the representation of the sample. Transpiration of single or several plants is used to calculate transpiration of large square plants accurately. Lu Zhenmin[6] believes that a certain rate of CO_2 comes into plants from the atmosphere to maintain plants' normal growth and development, while simultaneously water must be diffused into the atmosphere through pores from the plants, which requires the same amount of water supplied to the root accordingly. This is equivalent to the full amount of stomata transpiration, so that it establishes the transpiration model. Lu Zhenmin[7] established the field evapotranspiration estimation model; it can obtain field plants between evaporation by subtracting two models (Ev-T=E), which is meaningful to distinguish evaporation from transpiration. Kang Shaozhong[8] analyzed the influence of the evaporation force, effective soil moisture, and plants leaf area index on field evapotranspiration based on data from some sites on the Loess Plateau; found field

evapotranspiration amount when adequate water is supplied and restricted water is supplied; and also established the actual field evapotranspiration model on the basis of the water balance equation. The theoretical basis of all these methods is focused on their respective disciplines. Although evapotranspiration is a broad ecological processes, it is a comprehensive function of soil, air, and plant physiology, so it is inevitable that there are interdisciplinary differences and limitations between these methods. The SPAC water transport comprehensive study had a breakthrough until Philip presented a more complete soil - plant - atmosphere continuum (SPAC) concept. Nowadays, although there are a series of evapotranspiration calculation methods, each method is developed according to certain objects and conditions; it has not formed a perfect way and still uses two or more kinds of methods[9] to conduct numerous studies on evapotranspiration calculation.

In summary, nowadays people have a relatively weak understanding of substrate moisture migration in slope ecological protection areas. To this end, we should make vegetation concrete as a target for the study experiment, testing the basic characteristics of the substrate moisture. Carrying out ecological protection water evaporation at different temperatures, rainfalls, and different slope conditions is of great significance for ensuring water supply and vegetation growth. Meanwhile, it also provides a theoretical basis for expanding application of vegetation concrete that can be used in extreme environmental conditions.

ACKNOWLEDGMENT

This work was financially supported by the Natural Science Foundation of China (NO. 41202250).

REFERENCES

[1] Tang D.Y. & Cheng W.X & Hong J.L. 1984. Research Status and Prospects of evaporation. Ecographical Research: 84–97. China.
[2] Allen G, & Periera LS, & Raes D. F. 1998 Irrigation and drainage paper No.56:Crop evapotranspiration. ROME, Italy,:1–330.
[3] Xin X.Z. & Tian G.L. & Liu Q.H, 2003 Advances in quantitative remote sensing evapotranspiration. Journal of Remote Sensing: 233–240.
[4] Ming C. & Pan's Di. 1994 thermal infrared remote sensing satellite data to estimate evapotranspiration large area Advances in Water Science: 126–133.
[5] Brown, Rosenberg. 1985A resistance model to predict evapotranspiration and its application to a sugar beet field. Agron. J:341–347.
[6] Lu Z.M. & Zhang Y, 1987 computing crop water requirement and crop drying degree of discrimination method. Agricultural Meteorology: 57–59.

[7] Beijing Agricultural Ecology, Chinese Academy of Agricultural Experiment Station ecological research Beijing: China Meteorological Press, 1989:335–353.

[8] Kang SZ. 1985. A study of the mathematics model for the computation and prediction of evapotranspiration form the farmland. Northwestern Agricultural University: 30–35.

[9] Zhang J.S & Meng P.G. 2001. Methods of calculating evapotranspiration of plants. World Forestry Research:23–28.

Energy, Environment and Green Building Materials – Sheng (ed.)
© 2015 Taylor & Francis Group, London, ISBN 978-1-138-02718-3

Effecting analysis of the temporary traffic regulation on the traffic flow in urban expressway

M. Wang, J.F. Wang, R.X. Xiong & X.D. Yan

MOE Key Laboratory for Transportation Complex Systems Theory and Technology, Beijing Jiaotong University, Beijing, China

ABSTRACT: The effect of different traffic regulation modes on traffic flow in urban expressway and its time evolution laws are studied by the simulation method. A typical experiment segment is selected as the simulation model, and the influence of traffic regulation duration, the lane number of traffic regulation, the segment length of traffic regulation, and the distance between traffic regulation segment and on-ramp on traffic flow is explored. Based on the simulation results, the variation tendency of traveling time and queue length in different traffic regulation modes is analyzed, and the time evolution laws of queue length is studied in terms of traffic waves. The results indicate that traffic regulation duration and the lane number of traffic regulation have an obvious effect on traffic flow during the process of implementing traffic regulation.

KEYWORDS: Traffic engineering; Traffic regulation modes; Traffic regulation duration; Lane number of traffic regulation; Effect analyses

1 INTRODUCTION

Traffic regulation is a real-time traffic management mode. To better implement the cooperative management of local road networks and improve the travel LOS, traffic regulation has been considered one of the important strategies of urban active traffic management that interferes with traffic flow.

Many researchers studied the implementation strategies and traffic regulation impact on traffic flow. Cao et al. analyze how traffic regulation affects the traffic flow during traffic incidents and provide several macroscopic solutions for reference (Luo & Gan 2008, Du & Wang 2009). Due to its timeliness and randomness, traffic regulation might be induced to redistribute the traffic flow in a peculiar way and to cause a different influence degree on the nearby local road networks (Cheng et al. 2010, Dong et al. 2012, Chen et al. 2000). Some efforts are applied to study the special influence that traffic regulation has on traffic flow using different methods, such as mathematical model, microscopic simulation (Xie et al. 2004, Xu et al. 2011), and so on. Other researchers also studied the traffic flow operation characteristics and exhaust gas emission during the process of traffic regulation.

This paper studies the impact of different traffic regulation factors on traffic flow, including traffic regulation duration, lane number of traffic regulation, segment length of traffic regulation, and the distance between traffic regulation segment and on-ramp. The

artificial survey method and CCTV method are used to collect traffic flow data of experimental segments.

2 ANALYSIS OF THE EXISTING URBAN TRAFFIC REGULATION MODES

As an important measure in urban active traffic management, traffic regulation is widely used in metropolis in China. Table 1 compares the traffic regulation measures for major events of several cities.

Common traffic regulation measures include speed limit, changing lanes, temporary traffic interruption, lane closure, and so on, as shown in Table 1. The duration, road number, and road length of traffic regulation are vital factors for the implementation of traffic regulation.

3 THE SIMULATION

3.1 Simulation scenarios

The segment between Shawo Bridge and Sijiqing Bridge of West 4th Ring Rd in Beijing is selected as an experimental segment to establish the simulation scenarios.

The traffic regulation segment is about 1000 m away from the Shawo Bridge, as shown in Fig. 1. Point A is the position of on-ramp. Points B and C are the locations of ending point and starting point of traffic regulation, respectively.

Table 1. Comparison of traffic regulation measures of several cities.

City	Major events	Traffic regulation measures	
		similarities	differences
Beijing	The Olympic games, China-Africa Cooperation Forum	Vehicles are permitted to enter several certain segments in certain periods.	Olympic lanes Adopting different regulation for different vehicles
Shanghai	National Day		Adopting traffic regulations at different times and in different regions, Ramp closure
Guangzhou	The Asian Games		Permitted for passengers and vehicles
Xiamen	International hundred km race competition		Bi-directions or one-direction closure in several segments

Table 2. Corresponding parameters of experimental segment.

Parameters	Values
Starting point	Shawo Bridge
Ending point	Sijiqing Bridge
experiment segment length	About 8.5 km
Connected main roads	Lianshi road, Fuxing road, Fushi road, etc.
Large public places	301 hospital, Xijiao airport, Wukesong Stadium, etc.
Lane number	10 lanes both way
Direction of traffic	North-south

Table 3. Parameters set of simulation.

Parameters		values
The simulation time		3900s
Road length of regulation		500m
Regulation period		[600s, 1800s]
Interval of data collection		20s
Traffic flow (pc/h)	Main road of West fourth ring (S→N)	6800
	Lianhuachi west road (E→W)	2650
	Lianhuachi east road (W→E)	1115
	Zizhuyuanroad (E→W)	2189
	Xingshikou road (W→E)	1346
	Main road of West fourth ring (N→S)	6216
	on-ramp of Fuxing road	1464

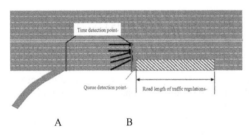

A   B

Figure 1. Simulation model of traffic regulation segment.

Table 2 shows the corresponding parameters of simulation scenarios segment.

3.2 Simulation parameter settings

There are mainly two parts of the parameter settings in the simulation model. The first part is the vehicle's composition, which includes car and bus. The other part is the expected speed distribution, in which the expected speed of car and bus is 50 km/h and 40 km/h, respectively, according to the investigation data. Other simulation parameter settings are shown in Table 3.

To collect the simulation result data, two detecting points are set up, as shown in Fig. 2. Simulation data gathered include vehicle travel time, the average queue length, and the biggest queue length.

4 ANALYSIS OF SIMULATION RESULT

4.1 Effect analysis of traffic regulation duration

Traffic regulation duration directly affects the queue length of traffic flow. Using the simulation model, the traffic regulation duration is set to 600s, 900s, 1200s, 1500s, and 1800s, respectively, and the impact of different traffic regulation duration on traffic flow is studied. Other parameters setting are shown in Table 4.

Table 4. Parameters setting of traffic regulation duration simulation.

Simulation parameters	values
the lane numbers of traffic regulation	2
position of the lane(in/out side)	inside
the distance between traffic regulation location and on-ramp(m)	500
the road length of traffic regulation(m)	500

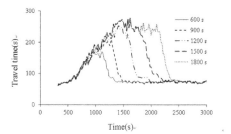

Figure 2. Variation curve of traveling time corresponding to different traffic regulation duration.

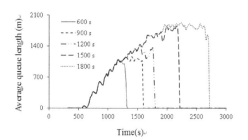

Figure 3. Variation curve of average queue length corresponding to different traffic regulation duration.

Figures 2 and 3 illustrate the variation curves of travel time of experiment segment and average queue length corresponding to different durations of traffic regulation. We can see that the curves have an increasing trend with the increase of traffic regulation duration. As the traffic regulation duration increases further, traveling time and average queue length will remain constant. It is explained that a certain proportion of vehicles from upstream of the traffic regulation segment would change their origin paths as the traffic regulation duration further continues.

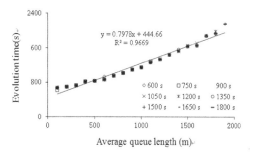

Figure 4. Time evolution curve of queue length in different traffic regulation duration.

Figure 4 illustrates the time evolution curve of queue length corresponding to different traffic regulation duration. We can see that the formation of a queue is approximately a uniform process, in which the speed of the traffic wave is approximately constant.

4.2 Effect analysis of lane number of traffic regulation

Figure 5 depicts the variation curve of average queue length corresponding to different lane number of traffic regulation.

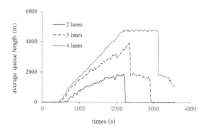

Figure 5. Variation curve of average queue length corresponding to different regulation lane numbers.

We can conclude that average queue length increases with the increase of lane number of traffic regulation, and the speed of traffic wave of queue formation appears to be an acceleration trend.

Figure 6 and Table 5 are the time evolution curve of queue length and fitting models corresponding to different lane number of traffic regulation, respectively.

Queue length of different lane number of traffic regulation has linear time evolution laws, and the slope of linear models is different, as shown in Fig. 6. The slope of the model becomes small with the increase of lane number of traffic regulation, which means the traffic wave of queue formation is quicker under the condition of 4-lane closure.

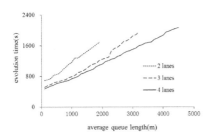

average queue length(m)

Figure 6. Time evolution curve of queue length in different lane number of traffic regulation.

Table 5. Time evolution laws models of queue length in different road number of traffic regulation.

Lane number of regulation	fitted equation	Slope	R^2
2lanes	$t=0.4896l+618.19$	0.4896	0.9689
3lanes	$t=0.4716l+367.98$	0.4716	0.9812
4lanes	$t=0.3711l+392.38$	0.3711	0.9978

4.3 Effect analysis of other parameters

Similar to earlier methods, the influence of traffic regulation segment length, distance between traffic regulation segment, and on-ramp on traffic flow is further explored. The corresponding results are shown as follows:

1 The traffic regulation segment length is set to 500m, 750m, and 1000m, with other things being equal. It is found that the factor has a small effect on the queue length of upstream traffic flow.
2 The distances between traffic regulation segment and on-ramp are set to 400m, 500m, 600m, 700m, 800m, 900m, and 1000m, respectively, to investigate their influences on traffic flow. The result shows distance interval changes.

5 CONCLUSION AND SUGGESTIONS

According to the results of simulation, some advice on the traffic regulation measures is proposed. The traffic regulation should be implemented in inside lanes. The traffic regulation duration, the segment length of traffic regulation, and the distance between the traffic regulation segment and the on-ramp should be set reasonably.

REFERENCES

DU, J.W & WANG Wei. 2009. Study on urban traffic control holding important activity. *Transportation Standardization*(24):101–104.
LUO, D. & GAN, R.H. 2008. The analysis to impact of traffic control measures during big event. *Traffic and Transportation*(H12):62–64.
CHENG, X.Q & TANG, R.X & ZHU, H. 2010. Simulation and data processing of highway traffic incident based on VISSIM. *Journal of Transportation Engineering and Information* 8(04):14–20. (in Chinese).
DONG, C.J & SHAO, C.F & MA, Z.L. 2012. Temporal-spacial characteristic of urban expressway under jam flow condition. *Journal of Traffic and Transportation Engineering* 12(03):73–79.(in Chinese).
CHEN, X.M & WANG, W & WANG, J.R. 2000. Link performance functions based on the traffic management and control. *Journal of Highway and Transportation Research and Development*:40–43. (in Chinese).
XIE, Y.CH & LI, X.H & FAN, Y.L. 2004. Urban road network model and assignment algorithm analysis under traffic control condition. *Journal of Highway and Transportation Research and Development* 21(2): 84–87. (in Chinese).
XU, J.M & LI, L & LIN, P.Q. 2011. Simulation-based evaluation for expressway traffic control strategies. *Highway*(02):83–87. (in Chinese).
LIU, F.T & Tang, Y. 2008. Simulation Study on the application of trafficflowcomplexity measure in traffic control. *Proceedings of the 4th International Conference on Wireless Communications, Networking and Mobile Computing,Dalian, China, 2008*:1–4.
CHOI, Jae-hyeongKimKi-ll.Performance evaluation of traffic control based on geographical information. *Proceedings of the IEEE InternationalConference on Intelligent Computingand Intelligent Systems, Shanghai, China,2009*:85–89.
WANG, Q.Y & XU, L.H & QIU, J.D. 2007. Research and realizationof the optimal path algorithm with complex trafficregulation in GIS. *Proceedings of the 2007 IEEE International Conference on Automation and Logistics, Jinan, China*:516–520.

Energy, Environment and Green Building Materials – Sheng (ed.)
© 2015 Taylor & Francis Group, London, ISBN 978-1-138-02718-3

Study on rural domestic sewage treatment modes of Ningbo

J. Wei, J. Yang & Y.L. Wang
Powerchina Huadong Engineering Corporation Limited, Hang Zhou, China

D.H. Hu
University of Windsor, Windsor, Ontario, Canada

ABSTRACT: The present situation about rural domestic sewage pollution and treatment was analyzed. Treatment models adaptable to Ningbo district were put forward, including centralized treatment model, decentralized treatment model, and transition model. This article analyzes the model selection method by influential elements, including population scale, convergent mode, terrain, geological conditions, and function areas. This article also sorts out the target villages accordingly, establishes mathematical models, and discusses the economical distance of centralized and decentralized treatment models and the control distance of average investment per user under the centralized mode. Finally, this article makes the selection of the treatment models for the 14 administrative villages, which includes 25 natural villages that are first taken into implementation.

KEYWORDS: Rural domestic sewage; Treatment models; Centralized treatment; Decentralized treatment; Ningbo

1 INTRODUCTION

1.1 Background

Following the path of China's rural areas' development, treatment process of domestic wastewater in rural area has been practiced and the experiences are being summarized and analyzed. The program is invested by The Word Bank and supported by the Ningbo Municipal Government as a part of "Construction of a New Socialist Countryside." The first 14 villages are selected from 21 villages of 5 countries in Ningbo as a result of water function and the opinions investigation of villagers who might be affected by the project.

1.2 Limitation of current treatment process

A few provinces have published several guidebooks, such as *The Technology And Practice of Zhejiang Rural Domestic Sewage Treatment* published by Zhejiang Province in November 2007 and *Rural Sewage Treatment Manual* published by Jiangsu Province in May 2008.

Thousands of methodologies are used to treat the rural domestic sewage water; however, in accordance with the rural economy situation, initial investment reduction, and the management simplicity, the treatment models usually could be classified as centralized and decentralized. And the treatment process could also be divided into two stages:

1 anaerobic stage: responsible for the removal of some SS and organic pollutants. Facilities usually include septic tank, digester tank, anaerobic tank, and so on.
2 biological treatment stage: includes aerobic tank and wetland treatment. It is responsible for the removal of organic pollutants, especially the nitrogen and phosphorus, containing organic compounds. The facilities usually include artificial wetland, detention pond, anoxic filter, and trickling filter.

The outflow from a rural wastewater treatment plant must reach the 2nd-class standard of the Town Wastewater Plant's requirement. The 1st-class A or the 1st-class B standard might be required according to different water functions. Here are several common treatment processes: a). Integrated processing equipment or highly purified tank, which could get the 1st-class standard of outflow. A/O is a common facility for this process. b). anaerobic + artificial wetland process, which could get the 2nd-class standard of outflow but will be influenced by season. c). the outflow from a digester tank can reach the 3rd-class standard, and only a few indexes could reach the 2nd-class standard. Some problems have been noticed during the design, construction, and utilization processes. d). other non-power facilities may only reach the 3rd-class standard.

For the areas near drinking water protection zones, additional treatments such as artificial wetland, land filter, and seepage wells will be needed at the outlet, according to the *China Water Pollution Prevention Law*, Rule 57.

1.3 *Objects and treatment features of this program*

According to the site observation, the villages in the first stage have these characteristics:

1 Good economic conditions as well as road hardening has been accomplished in some villages, which reduces the costs of excavation and restoration.
2 Water pollution is a big issue there; fecal sewage is usually treated by non-bottom septic tanks and infiltrated into the ground or collected by outdoor toilets and discharged to farmland. Domestic sewage is discharged freely to the environment, and it is a threat to the environmental conditions and public health.
3 Some villages are separated from each other by a wide distance; the treatment processes should be much more flexible.
4 Most of these villages are within the area of water source protection zones, and centralized discharge outlets are not applicable.

As a conclusion, it is necessary to conduct research and analysis in Ningbo rural areas in order to design better alternatives of wastewater treatment, to achieve the optimized economic and environmental benefits.

2 MODES

Four modes have been proposed according to the situation of Ningbo's rural areas: one centralized mode, two decentralized modes, and one transition mode. Single or multiple modes could be utilized according to the real conditions.

Mode 1. Combined wastewater system with centralized treatment process:

Waste water will be treated after collecting it by the gravity pipe network. This model has the best environmental performance, but it is only applicable for the villages with flattening, steep slopes, and centralized houses. Otherwise, the costs would be higher. These modes could be also divided into entire centralized model and part centralized model.

Mode 2: Sewage combined system with decentralized treatment process:

After the collection of wastewater, it goes into septic tanks for the pre-treatment and is then treated by the infiltration trench. This mode has a median cost but extra land area will be needed near the residential area, and it needs to satisfy the regulation of the groundwater level for the infiltration. This mode does not need pipelines, and it is the most applicable for the fertilizer-non-demanding villages.

Mode 3: Sewage diversion system with decentralized treatment process:

Fecal sewage from each house is discharged into an outdoor double-urn toilet; domestic wastewater will be treated by separated seepage wells. This mode has the lowest costs and is compared with mode 2; it has less land requirement but needs to satisfy the regulation of groundwater level for the infiltration. This mode fits the villages that are fertilizer demanding and have hard-to-lay pipelines.

Mode 4: Sewage diversion system with transition treatment process:

A few old residential areas exist in some villages and they need to be reconstructed. But reconstruction in rural areas is usually poorly organized and cannot ensure the lifetime of treatment facilities using the other 3 modes. Mode 4 proposes to increase the amount of outdoor toilets to collect the fecal sewage; domestic sewage will be treated by separated seepage wells. This mode has a low cost and satisfies both treatment requirement and longer lifetime facilities, which could be a transition mode for old residential areas.

3 VILLAGE CLASSIFICATION

According to the different features of villages, the models will be selected by the 5 different aspects of villages.

3.1 *Population*

The development of a village is determined by its population. There will be a higher requirement for time-enduring systems by villages of large population. In the meanwhile, the level of economic development of this type of village is comparatively high, which means advanced and motivating approaches can be adopted. On the contrary, smaller villages prefer the transition mode. According to the village classification standard, villages could be graded by status as basic-level villages and central-level villages, and they could also be graded by population as small, medium, and big villages.

Table 1. Grading by village scales.

Grading by scales	Villages	
Village level	Basic level	Central level
large	>300	>1000
medium	100–300	300–1000
small	<100	<300

3.2 *Aggregation mode*

The aggregation mode directly decides which mode would be the most appropriate: the centralized mode or the decentralized mode. There are 4 categories: linear aggregation, radial aggregation, planar aggregation, and scattered punctate.

3.3 *Terrain conditions*

The terrain conditions decide the direction of collection pipeline network and the address selection of the outlet treatment. The villages could be classified by the terrain conditions as (1). Plain type: Usually located in the plains region or on the foot of the hillside, the slope is less than 5%; (2). Valley type. Usually located on the hillside, most residences live by farming, the slope is usually greater than 10%, and houses are mostly step-like distributed, connected by steps with big elevation; (3). Half plain half valley: This is a type between plain and valley; the villages are half-plain type and half-valley type.

3.4 *Geological conditions*

The geological conditions decide the selection of treatment process at the outlet. Treatment process and area of occupied file change with the type of soil and groundwater level. According to the geological survey report, the soil conditions in the selected villages could be classified as follows: Sand, pebbly clay, silty clay, and the groundwater levels are between -0.3 ~ -4.0 m.

3.5 *Area function*

The outflow quality and discharge standard are determined by the environmental function of area. The villages could be classified as water resource protection zone and non-water resource protection zone by the different environmental functions.

4 DISCUSSION OF BOUNDARY CONDITIONS

We could choose the mode by the classification of treatment modes and villages, but in order to perform the quantitative comparison and selection, we need to discuss the boundary conditions that would be appropriate for each mode. Numerical models should be built with the data from construction cost information of Ningbo, Zhejiang Province. The results will be used as the theoretical basis for the selection of mode. Two boundary conditions have been researched as follows:

1 The economic distance for centralized mode and decentralized mode.

   The reason for conducting a boundary condition research is to quantitate the economic differences between centralized mode and decentralized mode utilized in the conditions of different distance between households. The average per household cost for the decentralized mode = Septic tank cost + Land infiltration cost + construction cost. For the centralized mode, the cost = Gravity pipeline cost + outlet treatment cost, and these costs may fluctuate depending on whether the road has been completely hardened or not. The results showed the following: First of all, when the road has been completely hardened, the decentralized mode would be more economic if the pipeline is longer than 18 meters per household; second, when the road has not been completely hardened, the decentralized mode would be more economic if the pipeline is longer than 25 meters per household.

2 The controlling distance for the centralized mode per household.

   When there is a limit of cost per household, sometimes we prefer the centralized mode so that it could be easier for management and maintenance and also exhibit good treatment effects; however, the decentralized mode is more economic. Compared with the similar construction costs, the average cost per household in Ningbo should be limited under 5000 RMB, and a controlling distance per household should be taken into consideration. Construction cost per household = (gravity pipeline cot + outlet treatment cost) / household number + single household + indoor renovation cost. It should also be discussed in two conditions: completely hardened road and non-completely hardened road: (1). For the completely hardened road, the outdoor pipeline should be no longer than 34 meters per household. (2). For the non-completely hardened road, this number should be less than 59 meters.

5 MODE SELECTION

Table 2 demonstrates the research of 14 administrative villages and 25 nature villages, as well as the selected mode for each.

6 CONCLUSION

The result shows that most of the villages are situable for the combined system with centralized mode, and some villages could also use the combination of 2–3 different modes. This research could contribute to the practice of wastewater treatment system in China's other rural areas.

Table 2. Statistics of sewage treatment models.

nature Village	Population Forecast	the village classification							Mode selection
		Village level	Aggregation mode	Terrain conditions	Geological conditions	Underground water level	Area function		
1	480	Big central	planar	Valley type	pebbly clay	1.6	Upstream of Baixi reservoir		1,2,4
2	94	Small basic	punctate	Valley type	pebbly clay	2.8			3
3	898	Medium central	planar	Valley type	silty clay	1.9			1,2,4
4	810	Medium central	planar	Plain type	pebbly clay	2.5	Downstream of Baixi reservoir		1
5	1888	Big central	planar	Plain type	pebbly clay	1.7	Upstream of Tingxia reservoir		1
6	1259	Big central	planar	Valley type	sand	4.0			1
7	399	Small basic	punctate	Valley type	pebbly clay	2.8			2
8	740	Medium central	planar	Half plain half valley	pebbly clay	2.8	Upstream of Hengshan reservoir		1,2
9	309	Small basic	planar	Half plain half valley	pebbly clay	2.8			1,2
10	205	Small basic	planar	Valley type	pebbly clay	2.8			1,2,4
11	399	Medium basic	planar	Valley type	silty clay and Round gravel	0.6			1
12	619	Medium basic	planar	Valley type	pebbly silty clay and Round gravel	1.5			1
13	257	Small basic	linear	Valley type	pebbly clay	1～3			1,2
14	155	Small basic	planar	Valley type	pebbly clay	1～3.1			2
15	336	Medium central	planar	Valley type	pebbly clay	0.3			1
16	944	Medium central	linear	Valley type	clay	0.3	Upstream of Siming Lake reservoir		1,2
17	671	Medium central	radial	Plain type	pebbly clay	0.4			1,2
18	357	Medium central	planar	Plain type	clay	1.2			1,2
19	526	Medium central	radial	Plain type	clay	2			1,2
20	315	Medium central	radial	Plain type	clay	2.1			1
21	234	Small basic	planar	Valley type	pebbly clay	0.6			1
22	206	Small basic	linear	Valley type	pebbly clay	1.5			1
23	1626	Big central	planar	Valley type	pebbly clay	5	Upstream of Xikou reservoir		1
24	1500	Big central	planar	Plain type	silty clay	0.3	Upstream of Datang Port reservoir		1
25	1206	Big central	radial	Plain type	silty clay	0.1			1

Energy, Environment and Green Building Materials – Sheng (ed.)
© *2015 Taylor & Francis Group, London, ISBN 978-1-138-02718-3*

Phenolic wastewater treatment by pulsed electrolysis with alternative current

X.M. Zhu, N.L. Nguyen, B. Sun, Z.R. Li & P.H. Zhan

College of Environment Science and Engineering, Dalian Maritime University, Dalian, China

ABSTRACT: Phenol is a major pollutant in wastewater. It has severe effects on human beings, in both the short term and long term. Various methods were used for removal of the phenol from wastewater such as adsorption, photodecomposition, activated carbon, Fenton process, electrocoagulation, and so on. Electrochemical treatment seemed to be a promising treatment method due to its higher effectiveness, lower cost, and so on. In order to remove the phenolic compounds in water, pulsed electrolysis method was used in this work. The possibility of using pulsed electrolysis with alternative current power (AC Power) to remove phenolic compounds from wastewater is researched by using the Ti/PbO_2 electrode. The removal rate of phenolic compounds is investigated in terms of various parameters, such as electrolysis time, voltage, frequency, and the flow rate. The energy consumption is also analyzed. Results show that the electrolysis method has optimal conditions by using the Ti/PbO_2 electrode according to the energy consumption results.

KEYWORDS: Phenolic Wastewater, Electrolysis, Pulsed AC Power

1 INTRODUCTION

The growing population, industrial development, and urbanization contribute to the deterioration of the environment. In recent years, due to the poor condition of surface waters, particular attention is paid to the quality of the aquatic environment. Serious threats to the quality and purity of water include phenols and its derivatives. Phenolic compounds were involved in many industries such as chemical synthesis, coke, refineries, manufacturers of resin, pharmaceuticals, pesticides, dyes, explosives, and herbicides, and they can also occur in their wastewaters. Phenolic compounds have been shown to be toxic to aquatic life at parts per million levels, and several phenolic have the ability to impart taste and odors to drinking water supplies and edible aquatic life at parts per billion levels [1].

Different methods are used for the removal of phenols, such as membrane technique, membrane-based solvent extraction, adsorption, and electrolysis [2–6]. The membrane technique needs to adjust the concentration of phenol to a low level; electro-coagulation technique is a process of destabilizing suspended, emulsified, or dissolved contaminants in an aqueous medium by introducing an electrical current into the medium; adsorption usually uses activated carbon, and it can be prepared from almost any carbonaceous material, but its removal rate is low; and electrolysis is the driving of a non-spontaneous chemical reaction by passing a direct electric current through an electrolyte, and the cathode oxidation can remove the

pollutants effectively. Nowadays, high-voltage direct electric current is not used due to reasons such as high energy consumption, it being dangerous, exhibiting high temperature, and so on [7–11].

In this work, the electrolysis process using alternative current pulsed power is considered as removing phenol from the solution. The pulsed power can decrease the energy consumption. Multiple parameters such as electrolysis time, voltage, frequency, and flow rate are attempted in order to obtain suitable conditions. Energy consumption analysis is also carried on to supply the parameters experiments.

2 EXPERIMENTAL

2.1 *Chemicals and instruments*

A phenolic solution was prepared by using an analytical grade of chemicals and dissolving them in distilled water. The energy for the experiment was served by a standard alternative current pulsed power supply.

The pH of the solution was measured by a digitally calibrated pH-meter (Shanghai Leici Xinjing Yiqi Co,Ltd , PHS-3C). The conductivity of the solution was measured by a conductivity meter (Shanghai Leici Xinjing Yiqi Co,Ltd , DDS-11A). The Chemical Oxygen Demand (COD) of the solution before and after the experiment was determined by the standard procedure with a COD analyzer (Lanzhou Lianhua Huanbao Keji Co. Ltd).

2.2 Experimental setup

The setup of the present work's experimental process is shown in Fig. 1. The size of the plexiglas's tank is 10 cm × 10 cm × 12 cm. The anode is made by Ti/PbO$_2$, and the main material of the cathode is ferrum.

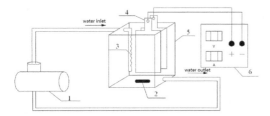

Figure 1.   Experimental Setup.
1, Circulation Pump, 2, Rotor, 3, Pole for Water Inlet, 4, Electrodes, 5, Plexiglass tank, 6, High Voltage Pulse Power Supply

First, 1 L of the phenolic wastewater solutions was placed into the electrolytic cell. The voltage was adjusted to a desired value by the standard alternative current power supply and the electrolysis was started. At the end of electrolysis, the solution was analyzed, and the electrodes were thoroughly washed with water to remove any residues on the surfaces.

3   RESULTS AND DISCUSSION

3.3   Effect of operation time

To explore the effect of operating time, the other experimental conditions were controlled as follows: Concentration of Phenol is 50 mg/L, Voltage is 30V, Concentration of NaCl is 3 g/L, Frequency is 3000 Hz, and Flow rate is 20 L/h.

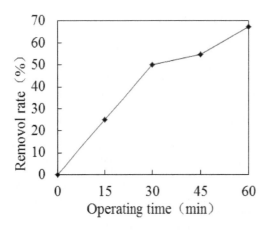

Figure 2.   Relationship between removal rate of the COD and the operating time.

As shown in Fig. 2, the removal rate of COD changed linearly with the operating time between 0 and 30 minutes. Then, COD removal rate increased slightly from 30 to 60 minutes. The reasons used to explain this phenomenon were that, first of all, phenol's degradation in the electrochemical process has many steps. In the beginning, they were just simple oxidizable reactions, but in the end, oxidizable reactions became complex, and the reaction speed was slower, thus making the removal rate of COD decrease obviously in the 0 to 30 minutes and slightly in the 30 to 60 minutes. Second, in the electrochemical oxidation process, a lot of substances were produced; these substances may cover the anode and slow down the generation speed of ·OH; thus, the removal rate of COD from 30 to 60 minutes slowly increased.

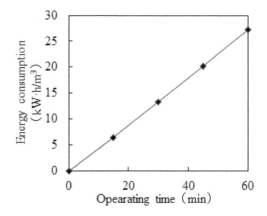

Figure 3.   Relationship between energy consumption and operating time.

The energy consumption for the electrolysis process was calculated. As seen in Fig. 3, it was very clear that the energy consumption and operating time had a linear relation.

3.4   Effect of voltage

Voltage is an important parameter in the electrolysis process. It can affect the current and the energy consumption. Fig. 4 showed the effect of voltage in the electrolysis process. As the voltage changed from 20V to 50V, the COD removal rate almost increased from 35% to 80% lineally, whereas the COD removal rate was maintained from 50V to 60V.

As seen in Fig. 5, the energy consumption was calculated, and the removal rate of COD was directly proportional to the energy consumption with the voltage changed from 30V to 50V. Then, the extra energy consumption did not increase the removal rate of COD. The COD removal rate was 74.2% with a

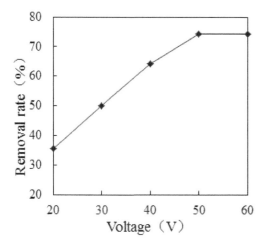

Figure 4. Relationship between voltage and COD removal rate.

50 V voltage, and the energy consumption was 40.75 kW·h/m³. The COD removal rate was 74.3% with a 65 kW·h/m³ consumption of energy. The optimal voltage was 50V in the experiment.

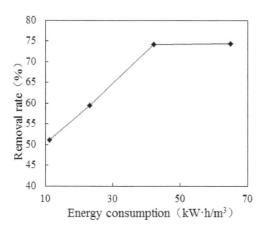

Figure 5. Relationship between energy consumption and COD removal rate.

3.5 Effect of frequency

In this work, we chose 3000 Hz and 500 Hz as the maximum and the minimum frequency values. It can be seen in Fig. 6 that with the increase of frequency from 500 Hz to 3000 Hz, the COD removal rate increased. They had a linear relation. According to the impact of the frequency on current, it is easy to know that the higher the frequency, the greater the current and the higher the COD removal rate. Figure 7 showed the effect of frequency on the energy consumption.

It can be realized that the energy consumption was directly proportional to the COD removal rate and the frequency. The results showed that the higher the frequency, the higher the current and the higher the energy consumption.

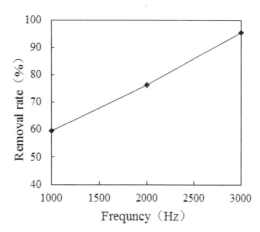

Figure 6. Relationship between frequency and COD removal rate.

By analyzing the results shown in Figs. 6 and 7, it can be concluded that, first, frequency directly affected the current; the higher the frequency, the better the treatment results. Second, the power supply with the constant voltage and variable current mode should select the frequency according to the actual experimental conditions and then find out the suitable terms; or maybe in constant current and variable voltage mode, with the voltage changing in a permission range, it could be easier to increase the frequency and to decrease the energy consumption.

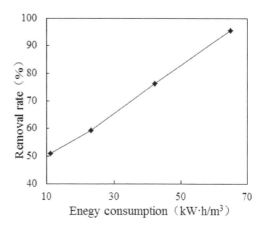

Figure 7. Effect of frequency on the energy consumption.

3.6 Effect of flow rate

The wastewaters flowing or not were important parameters in this experiment. Figure 8 showed the effect of wastewater flow on the COD removal rate. With the same experimental conditions except the flow rate of wastewater, the results showed that the flow rate could affect the COD removal rate. When the circulating water's speed was 20L/h, the COD removal rate was about 74% with 30 minutes operating time, which was better than the result with no flow rate. This result could be explained by the fact that the suitable flow rate could help improve the range of active species, which could increase the degradation of organic substances. If the flow rate was slow or zero, the speed of the active species transmission was also slow, and most of the energy in the reaction would be changed to thermal energy, thus decreasing electrolysis efficiency.

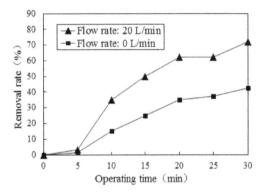

Figure 8. Effect of flow rate on the COD removal rate.

3.7 Conclusions

The results of this study have shown the applicability of electrolysis with an alternative current in the treatment of wastewater containing phenolic compounds by using the Ti/PbO$_2$ electrode. The removal rate of phenolic compounds was investigated in terms of various parameters, such as electrolysis time, voltage, frequency, and flow rate. The energy consumption was also analyzed. Results showed that the electrolysis method had optimal conditions by using the Ti/PbO$_2$ electrode according to the energy consumption results. A suitable reaction time was 30 minutes. The COD removal rate depended on the voltage, and simultaneously energy consumption would increase with voltage in the range of 20V to 50V. This was characterized by lower voltage, lower energy consumption, but worse removal result. The suitable voltage was 50V in this experiment. The frequency and the COD removal rate had a linear relation. According to the impact of the frequency on current, 3000 Hz was the optimal frequency. The COD removal rate increased with a suitable flow rate, and the flow rate 20L/h was better than no flow rate.

ACKNOWLEDGMENTS

This work is supported by the National Natural Science Foundation of China (Grant No. 41005079) and the Fundamental Research Funds for the Central Universities (3132014098).

REFERENCES

[1] X. Liu, Current situations and developments of industrial phenolic wastewater treatment techniques, *Chin. J. Ind. Water Treat.*, 1998, 18(2): 4–6.
[2] R. J. Bigda, Consider Fenton's chemistry treatment, *Chem Eng Prog.*, 1995, 91(21):62–66.
[3] F. J. Beltran, M. Gonzalez, F. J. Rivas, Fenton reagent advanced oxidation of polynuclear aromatic hydrocarbons in water, *Water Air Soil Pollution*, 1998,105 (3~4): 685–700.
[4] O. Laetitia, B. J. Jacques, A. D. Duprez, Catalytic waster oxidation of phenol and acrylic acid over Ru/C and Ru-CeO2/Ccatalysts, *Applied Catalysis B:Environmental*, 2000, (25): 267–275.
[5] H. Yu, G. W. Gu, Treatment of phenolic wastewater by sequencing batch reactor with aerated and unrated fills, *Waste Mange*, 1996,16(7): 561–566.
[6] D. W. Kirk, H. Sharifian, F. R. Foulkes, Anodic oxidation of aniline for wastewater treatment, *Appel. Electro-chem*, 1994, 39: 1857–1862.
[7] N. Mohan, N. Balasubramanian, C. Ahmed Basha, Electrochemical oxidation of textile wastewater and its reuse, *Joumal of Hazardous Materials*, 2007, 147(1–2): 644–651.
[8] C. Comninells, Electro catalysis in the electrochemical conversion/combustion of organic Pollutants for waste water treatment, *Electrochemical Acta*, 1994, 39(11): 1857–186.
[9] M. Panizza, C. Bocca, G. Cerisola, Electrochemical treatment of wastewater containing poly-aromatic organic pollutant, *Water Res.*, 2000, 34(9): 2601–260.
[10] D. Svetashova, Effect of electric current pulse shapen efficiency of electrochemical treatment of waters containing petroleum products, *Khimiya I Technologiya Vody*, 1992, 856–859.
[11] C. Comninellis, C. Pulgarin, Anodic oxidation of phenol for water treatment, *J. Appl Electrochem.*, 1991, 21:703–708.

Energy, Environment and Green Building Materials – Sheng (ed.)
© 2015 Taylor & Francis Group, London, ISBN 978-1-138-02718-3

Fundamental theory and impact factors of art city

Jie Zhao
Urban Design, Wuhan University, Hubei Province, China

ABSTRACT: China is in the midst of transformation, presenting characteristics such as rapid urbanization, cultural globalization, and social polarization, which pose as challenges to urban development and "identity crisis" of urban space. In 2007, China set up Art Committee of China Research Society of Urban Development, managing and controlling identity art activities and identity art space in the urban area in a centralized way. This article draws the definitions and indicators of various types of creative cities, innovation cities, eco-cities, and post-industrial cities, and it brings forward the concept of the art city, attempting to perform an exploratory study of the art city construction in China during the period of social transformation. With such hope to fill the gap in this field theoretically, this paper is expected to provide corresponding methods to build the art city, providing ten indicators and the Quantitative criteria to define which is of double significance in terms of both theory and practice.

KEYWORDS: Art city, Fundamental theory, Impact factors

1   INTRODUCTION

Domestic study on the art city is rare, and even the worldwide study on the topic is thin on the ground. The art city itself pursues a spiritual (art) altitude with material (city) as the carrier, which decides its complexity and diversity; the city is a complex and scientific macro-system, whereas the art is the synonym for variety. Similarly, existence of the art city needs to be tested and falsified in a long time, so it must be tested by history, proved by reality, and improved in the future, since both the art and the city will be greatly changed as time passes.

Currently, most first-tier and second-tier cities in China have met the basic needs of material progress, and they are currently yearning for a better and more enriched spirit world. Study on the art city can bring brand-new strength to these representative cities. Such strength can not only guide the city to seek and present its identity but also create and promote a new culture of its own, which will directly influence the urban planning and construction, hence to making the urban construction more rational, more oriented, and more sustainable. Such strength will also influence the ideas and image of the citizen, increasing the sense of belonging and the pride for city folks.

2   CONCEPT OF ART CITY

Throughout the world, the following cities have been recognized as art cities: New York stays the cutting-edge of art fashion, undoubtedly, having a dozen years of devotion and contribution to art; London maintains the time-honored elegance and modern aggressiveness, a talent exhibition place and a milestone for modern art; Berlin is a European cultural center with freedom and inclusiveness, being the initiating place for Avant Garde art; Los Angeles is one of the pioneering art centers in America with geniality, moderation, indolence, and leisure; Paris is refreshing and diverse, a city of romance with loftiness and constraints; Mexico city is a busy art market with coruscation and inclusiveness, a must-have destination for adherents of art fashion and collectors of artworks; Tokyo embraces classic architecture technology and modern top craftworks, a restrained and glaring city style; Madrid is the synonym for "post-Franco" era, with a strong smell of leisure and coziness permeating every corner of the city; Buenos Aires has extensive and profound cultural transplant and integration, following Argentina to be a more potential art market; and Miami is the most concerned showplace for modern art, an art tower with strong artistic air and a unique style. Each city has a strong artistic atmosphere, profound artistic culture, and rich art flavor.

We can figure out that the art city must be an organism of art expressive force and city functionality from the analysis and studies of these cities. In a broad sense, the art city is an epitomizer that is harmonious in general and rigorous in levels, possessing both material values and spiritual values, objective construction, and subjective evaluation. It embodies the large-scale and advanced socioeconomic level, accumulates the historical culture of a country or a nation, possesses an artistic space that can provide beauty and comfort to people, and forms a sustainable human settlement environment with cultural connotations. In a narrow sense, the art city links history with modern artistic city space and city functions, with the city as the basic carrier, the art as the sublimating method, human beings as the orientation, and the history as the bridge. The art city makes full use of cityscapes and resources, historical culture and features, artistic techniques, and expressive force to build a developable and sustainable city of unity and identity, which can guide the senses and ideas.

3 IMPACT FACTORS OF ART CITY'S FORMATION

3.1 Urban transportation pattern and environment

Serving as the traveling path and the art lifeline of a city, urban transportation, with various patterns—grid, radial, ring-radial, grid-ring-radial hybrid, and free style—is the first element to display its charm and allocate its resources. Besides, the traffic planting, road signs, pavement, road lighting, and infrastructure are optimal carriers of a city's artistry. When citizen and tourists roam in a city, they will judge the city's artistry by the direct perception that is brought by the city roads. In the art city, the road traffic system should employ the landscape design and the creative design mentality to embellish the city roads while meeting the traffic needs for roads to be unimpeded and smooth, hence making the road space artistic, interesting, and creative and avoiding simple and rough constructions.

Table 1. Traffic quantitative criteria.

Indicators	Excellent	Good	Medium	General	Poor
Average Transfer times	1–1.1	1.1–1.3	1.3.1.5	1.5–1.8	≥1.8
Air Pollution Index	0–50	50–100	101–200	201–300	>300
The average equivalent sound level of road traffic noise	≤68	68–70	70–72	72–74	≥74
Traffic landscape	Excellent	Good	Medium	General	Poor

3.2 Urban ecological corridor and park landscape

As an important aspect to keep and form landscape, cultural context, and features of a city, based on natural ecological conditions and zonal vegetation, urban afforestation should incorporate the folk customs, traditional culture, religion, and historical relics into the landscape greening to endow the urban green space system with territorial and cultural characteristics, hence making it identifiable and distinctive. It should be designed reasonably from an overall spatial layout to a partly greening landscape design, give prominence to the artistic merits, tap into city natural landscape resources, and study urban historical features in detail. The urban landscape building should take root in the national culture and regional culture. Meanwhile, eco-environmental construction of the city and landscape greening should gradually progress toward the realization of city biodiversity. With such a mentality of point-line-plane coordination from landscape ecology to design the urban landscape, the artificial landscape and natural resources inside the city can be combined organically, form an effective green ecological network, endow the mountains and rivers with cultural characteristics, incorporate artificial landscape into the natural environment, and achieve harmonious and integral construction of the urban ecosystem.

Table 2. Urban green space quantitative criteria.

Indicators		Standard
Built-up area Green coverage (%)		≥36%
Built-up area green rate (%)		≥31%
Park green area per capita	Construction land per capita≤80m²	≥7.50m²/p
	80≤m²Construction land per capita≤100m²	≥8.00m²/p
	Construction land per capita≥100m²	≥9.00m²/p
Minimum value of green space for each district		≥25%
Parkland service radius coverage (%)		≥70%
Number of Arboretum over 40hm²		≥1.00
Green penetration along River (%)		≥80%
Ecological and Landscape recovery rate of disposal land(%)		≥80%
Comprehensive evaluation value of urban landscape		≥8.00
Functional evaluation value		≥8.00
Landscape evaluation value		≥8.00
Cultural evaluation value		≥8.00
Trees protection rate (%)		≥95%
Water shoreline naturalization rate (%)		≥80%
Avenue promotion rate (%)		≥50%

3.3 *Urban public art space*

Urban public space is the place employed for conducting free activities for the citizen, providing a window for enjoying city life, experiencing cityscapes, displaying city personality, and appreciating charm of the city. Well-designed public space in general is unique, continuous and closed, attractive, easily accessible, identifiable, adaptable, and diverse. Urban public art space should be highly open, participatory, and comprehensive. Openness guarantees enough space for relevant art activities or adding artistic facilities, which are more suitable to construct a public art space because of open space and a good flow of people, such as squares, parks, and stations. Participatory public art space can lead people to spontaneously focusing on or appreciating the spatial artistry in use, such as artistic sculptures, art pieces, and art shows; enabling the art carriers in the space to communicate with people; and preventing them from falling into vulgar arts or making it above people. Finally, comprehensiveness is necessary as public art space design is expected to adapt to the esthetic views of various social communities. Thus, pertinent evaluations are expected to be made on the communities both mentally and emotionally. A good art space must consider the functions, humanity, and use of natural environment and construction materials in a comprehensive way, and it should reach the best effect on public art space though combining various design ideas and skills.

3.4 *Reconstructing historic urban quarters*

Urban history exists in everlasting mountains and rivers, in visible historical objects, in national customs, in literature, and, more importantly, in the memory of city people. Reflected in urban construction, it is the protection of historic urban blocks, which is not only at the material level but also at the spiritual level. What needs to be protected are original cultural accumulations, city memory, and original identity.

Table 4. Urban historical block quantitative criteria.

Indicators	Quantitative Analysis		
Spatial development direction	Border fractal dimension		
Geometric spatial scales	scale	Block area and depth	
		Block plane scale	Commercial Street Width: 3-6m Residential street width: 2-3.5m Node depth: 6-25m
	Volume rate		
	The flow of people		
Structural features	Space Syntax		

As for the building of art city, understanding the essence of protecting historic urban blocks meets the basic needs of people, corresponds to the general esthetic standards of city space, and better displays the indigenous culture and the distinctive personality of the city. Protection of the historic urban blocks will be favorable to promote the transformation of the previous simple space of aesthetic appeal into the high-quality city space of local characteristics, human touch, and artistry.

3.5 *Using urban waterfront space*

To some extent, unique landscapes of ancient cities resulted from the influence of landscape culture on mountains and water. Based on economic factors, urban construction factors, and political factors, the focuses of worldwide city development converge on the development of waterfront area these days. The development of the waterfront area in a city is diverse in ecology, collective in urban functional space, and high in urban landscape value.

The use of waterfront space in an art city is mainly embodied in aspects such as inheriting and continuing

Table 3. Urban public art space quantitative criteria.

Indicators			Weights(%)
The quality of urban public art space	Surrounding accessibility	Public Transport	5<x<8
		Walking System	5<x<7
		Point mark	3<x<5
	Landscape	Integrity	5<x<6
		Outline	3<x<5
		Scale	3<x<5
		Proportion	3<x<5
		Unobstructed line of sight	3<x<5
		Scene style and features	3<x<6
		Color	5<x<6
	Space	Land Nature stability	5<x<6
		Fitness of Use and planning	5<x<6
	Management	Green maintenance	5<x<6
		Environmental maintenance	5<x<6
	Culture contents	overall style	6<x<8
		Image recognition	3<x<5
		Cultural Activities	3<x<5

the urban texture, emphasizing the protection of waterfront ecological environment, and paying attention to the continuity of regional features and landscape characteristics as well as the metaphor of traditional symbols. Only until the waterfront area is improved, and the use of an inviting, continual waterfront space of human touch and ecological concerns is transformed, a city can prove to be an art city.

3.6 Urban street building space

As one major component of a city, streets are indispensable in connecting various functional areas of the city, serving as a major standard for whether a city is dynamic. Urban blocks may extend to any corner of the urban area, turning a loose urban framework into an organic network, whereas the concrete spatial entity of street space has a more powerful permeation and influence and lays a foundation for further manifestation of artistry of the city.

Design of street space in an art city should include the size of the street, constructions along the two sides of the street, alleyways, squares, parks, and iconic structures. Design methods and styles of these components of street space are the visible expressions of artistry of the city.

3.7 Design of urban architecture

The urban architecture is an important expression of urban civilization and a major indicator of the city's quality, which interprets the aesthetic appeal and cultural traits in artistic areas of the city as well as inviting stories and classic marks of the developing art city. During the construction of the art city, the urban architectural landscape brings out not only the strong visual impact but also the huge heart-shaking influence on people. Based on an accurate understanding of the city's historical and cultural contexts, regional architecture identity, and local folk customs, architectural design in the art city needs to respect the traditions, blaze new trails, improve the quality, and pursue harmony. While we are designing the architecture itself, its surrounding spatial environment should be coordinated and incorporated into the macro-constructions of the city, hence achieving such a harmonious state of the unity of human beings, the architecture, and the environment for the urban architecture in the art city.

3.8 Urban art and cultural representation programs

The art and culture of a city must be represented by specific structures that demonstrate not only the city as the art carrier but also the importance of the art to the city. Exhibition space of an art gallery in an

Table 5. Urban street building quantitative criteria.

Content			Requirements(m)
Scale	Street aspect ratio		1-2
	Building height along the street		<120
Continuity	Back line of buildings along the street		5<
	Eaves height of the first floor		4.5
Comfort	Facilities zone		1.5-2
	Walkways		4<
	Green belt	City Inner Ring	3<
		Inner Ring to Central Link	5<
		Central Link to the Outer Ring	10<
	Green and facilities	Street tree planting space	4-8
		Planting height	Trees	9-15
			Shrubs	0.6-1.2
		Street-lights	Spacing	5-20
			height	6-15
		Bus station spacing	Ordinary buses	500-800
			Rapid Transit	1500-2000
		Waste bins pitch		<100
		Rest seat pitch		300-500
		Accessible		Specification
	Parking and separation	Street parking	Parking time	<5h
		Vehicles and non-motor vehicles Separation		Increase the number of import and export

art city must display the cultural context from aspects of color reconstruction, and so on. In the meanwhile, skills that can display the regional culture should be presented in the space design, colors, and so on. A design philosophy is proposed that suits the cultural continuity, integration, development, and innovation of the art galleries; with tentative conclusions of speculative knowledge relevant to exhibition space design of art galleries, a new thinking and innovation in exhibition mode are developed based on the exhibition space design in modern art galleries to fully demonstrate the standards and the height of the art city.

3.9 Creating urban art and cultural industry

The urban art and cultural industry is the intersection point of culture and economy; such particularity distinguishes it from simple economic activities by paying attention to cultural quality, and from pure cultural sectors by pursuing economic benefits. The cultural industry of a city can provide spiritual and cultural goods to massive consumers on the one hand, and it can urge the city to improve its culture on the other hand. The art industry for a city is expected to have functions of information dissemination and entertainment while having art and cultural connotation.

3.10 Fostering atmosphere for artistic creation

The art city needs to have an atmosphere for artistic creation. Only by constant innovation can the city maintain its vitality and go further along the path of art. Some fundamental factors to keep this atmosphere are as follows: 1. Source of creation. It is the primary condition for art creating. It is also the city's artistic heritage, which includes features such as the natural environment, history and culture, custom, and folklore. 2. System of creation. With the institutions and systems that the creation needs, it is fundamental for fostering an atmosphere for art creation. Top institutions of art creation can provide corresponding talents, information, and the knowledge of art, and maintain a long-term and steady obtaining of art creation. And the system of creation serves as an umbrella for those institutions. A perfect system can stimulate artists' enthusiasm to create art products and guarantee the quality product as well as the fairness of art creation. 3. Environment for art creation. Besides the hardware, soft needs of creation can promote the necessity and security of the art creation from perspectives of ideas and policies, leading to the fostering of an artistic atmosphere in a healthy and smooth way.

4 CONCLUSIONS

Currently, the art city is a newly emerged concept. However, as time passes, more and more cities will reach a certain high level of economic development, and the identity of the city will be increasingly focused, so will the spiritual layers of the city. And more and more people will focus on how to reflect the city's cultural characteristics through city construction. Therefore, the idea of the art city can provide a new thinking and open a wider vision for planners inwardly. The art city is an expression of its culture and spirit. Its complicacy and comprehensiveness determine itself to be tested in a long time and

gradually change the urban features in a unique way. "The city can be used as a material for creation" "Art can be the source of originality to shape the urban landscape." Nowadays, the city has colorful artistic forms and a broader scope. The society requires larger-scale and more exceptional themes of city arts. Future art of the city asks us to shake off the bonds of traditional thoughts and techniques of creation, widen our vision, accept new art thoughts, assimilate the advanced thoughts and techniques abroad, and combine them with the essence of our traditional culture to build a variety of distinctive art cities reflecting national spirit and features of the times.

REFERENCES

1. Grigg, David.(1989). English agriculture : an historical perspective[M]. Oxford
2. Bailey, C. et al. (2004). Culture led urban regeneration and the revitalization of identities in Newcastle, Gateshead and the North East of England. International Journal of Cultural Policy, 10(1), 47–65.
3. Belfiore, E. (2004). Auditing culture. The subsidized cultural sector in the new public management. International Journal of Cultural Policy, 10(2), 183–202.Bell, D. (1976). The cultural contradictions of capitalism. New York: Basic Books. Bell, D., & Jayne, M. (2006). Conceptualizing small cities. In D. Bell & M. Jayne (Eds.),Small cities: urban experience beyond the metropolis (pp. 1–18). Abingdon, UK, New York: Routledge.
4. De Propris, L., Chapain, C., Cooke, P., Mac Neill, S., & Mateos-Garcia, J. (2009). The Geography of Creativity. London: NESTA.
5. Dixon, Tim. (2009). Urban land and property ownership patterns in the UK: trends and forces for change[J]. land use policy 26s:43–53.
6. EU Commission (2010a) EUROPE 2020. A Strategy for Smart, Sustainable and Inclusive Growth, Brussels: EU Commission, COM (2010).
7. Ginsburgh, V., & Throsby, D. (2006). Handbook on the economics of art and culture. Amsterdam: Elsevier.
8. Goodman, L. A. (1961). Snowball sampling. Annals of Mathematical Statistics, 32(1),148–170.
9. Hippel, E. Von (2005). Democratizing innovation. The MIT Press.
10. Home, Robert. (2009). Land ownership in the United Kingdom: Trends, preferences and future challenges[J].land use policy 26s:103–108.
11. Kagan, S., & Kirchberg, V. (Eds.). (2008). Sustainability: a new frontier for the arts and cultures, Frankfurt am Main: Verlag für Akademische Schriften.
12. Landry, C. (2000). The creative city: a toolkit for urban innovators. London: Earthscan. Landry, C. (2006). The art of city-making. London: Earthscan.
13. Lazzeretti, L., & Cinti, T. (2009). Governance-specific factors and cultural clusters: the case of Florence.vCreative Industries Journal, special issue on "The drivers and processes of creative industries in regions and cities", 2(1), 19–36.

14. Lloyd, R. D. (2006). Neo-Bohemia: art and commerce in the post industrial city. NewYork: Routledge. McGuigan, J. (2005). Neo-liberalism, culture and policy. International Journal of Cultural Policy, 11(3), 229–241.

15. Margheri, F., Modi, S., Masotti, L., Mazzinghi, P., Pini, R., Siano, S., & Salimbeni, R.(2000). SMART CLEAN: A new laser system with improved emission characteristics and transmission through long optical fibres. Journal of Cultural Heritage, 1(Suppl. 1), S119–S123.

16. Potts, J., Hartley, J., Banks, J., Burgess, J., Cobcroft, R., Cunningham, S., & Montgomery, L. (2008). Consumer co-creation and situated creativity. Industry and Innovation,15(5), 459–474.Power, D., &

Scott, A. J. (2004). Cultural industries and production of culture. London-New York: Routledge.

17. Chreiner, M. and Strlic, M. (Eds.), (2006) Handbook on the use of lasers in conservation and conservation science, COST G7.Storper, M., & Scott, A. J. (2009). Rethinking human capital, creativity and urban growth. Journal of Economic Geography, 9(2), 147–167.

18. Vergès-Belmin, V., Wiedemann, G., Weber, L., Cooper, M., Crump, D., & Gouerne, R.(2003). A review of health hazards linked to the use of lasers for stone cleaning. Journal of Cultural Heritage, 4(Suppl. 1), 33s–37s.Wasserman, S., & Faust, K. (1994). Social Network Analysis: Methods and Applications. New York: Cambridge University Press.

Energy, Environment and Green Building Materials – Sheng (ed.)
© 2015 Taylor & Francis Group, London, ISBN 978-1-138-02718-3

Study of spatial color reconciling in urban color planning—based on the thinking of urban color planning of Harbin

Z. Bian & Y.W. Zhang

Art Academy, Northeast Agriculture University, Harbin, China

ABSTRACT: Urban Color Planning is an important means of shaping the cultural characteristics of the city. This paper is mainly engaged in tailoring the standard of Harbin's color configuration system by analyzing Harbin city's regional and traditional color culture, creating the main theme of urban color, and combining the space. The experimental environment design is based on typical regional space, the visual testing is run by the cognitive experimental scale, subprojects are investigated by examining the contents of urban color introduction and color perception, and, finally, the color harmony style of the urban space is drawn with the statistical analysis of quantitative criteria.

KEYWORDS: city color; space combined; colors tone; color reconciling

1 INTRODUCTION

The acceleration and development of urban construction need a new, standardized, and rational new design pattern as a guide. People living in the urban city are now influenced by the psychological impact caused from the trend of building complex and the generation of international city patterns. The development of the city construction and the improvement of the urban life quality caused a series of new problems. Problems such as color pollution, disorganized design, and lack of spatial color showed up while the traditional urban landscape and the new urban city color were mixing up.

By studying this project, a method for color harmony of urban space is developed, to re-map the traditional characteristics of the city and culture and to help people know and understand the city's image. Taking the urban planning's lack of spatial color as a break point, we need to test with the virtual and reality experiential modes; create a clear regional spatial color characteristic; and make a better city image by planning the spatial color concept, inheriting the design research and techniques of the geographical and cultural color performance, and enhancing the city image's macroscopical visual experience.

2 METHODS

2.1 Experimental settings

The experiment was set in Guogeli Street, which is located in the central of Nangang District; it is an A class street in Harbin. From south to west, it starts from Wenchang Street and ends up at Yiman Street. Its total length is 2642 meters, the driveway is 12 to 21 meters wide, and the boardwalk on each side is 2 to 4 meters wide. Guogeli Street is very famous for its Russian style. In the street the Qiulin company stands, which is a classical Renaissance building; the Huayuan Elementary School was originally the residence of the Japanese General Consulate; the provincial Foreign Affairs Office's eclectic style architecture was also originally a consulate. The street is lined with many Baroque-style architectures, such as Lek Sellye, the Russian Catholic Church, the Russian folk merchandise trade blocks, the Russian River park with the surrounding musical fountain, the waterscreen film, local bars, and many sculptures and pavilions that show the Russian humanistic culture. This street is a large building complex that has shopping, leisure, entertainment, offices, restaurants, hotels, and many other functions. This experiment tested 98 people, including 46 men and 52 women, all of whom are passers from the street or from the surrounding buildings.

2.2 Scale setting

2.2.1 Investigation method

The visual test method is mainly about local regional color, the color of the streets, the color of buildings, and color perception. The contents are shown in the figure next. (Figure 1)

Using questionnaire, color card cognition, different material colors, and other means, we can examine subjects' perception of identifying the urban space environmental changes and spatial cognition ability under the conditions of color factors imported and, respectively, set the appropriate variables as shown in Table 1.

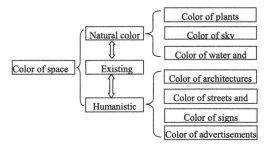

Figure 1. Survey content of color space.

2.2.2 *Quantitative criteria*

Corresponding the variables' order from Table 1, according to the examination content of the subjects, we can calculate the color variables by the color value assignment. Quantitative criteria values increase from negative to positive; the closer they get to a positive value, the smaller the error range, the more efficient the recognition effect, and also the higher the visibility of the spatial color (to the subjects). As shown in Table 2, the sampled color information was analyzed from the viewpoint of urban geography, plants, traditional architecture colors, modern architecture, and historical cultures and customs.

2.3 *Experimental site*

Subjects were tested without mutual interference or any hint in each experiment location (intersection, landmark buildings, and large building complexes). All the respondents were individuals; the male female ratio was close to 1:1; and the age range was 15–75. One of the experimenters was in charge of briefing, whereas another was writing down the data, and all others were either recording or supplying further questions.

2.4 *Data statistics*

According to the analysis of Harbin city's urban geographical and cultural environment space color identification, the control of CMYK color scheme is as follows:

Dark color 80>C>73 Gray color 58>C>15

Bright color25>C>2

Dark color 71>M>47 Gray color 39>M>6

Bright color 6>M>2

Dark color 71>Y>47 Gray color 39>Y>6

Bright color 6>Y>2

Dark color 71>K>47 Gray color 39>K>6

Bright color 6>K>2

The city's color tone is low brightness based (green, blue colors), with secondary high brightness colors (yellow-green, light green colors).

3 RESULT

3.1 *Natural color and humanistic color*

Natural color influences the spatial environment. Overall, 32% subjects considered that their identification of the whole regional environment was not affected by the color differences, and no recognition result changes occurred. Next, 68% subjects were able to find the color differences in the regional color space, and the identification time can effectively be shortened, of whom 51% can carry out a clear color control and find out all the differences.

By sampling experimental analysis, the urban planning spatial color and geographical landscape color were mainly influenced by sunlight and other weather factors. The characteristic of sunlight is a

Table 1. Research project and methods.

Total Project	Sub-projects	Examination Content	Empirical Approach	Variable Code
Natural color	Climatic conditions\ Local material	City culture \ Street culture \ Architectural culture\ landscape culture	Questionnaire\ Color card cognition	1
Spatial recognition	Color spatial recognition	Shape\quantity\ Constitution \Distance	Questionnaire\ Indentified cognition	2
	Differences in color direction	Identify direction\Locate direction\Search direction	Questionnaire\ direction\ Indentified cognition	3
Color performance	Color factor's influence on culture	Hue\Brightness\Cold and warm tones	Questionnaire\ Direction	4
	Visual balance of color factors	Disorganized design\ Color Pollution\Cultural character	Questionnaire\ Direction	5
Humanistic color	The environmental impact of color harmony	Natural color\ Humanistic color \Prompt design	Questionnaire\ Direction	6

Table 2. Variables and quantitative criteria.

Variable Code	Quantitative Criteria	Value Assignment	Variable Code	Quantitative Criteria, Sampling color information	Value Assignment
1	<3 <2 <1 Color space differences >1 >2 >3	−3 −2 −1 0 1 2 3	3	Dark color/Gray color/Bright color C:50 C:82 C:4 M:21 M:70 M:2 Y:20 Y:63 Y:2 K:0 K:35 K:0	−2 −1 0 1
			4	Dark color/Gray color/Bright color C:73 C:21 C:21 M:47 M:6 M:6 Y:47 Y:47 Y:47 K:3 K:3 K:0	−3 −2 −1 0 1 −3
2	<3 <2 <1 Color spatial cognition >1 >2 >3	−3 −2 −1 0 1 2 3	5	Dark color/Gray color/Bright color C:80 C:57 C:25 M:62 M:39 M:3 Y:83 Y:85 Y:70 K:37 K:4 K:0	−2 −1 0 1 2
			6	Dark color/Gray color/Bright color C:82 C:16 C:2 M:71 M:12 M:2 Y:69 Y:23 Y:2 K:64 K:0 K:0	−1 0 1

crucial element that decisively influences the urban spatial color. Using light and color principle, soft light increases color whereas strong light weakens color. Therefore, the color of Harbin appears a low saturation and high brightness color. Building materials' choices were limited by the climatic conditions; the weather also reflects different colors. Locals gradually formed a common understanding of spatial color in accordance with their own way of life. With the high and low temperature changes, they use the gentle and high brightness cool color or air color, decorated with gray, dark green, and green-blue as secondary color.

The experimental results show that the urban spatial color was affected by the regional, national characteristics and the cultural traditions. With green, light blue, lake blue, gray, white, and other colors, we inherent the urban culture.

3.2 Color recognition

Theoretically, the color recognition is based on the visibility, namely, the identification scale when one color overlays another. The scale of recognition is related with the urban spatial color, background color, brightness, purity, and hue.

The experiment shows that the color contrast between foreground and background colors has a direct impact on the characters' visual code. The search time for color coded characters is related to the foreground and background colors of the character. The color coded characters are more significantly contrasted than the non-color coded ones. Generally, the time of recognition increases while the color difference decreases between foreground and background colors. In terms of color brightness and warm-cold tones, the bright colors give a feeling of expanding or moving forward, whereas the dark colors give a feeling of reducing or moving backward. When it comes to color tones, bright warm colors have the sense of amplification, whereas low brightness warm colors show diminution; the specific recognition varies from different brightness levels[1]. The color tone was controlled to be warm in urban city design, such as yellow, light yellow, or milky white. The buildings tend to be beige (as representations: the Workers Cultural Palace, Youth Palace, Asia cinema, Dragon Apartment Building, the Provincial Government Building, and other city landmarks). Other buildings' color contains red, yellow, white, and other warm colors.

3.3 Spatial color cognition

In terms of spatial cognition, 86% subjects considered the larger the color areas are, the simpler the colored graphs are, the easier they could be remembered, and 59% of these have become accustomed to the simple color patterns with cognitional symbolizations.

Overall, 82% subjects believed the identification feature would be gradually weakened when the instructive color shows three continuous characteristics. By testing the hue within the same distance among red, orange, yellow, green, blue, purple, black, and white colors, 93% subjects considered that red, orange, and yellow have the highest recognition, and 33% of them have a certain color reference for the recognition of the GB color.

The experiment shows that the color areas, the simple graphics, and the spatial cognition are directly proportional. Similarly, red, orange, and yellow have the highest proportion during the hue test in the same distance. The interval size between foreground and background colors will affect its results; the higher the brightness of warm color, the easier the distance changes could be recognized. When the color areas were the same, most subjects would link the color areas with the background color or the surrounding colors. The brighter the surrounding colors are, the smaller the inner colors become. Therefore, audiences are sensitive to the color differences cognition.

3.4 Space color harmony

Color reconciling is about seeking a balance relationship between unity and contradiction. The color in space should not only be related to its own hue, brightness, and purity but also should be reconciled with the contrast from other different colors. Space color shows both the harmonic and contrastive quality in a regional environment, both of which are interdependent and mutually contradictory. Taking color theory as a foundation, urban architecture and other visual elements as reference, the color of the city could be simultaneously more diverse and unified, and the ambience of the city could be conveyed even better. Meanwhile, systematically designing the space color plays both an identifiable and a directional role in the urban planning.

The experiment shows that the color reconciliation is associated with two or more different colors; it is based on the reconciling of same or similar color tones, such as reconciling same color, same brightness colors, same chromaticity colors, and no colors. A small area of low brightness color reconciling gives a heavy and vigorous feeling; whereas large area of moderate color reconciling gives a mature, pleasurable feeling. Thus, the color reconciling is based on the color physiology; it gives a visual balance and a feeling of harmony. The reconciling relationship is as follows:

1 High chromaticity color contrast reconcile as a master– slave relationship, as well as the area and the graphic language.

2 The reconciling of contrast achieve harmony by separating the matching colors from the complementary colors.
3 Changing the brightness and chromaticity of two complementary colors to reconcile.
4 Under certain circumstances, two complementary colors are able to mix to reconcile.
5 Complementary colors can use gradient colors or compartmental colors to reconcile.

4 DISCUSSION

According to the results of this experiment, the urban color planning should consist of historical culture, natural environment, color aesthetics, color guidance system, and people's living habits; it is an effective means to create a diverse, harmonious, and unified city. City color planning of Harbin highlighted the research based on the color theory of this experiment. In order to create and plan urban style, we should use urban color impression as a guide and take space color reconciling as a means, to plan the urban color fully integrated with the local practice, strengthen the pertinence and practicability of the urban color constructing system, and contribute to a new exploration of urban color planning.

ACKNOWLEDGMENT

This research was supported by the Planned Project of Heilongjiang Province on Art Science (No. 12D005).

REFERENCES

[1] Shuwen Shi. Architectures' Environmental Color Design [M], Beijing, China Construction Industry Press. 1991.
[2] Aysu Baskaya, Christopher Wilson, Yusuf Ziya; zcan. Wayfinding in an unfamiliar environment: different spatial set tings of two polyclinics [J]. Environment and Behavior, 2004, 36(6):839.
[3] Peihong Zhang. The crowd evacuation rules during building fires. [D]. Ph D Dissertation.
[4] Leiqing Xu, Bo Huang, Zhong Tang. The affection on the sense of direction and way finding brought by the spatial differences in Gestalt Space. Academic Journal Of Tongji University. 2009. 2.
[5] Jacques Besner, A Master Plan or A Regulatory Approach for the Urban Underground Space Development, The Montreal Case (translated by Bo Zhang), International Urban Planning, 2007 (22):16~20.
[6] Rong Cheng. Study on city color planning management [D]. Zhong Nan University, 2010.

Energy, Environment and Green Building Materials – Sheng (ed.)
© 2015 Taylor & Francis Group, London, ISBN 978-1-138-02718-3

Influence factors of fluorine concentration in groundwater in Gaomi area, Shandong province, China

J.G. Feng & Z.J. Gao
Shandong Provincial Key Laboratory of Depositional Mineralization & Sedimentary Minerals, Qingdao, P.R. China
College of Earth Science and Engineering, Shandong University of Science and Technology, Qingdao, P.R. China

J.F. Duan
Hydro-geology Bureau of China National Administration of Coal Geology, Handan, P.R. China

M.J. Shi
College of Earth Science and Engineering, Shandong University of Science and Technology, Qingdao, P.R. China

ABSTRACT: High fluorine in shallow groundwater in Gaomi area is typical in Shandong province. Analyses studied the influence of groundwater exploitation, rainfall, evaporation, geological lithology, hydrodynamic conditions, and other factors on fluorine concentration. Fluorine concentration in shallow groundwater is impacted by a special local condition.

KEYWORDS: High Gluorine Groundwater; Influence Factors; Gaomi

1 INTRODUCTION

High fluorine groundwater in Gaomi is well known in China (Sun 1984, Chen et al. 2006, Chen et al. 2009, Dai et al. 2012). High fluorine groundwater generally formed in certain areas. Analyzing the influence factors is important for researching the high fluorine groundwater in Gaomi area.

The high fluorine groundwater is distributed in Damoujia, Jiangzhuang, Renhe, Xiazhuang, and Heya town, which constitute a total area of about 793 km².

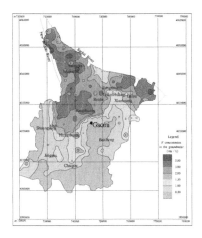

Figure 1. Regional distribution of fluorine concentration in shallow groundwater of Gaomi area.

2 RELATIONSHIP BETWEEN GROUNDWATER EXPLOITATION AND FLUORINE

The sanitation and antiepidemic station of Gaomi in Shandong province has resulted in the fluorine water being monitored continuously from the time of taking water control measures in 1985 until 16 years later (Ge et al. 2002). A part of the monitoring results is shown in Table 1.

Table 1. Fluorine concentration changes of drinking water in Gaomi from 1986 to 2001.

Town		Before taking control measures					After taking control measures			
		1986	1988	1990	1992	1994	1995	1997	1999	2001
Da moujia	F⁻ (mg/L)	1.18	1.30	1.50	1.50	1.61	0.98	1.01	0.99	1.03
	exploitation (m³)	108000					87703			
Jiang zhuang	F⁻ (mg/L)	0.98	1.02	1.10	1.15	1.25	1.06	1.09	1.05	1.04
	exploitation (m³)	144000					117684			

From 1985 to 1994, fluorine concentration in drinking water increased without taking control measures and it was reduced after 1995. It shows that hydrodynamic conditions changed by groundwater exploitation; this promoted the leaching and dissolution of fluorine in solid medium of high fluorine area and increased the fluorine concentration in groundwater.

3 RAINFALL, EVAPORATION, AND FLUORINE

The ref. (Ge et al. 2002) lists the rainfall, evaporation, and fluorine concentration of 10 sources of drinking water (30 samples per year). A part of the data extracted is depicted in Table 2.

Table 2. Rainfall, evaporation, and fluorine concentration of 10 sources of drinking water from 1985 to 1994 in Gaomi.

year	average concentration of fluorine (mg/L)	Rainfall (mm)	Evaporation (mm)
1985	0.92	442.4	1749.3
1986	0.95	453.3	1681.9
1987	0.81	730.1	1449.1
1988	1.12	466.7	1676.4
1989	1.20	486.4	1799.3
1990	1.21	465.7	1658.2
1991	1.23	476.5	1659.1
1992	1.28	454.4	1756.4
1993	1.18	656.3	1654.2
1994	1.45	432.1	1865.4

Table 2 shows that the average fluorine concentration of 10 sources of drinking water is on the rise from 1985 to 1994. Rainfall also shows a trend of fluctuation rise at the same time. If in a given year rainfall is on the high side, average fluorine concentration is lower.

Fluorine concentration of Gaomi faced a rising trend from 1985 to 1994, and evaporation had the same change law at the same time. We need to explain that evaporation plays an important role in high fluorine groundwater formation of the area.

Caixia Li, the senior engineer of 4th geology and mineral resources exploration of Shandong provincial, analyzed the hydrological and geochemical characteristics in high fluorine areas in Gaomi in 2008 (Li 2008). Figure 2 reflects that the F⁻ concentration of groundwater in the two sample points has a similar variation, but the change law of rainfall is more different.

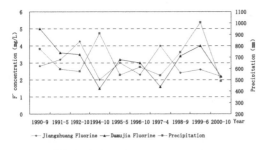

Figure 2. Fluorine concentration in groundwater and rainfall changes.

Fuorine concentration in groundwater and the rainfall in Jiangzhuang have the opposite changing

trend; fluorine concentration is relatively low during the large rainfall month. It indicated that rainfall can dilute the fluorine concentration in Jiangzhuang. The fluorine concentrations in groundwater and rainfall in Damoujia have a similar change tendency. It is because in Damoujia, which is far from the source of fluorine, the influence of other factors on the fluorine concentration in water increased; the comprehensive performance is the fluorine concentration; and the rainfall has the same change tendency. The abnormal law from October 1992 to October 1994 is believed to be caused by a lack of monitoring data in 1993.

Fluorine concentration and evaporation in Jiangzhuang have the same change tendency. Fluorine concentration is relatively high during the large evaporation month. This illustrates that evaporation has a great influence on fluorine concentration in Jiangzhuang. The fluorine concentration and the evaporation in Damoujia have opposite changing trends; the reason is Damoujia is far from the source of fluorine and the influence of other factors increased in the fluorine concentration.

4 GEOLOGICAL LITHOLOGY AND FLUORINE

The source of fluorine in shallow groundwater in Gaomi area is mainly endogenous. Fluorine mainly originates from the dissolution of containing fluorine minerals in rocks and soil and migrates by water flow. The area of high fluorine is located in Jiaolai basin; formation lithology is Mesozoic Cretaceous sandstone, conglomerate, mudstone, volcano rock, and volcano clastic rock. All the Mesozoic strata rocks contain fluorine. The Wang group of sedimentary clastic rocks have not only high fluorine concentration but also higher soluble coefficient, which is the main source of fluorine in that area. The total fluorine concentration of Wang group clastic rocks was widely distributed in high fluorine areas between 540 and 600 ppm, and its fluorine soluble coefficient is greater than 3 percent. The upper strata of Wang group was mostly covered by the quatemary loose rocks, mainly divided into Heituhu group and Yihe group, and the soil total fluorine concentration was much more than the world average value (according to the data from China's environmental monitoring station, there were 4093 different samples of soil (A layer), the average of fluorine concentration is 478 mg/kg, the minimum is 50 mg/kg, the maximum is 3467 mg/kg, and the 95 percent fiducial interval range is from 191 to 1012 mg/kg(CNEMC 1990)). The fluorine concentration of 60-cm depth soil reached 2266.7 ppm, and the soluble fluorine was 21.6 ppm in Zhoujiazhuang village of Kangzhuang town. Both the quatemary sediments and the underlying bedrock were leached by rain or dissolved by groundwater; there was some fluorine transfer from solid phase to liquid phase, eventually leading to the fluorine elements being enriched in groundwater.

5 HYDRODYNAMIC CONDITION AND FLUORINE

Hydrodynamic condition is also an important factor of influence for fluorine concentration in groundwater. In general, the better the hydrodynamic conditions of groundwater, the easier it is to lose fluorine and the lower the fluorine concentration in groundwater. In Lixianzhuang village, located in Jiangzhuang town in the high fluorine area of northern Gaomi, the F- concentration in groundwater was far lower than the surrounding villages in every survey. An ancient river was found between Lixianzhuang village and Wangjiasi village through the survey. There exists a thick sand layer about 10-m depth underground from Lixianzhuang to the north boundary of Gaomi (figure 3). In the villages around Lixianzhuang, the thickness of the sand layer suddenly decreased and even disappeared to the area of Renhe town. These stratum conditions led groundwater flow to become smooth near Lixianzhuang; hydrodynamic condition was very good, and fluorine concentration was mainly below 1 mg/L.

The fluorine concentrations in groundwater in Beilijia village, Anjia village, and Tanjia village are higher than in others, and the maximum is more than 15 mg/L, but close to Huaijia village in the north it is less than 1 mg/L. A geological survey found that the southern regional (from Beilijia village to Tanjia village) strata do not contain sand. The thickness of the sand layer gradually increased from south to north, and the sand layer increased to 3–5 m in Huaijia village. The groundwater hydrodynamic condition is obviously strengthened by the result, which is a decrease in fluorine concentration in groundwater of Huaijia village area.

6 CONCLUSIONS

1 High fluorine groundwater is mainly distributed in the central and northern areas of Gaomi.
2 The amount of groundwater exploitation, rainfall, and evaporation have a significant influence on the forming of high fluorine groundwater.
3 Hydrodynamic conditions have important influence on the enriching and migration of fluorine in groundwater.

ACKNOWLEDGMENTS

This study was supported by SDUST Research Fund (No: 2012KYTD101) and the National Natural Science Foundation of China (41102149).

REFERENCES

Chen Peizhong, Yun Zhongjie, Ma Aihua, et al. 2006. Survey Analysis on Prevalent Status of Endemic Fluorosis in Shandong Province . *China Preventive Medicine*, 7(4): 254–256.
Chen Peizhong, Yun Zhongjie, Bian Jianchao, et al. 2009. Analysis on Surveillance of Endemic Fluorosis in Shandong Province in 2007. *China Preventive Medicine*,10(7):570–573.
China National Environmental Monitoring Centre(CNEMC). 1990. *China's Soil Element Background Values*. Beijing: China Environmental Science Press.
Dai Jierui, Wang Cunlong, Pang Xugui, et al. 2012. The evaluation of the geochemical environment quality of shallow underground water in east Shandong province. *Geophysical & Geochemical Exploration*, 36(2): 277–282.
Ge Xiangjin, Jiang Yuting, Tang Yizhu, et al. 2002. Water Fluoride Monitoring and the Reason of Rise for Water Defluoridation Source of Water in Gaomi. *Literature and Information of Preventive Medicine*,8(5):556–557.
Li.Caixia 2008. Hydrogeochemical characteristics of high-fluorine groundwater in the Gaomi area, Shandong, China. *Geological Bulletin of China*, 27(5):689–699.
Sun Zhuyou. 1984. Endemic Fluorosis Research in Shandong Province.*Shanghai Environmental Science*. 3(6):18–20.

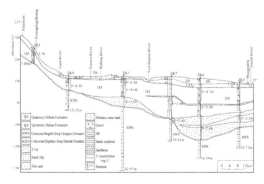

Figure 3. Hydrogeological section from Xiejia Village to Wangganba.

Energy, Environment and Green Building Materials – Sheng (ed.)
© *2015 Taylor & Francis Group, London, ISBN 978-1-138-02718-3*

Study on the training of safety signs in coal mines

Wei Jiang
Faculty of Resources & Safety Engineering, China University of Mining & Technology, Beijing, China

Ya-Nan Liu
Faculty of Resources & Safety Engineering, China University of Mining & Technology, Beijing, China
Binzhou Administration of Work Safety, Shandong, China

Shang-Hong Shi
Faculty of Resources & Safety Engineering, China University of Mining & Technology, Beijing, China

ABSTRACT: Training can improve people's understanding of safety signs. The influence of information content of safety signs on the experimental subjects to get the learning times and the reaction time to a certain standard is studied based on the fixed learning frequency. Conclusions are drawn as follows. When the accuracy reaches 100, learning frequency increases with the increase of information content of safety signs, namely the learning frequency for safety signs of high information content is 4, medium information content is 3.3, and low information content is 1.7. So does the reaction time. The reaction time for safety signs of high information content is 1271 ms, of medium information content is 1133 ms, and of low information content is 1009 ms. The accuracy decreases with the increase of information content of safety signs.

KEYWORDS: Coal mines, Safety signs, Training

1 INTRODUCTION

Safety signs are used for expressing specific safety information, which consists of graphic symbols, safety colors, geometric shapes, or geometric borders or texts[1]. The last defense for safety precaution is a safety sign whose failure may lead to the occurrence of safety accidents[2]. Therefore, studying the safety signs more intensively and comprehensively and increasing their effectiveness are of great significance for accident prevention.

With the deepening of the study, research scope of safety signs is also expanding[3,4]. Coonetilleke et al pointed out that when the users saw the signs first, although they might have guessed the meaning of the signs, training can improve the cognitive processes of understanding the meaning of the signs[5]. Previous studies show that training can enhance people's understanding of the meaning of the signs remarkably[6]. The safety signs regulated in Safety Signs for Coal Mines are used as the experimental materials to study the training frequency and reaction time to completely memorize the safety signs of different information content.

2 TRAINING

1 Directions of the experiment. The influence of information content of safety signs on the experimental subjects to get the learning times and the reaction time to a certain standard is studied based on the fixed learning frequency.

2 Experimental methods. Present while using the computer to fill in the questionnaire method for testing. Artificially control safety signs pictures on a computer screen presentation and time. Test participants in the experiment one by one. The main trial records each person's situation related to the amount of information on signs in a table; mark the impact of processing experiment, and use mathematical and statistical tools for statistical analysis of the experimental results.

3 Subjects. The experiment chose twenty 24–30-year-old graduate students as subjects, including 10 boys and 10 girls; all subjects had normal vision or corrected vision and had no experience of training in safety signs.

4 Experimental material. A formula was used to calculate the average amount of information sources to determine the 74 safety signs amount of information in the "mine safety signs" (GB14161-2008) (22 prohibition signs, 19 warning signs, 12 instruction signs, and 21 tips mark), followed by a mine safety signs uncertainty measurement; whereas mine safety signs were distinguished by the degree of comprehensibility difficulty. We got 28 zero information mark mine safety signs, 14 0-1 information

mark mine safety signs, 13 1-2 information mark mine safety signs, and 19 mine safety signs' amount of information mark that is greater than 2. The information mark for 0-1 randomly picked 10 as the minimum amount of information for the low information group; from the amount of information for 1-2, it randomly picked 10 as the amount of information group, randomly picked from the amount of information mark being greater than 2, and randomly picked 10 as the high amount of information mark for a group. Safety signs in these 30 pictures were selected and displayed to be tested on a computer monitor screen.

5 Laboratory equipment. A computer has a 23-inch monitor. Experimental subjects are away from the monitor by about 50 cm, and safety signs on the display size are 30 cm × 30 cm. The first stage is the learning stage; the laboratory interface simultaneously shows signs, safety signs pictures and names. The second stage is the testing phase; the experimental interface presents only safety signs pictures, no longer showing the sign's name.

3 TRAINING RESULTS ANALYSIS

3.1 *Number of training and learning analysis*

Statistical tools spss 19.0 statistical software for data analysis is used, mainly applying GLM (General Linear Model) repeated-measures analysis.

GLM repeated measurement is the process of advanced analysis and repeating the same dependent variable measure. It can be a repeated measure under the same conditions, and it can also be repeated measure under different conditions[7, 8].

1 Analysis of the information size impact on the study times

Research for the single factor experiment, arrange the experimental results, and obtain 100% correct number form of learning, each of which was tested for different levels of safety signs after the amount of information to be learned. Everyone corresponds to a set number of a learning record, recording a total of 20 group study times (see Table 1). The average difference of 20 sets of data was analyzed for repetitive measure analysis of variance.

As can be seen from Table 1, with the amount of information increasing, the number of learning increases in order to achieve a certain accuracy standard.

Table 2 shows the amount of information for the three different learning standard safety signs to reach 100% accuracy rate of learning times descriptive statistical analysis. These include the number of samples to be tested (N), the average number of learning (Mean), and standard deviation (Std. Deviation). Subjects were 20, and standard deviation reflects the degree of dispersion

Table 1. Amount of information on signs marks the impact of the experimental stages of the learning process.

Sequence number	gender	Learning frequency		
		Low information content	Low information content	Low information content
1	male	2	3	3
2	male	1	4	4
3	male	2	3	4
4	male	2	2	4
5	male	1	4	4
6	male	2	3	4
7	male	1	3	4
8	male	1	2	5
9	male	2	4	4
10	male	2	4	4
11	female	2	3	4
12	female	2	4	4
13	female	2	3	4
14	female	1	3	5
15	female	1	3	3
16	female	2	4	4
17	female	1	3	4
18	female	3	4	4
19	female	2	3	3
20	female	2	3	4

Table 2. Descriptive statistics.

	N	minimum	maximum	Mean value	Standard deviation
Low information content	20	1.00	3.00	1.7000	.57124
Medium information content	20	2.00	4.00	3.2500	.63867
High information content	20	3.00	5.00	3.9500	.51042
Effective N (listing status)	20				

of data. Analyzing data in the table, the wnumber of higher information safety signs learning were the most concentrated, followed by a low amount of information and a medium amount of information learning times. From the average number of learning, we can get a preliminary analysis, that is, at the rate we reached a 100% correct premise, low information level of safety signs are more easily recognized and remembered; with increasing levels of information, safety signs are more difficult to identify and remember.

Table 3 consists of multiple test results, such as the F-test Pillai's Trace test methods, Wilks' Lambda test method, Hotelling's Trace test methods, and Roy's Largest Root test method probability Sig. Value 0.000 is less than 0.05; considering to be within the group of factors, the amount of information has a significant effect (F = 61.992, p <0.05), that is, there is 95% certainty that the group has a significant influence on the effect of the number of learning differences, indicating that the amount of information for safety signs to reach the level of 90% correct rate of learning times has a significant effect. This indicates that the level of information for safety signs reaching a 100% correct rate of learning times is significantly affected. Similarly, in four of the test results, P values between gender and the amount of information that are greater than 0.05 are 0.89, indicating no interaction between gender and the amount of information and that the number of sex influence learning on the 100% correct rate for safety signs was not significant.

Table 3. Multivariate testing.

Effect		Value	F	Suppose df	Error df	Sig.
Intercept	Pillai's trace	.921	61.992[a]	3.000	16.000	.000
	Wilks' Lambda	.079	61.992[a]	3.000	16.000	.000
	Hotelling's trace	11.624	61.992[a]	3.000	16.000	.000
	Roy's largest root	11.624	61.992[a]	3.000	16.000	.000
Gender	Pillai's trace	.037	.207[a]	3.000	16.000	.890
	Wilks Lambda	.963	.207[a]	3.000	16.000	.890
	Hotelling's trace	.039	.207[a]	3.000	16.000	.890
	Roy's largest root	.039	.207[a]	3.000	16.000	.890

3.2 Training reaction time results analysis

Research is a single factor test. Finish the experimental results and obtain each of the different levels of information security mark after testing an average response time and maintaining accuracy of records in Table 4. Everyone corresponds to a group, there is an average response time and accuracy, there was a total of twenty groups of test records, the average data for the twenty-group differences were repeated measurements, (the reaction) signs mark the amount of information on the impact of processing the experimental testing phase analysis of variance, and gender is a between-group variable. All these are used to get a series of statistical analysis.

Table 4. Average reaction time and accuracy record sheet.

Sequence number	gender	Reaction time (ms)		
		Low information content	Medium information content	High information content
1	male	1001.25	1224.38	1343.05
2	male	1011.05	1221.04	1245.07
3	male	957.26	1108.15	1130.08
4	male	987.34	1013.17	1165.87
5	male	1048.98	1269.52	1334.56
6	male	987.34	1046.79	1456.97
7	male	1004.28	1165.65	1268.04
8	male	978.29	1098.36	1224.63
9	male	956.14	1124.68	1246.05
10	male	1122.05	1234.04	1391.87
11	female	1099.26	1106.45	1232.68
12	female	973.14	1003.19	1265.19
13	female	1021.00	1276.52	1465.06
14	female	925.34	1046.09	1256.45
15	female	1176.28	1235.65	1284.04
16	female	934.26	1105.36	1304.63
17	female	987.14	1122.68	1233.05
18	female	1012.05	1106.04	1185.87
19	female	1054.26	1136.45	1192.61
20	female	956.14	1023.14	1205.17

Table 5 consists of three different groups of information security flag test reaction time and accuracy of the descriptive statistical analysis results. These include the number of samples to be tested (N), the mean (Mean), and standard deviation (Std. Deviation). Among them, the number of samples to be tested (N) of 20, standard deviation (Std. Deviation) reflects the degree of dispersion of the data. From the analysis of the data in Table 5, the average response time is shown to be the most concentrated in the low information group, followed by the medium information

Table 5. Descriptive statistics.

	N	minimum	maximum	Mean value	Standard deviation
Low information content	20	925.34	1176.28	1009.6425	63.77753
Medium information content	20	1003.19	1276.52	1133.3675	85.72043
High information content	20	1130.08	1465.06	1271.5470	89.62617
Effective N (listing status)	20				

group. The average response time is relatively found in the most high information group and is not focused. Low information level product safety signs have the shortest response time: As the level of the constant increases in the amount of information being tested for safety signs, it will gradually increase the average response time.

Data analysis is conducted based on Table 6, where the value of the F-test probability P Pillars Trace test methods, Wilks' Lambda test methods, Hotelling's Trace test methods, and Roy's Largest Root test methods are 0.000 less than 0.05; the amount of information considered to be within the group of factors has a significant effect ($F = 209.619$, $p < 0.05$), that is, 95% certainty that the effector group has a significant influence on the average response time difference, indicating thst the average response time for the level of the amount of information is significantly affected. Similarly, the results of four test P values between gender and information are 0.648 greater than 0.05, indicating no interaction between gender and the amount of information, that is, the average response time for sex had no significant effect.

Table 6. Multivariate testing.

Effect		Value	F	Suppose df	Error df	Sig.
Intercept	Pillai's trace	.975	209.619a	3.000	16.000	.000
	Wilks' Lambda	.025	209.619a	3.000	16.000	.000
	Hotelling's trace	39.303	209.619a	3.000	16.000	.000
	Roy's largest root	39.303	209.619a	3.000	16.000	.000
gender	Pillai's trace	.095	.561a	3.000	16.000	.648
	Wilks Lambda	.905	.561a	3.000	16.000	.648
	Hotelling's trace	.105	.561a	3.000	16.000	.648
	Roy's largest root	.105	.561a	3.000	16.000	.648

4 CONCLUSION

In summary, the amount of information for different levels of safety signs learning and test experiments initially proved to be an effective indicator of the amount of information used to evaluate the safety mark of quality. A safety sign has its own graphics encoding, and users need to learn to master it after its design implications.

1 According to the amount of information based on the experimental results for different levels of safety signs learning phase of the experiment, to reach some understanding with the correct rate, number learning times increase with the amount of information increase. The amount of information and different levels of safety signs learning number are significantly different. Safety signs used to learn the number of high information content is 4 times, the amount of information to learn the number of safety signs is 3.3 times, and the number of the low amount of information to learn safety signs is 1.7 times. For the high level of information security signs, due to its ability to pass on information and to be useful to more users, these safety signs have their own definition. There is an uncertain answer, so this allows the user to easily produce safety signs ambiguity, not the reminiscent meaning they want to express. Therefore, with the amount of information for higher levels of safety signs, the user must master these icons and needs to go through more learning times.

2 According to the amount of information based on the experimental results for different levels of safety signs test phase of the experiment, the different levels of information security mark on the reaction time were significantly different. With increasing the amount of information, the reaction time increases. High information content, medium information content, and low information content groups are different in reaction time. High information safety signs reaction time is 1271 ms, medium information safety signs reaction time is 1133 ms, and low information safety signs response time is 1009 ms; these explain the value of information and that it can affect the user response for the processing of safety signs. By the difference in the reaction time, the learning can be described by using the same standard; the user has faster processing speed for low information than for high information. For the large amount of information safety signs, because the user's processing speed is slow, if it is in an emergency situation, there is no chance to react and the user is prone to accidents.

3 According to the accuracy of the experimental results from the test phase of the experiment, the different levels of information security signs exhibit significant differences. With an increasing amount of information, the correct rate is gradually decreased. That is, under the same premise of learning standards, the user has deeper memories for safety signs of low information content. The main features of these lower levels of information security signs are that they are intuitive, and they are able to inspire people to think of their proper meaning.

Author: Jiang Wei (1982-), female, Faculty of Resources & Safety Engineering, China University of Mining& Technology (Beijing), a lecturer and doctor, engaged in research on safety culture, behavioral safety, and training and safety signs. Tel: 13426323576, E-mail:jiangwei678@126.co

Detailed address: Room 404, Comprehensive Building, Faculty of Resources & Safety Engineering, China University of Mining& Technology (Beijing), Ding No.11 Xueyuan Road, Haidian District, Beijing.

ACKNOWLEDGMENT

Study Supported by the Fundamental Research Funds for the Central Universities (Item Number: 2013QZ02).

REFERENCES

[1] State Administration of Work Safety (SAWS). GB 2894 safety signs and the use of guidelines [S]. Standards Press of China, 2009.

[2] HU Yi-cheng, ZHOU Xiao-hong, WANG Liang. Evaluating effectiveness of safety signs on building site [J]. China Safety Science Journal, 2012, 22 (8): 37–42.

[3] HU Yi-cheng, ZHOU Xiao-hong, WANG Liang. Investigation into recognizability of safety signs:Sign features ad user factors [J]. China Safety Science Journal,2013, 23 (3):16–21.

[4] Lesch M F. A comparison of two training methods for improving warning symbol comprehension[J]. Applied Ergonomics, 2008,39(2):135–143.

[5] Goonetilleke R S, Martins Shih H, Kai On H,et al. Effects of training and representational characteristics in icon design[J].International Journal of Human-Computer Studies,2001,55(5):741–760.

[6] Alan H S Chan, Annie W Y Ng. Effects of sign charac-teristics and training methods on safety sign training effectiveness[J].Ergonomics,2010,53(11):1325–1346.

[7] SHI Wen-wen.SPSS 19.0 statistical analysis from entry to the master [M]. Tsinghua University Press, 2012.

[8] WU Song, PAN Fa-ming. SPSS statistical analysis collection [M]. Tsinghua University Press, 2014.

Energy, Environment and Green Building Materials – Sheng (ed.)
© 2015 Taylor & Francis Group, London, ISBN 978-1-138-02718-3

Constitution analysis of pollution sources in urban rivers of Hangzhou

Jin Yang & Jun Wei
Powerchina Huadong Engineering Corporation, China

Min Yuan & Yin Lu
The Renovation Construction Center of Urban River in Hangzhou, China

Feng-Fei Chen
Powerchina Huadong Engineering Corporation, China

ABSTRACT: Describing the characteristics of Hangzhou city rivers, introducing the water environmental status of Hangzhou urban rivers, and establishing the model of non-point source pollutant within the city-surrounding highways in Hangzhou city by utilizing SWMM software, this research has analyzed the runoff pollution by applying the model method and measured values comprehensively; simultaneously, the waste water without interception, the tail water discharged for standards, the combined sewer overflow, the atmospheric deposition, surface rainfall, and the river sediment pollution have been calculated quantitatively. Based on the analysis of the Constitution of pollution sources, the research indicates the direction for the next stage in improvement work of the Hangzhou city rivers.

KEYWORDS: Hang Zhou; River channel; Water environment; Runoff pollution SWMM; CLC number: X705

1 INTRODUCTION

Hangzhou, the provincial capital of Zhejiang Province, located in the southeast of China, is one of China's Seven Ancient historical Capital Cities, and it has a reputation of "Paradise cities." Water is the soul of Hangzhou. Currently, there are 1449 water courses in Hangzhou City with a total length of 7453 kilometers, among which 479 are within the urban area with a length of 1021 kilometers.

The characteristics of Hangzhou Water System are as follows: (1) It is obviously affected by artificial regulation, graded water system, and divided area. There are 82 pump houses and 83 sluices in urban areas, dividing the water system into five districts and five degrees of water level; (2) The river network is crowded, and the water surface ratio is large. The river network density reaches 1.2km/km², and the water surface ratio is about 14%. According to the "The guidelines of urban water system planning," the water surface ratio belongs to a category 1 partition. (3) The problem of flood and water quality still exists. The change of underlying surface due to urbanization results in water logging in the urban area; however, the development of the economy results in pollution of urban rivers. (4) The features of water culture and water landscape are prominent. The development history and the special status of Hangzhou thrive on water, which enables the water to carry too much of cultural landscape implication. West Lake and the Grand Canal have been listed in the world cultural heritage, whereas the Qiantang River and West Brook are also historical interests.

2 CURRENT STATUS OF WATER POLLUTION

According to the monitoring of 6 routine water quality indexes, ammonia and nitrogen, total phosphorus, COD_{Mn}, DO, transparency, and water temperature, in 137 sections of 100 urban river areas and two lakes in Hangzhou, the water that is less than V still occupies a larger proportion.

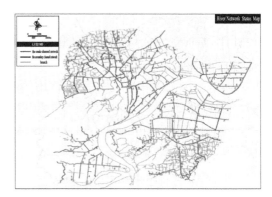

Figure 1. Figure of current channel distribution.

The eutrophication evaluation of 50 sections in the urban river was defined by the comprehensive nutritional status index; among them, 9.68% is in mild eutrophication state, 6.45% is in medium eutrophication state, and 83.87% is in hyper-eutrophic state. Based on the earlier analysis, after years of renovation, the water quality of Hangzhou city river is still not optimistic and needs further improvement.

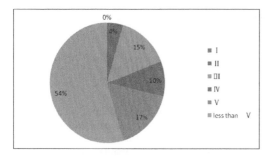

Figure 2. Diagram of various types of water distribution ratio.

3 CONSTITUTION ANALYSIS OF POLLUTION SOURCES

Around the Water Environment of Rivers in Hangzhou, recently, there is still lack of quantitative analysis on the relevant constitution of pollution sources. By data collection and field survey, the pollution sources can be preliminarily divided into six parts: the runoff pollution, the atmospheric deposition and surface rainfall pollution, the waste water pollution without interception, the pollution of tail water discharged for standards, the combined sewer overflow pollution, and the river sediment pollution.

3.1 Analysis of annual runoff pollution load

Two methods were used to analyze the annual runoff pollution load.

Method 1: the model method

Calculating runoff pollution load of area within the city-surrounding highways by selecting America EPA SWMM software, the main steps of the model method are as follows:

1 The GIS analysis of land uses a planning map

Using ArcGIS software, GIS analysis is carried out for the land use planning map of the area of 781 square kilometers within the city-surrounding highways. According to the different land use types, this involves setting different proportions of the runoff coefficient and the pervious area, and inputting the attribute of the model catchment area.

2 The basic parameter definition model

This mainly includes terrain slope, manning coefficient and depression storage, and infiltration select by using the GREEN_AMPT model.

3 Inputting natural rainfall concentration

Concentration of pollutants contained in natural rainfall is measured in values.

In the model selection, the index of TSS, TN, TP, and COD is 30 mg/l, 3 mg/l, 0.1 mg/l, and 10 mg/l in natural rainfall.

4 Selecting the typical mean flow year from annual rainfall series at the age of 30 years, and inputting the 5min interval rainfall data in the typical year into the model.

5 Adopting the pollution load module of SWMM

Tables 1. Concentration of pollutants contained in natural rainfall in Hangzhou City. (mg/l)

Index	TSS	TN	TP	COD
numerical value	25–43	2.83–6.25	0.038–0.250	< 20

software, we can calculate the total pollution load level year in mean flow year. (Table 2)

Method 2: The measured runoff concentration estimated by EMC method

Through choosing different functional areas in urban area, we can measure the runoff water quality of different precipitation and recommend the average pollutant concentration of runoff pollution as the estimated value. According to the recommended value, this paper calculated the total amount of Xiacheng district runoff pollution in the case of average annual rainfall.

Attention should be paid to the fact that both the wash off and buildup parameters of the model method and the measured data of the EMC method are the city built-up area monitoring data, and the concentration of those data is obviously higher than that in the undeveloped plots and natural mountain landscape. But there is still a considerable part of the area in the undeveloped plots and natural mountain landscape within the city-surrounding highways; if the city built-up area rain data are applied to the entire region, runoff pollution load calculation is on the high side. Therefore, we calibrate the two methods according to the proportion of the current built-up land area accounting for the total area within the city-surrounding highways.

3.2 Analysis of the load of the wastewater pollution without interception

According to the statistical data, the average sewage interception rate of urban area within the city-surrounding highways is 78%. Simultaneously, according to the wastewater discharge statistics in drought

season of the river regulation that has been implemented in the rain and sewage diversion storm inlet
on both sides, runoff accounts for about 10% of total
amount of sewage, that is to say, even if the diversion transformation has been more perfect, there is
still a part of mixed wastewater discharged by the
storm inlet, and the percentage was roughly 10%.
According to the statistics of other domestic cities,
the 10% sewage mixed connection and infiltration
ratio is the lower limit in general. Therefore, through
the earlier analysis, the untreated sewage flowing into
the river included not only the proportion of sewage
interception rate statistics but also about 10% of the
amount of sewage discharged that mixed into rainwater piping systems. Therefore, the total amount of
sewage that goes uncollected and that is discharged
directly into rivers should be around 32%. Currently,
the total amount of sewage discharge is 150million
m^3/d in Hangzhou city area; then according to a non-
interception ratio of 32%, we can get the sewage quantity of no closure; then, the load of the wastewater
pollution without interception can be calculated by the
water quality data of current sewage plant operation.

3.3 Analysis of the load of the pollution of tail water discharged for standards

In Hangzhou urban area, the main sewage treatment
plant outfalls all discharge into the Qiantang River.
There is no discharge into the urban river; therefore,
the load of the tail water pollution drainage standard
discharged from those sewage treatment plants is
not counted. Nevertheless, the tail water of part of
the sewage treatment plant and tail water outfall distributed outside the city limits flowed into the main
urban area by the river. This part of the sewage plant
should consider their tail water pollution load, where
this part of the load should be calculated based on the
amount of tail water and the tail water standards. The
total amount of this part of the sewage is 11 millon
t/d, which executed level A standard, and N/P reached
the standard of class III.

3.4 Analysis of the combined sewer overflow pollution load

According to "The Special Planning of the Sewage
Treatment Project in Hangzhou," the Middle East River
area and the Huansha canal area are still predominantly
the interception combined system, with an area of
around 4 square kilometers. In addition, the downtown
area of Hangzhou city has nearly 50 residential areas
of about 4.8 square kilometers for the combined
system. Chengxiang town of about 18 square kilometers in Xiaoshan old city for the combined system
and, consequently, service area of the combined system within the city limits is 26.8km². According to the
related research, for the combined sewer system, there

is a mathematical relationship between the amount of
sewage overflow on unit area per year (mm/a) and the
dry season sewage quantity and interception ratio. The
relation formula is as follows:

$$H0=[-C1(nQd)^{1/2}+C2]^2 \qquad (1)$$

 n——interception ratio;
 Qd——the dry season sewage quantity, m^3/skm²;
 H0——the amount of sewage overflow unit area
per year, mm/a.

There is a proportional relationship between the
coefficient C1, C2 and the annual runoff amount,
according to the runoff coefficient and the average
annual rainfall of Hangzhou, coefficient C1, C2 value
can be obtained. Then, according to the interception
ratio, the dry season sewage quantity of combined
drainage system can help calculate the water volume
per unit area of the combined sewer overflow; multiplying by the combined sewer service area and the
average water quality combined sewer overflow, we can
finally get the combined sewer overflow pollution load.

3.5 Analysis of the load of the atmospheric deposition and surface rainfall pollution

In Hangzhou region, the load intensity of atmospheric
deposition pollution of ammonia and nitrogen, TN, TP
is 0.054 t/(a.km²), 0.18 t/(a.km²), and 0.014 t/(a.km²);
it can help calculate the pollution load of atmospheric
deposition contribution on river water quality. The
surface rainfall is calculated by natural rainfall concentration (Table 1).

3.6 Analysis of sediments pollution substrate pollution load

According to sampling analysis of Hangzhou River,
the percentage of organic matter of river sediment
ranged from 0.88% to 4.67%, with an average of
2.43%. From the overall trend, the order of the average content of organic matter in the Grand Canal
water system is 2.672%> Shangtang river water
system is 2.654%> south of the Yangtze River water
system is 2.09%> Xiasha water system is 1.885%>
Shangsi water system is 1.79%. According to the survey, the percentage of organic matter of the Qiantang
river sediment ranged from 1.158% to 3.419%, with
an average of 1.45%, being obviously lower than
Hangzhou urban rivers. In addition, according to the
test of sediments effect on overlying water contaminant release in the case of static, moderate disturbance
and strong disturbance in the canal area water system,
the concentration of COD, NH_3-N, TN, and TP in
the overlying water would be increased by an average of 1.15mg/l, 0.08mg/l, 0.1mg/l, and 0.008mg/l.
The contaminant release values were amended by the

proportion of the average content of organic matter of 2.672% in the canal area water system and the average content of organic matter of 2.43% in the whole city area; the release concentration of COD, NH3-N, TN, and TP of sediments effect on overlying water within the city limits would implemented 1.05mg/l, 0.073mg/l, 0.091mg/l, and 0.0073mg/l. According to the external water transfer amount and the total runoff flowing into the river in the study area, the river sediment pollution load can be obtained by the earlier concentration value.

4 CALCULATION RESULTS

According to the earlier analysis, the proportion of main output of water environment pollution source in urban rivers of Hangzhou(Tables 5) under the current condition could be calculated through programing calculation procedures and inputting all the model parameters. Analyzing COD and TN, two indexes, thanks to all kinds of pollution, source detection indexes differ.

Tables 3. Main output of water environment pollution source in urban rivers of Hangzhou.

pollution source	COD (t/a)	TN (t/a)
1 Annual runoff pollution load	54202.2	3244.2
2. The load of the wastewater pollution without interception	65195.0	7334.4
3. The load of the pollution of tail water discharged for standards	2007.5	40.2
4. The combined sewer overflow pollution load	6800.0	582.9
5. The load of the atmospheric deposition and surface rainfall pollution	1235.5	24.7
6. The river sediment pollution load	2826.0	245.7

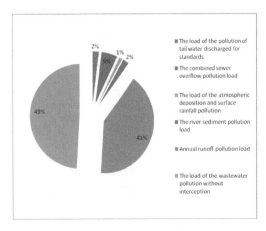

Figure 3. COD constitution of water environment pollution sources in urban rivers of Hangzhou.

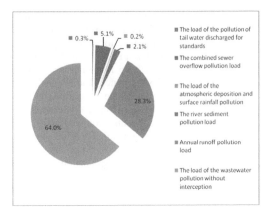

Figure 4. TN constitution of water environment pollution sources in urban rivers of Hangzhou.

5 CONCLUSION

1 Currently, the main source of water environment pollution sources in urban rivers of Hangzhou comprises the wastewater pollution without interception, the runoff pollution, and the combined sewer overflow pollution. Some control measures aiming at the earlier three kinds of pollution sources should be carried out.

2 Runoff pollution has constituted the second pollution source of river water environment in Hangzhou under the current condition. Along with the sewage interception pipeline and the separation of rain and sewage, the proportion of the wastewater pollution without interception and the combined sewer overflow pollution will gradually decline, whereas the proportion of runoff pollution will be increasing day and day, then becoming the greatest limiting factor of water environment estimate in Hangzhou urban river. Consequently, it is necessary to fully consider the control efforts of runoff pollution.

REFERENCES

[1] Hydro China huadong engineering corporation limited. The technology research and first phase of the demonstration subject report of water environment treatment methods in Hangzhou city rive. [R].2014
[2] Hangzhou city comprehensive transportation research center. Study on present situation Investigation and Technological Measure Schemes for Functional Improvement of Sewage Interception Pipe in Hangzhou Riverside[R]. 2009
[3] Hangzhou city comprehensive transportation research center. Study on the Establishment of Low-carbon City Storm System——Based on Low-impact Development. [R]. 2011

[4] Hangzhou urban planning and design institute. The Special Planning of the Sewage Treatment Project in Hangzhou[R]. 2004

[5] Hangzhou city canal group. A Comprehensive Protection Scheme for Water Environment of the Hangzhou urban section of the Beijing-Hangzhou Canal. [R]. 2013

[6] Hangzhou city river Supervision Center. A Research on River Dredging Technology in Hangzhou. [R]. 2009

[7] Hangzhou research academy of environmental sciences. Demonstration application research on the engineering of water environment treatment and ecological restoration universal technology and key link of Hangzhou city river. [R]. 2013

[8] Hangzhou city construction committee. Hangzhou city construction Yearbook.[M]. 2013

[9] Hangzhou urban planning and design institute. The Urban Rainwater Planning in Hangzhou. [R]. 2004

Energy, Environment and Green Building Materials – Sheng (ed.)
© 2015 Taylor & Francis Group, London, ISBN 978-1-138-02718-3

Thermal effects on the geometric nonlinearity of beam structures

J. Duan & Y.G. Li
China State Construction Technical Center, Beijing, PR China

ABSTRACT: This paper develops a geometric nonlinear beam element with the thermal effects taken into account. The elemental governing equation is obtained by degenerating the three-dimensional continuum mechanics to one-dimensional forms based on the plane section hypothesis of beam element. In the deduction, only the nonlinear effects of axial force are taken into account. The explicit expressions of the stiffness matrices and the load vectors are presented by integration along the cross-section and longitudinal direction. And finally, a numerical example calculated by the present beam element model is presented and compared with that of ABAQUS solid element model. The two results are in good agreement, demonstrating the validity and accuracy of the present method.

KEYWORDS: geometric nonlinear analysis, beam element, thermal effect, FEM

1 INTRODUCTION

Geometric nonlinear analysis is a classic problem that has drawn much attention around the world. According to the literature [Timoshenko, 1953], both Jacob Bernoulli (1654–1705) and Euler (1707–1783) invested a lot of effort in the geometric nonlinear problem centuries ago. However, due to the complexity of nonlinear analysis, most of the engineering problems are analyzed and solved by the small deformation theory, which was modified and improved by Navier (1785–1836) in the 19th century. This situation continued until the middle of the previous century when the finite element method (FEM) was proposed by Clough (1960).

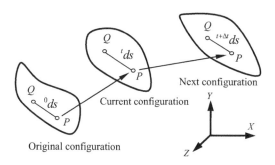

Figure 1. Illustration of geometric nonlinearity.

Figure 1 presents a motion mode of solid block for the geometric nonlinear analysis. In Fig. 1, the original configuration represents the undeformed configuration at time 0, the current configuration represents the deformed configuration at time t, and the next configuration represents the unknown deformed configuration at time $t + \Delta t$. According to the selection of different configurations as reference, the analysis of geometric nonlinear problem could be classified into three methods [Yang et.al, 2003]: total Lagrange format (T.L.), updated Lagrange format (U.L.), and Eulerian format. The T.L. and U.L. format could be collectively called generalized Lagrange format (G.L.), and the known configuration at time 0 and t is selected as reference, respectively. However, the Eulerian format chose the unknown configuration at time $t + \Delta t$ as reference, which leads to the inconvenience in calculation. Thus, as pointed out by Belytschko et.al (2000), the U.L. and T.L. format are the most popular methods in the geometrical nonlinear analysis whereas the Eulerian format is rarely adopted.

For the geometric nonlinear analysis of beam structure, Belytschko & Hsieh (1973) presented a co-rotational method; Argyris et.al (1979) presented a natural method; and Bathe & Bolourchi (1979) presented a U.L. format degenerated from the three-dimensional continuum mechanics. In all of the three methods, the displacements of each element are subdivided into rigid body displacements and elastic deformations. Based on the earlier idea, Duan & Li (2014a, 2014b, 2014c) present a beam element for the geometric nonlinear static and dynamic analysis.

In this paper, the thermal effect is taken into account in the geometric analysis of beam structure. The elemental governing equation is obtained by degenerating the three-dimensional continuum mechanics to one-dimensional forms based on the plane section hypothesis of beam element. In the deduction, only the nonlinear effects of axial force are taken into account. The explicit expressions of the stiffness matrices and the load vectors are presented by integration along the cross-section and longitudinal direction. Finally, a numerical example calculated by the present beam element model is presented and compared with that of ABAQUS solid element model. The two results are in good agreement, demonstrating the validity and accuracy of the present method.

2 THERMAL EFFECTS ON BEAM ELEMENT

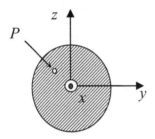

Figure 2. Illustration of a beam cross-section.

With the plane-section hypothesis of beam element, the incremental displacements of an arbitrary point P on the beam cross-section, shown in Figure 2, can be expressed in terms of the axial displacement $_tu$ at centroid, the transversal displacements $_tv$ and $_tw$, and the rotational angle $_t\theta_x$ as follows:

$$
\begin{aligned}
_tu^p &= {}_tu - y\frac{\partial_tv}{\partial x} - z\frac{\partial_tw}{\partial x} \\
_tv^p &= {}_tv - z\,_t\theta_x \\
_tw^p &= {}_tw + z\,_t\theta_x
\end{aligned}
\tag{1}
$$

where t represents that the earlier incremental displacements are measured in the current configuration at time t (see Fig. 1).

With the axial displacement $_tu$ at centroid and the rotational angle $_t\theta_x$ discretized using linear interpolation function, while the transversal displacements $_tv$ and $_tw$ are discretized using Hermit interpolation function [Wang, 2003], Eq. 1 could be represented as follows:

$$
\begin{bmatrix} _tu^p & _tv^p & _tw^p \end{bmatrix}^T = {}_t\bar{\mathbf{H}}_t\mathbf{a}^e
\tag{2}
$$

where

$$
_t\bar{\mathbf{H}} =
\begin{bmatrix}
N_1 & -yN_{1,x}^a & -zN_{1,x}^a & N_1 & zN_{1,x}^b & -yN_{1,x}^b \\
0 & N_3 & 0 & -zN_1^a & 0 & N_1^b \\
0 & 0 & N_3 & yN_1^a & -N_1^b & 0 \\
N_2 & -yN_{2,x}^a & -zN_{2,x}^a & N_2 & zN_{2,x}^b & -yN_{2,x}^b \\
0 & N_5 & 0 & -zN_2^a & 0 & N_2^b \\
0 & 0 & N_5 & yN_2^a & -N_2^b & 0
\end{bmatrix}
\tag{3}
$$

For the beam element, only the axial strain, $_t\varepsilon_{xx}$, and the transverse shear strain, $_t\gamma_{xy}$ and $_t\gamma_{xz}$, should be taken into account. These strains could be decomposed into the linear part and the nonlinear part as follows:

$$
\begin{aligned}
_t\varepsilon_{xx} &= {}_te_{xx} + {}_t\eta_{xx} \\
_t\gamma_{xy} &= {}_te_{xy} + {}_t\eta_{xy} \\
_t\gamma_{xz} &= {}_te_{xy} + {}_t\eta_{xy}
\end{aligned}
\tag{4}
$$

where $_te_{xx}, {}_te_{xy}$, and $_te_{xy}$ represent the linear strain and could be expressed as follows:

$$
\begin{aligned}
_te_{xx} &= \frac{\partial_tu^p}{\partial x} \\
_te_{xy} &= \frac{\partial_tu^p}{\partial y} + \frac{\partial_tv^p}{\partial x} \\
_te_{xy} &= \frac{\partial_tu^p}{\partial z} + \frac{\partial_tw^p}{\partial x}
\end{aligned}
\tag{5}
$$

and $_t\eta_{xx}, {}_t\eta_{xy}$, and $_t\eta_{xy}$ represent the nonlinear strain and have the following expression:

$$
\begin{aligned}
_t\eta_{xx} &= \frac{1}{2}\left[\left(\frac{\partial_tu^p}{\partial x}\right)^2 + \left(\frac{\partial_tv^p}{\partial x}\right)^2 + \left(\frac{\partial_tw^p}{\partial x}\right)^2\right] \\
_t\eta_{xy} &= \frac{\partial_tu^p}{\partial x}\frac{\partial_tu^p}{\partial y} + \frac{\partial_tv^p}{\partial x}\frac{\partial_tv^p}{\partial y} + \frac{\partial_tw^p}{\partial x}\frac{\partial_tw^p}{\partial y} \\
_t\eta_{xy} &= \frac{\partial_tu^p}{\partial x}\frac{\partial_tu^p}{\partial z} + \frac{\partial_tv^p}{\partial x}\frac{\partial_tv^p}{\partial z} + \frac{\partial_tw^p}{\partial x}\frac{\partial_tw^p}{\partial z}
\end{aligned}
\tag{6}
$$

Selecting the current configuration at time t as the reference configuration and using the principle of virtual work, the following equation could be obtained [Zienkiewicz & Taylor, 2005]:

$$
\int_0^l \iint_A \left({}_t^{t+\Delta t}\sigma_{xx}\delta_t\varepsilon_{xx} + {}_t^{t+\Delta t}\sigma_{xy}\delta_t\gamma_{xy} + {}_t^{t+\Delta t}\sigma_{xz}\delta_t\gamma_{xz} \right) dydzdx = {}^{t+\Delta t}R
\tag{7}
$$

where $_t^{t+\Delta t}R$ represents the external virtual work at the current time step, $_t^{t+\Delta t}\sigma_{xx}$ represents the axial stress, and $_t^{t+\Delta t}\sigma_{xy}$ and $_t^{t+\Delta t}\sigma_{xz}$ represent the shear stresses along the direction y and z, respectively; see the following equation for details:

$$_t^{t+\Delta t}\sigma_{xx} = {}_t^t\tau_{xx} + E\left({}_t\varepsilon_{xx} - \alpha\left({}^{t+\Delta t}T - {}^tT\right)\right)$$

$$_{t+\Delta t}^{t+\Delta t}\sigma_{xy} = {}_t^t\sigma_{xy} + G {}_t e_{xy} \qquad (8)$$

$$_{t+\Delta t}^{t+\Delta t}\sigma_{xz} = {}_t^t\sigma_{xz} + G {}_t e_{xz}$$

where α is the coefficient of thermal expansion; $^{t+\Delta t}T$ and ^tT represent the temperature at time $t + \Delta t$ and t, respectively.

By integrating Eq. 7 along the cross-section and longitudinal direction, the following equation could be obtained:

$$\left({}_t^t\mathbf{K}_L + {}_t^t\mathbf{K}_G\right){}_t\mathbf{a}^e = {}_t^{t+\Delta t}\mathbf{P}_e + {}_t^{t+\Delta t}\mathbf{P}_T - {}_t^t\mathbf{P} \qquad (9)$$

where $_t^t\mathbf{K}_L$ and $_t^t\mathbf{K}_G$ represent the linear and nonlinear stiffness matrices measured in current configuration. $_t^t\mathbf{K}_G$ can be expressed as follows:

$$_t^t\mathbf{K}_G = \frac{_t^tF_x}{l}\begin{bmatrix} {}_t^t\mathbf{K}_G^{11} & {}_t^t\mathbf{K}_G^{12} \\ {}_t^t\mathbf{K}_G^{21} & {}_t^t\mathbf{K}_G^{22} \end{bmatrix} \qquad (10)$$

where $_t^tF_x$ is the elemental axial force and ^tl is the current elemental length. They could be calculated by the following equations:

$$_t^tF_x = EA\left(\frac{{}^tl - {}^0l}{{}^0l} - \alpha\left({}^{t+\Delta t}T - {}^0T\right)\right) \qquad (11)$$

$$\left({}^tl\right)^2 = \left(l + \left({}_0^tu_2 - {}_0^tu_1\right)\right)^2$$
$$+ \left({}_0^tv_2 - {}_0^tv_1\right)^2 + \left({}_0^tw_2 - {}_0^tw_1\right)^2 \qquad (12)$$

In Eq. 9, $_t^{t+\Delta t}\mathbf{P}_e, {}_t^{t+\Delta t}\mathbf{P}_T$, and $_t^t\mathbf{P}$ represent the external force, the thermal force, and the internal elemental force, respectively. $_t^{t+\Delta t}\mathbf{P}_T$ and $_t^t\mathbf{P}$ could be calculated by Eq. 13 and Eq. 14, respectively.

$$\left[{}_{t+\Delta t}^{t+\Delta t}\mathbf{P}_T\right]_i = \mathbf{0} \qquad (i \neq 1,7)$$

$$-\left[{}_t^t\mathbf{P}_T\right]_1 = \left[{}_t^t\mathbf{P}_T\right]_7 = EA\alpha\left({}^{t+\Delta t}T - {}^0T\right) \qquad (13)$$

$$\left[{}_{t+\Delta t}^{t+\Delta t}\mathbf{P}\right]_i = \left[{}_t^t\mathbf{P}\right]_i + \left[{}_t^t\mathbf{K}_L\right]_{ij}\left[{}_t\mathbf{a}^e\right]_j \qquad (i \neq 1,$$

$$-\left[{}_t^t\mathbf{P}\right]_1 = \left[{}_t^t\mathbf{P}\right]_7 = {}_t^tF_x \qquad (14)$$

It should be emphasized that the calculation of internal force, $_t^t\mathbf{P}$, is based on the criterion of rigid body test presented by Yang & Chiou (1987), which could be stated as follows: For an element initially acted on by a set of nodal forces in equilibrium at an intermediate step of nonlinear analysis, this set of nodal forces must rotate after the rigid body rotation with no change in their magnitudes, thereby ensuring equilibrium of the element in the rotated configuration.

3 NUMERICAL EXAMPLE

A clamped-clamped beam with a thin-walled circle cross-section is shown in Fig. 7. It is subjected to a concentrated load, $P = 1KN$, at the middle point together with a uniform temperature rise ΔT from 0K till 100K. The purpose of this numerical simulation is to investigate the temperature effect on geometric nonlinear analysis of beam structure. Fixing the constant middle load, the largest deflection of the beam versus the uniform temperature rise is shown in Fig. 7. The present results are in good agreement with the results form ABAQUS solid element model. Obviously, the deflection of the beam will increase as the temperature increases. That is to say, the temperature rise will soften the structural stiffness of the clamped-clamped beam.

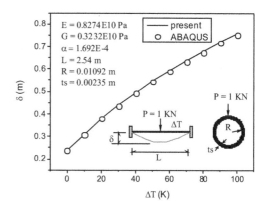

Figure 3. Results of the present beam element model compared with that of ABAQUS solid element model.

4 CONCLUSIONS

In this paper, the thermal effect is taken into account in the geometric analysis of beam structure. The elemental governing equation is obtained by degenerating the three-dimensional continuum mechanics to one-dimensional forms based on the plane section hypothesis of beam element. In the deduction, only the nonlinear effects of axial force are taken into account. The explicit expressions of the stiffness matrices and the load vectors are presented by integration along the cross-section and longitudinal direction. Finally, a numerical example calculated by the present beam element model is presented and compared with that of ABAQUS solid element model. The two results are in good agreement, demonstrating the validity and accuracy of the present method.

REFERENCES

Argyris J. H., Balmer G., Doltsinis J. S., Dunne P. C., Haase M., Klieber M., Malejannakis G. A., Mlejenek J. P., Muller M. & Scharpf D. W. 1979. Finite Element Method: the Natural Approach. *Comp. Meth. in Appl. Mech & Engng* 17/18: 1–106.

Bathe K. J. & Bolourchi S. 1979. Large Displacement Analysis of Three-Dimensional Beam Structures. *Int. J. Num. Mech. Engng.* 14: 961–98.

Belytschko T. & Hsieh T. 1973. Non-linear Transient Finite Element Analysis with Convected Coordinates. *International Journal for Numerical Methods in Engineering* 7: 255–271.

Belytschko T., Liu W. K. & Moran B. 2000. *Nonlinear Finite Elements for Continua and Structures.* New York: Wiley.

Clough R. W. 1960. The Finite Element Method in Plane Stress Analysis. *Proceedings of the Second ASCE Conference on Electronic Computation.* Pittsburgh, PA.

Duan J. & Li Y.G. 2014a. A beam element for geometric nonlinear dynamical analysis. *Advanced Materials Research* 919–921: 1273–1281.

Duan J. & Li Y.G. 2014b. A Large Rotation Matrix for Nonlinear Framed Structures, Part 1: Theoretical Derivation. *Advanced Materials Research* (to be published).

Duan J. & Li Y.G. 2014c. A Large Rotation Matrix for Nonlinear Framed Structures, Part 2: Numerical Verification. *Advanced Materials Research* (to be published).

Timoshenko, S. P. 1953. *History of Strength of Materials.* New York: McGraw-Hill.

Wang X.C. 2003. *Finite Element Methods.* Beijing: Tsinghua University Press (in Chinese).

Yang Y. B. & Chiou H. T. 1987. Rigid Body Motion Test for Nonlinear Analysis with Beam Elements. *Journal of Engineering Mechanics* 113(9): 1404–1419.

Yang Y. B., Yau J. D. & Leu L. J. 2003. Recent Developments in Geometrically Nonlinear and Postbuckling Analysis of Framed Structures. *Appl Mch Rev* 56(4): 431–449.

Zienkiewicz O.C. & Taylor R.L. 2005. *The Finite Element Method for Solid and Structural Mechanics* (6th edition). Oxford: Elsevier Butterworth-Heinemann.

5 APENDIX: STIFFNESS MATRICES

The camera-ready copy of the complete paper printed on a high resolution printer on one side of the paper as well as two copies of the paper should be sent to the editor after receiving the final acceptance notice.

$$
{}_t\mathbf{K}_G = \frac{{}_t^tF_x}{l}\begin{bmatrix} {}_t^t\mathbf{K}_G^{11} & {}_t^t\mathbf{K}_G^{12} \\ {}_t^t\mathbf{K}_G^{21} & {}_t^t\mathbf{K}_G^{22} \end{bmatrix} \tag{A.1}
$$

$$
{}_t^t\mathbf{K}_G^{11} = \begin{bmatrix} 1 & 0 & 0 & 0 & 0 & 0 \\ & \frac{6}{5}+\frac{12I_z}{Al^2} & 0 & 0 & 0 & \frac{l}{10}+\frac{6I_z}{Al} \\ & & \frac{6}{5}+\frac{12I_y}{Al^2} & 0 & -\frac{l}{10}-\frac{6I_y}{Al} & 0 \\ & & & \frac{J_x}{A} & 0 & 0 \\ & & & & \frac{2l^2}{15}+\frac{4I_y}{A} & 0 \\ & & \text{sym} & & & \frac{2l^2}{15}+\frac{4I_z}{A} \end{bmatrix} \tag{A.2}
$$

$$
{}_t^t\mathbf{K}_G^{12} = \begin{bmatrix} 1 & 0 & 0 & 0 & 0 & 0 \\ 0 & -\frac{6}{5}-\frac{12I_z}{Al^2} & 0 & 0 & 0 & \frac{l}{10}+\frac{6I_z}{Al} \\ 0 & 0 & -\frac{6}{5}-\frac{12I_y}{Al^2} & 0 & -\frac{l}{10}-\frac{6I_y}{Al} & 0 \\ 0 & 0 & 0 & -\frac{J_x}{A} & 0 & 0 \\ 0 & 0 & \frac{l}{10}+\frac{6I_y}{Al} & 0 & -\frac{l^2}{30}+\frac{2I_y}{A} & 0 \\ 0 & -\frac{l}{10}-\frac{6I_z}{Al} & 0 & 0 & 0 & -\frac{l^2}{30}+\frac{2I_z}{A} \end{bmatrix} \tag{A.3}
$$

$$
{}_t^t\mathbf{K}_G^{21} = \left[{}_t^t\mathbf{K}_G^{12} \right]^{\mathrm{T}} \tag{A.4}
$$

$$
\left({}_t^t\mathbf{K}_G^{22}\right)_{ij} = \left({}_t^t\mathbf{K}_G^{11}\right)_{ij} \quad \left(i, j = 1,2,3; \quad i = j\right) \tag{A.5}
$$

$$
\left({}_t^t\mathbf{K}_G^{22}\right)_{ij} = -\left({}_t^t\mathbf{K}_G^{11}\right)_{ij} \quad \left(i, j = 1,2,3; \quad i \neq j\right) \tag{A.6}
$$

Energy, Environment and Green Building Materials – Sheng (ed.)
© *2015 Taylor & Francis Group, London, ISBN 978-1-138-02718-3*

Material balance calculation and application of modified step-feed process

Q. Chen
State Key Laboratory of Urban Water Resource and Environment, Harbin Institute of Technology, Harbin, China

W. Wang
College of Civil and Architectural Engineering, Heilongjiang Institute of Technology, Harbin, China

C.X. Wang
Guangzhou Municipal Engineering Design and Research Institute, Guangzhou, China

Y.Z. Peng
State Key Laboratory of Urban Water Resource and Environment, Harbin Institute of Technology, Harbin, China

X. Zhang & C.H. Shao
College of Civil and Architectural Engineering, Heilongjiang Institute of Technology, Harbin, China

ABSTRACT: The modified step-feed process was used to treat low COD/TN domestic sewage, and equations of material balance calculation were reinstituted. Under the condition of Hydraulic Retention Time (HRT) equals 10h, Sludge Retention Time equals 10d to 15d, $Q_{anaerobic}:Q_{anoxic\,2}:Q_{anoxic\,3}$ equals 4:3:3, and $V_{anaerobic\,zone}:V_{total\,anoxic\,zones}:V_{total\,oxic\,zones}$ equals 4:9:9; the removal effects of pollutants were studied. The results showed that the effluent concentrations of COD, NH_4^+-N, TN, and PO_4^{3-}-P of the process were, respectively, 46.08mg/L, 0.45 mg/L, 17.76mg/L, and 2.61mg/L. According to the results of material balance calculation, effective carbon source utilization rate reached 62.8%, and the fundamental reason of the low removal rate of TN and PO_4^{3-}-P was the lack of carbon source.

KEYWORDS: Step feed; Low COD/TN; Material balance calculation

1 INTRODUCTION

Because of the constraint of the process form, traditional biological nitrogen and phosphorus removal process cannot exert a good treatment effect toward low COD/TN domestic sewage. Instead of the inlet water form of simple point, the water inflows in the anaerobic and anoxic zones in the modified step-feed process, it makes this process have many advantages such as short HRT and high MLSS. Since the raw water goes through anaerobic and anoxic zones first, effective utilization of carbon source will be conducted first to decrease adverse effects owing to the lack of carbon source. Material balance calculations of the modified UCT step-feed process have been put forward by Ge. Cao used the modified A/O process to treat low COD/TN sewage and put forward the material balance calculation equations. In this paper, new

material balance calculation equations were set up on the basis of the equations that were founded by predecessors. The transformation of pollutants in every reaction region was studied by this series of equations to provide a basis for the subsequent parameter optimization.

2 TEST PART

2.1 *Test device*

A schematic diagram of the modified step-feed process is shown in Figure 1. The total reactor volume is 100L, and the effective volume is 67L. The reactor is divided into 7 compartments: the anaerobic zone, anoxic zone 1, oxic zone 1, anoxic zone 2, oxic zone 2, anoxic zone 3, and oxic zone 3. The two-sedimentation tank is a vertical flow type and its volume is 44L.

Table 1. Raw water quality.

Item	COD	NH_4^+-N	NO_2^--N	NO_3^--N	TN	PO_4^{3-}-P	C/N
Range/mg·L^{-1}	175.8~344.5	43.4~55.6	0	0~1.04	53.3~66.33	2.55~4.45	3.16~5.81
Average/mg·L^{-1}	244.6	48.55	0	0.55	57.66	3.73	4.22

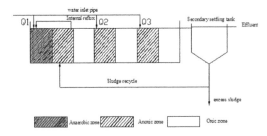

Figure 1. Schematic diagram of modified step-feed process.

2.2 Inoculation sludge and raw water quality

Inoculation sludge was taken from Harbin Wenchang sewage treatment plant, and the sludge activity was good after acclimation. Raw water comes from the staff quarters in Harbin Institute of Technology; raw water quality is shown in Table 1.

2.3 Water quality index and analysis method

NO_3^--N:Thymol spectrophotometry;NO_2^--N:N-(1- naphthyl)-ethylenediamine spectrophotometry; NH_4^+-N:Nessler's reagent spectrophotometry; TN: TOC-VCPN TN Meter;PO_4^{3-}P:SnCl$_2$ spectrophotometry; COD:5B-3(C) type COD quick-analysis apparatus (LianHua technology, China);MLSS: Filter paper weighing method; MLVSS: Muffle furnace burning method;pH, ORP, DO and temperature: WTW 3420 online tester (WTW Company, Germany).

3 REDUCTION OF MATERIAL BALANCE CALCULATION EQUATIONS

Based on the documents [2], [3], combining process characteristics with analysis purposes, material balance calculation equations were deduced as follows:

Anaerobic zone
$$Q_1 \cdot S_{0,x} + Q_{ir} \cdot S_{A1,x} = (Q_1 + Q_{ir}) \cdot S_{An,x} + \Delta S_{An,x} \quad (1)$$

Anoxic zone 1
$$(Q_1 + Q_{ir}) \cdot S_{An,x} + Q_{sr} \cdot S_{sr,x} = Q_{ir} \cdot S_{A1,x} + (Q_1 + Q_{sr}) \\ \cdot S_{A1,x} + \Delta S_{A1,x} \quad (2)$$

Oxic zone 1
$$(Q_1 + Q_{sr}) \cdot S_{A1,x} = (Q_1 + Q_{sr}) \cdot S_{O1,x} + \Delta S_{O1,x} \quad (3)$$

Anoxic zone 2
$$(Q_1 + Q_{sr}) \cdot S_{O1,x} + Q_2 \cdot S_{0,x} = \left(\sum_{i=1}^{2} Q_i + Q_{sr} \right) \cdot S_{A2,x} \\ + \Delta S_{A2,x} \quad (4)$$

Oxic zone 2
$$\left(\sum_{i=1}^{2} Q_i + Q_{sr} \right) \cdot S_{A2,x} = \left(\sum_{i=1}^{2} Q_i + Q_{sr} \right) \cdot S_{O2,x} + \Delta S_{O2,x} \quad (5)$$

Anoxic zone 3
$$\left(\sum_{i=1}^{2} Q_i + Q_{sr} \right) \cdot S_{O2,x} + Q_3 \cdot S_{0,x} = \left(\sum_{i=1}^{3} Q_i + Q_{sr} \right) \cdot S_{A3,x} \\ + \Delta S_{A3,x} \quad (6)$$

Oxic zone 3
$$\left(\sum_{i=1}^{3} Q_i + Q_{sr} \right) \cdot S_{A3,x} = \left(\sum_{i=1}^{3} Q_i + Q_{sr} - Q_{dis} \right) \cdot S_{O3,x} \\ + \Delta S_{O3,x} \quad (7)$$

Secondary settling tank
$$\left(\sum_{i=1}^{3} Q_i + Q_{sr} - Q_{dis} \right) \cdot S_{O3,x} = Q_{sr} \cdot S_{sr,x} + \left(\sum_{i=1}^{3} Q_i - Q_{dis} \right) \\ \cdot S_{eff,x} + \Delta S_{sst,x} \quad (8)$$

$$S_{nitrogen\ assimilation} = 0.1293X_v \cdot Q_{dis} \quad (9)$$

$$TN_{inf} = Q_{inf} \cdot S_{TN,\ inf} \quad (10)$$

$$TN_{eff} = Q_{eff} \cdot S_{TN,eff} \quad (11)$$

$$COD_{inf} = Q_{inf} \cdot S_{COD,inf} \quad (12)$$

$$COD_{eff} = Q_{eff} \cdot S_{COD,eff} \quad (13)$$

Q_1, Q_2, Q_3 : Influent flow of the anaerobic zone, anoxic zone 2, anoxic zone 3, m^3/d; Q_{ir}, Q_{sr}, Q_{dis}: Internal reflux flow, Sludge recycle flow, Sludge discharge flow in oxic zone 3,m^3/d; Q_{inf}, Q_{eff} : Total flow of influent, effluent, m^3/d;

$S_{0,x}$, $S_{sr,x}$, S_{eff}: Concentration of pollutants in raw water, sludge recycle, effluent, mg/L; $S_{An,x}$, $S_{A1,x}$,

$S_{A2,x}$, $S_{A3,x}$: Concentration of pollutants in anaerobic zone, anoxic zone 1, anoxic zone 2, anoxic zone 3, mg/L; $S_{O1,x}$, $S_{O2,x}$, $S_{O3,x}$: Concentration of pollutants in oxic zone 1, oxic zone 2, oxic zone 3, mg/L; $\Delta S_{An,x}$, $\Delta S_{A1,x}$, $\Delta S_{A2,x}$, $\Delta S_{A3,x}$: Consumption of pollutants in anaerobic zone, anoxic zone 1, anoxic zone 2, anoxic zone 3, g/d; $\Delta S_{O1,x}$, $\Delta S_{O2,x}$, $\Delta S_{O3,x}$, $\Delta S_{sst,x}$: Consumption of pollutants in oxic zone 1, oxic zone 2, oxic zone 3, two-sedimentation tank, g/d; TN_{inf}, TN_{eff}: Amount of TN in influent, effluent, g/d; COD_{inf}, COD_{eff}: Amount of COD in influent, effluent, g/d; $S_{TN,inf}$, $S_{TN,eff}$: Concentration of TN in influent, effluent, mg/L; $S_{COD,inf}$, $S_{COD,eff}$: Concentration of COD in influent, effluent, mg/L; $S_{nitrogen\ assimilation}$: Removal amount of nitrogen by assimilation, g/d; X_v: Mixed Liquor Volatile Suspended Solids, mg/L; 0.1293: Percentage of nitrogen in sludge.

4 RESULTS AND DISCUSSION

4.1 Removal performance of COD

Average effluent concentration of COD in system was 46.08 mg/L, and the corresponding removal rate was 80.5%. So COD could be efficiently removed. Process and results of material balance calculation are shown in Tables 2 and 3. The table showed that the effective utilization rate of carbon source reached

62.8% and was much higher than the 45.6% of ordinary UCT processes. Only 19.2% of COD was removed in oxic zones; therefore, the modified step-feed process could realize high-level utilization of carbon source.

4.2 Impact on nitrogen removal effect

Average effluent concentration of NH_4^+-N in system was 0.45mg/L, and the corresponding removal rate was 99.1%. Material balance calculation process of NH_4^+-N is shown in Table 4. The table showed that some of NH_4^+-N could be removed in anaerobic zone and anoxic zones, and NH_4^+-N could be completely removed in oxic zones.

Material balance calculation process of NO_x^--N is shown in Table 5. The majority of denitrification took place in the first A^2/O process, and the corresponding phase efficiency was 65.5%. NO_x^--N produced in the oxic zones could not be denitrified completely in anoxic zones due to the low COD/TN; in the meanwhile, NO_x^--N would be accumulated and be discharged along with the effluent.

Material balance calculation results of nitrogen were shown in Table 6. The table showed that the proportion of nitrogen removal by simultaneous nitrification and denitrification (SND) and assimilation was 26.2% and 13.8%, so SND and assimilation played important roles in TN removal.

Table 2. Material balance process of COD.

Reaction zone	Flow / m³·d⁻¹	Influx /g·d⁻¹	Outflow /g·d⁻¹	Removal amount /g·d⁻¹	Phase influx /g·d⁻¹	Phase outflow /g·d⁻¹	Phase efficiency /%
Anaerobic zone	0.228	27.004	16.847	10.197	23.52	12.412	43.4
Anoxic zone 1	0.391	24.065	25.767	-1.702			-7.2
Oxic zone 1	0.228	15.025	12.412	2.613			11.1
Anoxic zone 2	0.277	24.701	15.991	8.71	24.701	14.883	35.3
Oxic zone 2	0.277	15.991	14.883	1.108			4.5
Anoxic zone 3	0.326	27.172	18.677	8.495	27.172	14.535	31.3
Oxic zone 3	0.326	18.677	14.535	4.142			15.2
Secondary settling tank	0.321	14.535	14.168	0.367	14.535	14.168	2.5

Table 3. Material balance results of COD.

Item	Influx / g·d⁻¹	Consumption/ g·d⁻¹				
		Anaerobic zone / g·d⁻¹	Anoxic zone / g·d⁻¹	Oxic zone / g·d⁻¹	Secondary settling tank / g·d⁻¹	Effluent / g·d⁻¹
content	40.88	10.197	15.503	7.863	0.367	6.950
proportion	100%	24.9%	37.9%	19.2%	0.9%	17.0%

Table 4. Material balance process of NH_4^+-N.

Reaction zone	Flow /$m^3 \cdot d^{-1}$	NH_4^+-N					
		Influx /$g \cdot d^{-1}$	Outflow /$g \cdot d^{-1}$	Removal amount /$g \cdot d^{-1}$	Phase influx /$g \cdot d^{-1}$	Phase outflow /$g \cdot d^{-1}$	Phase efficiency /%
Anaerobic zone	0.228	4.956	4.530	0.426	3.109	0	13.7
Anoxic zone 1	0.391	4.530	4.43	0.1			3.2
Oxic zone 1	0.228	2.583	0	2.583			83.1
Anoxic zone 2	0.277	2.344	1.058	1.286	2.344	0	54.9
Oxic zone 2	0.277	1.058	0	1.058			45.1
Anoxic zone 3	0.326	2.344	1.330	1.014	2.344	0	43.3
Oxic zone 3	0.326	1.330	0	1.330			56.7
Secondary settling tank	0.321	0	0	0	0	0	0

Table 5. Material balance process of NO_x^--N.

Reaction zone	Flow / $m^3 \cdot d^{-1}$	NO_x^--N					
		Influx /$g \cdot d^{-1}$	Outflow /$g \cdot d^{-1}$	Removal amount /$g \cdot d^{-1}$	Phase influx /$g \cdot d^{-1}$	Phase outflow /$g \cdot d^{-1}$	Phase efficiency /%
Anaerobic zone	0.228	0.6787	0.025	0.654	2.485	3.890	26.3
Anoxic zone 1	0.391	2.444	1.47	0.974			39.2
Oxic zone 1	0.228	0.857	3.890	-3.033			—
Anoxic zone 2	0.277	3.939	2.922	1.017	3.939	4.504	25.8
Oxic zone 2	0.277	2.922	4.504	-1.582			—
Anoxic zone 3	0.326	4.553	3.638	0.915	4.553	5.274	20.1
Oxic zone 3	0.326	3.638	5.274	-1.636			—
Secondary settling tank	0.321	5.274	4.712	0.562	5.274	4.712	10.7

Table 6. Material balance results of nitrogen.

Item	Influx	Outflow			
		Denitrification	SND	Effluent	Assimilation
Content /$g \cdot d^{-1}$	9.050	4.122	1.306	2.37	1.252
Proportion/%	100%	45.5%	14.4%	26.2%	13.8%

4.3 Impact on PO_4^{3-}-P removal effect

Average effluent concentration of PO_4^{3-}-P in system was 2.61 mg/L, and the corresponding removal rate was 31.8%. Plenty of NO_x^--N was denitrified in anaerobic zone, and little NO_x^--N still appeared in effluent of anaerobic zone. The phosphorus release and uptake rates of PAOs were inhibited due to the lack of carbon source in anaerobic zone, so the removal effect of PO_4^{3-}-P was poor and the concentration of PO_4^{3-}-P in effluent was high.

5 CONCLUSION

In order to further study the transformation of pollutants in the system, the method of material balance calculation was studied and suitable material balance calculation equations were obtained. The results showed that effective carbon source utilization rate reached 62.8%, and the low removal rate of TN and PO_4^{3-}-P resulted from the lack of carbon source in raw water. The actual processing effect and the results of material balance calculation could be combined to seek the optimal control parameters in practical engineering application.

ACKNOWLEDGMENTS

This work was financially supported by the National Natural Science Foundation of China (NSFC) (NO. 51208185), the Heilongjiang Province Natural Science Foundation (No. QC2011C018), and the Heilongjiang Province Ordinary Colleges and Universities Young Academic Backbone Support Plan (1251G053). The corresponding author is Q. CHEN.

REFERENCES

[1] Cao,G.H.et al.2009. The problems and solutions of conventional biological nitrogen and phosphorus process. *Technology of Water Treatment* 35(03):102–106.

[2] Cao,G.H.et al.2013. Biological nutrient removal by applying modified four step-feed technology to treat weak wastewater. *Bioresource Technology* 128(0): 604–611.

[3] Ge,S.J.et al.2010. Performance and material balance of modified UCT step feed enhanced biological nitrogen and phosphate removal process. *Journal of Chemical Industry and Engineering* 61(4):1009–1017.

[4] Henze, M.et al.1995. Wastewater treatment: biological and chemical processes. Berlin:Springer-Verlag.

[5] Peng,Y.Z.et al.2011. Enhanced nutrient removal in three types of step feeding process from municipal wastewater. *Bioresource Technology* 102(11):6405–6413.

[6] Zhu,G.B.et al.2009. Performance and optimization of biological nitrogen removal process enhanced by anoxic/oxic step feeding. *Biochemical Engineering Journal* 43(3):280–287.

Energy, Environment and Green Building Materials – Sheng (ed.)
© 2015 Taylor & Francis Group, London, ISBN 978-1-138-02718-3

Study of phase change material based on solar building

W.H. Li, Ch.W. Zhang, M.Y. Zhang, H.T. Feng & Y. Liu
Army, Tieling, Liaoning, China

ABSTRACT: With the continued advancements of energy consumption, solar applications started becoming more and more. The characteristics of solar building are that they are good, of low cost, and easy to maintain. This paper summarizes the characteristics of the distribution of solar energy, passive solar house, and active solar house work principle. The use of phase change material to store a sufficient period of solar heat and release heat at night or during rainy weather is proposed. The study of portable phase change material storage tank module that is suitable for solar building has broad development prospects.

KEYWORDS: Phase change material; Storage; Solar building

1 INTRODUCTION

Our country is located in the east of Asia Europe continent, covers an area of 9.6×10^6 square kilometers, with a vast territory, and has very rich solar energy resources; the annual average of solar radiation is 590kJ/ (cm² • a). The solar radiation resources are influenced by climate and geographical environment, and their distribution is obviously regional.

The general situation of solar energy resource distribution is that the western is more than the eastern, and the southwestern is mostly less than the north. In addition to Tibet and Xinjiang, a low latitude area is less than a high latitude area. According to the annual size of solar radiation measured by China Meteorological Department, four solar radiation resources are generally classified in China [1].

2 SOLAR BUILDING CLASSIFICATION

Solar building can be divided into three categories: The first category is the passive solar house; it makes use of the direction of building, the structure, layout, and related materials for collection, storage, and distribution of solar building. The second category is the active solar house; it is the building of solar air-conditioning and a heating system composed by a solar heat exchanger, a fan, a pump, and a radiator. The third category is intended to add solar cell application on the active solar house, to provide heating, air conditioning, and lighting that are fully able to meet these requirements; it is called "zero energy house."

3 PASSIVE SOLAR HOUSE

3.1 Basic principle of passive solar house

The passive solar house lets the sun directly shoot into it and naturally makes use of it [2] (Figure 1). It does not require additional solar heating systems (collector, pipeline, and so on), and the building itself is a solar thermal system. The most simple and the basic working principle of the passive solar house is to let the sunshine through the south glass (collector area) of the building into the interior; then, the storage element, such as brick, adobe, stone, and furniture, absorbs the solar energy and transforms it into heat, thus heating the room. The main principle is the "greenhouse effect," that is, the shortwave solar radiation can go through the glass and the long wave infrared heat radiation cannot go through the glass. Once the solar energy gets through the glass, and it is absorbed by the material, the thermal radiation from these materials will not be returned to the outside through the glass, and it is limited to use in the room. Therefore, the three aspects of the solar house are heat gather, heat storage, and heat preservation.

3.2 Heat storage mode of the passive solar house

The heat storage mode of the passive solar house is mainly used in the building envelope walls or windows, or a simple flat device is used as a collector. The heat storage mode can be divided into the following: direct benefit, heat storage wall, attached sunspace, and heat storage roof. The most commonly used as well as the most economical and practical is direct benefit, heat storage wall, the hybrid type of direct benefit, and heat storage wall [3, 4].

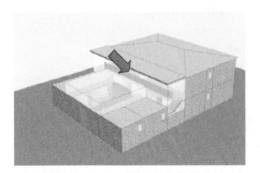

Figure 1.   Basic working principle of passive solar housing.

3.2.1   *Direct benefit type (solar window) passive solar house*

The direct benefit passive solar house is intended so that the sunlight can directly heat it. During the day, the sun through the wide surface of the south window glass shines to heat storage objects such as wall, floor, and furniture. During the night, stored heat is released by convection, radiation, and conduction, which maintain the room temperature at a certain level. It not only sets a higher thermal efficiency but is also easy to operate and has the lowest cost. But its drawback is that the reaction to the changes of the outdoor environment is more sensitive, and it is not conducive to the stability of the indoor temperature.

3.2.2   *Heat storage wall-type passive solar house*

The sun first shines into the dark storage wall with the glass cover, which is located between the sun and the housing, and then heat enters indoors through two ways:

1   The outer surface of the storage wall absorbs solar radiation, then the heat is transferred to the inner wall by conduction, and finally the heat is exchanged with the indoor air by radiation and convection, thus heating up the room.
2   The interlayer air between the glass and the wall rises after being heated by the outer surface of the wall. The hot air enters the room, through the ventilation hole upper wall; simultaneously, the indoor cold air moves through the vent into the interlayer through the bottom of the wall. Then, convective circulation transits hot air into the room.

Storage wall can not only use the traditional building materials but also use the phase change materials of high density [5]. According to the requirements range of indoor temperature, phase change materials of reasonable phase transition temperature are selected, to achieve a more comfortable environment.

3.2.3   *Attached sunspace type passive solar house*

It is a development form of heat storage wall in which the air interlayer between glass and wall is widened, forming an available space. Through a part of the windows, sunshine has direct exposure in the heating room. A part of the sun is absorbed by the ground and public wall, the heat comes into the heating room by hot air circulation, and heat conduction plays the role of solar heating. The relationship between the attached sunspace and room is flexible; it can be a brick wall, and we can also set the glass window or set a masonry wall with doors and windows.

It is unfavorable to set the sun room protruding from the walls, since it would block the rest of the south wall, to influence the absorption of the heat of the sun. The exposed area of the glass wall cannot be too large; otherwise, it will affect the heat storage of the wall [6] as shown in Fig. 2.

The heat collecting area is the largest in Fig. 2 (c), the collector area is the smallest in Fig. 2 (a), and collector area is at the center in Fig. 2 (b). But considered from the heat storage performance, storage performance of Fig. 2 (a) is the best, storage performance of Fig. 2 (b) is the second, and storage performance of Fig. 2 (c) is the worst. Ordinary glass can not only gather a lot of heat through the solar radiation but also absorb the long wave radiation above it; the absorption rate is more than 84%.

Because of the high thermal conductivity of glass, the absorbed long wave radiation can be easily achieved on the other surface by heat conduction and then be dissipated by long wave radiation or convection mode. This requires an adequate regenerator for

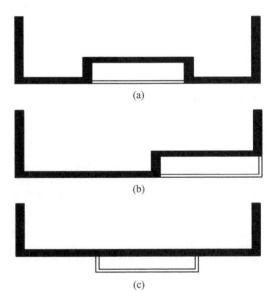

Figure 2.   Sun room setting type.

storage heat, to achieve the heating requirements. At the same time of heat collection, it also has three surface storage walls; the heat can be stored well, thus maintaining indoor temperature stability.

3.2.4 *Heat storage roof-type passive solar house*

The heat storage roof-type passive solar house has two kinds of design schemes: Heat storage equipment can be a plastic bag full of water or phase change heat storage materials [7]. The sun room is suitable to that condition, in which heating load is not high in winter and cooling is used in summer. But the roof needs a strong bearing capacity; heat insulation cover operation is troublesome.

4 ACTIVE SOLAR HOUSE

Active solar house is a solar construction that is composed of solar heat exchanger, hot water tank, pump, radiator, controller, and heat storage device. It is the same with the passive solar house, where retaining structure has good heat insulation performance. Active solar house working principle can be divided into the heat collecting circuit, heating circuit, and living hot water circuit [8], as shown in Fig. 3.

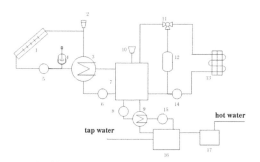

1-collector; 2- exhaust valve; 3- heat exchanger; 4- storage;5, 6, 8, 14, 15 - circulating pump; 7- storage tank; 9- hot water exchanger; 10- decompression valve; 11- electric valve; 12- auxiliary heat source; 13- indoor radiator; 16- water preheating tank; 17- auxiliary heating water tank

Figure 3.   Active solar house principle.

4.1 *Heat collecting circuit*

The collector loop mainly comprises a heat collector, storage, a collector heat exchanger, and a circulating pump. It adopts a differential control. A temperature sensor is installed on the collector heat absorption plate close to the heat transfer medium outlet, and another temperature sensor is installed on the heat storage tank bottom near the collecting loop reflux outlet. When the first sensor temperature is 5 ~ 10 °C higher than the second sensors temperature, pump 5 is opened. In this case, the fluid from the reservoir

moves by pump into the collector, and the heat from the air moves through the collector replacement into the storage. On the contrary, the heat pump is closed when the difference is 1 ~ 2 °C between the hot water storage tank outlet temperature and the heat collector heat plate temperature. In this case, the water in the collector flows into the container for deicing.

The storage device must be heated or closed to prevent freezing temperature. Due to the efficient energy from the sun possibly exceeding the amount of water, the system should set the temperature control device. When the hot water storage tank temperature exceeds a certain limit, the heat collecting circulating pump will automatically shut down.

4.2 *Heating circuit*

The heating circuit mainly includes the thermal storage tank, radiator, the auxiliary heat source, electric valves, and other components. Heating loop is a heat medium loop in the room. One of the temperature sensors placed in the outlet of the storage tank, the temperature-sensitive element measures indoor temperature. When the indoor temperature is reduced, the temperature of the heat storage water tank is high and reaches a certain value; the auxiliary heater is turned off, providing heat from a heat storage tank.

The other temperature sensor is installed in the return pipe heating loop. If the indoor temperature continues to decline, and the first sensor readout temperature is lower than the second, the heat storage water tank heat cannot meet the load requirements, the electric valve should cut off the contact between the hot water storage tank and the system, the hot water storage tank goes out of the loop, and then heating takes place by an auxiliary heater.

4.3 *Listing and numbering*

The living hot water circuit includes hot water heat exchanger, the preheating water tank, the auxiliary heating water tank, and pump parts. Tap water flows into the preheating water tank through the heat exchanger, and after being preheated the water circulation to the auxiliary heating water tank takes place from the top of the preheating tank. The water in the auxiliary heating water tank is heated till the desired temperature, for use in every room. Any domestic hot water system must use the thermostat valve or other methods, to ensure hot water temperature is appropriate.

5 DESIGN OF STORAGE TANK

The heat storage tank is the main component of active solar house that is used to store solar heat. It has a great influence on the use of the active solar house, Heat storage being on time or not results in storage quantity being appropriate or not. The shape,

(a) Energy storage unit after charge

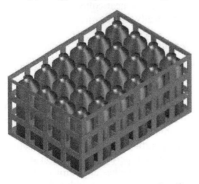

(b) Storage box filled with energy storage unit without cover

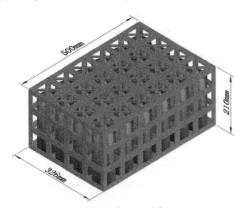

(c) Complete energy storage box module

Figure 4. Schematic diagram of accumulator module with detailed size and composition.

arrangement, layout for heat and hot water storage tank, storage tank, and insulation materials have relation to the radiation of the heat storage tank [9].

In order to store more heat for a longer time, both water and phase change materials can be used as storage medium. Because phase change materials have a big heat storage density, they can increase the storage amount when used as storage medium in the heat storage tank. We can select quantity and volume of the storage module according to needs. The size of the energy storage module is as follows: The length

is 500mm, the width is 328mm, and the height is 210mm.

There are 24 energy storage units in one box, and these are arranged in a monolayer. There are small protrusions on the bracket to fix the energy storage unit. The box cover adopts embedded design; after installed phase change of energy storage materials, the box cover is inserted into the box body. This method can be convenient for assembling and replacing the energy storage unit. The storage box module is shown in Fig. 4.

6 CONCLUSIONS

Solar energy is inexhaustible, and being a new energy it has good economic benefits. The use of solar energy is one of the important measures for saving energy and protecting the environment. Solar building is a kind of ecological construction; it is clean, safe, and comfortable; and it has no pollution but a very strong vitality. The sun room can directly use solar radiation energy for heating, water heating, and cooling, to improve living conditions and the indoor temperature, save energy, and improve the ecological environment; this has a very important significance. The heating and hot water supply system of active solar house can be automatically adjusted according to need and it can provide a comfortable indoor environment. The active solar house has amplitude application prospects.

REFERENCES

[1] Hao Gaihong. Study on thermal performance of a solar building and its parts [D]. Taiyuan: Taiyuan University of Technology, 2002.
[2] Chen Qianyi, Li Wei, Lin Geng. Numerical simulation and optimization of rural passive solar house dynamic energy consumption [J]. Anhui Agricultural Sciences, 2011, 39 (33):20908–20910.
[3] Chen Yu, Li Zhenmao. Design of low temperature architecture technology [J]. Cold passive solar housing area, 2005, (5):93–94.
[4] He Zinian. Passive solar house [J]. renewable energy, 2005, (5):84–86.
[5] Chen Chao, Liu Yuning, Guo HaiFeng, Zhou Wei. A preliminary study on application of composite phase change wall in passive solar house [J]. Journal of building materials, 2008, 11 (6):684–689.
[6] Jin Hong, Zhou Chunyan. Energy saving solar rural houses in cold areas combined analysis [J]. Journal of Harbin Institute of Technology, 2008, 40 (12): 2007–2010.
[7] Wang Lei, Jiang Shuguang. The adaptability and economy of passive solar house of [J]. Construction economy, 2010 (12):83–86.
[8] Zhou Yan, Xie Junlong, Shen Guomin. Application of the active solar house [J]. Building heat ventilation air conditioning, 2003 (1):17–19.
[9] Wang Jinping, Tang Runing. The economy analysis of solar house with seasonal heat storage system [J]. Energy saving technology, 2007, 23 (2):169–171.

Energy, Environment and Green Building Materials – Sheng (ed.)
© 2015 Taylor & Francis Group, London, ISBN 978-1-138-02718-3

2D Numerical simulation of flow field around bridge piers in Bend River based on FLUENT

W.C. Pan & M.J. Liu
School of Navigation, Wuhan University of Technology, Wuhan, China

ABSTRACT: 2-D flow field around bridge piers on Bend River is numerically simulated based on FLUENT software in order to grasp the flow field conditions near the bridge piers. The varied flow field of bridge piers setting and the effect scope of ships passing through in Bend River have been analyzed. The paper used RNG k-ε method to simulate different conditions. The references of pier setting in bend water area and safety measures for ship navigation have been studied. This is useful to offer an important reference for ship safety across bridge areas.

KEYWORDS: FLUENT; Numerical simulation; Bridge area; Safety measures

1 INTRODUCTION

Yangtze River is the main artery of Chinese inland waterways transport. It is necessary to understand the velocity and direction of water flow and other hydrological conditions except the channel scales in order to overcome the effects of flow field by using engine and rudder when sailing in the River. It is an important guarantee of navigation safety for ships sailing to the chosen track, especially in the Bend River of the Yangtze River with piers setting in navigation restricted water areas.

Currently, more than 120 bridges are being built along the Yangtze River. With the development of the main traffic arteries and port demands along the Yangtze River, the bridge construction is gradually increased. The bridges are expected to be more than 200 in the year 2020. The city port of the bridge river area is concentrated with approximately 15 km a bridge. The straight river and Bend River had the same proportion in the Yangtze River; most straight water areas had built bridges in accordance with the standard requirements; and consequently, it is necessary to consider building bridges in Bend River. Some city roads were located in the bend area with a large distance around them; however, this was not convenient and increased the bridge construction cost. Therefore, the demand of bridge constructions in Bend River is more prominent in need. Bend River will be the required water area for bridge building.

The piers setting greatly narrowed the channel width in the river with the bridge building, and its influence is more evident in Bend River. Simultaneously, the velocity distribution is extremely uneven and the flow field is inordinate by the influence of bend-rushing current in Bend's water. The flow regime of flow field after the piers building becomes more special. Therefore, shipping transportation in the Yangtze River had brought many restrictions and obstacles (Hu Xuyue et al. 2011). In order to alleviate the risk, it is necessary to master the piers influence and propose mitigation measures in the research on the hydrology surrounding bridge piers in Bend River.

Based on CFD research, Luo et al. (2014) designed the width clearance requirement and current speed limits for navigation safety by using a three-dimensional flow field. Based on the CFD code Fluent, Gan et al. (2014) adopted quasi-steady and unsteady methods to numerically simulate the viscous flow field around a ship navigating in the bridge area. Through the establishment of a three-dimensional flow model, Lv (2012) simulated the flow characteristics of a continuous curve. B. Yulisitiyanto (1998) used the two-dimensional shallow water equations to solve the flow around a circular cylinder and surface wave. Fang et al. (2011) simulated the instantaneous flow field by using a mathematical model and discussed the relevant hydraulic characteristics. Zu (2006) used the method of theoretical analysis and experimental study in his article. The characteristics and the width of the turbulence around the round-ended pier were studied. This paper researches the effect of different factors on the flow field surrounding the bridge area by using 2D numerical simulation, and analysis safety navigation range and control requirements of corresponding waters after piers setting.

2 EXPERIMENTAL MODEL PARAMETERS

This paper studies the flow field after piers setting on different conditions of piers spacing, and velocity by using numerical tank simulations based on FLUENT. The 1050m typical curvature radius of lower reaches of Yangtze River is numerically simulated. The width of the experimental pond is 300m with two 30m radius piers. Two typical situations, 120m and 180m, are taken into consideration as piers spacing. The direction of initial flow that is uniform follows the X axis. The initial velocity has 2 and 2.5m/s, which are two conditions. The inland navigation vessels have shallow draught load and an unobvious influence of bottom current, so this paper conducts research on the surface flow in Bend River. Therefore, 4 kinds of experimental simulations have been analyzed and then the velocity distribution surrounding the pier waters is taken out. The experiment simulation model selected the entrance flow on the right side of Bend River, as shown in Fig. 1.

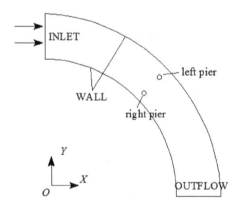

Figure 1. Bend River calculation model.

The flow field has been solved by using finite volume method based on FLUENT software. Triangle mesh division was selected in 2D geometric models. Double precision solver has been used in the numerical process. The fluid velocity entrance boundary is Velocity-inlet, the air outlet boundary is Outflow, and the wall boundary condition is Wall. This paper selected 2nd-order implicit as solving mode, and its operational environment set operating conditions. In the process of solving the parameters, RNG k-ε was chosen according to the practical considerations. The control parameters used Second-Order Upwind.

3 FLOW FIELD SIMULATION

Gambit software is used to model and mesh the experimental pond. In order to analyze the bridge

piers' flow field interference accurately, the grids increased density nearing the bridge axis. 1050m curvature radius has been studied in 120m and 180m piers spacing with the initial velocity of 2.0m/s and 2.5m/s. The simulation results of flow field distribution after the bridge piers setting that was based on FLUENT are shown in Figs. 2–5.

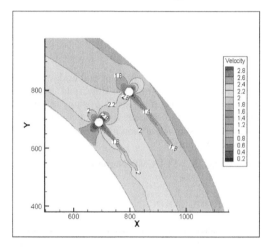

Figure 2. Flow velocity distribution of 120m piers spacing at the initial velocity of 2.0m/s.

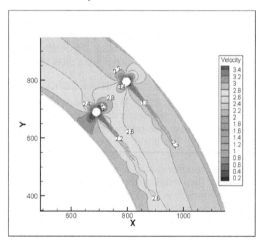

Figure 3. Flow velocity distribution of 120m piers spacing at the initial velocity of 2.5m/s.

The results of 120m piers spacing show that ① when the initial velocity is 2.0m/s, the flow velocity of Bridge area is in the range of 0.003–4.003m/s. The pier velocity peaks at 4.003m/s on the right pier and below 4.00m/s on the left pier. The right pier side velocity distribution is significantly higher than the left side, and it is obviously reduced in the middle side. The velocity distribution has an increasing trend near the

150

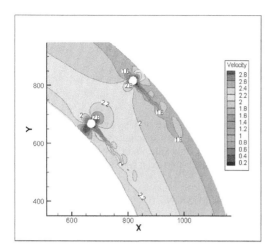

Figure 4. Flow velocity distribution of 180m piers spacing at the initial velocity of 2.0m/s.

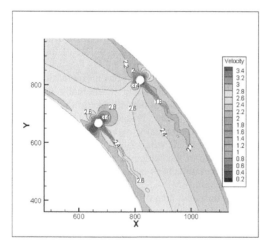

Figure 5. Flow velocity distribution of 180m piers spacing at the initial velocity of 2.5m/s.

right pier, and it gradually decreases from the middle to the left pier in the 50 meters upstream of the bridge axis. The peak velocity is about 2.11m/s. The velocity distribution gradually decreases from the right pier to the left side in the 100 and 200 meters upstream of the bridge axis, and its peak velocity is about 2.1m/s; ② when the initial velocity is 2.5m/s, the flow velocity of Bridge area is in the range of 0.012–5.031m/s. The pier velocity peaks at 5.031m/s on the right pier and is below 4.4m/s on the left pier.

For 180m piers spacing: ① When the initial velocity is 2.0m/s, the flow velocity of Bridge area is in the range of 0.008–4.190m/s. The pier velocity peaks at 4.190m/s on the right pier and is below 3.5m/s on the left pier; ② when the initial velocity is 2.5m/s, the

flow velocity of Bridge area is in the range of 0.013–5.262m/s. The pier velocity peaks at 5.262m/s on the right pier and is below 4.5m/s on the left pier.

4 STATISTICAL ANALYSES OF FLOW FIELD RESULTS

4.1 Simulation data analysis

In conclusion, the velocity distribution was obtained from FLUENT data based on Tecplot software. The velocity distribution of 120m piers spacing with different entrance velocity and different curvature radius in the distance of 0, 15, 30, 45, 60, 75, 90, 105, and 120 m from the right pier according to pier axial is shown in Fig. 6. The velocity distribution of 180m piers spacing with different entrance velocity and different curvature radius in the distance of 0, 22.5, 45, 67.5, 90, 112.5, 135, 157.5, and 180 m from the right pier according to pier axial is shown in Fig. 7. Table 1 shows the velocity in the 50, 100, and 200 meters upstream of the bridge axis.

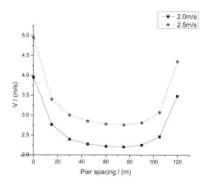

Figure 6. Velocity distribution of 120m piers spacing with different entrance velocity and different curvature radius.

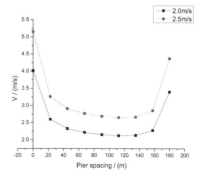

Figure 7. Velocity distribution of 180m piers spacing with different entrance velocity and different curvature radius.

Table 1. Velocity distribution in the 50, 100, and 200 meters upstream of the bridge axis.

Upstream of the bridge axis (m)	Velocity from the right pier side(m/s)				
	0	30	60	90	120
50	2.01	2.11	2.09	2.01	1.82
100	2.1	2.06	2.01	1.96	1.89
200	2.11	2.05	1.99	1.94	1.89

The simulation data shows the following: The velocity surrounding the piers increases with the pier spacing widened, but its difference is not obvious. Right pier velocity is obvious more than the left side. Because of the flow obstruction from the pier, the velocity on both piers sides becomes high. It makes superimposing water, oblique flow in front of bridge piers. Therefore, strong push flow appears from the pier of both sides. Figures 2–5 show that piers turbulent area increases with the pier spacing widened, and the right pier bottom velocity of turbulent area is larger than the left pier side. From Fig. 6 and Fig. 7, we can see that the flow velocity decreased from two pier sides, and the minimum value on the middle side was higher than the initial velocity. Simultaneously, its corresponding velocity point becomes smaller when its pier spacing is widened in addition to the pier side.

4.2 Fitting and simulation

According to the fitting of simulation data, the relationship between piers axis distance from right pier and corresponding velocity has been obtained in different pier spacing and initial velocity. The velocity distribution on bridge axis can be calculated by Eq. 1–4, and the relationship determines its maximum velocity region.

Equations 1–2 express the relationship between 120m and 180m piers spacing between velocity and distance of piers axis from right pier side at the initial velocity of 2.0m/s.

$$v = 7\,e^{-5}x^5 + 0.005x^4 - 0.123x^3 + 0.972x^2 - 3.291x + 6.388, R^2 = 0.998 \tag{1}$$

$$v = 0.008x^4 - 0.178x^3 + 1.307x^2 - 4.138x + 6.997, R^2 = 0.994 \tag{2}$$

Equations 3–4 express the relationship between 120m and 180m piers spacing between velocity and distance of piers axis from right pier side at the initial velocity of 2.5m/s.

$$v = 0.009x^4 - 0.182x^3 + 1.352x^2 - 4.399x + 8.154, R^2 = 0.997 \tag{3}$$

$$v = 0.011x^4 - 0.24x^3 + 1.759x^2 - 5.551x + 9.149, R^2 = 0.994 \tag{4},$$

where x is distance of piers axis from right pier in 120 and 180 m piers spacing; v is the corresponding velocity from right side, and flow direction is perpendicular to the bridge axis; and R is fitting coefficient.

5 CONCLUSION

This paper calculated the effect of various factors on the flow field surrounding bridge area of the right side Bend River by using numerical simulation.

1 The current velocity of right pier side is obviously larger than the left side, and right pier flow field has the greater influence area. The influence range of flow field is increased with an improvement in the flow velocity. The flow field overlapping phenomenon is seen in the middle of the bridge axis.
2 The velocity distribution has an increasing trend near the right pier, and it gradually decreases from the middle to the left pier in the 50 meters upstream of the bridge axis. The velocity distribution gradually decreases from the right pier to the left side in the 100 and 200 meters upstream of the bridge axis. It is the pivotal flow field that influences safety of ships sailing through bridge areas.
3 The turbulent area of the right pier is longer than the left pier side, and piers turbulent area decreases with the pier spacing widened at different initial velocity.

The flow condition is complicated when piers are setting in Bend River. It creates adverse effects when the ship sails safety through the bridge waters. The mariner on duty must properly close the vessel to the right side before the vessel gets into the bridge area. It is necessary to use engine and rudder for safety navigation when sailing in the bridge area to overcome the effect of push flow and bend-rushing current.

REFERENCES

Hu Xuyue, Zhang Qingsong, Ma Lijun. 2011.Numerical simulation of influence of transition section on continuously curved channel flow, *Advance in Waterscience*: 2011.11(6): 851–858.

Lun Weilin,Gun Langxiong,Zou Zaojian. 2014.Flow Field in Vicinity of Pier and its Effect on Navigating Ship, *Navigation of China*: 66–70.

Gan Langxiong, Zou Zaojian, Xu Haixiang. 2014. Calculation and Analysis of the Hydrodynamic Forces on a Ship Navigating in Bridge Area, *Journal of Ship Mechanics*: 613–622.

Lv Linkai. 2012. Research on numerical simulation of the bending fairways widened and the analysis of navigation safety, *Changsha University of Science&Technology*.

Yulisitiyanto B, Zech Y, Graf W H. 1998. Flow around a cylinder: shallow-water modeling with diffusion-dispersion, *Journal of Hydraulic Engineering*: 419–429.

Fang Sensong. 2011. Study on Flow Characteristics Around Continuous Arrangement of Pier, *Changsha University of Science&Technology*.

Zu Xiaoyong. 2006. Experimental study of turbulent flow width and the characteristic of the turbulence around round-ended pier, *Changsha University of Science&Technology*.

Energy, Environment and Green Building Materials – Sheng (ed.)
© *2015 Taylor & Francis Group, London, ISBN 978-1-138-02718-3*

Research on control system of multidimensional dynamic indoor environment

F.Z. Nian & D.W. Shao

College of Computer and Communication, Lanzhou University of Technology, China

ABSTRACT: Since "12 FYP" of China, the target of energy saving and emission reduction is higher and clear. In this paper, a system that possesses user feedback and integrated indoor environment control was developed based on the Internet of things. On the premise of not reducing the comfort of the user, the system is aimed at energy saving. First, we considered the open question of energy saving and emission reduction in China; the shortcomings of corresponding methods were also investigated. Then, a test bed of multi-dimensional dynamic indoor environment was established based on the concept of integrated control. The software part includes the clients of mobile phone and computer. The hardware includes the sensors, single chip, and all kinds of energy-consuming devices. Finally, the experiments indicate that the proposed method is effective.

KEYWORDS: Energy saving, Thermal comfort, Integrated control, Feedback

1 INTRODUCTION

Saving energy and reducing greenhouse gas emissions has become the consensus for many countries around the world. In the "12 FYP" of China, it provides detailed energy-saving emission -reduction targets: Non-fossil energy in primary energy consumption reached 11.4%, per unit of GDP energy consumption by 16%.[1]

Based on the background just described, energy-saving and emission reduction research has become a new hot topic [2]. SufengWang studied the investment and the emission reductions of the enterprise in a complex environment[3]. By collecting and collating a large number of scholarly researches on energy conservation [4], LingxiCai and LiliFan analyzed and extracted the three evaluation dimensions for the policies of urban energy conservation, which are the economic growth structural adjustment and innovation development. From the existing research results, energy-saving and emission reduction research are from a policy or the effect of implementation aspects, but the research on energy-saving program design, development, and implementation affecting the evaluation system are rarely a concern. After studying a number of energy-saving design concepts and programs, this paper especially pays attention to the issue of emission reduction in the daily operation of the building process and provides an effective solution.

2 SITUATION INDOOR ENVIRONMENTAL CONTROL AND RESEARCH

2.1 *Current situation of indoor environment quality control*

In the current adjusting of indoor environmental quality, sound, light, and heat air quality control everything in their own way and do not communicate. If all kinds of indoor environmental control equipment are integrated and optimized, we can further reduce the energy consumption of the indoor environment.

Based on indoor environmental control strategies, for example, the indoor temperature and humidity control system form is mainly divided into two categories. One is a control scheme without user involvement, and this system cannot satisfy the individual requirements of some people; the other is the user involvement[5]. Users enter the room temperature set-point to control the running way in which air conditioning causes heat or cold, and so on so as to result in waste of energy.

2.2 *Study of the indoor environment control system*

Based on the idea of complained indoor environment integrated control, a more personalized control system, indoor environment intelligent control system, was introduced in this paper. The earlier problems were solved successfully. The system integrates the various control devices for optimal control, to better satisfy the comfort aspects of the user. Meanwhile, the control system has developed a user-friendly interface HMI for accepting user complaints so that we can control the will of the people in the control loop and enable intelligent control.

To achieve multidimensional dynamic indoor environment control system based on complaints concept, we need to develop a test bed that uses integrated control equipment and optimizes the environmental regulation on the basis of the user feedback. According to this requirement, the test bed is divided into four levels for design, which include data acquisition and device control layer, storage layer, functional layer, and interaction layer. Technical architecture for the control system is shown in Fig. 1:

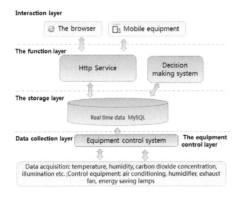

Figure 1. Technical architecture for the control system.

The bottom of the system is the data acquisition and equipment control layer, which collect the current real-time indoor environmental quality data through sensors, simultaneously accept and implement decisions to adjust the upper device control commands issued[6].

The storage layer is the core of the whole system, which records generated data of the whole system at runtime.

The function layer is responsible for providing business support between the storage layer, the interaction layer, and the equipment control layer [7]. In addition, the functional layer of the decision-making system provides an intelligent control strategy to achieve an automatic control system.

Interaction layer provides the user's interaction with the system.[8] We design the user's intuitive feelings into the interface, such as cold, hot, dry, wet, dizzy, dark, stuffy, and so on, which can more

accurately receive feedback from users in order to achieve accurate regulation.

3 THERMAL COMFORT EVALUATION AND ANALYSIS OF THE EXPERIMENTAL DATA

This paper will deduce two indicators by the equation of thermal comfort to evaluate the indoor thermal environment and the effect of the system regulating the indoor environment, which are predictive values of PMV and predicted average voting percentage dissatisfied PPD [9].

3.1 Human body heat exchange with the environment and Fanger's thermal comfort equation

From the viewpoint of thermodynamics, the heat exchange between humans and the environment follows the first law of thermodynamics. Therefore, the obtained energy subtracting the loss of human energy would mean the accumulation of human energy, that is, the difference between the body and the net heat gain H for human regenerative heat dissipation rate of S [11].

$$S = H - C_{res} - E_{res} - E_{dif} - E_{rsw} - (C+R)(W/m^2)$$

In the formula:

S — Body heat storage rate;

H — Human net heat gain;

E_{dif}—Proliferation of skin evaporative heat loss;

E_{rsw}—Heat loss due to sweating body close to the skin surface in comfortable conditions/ state;

E_{res}—Latent heat loss during respiration;

C_{res}—The sensible heat loss during respiration.

C—The surface of the skin into the surrounding air convective heat loss;

R—By the radiation heat loss to the environment of the skin surface;

Human net heat gain H is the amount of the difference between the internal body metabolism and body made M W external power, namely H = M-W. When people are still moving slightly, W does not count, so H = M. Along with skin temperature and perspiration while maintaining a comfortable state, S = O. Otherwise, S ≠ 0, S > 0 represents the body from the environment being too hot, resulting in thermal; S <0, represents the body's ability to dissipate heat to the environment, resulting in a cold feeling.

After lots of experiments and research, Danish scholar Fanger put the heat balance equation with absorption and radiating heat expressions into S = O. It can obtain the following thermal comfort equation:

$$
\begin{aligned}
L = H - &3.054\left(5.765 - 0.007H - P_a\right) \\
&- 0.42\left(H - 58.15\right) \\
&- 0.0173M\left(5.867 - P_a\right) \\
&- 0.0014M\left(34 - t_a\right) - f_{cl}h_c\left(T_{cl} - t_a\right) \\
&- 3.96 * 10^{-8}f_{cl}\left(T_{cl}^4 - T_{mrt}^4\right)
\end{aligned}
$$

In the formula:

H—Human net heat, H = M (1-η), and η- body's mechanical efficiency;

P_a—Air water vapor partial pressure;

t_a—Air temperature;

f_{cl}—Clothing area factor;

h_{cl}—Convective heat transfer coefficient;

T_{mrt}—The average temperature of the outer surface of the body dress;

T_c—The average radiation temperature of the environment;

L — Human thermal load: If L = 0, it shows that thermal comfort conditions are met, and the body of the surrounding environment feels comfortable; otherwise, it will feel uncomfortable, such as in the winter, where $t_{a2}<t_{a1}$, along with the corresponding $\Delta L = L_1 - L_2 < 0$. As the temperature dropped, ΔL got smaller, therefore feeling more cold.

3.2 Statistical analysis of experimental data

In order to examine the moderating effect of indoor environmental quality control system, we carry out experiments for a period of six days; every experiment is from 8:30 am to 17:30 pm. Fifteen volunteers participate in our experiment. PMV-PPD thermal comfort evaluation index is mainly affected by air temperature, mean radiant temperature, relative humidity, air velocity, the body's metabolism rate, and the combined effect of the dress situation. The following data are acquired by a least-squares fit of the laboratory's items in the experimental period.

Figure 2 shows the trends of indoor temperature; Fig. 3 shows the change of the indoor humidity. The indoor wind speed trend is displayed in Fig. 4. Due to differences in the location and placement of the sensor itself, the measurement data are slightly different, but the overall trend is the same and does not affect the calculation of our results.

3.3 Evaluation of thermal comfort regulation

PMV index is a measure of the scale of thermal sensation, and this index value is generally within a range from -3 to +3 changes; PMV=0

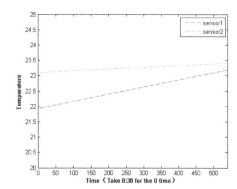

Figure 2. Trends of indoor temperature.

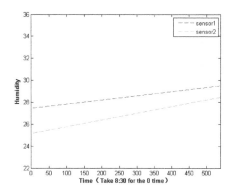

Figure 3. Trends of the indoor humidity.

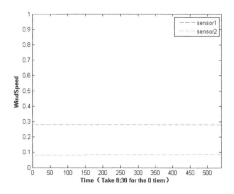

Figure 4. Trends of the indoor wind speed.

represents the thermal neutral (namely thermal comfort); PMV<0 indicates cold feeling; and PMV>0 corresponds to thermal sensation. Since PMV comfort index represents the vast majority of people in the same environment, but because of physiological differences between people, a few people are not satisfied with the thermal environment. Therefore, PPD index is used to indicate the percentage of people dissatisfied with the thermal environment [12].

According to statistics, every day, the experimenter is intended mainly to sit reading, and their metabolism M is $58.14m^2$. According to the general computing habits, the amount of clothes is set to 0.155 m^2K/W, the indoor temperature is set at 21.9–23.4°C, the indoor humidity is between 25.1% and 29.2%, and the air flow rate is less than 0.3m/s. Meanwhile, we assume the indoor air temperature and mean radiant temperature are equal. Therefore, the experimental PMV value is -0.48, and PPD value is 12%; in other words, the indoor environment feels comfortable and acceptable to the human body.

4 EVALUATION OF ENERGY SAVINGS

Under the premise of guaranteeing indoor comfort, we counted indoor environmental devices work time. According to statistics, the fans work 6 hours 4 minutes 45 seconds per day, and air conditioning daily working time is 2 hours 33 minutes 30 seconds. Because of northwest dry weather, we use two rated power of the humidifying 300ml/h of the humidifier that cannot meet the needs of the humidifier; the humidifier daily working time is 9 hours. Equipment uptime chart is depicted next. Figure 5 depicts the trends of the ventilation fans working time. Figure 6 shows the trends of the air conditioning working time.

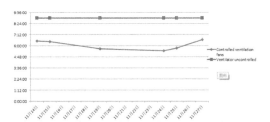

Figure 5.   Trends of ventilator working time.

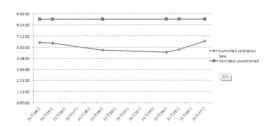

Figure 6.   Trends of air conditioning working time.

During the experiment, the air conditioning input power we used was 2219W, and fan input power was 29W. According to calculation, under the premise of ensuring the comfort and exhaust fan working full-time, the system saves about 14.4 kWh per day, which is compared with air conditioning. Every unit of electricity saved is equivalent to saving 0.361kg of standard coal and reducing 1.324kg of carbon dioxide emissions. Then, the multidimensional dynamic indoor environment control system can save 5.2kg of standard coal each day and reduce 19kg of carbon dioxide emissions. Of course, this test only examined the indoor temperature and humidity, carbon dioxide, and other environmental indicators. The control equipment only include air conditioners, humidifiers, and exhaust fan. With the addition of more environmental quality factors and more equipment, the energy efficiency of the system will be even more impressive.

5 CONCLUSION

During the national "12 FYP" and the rapid development of China's city construction, according to the actual demand of energy saving and emission reduction, we combined with the Internet of things thought to provide a solution for the office buildings and residential areas indoor electric equipment integrated to control. From the experimental results, we can observe that the proposed design not only ensures a comfortable indoor environment but also reduces energy consumption in buildings transformed in daily use.

ACKNOWLEDGMENTS

This research is supported by the National Natural Science Foundation of China (No: 61263019), the Program for International S&T Cooperation Projects of Gansu province (No.144WCGA166), and the Doctoral Foundation of LUT.

REFERENCES

[1] Building Energy Research Centre of Tsinghua University, China Building Energy Efficiency Annual Report 2012. China Architecture and Building Press.
[2] Guoxing Zhang & XiulinGao. Effectiveness of energy conservation policies and measures [J].East China Economic Management, 2014.
[3] SufengWang. Chinese carbon emissions reduction mechanism of the initial allocation and [D]. Hefei University of Technology, 2014.
[4] LingxiCai&Lili Fan &Yanghong Xian Urban dimension of energy saving policy evaluation research [J]. Ecological and economic, 2014.

[5] Jun Luqiu networking technology in the wisdom of life and smart home networking technology 1, NO 9 (2011):57–59.

[6] Jianchun Zhou & Min Qian & WenshiLi. Temperature Measurement System Based on MCU and PC serial communication[J] Communication Technology, 2011, 44 (5):157–159.

[7] Xiangjian Wu & Hui Wang, & JinkeCai.Small household environmental control systems. Modern electronic technology 33, NO 018 (2010):38–40.

[8] Congcong Liu. Intelligent living environment comfortable control method [D]. Shandong Jianzhu University, 2012.

[9] Bluyssen, Philomena M. The Indoor Environment Handbook: How to Make Buildings Healthy and Comfortable. Routledge, 2013.

[10] Andersen, Rune Vinther, JørnToftum, Klaus Kaae Andersen and Bjarne W Olesen.Survey of Occupant Behaviour and Control of Indoor Environment in Danish Dwellings. Energy and Buildings 41, no. 1 (2009):11–16.

[11] Xiaolin XU & BaizhanLI, & Mingzhi.Luo Affect indoor thermal environment on human comfort analysis [J].

[12] Huamei Zheng.Preliminary PMV Nomograph method and formulas for solving the winter heating [J]. Building Energy & Environment, 2007, 26 (5):74–78.

Energy, Environment and Green Building Materials – Sheng (ed.)
© *2015 Taylor & Francis Group, London, ISBN 978-1-138-02718-3*

Research on the FLUENT simulation of the single aluminum pore growth

M. Zhang & C.J. Chen
School of Mechanical and Electrical Engineering, Laser Processing Research Center, Soochow University, China

X.N. Wang
Shagang School of Iron and Steel, Soochow University, China

ABSTRACT: The purpose of this paper is to use fluent simulation to study the single aluminum pore growth. In this paper, the flow of molten aluminum on the surface of the bubble is simulated to reflect the effect parameters on foaming process. The flowing of molten aluminum was researched in two conditions: During the flowing of molten aluminum when the shape of bubbles was spherical, the viscosity of the molten aluminum was changed; when the shape of bubbles was spherical, ellipsoid, and dodecahedron, the viscosity of aluminum molten was constant. It was concluded that during the foaming process when the aluminum foam bubbles were spherical in shape, with the decrease of the aluminum melt viscosity, the stability of the bubbles was reduced; when the viscosity (0.00308kg/ms) of molten aluminum was unchanged, dodecahedron bubbles were more stable than ellipsoidal and spherical bubbles. The conclusions coincided with experimental results.

KEYWORDS: single aluminum pore; fluent; viscosity; stability

1 INTRODUCTION

Aluminum foam is a promising functional and structural material containing a large pore structure, and it has excellent thermal, mechanical, and acoustic performance. There are various approaches for producing aluminum foams, such as foaming of melts with blowing agents, foaming of melts by gas injection, foaming of metal powder compacts, and so on. The powder metallurgy manufacturing route offers the possibility of creating foam having a near-net shape without the need for the addition of abrasive stabilizing particles [1,2]. Although the base powders can be expensive, this technique can eliminate the need for extensive machining operations. The production of foams by using powder metallurgy was developed some years ago and has ever since been refined [3]. As technology moves forward and new potential applications emerge, such as in aluminum sandwich panels, research is still required in order to obtain aluminum foams with the desired properties. Limiting factors such as non-uniformity of pores and poor overall foaming reproducibility have plagued the process for years. Significant research has been done on the processing experiment [4,5]. This might be due to the difficulties in experimental observation and measurement because of the nontransparency and high temperature of the metallic melt. On the other hand, numerical simulation has become a very important and powerful tool for investigating and understanding complex flow phenomena in science and engineering.

Over recent years, more and more efforts have been made to investigate metal melt flows[6,7], both one phase and two phases, by numerical simulations.

However, fluid dynamics in a bubble-melt two-phase system is highly complex. Hong Liu and Maozhao Xie [7] have done some research especially on the stirring foaming process. Unfortunately, no open publication has been found working on the influence of processing variables on the foam cell structure of powder metallurgy manufacturing using laminar model. So, in this paper, the flow of molten aluminum on the surface of the bubble is simulated to reflect the effect parameters on the foaming process.

2 EXPERIMENTS

2.1 *Foaming experiment of aluminum foams*

Since the details of the experimental procedure and results have been reported elsewhere [8], here we present only a brief description. Aluminum and Si elemental powders were mixed with 1.0 wt% TiH2 having the following compositions: Al–12 wt%Si. The powders were mixed in a tumbling mill for 30 min along with alumina spheres to produce uniform mixed powders. The surface of two steel plates was degreased, descaled, and roughened in order to provide a fresh surface of contact for the powders. After the foaming agent was uniformly distributed within the matrix powders, the two plates were stacked and 1Wt. % TiH2 alloy powder was put between the plates.

The plate was roll-bonded with a draft percentage of more than 60% reduction at room temperature. A well-bonded precursor was successfully obtained. A high-temperature foaming test was performed in the atmosphere using a silicon carbide heating furnace. After reaching the foaming temperature of 620°C, the specimen was held isothermally for Δt, and then cooled down to room temperature. An aluminum foam sandwich structure was obtained (Fig.1).

Figure 1. Aluminum plate/aluminum foam core sandwich structure.

3 PHYSICAL AND MATHEMATIC MODEL

For the sake of simplicity, the following assumptions were introduced into the numerical model:

(a) The foaming tank is adiabatic, so there is no heat loss to the environment. Furthermore, the heat transfer rate between the melt and gas bubbles was infinite; consequently, the pores have the same temperature as the aluminum melt. Therefore, no expansion exists in the bubbles due to temperature variation and then the energy equation is not needed anymore.

(b) Inside the grid cell, the gas and liquid have the same pressure.

In this paper, during the process of producing foam aluminum, the flow of molten aluminum on the surface of the bubble is simulated to reflect the stability of foam. The following two aspects were discussed:

(1) The shape of the bubble was constant, the bubbles were spherical, and the viscosity of the molten aluminum was changed: 0.0045kg/ms, 0.00321kg/ms, 0.00308kg/ms, 0.00301kg/ms, and 0.00292kg/ms.

(2) Viscosity was constant at 0.00308kg/ms; the bubbles were spherical, ellipsoid, and dodecahedron in shape.

3.1 Cell modeling

According to Fig. 1, the structure of the cell was simulated as sphere, ellipsoid, and regular dodecahedron exemplified in Fig. 2a, b, and c. It can be seen that the real structure of the cells (Fig.1) has some similarities with the structure model.

3.2 Numerical procedure

Gas and liquid properties are set as follows: density of aluminum melt: q_l = 2650kg/m3, H2

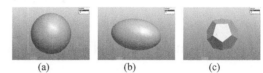

Figure 2. Aluminum plate/aluminum foam core sandwich structure.

(a) spere; (b) ellipsolid; (c) odecahedron

density q_g = 0.0899 kg/m3, and viscosity of H2 μ_l= 1e-05 N·S/m2; viscosity of molten aluminum at a different temperature is shown in table 1. The three-dimensional two-phase CFD model described earlier was solved using the commercial flow simulation software FLUENT13.0 as well as a number of user-defined routines. In this model, H2 was the suppressed body, so only aluminum flows were studied in this paper. Meshing method was tetrahedrons, and grid growth pattern was Patch Independent. The minimum limit size was set at 0.05mm. Unfortunately, very few experimental data are available for a systematic comparison of predicted results with measurement. Therefore, we restrict the scope mainly to qualitative discussions of the predicted results. The bubble-melt two-phase flows in the tank are described with a multiphase volume of the fluid model. In this model, H2 was the suppressed body, so only aluminum flows were studied in this paper.

Table 1. Viscosity of molten aluminum at a different temperature.

temperature/ °C	600	610	620	630	640
viscosity/ (N·S/m²)	0.0045	0.00321	0.00308	0.00301	0.00292

For the discretization and solution of variables, a first-order upwind scheme is applied to the equations of momentum, laminar kinetic energy, and the volumetric fraction equations for gas and liquid, which are solved with the Geo-Reconstruct scheme. The simple algorithm for pressure–velocity coupling is used for numerically solving the model equations.

Figure 3. Meshing in workbench.

(a) sphere; (b) ellipsolid; (c) odecahedron

4 RESULTS AND DISCUSSION

4.1 *Flow characteristics of liquid phase*

Bubble surface contours pressure distribution is substantially the same in a different viscosity, so in this paper only the pressure contours of viscosity 0.0045 N·S/m² are shown in Fig. 4. But in a different viscosity, the variation of pressure is plotted as shown in Fig. 4. The simulation results show that with a decrease in the aluminum melt viscosity, the surface bubble pressure decreases. This is to say, with the surface pressure of the bubble decreasing, the flow resistance of aluminum melt is reduced, the trend of bubble growth increases, and the bubble grows and bursts easily. So during a certain temperature, the melt viscosity decreases, surface pressure of the bubble decreases, and the bubble stability also decreases.

Typical predicted results for the aluminum liquid phase velocity distribution are shown in Fig. 5. It can be seen that in different viscosities, the aluminum melt surface of the bubble velocity vector distribution is roughly the same, but the maximum and minimum velocity change with the increase of viscosity, as plotted in Fig. 6.

In Figs. 5 and 6, the surface velocity of the sphere is not the same, the speed of the local area is large,

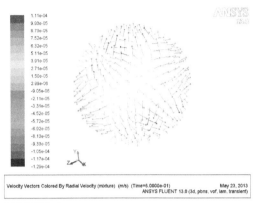

(a) 0.0045kg/ms

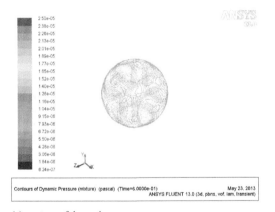

(a) contour of dynamic pressure

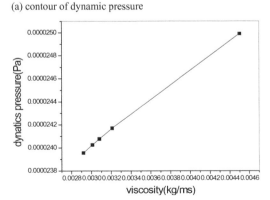

(b) change of pressure

Figure 4. Contour of dynamic pressure and change of pressure in different viscosity.

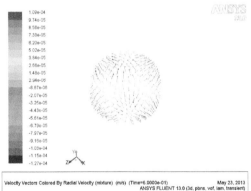

(b) 0.00321kg/ms

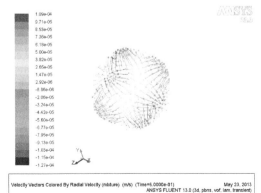

(c) 0.00308kg/ms

163

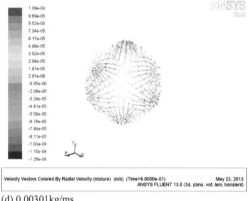

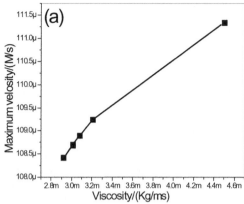

(d) 0.00301kg/ms

(a) maximum

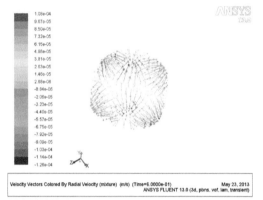

(e) 0.00292kg/ms

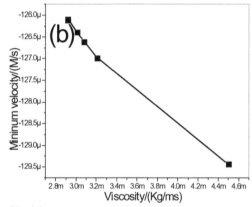

(b) minimum

Figure 5. Velocity distribution of liquid in H2 surface in different viscosity.

Figure 6. Maximum and minimum velocity of the bubble surface in different viscosity of molten aluminum.

and the speed is very small in the local area. With the decrease in viscosity, the bubble surface velocity variation gets smaller and smaller.

Velocity difference on the surface of the bubble causes spherical bubble growth to become unstable, with the direction toward the ellipsoidal. With the increase in viscosity of molten aluminum, the bubble surface speed variation is large, the growth trend from spherical to ellipsoidal is big, and the spherical can quickly grow into the desired ellipsoidal and form a better bubble stability. On the contrary, as the aluminum melt viscosity decreases, the velocity of the bubble surface variation is smaller, the trend of the sphere bubble growing into the ellipsoidal bubble becomes slower, and so the bubble is not instable.

Anyway, during the definite temperature range, with the decrease of molten aluminum viscosity, the stability of the bubble has the trend of decreasing. It is obtained from theory that bubbles are separated by the liquid membrane in molten aluminum. At the

juncture of the bubble, the fluent of the liquid film surface will flow onto the junction. Reduced viscosity will enhance the molten aluminum flow and weaken the strength of the liquid film; bubble film will be ruptured easily; small bubbles will grow and merge easily; and bubble stability will become poor. So simulation was in accordance with experimental results.

4.2 Effect of bubble shape on the stability

According to preliminary experimental research, it was shown that the foaming process was more stable at 620°C. So, it was set to a temperature of 620°C, and the viscosity was 0.00308kg/ms. The effects of the different shapes of bubbles on the aluminum melt flow was analyzed as follows. The pore was, respectively, spherical, ellipsoid, and odecahedron in shape, according to the experimental results depicted in Fig.1.

Contours of dynamic pressure and velocity vector of three different shapes of bubbles are shown in Fig. 7 and Fig. 8.

In Fig. 6, it was shown that the dynamic pressure of molten aluminum on the spherical bubble surface is roughly uniformly distributed. On the ellipsoidal bubble surface, the dynamic pressure is less distributed on the long axis and more distributed on the short axis; whereas on the dodecahedron bubble surface, the dynamic pressure is mainly concentrated on the edge. The dynamic pressure of the spherical bubble is the smallest, followed by the ellipsoidal bubble and the dodecahedron bubble is the largest. The larger the pressure is, the larger the surface tension between the two phases is. So the ability to prevent air bubbles from growing stronger indicates that the bubbles have good stability.

In Fig. 7, it is shown that velocity distribution in the spherical bubble is relatively uniform; velocity distribution in the ellipsoidal bubble is relatively small on the long axis and relatively large around the short axis; and velocity distribution at the edge of the dodecahedron bubble is the largest.

It can be illustrated that the bubbles grow from generating into contact with each other. Because of the surface tension, bubbles grow into spherical shape and grow slowly and in the draining effect, the pore wall becomes thinner, Finally, there is only a thin film between two bubbles. Under the pressure, pores change into polyhedral foam.

Thus, simulation results show that regular polygon bubbles have better stability than ellipsoidal bubbles, and the stability of ellipsoidal bubbles is better than that of spherical bubbles.

The earlier simulation results coincide with experimental results[9].

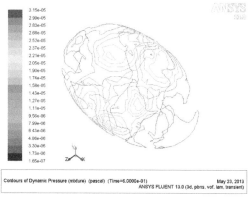

(b) ellipsolid

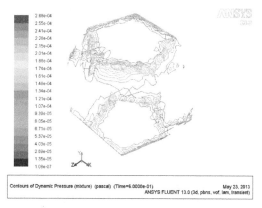

(c) odecahedron

Figure 7. Dynamic pressure contours of different-shaped bubbles at 620°C.

CONCLUSIONS

Based on the earlier simulation analysis, two following conclusions were obtained:

1 During the foaming process when the aluminum foam bubbles were of the same spherical shape, with the decrease of the aluminum melt viscosity, the stability of bubbles was reduced;

2 When the viscosity (0.00308kg/ms) of molten aluminum was unchanged, dodecahedron bubbles were more stable than ellipsoidal and spherical bubbles.

ACKNOWLEDGMENTS

This work was supported by the National Nature Science Foundation of China (grant No. 51104110), the fund of the State Key Laboratory of Advanced Processing and Recycling of Non-ferrous Metals, and the Lanzhou University of Technology and

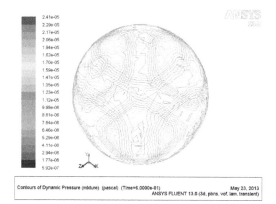

(a) sphere

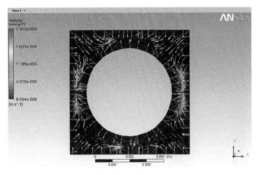

(a)

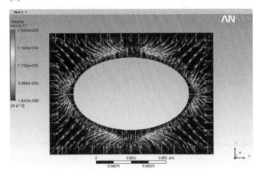

(b) ellipsolid

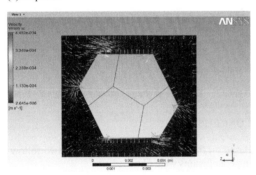

(c) odecahedron

Figure 8. Velocity vector of different-shaped bubbles at 620°C.

Suzhou Science and Technology Bureau (grant No. SYG201231).

REFERENCES

[1] A.R. Kennedy, Effect of compaction density on foamability of Al–TiH$_2$powder compacts[J]. Powder Metallurgy 2002, 45(1):75–79(5).

[2] F. Baumgärtner, I. Duarte, J. Banhart. *Industrialisation of powder compact foaming process[J]*. 1st International Conference on Metal Foams and Porous Metal Structures, 14-16 June 1999, Bremen, Germany [] Advanced Engineering Materials 2(4), 168–174 (2000).

[3] M. Lafrancea,*, M. Isaca, F. Jaliliana, K.E. Watersa, R.A.L. Drewb. The reactive stabilization of Al–Zn foams using a powder metallurgy approach [J]. Materials Science and Engineering A, 2011, 528A: 6497–6503.

[4] Peter Schäffler, Walter Rajner. Process stability in serial production of aluminium foam panels and 3D parts [J], Advanced Engineering Materials, 2004, 6(6):452–453.

[5] ZHANG Min, ZU Guo-yin,YAO Guang-chun. The effect of Mg addition on the stability of foams in preparation of foam aluminum sandwich [J]. Journal of Functional Materials, 2007, (4):576–579.

[6] Weigang Xu, Hongtao Zhang, Zhenming Yang, Jinsong Zhang.Numerical investigation on the flow characteristics and permeabilityof three-dimensional reticulated foam materials[J]. Chemical Engineering Journal 140 (2008) 562–569.

[7] Hong Liu, Maozhao Xie, KeLi, Deqing Wang bThree-dimensional CFD simulation of bubble–melt two-phase flow withair injecting and melt stirring[J]. International Journal of Heat and Fluid Flow 32 (2011) 1057–1067.

[8] ZHANG Min, YAO Guang-chun, ZU Guo-yin.. Research on preparation of aluminum foam sandwich and steel plate/foam core interfacial microstructure[J]. JOURNAL OF FUNCTIONAL MATERIALS, 2006, 37(2):281–283. (In Chinese)

[9] ZHANG Min, YAO Guang-chun. Study on the foaming process of the foam aluminum sandwich[J]. JOURNAL OF FUNCTIONAL MATERIALS, 2008, 39(4): 596–599. (In Chinese)

Energy, Environment and Green Building Materials – Sheng (ed.)
© 2015 Taylor & Francis Group, London, ISBN 978-1-138-02718-3

Exploration for geographic information of an ancient village based on GIS technique—Da Liangjiang Village in Jingxing County of Hebei Province as an example

C.Y. Wang & G.Q. Li
Agriculture University of Hebei, Baoding, Hebei, China

Y. Yang
National Computer Network Emergency Response Technical Team/Coordination Center of China Shanghai Branch, Pudong, Shanghai, China

ABSTRACT: This paper took Da Liangjiang Village in Jingxing County of Hebei Province as an example, and it investigated the method and technique of geographic information of ancient village by using GIS. As an important part of the protection of ancient village, this research applied data analysis, spatial analysis, and 3D visual function of GIS technique; it also conducted a deep analysis of the ancient wisdom in site selection, development, and essence of places from the angles of Planning, Architecture, and Science of Human Settlements. All this is done with the aim of recording and preserving the precious values of ancient village, which offers beneficial reference and experience for planning and construction in both the present and future.

KEYWORDS: GIS; ancient village; data analysis

As the common wealth of mankind, ancient village is a complex of tangible cultural heritage and intangible culture heritage, which concentrates on human beings' labor of thousands of years and infinite wisdom as well as contains a long history and profound culture. From the protection status of ancient village, the protective range contains not only tangible cultural heritage, such as planning of the village, various types of construction, and rare plants and ancient trees, but also intangible cultural heritage, such as folk custom, folk belief, and traditional crafts. However, with the development of modern society, the lifestyle and mode of production has been greatly changed; this effect is comprehensive and results in the absolute crashing of ancient village. It is an important issue for planners and architects to deal with the protection and research of ancient village.

1 BACKGROUND

With the development of technology, GIS has been frequently applied in protection and practice of ancient village as a new technique.

Currently, the mainstream of ancient village preservation by using GIS places emphasis on establishing a database, including graphic database, attribute database, and image database, and finally establishing

a protective system according to the requirement of planning, preservation, and management. On the whole, the application of GIS technique in the protection of ancient village places emphasis on information collection and archiving, but it ignores the traditional methods of data processing and analysis from the angle of Planning and Architecture. This

Figure 1.

thesis will discuss how to analyze the geographic information of ancient village by using GIS technique in order to obtain direct analysis results; excavate the content of ancient Planning, Architecture, and Science of Human Settlements that lie hidden in geographic information of ancient village and its surrounding regions; and find out ancient wisdom of old village in the process of construction and development. All this makes preservation of ancient village a beneficial task.

2 PROFILE OF DA LIANGJIANG VILLAGE

Da Liangjiang Village (Figure 1) is located on the hiterland of Taihang Mountain, which is situated on the borders between Shanxi and Hebei Provinces. Because of the mountain, a natural barrier that makes poor communication, the village is protected from the destruction of wars during hundreds of years, and its ancient appearance is entirely preserved. The pattern of the village shows the architectural feature of the folk house of the northern mountain area. Constructions in the village reveal evident folk house style of Shanxi province and also contain the features of quadrangle dwellings in Beijing and Hebei province. Besides, the surroundings of the village have a graceful environment that together constitute the natural landscape and historical culture.

By 2007, the research team of the department of architecture, College of urban and rural construction in Agricultural University of Hebei, had finished detailed surveying and mapping, and it drew up the protection planning. As one of the most complete ancient villages, Da Liangjiang Village was honored as the second batch of famous historical and cultural villages in Hebei in 2008, and the fifth batch of famous historical and cultural villages in China in 2010.

During the process of surveying and mapping, a lot of geographic information data was collected, which is the research basis of this thesis.

3 RESEARCH PROCESS

This thesis chose ArcGIS of ESRI as research tool, and it analyzed the topographic map that was calibrated according to the data collected during mapping. Contour method and elevation point method were chosen to input data so as to gain accurate results.

There were two methods of surface analysis and profile analysis. According to the situation, data were processed mainly by surface analysis and in addition to it, profile analysis.

3.1 *Mapping check*

Before conducting an analysis, mapping data need to be checked. First, we should conduct a pre-analysis of CAD files. We need them to generate topographic data and to check out whether there are outliers (Figure 2); then, we should use GPS receiver to locate the outliers for secondary mapping for supplementary survey and checking. After that, we should use this method again until there are no outliers.

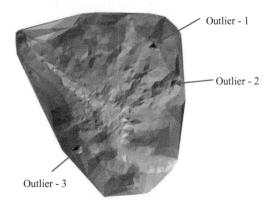

Figure 2.

3.2 *Surface analysis*

The topographic data in CAD files that had been checked contained continuous contour lines and discontinuous elevation points. Using the auto-analysis function of Sketchup, we obatined a series of discontinuous mesa, but the real topography changes continuously, which means there are countless points on the surface; so they can be measured and recorded. The surface model of GIS allows the users to record the surface information in the software. This surface model collects the points from different locations and then uses interpolation to get a simulation of the surface and approximate fitting. In a word, using limited points to speculate the unknown points, we obtained the whole surface. By analyzing the curved face trend and using different colors, we can directly know the changes in geographic information.

This thesis mainly uses TIN trigonometric analysis and Kriging analysis:

TIN triangular mesh is generated by vector data, which could be points, lines, or mixed data element. One or more vector data that need to contain elevation values could be chosen to form TIN surface. The TIN surface that has already been created can continuously add some elements or features in order to change its surface or improve the precision. In addition, the TIN triangular mesh can exactly show the changing process of rock mass.

The assumption of Kriging analysis is that the distance and direction among sampling points must reflect some spatial correlation and explain spatial difference. Kriging model can use mathematical function to fit the particular points or all the points in a given search radius and to estimate the value of every point. This method is applied to the situation where the data have a deviation in distance and direction, so the surface generated is smoother and more continuous. Because the surface generated by Kriging is smoother, it is used to analyze the slope of topography, which can be combined with the rule of construction slope; thus, it can simply analyze the area appropriate for a building.

We can use the data to output 4 kinds of maps: altitude analysis, aspect analysis, slope analysis, and hillshade analysis. They can show the difference between the two analysis results through contour method and elevation-point method.

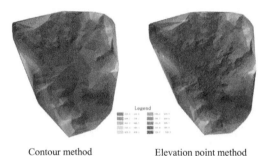

Contour method   Elevation point method

Figure 3.

Altitude analysis (Figure 3): We used TIN triangular mesh to finish the analysis. The blue points are the lowest elevation points, and the brown ones are the highest; elevation values change gradually between them. It is observed that the topography of this area is a mountain valley. This ancient village is located along the northeastern slope of the lower middle part, which is a reasonable choice for protecting from floods and mountain torrents and simultaneously convenient to build. Simultaneously, the contour method compared with the elevation method outputs more platforms and more evidently reflects the variation trend but has lower accuracy.

Aspect analysis (Figure 4): We used TIN triangular mesh to finish this analysis. The aspect reflects whether the terrain is suitable for receiving sunshine. From the output, we can observe that this village has a better aspect that could be beneficial for receiving sunshine. Simultaneously, the contour method compared with the elevation method has a deviation when calculating the aspect.

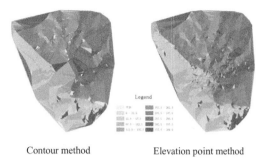

Contour method   Elevation point method

Figure 4.

Slope analysis (Figure 5): We used Kriging analysis to show the gentle variation trend of the slope. The slope reflects the steepness of slopes. According to the national regulation, the slope less than 25% can plan and construct, which means less 14.04° using acrtan 0.25. We set 0 to 14.04° as the first sector that was painted by bright color, so the bright area in the graph means it is suitable for planning and construction. From the output result, this valley has a rich area for planning and construction without regard to ecological factors and geologic structure. The output of the elevation method can obviously show the boundary in the northern part due to the slope, and the output of the contour method can show the variation trend of the slope, but it has lower accuracy.

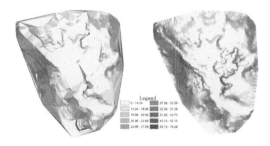

Contour method   Elevation point method

Figure 5.

Hillshade analysis (Figure 6): We used calculus of interpolation directly. Regardless of contour method and elevation method, both obviously show the hillshade situation. From the analysis result, this area is basically unaffected by hillshade and has good sunshine, which is suitable for living and producing.

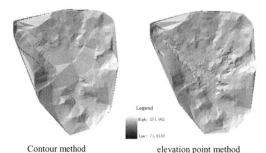

Legend

High: 254.962

Low: 71.8139

Contour method   elevation point method

Figure 6.

3.3 Profile analysis

After surface analysis, we got only the slope trend of this region, which is a sketchy result. We need to process profile analysis in order to analyze the cross-section of the critical area and take the accurate slope of the valley. First, we need to choose a representative cross-section by choosing cutting line (1–5) in the characteristic region.

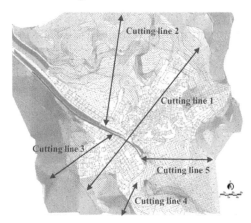

Figure 7.

Cutting line 1 is medial axis from ancient village to the new one, cutting line 2 is the line from the mountain of the northern part to the new village, cutting line 3 is the line from the mountain of the southwestern part to the new village, cutting line 4 is the line from the mountain of the southern part to the southern entrance area of this village, and cutting line 5 is the line from the mountain of the eastern part to the ancient village.

We put the topographic data and cutting line input to ArcGIS and obtained the output: Cutting line 1 has 938 results and it chooses an operating polygonal line; cutting line 2 has 649 results; cutting line 3 has 344 results; cutting line 4 has 196 results; and cutting line 5 has 363 results. All the results are shown in Figures 8 to 12.

In Figure 8, the left is the beginning of the northern cutting line and the right is the ending of the southern cutting line. The polygonal line obviously shows that the location of ancient village in sector 150 to 200 has a moderate slope, which is clearly better than the southwestern slope where the new village is located. It verifies the correctness of the site selection of ancient village.

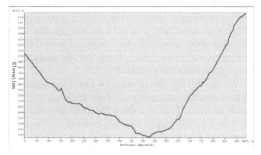

Figure 8.

In Figure 9, the left is the beginning of the northern cutting line and the right is the ending of the southern cutting line. From the polygonal line, it shows the variation trend of the slope from northern mountain to the new village.

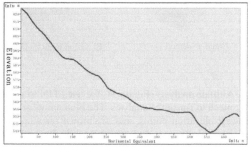

Figure 9.

In Figure 10, the left is the beginning of the cutting line of southwestern mountain and the right is ending of central section of new village. From the polygonal line, it shows the variation trend of the slope from southwestern mountain to new village.

In Figure 11, the left is the beginning of the cutting line of southern mountain and the right is southern entrance of village area. From the polygonal line, it shows the variation trend of the slope from southern mountain to southern entrance of village area. It shows that the topside of the slope is gentle, while the slope near the village is sharp.

In Figure 12, the left is the beginning of the cutting line of eastern mountain and the right is the direction of ancient village. From the polygonal line, it shows the variation trend of the slope from the eastern

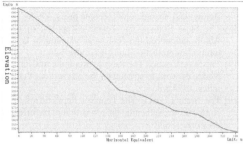

Figure 10.

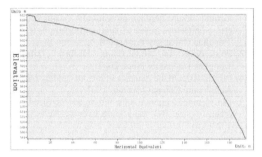

Figure 11.

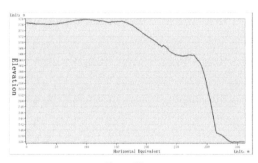

Figure 12.

mountain to ancient village. It shows that the topside of the slope is gentle fluctuant, while the slope near the village is sharp, which is close to right angle.

4 COMPREHENSIVE ANALYSIS OF GEOGRAPHIC INFORMATION

From the analysis of surface and profile, we can give a relatively accurate conclusion about the physical geography information of this region.

The whole village is located in a mountain valley; the valley bottom is long and narrow and has a flat terrain, which is a small flood plain. This valley is surrounded by mountains, and the elevation increases gradually from the bottom but does not

change sharply. Therefore, the valley has a good sunshine condition, which is beneficial for producing activities.

The topographic feature from the bottom to the slope is that overall the slope of the eastern mountain is gentler, whereas that of the western mountain is steeper. The northern, northeastern part of the valley bottom has a gentle slope. Profile 2 shows that the northeast to west-southwest slope is long and gentle, whereas the northwestern part of the mountain has a steep slope. In the middle of the valley bottom, from northeast to southwest there is a gentle slope; whereas from southwest to northeast, there is a steeper slope. Profile 1 shows that the northeastern slope is long and gentle, whereas the southwestern one is short and steep. The variation trend of the slope southwestward of ancient village is from sharp to gentle in the western mountain to the valley bottom. Profile 3 has an obvious inflection point where there is 160m relative displacement toward the cutting point, and it has a gentle slope from the range 160m to 280m. In the southern part of the valley bottom, shown in profile 5, there appears to be a wavy terrain. On the southwestward side, there is a steep slope (profile 4); the terrain from the bottom is gradually uplifted, afterward appears as a short mesa, is rapidly uplifted, and then suddenly dropped.

Site selecting analysis of planning and construction: According to the geographic information of this region, two areas are suitable for building: One is the northern and northeastern slope in the north, and another is the slope from the northeast to southwest in the central part. However, the northern part of the mountain was the spring channel, and a damn and protection project were built there; therefore, the slope in the central part from the northeast to southwest is the best choice, where the ancient village is located. Besides, the southwestern slope in the central part, where there is a gentle and moderate slope, is the secondary choice for construction that could be chosen to build the new village. Other areas, such as the western and southwestern slope, that have a sharp slope or are close to the bottom are unsuitable for planning and building. The eastern and southeastern slope could be used for terrace cultivation, and these are steeper and more gradual.

Using modern technology, we can fully understand the geographic information of the target area and even restore the topography. From this research, we found out that the location of the ancient village is surprisingly universal with the answer calculated by digital technology. We can even conclude that the range and texture of ancient village are also exactly the same with the texture formed by the spontaneous change of the slope. We cannot help thinking about how those ancient people chose their ideal homeland without advanced science and technology. Backed by the mountain and fronting the road, the site is selected in a gentle slope where it is convenient for transportation and without the damage of floods. This is the enlightenment from the ancient wisdom during research of this village.

REFERENCES

[1] Zhang Fan, Historical and cultural protection countermeasures in the development of cities, Southeast University Press, 2006.

[2] Wang Jun, Nanjing ancient buildings in the old city reconstruction, 2006.11.10.

[3] Ruan Yisan,Wang Jinghui,Wang Lin, Historical and cultural city protection theory and planning, TONGJI University Press, 2004.

[4] Luo Zhewen, Ancient Chinese architecture [M], Shang HaiGuji Press, 2001.10.

[5] Yang Bingde, Chinese modern city and architecture [M], China architecture&building press, 1993.10.

[6] Contemporary China Beijing (B)[M], Contemporary China Publishing House, 1993.86.

[7] Yang Dongping, Urban monsoon: The cultural spirit of Beijing and Shanghai [M], People press, 1994.

[8] The State Council The People's Republic of China, About announced the first batch of national key cultural relics protection units list, 1961.3.4.

[9] The State Council The People's Republic of China,The Notice of work to strengthen the protection of cultural relics, 1974.8.8.

[10] The State Council The People's Republic of China, Notice on strengthening the protection of historical relics work, 1980.5.7.

[11] The State Council The People's Republic of China, About the second batch of national key cultural relics protection units, 1982.2.

[12] Xu Subin, The comparative study of east Asian architectural culture heritage protection[A], Chinese modern architecture research and conservation (B)[C], Tsinghua University Press, 2001.

[13] Proceedings of Architectural History (15th)[M], Tsinghua University Press, 2002.

[14] MOHURD, On the issuance of the outstanding modern architecture Council Minutes, 1991.7.2.

[15] Fan Kening, Lingnan authentic arcade civilized road reconstruction [N], Yangcheng Evening News, 2000, 3, 23(A5).

Energy, Environment and Green Building Materials – Sheng (ed.)
© *2015 Taylor & Francis Group, London, ISBN 978-1-138-02718-3*

Research on a sewage treatment system with low operation cost applied in rural areas

C.Y. Wang & G.Q. Li
Agriculture University of Hebei, Baoding, Hebei, China

N.Z. Wang
Hebei Arts and Design Academy, Baoding, Hebei, China

ABSTRACT: With the development of society, people's demand on water environment improves obviously in Chinese rural areas. For example, every administrative village should establish sewage treatment system according to requirements on Planning and Design Technical Guidance on Rebuilding and Improving Rural Appearance in Hebei province (2014). This paper tries to research a cheap solving scheme based on sewage treatment technology of combined constructed wetland, which is sewage treatment system of low operation cost characterized by low construction cost, low maintain cost, abroad range of application, high disposal efficiency, strong compositionality resistance and suitable for popularization in rural areas etc. It is found that the effluent's concentration range of main pollutant materials COD, BOD, SS, NH and TP of domestic sewage after disposal are 16-32mg/L, 7.8-14.3mg/L, 3.3-6.7mg/L, 6.1-11.7mg/L, 0.21-0.41mg/L respectively. Water quality of effluent is superior to various national standard of sewage recycling currently in effect. The design scheme completed by this research has received national patent (patent number: ZL 201010115523.2, ZL 201020120887.5, ZL 201010115495.4,ZL 20101.115531.7,ZL 201020120813.1,ZL 201020120904.5).

KEYWORDS: rural areas, sewage treatment, low cost

In recent years, with the development of the economy, treatment of water pollution in rural areas has become increasingly important. All local governments issue relevant requirements on sewage treatment in rural areas in succession. Technically, sewage treatment system of traditional industrialization has achieved distributed miniaturization, and related technology has matured. But in the process of planning, construction, and practice, we found that it is extremely difficult to popularize sewage treatment facilities at the village level. The problems are construction costs and operating costs fundamentally.

Taking the planning and design project of rebuilding and improving Sun Guzhuang village's appearance in Dingxing Country, the author participated, for example, in Hebei province putting forward the requirement of establishing a sewage treatment station for each village, and we also made the plan and design of a sewage treatment station in plan formulation. But in practical terms, investing more than 400,000 yuan to build a small sewage treatment station is quite unreasonable for a medium-scale plain village with less than 400 households and where each person's annual per capital income is less than 5000 yuan. In addition, throughout the service cycle of 20 years, each villager has to undertake processing cost

of about 0.5 yuan per ton water. From inquiring the situation of investigation research on villagers, we can conclude that villagers prefer spending the same money on road building and landscaping rather than on sewage treatment facilities with such high construction and operating cost. Therefore, according to the current situation in rural China, regardless of construction cost or operation cost of sewage treatment technology, traditional industrialization cannot meet the actual needs in the countryside in our country. And there is an urgent need for a new kind of solution.

1 BACKGROUND

The real sewage treatment started in the late 1990s, which pays great attention to sewage treatment technology in the city. However, with a large amount of power consumption, the traditional sewage treatment system disposes pollutants at the cost of energy and money consumption, which cannot combine energy saving with emission reduction and cannot achieve low carbon of water purification. Such a traditional method transformed environment destruction by pollutants to ecological system destruction by energy consumption. It is acceptable to utilize such technology in the city to dispose high-density sewage, but its high operation cost decides that it is inappropriate to apply it in rural areas. According to the Planning and Design Technical Guidance on Rebuilding and Improving Rural Appear-ance in Hebei province(2014), the low-cost sewage treatment technology recommended by Hebei province governments includes a kind of stabilization pond and a kind of constructed wetland. Among this, stabilization pond scheme is quite outdated, and it will bring problems such as underground water pollution, breeding mosquitoes, and releasing peculiar smells. Represented by sewage treatment technology of constructed wetland, the technology of sewage ecological treatment has the characteristics of high efficiency, low investment, low operation cost, and low maintenance technology. The technology receives more and more attention in the world. However, the shortages limiting its application are mainly large areas occupied and not ideal sewage treatment efficiency. As for the actual situation of sewage treatment in rural areas, it is alright to utilize integrated technology of combined constructed wetland under the scheme of sewage treatment of con-structed wetland in order to realize ideal treatment effects based on cost control and to achieve a recognized balance in the market.

2 TECHNOLOGY CHOICE AND OVERALL STRUCTURE OF SEWAGE TREATMENT SYSTEM

Looking at the current sewage treatment technology, vertical flow wetland can purify inorganic pollutant efficiently whereas the capability of disposing organic pollutant is insufficient. Besides, the horizontal undercurrent wetland is suitable for disposing organic and heavy metal pollutant whereas the oxygen supply is insufficient and nitrification is weak. Therefore, this research tries to design a wetland system of vertical flow and horizontal flow that is characterized by high sewage purification efficiency per unit area and sufficient oxygen supply.

Overall structure (see Figure 1) is a vertical reaction pond of combined constructed wetland consisting of vertical-flow and horizontal-flow constructed wetland. Mixed matrix is filled in the pond, and hyper accumulator plants are planted. The roughly treated sewage will be introduced into the reaction system of combined constructed wetland. The combined constructed wetland will dispose the inorganic, organic, and heavy mental pollutant in sewage and eliminate the pollutant to the furthest extent. In addition, all current sewage treatment equipment can be used as a pre-treatment system of the reaction pond so as to realize technology upgrading in total by seamless connection. This characteristic makes this system

have very strong flexibility in the field of improving and rebuilding traditional technology.

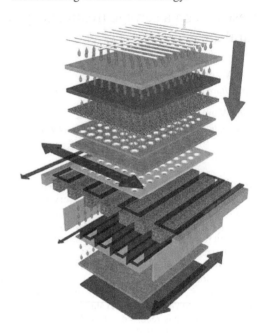

Figure 1.

3 CONSTRUCTION OF SEWAGE TREATMENT SYSTEM

The technical solution adopted in this system is as follows: One is a combined constructed wetland system of vertical flow and horizontal flow, whose key technology lies in the fact that the vertical-flow wetland pond is upper, and the horizontal-flow wetland pond is lower. The water treated by the vertical-flow wetland pond flows into the horizontal undercurrent wetland pond with the help of setting up the backflow pond. From top to bottom, the added substrate in the vertical-flow wetland pond consists of coarse sand, artificial mixed layer, gravel area, and artificial mixing layer with shallow rooted plants. The water distributor is equipped on the top of the vertical- flow wetland pond; the bottom is equipped with sluice channel; the upper part of the sluice channel is a porous plate; and the bottom part of the sluice channel is a water proofer. Both sides of the vertical-flow wetland pond are gravel areas, and the middle is the matrix region. From top to bottom, its matrix region is a coarse sand layer, an artificial mixed layer, and coarse sand. The set diversion plate divides the matrix region into S-shaped water channels, an artificial mixing layer with deep-rooted plants. The bottom is equipped with sluice channel, the bottom of the sluice channel is the water proofer, and the top of the sluice channel is located in the outlet of gravel areas and a set porous plate. The sewage pump is set in reflux pond; the sewage pump is connected with the gravel area of horizontal-flow undercurrent wetland pond via delivery pipe; and leak-off pipes are set between vertical-flow wetland pond and horizontal-flow wetland pond.

The water proofer of the vertical-flow wetland pond consists of plastic materials, the cavity that can be created through by deep-rooted plants will be preset on the water proofer, and the cavity inside is filled with sealing material to prevent leakage.

As a further improvement, the micro-porous burst pipes are set at the bottom of the matrix region of the horizontal-flow wetland pond mentioned earlier; these pipes are connected with the underwater micro-porous burst engine that is set in the reflux pond.

The artificial mixed layer consists of vermiculite, zeolite, and coal cinder, and its thickness is at least 20 cm with a dense upper and loose bottom.

The particle size of coarse sand in vertical-flow wetland pond and horizontal-flow wetland pond is 2 to 7 mm, the particle size of coarse sand in vertical-flow wetland pond is 4–8 cm, at one side of the horizontal-flow artificial wetland pond's grave area is the water inlet, at the other side is the water outlet, the particle size of coarse sand in the inlet is 8–15 mm, and the particle size of coarse sand in the outlet is 2– 4 cm.

The beneficial effect produced by this system and technology scheme lie in the following: The integrated application of the two constructed wetlands not only effectively cleans the entropic inorganic elements of sewage but also effectively decomposes and adsorbs the organics and heavy metals of sewage. The purification of sewage in multiple direction and flow with enough clarification time can make better use of river closure, filtration, adsorption of matrix, and so on, to improve the efficiency of purification, and the vertical flow is not easily jammed. The adoption of up and down vertical composite structure effectively reduces floor area; a further improvement is that the underwater micro porous burst engine will conduct microburst by delivering oxygen from micro-porous burst pipes to the horizontal-flow undercurrent pond, which add oxygen content effectively, improve the purification efficiency on the reaction of plant roots and microbial flora, efficiently deal with and absorb heavy metal, and decompose organic pollutants. After the sewage's flowing to the underflow wetland pond, organic pollutants, heavy metal, as well as inorganic pollutants will be reduced a lot.

4 INSTRUCTIONS OF PROCESSING FLOW

Further details on the system will be explained based on the combination of design map and concrete implementation methods.

First, all sewage will be pretreated to filter large floaters and suspended matter, reduce the scum produced in subsequent processing, and ensure the normal operation of sewage treatment equipment. The pretreated sewage will flow into the vertical-flow wetland pond 25 through sewage pipe 2, and the flow rate can be controlled by valve. Sewage evenly flows into the vertical-flow wetland pond 25, vertically downward through the coarse sand layer8, artificial mixed layer 5, gravel layer 6, and enrichment plant 4 of shallow-rooted system. Sewage adsorbs a part of heavy metal pollutants by the vertical infiltration layers of substrate. The nitrification ability of roots and Microbial flora growing in the substrate focuses on removing inorganic matter such as nitrogen, ammonia, and so on; suspended solid content in the sewage is dealt with by vertical-flow wetland pond and is sharply reduced. Finally, it flows into slicing sink 9 of the vertical-flow wetland pond through the Permeable plate 27 on the wall of slicing sink 9 in the vertical-flow wetland pond 25, then via tap pipe 10, and next to reflux pond 12.

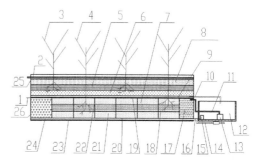

Figure 2.

Sewage pump 13 and diving aerator 15 are set in reflux pond 12. Sewage dealt with by vertical-flow wetland pond 25 influxes into reflux pond 12 through tap pipe10. The sewage in reflux pond 12 will be retransfused to the horizontal-flow undercurrent wetland pond 26 located below vertical-flow wetland pond 25 through sewage pump 13 and counter flow pipe 11. Flow speed is controlled by a valve, and horizontal-flow undercurrent wetland pond is located beneath the surface, which is beneficial to the heat preservation for the north in winter. Diving aerator 15 supplies oxygen for horizontal undercurrent wetland pond 26 by microspore aerator 16. Sewage first passes the gravel area 17 of outlet; then, under the guidance of guide water separator 22, it flows in the vertical and horizontal direction in the horizontal undercurrent wetland pond 26, which can increase evolutionary time and improve the efficiency of purification. Next, it inflows into the gravel area 30 of outlet at the end, affluxes into the slicing sink 24 of

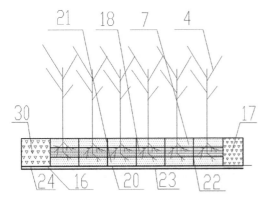

Figure 3.

horizontal undercurrent wetland pond via the porous plate below the gravel area 30, and is discharged through tap pipe 14. The discharged water can be recycled. When the sewage passes through the horizontal undercurrent wetland pond, organic pollutants, heavy metal pollutants, as well as inorganic pollutants will be reduced a lot.

Both sides of the horizontal-flow wetland pond are gravel areas: On one side is the inlet, on the other side is the outlet, and the unfilled area encircled by water guide plate at the gravel areas of the outlet is just filled with some gravels. Addition of some gravel to the outlet can effectively prevent congestion and drain water much better. A porous plate is set under the gravel area 30 of the outlet, so that water can flow into sluice channel 24, and water proofer 23 is set on the top of sluice channel 24 except for the permeable plate parts. From the top to the bottom, the substrate consists of coarse sand layer 7, artificial mixed layer 18, coarse sand layer 21, and artificial mixed layer 18. It blends with Zeolite, vermiculite, and coal cinder. The set diversion plate 22 divides water into S-shaped water channels, which can increase the treatment effect of sewage. In a horizontal undercurrent wetland pond, plant-microbiota ecosystem will be formed, a lot of superior micropopulations will appear in the Matrix, and root systems can further decompose organic pollutants and heavy metal pollutants relying on the microbial flora. Matrix and micro-porous aerator 16 are set in horizontal undercurrent wetland pond, micro-porous aerator 16 is porous network PVC air duct, and the aeration is conducted through micro-porous aerator 16 by diving aerator. Commonly used horizontal-flow wetland pond mainly relies on the plants translocating to compensate oxygen vertically, the amount of compensation is small, and vertical-flow layer itself in this system has already increased the vertical amount penetration of oxygen. Then, making use of reflux pool and a micro-aeration device, which can efficiently increase the oxygen content of the system, we

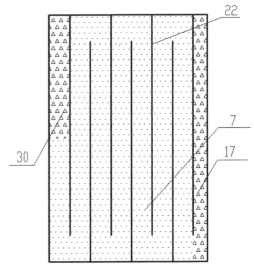

Figure 4.

improve the reaction purification efficiency of roots and microbial flora, efficiently absorb heavy metal pollutants, and decompose organic pollutants. When the sewage passes through subsurface wetland pond, organic pollutants, heavy metal pollutants, as well as the inorganic pollutants will be reduced a lot.

Shallow-rooted plants 3 are planted in vertical-flow wetland pond 25. The plant has functions

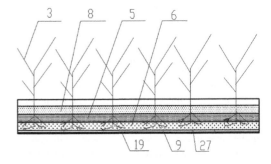

Figure 5.

of enrichment. Because the vertical-flow wetland pond is above the ground, most of the plants we choose are shallow-rooted plants. They have strong processing capability against inorganic pollutant and scenery effects such as rush, water lily, willow herb, and mosaic canna. Sewage flows into vertical-flow wetland pond 25 evenly through sewage pipe 2 and vertically downward through the coarse sand layer 8, artificial mixed layer 5, and gravel layer 6. Sewage adsorbs a part of heavy metal pollutants by the vertical infiltration layers of substrate, nitrification ability of roots, and Microbial flora growing in the substrate focuses on removing inorganic matter such as nitrogen, ammonia, and so on. Suspended solid content in the sewage dealt with by vertical-flow wetland pond sharply is reduced, and it finally flows into slicing sink 9 of the vertical-flow wetland pond through the Permeable plate 27 on the wall of slicing sink 9 in the vertical-flow wetland pond, then via tap pipe 10, and finally next to reflux pond 12. At the bottom of sluice channel, water proof level 19 is set to prevent the water at disposal from flowing into the horizontal-flow wetland pond directly. To ensure the growth of plants, it is necessary to preset cavity and flexible plastic water tight packing among plant stalks such as plastics that not only can prevent sewage from circulating among ponds but also can ensure the normal growth of plants without affecting efficiency. The porous disc touches coarse sand of a big size directly so that it is difficult to block and there is no need for special handling.

By the earlier engineering design, this system not only can conduct high-efficiency disposal of inorganic eutrophication elements in sewage but also can absorb organics and heavy metal pollutant in sewage sufficiently and effectively and improve sewage purification efficiency per unit area.

5 ANALYSIS OF PURIFICATION AND ECONOMY EFFICIENCY

In order to examine the actual operation effects of combined ecological sewage treatment and recycling system, we establish a laboratory pilot-plant system, as shown in Table 1. We extract swage with 6 times in Sunjiazhuang village to examine the feasibility of the verification system.

Disposing and analyzing experimental data of building analog system, we find that adopting the sewage treatment and recycling system based on combined constructed wetland can save 10% of establishing cost and reduce a half to one-third operation staff compared with traditional sewage treatment technology. In addition, by the sewage treatment and recycling system based on combined constructed wetland, power consumption will decrease to less than 0.1Kwh per ton; direct operation cost will decrease from 0.7–0.8 yuan to 0.2–0.3 yuan, and more than 80% of sewage can realize recycling after treatment. And it can save a large amount of cost in the long run.

6 CONCLUSION

Popularization of the sewage treatment system in rural areas not only can alleviate the current severe water pollution and water shortage in our country but also can improve living conditions in rural areas

Table 1. Experimental data of domestic sewage.

Domestic sewage Unit: (mg/L)		COD	BOD	SS	NH	TP
Domestic Sewage outlet collection and test	1	157	125	73.7	27.8	3.9
	2	210	170	96.4	37.2	5.8
	3	178	148	78.5	31.4	4.5
	4	190	159	87.4	32.8	4.9
	5	169	139	70.6	28.5	4.3
	6	240	198	117.6	41.4	6.2
Introduce treatment system, and examine collection tank after treatment	1	16	7.8	3.3	6.1	0.21
	2	24	12.3	4.9	9.1	0.31
	3	19	9.2	3.8	7.0	0.26
	4	21	9.8	4.2	7.3	0.28
	5	18	8.6	3.6	6.5	0.25
	6	32	14.3	6.7	11.7	0.41

to a great extent. The key is to research and develop a kind of sewage disposal system that is suitable in rural areas, which can gain better treatment results at a low cost. The ecological treatment of sewage at a low cost in rural areas will contribute a lot to realizing our country's goal of circular economy development; result in energy saving and emission reduction; and produce huge social benefit.

As a kind of ecological treatment technology against sewage disposal in rural areas, we think this is quite practical through designing, fabricating, arguing, and proving. Compared with traditional technologies, it gains better treatment effects and improves the utilization ratio of water resources. Meanwhile, it is much cheaper and easier to realize rather than new technology. Its characteristics of flexibility and convenience are quite favorable to mass market popularization and application.

REFERENCES

[1] Gao Tingyao, Gu Guowei, Zhou Qi Higher Education Press, Water Pollution Control Engineering (a&b),the Third Version.
[2] Ren Zhouyu, Chen Zhongzheng, Li Tianrong, City Water Supply and Drainage, China Building Industry Press, the Second Version.

[3] Zhou Qunying, Gao Yaoting, Environmental Engineering Microbiology,Higher Education Press, the Second Version.

[4] Zhou Qunying, Gao Yaoting, Environment Equipment-Principle & Design & Application, Higher Education Press, the Second Version.

[5] Li Keguo, Wei Guoyin, Zhang Baoan, Environment Economics, China Environmental Science Press.

[6] Zhou Lv, Economy and Cost Management of Environmental Protection Engineering Technology, Environmental Science and Engineering Center of Chemical Industry Press.

[7] Water Pollution Control Commitee of China Association of Environmental Protection, Industry development report of China's Sewage Treatment Industry in 2009[J] Water Industry Market, 2010.

[8] Xinxiang Reseach group in Ecological Environmental Research Center of Natural Sciences Chinese Academy of Sciences, Analysis of China Environmental Protection Industry and Investment Advisory Report[J].

[9] Office file of environmental protection leading group in hebei province (JiHuan LingBan [2009] No. 8).

[10] Consulting Report of Industry Chain Competition Trend of Sewage Treatment Equipment in 2010 and Integration Strategy of Corporate Iinternal and External Resources[R]. 2009.

[11] Integrated Pollutant Discharge Standard. On January 1, 1998.

[12] National Sewage Recycling - Water Quality for Industries.2005.

Energy, Environment and Green Building Materials – Sheng (ed.)
© *2015 Taylor & Francis Group, London, ISBN 978-1-138-02718-3*

Numerical studies of the oxidation process of Hg by ozone injection

Bo Li & Jin-Yang Zhao
Electric Power Planning & Engineering Institute, China

Jun-Fu Lv
Key Laboratory for Thermal Science and Power Engineering, Ministry of Education, Department of Thermal Engineering, Tsinghua University, China

ABSTRACT: The numerical simulations of oxidation process of $Hg/NO/O_3$ mixtures were conducted with CHEMKIN computer package. Plug Flow Reactor (PFR) computer code was used to predict the oxidation process of $Hg/NO/O_3$. A simplified kinetic mechanism that includes 11 species and 14 reactions was used here. The effects of molar ratio of O_3/NO and temperatures were assessed, and a further insight of the chemical controlling process was obtained by sensitivity analysis. The results revealed that the optimized oxidation conditions by ozone for Hg in the $Hg/NO/O_3$ mixtures are that molar ratios of O_3/NO are larger than 1.4 and temperature is higher than 150 °C. Furthermore, the sensitivity analysis showed that NO_3 is an important product to promote the oxidation process of Hg; NO_3 can help Hg transform into HgO via reaction $NO_3+Hg=HgO+NO_2$.

1 INTRODUCTION

Air pollution has become one of the most serious environmental problems all over the world. Fine particles, SO_2, NOx, and mercury are included in the flue gas of coal-fired power plants. As the emission standard of China is strengthened in recent years, more and more studies are focused on the new technologies on pollutant control in order to increase the overall efficiency of removal [1, 2].

Coal combustion is the main source of the atmospheric mercury emissions from human activities worldwide. China's new emission standard for thermal power plants (GB 13223-2011) includes a limit on the amount of mercury ($0.03mg/m^3$) that can be emitted from coal-fired power plants. Water-soluble oxidized species of mercury are more easily absorbed in scrubber systems. Wet oxidation method is an option for simultaneously removing two or more pollutants such as NO, SO_2, and Hg [3]. Various oxidation agents are used for oxidizing NO, SO_2, and Hg in this method, such as chlorine dioxide, sodium chlorite, hydrogen peroxide, and ozone. Ozone is a stable molecular and promising oxidant in practical applications. The mercury is able to be oxidized to mercury oxide, which can be then removed by wet scrubbing. O_3 oxidation has been previously investigated in many studies. Wang et al. [4] numerically investigated the oxidation characteristics of mercury by ozone under different temperature and mole fractions; results indicated that the O_3/NO mole fraction should be kept at 1.2 at least to ensure that all the Hg

are transformed into HgO in the oxidation process. Jiang et al. [5] developed a new mechanism to predict the oxidation behavior of mercury by ozone. In addition, other studies [6-7] indicated that the concentration of HCl in the gas mixtures significantly affects Hg conversion from Hg^0 to HgO.

To gain further insights of the effect of nitric oxide on the oxidation process of mercury by ozone, the numerical studies were conducted in this research. The effects of different temperature and molar ratios of O_3/NO on the conversion rate of Hg to HgO were considered; further insights of the chemical controlling process were obtained by sensitivity analysis; and the intermediate species in the oxidation process were also discussed.

2 SIMULATION STRATEGY

The kinetic mechanism used in this study was from Ref. [5], which includes 11 species and 14 reactions. The reactions and related parameters were listed in table 1.

The present kinetic calculations were performed using the CHEMKIN computer package. Plug Flow Reactor (PFR) computer code was used to predict the oxidation process of $Hg/NO/O_3$. The PFR model describes the steady state; the tube flow reactor that can be used for process design, optimization, and control. The mixing in the axial flow direction is ignored but perfect mixing in the directions transverse to this is taken into account in PFR models. Thermodynamic curve fits were obtained from the National Institute of Standards and Technology chemical species database [8] and CHEMKIN database [9].

Table 1. Key parameters in kinetic mechanism from Ref. [5].

No.	Reactions	A cm³/mol-sec	β	Ea kcal/mol
1	$O_3+NO=O_2+NO_2$	2.59E+12	0.0	3180
2	$O_3=O+O_2$	7.6E+12	0.0	24400
3	$O_3+O=2O_2$	4.82E+12	0.0	4094
4	$NO+O=NO_2$	5.62E+15	0.0	-1161
5	$NO_3+NO_3=2NO_2+O_2$	2.08E+13	1.0	13345.8
6	$NO_3+NO=NO_2+NO_2$	1.08E+13	0.0	-219.0
7	$O_3+Hg=O_2+HgO$	5.08E+07	0.0	2796.0
8	$O_3+NO_2=O_2+NO_3$	8.43E+10	0.0	4870
9	$O+O+M=O_2+M$	1.89E+13	0.0	-1788
10	$O+NO_2=NO_3$	2.9E+21	-2.0	0.00
11	$NO_3+O=NO_2+O_2$	6.03E+12	0.0	0.00
12	$NO_3+O_3=NO_2+2O_2$	2.08E+13	1.0	16458.3
13	$NO_2+NO_3=N_2O_5$	3.87E+11	0.2	0.00
14	$NO_3+Hg=HgO+NO_2$	1.49E+14	0.0	6249

3 RESULTS AND DISCUSSION

Figure 1 shows that the Hg conversion rate of molar ratio of O3/NO is 1 at different temperatures ranging from 100 °C to 250 °C. It can be seen from the figure that as the temperature is increased, the Hg conversion rate is increased. The conversion rate of Hg is 12.9% even at temperature T = 250 °C. Results of figure 1 indicate that at a molar ratio of O_3/NO, $(O_3/NO)_{\text{mole fraction}} = 1$, it is difficult for Hg to be transformed into oxidation state.

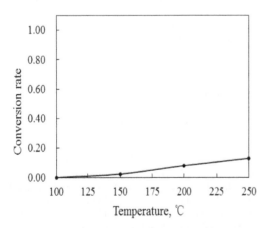

Figure 1. Hg conversion rate of molar ratio of O_3/NO is 1 at different temperatures.

Figure 2 shows that the Hg conversion rate of molar ratio of O_3/NO is 1.2 at different temperatures ranging from 100 °C to 250 °C. It can be seen from the figure that as the temperature is increased, the Hg conversion rate is increased. The conversion rate of Hg is 100% at T = 250 °C. The result of figure 2 shows that at a higher molar ratio of O_3/NO, the Hg conversion is higher; this conclusion can also be confirmed by Figs. 3 and 4.

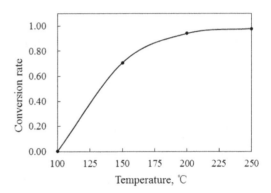

Figure 2. Hg conversion rate of molar ratio of O3/NO is 1.2 at different temperatures.

Figures 3 and 4 depict the Hg conversion rates of molar ratios of O_3/NO, which are 1.4 and 1.6, respectively, at different temperatures ranging from 100 °C to 250 °C. It can be seen from the two figures that as the temperature is increased, the Hg conversion rate is increased. The conversion rate of Hg is 100% at temperature T = 250 °C. Furthermore, in Figs. 3 and 4, the Hg conversion rates are nearly 100% at T = 150 °C; this temperature is also in the optimized temperature range of NO oxidation by O_3[5].

Based on the earlier discussion, at different $(O_3/NO)_{\text{mole fraction}}$s, Hg can be oxidized into HgO by O3; however, at lower $(O_3/NO)_{\text{mole fraction}}$, the oxidation effect of Hg is not obvious. However, at $(O_3/NO)_{\text{mole fraction}}$ is larger than 1.4, the oxidation rate is increased significantly as the temperature varies. Therefore, in the multi-pollutant control of coal-fired power plant, with regard to the applications of O_3, the $(O_3/NO)_{\text{mole fraction}}$ is suggested to be kept larger than 1.4. Hence, nearly all the Hg can be transformed into HgO at a temperature of 150 °C.

To obtain further insights of the chemical controlling process of the Hg oxidation process, the sensitivity analyses were conducted. Figure 5 shows the ranked logarithmic sensitivity coefficients of Hg formation. The simulation conditions are $(O_3/NO)_{\text{mole fraction}} = 1.6$, T = 150 °C. The negative value represents the Hg consumption, and the positive value represents Hg formation.

It can be seen from Fig. 5 that reaction 1, reaction 2, reaction 6, reaction 8, and reaction 14 are dominant in controlling the overall reactivity. The most important

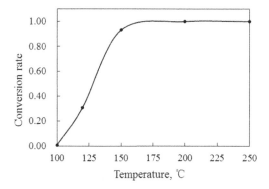

Figure 3. Hg conversion rate of molar ratio of O3/NO is 1.4 at different temperatures.

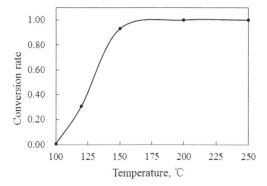

Figure 4. Hg conversion rate of molar ratio of O3/NO is 1.6 at different temperatures.

reaction for the Hg oxidation is reaction 14. It can be concluded that NO_3 is responsible for the oxidation of Hg. For $(O_3/NO)_{mole\ fraction}$s that are larger than 1.0, NO can be oxidized into NO_3 by the excess O3, and then NO_3 can help Hg transform into HgO via reaction 14.

Figure 6 shows the mole fractions of intermediate species as the variations of residence time. The simulation conditions are $(O_3/NO)_{mole\ fraction} = 1.6$, T = 150 °C. It can be seen from this picture that as the residence time is increased, O_3 and Hg are decreased. NO_2 is first increased and then decreased. NO3 starts being generated at 0.5 second; in the meanwhile, HgO starts being formed. At the end of the oxidation process, all the O_3 is consumed and Hg is transformed into HgO. The observation in Fig. 6 is consistent with Fig. 5, which confirmed that NO_3 is the important specie in the oxidation process of Hg in Hg/NO/O_3 mixtures. Therefore, $(O_3/NO)_{mole\ fraction}$s that are larger than 1 should be chosen in practical applications to ensure that all the Hg is transformed into HgO.

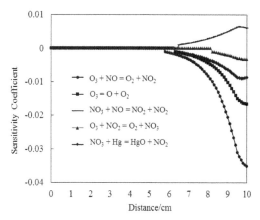

Figure 5. Ranked logarithmic sensitivity coefficients of Hg formation.

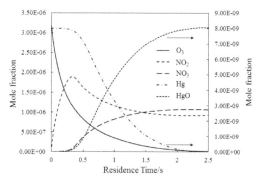

Figure 6. Mole fractions of intermediate species as the variations of residence time.

4 CONCLUSIONS

In this paper, the numerical simulations of oxidation process of Hg/NO/O_3 mixtures were conducted with CHEMKIN computer package. Plug Flow Reactor (PFR) computer code was used to predict the oxidation process of Hg/NO/O_3. A simplified kinetic mechanism that includes 11 species and 14 reactions was used here. The effects of molar ratio of O3/NO and temperatures were assessed; further insights of the chemical controlling process were obtained by sensitivity analysis.

The results indicated that at different $(O_3/NO)_{mole\ fraction}$s, Hg can be oxidized into HgO by O_3; however, at lower $(O_3/NO)_{mole\ fraction}$, the oxidation effect of Hg is not obvious. But at $(O_3/NO)_{mole\ fraction}$ is larger than 1.4, the oxidation rate increased significantly as the temperature varied. Therefore, in the multi-pollutant control of coal-fired power plant, with regard to the applications of O_3, the (O_3/NO) mole fraction is suggested to be kept at larger than 1.4. Hence, the Hg can be transformed into HgO at a temperature of 150 °C. $NO_3 + Hg = HgO + NO_2$ is a key reaction for the oxidation of Hg, and NO_3 can help Hg transform into HgO via this reaction.

REFERENCES

[1] E. H. Chui, H. Gao, A. J. Majeskiet et al., Performance improvement and reduction of emissions from coal-fired utility boilers in China, Energy for Sustainable Development, 2010, 14: 206–212.

[2] A. M. Carpenter, Advances in multi-pollutant control, IEA report, 2013.

[3] C. Sun, N. Zhao; Z.K. Zhuang et al. Mechanisms and reaction pathways for simultaneous oxidation of NOx and SO2 by ozone determined by in situ IR measurements, Journal of Hazardous Materials, 2014, 274: 376–383.

[4] Z. H. Wang, Mechanism study on multi-pollution control simultaneously during coal combustion and direct numerical simulation of reaction jets flow. Ph.D thesis, Zhejing, China, 2005.

[5] S. D. Jiang, Experimental and mechanism study on multi-pollutants control by ozone and active molecule. Ph.D thesis, Zhejing, China, 2010.

[6] L. S. Wei, J. H. Zhou, Z. H. Wang et al. Kinetic modelling of homogeneous low temperature multi-pollutant oxidation by ozone: The importance of SO and HCl in predicting oxidation. Journal of Zhejiang University SCIENCE A, 2006, 7: 335–339.

[7] M. H. Xu, Q. Yu, C. G. Zheng et al. Modeling of homogeneous mercury speciation using detailed chemical kinetics, Combustion and Flame, 2003, 132: 208–218.

[8] Mallard, W.G.; Westley, F.; Herron, J.T. et al. NIST Chemical Kinetics Database, 1998.

[9] Kee, R.J.; Rupley, F.M.; Miller, J.A. et al. CHEMKIN-PRO Release 15083, Reaction Design, Inc, SanDiego, CA, 2008.

Energy, Environment and Green Building Materials – Sheng (ed.)
© 2015 Taylor & Francis Group, London, ISBN 978-1-138-02718-3

Studies on image processing technology of chalkiness detection in early indica rice

P. Cheng, Q. Liu, Y. Su & D.S. Cheng
College of Biological Science and Technology, Hunan Agricultural University, China

ABSTRACT: A computer image processing system Chalkiness2.0 has been developed for detection of chalkiness characteristics of early indica rice, which includes chalkiness degree, shape of grain, ratio of chalky grain, and so on. In this process, based on iterative adaptive threshold segmentation and polar processing, a single kernel image from the mass grain images is extracted using gray transformation and adaptive median filter for denoise, and then the parameters of chalkiness are examined. The results show that with only once image processing, all the related chalkiness indexes can be exported automatically. The outputs are more than 97% correlated with those of the manual detection, thus chalkiness of the early indica rice can be detected effectively.

KEYWORDS: Early indica rice; Chalkiness; Image Processing; Rapid Detection

1 INTRODUCTION

Detection of grain quality factors has been one of the hot topics in recent years, and how to detect grain quality factors quickly and accurately is becoming more and more important (Gao Y et al.2009). Grain quality factors include many indicators, and higher chalkiness is one of the major limitation factors of early indica rice quality (Zhou LJ et al.2009). The traditional method for grain chalkiness detection, which includes tedious steps, uses manual sampling and measurement and visual observation, and it is slow, labor intensive, subjective, and poorly consistent. As a result, three-dimension image processing system and video microscopy were preliminarily applied in the method of grain chalkiness detection (Hou CY et al. 2001, Lei DY et al. 2008, Zeng DL et al. 2001), and these methods have significantly reduced the subjectivity and increased the objectivity. However, due to the damage in grain detection and the high costs, it is necessary to establish a new objective, as well as cheap and accurate chalkiness detection system.

2 MATERIALS

In this experiment, three varieties of early indica rice, including FB, F11, and V20B, were used as materials. One hundred rice grains were randomly selected from these varieties, respectively, and placed on the scanner board. The preview area was adjusted to the appropriate range, and then the images were obtained by scanning and read for image processing.

3 METHODS

3.1 Preprocessing of images

Image preprocessing is especially important, because later work will not be able to start if this is not properly handled. The purpose of preprocessing is to decrease the stigma, noise, and various factors interfering in grain images, improve image quality, and ensure the accuracy of subsequent analysis. Thus, it will be helpful to image feature extraction and pattern identification after preprocessing (Shi ZX et al. 2008).

3.1.1 Transformation of grayscale

Because the original image was 24-bit true color images, which included red, green, and blue color channels, it will be converted to a grayscale image using algebraic sum of the basic color value in order to reduce the calculation amount. The conversion formula is as follows:

$$Y = 0.299 \times R + 0.587 \times G + 0.114 \times B \qquad (1)$$

Pixel gray value of the grayscale image has characters of discontinuity and similarity; that is, the image has grayscale similarity within the grain area but grayscale discontinuity at the boundary area. Therefore, the system will first find out all the connected components of the gray images, and then assign the same connected components with the same mark, but different areas with different marks to distinguish the grain identification.

3.1.2 *Denoising of images*

The adaptive median filter algorithm was applied for the image denoising (Toprak A et al.2006). First, noises were detected from the image of all areas according to the noise pollution situation, then the size of filter window was determined, and the detected noise points were filtered accordingly. Inside the window, the pixel grayscale values $x(i,j)$ were selected to sort out the several specific directions, and each of the median values was weighted(W_k) after sorting. Then, the pixel grayscale values of the window center were replaced by operation results using a formula (2) expressed. The experimental results showed that outlines of images were clearer after filtering, and granular noise was well suppressed. As a result, the local details of grain characteristics were well retained.

$$y(i,j) = \sum_{i=i-1}^{i+1} \sum_{j=j-1}^{j+1} w_k x(i,j) \tag{2}$$

3.1.3 *Segmentation of images*

The purpose of image segmentation is to separate the effective information from the target for subsequent analysis and processing (Zhang L et al.2010). The specificity of each grain was identified because of the individual characteristic differences, the values of feature-point thresholds were integrated, and finally an average value was taken as a new threshold. The image was divided into foreground and background according to the new threshold function switch. After several times of repeats, the function switch would not change and converge to a stable threshold; then the stable threshold was employed as the final threshold and used for the image segmentation.

3.1.4 *Extraction of single grain kernel image*

A single grain kernel image was extracted from grain images of binarization for the subsequent processing. In this process, the scanning method and the eight-neighbor connectivity method were employed to mark the grain kernel area(Chen BS et al.2006). After the grain binary image area was marked, the area of grain was segmented from binary image and the positional information of grain outlines (row, column) was reflected in the original grayscale image. Then, a single grain kernel image was segmented by logical operation in groups of grain images (Fig 1).

Figure 1. Single grain kernel image.

3.2 *Detection of chalkiness degree*

Grain kernel chalkiness, which is the opaque portion in grain endosperm, can be divided into white belly, white core, and white back according to its location. For detection of chalkiness degree, the background area was removed by the system image preprocessing; then, the shape of grain area, polar coordinates of the boundary point, and the initial direction were calculated. Next, the chalky area was extracted from the grain area after second threshold processing conducted on them. The chalky magnitude P_1 and P_2 within the two fan areas on both sides of $[50°, 160°]$, $r \in [0, r \times 9/10]$ in the initial directions were calculated using formula (3), and the maximum value will be taken as grain chalky magnitude P. If it is chalky grain, the value is relatively large, but if it is standard grain, the value is smaller. Whether it is chalky grain or not, it is determined by the chalky magnitude threshold value *ThdP* (set up by the experimental data).

$$\begin{cases} x_{ij} = r_i \times \cos\theta_j + C_x \\ y_{ij} = r_i \times \sin\theta_j + C_y \end{cases} \tag{3}$$

$r_i = 1,2,3,\ldots,[r]$ ("[]": integer);
$\theta_j = \theta_1, (\theta_1 + 1°), (\theta_1 + 2°), \ldots, \theta_2$

3.3 *Detection of chalky grain rate*

For detection of chalky grain rate, the binary image was obtained by the system threshold processing and calculated by open operation of mathematics to eliminate granular noise from the binary image. After the area was marked and the total number of grains was calculated, the area marked images and chalky area images extracted were calculated and operated. Then, the number of chalky grains was determined according to the marked chalky area. At last, the chalky grain rate was calculated as follows: the number of chalky grains divided by the total number of grains. The flow chart of program for chalkiness analysis system (Chalkiness2.0) is shown in figure 2.

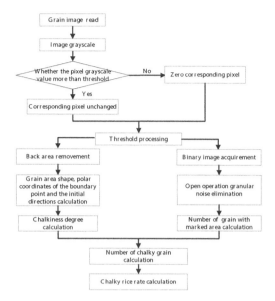

Figure 2.   Flow chart of program for chalkiness analysis.

3.4 *Detection of grain shape*

The ratio of grain length to width, which is one of the characters of grain shape, is shown in Fig. 3. Because grain images are not axisymmetric, the length between the two farthest points is taken as the grain length and the straight line between these two points is not necessarily the buttock line of the grain. Therefore, the distance from each side of the outline to the grain length line is not equal. If a perpendicular line were drawn to the grain length line, the line segments will be crossed with grain outlines on two points. After calculating the distance from each points to the grain length line, respectively, the grain width could be obtained and thus the ratio of grain length to width could be calculated.

Then, the system will automatically find out the maximum distance between the two points in grain outlines; their coordinates are recorded as $(x_1, y_1),(x_2, y_2)$; and the distance between the two points is taken as grain length d_1:

$$d_1 = \sqrt{(x_2 - x_1)^2 + (y_2 - y_1)^2} \qquad (4)$$

The equation of the grain length line is $Ax+By+C=0$, and the distance from each side of grain outline to grain length line is calculated, respectively, using point to straight line distance formula:

$$r = \frac{|Ax_0 + By_0 + C|}{\sqrt{A^2 + B^2}} \qquad (5)$$

Thus, the value of r_1 and r_2 can be calculated and the grain width $d_2=r_1+r_2$

4   RESULTS AND ANALYSIS

In order to compare the computer image processing method with the manual detection method in chalkiness characters analysis, three varieties (i.e. FengYuan B, MiYang 46 and V20B) of rice were randomly sampled and analyzed, respectively, by five experienced chalkiness analysts according to the standard of GB/T17891-1999. Simultaneously, the same three varieties of rice were randomly sampled and analyzed, respectively, by a scanner and computer image processing system Chalkiness2.0 using the set parameters. The experiments were repeated thrice, and the results are shown in table 1.

Figure 3.   Length and width of grain.

The experimental results show the condition when FengYuan B was used as a sample to adjust and optimize the parameters of Chalkiness2.0; the obtained values of chalkiness character are consistent with the reference values of the national standard. Then, Chalkiness2.0 with the optimized parameters was used for chalky detection of MiYang 46 and V20B; the results of each index were also consistent and reliable. Moreover, on comparing with the results of the manual detection method, the ratio of length to width, chalky grain rate, chalkiness degree, coefficient variation, and standard deviation were smaller with good repeatability. Although the values of chalky indexes were also close to the reference values when the manual detection method was applied, their standard deviation and coefficient variation were significantly higher than those of the computer image detection system, within which the variation coefficient of chalky grain rate and chalkiness degree were even twice more than the corresponding results of the image processing method (Chalkiness2.0). These experimental data suggest that the image processing method is more accurate and repeatable in chalkiness characters analysis than the traditional visual observation method.

5 DISCUSSION

It was shown that through the proper threshold selected, the similarity of the results of grain chalkiness degree, chalky grain rate, and grain shape calculated by the image processing system Chalkiness2.0 is more than 97% to that of the manual detection. As a result, the system has a strong advantage in the mass screening of breeding material, germplasm resources evaluation, and genetic group analysis because of its accuracy, consistency, and reliability. According to the analogy between chalkiness and RLA, a method for measuring the relative lesion area on plant leaves (RLA) using Chalkiness1.0 was proposed by ZhengYan (Zheng Y et al.2008); this suggests that the advanced present system (Chalkiness2.0) can be applied to other relative studies and has a promising prospect.

The preliminary analysis of chalkiness character of different varieties showed that the results detected by Chalkiness2.0 were affected by grain quantity, grain type, and grain separation degree. Thus, further verification for different types of rice varieties and adjustment of the corresponding algorithm is needed. It is also important to establish standard samples in order to achieve detection consistency in all kinds of environment. Meanwhile, this system needs to be further improved in three aspects: (1) Combining the computer vision technology with neural network technology, comprehensive results should be obtained after the weight of grain indexes is increased; thus, the quality of grain appearance is classified (Yin YG et al.2009). (2) The corresponding rice database should be established according to the characteristics of regions and variety in order to promote the evaluation of rice quality both scientifically and automatically. (3) The detection accuracy of dynamic images should be improved to meet the demand of online detection.

Table 1. Comparison of chalkiness character between computer image processing and manual detection.

Detect method	Statistic parameter	FB			F11			V20B		
		Chalky grain rate	Chalkiness degree	Length /width ratio	Chalky grain rate	Chalkiness degree	Length / width ratio	Chalky grain rate	Chalkiness degree	Length /width ratio
Computer image method	Mean	89.37	27.03	3.20	83.79	23.12	2.14	54.13	15.42	1.82
	SD	1.65	0.75	0.05	1.85	0.84	0.04	2.39	1.20	0.08
	CV	1.87	2.66	1.55	2.24	3.62	1.82	4.60	7.78	4.36
Manual method	Mean	92.92	28.18	3.67	85.64	23.88	2.28	57.88	16.56	1.85
	SD	2.61	1.87	0.16	3.24	1.75	0.14	4.44	2.22	0.14
	CV	2.79	6.74	3.48	3.77	7.33	4.45	7.62	13.7	7.40

(SD:Standard Deviation, CV:Coefficient Variation).

REFERENCES

Gao Y, Zhou JP, Dai QG, Jiang N. 2009. Consideration and strategy about methodology and technology in measurement of rice quality factors.*Journal of the Chinese Cereals and Oils Association*. 24(12):158–161.

Zhou LJ, Jiang L, Zhai HQ, Wan JM. 2009. Current status and strategies for improvement of rice grain chalkiness. *Hereditas*. 31(6):563–572.

Hou CY, Seiichi O, Yasuhisa S, Yoshinori K, Toru T, Kenichi K,Toshiro H, Gabsoo D. 2001. Application of 3D-microslicing image processing system in rice quality evaluation. *Transaction of the CSAE*. 5(3):92–95.

Lei DY, Xie FM, Xu JL, Chen LY. 2008. QTL mapping and epistasis analysis for grain shape and chalkiness degree of rice.*Chinese J Rice Sci*. 22(3):255–260.

Zeng DL, Teng S, Qian Q, Kunihiro Y. 2001. The application of video microscopy in rice chalkiness study. *Scientica Agricultura Sinica*. 34 (4):451–453.

Shi ZX, Cheng H, Li JT, Feng J. 2008. Characteristic parameters to identify varieties of corn seeds by image processing. *Transaction of the CSAE,*. 24(6):193–195.

Toprak A, Guler I. 2006. Suppression of impulse noise in medical images with the use of fuzzy adaptive median filter. *Journal of Medical Systems*. 30(6):465–471.

Zhang L, Ji Q. 2010. Image segmentation with a unified graphical model. *IEEE Trans Pattern Anal Mach Intell*. 32(8):1406–1425.

Chen BS. 2006. A new algorithm for binary connected components labeling.*Computer Engineering and Applications*. 42(25):46–47.

Zheng Y, Wu WR. 2008. A method for measuring relative lesion area on leaves using a rice chalkiness ratio analysis software. *Scientia Agricultura Sinica*. 41(10):3405–3409.

Yin YG, Ding Y. 2009. Rapid method for enumeration of total viable bacteria in vegetables based on computer vision. *Transaction of the CSAE*. 25(7): 249–254.

Energy, Environment and Green Building Materials – Sheng (ed.)
© *2015 Taylor & Francis Group, London, ISBN 978-1-138-02718-3*

Properties and coke deactivation of Pt-Sn-Na/ZSM-5 catalysts for propane dehydrogenation to propylene

C. X. Wang & Z.Q. Wang

College of Chemical Engineering, Changchun University of Technology, China

ABSTRACT: The Pt-Sn-Na/ZSM-5 catalysts were prepared by the sequential impregnation method with different contents of Na for propane dehydrogenation to propylene. By using XRD, H_2-TPR, and NH_3-TPD, the influence of Na and Sn on the Pt(Sn) / ZSM-5 catalysts was studied with regard to their physicochemical properties. The L acid centers on the catalyst surface may promote the yield of propylene, and the equivalent metal Sn and Na oxides of Pt^{4+} and Pt^{2+} take part together in the reaction process. The catalyst structure and dispersion of metals have not been destroyed before and after the reaction. The addition of Sn and Na can decrease the coke in the catalysts.

KEYWORDS: Propane ; Propene ; Pt-Sn-Na/ZSM-5; Coke deactivation; Properties; Characterization

1 INTRODUCTION

Propylene is an important basic organic chemical raw material for the production of polypropylene, acrylonitrile, butanol, octanol, propylene oxide, and isopropyl alcohol, and the demand of the downstream products for propylene is growing [1]. How the abundance of the inexpensive propane is utilized and converted into the shortage and high value-added propylene in the market through the catalytic dehydrogenation is a problem and currently a focus of research[2].

In this paper, the catalysts are prepared through continuous impregnation method and the content of Na is changed to study the effect of additives and their actions both on each other and on the catalysts, so as to obtain a new kind of catalyst loaded on HZSM-5 zeolite and make it have a higher activity, selectivity, and stability, which will be applied to the propane dehydrogenation in industry.

2 EXPERIMENTAL SECTION

2.1 *Reagents*

$H_2PtCl_6 \cdot 6H_2O$ (Guoyao Chemical Reagent Co., AR), $SnCl_4 \cdot 5H_2O$ (Xilong Chemical Co., AR), NaOH (Tianjin Chemical Reagent Co.AR), HZSM-5 zeolite of Si/Al = 45 (specific surface area ≥200m²/ g, bulk density 0.147g / mL, crowded bar, Nankai University Chemical Plant). Propane gas (Dalian Date Gas Co., 8- liter capacity, purity 99.95%). The reactions were finished in an MRT-M0102BG micro-catalytic reactor made in Spaceflight Century Star Technology Co., LTD. (Beijing). The reaction conditions were that catalyst loading was 2.0g, pressure was 0.1MPa, temperature was 590°C, molar ratio of H_2 to propane was 0.05, and space velocity of gas was 3.0h⁻¹.

2.2 *Catalyst preparation*

The catalysts were prepared by the sequential impregnation method. The preparation procedure of the three-metal catalysts was that H_2 $PtCl_6 \cdot 6H_2O$ and NaCl were dissolved in deionized water, respectively, and $SnCl_4 \cdot 5H_2O$ was dissolved in ethanol, in order to obtain the three kinds of solution. The ZSM-5 zeolite was first impregnated into the sodium ion solution with the same volumes for 12h, evaporated to dryness in a water bath at 80°C, and dried overnight at 120°C. Then, the samples were heated at a 10°C/ min rate to 500°C and calcined in air for 4h.The catalysts were obtained with the mass fraction of Na being 0, 0.5, 1.0, 1.5, and 2.0%, respectively.

2.3 *Catalyst characterization*

Temperature-programmed reduction (TPR) was measured with the catalyst characterization apparatus made in PengXiang Technology Co., LTD (Tianjin). Before the TPR experiments, the catalysts were dried in flowing N_2 at 500°C for 1h. 5% H_2/N_2 was used as the reducing gas at a flow rate of 40 mL/min. The rate of temperature rise in the TPR experiment was 10°C/ min till 800°C.

Temperature-programmed desorption (TPD) was measured with the same apparatus as that of H₂-TPR. TPD measurements were carried out using ammonia as a sorbent. About 0.15g of catalysts was placed in a quartz reaction tube under N_2 purge , and the heating rate was 10°C/ min to 500°C; then, the surface of the sample was cleaned by N_2 and it was cooled down to room temperature.

3 RESULTS AND DISCUSSION

3.1 *Catalytic performance*

Figure 1 depicts the catalytic properties of the Pt-Sn-Na/ZSM-5 catalysts for the dehydrogenation of propane. As can be seen from Fig. 1, the propane conversion decreases from initial 0.34 to 0.23 and the selectivity of propylene increases from 0.82 to 0.95 with the catalyst-used time being within 500min.

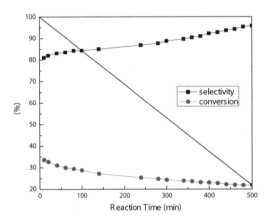

Figure 1. Catalytic performance of Pt-Sn-Na (1.0%)/ ZSM-5 catalyst for propane dehydrogenation.

3.2 *H₂-TPR of catalysts*

Figure 2 shows that the Pt/ZSM-5 catalyst has two reduction peaks that are strong at 424°C and weak at 212°C, which are consistent with the literature data[3]. After adding Sn, the peak at 212°C moves to the higher temperature and becomes stronger whereas the peak at 424°C also moves to the higher temperature and becomes weaker. After adding Na, the two peaks move first to the higher temperature and then slightly return. When the Na content is 1.0%, the two peaks at 250°C and 420°C, respectively, are alike in the area[4]. This indicates that there are two kinds of metal oxide and they possess equivalent active points, which interact with each other and promote the conversion of propane to propylene on the surface of the catalyst.

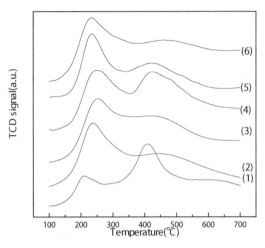

Figure 2.   H₂-TPR profiles of the different catalysts.

(1) Pt/ZSM-5; (2) Pt-Sn/ZSM-5; (3) Pt-Sn-Na(0.5%)/ ZSM-5;(4)Pt-Sn-Na(1.0%)/ZSM-5;(5)Pt-Sn-Na(1.5%)/ ZSM-5 (6) Pt-Sn-Na(2.0%)/ZSM-5

3.3   *NH₃-TPD of catalysts*

Acidity of the catalysts can significantly affect their dehydrogenation performance. The acidity of catalysts can be measured by NH₃-TPD. NH₃-TPD spectrum of the catalysts is shown in Fig. 3. The Pt/ZSM-5 catalysts possess two peaks: one at 198°C, which is strong, and another at 446°C, which is weak[5]. The addition of Sn makes the high temperature peak become weaker, whose position does not change. When the loading amount of adding Na is from 0.5% to 2.0%, the amount of strong acid on the catalyst surface successively decreases and moves to the lower temperature, which makes the two peaks combine into a large peak at the low temperature and become stronger.

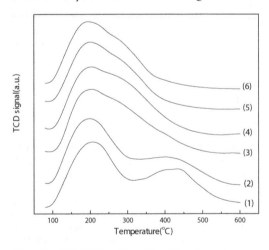

Figure 3.   NH₃-TPD profiles of the different samples.

(1) Pt/ZSM-5; (2) Pt-Sn/ZSM-5; (3) Pt-Sn-Na (0.5%)/ZSM-5;(4)Pt-Sn-Na(1.0%)/ZSM-5;(5)Pt-Sn-Na(1.5%)/ZSM-5(6); Pt-Sn-Na(2.0%)/ZSM-5

on the catalyst surface do not change before and after the reactions, and the catalyst structure has not been destroyed.

4 CONCLUSION

Under the optimum conditions (the mass of catalyst = 2.0g, pressure = 0.1MPa, temperature = 590°C, mole ratio of H_2 to propane = 0.05 , the gas space velocity = $3.0h^{-1}$), the conversion of propane obtained was 27.96% and the selectivity of propylene was 95.31% over the catalyst Pt0.5%-Sn1.0%- Na1.0%/ZSM-5.

The results show that the active ingredient and additives can be highly dispersed on the carrier surface of ZSM-5; the addition of Sn and Na may not only maintain the activity of catalysts but also increase the selectivity of the main product. The L acid centers of the catalyst surface may promote the yield of propylene, and the equivalent metal Sn and Na oxides of Pt^{4+} and Pt^{2+} take part together in the reaction process of propane dehydrogenation to propene. The Pt particles

REFERENCES

[1] Yu Jiao propylene market supply and demand status quo and development trend [J] contemporary petrochemical, 2004, 12 (9):35–38.
[2] Feng Jing, Zhang Mingsen, Keli, Zhang Yuan, a propane dehydrogenation catalyst research progress [J] Industrial Catalysis, 2011, 19 (3):8–13.
[3] Lin Li-Wu, Yang Wei Shen, Wu Rongan.PtSn-Al2O3 supported catalyst propane dehydrogenation performance improvement [J] petrochemicals, 1992, 21: 511–515.
[4] TENG Jia Wei, Zhao Guoliang, Xie in the library, Yang Weimin increasing propylene olefin catalytic cracking catalyst [J] petrochemicals, 2004, 33 (2): 100–103.
[5] You Tingxiu , propane dehydrogenation propylene oxidation kinetics of reactions in the macro-jun [J] by the natural gas industry, 2004, 29:51–56.

Energy, Environment and Green Building Materials – Sheng (ed.)
© 2015 Taylor & Francis Group, London, ISBN 978-1-138-02718-3

Turbidity variation in a detention pond after storing runoff

C. Tang & T. Y. Guo

Department of Soil and Water Conservation, National Pingtung University of Science and Technology, Taiwan

ABSTRACT: The Pond of Flood Detention and Sedimentation (PFDS) could be created into an ecological pond, and its spatial variation of aqua quality should be analyzed for the reference to soil and water conservation of practice and maintenance. The experimental site was a flood detention pond and was located in Jing-Se Lake, the National Pingtung University of Science and Technology, Taiwan. The water temperature and turbidity were measured from January 29, 2013 to April 2, 2014. Overall, 24 investigations had been performed for an analysis. The results show that the turbidity difference near the outlet was less than the inlet in PFDS with 3.5-65.2 FTU. The flow velocity of PFDS became slower without rainfall; then, the turbidity range and mean value were between 15 to 55.6 FTU and 32.5 FTU; in the meanwhile, the aquatics migrated outlet of pond could increase turbidity. The turbidity where the water depth was more than 1 meter would appear four times of the normal situation.

KEYWORDS: Flood detention pond; Water temperature; Turbidity

1 INTRODUCTION

The ground surface runoff during the storm should be retained temporarily in a detention pond for the slopeland stability (Huang, 2012). When the flood detention pond could reduce the runoff discharge and peak flow and restrict sediment from soil erosion, then water quality of runoff could be clarified (Huang, 2001). The flood detention pond can be constructed into a steady and suitable ecological environment. The lowest limitation for the ecological pond of the aquatic organisms was water quality. The soil erosion would be carried by runoff and into the detention pond; it could change the water quality. All further research and analysis was necessary. The turbidity of storage runoff was the most apparent change, and it was adopted as the index for water quality. Thus, it should be feasible for the detention pond.

The turbidity was a reliable index for the aquatic environment, water resource conservation, and waterworks operation (USGS, 2003). The observational station of turbidity was commonly used to collect a long period of accumulated data (Liu *et al.*, 2008). Besides, turbidity indicated the sediment concentration of public water utility, which was used for the apparent linear correlation (Hsu *et al.*, 2007). The different river basin caused the difference between turbidity and sediment concentration, and the turbidity could not comprehensively interpret the aquatic quality, regardless of whether it had already been an initial assessment or not (Randerson *et al.*, 2005).

The soil, organic matters, and mineral substance will flow with runoff into the detention pond, and they will then directly increase the turbidity. The time and spatial distribution of the rainfall would influence the amount of soil erosion and the distribution of non-point source pollution; hence, rainfall would affect the value of the turbidity (Lo *et al.*, 2010). It meant there was an apparent correlation between rainfall and turbidity (Liao *et al.*, 2011). Simultaneously, the difference of geological environment and the rainfall pattern in distinct areas would also result in the variation of the sediment concentrations (Lin, 2002). The measurement of the turbidity in the pond would proceed with before and after the rainfall, and then we could realize the turbidity variation in the pond of flood detention and sedimentation.

2 MATERIALS AND METHODS

The experimental site was located in front of Jing-Se Lake, the comprehensive district of soil and water conservation of National Pingtung University of Science and Technology, Taiwan. A levee was set in one of the five distances along the east-west axis of the lake area as well as a trapezoid pond of flood detention and sedimentation (PFDS), which would be constructed with 2500m^2 enclosed area. The PFDS could delay the peak flow after accumulated surface runoff from upstream of the watershed. The mean and deepest depths of pond were 0.6m and 1m. The PFDS constructed a porous circumstance for biological growth, the shore had planted terrestrial and waterfront plants to increase levee stability, and the bottom base was paved in fine texture soil. This pond could delay and store the peak flow, and in order to retain the water level it had provided supplements from the aquaculture pond's effluent water. All the turbidity observation of droughty and rainy seasons was carried out in 24 times investigation and included eight times of full-day turbidity observation from January 29, 2013 to April 2, 2014; it had twenty available data to analyze.

To realize the spatial turbidity variation in the PFDS, sixteen observational sites were set with a mesh distance of 10m and two observational sites in the inlet and outlet. The water turbidity sensor (JFE, ACLW-USB, NO.249, Advantech Co., Japan) was used, and the unit adopted FTU (ISO7027, Formazine Turbidity Unit), which could be measured by the scattering proportion of direct light with suspending particle in water. The simple floating raft could consist of six plastic 2L-bottles with 30 cm height, and it set the turbidity sensor in the central tube of raft into the water. This would sink at a 20-cm depth and prevent shift from disturbing the bottom base. There was a small hill close to the western side of the pond; it could directly decrease solar radiation and affect the turbidity of the pond after noon. Then, the measuring period was decided from 3pm to 6pm. Besides, turbidity of the inlet, outlet, and middle side would precede continuous observations during the whole day.

3 RESULTS AND DISCUSSION

First, the result of the turbidity in different sampling points in the pond of flood detention and sedimentation (PFDS) was summarized in Table 1. To compare the numerical difference of turbidity of inlet (θ_i) and outlet (θ_o) in Table 1, we could see that the condition of increasing and decreasing tendencies was 6 times at both sampling sites. The θ_i greater than θ_o was about 1.8~24.8 FTU in increasing tendency, and the θ_i less than θ_o was about 3.5-65.2 FTU in decreasing tendency.

Preliminarily inferred PFDS had the ability to decrease the turbidity of storage water. The θ_i was mainly greater than θ_o had appeared from February to March, for the effluent from the aquaculture pond was used in the supplement to keep the dead water level in the pond during droughty season, and there were many nutritive substances that easily aggregated the activities of aquatic organisms at the inflow area. The relate turbidity observations could be completed in a few days after the rain event during rainy season, and the θ_i could still be less than θ_o, about 2.1–11.5 FTU, from May to July. This meant that the rainfall would directly disturb the water turbidity of PFDS. On comparing the turbidity difference of different sampling sites, we could find out that the region with lower turbidity appeared at sampling sites 2, 3, and 4, which were near the north side of the pond. There was no sheltering effect by the taller canopy; it could decrease aquatic activities and then beneficially decreased the water turbidity. In addition, the region with higher turbidity appeared at sampling sites 5, 11, and 13. Especially based on the observation, one could infer that the region around the sampling site 5 frequently passed by waterfowl; it even appeared that the highest turbidity was equal to 118.3 FTU during the observational period. The sampling site 11 was near the central region of PFDS, where the water depth was quite shallow, and the alluvial sedimentation could often exist, resulting in higher turbidity. The sampling site 13 was close to the inlet of PFDS since the region was apparently influenced by inflow, and that of turbidity tendency would be increased. Further, the mean turbidity at sampling sites 8, 11, 12, 14, 15, and 16 located at the eastern and southern regions of the pond was more than 30 FTU; those with a higher turbidity would be induced by shelter of a taller canopy and more biological activities. Statistical results of turbidity had shown that the range and average were 15–55.6 FTU and 32.5 FTU from January to May, 12.3–32.8 FTU during November, and 19.2 FTU from July to September, respectively.

Twelve isopleths had used turbidity at different sampling sites for interpreting spatial variation of PFDS (Figure 1). As shown in Figure 1a, the lack of rainfall in January only recharged the effluent water from the aquaculture pond to remain the dead water level. Besides, the observational time was after 15:30; the shelter effect by hilly topography at the west side

Table 1. Data of turbidity measurement in the pond of flood detention and sedimentation (Jan. 29, 2013–Apr. 2, 2014).

Site	Date												Average
	2013											2014	
	Jan 29	Feb 4	Feb 19	Feb 25	Mar 22	May 7	May 31	Jul 15	Sep 4	Sep 29	Nov 3	Apr 2	
1	16.8	13.5	35.3	34.3	---	45.3	37.8	12.9	10.6	30.1	13.7	13.7	24.0
2	12.7	16.0	33.7	31.5	47.5	29.0	36.8	13.2	10.8	26.9	15.0	14.6	24.0
3	15.4	16.9	32.2	27.7	28.5	32.6	23.4	13.1	9.8	18.8	20.1	16.3	21.2
4	20.9	24.5	31.8	34.1	29.1	38.7	30.2	14.1	10.2	18.1	24.3	19.0	24.6
5	14.9	18.9	---	38.6	24.3	40.8	---	11.8	14.6	44.7	11.9	13.8	23.4
6	17.9	17.4	31.5	40.9	39.4	39.5	23.2	12.5	21.1	37.4	14.8	17.8	26.1
7	12.3	17.6	30.8	52.1	33.7	41.5	34.9	13.7	12.5	33.6	18.0	19.4	26.7
8	---	25.6	48.6	80.2	54.6	51.1	65.8	9.9	12.9	55.0	16.4	16.2	39.7
9	16.5	23.9	37.1	54.5	49.1	52.4	25.2	19.3	14.5	18.3	15.1	14.9	28.4
10	17.5	18.0	40.4	52.6	51.3	44.7	26.8	20.8	11.8	18.1	15.9	14.0	27.7
11	14.4	18.7	36.6	61.8	---	35.2	52.2	---	12.2	35.3	17.4	12.4	29.6
12	17.8	19.6	37.6	71.2	57.6	51.4	86.0	9.5	13.6	44.8	22.5	15.3	37.2
13	20.5	17.1	24.9	76.4	40.5	49.3	---	11.8	11.3	27.0	33.8	12.7	29.2
14	22.4	19.9	27.5	94.1	67.2	47.8	56.9	9.0	10.1	35.8	11.7	12.1	34.5
15	20.6	19.0	30.7	86.0	49.0	48.2	73.3	12.4	10.8	41.7	13.4	14.1	34.9
16	13.1	18.5	34.1	78.2	59.1	48.9	69.1	10.0	10.8	---	13.9	15.1	33.7
Inlet	15.4	15.3	---	30.6	---	40.0	51.8	8.3	15.9	42.1	18.0	13.7	25.1
Outlet	15.3	18.0	38.8	55.4	36.2	51.5	54.7	10.3	11.9	30.4	14.5	15.5	29.4
θ	16.7	18.8	34.5	55.6	44.5	43.8	46.8	12.3	12.5	32.8	17.2	15.0	Σ 28.9
Tw	24.5	28.1	28.4	27.5	30.2	29.7	33.2	32.9	30.3	31.3	27.6	25.5	

---:No data; θ: Mean turbidity; Tw: Water temperature

of the pond could decrease the solar radiation and cause the weak eastern downhill air flow. A similar tendency was also shown in Figure 1b and 1c. It indicated that the turbidity in the east region of the pond was higher than that of the west region, which was about 5–20 FTU.

The increasing tendency of turbidity was shown in Figure 1d, and that turbidity increased more than 4 times without rainfall. This phenomenon might be due to the great quantity of discharge with many nutritive substances that would drain into the pond. The activity of aquatic organisms gathered around in the southern and south-east side along the water flow,

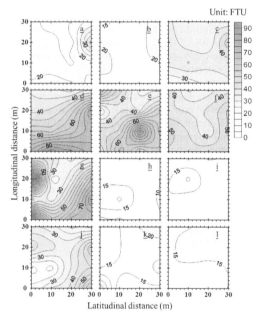

Figure 1. Spatial distribution of turbidity in the pond of flood detention and sedimentation for observational period at (a) 01/29/13, 15:36-16:36, (b) 02/4/13, 15:35-16:35, (c) 02/19/13, 15:20-16:20, (d) 02/25/13, 15:50-16:50, (e) 03/22/13, 16:49-17:49, (f)05/07/13, 16:16-17:16, (g) 05/31/13, 15:38-16:38, (h) 07/15/13, 16:07-17:07; (i) 09/04/13, 15:12-16:12; (j) 09/29/13, 15:41-16:41; (k) 11/3/13, 15:26-16:26; (l) 04/02/14, 15:19-16:19.

and this could make the turbidity possess an increasing tendency. A similar situation was also seen in Figure 1f and 1g. From both figures, we could see that the turbidity of the northwest region appeared at 115 FTU, which was quite a higher situation during the observational period, and by contrast we found that the waterfowl frequently passed for hunting aquatic organisms. From Figure 1e, water depth at the southeast region was deeper, and the turbidity was more

than 90 FTU. The turbidity value in the outlet region was twice higher than the inlet region, since the practical observation could illustrate that the fish in the pond had started reproducing in March, and this would disturb the sediment at the bottom of the pond. The turbidity of every sampling site in Figure 1h, 1i, 1k, and 1l had shown all the values to be less than 30 FTU. In the meanwhile, this could reach the lowest drinking water standard, and it meant the storage water was clear enough. On comparison with weather conditions for the four observations, we could find that the rainfall occurred before less than 12 hours to start the observation, and there were 16 mm, 12 mm, 16.5 mm, and 1.4 mm rainfall for each observation. It indicated that the outside surface runoff would flow into the pond, increased the water level and flow velocity, and was beneficial to remove the suspended matter in the water. Finally, in Figure 1j, it was shown that the spatial variation of turbidity was similar to Figure 1d. Since this observation was performed before 5 days of the sampling date, the 81.5 mm rainfall caused by the typhoon circulation had discharged runoff into pond, and it disturbed the sediment in the pond. The increasing tendency of turbidity appeared.

According to Figure 1, three regions with higher turbidity in the PFDS are selected, including middle side, outlet, and inlet. Based on eight serial turbidity observations and contrast of water temperature variation with diurnal cycle, three observations were determined on January 17, February 17, and March 4, 2014 to analyze and illustrate the daily turbidity variation at the middle side, outlet region, and inlet region of PDFS (Figure 2).

The first daily turbidity observation was made on January 17–18, 2014. The effluent water of aquaculture pond flow into the middle site of PFDS would decrease its velocity by the larger receiving section (Figure 2a). The turbidity maintained 10 FTU from 18:00 on January 17 to 3:00 on January 18, 2014, and the turbidity decreased to 8 FTU on the second day from 3:00 to 6:00. The decrease in value might be a result of lower inflow from the aquaculture pond. The turbidity showed the increasing tendency with 3–4 FTU, due to the water temperature being only 18°C; this was associated with less activities of organisms in the pond. The turbidity increased with 9-41 FTU on the second day at 15:00, and the water temperature reached 24°C, which could result in larger biological activities in the PFDS. The turbidity variation was related with inflow amount for the remaining dead water level of PFDS. The lower water temperature could decrease biological activities and turbidity. In the second observation (Figure 2b), the water temperature near the outlet region was around 22–28 °C from the next 24 hours of the sampling day starting from 18:00 on February 17, 2014. The water depth in the outlet region was more than 1m, in which

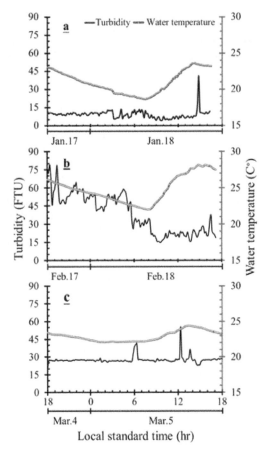

Figure 2. Daily variation of turbidity and water temperature in middle site, outlet, and inlet at the pond of flood detention and sedimentation on (a) Jan.17, (b) Feb.17, and (c) Mar. 4, 2014.

a relatively higher thermal capacity induced the higher water temperature. Hence, the turbidity decreased whenever the water temperature was lower. The water temperature and turbidity reached the lowest value on the next day at 7:00 and 9:00. It also showed that turbidity decreased to a certain value, and the water temperature would lag to decrease for 2 hours. The turbidity remained in the range of 15–30 FTU, and the PFDS would be limpid due to the decrease of effluent water from the aquaculture pond in the day time. Besides, the turbidity range was 45–80 FTU and the muddy water was caused by the activities of aquatic organisms from 18:00 to 19:00 on the sampling day. It indicated that the water depth of the outlet region was more than 1m and was beneficial for biological activities in the pond; then, the resulted turbidity was four times higher than the normally steady situation. In the third observation (Figure 2c) on March 4, 2014, the

water temperature near the inlet region was 22–25°C, which was close to the normal water temperature, and the increasing ranges of turbidity were 15, 30, and 7 FTU at 6:00, 12:00, and 13:30, respectively. The phenomenon might be due to the more foraging by aquatic organisms at 6:00. The turbidity was significantly increased by the addition of feeding stuff and decreasing the amount of outflow from the aquaculture pond. This could induce the biological aggregation for moving to the shallower inlet region.

4 CONCLUSIONS

Summarizing the results and discussion, we could obtain the following conclusions:

The turbidity of outlet region was less than inlet region with 3.5–65.2 FTU range in the pond of flood detention and sediment. No rainfall would decrease flow velocity in the pond, and the more aquatic organisms were removed to the outlet region of the pond; then, the turbidity at those regions increased. The heavy rain could result in a great amount of runoff, and it flowed into the pond, disturbing the alluvial substances. Then along the flow direction reaching the outlet for turbidity, variation would be increased along with its distance. Maintaining the dead water level in the pond would change the water temperature, activity of aquatic organisms, and turbidity. The region with a 1m water depth was beneficial for the activity of aquatic organisms, and turbidity was four times greater than that of a usually steady situation.

REFERENCES

Hsu, Y.S., J.F. Cai, C.M. Wei & H. P.Huang. 2007. Rating relations between turbidity and suspended solids concentration of reservoir sediment -a case study of Shi-Men reservoir. *Journal of Chinese Agricultural Engineering.* 53(1):62–71.

Huang, C.C. 2001. A study of the effectiveness of rainwater catchments on flood control in slope land area—A case study in Hsichih. *Master thesis of the Graduate school of Earth Science, Chinese Culture University.*

Liao, C.S., C.L. Chang, S.L. Lo, C.Y. Hu & J. L. Ma. 2011. The correlation analysis of Taiwan's main water treatment plants between upstream of the water turbidity and hydrological characteristics. *Journal of Taiwan Agricultural Engineering,* 57(2):78–86.

Liu, C. H., Y.S. Hsu, M.C. Lee & S.M. Lin. 2008. Characteristics of discharge, suspended sediment concentration, and turbidity at the Gao-Ping river weir. *Journal of Chinese Soil and Water Conservation,* 39(3):345–353. (in Chinese with English abstract)

Lin, Y.S. 2002. Relationship between heavy rainfall and water turbidity in the watershed of Shihmen Reservoir.

Master thesis of the Department of Applied Geology, National Central University.

Lo, S.L., C.C. Sung, C.S. Lai, C.Y. Hu, C.L. Chang & Y.C. Huang. 2010. Establishment of emergency response information management system for public water supply and technique for high turbidity raw water treatment (2/2). *Water Resources Agency, Ministry of Economic Affairs Subsection, National Taiwan University execution.*

US Geological Survey. 2003. National field manual for the collection of water-quality data: US Geological Survey Techniques of Water-Resources Investigations. *Book 9. Chaps.* A1–A9.

Randerson, T.J., J.C. Fink, K.J. Fermanich, P. Baumgart & T. Ehlinger. 2005. Total Suspended Solids-Turbidity Correlation in Northeastern Wisconsin Streams. AWRA–Wisconsin Section Meeting.

Energy, Environment and Green Building Materials – Sheng (ed.)
© 2015 Taylor & Francis Group, London, ISBN 978-1-138-02718-3

Experimental study on kitchen garbage hydrolysis conditions

G.S. Cheng & Y. Zhao

National Engineering Laboratory for Biomass Power Generation Equipment, North China Electric Power University, Beijing, China

C.Y. Luo

School of Geography and Life Sciences, Bijie University, Bijie, China

ABSTRACT: Kitchen garbage is a rich-nutrition resource and it is suitable for saccharification and fermentation. The glucoamylase and amylase were used to hydrolyze kitchen garbage in order to improve the hydrolysis efficiency of kitchen garbage. The experimental results show that the reducing sugar content of the kitchen garbage is greatly improved by hydrolysis with glucoamylase and amylase. The best processes are as follows: enzymes activities ratio (7.5:1), pH (4.5), temperature (55°C), and time (5h). The lipase is helpful to the further hydrolysis of the kitchen garbage, and the crude fat content decreases greatly when lipase is added to the kitchen garbage. The reducing sugar content increases by 5.17%, and the crude fat content decreases by 63.96%.

KEYWORDS: Kitchen Garbage; Hydrolysis; Fermentation; Reducing Sugar

1 INTRODUCTION

Kitchen garbage is a rich-nutrition resource due to its composition of sugar, lipids, proteins, cellulose, and other compounds. But using proper treatment methods for kitchen garbage is very important because improper treatment methods will bring pollution and become a potential hazard affecting the food safety and ecological safety [1–3]. Kitchen garbage treatment technology can be divided into non-biological treatment and biological treatment technology according to the current processing media. Non-biological treatment technology mainly refers to the traditional waste treatment methods, such as incineration [4], landfill [5–6], and so on. Biological treatment technology mainly includes the anaerobic digestion [7–9] and aerobic composting [10–11]. Biological treatment technology, such as saccharification and fermentation, is a more promising technology that provides resource utilization efficiency than non-biological treatment technology, but it needs to be made more effective.

This study is an important part of the development and utilization of kitchen garbage recycling. The main objective was to identify and select suitable hydrolysis reaction conditions of kitchen garbage; make complete use of various nutrients (cellulose, starch, crude fat, crude protein, and so on) through a series of hydrolysis experiments with glucoamylase and amylase. Reducing sugar content was measured by 3,5-dinitrosalicylic acid method and compared under different hydrolysis reaction conditions and with composite enzymes. It will provide the theoretical

basis for kitchen garbage hydrogen fermentation, methane fermentation, alcohol fermentation, and lactic acid fermentation process, and provide the basis for harmless kitchen garbage, resource recycling, and reduction.

2 MATERIALS AND METHODS

The garbage sample was collected from the second student public canteen of Bijie University. The kitchen garbage sample was stored in a refrigerator (the temperature is under -18 °C) after preliminary sorting and crushing. The compositions of kitchen garbage were analyzed by Anthrone colorimetric method, 3,5-dinitrosalicylic acid method, and Soxhlet distill method. The contents of total sugar, reducing sugar, and crude fat are 14.72%, 2.56%, and 8.74%, respectively. The analysis results show that carbohydrate content is high and there are sugars, starch, and crude fat in the kitchen garbage. The kitchen garbage is a rich-nutrition resource and is suitable for further fermentation production. The glucoamylase, amylase, and lipase used to hydrolyze kitchen garbage were bought from Zhangjiagang Jin Yuan Biological Chemical Co., Ltd and Heshibi Biotechnology Co., Ltd. The composite enzymes activities of glucoamylase, amylase, and lipase are 50000u/g, 10000u/g, and 10000u/g, respectively.

The kitchen garbage hydrolysis samples were prepared through the following steps: 1) the 100g kitchen garbage was put out of the refrigerator and put into the experimental container; 2) acid (3.0%HCl) was added to the kitchen garbage and the pH of kitchen garbage was regulated; 3) the enzymes were added to the container to hydrolyze the kitchen garbage; and 4) the 4% NaOH was added to the container to terminate the biochemical reaction after a period.

3 RESULTS AND ANALYSIS

3.1 Effect of pH on kitchen garbage hydrolysis

The initial pH values of 100g kitchen garbage samples are regulated to 3.5, 4.0, 4.5, 5.0, and 5.5, respectively. The hydrolysis experiments of the 100g kitchen garbage samples after regulating pH were performed to investigate the effect of pH on kitchen garbage hydrolysis under the conditions of temperature (55°C) for 5h with glucoamylase (1500U). The experimental results are shown in Fig. 1.

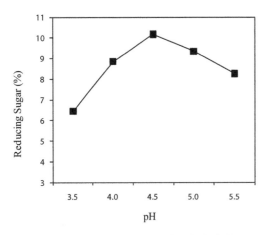

Figure 1.   Effect of pH on kitchen garbage hydrolysis.

It can be seen from Fig. 1 that with the increase of initial pH in the kitchen garbage sample, the reducing sugar increased at the beginning and decreased when the initial pH value exceeded 4.5. The kitchen garbage hydrolysis efficiency is best when the initial pH value of kitchen garbage is 4.5 than the hydrolysis efficiency under other initial pH. From the earlier experimental results, we can draw the conclusion that the glucoamylase activity is the highest when the initial pH value of kitchen garbage is 4.5.

3.2 Effects of glucoamylase dosage on kitchen garbage hydrolysis

The 100g kitchen garbage samples were used to hydrolyze under the conditions of initial pH value (4.5), temperature (55°C), time (5h), and glucoamylase dosage (500U, 1000U, 1500U, 1800U, and 2000U). The experimental results are shown in Fig. 2.

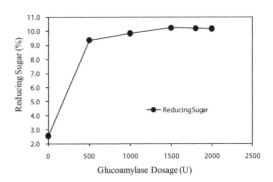

Figure 2.   Effect of glucoamylase dosage on kitchen garbage hydrolysis.

It can be seen from Fig. 2 that the kitchen garbage hydrolysis efficiency is better under the conditions of initial pH (4.5), temperature (55°C), time (5h), and glucoamylase dosage (1500U). The reducing sugar content reaches 10.25% under the condition of glucoamylase dosage (1500U), higher than 10.21%, 10.18% under the condition of glucoamylase dosage of 1800U and 2000U, respectively. A very high glucoamylase dosage is not good for the kitchen garbage hydrolysis, which is that the complex reaction of glucose may be induced when the glucoamylase dosage is overdose.

3.3 Effect of amylase dosage on kitchen garbage hydrolysis

The hydrolysis experiments of 100g kitchen garbage samples were performed to investigate the effect of amylase dosage on kitchen garbage hydrolysis under the conditions of initial pH (4.5), temperature (55~) for 5h, glucoamylase dosage (1500U), and amylase dosage (60U, 120U, 200U, 230U, and 250U), respectively. The kitchen garbage hydrolysis efficiency with glucoamylase and amylase is compared with the kitchen garbage hydrolysis efficiency while using only glucoamylase. The experimental results are shown in Fig. 3 and Fig. 4.

It can be seen from Fig. 3 and Fig. 4 that with the increase of amylase dosage, the reducing sugar content increased at the beginning, but when the amylase dosage was overdose, the reducing sugar content slightly decreased. The kitchen garbage hydrolysis efficiency is better with glucoamylase dosage (1500U) and amylase dosage (200U) under the conditions of initial pH (4.5), temperature (55°C) for 5h. The reducing sugar content reaches 11.28%

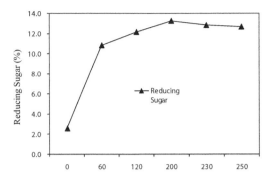

Figure 3. Effect of amylase dosage on kitchen garbage hydrolysis.

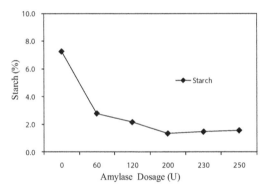

Figure 4. Effect of amylase dosage on starch hydrolysis.

under the condition of amylase dosage (200U), higher than 10.84%, 10.59% under the condition of amylase dosage (230U, 250U), respectively. The very high amylase dosage is not good for the kitchen garbage hydrolysis, which may be that the complex reaction of glucose will be induced when the amylase dosage is overdose. The amylase dosage (200U) is appropriate for 100g kitchen garbage hydrolysis. The experimental results from Fig. 4 also show that the starch content decreases greatly with amylase, and the content of starch in kitchen garbage decreases from 7.26% to 1.35%.

3.4 Effect of lipase dosage on kitchen garbage hydrolysis

The hydrolysis experiments of 100g kitchen garbage samples were performed to investigate the effect of lipase dosage on kitchen garbage hydrolysis under the conditions of initial pH (4.5), temperature (55°C) for 5h, with glucoamylase dosage (1500U) and lipase dosage (800U, 1600U, 2500U, 2800U, and

3000U), respectively. The kitchen garbage hydrolysis efficiency with glucoamylase and lipase is compared with that of using only glucoamylase. The experimental results are shown in Fig. 5.

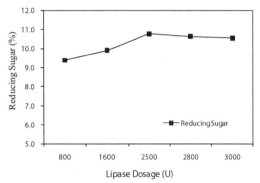

Figure 5. Effect of lipase dosage on kitchen garbage hydrolysis.

It can be seen from Fig. 5 that the kitchen garbage hydrolysis efficiency with lipase is better under the conditions of initial pH (4.5), temperature (55°C) for 5h, and glucoamylase dosage (1500U). The reducing sugar content reaches 10.78% under the condition of lipase dosage (2500U), higher than 10.64%, 10.55% under the condition of lipase dosage (2800U, 3000U), respectively. The very high lipase dosage is not good for the kitchen garbage hydrolysis, which may be that the complex reaction of glucose will be induced when the kitchen garbage is hydrolyzed to a certain extent. The lipase dosage of 2500U is appropriate for 100g kitchen garbage hydrolysis. The experimental results show that the lipase is helpful to the further hydrolysis of the kitchen garbage, and the crude fat content decreases greatly when lipase is added to the kitchen garbage. The reducing sugar content increases by 5.17% (from 10.25% to 10.78%), and the crude fat content decreases by 63.96% (from 8.74% to 3.15%).

4 CONCLUSIONS

The components of kitchen garbage were analyzed and the glucoamylase, amylase, and lipase were used to hydrolyze kitchen garbage. The experimental results show that carbohydrate content is high and there are cellulose, starch, and crude fat in the kitchen garbage. The kitchen garbage is suitable for further fermentation production. The optimum conditions of kitchen garbage are as follows: enzymes activities ratio (7.5:1) (glucoamylase: amylase), temperature (55°C), and pH (4.5) for 5h. The lipase is helpful to the further hydrolysis of the kitchen garbage, the crude fat content decreases greatly when lipase is added to the kitchen garbage, the reducing sugar content increases by 5.17%, and the crude fat content decreases by 63.96%.

ACKNOWLEDGMENTS

This work was financially supported by the National Science and Technology Support Program (2013BAG25B03), the Fundamental Research Funds for the Central Universities (12MS42, 12QN12), Bijie cycle Economic Research Special Fund (ZK011), Social Development Technology Research Plan of Guizhou Province (SY20113077), and Beijing Higher Education Young Elite Teacher Project (YETP0712).

REFERENCES

[1] Wenjun Zhao, Xiaohong Sun, Qunhui Wang, et al. Lactic acid recovery from fermentation broth of kitchen garbage by esterification and hydrolysis method. Biomass and Bioenergy, 2009, 33(1):21–25.

[2] Guishi Cheng,Qunhui Wang,Xiaohong Sun, et al. Experimental study on concentration of ammonium lactate solution from kitchen garbage fermentation broth by two-compartment electrodialysis. Separation and Purification Technology, 2008, 62(1):205–211.

[3] Aikaterini Ioannis Vavouraki, Evangelos Michael Angelis, Michael Kornaros. Optimization of ther-mo-chemical hydrolysis of kitchen wastes. Waste Management, 2013, 33(3):740–745.

[4] Zaili Zhang, Ling Zhang, Yaliang Zhou, et al. Pilot-scale operation of enhanced anaerobic digestion of nutrient-deficient municipal sludge by ultrasonic pretreatment and co-digestion of kitchen garbage. Journal of Environmental Chemical Engineering, 2013, 1(1-2):73–78.

[5] Yongjin Park, Feng Hong, Jihoon Cheon, et al. Comparison of thermophilic anaerobic digestion characteristics between single-phase and two-phase systems for kitchen garbage treatment. Journal of Bioscience and Bioengineering, 2008, 105(1):48–54.

[6] Wen-Tien Tsai. Management considerations and environmental benefit analysis for turning food garbage into agricultural resources. Bioresource Technology, 2008, 99(13):5309–5316.

[7] Hongzhi Ma, Qunhui Wang, Wenyu Zhang, et al. Optimization of the Medium and Process Parameters for Ethanol Production from Kitchen Garbage by Zymomonas Mobilis[J]. International Journal of Green Energy, 2008, 5(6):480–490.

[8] Hironori Yabu, Chikako Sakai, Tomoko Fujiwara, et al. Thermophilic two-stage dry anaerobic digestion of model garbage with ammonia stripping. Journal of Bioscience and Bioengineering, 2011, 111(3):312–319.

[9] Ahmed Tawfik, Mohamed El-Qelish. Key factors affecting on bio-hydrogen production from co-digestion of organic fraction of municipal solid waste and kitchen wastewater. Bioresource Technology, 2014, 168:106–111.

[10] Zhentong Li, Hongwei Lu, Lixia Ren, et al. Experimental and modeling approaches for food waste composting: A review. Chemosphere, 2013, 93(7):1247–1257.

[11] S.Kanazawa, Y. Ishikawa, K. Tomita-Yokotani, et al. Space agriculture for habitation on Mars with hyper-thermophilic aerobic composting bacteria. Advances in Space Research, 2008, 41(5):696–700.

Energy, Environment and Green Building Materials – Sheng (ed.)
© *2015 Taylor & Francis Group, London, ISBN 978-1-138-02718-3*

Study of uniaxial constitutive model of concrete in code and development in ABAQUS

H. Qi, Y.G. Li, X.M. Chen, J. Duan & J.Y. Sun
Technical Center, China State Construction Engineering Corporation Ltd., Beijing, China

ABSTRACT: A damage uniaxial model of concrete is given in "Code for design of concrete structures (GB50010-2010)," which is developed in ABAQUS via UMAT (user material). In this paper, the compression skeleton curve of the model in new code is studied and compared with Mander model and Guo-Zhang model, and the damage development of the model is studied in a new code. The complete cycle curve is not deduced from the model in the new code, such as the unloading curve after tension loading and the tension loading after compression loading. In this paper, the model in the new code is replenished to improve all the loading paths.

KEYWORDS: Code for design of concrete structures; Uniaxial constitutive model of concrete; ABAQUS; UMAT

1 INTRODUCTION

Concrete uniaxial constitutive model is deduced from the code (GB50010–2010) in the form of constitutive damage; however, the complete cycle curve is not deduced with the missing part of the unloading curve after tension loading and tension loading after compression loading. The whole cycle curve is completed in this article based on damage theory, and the completed constitutive model is developed in ABAQUS using subprogram UMAT. The damage development rule is studied, and the difference between the skeleton curve of concrete uniaxial constitutive model in the code and the skeleton curve of concrete uniaxial constitutive model in the old code is compared. In a word, it is meaningful for the reorganization and application of the constitutive model in the code through the study and improvement of concrete uniaxial constitutive model.

It is found (Qi et.al 2011) that the calculation result of Mander model and Guo-Zhang model meets well with the experimental result, and in this paper, the constitutive model in the code is compared with the two classical models in particular.

2 CONCRETE UNIAXIAL CONSTITUTIVE MODEL

2.1 Concrete constitutive model of code

The uniaxial tension stress–strain curve of concrete (GB50010–2010) is as follows:

$$\sigma = (1-d_t)E_c\varepsilon$$

$$d_t = \begin{cases} 1 - \rho_t[1.2 - 0.2x^5] & x \le 1 \\ 1 - \dfrac{\rho_t}{\alpha_t(x-1)^{1.7} + x} & x > 1 \end{cases}$$

$$x = \frac{\varepsilon}{\varepsilon_{t,r}}$$

$$\rho_t = \frac{f_{t,r}}{E_c\varepsilon_{t,r}}$$

The uniaxial compression stress–strain curve of concrete (GB50010–2010) is as follows:

$$\sigma = (1\text{-}d_c)E_c\varepsilon$$

$$d_c = \begin{cases} 1 - \dfrac{\rho_c n}{n-1+x^n} & x \le 1 \\ 1 - \dfrac{\rho_c}{\alpha_c(x-1)^2 + x} & x > 1 \end{cases}$$

$$x = \frac{\varepsilon}{\varepsilon_{c,r}}$$

$$n = \frac{E_c\varepsilon_{c,r}}{E_c\varepsilon_{c,r} - f_{c,r}}$$

$$\rho_c = \frac{f_{c,r}}{E_c\varepsilon_{c,r}}$$

Under cycling loading, the unloading and re-loading stress path of compression concrete is determined by the following equation:

$$\sigma = \mathbf{E}_r\left(\varepsilon - \varepsilon_z\right)$$

$$\mathbf{E}_r = \frac{\sigma_{un}}{\varepsilon_{un} - \varepsilon_z}$$

$$\varepsilon_z = \varepsilon_{un} - \frac{(\varepsilon_{un} + \varepsilon_{ca})\sigma_{un}}{\sigma_{un} + E_c\varepsilon_{ca}}$$

$$\varepsilon_{ca} = \max\{\frac{\varepsilon_c}{\varepsilon_c + \varepsilon_{un}}, \frac{0.09\varepsilon_{un}}{\varepsilon_c}\} \cdot \sqrt{\varepsilon_{un}\varepsilon_c}$$

2.2 Mander tension stress–strain curve (Mander et.al 1988)

$$\sigma = \frac{f_{cc}xr}{r - 1 + x^r}$$

in which: $x = \dfrac{\varepsilon}{\varepsilon_{cc}}$, $r = \dfrac{E_0}{E_0 - E_{sec}}$.

The relationship between unloading point strain and residual plastic strain is (Mander et.al 1988) as follows:

$$\varepsilon_z = \varepsilon_{un} - \frac{(\varepsilon_{un} + \varepsilon_{ca})\sigma_{un}}{\sigma_{un} + E_c\varepsilon_{ca}}$$

$$\varepsilon_{ca} = \max\{\frac{\varepsilon_c}{\varepsilon_c + \varepsilon_{un}}, \frac{0.09\varepsilon_{un}}{\varepsilon_c}\} \cdot \sqrt{\varepsilon_{un}\varepsilon_c}$$

2.3 Guo-Zhang model (Guo and Shi, 1982)

$$y = \begin{cases} \alpha_{ac}x + (3 - 2\alpha_{ac}) \cdot x^2 + (\alpha_{ac} - 2) \cdot x^3 & x \leq 1.0 \\ \dfrac{x}{\alpha_{dc}(x-1)^2 + x} & x > 1.0 \end{cases}$$

where $x = \varepsilon / \varepsilon_{cc}$; $y = \sigma / f_{cc}$; f_{cc} is confined concrete compressive strength; $f_{cc} = (1 + 0.5\lambda_v)f_c$; and ε_{cc} is confined concrete peek compressive strain, $\varepsilon_{cc} = (1 + 2.5\lambda_v)\varepsilon_c$.

Guo-Zhang model: The relationship between unloading point strain and residual plastic strain is as follows[7]:

$$\frac{\varepsilon_{pl}}{\varepsilon_0} = 0.31 \cdot (\frac{\varepsilon_{un}}{\varepsilon_0})^{1.6}$$

3 DEVELOPMENT AND STUDY OF CONSTITUTIVE MODEL OF CODE

The constitutive model of code is re-developed in ABAQUS, the calculation result of which is compared with Guo-Zhang model; the result is as follows (the conclusion is suitable to all kinds of strength of concrete material, C35 is taken into calculation in this paper):

3.1 Study of compression skeleton curve

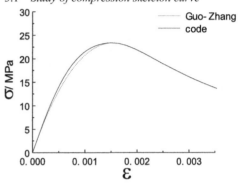

Figure 1. Comparison of Guo-Zhang model and model of code.

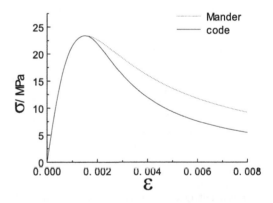

Figure 2. Comparison of Manderand model of code.

It can be seen that both the ascent stage and descent stage are different from Mander model. The ascent stage of model of code is in accordance with Mander model, and the descent stage is in accordance with Guo-Zhang model.

3.2 Calculation comparison of compression unloading

It can be seen from what has been said earlier that the calculation result of plastic strain of code model is in accordance with Mander model, and as to the

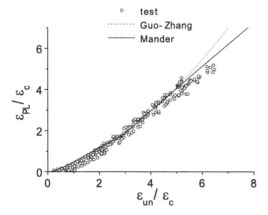

Figure 3. Comparison of Guo-Zhang model and model of code.

hoop-confined component [Mander, 1988; Guo and Zhang, 1982], the calculation result and experimental result of plastic strain of Guo-Zhang model and Mander model are compared as follows:

The relationship between residual plastic strain and unloading strain of Mander model and Guo-Zhang model is as shown in Fig. 3, from which it can be seen that when $\varepsilon_{un}/\varepsilon_0 < 6$ the models are not quite different from each other, and they fit the experimental result (Guo and Zhang, 1982) well, when $\varepsilon_{un}/\varepsilon_0 > 6$ Mander model fits the experiment result better.

3.3 Study of the damage development of constitutive model in the code

1) Study of compression damage development

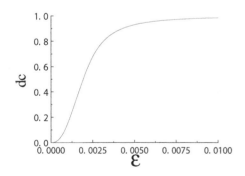

Figure 4. Compression damage.

From the figure, it can be seen that the compression damage develops from 0 to 1 slowly at the beginning, then quickly later, and turning slowly when close to 1, which fits the damage theory.

2) Tension damage development

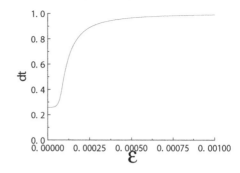

Figure 5. Tension damage.

From Figure 5, it can be seen that the tension damage development rule is in accordance with compression damage, but the model is of some tension damage at the beginning; in other words, there is initial tension damage of the material, which can be also deduced from equation d_t. When the compression strain is 0, it can be deduced from chapter 1.1 that

$$d_t = 1.2\rho_t = 1.2\frac{f_{t,r}}{E_c\varepsilon_{t,r}} \neq 0$$

4 HYSTERESIS RULE COMPLEMENT

4.1 Tension unloading hysteresis rule complement

Both Mander model and Guo-Zhang model are short of model tension hysteresis curve; the code just introduces the tension skeleton curve without the tension unloading curve or the tension loading after the compression loading curve. The Ten-Zhou model is considered in paper 4. However, in this paper, it is believed that the hysteresis rule of Ten-Zhou model is too complicated, and needs to be simplified in the engineering application. A very complicated consideration may reduce the practicability of the constitutive model, in view that the tension performance is far weaker than the compression performance and slightly affects the result. In this paper, it is considered that the tension may not produce plastic deformation, based on which the model tension part of the code is completed; the result is shown in Fig. 6.

4.2 Completion of hysteresis rule of tension loading after compression loading

It is considered in this paper that, when the concrete is on tension loading after compression loading, the former tension and compression damage affect the

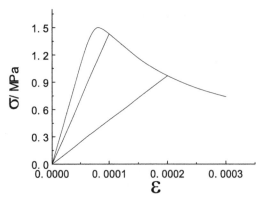

Figure 6. Tension hysteresis curve.

constitutive tension, based on which the tension damage equation is put forward as follows:

$$d_{t'}=d_t+\left[(1-E_r/E_c+d_{t0}\times w)/(1+d_{t0}\times w)\right]\times(1-d_t),$$

In which d_t is the tension damage of the tension loading, $1-E_r/E_c$ is the compression damage of compression stiffness degradation, d_{t0} is the former tension damage, and w is the parameter.

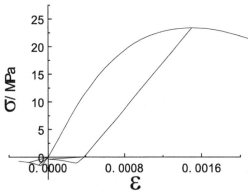

a) Calculation result of the model

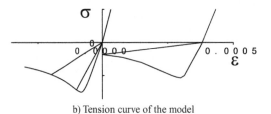

b) Tension curve of the model

Figure 7. Tension hysteresis curve of the model.

Taking $w=0.2$, and taking the earlier equation as the equation of the tension skeleton curve of the code, the calculation result of tension loading after compression loading of the model is shown in Figure 7.

From Figure 7, we can see that when there is tension loading after compression loading, the initial stiffness is smaller than the unloading stiffness, owing to the effect of former tension stiffness degradation.

5 CONCLUSIONS

The concrete uniaxial constitutive model is developed in ABAQUS, the tension and compression skeleton curve are studied, and the damage development rule is concluded as follows:

1 The ascent stage of the compression skeleton curve of the code is in accordance with Mander model;
2 The descent stage of the tension skeleton curve of the code is in accordance with Guo-Zhang model;
3 The plastic deformation equation of the code is in accordance with Mander model, and it fits well with the experimental result of the hoop-confined reinforced concrete components;
4 The initial tension damage value of the code is not 0, and there is initial damage of the concrete material;
5 The constitutive model of the code is completed based on damage theory, which covers all the uniaxial loading paths.

REFERENCES

GB50010–2010, 2010. Code for design of concrete structures. *Beijing: China Architecture & Building. Press.*
Hu Qi, Yungui Li, Xilin Lv. 2011. A practical confined concrete constitutive model under uniaxial hysteresis load. *Engineering Mechanics, 28(9): 95–102.*
Mander, J.B., Priestly, M.J.N., Park, R. 1988. Theoretical Stress—Strain Model for Confined Concrete. *Journal of Structural Engineering, ASCE, 114(8): 1804–1826.*
GUO Zhenhai, ZHANG Xiuqin. 1982. Experimental study on complete stress2strain curves of concrete. *J Building Structures, 3(1): 112. (in Chinese)*
Guo Zhenhai, Shi Xudong. 1982. Reinforced concrete theory and analyse. *beijin: Tsinghua University Press, 2003.*

Energy, Environment and Green Building Materials – Sheng (ed.)
© *2015 Taylor & Francis Group, London, ISBN 978-1-138-02718-3*

An ecosystem-level ecological risk assessment framework for lake watershed: A case study of upstream Taihu lake watershed

N. Li
Civil and Environmental Engineering School, University of Science and Technology Beijing, Beijing, China

J.C. Qi
College of Architecture and Environment, Sichuan University, Chengdu, Sichuan, China

R.R. Han, B.H. Zhou & Y.Y. Huang
Civil and Environmental Engineering School, University of Science and Technology Beijing, Beijing, China

ABSTRACT: Ecosystem-level regional Ecological Risk Assessment (ERA) has been a focus of research in recent years. In this study, based on ecosystem-level ERA frameworks of watershed developed by previous researches, a lake watershed ecosystem-level ERA model is built by establishing relations among Risk Sources (RS), Threatening Factors (TF), Ecosystem Indicators (EI), Ecosystem Functions (EF) and Ecosystem Services (ES), with ES as assessment endpoints. The model is applied in upstream Taihu Lake watershed, and result shows that the region of Wuxi-Jiangyin has the highest ecological risk, while Changxing-Anji region the lowest. Effects caused by industry, agriculture and aquaculture are relatively more severe as risk sources in the overall area, while COD, nutriment, heavy metal, Persistent Organic Pollutants (POPS) and hydrological variation are main TF in the regions assessed.

KEYWORDS: lake ecosystem, ecological risk assessment, ecosystem-level, Taihu Lake watershed

1 INTRODUCTION

From the perspective of overall ecosystem, Ecological Risk Assessment (ERA) could be defined as "the process to evaluate the possibility of the formation, or the likelihood of formation, of the negative ecosystem effects, when the ecosystem or its components are exposed to one or multiple anthropic stressors" (Hope, 2006). ERA usually contains three stages, which are Problem Formulation, Analysis and Risk Characterization (Cormier et al., 2000). Comparing with ERA in limited areas, regional ERA focuses more on the effect caused by configurations of spatial factors, due to the features and complexity of the regions assessed (Hunsaker et al., 1990).

Ecosystem-level regional Ecological Risk Assessment (ERA) has been a focus of research in recent years. Ecosystem-level based ERA pays more attention to interactive changes between organism and environmental factors under the stress of certain hazardous circumstances, since ecosystem is an interactive integrated dynamical synthesis of organism and inorganic environment. Multi Criteria Decision Making (MCDM) and Geographic Information System (GIS) are integrated into the ERA framework to conduct research in division of wetland ecosystem and ERA methodology (Malekmohammadi and Blouchi, 2014).

Meanwhile, ERA has been turning to integrated ERA modes with multi-stressors, multi-receptors and multi-relations. Several studies have been conducted, as a regional ERA was conducted in Androscoggin River with Relative Risk Model (RRM) (Landis et al., 2006) and an ecosystem-level river ERA model was built and applied in the watershed of Yellow River (Zhao and Zhang, 2013). The river ERA model build relations between various risk components, including sources, stressors, ecosystem indicators and assessment endpoints, which are Ecosystem Services. In this study, an ecosystem-level ERA model for lake watershed is built, based on the river ERA model developed by Zhao and Zhang and results of a further research of lake watershed.

2 METHODOLOGY AND DATA

2.1 Methodology

A lake watershed ERA framework is built in this study, after a thorough research and analysis of the development process of ecological risks on ecosystem level. Risk Sources (RS) generate various Threatening Factors (TF), which may cause effects to certain structures and processes to the ecosystem.

Complex internal relations within the ecosystem transfer the effects to other related factors, which causes imbalance of the Ecosystem Functions (EF), and eventually changes the Ecosystem Services (ES) people gain from the ecosystem. Consequently, ES are selected as the assessment endpoints, which are shown in Table 1.

Table 1. Ecological services list.

Code	ES
ES1	Water Supply
ES2	Aquatic Production
ES3	Shipping
ES4	Water Conditioning
ES5	Climate Regulation
ES6	Water Quality Purification
ES7	Biological Diversity
ES8	Carbon Fixation and Oxygen Release
ES9	Sediment Epeirogenic
ES10	Entertainment and Tourism

The lake watershed ERA model built in this study is based on RRM. RRM has been adopted in previous ERA researches. Landis and Wiegers built RRM to solve the problem of risk assessment of multiple risk sources in different scales of space and time. It was successfully used in landscape and regional ERA, e.g., Angela M.Obery's risk assessment in Codorus lake watershed, and R.E.Bartolo in tropical river of Australia (Landis, 2004, Bartolo et al., 2012, Obery and Landis, 2002). RRM introduced exposure "factors" and effect "factors" between risk sources and the ecosystem, and between ecosystem and assessment endpoints. Quantification and half-quantification problems of large scale risk assessment could be resolved to a certain extent with calculated rankings of risk sources, ecosystem and assessment endpoints, which are determined with ranking standards (Suter, 1990).

First, data of all the risk sources are normalized in regions to be assessed with RRM methodology, and the Source Rank Matrix (SRM) is built, which is expressed as Eq.1:

$$SRM^i = \begin{pmatrix} rsr_1^i \\ rsr_2^i \\ \cdots \\ rsr_{k-1}^i \\ rsr_k^i \end{pmatrix} \qquad (1)$$

where i and j represent regions to be assessed and risk source indicators respectively, and rsr_j^i represents rankings of relative strengths of risk source j in region i.

Second, ecosystem resilience is quantified. Data of ecosystem indicators are transformed into relative state rankings representing the habitat condition of the ecosystem, to build Ecosystem Habitat Matrix (EHM), representing states of the ecosystem indicators. The ranking criteria for all EI are shown in Table 2. Resilience of ecosystem is closely related to its state, as the higher rankings of state are, the better of the state of the ecosystem, and the stronger of the ecosystem resilience. Thus EHM and Habitat Rank Matrix (HRM) are built as Eq.2:

$$EHM^i = \begin{pmatrix} a_1^i \\ a_2^i \\ \cdots \\ a_{h-1}^i \\ a_h^i \end{pmatrix} \qquad HRM^i = \begin{pmatrix} ei_1^i \\ ei_2^i \\ \cdots \\ ei_{r-1}^i \\ ei_h^i \end{pmatrix} \qquad (2)$$

where both EHM and HRM contain h columns and one row, in which i represents regions to be assessed, a_h^i in EHM and ei_h^i in HRM represent rankings of EI h for states and resilience respectively.

Third, relations of ecosystem components, including between RS and TF, between TF and EI, among all the EI and between EI and ES are established to describe the ecosystem-level risk development process with RRM ranking methods. Ranking criterion of relation between RS and TF are showed in Fig.1-a, results of which constitute Source-Stressor Matrix (SSM). Relation between TF and EI is determined by whether a TF is capable of influencing an EI and the strength of the effect. The ranking criterion is as shown in Fig.1-b, results of which constitute Stressor-Ecosystem Indicator Matrix (SEM). Relations of EI within the ecosystem are in the forms of substances and energy, in which the quantification process is finished, as shown in Fig.1-c, to form Among Ecosystem Indicators Matrix (AEM).

Ecosystem Indicator-Ecosystem Function Matrix (IFM) is built to express the relation between structure and process of ecosystem and EF, as Eq.3:

$$IFM = \begin{pmatrix} a_{11} & \cdots & a_{1f} \\ \vdots & \ddots & \vdots \\ a_{h1} & \cdots & a_{hf} \end{pmatrix} \qquad (3)$$

where the columns and rows represent EI and EF respectively, while a_{hf} represents whether EI h relates to EF f.

Table 2. Ranking criteria of EI.

EI	Ranking Criteria	Note	Source
pH	6~9 as 3,others as 0	Normal pH value is a feature for healthy ecosystem. Over high or low pH may jeopardize ecosystem.	Environmental Quality Standards for Surface Water (GB3838-2002)
DO	<2 as 0, 2~3 as 1, 3~6 as 2, >6 as 3	Low DO risk life of aquatic organism, which may jeopardize the ecosystem.	
TN	>2 as 0, 1.5~2 as 1, 0.5~1.5 as 2, <0.5 as 3	Over high TN concentration causes water body eutrophication, which leads to imbalance of the ecosystem.	
TP	>0.2 as 0, 0.1~0.2 as 1, 0.025~0.1 as 2, <0.025 as 3	Over high TP concentration causes water body eutrophication, which leads to imbalance of the ecosystem.	
COD	>30 as 1, 15~30 as 2, <15 as 3	—	
NTU	Qualitative Determination	No standard for water quality could be referenced.	—
IBI	Qualitative Determination	—	—
Ammonia Nitrogen	>1.5 as 1, 0.5~1.5 as 2, <0.5 as 3	—	Environmental Quality Standards for Surface Water (GB3838-2002) Environmental Quality Standard for Soils
Heavy Metal	<90 as 3, 90~300 as 2, 300~400 as 1	With the amount of Chromium in the bottom mud.	
Water Temperature	Within normal range as 3	Water temperature in the normal range is a feature for healthy ecosystem.	
Water Quantity	Proportion of average quantity in recent years, >90% as 3, 60%~90% as 2, 10%~60% as 1, <10% as 0	Over low water quantity may cause negative effect on operation of the ecosystem.	Determined with Tennant method.
EI	Ranking Criteria	Note	Source
Chlorophyll a	Qualitative Determination	Over high and over low are both negative effects.	—
Aquatic Species	Qualitative Determination	Characterized with numbers of species in a certain food chain.	(Weijters et al., 2009)
Lake Area Rate	Proportion of average rage in recent years,>90% as 3, 60%~90% as 2, 10%~60% as 1, <10% as 0	Degeneration of lake may jeopardize the lake ecosystem.	—
Dam	Qualitative Determination	It is thought that the more dams, the more lake connectivity is affected.	—

Function-Service Matrix (FSM) is built based on EF and ES of lake watershed, as is expressed in Eq.4:

$$FSM = \begin{pmatrix} b_{11} & \cdots & b_{1e} \\ \vdots & \ddots & \vdots \\ b_{f1} & \cdots & b_{fe} \end{pmatrix} \tag{4}$$

where the columns and rows represent EF and ES respectively, while b_{fe} represents whether EF f relates to ES e. The criterion of "Yes as 1 and No as 0" is adopted for building IFM and FSM, as Fig.1-d shows.

Consequently, Ecosystem Indicator-Ecosystem Service Matrix (EEM) could be built with IFM and FSM, as is expressed in Eq.5:

$$ISM = IFM \times FSM = \begin{pmatrix} c_{11} & \cdots & c_{1e} \\ \vdots & \ddots & \vdots \\ c_{h1} & \cdots & c_{he} \end{pmatrix} \tag{5}$$

where the columns and rows represent IE and ES respectively, while c_{he} represents the relations between EI h and ES e.

After quantifying the risk components and their relations, the effect of ES caused by RS in a watershed could be calculated with the matrixes built. Final scores of effect caused by TF released by RS through complex processes in watershed i could be calculated with Eq.6:

$$impact_i = \sum_j \sum_k \sum_h \sum_e (SRM_{ij} \times SSM_{jk} \times SEM_{km}$$

$$\times HRM_{ih} \times AEM_{hh} \times EEM_{he}) \tag{6}$$

where $impact_i$ represents overall ecological risk in watershed i. $\sum_j SRM_{ij} \times SSM_{jk}$, $\sum_j \sum_k SRM_{ij} \times SSM_{jk} \times SEM_{kh}$ and $\sum_j \sum_k SRM_{ij} \times SSMjk \times SEM_{kh} \times HRM_{ih} \times AEM_{hh}$ represent the cumulative strength of TF k generated by all the RS, integrated effect of EI h caused by all the TF, and that with ecosystem resilience and internal relations taken into consideration in region i.

2.2 Data source

In this study, the ecosystem-level ERA model for lake watershed is used in upstream Taihu Lake watershed. Data related to RS and EI of 2012 are collected to support the assessment, mainly from the statistical yearbooks, reports and database of the provinces and cities involved, and the reports and bulletin published by relevant management organizations, including "2012 Taihu Health Report" (Taihu, 2012), "China's Water Resources (2012)" (China, 2012), "Annual Report for Flood and Typhoon in Taihu Lake Watershed" (China, 2013a) and "Report for Water Resources Quality of Taihu Lake Watershed Provincial Boundary (2013)" (China, 2013b), etc.

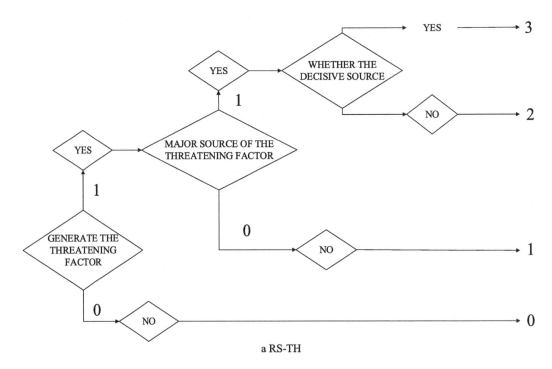

a RS-TH

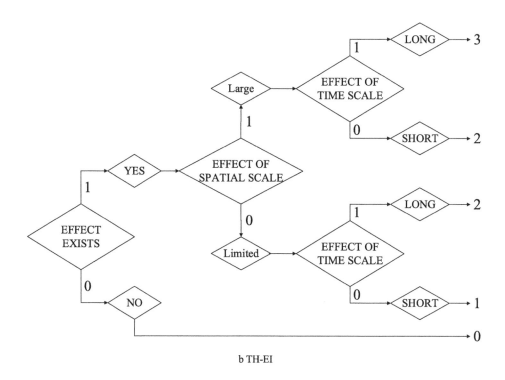

b TH-EI

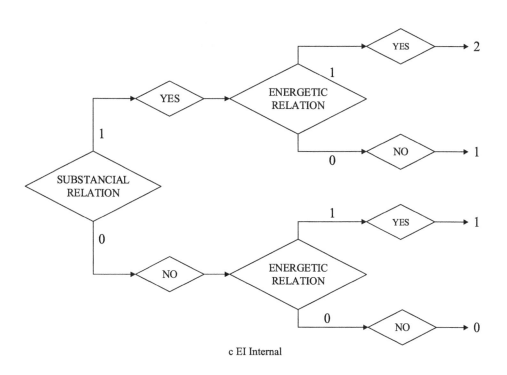

c EI Internal

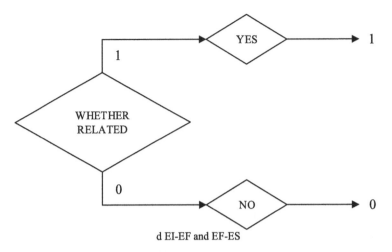

d EI-EF and EF-ES

Figure 1. Ranking criteria for relations of risk components.

3 RESULT

3.1 *Division of regions*

The ERA model is applied to upstream Taihu Lake watershed. Upstream Taihu Lake watershed lies in southwest-west-northwest of Taihu area, including four main drainage systems of Taoge, Tiaoxi, Nanhe and Yanjiang, in which Yanjiang is regarded as upstream because the amount of pollutant is large, even though it contains both inflowing and outflowing rivers. The criterion of region division takes both drainage systems and administrative districts into consideration, which divides the whole area into five assessment units: R1 Nanhe (Liyang-Yixing), R2 Taoge (Changzhou-Jintan), R3 Yanjiang (Wuxi-Jiangyin), R4 West Tiaoxi (Changxing-Anji) and R5 East Tiaoxi (Huzhou-Deqing).

3.2 *Assessment results*

The final scores of all the risks of ES, RS and TH are shown in Table 3.

With the ecosystem-level ERA model for lake watershed, scores representing overall ecological risks caused by RS of upstream Taihu Lake are calculated, as shown in Fig.2.

As Fig.2 shows, the risk scores for five assessed regions are: R3 397030, R2 312922, R1 290662, R5 295222 and R4 237922.

Scores of all the RS in the assessed regions are shown in Fig.3. It demonstrates that RS4 (Industry), RS5 (Agriculture), RS7 (Aquaculture), RS9 (Domestics) and RS10 (Water Conservancy Establishments) are comparatively strong sources. High scores include: RS4 77400 in R3; RS5 74832 in

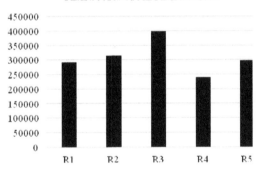

Figure 2. Risk scores of all risk sources.

R3; RS7 62856 in R2, and 47420 in R3; RS9 60900 in R3; RS10 59088 in R4.

Results of all the TH in the regions concerned are illustrated in Fig.4. It could be concluded that TF1 (COD), TF2 (Nutriment), TF3 (Heavy Metal), TF4 (Persistent Organic Pollutants, POPS) and TF5 (Hydrological Variation) are main TF in these regions. TF mentioned above scores higher in R3, which are 70700 (TF1), 54780 (TF2), 46200 (TF3), 46200 (TF4) and 80142 (TF5), respectively. TF5 scores relatively high in all the regions assessed, which are 81600 in R1, 67104 in R2, 80142 in R3, 97048 in R4 and 87987 in R5.

4 DISCUSSION

In the overall risks of lake watershed ecosystem, risks in R3 (Wuxi-Jiangyin, Yanjiang drainage system, north) caused by RS is the strongest, R2

(Changzhou-Jintan, Taoge drainage system, northwest) comes the second, while R4 (Changxing-Anji, West Tiaoxi, southwest) has the lowest ecological risk by RS. R3 is mostly influenced by sources of Industry, Agriculture and Domestics, while R2 mainly suffers from Industry and Aquaculture. Results show that R3 and R2 should be focused in ecological risk management in upstream Taihu Lake watershed, and high-risk sources of Industry, Agriculture and Domestics should be paid more attention.

In the severity of various sources, Industry, Agriculture and Aquaculture are comparatively more hazardous risk sources in the overall upstream Taihu Lake watershed. Industry has relatively more intensive impact in R2 and R3. Aquaculture are strongly jeopardizing R2, R3 and R4. Meanwhile, R3 suffer from Agriculture and Domestics more than any other regions, and R4 is comparatively more effected by Water Conservancy Establishments. Therefore, high-risk sources including Industry, Agriculture and Aquaculture should be emphasized in lake watershed ecological management on a macro level, and specific measures of management should be proposed according to different situations of all the regions.

From the perspective of the TF, COD, Nutriment, Heavy Metal, POPS and Hydrological Variation are main TF in this area, in which Hydrological Variation are of comparatively intensive threat to the whole upstream watershed. Results show that the proportions of threats in R2 and R3 are similar, only the threat caused by TF in R3 are stronger. Attention of ecological management should be paid to the controlling of COD, Nutriment, Heavy Metal and POPS, together with the Hydrological Variation, to minimize the threats caused by TF.

5 UNCERTAINTIES

In the process of lake watershed ERA, quantification of relations between ecosystem risk components is the key to the quantification of ecosystem risk. Uncertainties in final results may be caused due to the limitation of the quantification standard and quality of data collected.

To quantify risk components, certain quantification standards are used to rank risk components to numbers with data of the characteristics. Selection of the standards could have direct impact on the result. Environmental Quality Standards for Surface Water (GB3838-2002) is used in common monitoring objectives, while criteria based on quantitative and qualitative analysis of previous studies are adopted in some other indicators, which may cause uncertainty to a certain extent. In the procedure of ranking matrix with RRM method, data with reliable data sources, e.g. COD, is regarded as with low uncertainties, while for some indicators, some of which require integrated characterization with data of multiple monitoring objects, are regarded as of medium reliability with high uncertainties. Qualitative analysis is conducted to demonstrate relations between risk components, while leads to relatively high uncertainty.

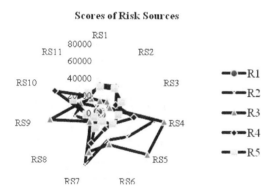

Figure 3. Scores of RS in assessed regions.

6 CONCLUSION

An ecosystem-level ERA model is built based on previous researches about ERA in system level. Relations involving Risk Sources, Threatening Factors, Ecosystem Indicators, Ecosystem Functions and Ecosystem Services are established, with Ecosystem Services as assessment endpoints, in order to provide quantified support to ecosystem management for lake watershed. A case study of upstream Taihu Lake watershed is conducted to

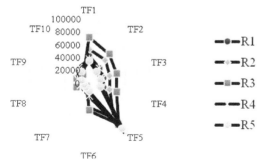

Figure 4. Scores of TF in assessed regions.

Table 3.   Risk scores of ES, RS and TH.

	R1	R2	R3	R4	R5
ES1	4662	4284	6676	4540	5144
ES2	51518	57184	69350	40660	51696
ES3	4662	4284	6676	4540	5144
ES4	14268	13882	18804	11762	14560
ES5	57700	61230	80432	48872	59166
ES6	4662	4284	6676	4540	5144
ES7	65014	73420	92356	54140	66020
ES8	4772	5089	2350	3144	3016
ES9	72000	76278	94532	56288	72206
ES10	16176	18076	21528	12580	16142
Total	290662	312922	397030	237922	295222
SR1	2088	12064	13366	32088	29160
SR2	2276	13662	15044	35916	32292
SR3	2880	11288	12752	20696	28116
SR4	2342	66192	77400	40636	22278
SR5	4520	43044	74832	19726	20566
SR6	2516	33138	38634	20200	11180
SR7	6120	62856	47420	37052	18634
SR8	1016	14274	16248	4260	8752
SR9	3528	30348	60090	15882	16378
SR10	4224	24046	26032	59088	35352
SR11	944	7982	19016	22308	26124
ST1	41136	52018	70700	36340	33876
ST2	28152	43440	54780	27280	21000
ST3	24060	34232	46200	22720	18942
ST4	24060	34232	46200	22720	18942
ST5	81640	67104	80142	97048	87984
ST6	21420	37776	42796	20032	13260
ST7	9312	6468	6834	14832	8496
ST8	11142	10476	21470	14080	8148
ST9	11572	16048	21470	18880	12844
ST10	20952	17280	15876	28776	20800

demonstrate the appliance of the model. Results show that region of Wuxi-Jiangyin (Yanjiang) has the highest risks caused by sources and stressors, with the score of 397030, while Changxing-Anji (West Tiaoxi) has the lowest, with the score of 237922. Industry, Agriculture and Aquaculture are main sources in the overall assessed regions. COD, Nutriment, Heavy Metal, POPS and Hydrological Variation are major Threatening Factors in this area, in which Hydrological Variation has strong ecological risk impact on all the regions of upstream Taihu Lake watershed. Prior indicators of Risk Sources and Threatening Factors should be more focused, to minimize the hazardous risks brought by the sources and stressors with effective management measurements. Meanwhile, an integrated consideration with other sources and stressors should be taken to guarantee the sustainable healthy development of the ecosystem.

The research was supported by Major Science and Technology Program for Water Pollution Control and Treatment (No. 2012ZX07503-003-005). Corresponding author: B.H.ZHOU, Email: zhoubeihai@sina.com, Tel: +86 10 62334821.

REFERENCES

Bartolo, R. E., Dam, R. a. V. & Bayliss, P. 2012. Regional Ecological Risk Assessment for Australia's Tropical Rivers: Application of the Relative Risk Model. *Human and Ecological Risk Assessment: An International Journal*, 18, 16–46.

China 2012. China's Water Resources. Shanghai, China: The Ministry of Water Resources of the People's Republic of China.

China 2013a. Annual Report for Flood and Typhoon in Taihu Lake Watershed. Shanghai, China: The Ministry of Water Resources of the People's Republic of China.

China 2013b. Report for Water Resources Quality of Taihu Lake Watershed Provincial Boundary. Shanghai, China: The Ministry of Water Resources of the People's Republic of China.

Cormier, S. M., Smith, M., Norton, S. & Neiheisel, T. 2000. Assessing ecological risk in watersheds: A case study of problem formulation in the Big Darby Creek watershed, Ohio, USA. *Environmental Toxicology and Chemistry*, 19, 1082–1096.

Hope, B. K. 2006. An examination of ecological risk assessment and management practices. *Environment International*, 32, 983–995.

Hunsaker, C. T., Graham, R. L., Suter, G. W., O'neill, R. V., Barnthouse, L. W. & Gardner, R. H. 1990. Assessing ecological risk on a regional scale. *Environmental Management*, 14, 325–332.

Landis, W. G. 2004. *Regional Scale Ecological Risk Assessment*, CRC Press.

Landis, W. G., Chen, V., Pfingst, A. & Kushima, G. 2006. Androscoggin River Watershed Ecological Risk Assessment. *National Council for Air and Stream Improvement Grant 1–56189*. Western Washington University.

Malekmohammadi, B. & Blouchi, L. R. 2014. Ecological Risk Assessment of Wetland Ecosystems Using Multi Criteria Decision Making and Geographic Information System. *Ecological Indicators*, 41, 133–144.

Obery, A. M. & Landis, W. G. 2002. A Regional Multiple Stressor Risk Assessment of the Codorus Creek Watershed Applying the Relative Risk Model. *Human and Ecological Risk Assessment: An International Journal*, 8.

Suter, G. W. 1990. Endpoints for regional ecological risk assessments. *Environmental Management*, 14, 15.

Taihu 2012. Taihu Health Report 2012. Shanghai, China: Taihu Basin Authority.

Weijters, M., Janse, J., Alkemade, R. & Verhoeven, J. 2009. Quantifying the effect of catchment land use and water nutrient concentrations on freshwater river and stream biodiversity. *Aquatic Conservation: Marine and Freshwater Ecosystems*, 19, 104–112.

Zhao, Z. & Zhang, T. 2013. River Risk Assessment Based on Ecosystem Level. *China Environmental Science*, 516–523.

Energy, Environment and Green Building Materials – Sheng (ed.)
© 2015 Taylor & Francis Group, London, ISBN 978-1-138-02718-3

Study of preconditioning methods for flows at all speeds

L. Li, G.P. Chen, G.Q. Zhu & W. Zhang
Southwest University of Science and Technology, Mianyang, China

ABSTRACT: Based on multi-component Navier-stokes equations, Weiss–Smith matrix preconditioning method is implemented within pseudo-time derivative term. AUSM+-up family schemes and LU-SGS implicit iterative method were used to solve chemical reacting flows at all speeds and were compared with experimental data and theoretical value. Through a comparison of calculation with available numerical and experimental data and theoretical value in literature, the results showed that the preconditioning algorithm can be applied efficiently to the reacting flows at all speeds. All these works built foundations for further application of aeroengine.

KEYWORDS: Preconditioning; AUSM+-up; LU-SGS; All speeds flow; Chemical flow

1 INTRODUCTION

It is becoming more and more popular that the algorithm for compressible fluid is improved through using precondition in the past twenty years, which can be used to calculate the number of low Mach flow[1][2]. First, some problems are mixed problems of compressible and incompressible flows, that is a part of the flow field is low Mach number and the others are compressible flow. A typical example is the wing at high angle of attack. Second, density of the low-speed flow variations of surface heat transmission and the amount of adding heat may also be compressible; for example, low combustion problems and heating gas through electromagnetic radiation, such as using of laser, solar, or microwave to achieve space propulsion advanced technology. Last, engineers like using existing compression flow software to calculate flow field of all speed ranges; this can be avoided using a number of different calculation software. In recent years, research and application of precondition method developed fast in the domestic realm, but there are still certain gaps abroad, and they are less used for chemical reactions. Based on the original variables and using the finite difference numerical to simulate flow field of rotor in hover, the precondition method without chemical reaction is proved through using experimental data from Xiao Zhong-yun[3]. A uniform method is used for unsteady flows at arbitrary mach number from low flow to high flow of a two-dimensional model proposed by Ou-Ping[4]. N-S equations with precondition method from Xia Shu-Ning were applied to 2D flows at all speeds without a chemical reaction. The results showed that the precondition method can improve the convergence speed and precision of calculation[5]. Finite difference method and LU-SGS implicit iterative method were used to solve the flow equations and species transport equations in a fully coupled manner. Computations were carried out in reacting shear layer and shock ignition, which proved that the methods and programs could be used in chemical non-equilibrium flow from Pan-Sha[6]. An effective algorithm for simulating the reacting flows over a wide range of Mach numbers was developed using the preconditioned AUSM+ scheme from Liu-Chen[7]. To use AHL3D, massively parallel software that the software can simulate is used in the two-dimensional or three-dimensional, steady or unsteady, perfect gas, or chemical non-equilibrium flow in this paper. Calculation of control equation was done with the help of the three-dimensional flow with the chemical reaction, and the fuel has been assumed to be gaseous. Turbulence model used kW TNT double-equation model. AUSM+-up splitting was used to inviscid format[8]. Equations are solved by using LU-SGS method. The software for the scramjet combustor has been used very extensively, and its reliability has been universal authentication[9–10]. Precondition method was used through adding precondition matrix on the original control equation of time and numerical simulation perfect gas of the full speed range, and the results were compared with the experimental and theoretical data. All the simulation results indicate that the development of the algorithm can be used to simulate chemical reaction flow field of full range.

2 CONTROL EQUATION

Control equation of three-dimensional N-S equation of precondition with components form is as follows:

$$\Gamma_P \frac{\partial Q_P}{\partial \tau} + \frac{\partial Q}{\partial t} + \frac{\partial F}{\partial x} + \frac{\partial G}{\partial y} + \frac{\partial E}{\partial z} = \frac{\partial F_v}{\partial x} + \frac{\partial G_v}{\partial y} + \frac{\partial E_v}{\partial z} + S \tag{1}$$

Γ_P is precondition matrix.

$Q_P = (P, u, v, w, T, c_i|_{i=1,ns-1})^T$ is the original variable.
$Q = (\rho, \rho u, \rho v, \rho w, \rho E_t, \rho_{ic}|_{i=1,ns-1})^T$ is conservation variable.

F, G, E is the inviscid flux; F_v, G_v, E_v is the viscous flux; S is the source; and u, V, and w are X, y, and z direction velocity, respectively; ρ, ρ_{ci} is the gas density and component mass fraction, respectively, assuming the component number is ns; $E_t = e + \frac{1}{2}(u^2 + v^2 + w^2)$ is the total internal energy of gas, and e is the thermodynamic energy.

Precondition matrix obtained through the integration of Weiss[11] and Ding[12] is as follows:

$$\Gamma_P = \begin{bmatrix} \Theta & 0 & 0 & 0 & \rho_T & \rho_{c1} & \cdots & \rho_{c(ns-1)} \\ \Theta u & \rho & 0 & 0 & \rho_T u & \rho_{c1} u & \cdots & \rho_{c(ns-1)} u \\ \Theta v & 0 & \rho & 0 & \rho_T v & \rho_{c1} v & \cdots & \rho_{c(ns-1)} v \\ \Theta w & 0 & 0 & \rho & \rho_T w & \rho_{c1} w & \cdots & \rho_{c(ns-1)} w \\ \Theta H - 1 & \rho u & \rho v & \rho w & \rho_T H + \rho C_p & \rho_{c1} H + \rho H_{c1} & \cdots & \rho_{c(ns-1)} H + \rho H_{c(ns-1)} \\ \Theta c_1 & 0 & 0 & 0 & \rho_T c_1 & \rho + c_1 \rho_{c1} & \cdots & c_{ns-1} \rho_{c(ns-1)} \\ \vdots & 0 & 0 & 0 & \vdots & \vdots & \cdots & \vdots \\ \Theta c_{ns-1} & 0 & 0 & 0 & \rho_T c_{ns-1} & c_1 \rho_{c1} & \cdots & \rho + c_{ns-1} \rho_{c(ns-1)} \end{bmatrix} \tag{3}$$

in which $\Theta = \left(\frac{1}{U_r^2} + \frac{\gamma-1}{c^2} \right)$

U_r is reference speed.

Different definitions of U_r are studied by Venkateswaran [13]; using the local maximum value of precondition methods, we have achieved relatively good results. Therefore, this paper uses the local maximum precondition method.

$$U_r^2 = \min\left(\max\left[\bar{V}^2 \Big|_{neighbors} \right], c^2 \right) \tag{4}$$

According to literature [3], preconditions on viscous characteristic value have an impact, but the impact is not big. In order to simplify the calculation, viscous characteristics values of no-precondition or precondition are the same in this paper.

2.1 AUSM+-up format

AUSM+-up was first proposed in 2001 by Meng sing-Liou [14]. The pressure fluctuation phenomenon of low-speed region of the AUSM+ format has been solved through modifying the pressure flux and numerical velocity and extending the compressible formula to the low Mach number flow. Comparable results are achieved with AUSMDV from Roe and Godunov. Some improvements were made in 2003 by Meng sing-Liou[15]; a simplified calculation formula was obtained along with a more convenient AUSM+-up.

$$F_{i+1/2,j,k} = S a_{i+1/2,j,k} \left[M_{i+1/2,j,k}^+ \begin{pmatrix} \rho \\ \rho u \\ \rho v \\ \rho w \\ \rho H \\ \rho_s \end{pmatrix}_L \right.$$
$$\left. + M_{i+1/2,j,k}^- \begin{pmatrix} \rho \\ \rho u \\ \rho v \\ \rho w \\ \rho H \\ \rho_s \end{pmatrix}_R \right] + \begin{pmatrix} 0 \\ n_x P_{i+1/2,j,k} \\ n_y P_{i+1/2,j,k} \\ n_z P_{i+1/2,j,k} \\ 0 \\ 0 \end{pmatrix} \tag{5}$$

$S = \sqrt{n_x^2 + n_y^2 + n_z^2}$ is the grid surface area, and $H = E + p/\rho$ is total enthalpy; in order to write conveniently, we should omit the subscript jk.

$$M_{i+1/2}^\pm = \frac{1}{2}(M_{i+1/2} \pm |M_{i+1/2}|) \tag{6}$$

$$M_{i+1/2} = \mu^+(M_{i+1/2}^L) + \mu^-(M_{i+1/2}^R) \tag{7}$$

$$\mu^\pm = \begin{cases} \pm\frac{1}{4}(M \pm 1)^2 \pm \beta(M^2 - 1)^2 & |M| < 1 \\ \frac{1}{2}(M \pm |M|) & |M| \geq 1 \end{cases} \tag{8}$$

$-\frac{1}{16} \leq \beta \leq \frac{1}{2}$, $\beta = \frac{1}{8}$ is suggest by Liou.

$$M_{i+1/2}^{L,R} = \frac{\tilde{n} \cdot \vec{V}_{i+1/2}^{L,R}}{a_{i+1/2}} \quad (9)$$

$$a_{i+1/2} = \frac{1}{2}(a_L + a_R) \quad (10)$$

$$M_{i+1/2} = \mu^+(M_{i+1/2}^L) + \mu^-(M_{i+1/2}^R)$$
$$- \frac{K_p}{f_a} \max\left(1 - \sigma \overline{M}^2, 0\right) \frac{p_R - p_L}{p_R + p_L} \quad (11)$$

$1 \le K_p \le 1, \sigma \le 1$.
$K_p = 0.25, \sigma = 1$ is suggested by Meng-sing Liou[15].

$$f_a(M_0) = M_0(2 - M_0) \in [0,1]$$

$$M_0^2 = \min\left(1, \max\left(\overline{M}^2, M_\infty^2\right)\right) \in [0,1]$$

$$\overline{M}^2 = \frac{\left[\left(u^2 + v^2 + w^2\right)_L + \left(u^2 + v^2 + w^2\right)_R\right]}{2a_{1/2}^2}$$

$$\alpha = \frac{3}{16}(-4 + 5f_a) \in \left[-\frac{3}{4}, \frac{3}{16}\right]$$

$0 \le K_u \le 1$, $K_u = 0.75$ is suggested by Meng-sing Liou[15].

3   RESULTS AND DISCUSSION

3.1   *Example for supersonic*

Calculation model uses a single groove; there are four periods of expansion [10] as shown in Fig. 1. The cross-sectional area of the entrance is 50 mm, and the total length is 1700 mm.

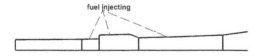

Figure 1.   Schematic of a scramjet combustion.

Flow conditions: Mach is 4, total temperature is 937 K, the total pressure is 0.8MPa, the corresponding isolator entrance Mach is 2, static temperature is 530K, and pressure is 0.1MPa. Calculations and tests are polluting air and adiabatic wall.

In this paper, the numerical simulation is two-dimensional simulation, heat effect cannot be neglected in actual three-dimensional simulation [10], and two-dimensional numerical simulation of cold flow is feasible as well as suitable[16]. Numerical simulation and experimental results in this paper also validate this conclusion. Figure 2 gives the flow of wall pressure curve graph; cold flow calculation data are compared well with the experimental data well, and there are no precondition results, because high-speed flow program equation values automatically return to the precondition before closing the form, precondition. Figure 3 gives the flow Mach number images. This example shows the feasibility of the numerical simulation of cold flow conditions.

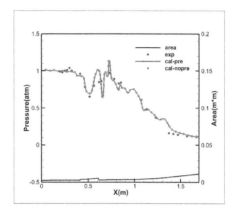

Figure 2.   Wall pressure in the cold flow.

Figure 3.   Mach contour in cold flow.

3.2   *Example of NACA0012*

A simulation was used for inviscid and viscous flow around an NACA0012 airfoil, by O structured grid and surface grid distance being 0.001. Double precision is used for calculation. Figures 4 and 5 depict surface pressure coefficients and pressure contours, respectively, of Mach number 0.01; the angle of attack of 0° is inviscid. Figures 6 and 7 depict surface pressure coefficients and pressure contours, respectively, of Mach number 0.05; the angle of attack graph 5° is inviscid. Figure 8 depicts surface pressure coefficients of Mach number 0.3, the angle of attack 3.59°,

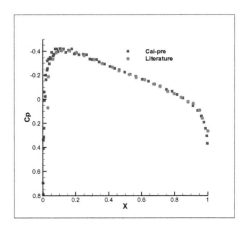

Figure 4.　Surface pressure coefficients of NACA0012.

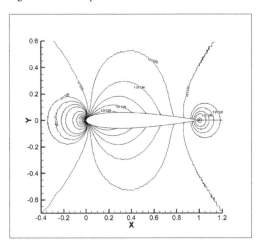

Figure 5.　Pressure contours of NACA0012.

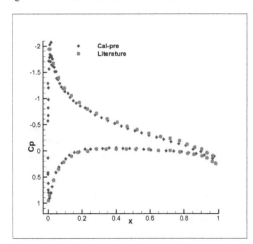

Figure 6.　Surface pressure coefficients of NACA0012.

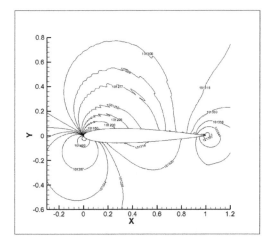

Figure 7.　Pressure contours of NACA0012.

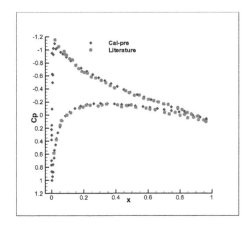

Figure 8.　Surface pressure coefficients of NACA0012.

and Re is 8.5×10^5. It was seen from Fig. 4, Fig. 6, and Fig. 8 that wall surface pressure coefficient and potential flow solution agree well with the feasibility and the precondition method to calculate low speed in a cold.

3.3 Blunt body turbulent combustion

The SANDIA examples are axisymmetric bluff body turbulent combustion model, ethylene / air had 7 chemical components, and 3-equation model. Bluff body burner structure was schematically shown in Fig. 9. Table 5 gives the concrete experimental condition. The bluff body was a way in which the flame could be stabilized. The gas passes behind a bluff body and forms a stable recirculation zone by eddy composition and double vortex structure. Calculated

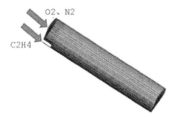

Figure 9. Schematic of SANDIA.

Table 5. Simulation conditions of SANDIA.

	oxidant	fuel
U(m/s)	40	118
P(Pa)	101325	101325
T(K)	300	300
C2H4	0	1.0
O2	0.21	0
N2	0.79	0

Figure 10. Streamlines of SANDIA.

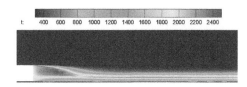

Figure 11. Temperature contour of SANDIA.

streamlines were shown in Fig 10; these reflect the double vortex structure. The internal recirculation combustion products were sufficient; the fresh gas recirculation zone adjacent to the free flow was constantly sucked into the recirculation zone, mixed, and burned. The highest temperature of the reflow zone is depicted in Fig. 11.

4 CONCLUSION

Based on the time used with the low-speed precondition technology from AHL3D, massively parallel software has been used for calculation of scramjet. The method of solving a suitable set for the development of flow field and speed scale has been carried out through the examination and verification of various complete gases and chemical non-equilibrium flow of classical examples. The results show that the method can simulate the perfect gas and chemical non-equilibrium flow field perfectly and the algorithm can be extended to the next step on the aircraft engine.

REFERENCES

[1] E. Turkel, R. Radespiel and N. Kroll, Assessment of preconditioning methods for multidimensional aerodynamics[J]. Computers & Fluids, Vol. 26, No. 6m pp. 613–634, 1996.
[2] Y.H Choi and C. L. Merkle.The application of preconditioning in viscous flows[J]. Journal of computational physics, Vol. 105, pp. 207–223, 1993.
[3] Xiao Zhong yun. Investigation of Computational ModelingTechniques for Rotor Flow fields[D]. Airbreathing Hypersonics Research Center of China Aerodynamics Research and Development Center, 2007.
[4] Ou Ping.A Uniform Method for Unsteady Flows at Arbitrary Mach Number[J].chinese journal of computational of physics, 2007, 4.
[5] Xia ShuNing, Wu Zhong Cheng, Zhu Zi Qiang. Research on Application of pretreatment method in full field velocity in the flow field[J].Chinese Journal of Computational Mechanics, 2011, 4.
[6] Pan Sha, Li Hua, Fan Xiao Qiang etc.Application of preconditioning methods in numerical simulation for chemical non-equilibrium flows[J].Acta Aerodynamica Sinica, 2004, 11.
[7] Liu Chen, Wang Giang Feng, Wu Yi Zhao.Application of preconditioned AUSM + scheme on numerical simulation of reacting flows at all speeds[J]. Journal of Aerospace Power 2009, 8.
[8] Meng-sing Liou. A further development of the AUSM+ scheme towards robust and accurate solutions for all speeds. AIAA paper No. 2003–4116, 2003.
[9] Zhao Hui Yong. Parallel Numerical Study of Whole Scramjet engine[D]. chinese journal of computational of physics, 2005.
[10] TIAN Ye, LE Jia-ling, YANG Shun-hua, et al. Numerical study on air throttling influence of flame stabilization in the scramjet combustor [J], 2013, 34(6).
[11] J. M. Weiss and J. Y. Murthy, Computation of reacting flow fields using unstructured adaptive meshes, AIAA paper No. 95–0870, 1995.
[12] Ding Li, Venkateswaran Sankaran, Jules W. Lindau and Charles L. Merkle, A unified computational formulation for multi-component and multi-phase flows[J]. AIAA paper No. 2005–1391, 2005.
[13] S. Venkateswaran, D. Li and C. L. Merkle, Influence of stagnation regions on preconditioned solution at low speed[J]. AIAA paper No. 2003–435, 2003.
[14] Meng-sing Liou, Ten years in the Making –AUSM-family, AIAA paper No. 2001–2521, 2001.

[15] Meng-sing Liou, A further development of the AUSM+ scheme towards robust and accurate solutions for all speeds[J]. AIAA paper No. 2003–4116, 2003.

[16] Jeong Y C, Jinhyeon N, Jong R B, et al.Numerical Investigation of Combustion/Shock-train Interactions in a Dual-Mode Scramjet Engine [R]. AIAA 2011–2395.

[17] JACK R EDWARDS, CHRISTOPHER J Roy. Preconditioned multigrid methods for two-dimensional combustion calculations [J]. AIAA Journal, 1998, 36(2):1852192.

Energy, Environment and Green Building Materials – Sheng (ed.)
© 2015 Taylor & Francis Group, London, ISBN 978-1-138-02718-3

Finite element simulation and optimization design of biodegradable seedling pot

Xing-Dong Liu
Chinese Academy of Agricultural Mechanization Sciences, Beijing, China
College of Mechanical Engineering, Jiamusi University, Jiamusi, China

Shu-Jun Li
Chinese Academy of Agricultural Mechanization Sciences, Beijing, China

Lu-Jia Han
China Agricultural University, Beijing, China

Quan-Rong Jing, Dao-Yi Li & Zhao-Yang Qiu
Chinese Academy of Agricultural Mechanization Sciences, Beijing, China

ABSTRACT: The design and preparation for the biodegradable seedling pot has been carried out. Based on the results of mechanical performance test of the biodegradable seedling port, the finite element model has been established. Through the simulation analysis, stress distribution, and deformation law of the seedling pot body, it has also been concluded under the action of external pressure, comparing the results with experimental research, that the easy damaged parts of the biodegradable seedling pot under outer loading can be determined. Adjusting the preparation process accordingly, the optimized shape design scheme can be achieved.

KEYWORDS: Biodegradable seedling pot; Finite element simulation; Optimized design

1 INTRODUCTION

Nowadays, a large amount of waste plastics bring "white pollution" because of their degradability, which seriously pollute the environment and endanger people's health; therefore, the biodegradable polymer materials become one of the key focuses. Currently, the second big application field of biodegradable polymer materials is in agriculture; the biodegradable polymer materials can become compost through the organic degradation process under appropriate conditions; and the degradation products are not only beneficial to plant growth but also improve soil environment.

To solve the earlier problems of plastic containers, two types of biodegradable seedling containers with plant fiber and paper have been developed in recent years[1]. The current studies of the biodegradable seedling container mainly focused on the materials choice and production. Mingwei Shen and Feilin Hao [2] researched the feasibility of the production of the biodegradable seedling pot using the invasion plants "water hyacinth" as raw material. Youzhi Wang, etc. [3] used the activated sludge to produce the biodegradable nutrients seedling pot.

Till date, the application study of the biodegradable nutrition seedling containers in the actual production is still less; it has not yet formed a complete system, especially the application research in large-scale production is rare. Wheat bran and rice husk powder were the matrix materials adopted in the study, whereas melamine resin and polyvinyl alcohol were used as adhesives. The choice of materials can fully realize biodegradation of the seedling pot. Simultaneously, the material design with different formulations has been realized, thus optimizing the performance and strength of the seedling pots, and with the help of finite element analysis software, the shape optimization design has also been realized.

2 MATERIALS AND FORMULATIONS

Four different formulations (Table 1) of the seedling pot were adopted for the product design in the study. The specific proportions of materials and chemicals were shown in Table 1. Wheat bran and rice husk powder were the seedling pot matrix, whereas melamine resin and polyvinyl alcohol were used as adhesive, which can be considered natural degradation. A5 in

Table 1 was melamine resin, and the chemical name was melamine formaldehyde resin. PVA was polyvinyl alcohol, and it can be used as emulsifying stabilizer to manufacture water-soluble adhesives. EBS in Table 1 was acid ethylene bistearic amide; it can be added in the adhesive and wax to improve the anti-caking and demolding effect. DCHP is dicyclohexyl phthalate, and it is used as a secondary plasticizer. RCOO- Na is hard acid sodium, and it can be used as a corrosion inhibitor.

2.1 Preparation technology

Rice husk powder, wheat bran, adhesives, demolding agent, and so on were blended together in proportion for later use. Adopting secondary dry hot pressing molding process, the processing conditions were 180 ℃ mold temperature for the upper mold; the temperature of the upper mold was 15 ℃ higher than that of the lower mold; the maintaining pressure was 165 MPa; and the holding time was 3 seconds. The holding pressure and time was maintained the same in all tests. At the same formulation, 50 seedling pots specimens were produced (the specimens that were taken were specimens 1, 2, 3, and 4 given next); through cooling at room temperature and aging for 1 day, subsequent performance tests can be conducted.

2.2 Measurement of elastic modulus and poisson's ratio

The determining mechanical parameters of seedling pots were the basis of the mechanical properties analysis. Before the finite element analysis, it is necessary to determine the elastic modulus and poisson's ratio parameters of the degradable material. The seedling pots were cut into 20×100 mm rectangle specimens. In order to determine the poisson's ratio and elastic modulus of materials, the strain gages were glued on the horizontal and vertical position of the rectangular specimens. Under the constant temperature, the specimen was fixed in the XL2118C type force and strain comprehensive parameter tester (seen in Fig. 1), with load at a low 0.5 mm/m speed, under the action of tensile load. According to Eqs. 1 and 2, the test results of elasticity modulus and poisson's ratio obtained are shown in Table 2.

The calculation equations of poisson's ratio μ and elasticity modulus E are as follows:

$$\mu = \left| \frac{\varepsilon_h}{\varepsilon_z} \right| \qquad (1)$$

$$E = \frac{\sigma}{\varepsilon_z} = \frac{F}{A\varepsilon_z} \qquad (2)$$

where $\varepsilon_z, \varepsilon_h$ are vertical strain and horizontal strain for specimen; F is the tensile load; and A is the cross-section area of the specimen.

a) Clamping of specimen

b) Measuring instrument

Figure 1. Measurement of elastic modulus and poisson's ratio.

Table 1. Formulations design.

Formulation	Wheat bran	Rice husk powder	A5	PVA	EBS	RCOO-Na	DCHP
1	30%	56%	5%	5%	1.5%	1.5%	1%
2	56%	30%	5%	5%	1.5%	1.5%	1%
3	56%	30%	7.5%	2.5%	1.5%	1.5%	1%
4	30%	56%	2.5%	7.5%	1.5%	1.5%	1%

Table 2. Experiment value of poisson's ratio and Elastic modulus.

Specimen	Maximum tensile load (N)	Poisson's ratio	Elastic modulus (KPa)
1	52	0.400	3.100
2	74	0.404	4.800
3	40	0.410	2.800
4	45	0.412	4.500

2.3 Static compression performance experiment

A further study of the crack damage can be based on the following static compression experiment. This study can also provide the basis for the establishment of the mechanical model of plant straw and the finite element analysis of the seedling pots.

Using electronic universal testing machine WDT-10 produced by the Tianshui Hongshan testing machine Co., LTD to perform the statics compression experiments, the maximum static compression loads of the specimens 1, 2, 3, and 4 were 963 N, 1297 N, 769 N, and 918 N, respectively. Specimen 2 had optimal compression performance, so Formulation 2 had the optimal proportions of the raw materials. The typical curves of stress and displacement were as shown in Fig. 2. The final compression damage can be observed in Fig. 3.

From Fig. 2, it can be seen that biodegradable material showed obvious brittle characteristics in compression properties. The cracks on the top can be observed first under certain pressure. When the compression load reached the peak load, the seedling pot endured the whole damage, and it cannot be used anymore. The continue increase of compression load after the peak load can be explained as the remaining part not being damaged, but it is of no use for the total compression ability.

Some cracks occurred in the top pot first, but the final destruction position of the specimen was the pot bottom by experimental observation (the seedling pot bottom and seedling pot body spined off each other). From the separation destruction phenomenon of the variable cross-section, with the dominant factors influencing the compression strength, eventually damage was still the local extreme stress of the variable cross-section at the bottom of the seedling pot.

3 FINITE ELEMENT ANALYSIS OF SEEDLING POT

Because the shape of the seedling pot was irregular, it was difficult to directly determine the inner stress distribution under the outer load. In order to further

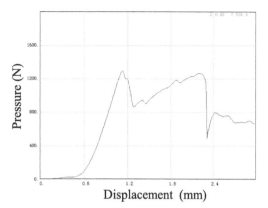

Figure 2. Compression–deformation curves.

Figure 3. Image of compression damage.

understand its stress distribution principles and the damage causes, the finite element analysis was carried out[4].

ANSYS software has been used as the analysis tool and according to the actual shape of the seedling pot to build the 3D model. Using the poisson's ratio and elastic modulus measured in Table 2 as the initial parameters, meshing the 3D model, 100MPa pressure was loaded on the top of the 3D model; the bottom was fixed in this condition. The analysis results can be observed in Fig. 4. Figure 4. a) shows the total deformation, and Fig 4. b) shows the changing tendency of the equivalent stress.

The maximum pressure values were 76.6 MPa, 69.3 MPa, 74.6 MPa, and 70.5 MPa, respectively, for the four formulations as shown in Table 1. The values showed that specimen 2 had the lowest stress values under the same pressure, which testified that specimen 2 can endure larger pressure in the same condition; so this formulation was the best for compression performance. This conclusion can be also proved by the actual compression test.

According to the actual compression test, the pressure was loaded on the bottom of the seedling pot and the top was fixed in the following finite element analysis. The maximum load can be achieved at the edge of the bottom of the pot as in Fig 5. a), where just in the position, the damage occurred. When the specimen was under external pressure load, the drum shape occurred and formed a tensile stress area, so the tensile stress was concentrated at the edge of the bottom of the pot and then the part was broken, causing the damage of the specimen. The bottom edge of the pot should be strengthened in the optimization design. The structural design cannot be ignored with regard to the overall compressive strength of the seedling pot. Therefore, reducing the variations in cross-section was appropriate and can be hoped to greatly increase the compression deformation resistance. Changing the pot structure, that is increasing the curvature radius of the pot bottom, the simulation analysis has been done to obtain the extreme stress variation. The curvature radius of the pot bottom of the original design was 0 mm, and the changed curvature radius was 100mm. The results can be observed in Fig. 5. The maximum equivalent stress can be reduced from 21.29 MPa to 20.46 MPa in Fig. 5 a) and b), where the total reducing ratio was 4.036%. So this is an effective way for improving the total compression strength for the future structure design of the seedling pot.

4 CONCLUSIONS

Through a comparison of the experimental research of different formulas, the optimal compression properties can be achieved by formulation 2, and it can be selected as the product formulation. According to the results of the finite element simulation, the shape optimization of the seedling pot can be improved in future design, which can strengthen the weak position; thus, the production can be more durable.

ACKNOWLEDGMENTS

This project was supported by the National science and technology plan project in rural areas for the twelfth five-year (2012BAD32B02) and the Science and technology research project of Heilongjiang province education department (12541832).

a) Total deformation

b) Equivalent stress

Figure 4. Loading on the top of the seedling pot.

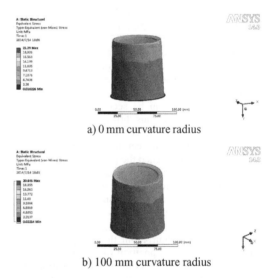

a) 0 mm curvature radius

b) 100 mm curvature radius

Figure 5. Equivalent stress of different curvature radius.

REFERENCES

[1] Huang X.M., Tang H., Xu S.G., 2001. Development of Biodegradable Seeding Containers. *Technical Textiles* 19(127):18–20.

[2] Shen M. W., Hao F. L. 2006. Investigation into the Use of Dried W ater Hyacinth as a Seed Germination Medium. *Journal of Agro-Environment Science* 25(1):258–260.

[3] Wang Y.Z., Du C.J. 1995.Studies on the Manufacture of Multi—Nutrtent Seedling Raising Pot, Its Physical and Chemical Properties. *Heilongjiang Agricultural Science* 18(1):18–20.

[4] Zhao D., Sun Y.L., Zhao X.J. 2002. Computer-aided simulation of plant stalk die compaction processes in cup-shaped mold. *Journal of Beijing Forestry University* 24(5):208–210.

Energy, Environment and Green Building Materials – Sheng (ed.)
© 2015 Taylor & Francis Group, London, ISBN 978-1-138-02718-3

Design of robust sliding mode observer for a class of linear system with uncertain time delay: A linear matrix inequality method

F.X. Xie, L. Chen & A. Zhou
Civil and Transportation College, Hohai University, Nanjing, Jiangsu, China

ABSTRACT: This paper deals with the design of robust observers for a kind of linear system with time-varying input time delay, which is still an unsolved problem in the field of structural control. A class of novel robust sliding mode observer with time-varying input delay was developed from the basis of coordinate transformation. The delay dependent conditions formulated by linear matrix inequalities (LMI) were proposed to find the gains of the proposed observer. The dynamic properties of the robust observers were discussed and proved by Lyapunov-Krasovskii method. Finally, a numerical example showed both the feasibility and efficacy of the proposed observer.

KEYWORDS: robust sliding mode observer; input time delay; variable delay; LMI

1 INTRODUCTION

Input-time delay is a common phenomenon in control systems. The transmission of the signals and the mechanical facilities of actuator operation both introduce input time delay. The control input may cause stability and efficiency problems in control systems problem to some extent. Meanwhile, most of the input delays are difficult to measure as they vary constantly. How to obtain state information from the outputs of the delay system is an important issue for both linear and non-linear systems. Sliding mode observers for systems with delays in the state variables have attracted considerable interest over the years and several design methods have been proposed (Edwards & Spurgeon 1998, Tan & Edwards 2010). The problem of asymptotic observers has been considered and various methods have been proposed for the design of full-order observers (Wu & He 2008). Some observers for delayed systems were proposed by several authors (Liu 2002, Spurgeon 2008, Richard 2003). However, most of these designs of sliding mode observers were based on the assumption that the value of the delay is known, or could be measured.

In the real applications, to simplify the complexity of the design of a sliding mode observer, the time delays are assumed to be known and invariant, which are not necessarily true in reality. There are only a few cases in which the observers are independent of the precognition of state or input delays (Choi & Chung 1996, Choi & Chung 1997, Wu, He, She, & Liu 2004, Fridman, Shaked, & Xie 2003). These approaches focus on linear systems and

guarantee the H_∞ performances of the control systems based on LMI formulations. Many sufficient delay independent stability conditions are proposed in these literatures cited. However, the research on the design of robust sliding mode controller with unknown input delay has obtained much less concentrations.

2 PRELIMINARIES AND ASSUMPTIONS

A linear system with state and input delays is considered:

$$\begin{cases} \dot{x}(t) = Ax(t) + Bu(t) + B_h u(t - h(t)) + E\omega(t) \\ y(t) = Cx(t) + D\omega(t) \\ u(t) = \varphi(s), \forall s \in [-h_m, 0] \end{cases} \tag{1}$$

where $x \in \mathbb{R}^n, u \in \mathbb{R}^m, y \in \mathbb{R}$ are the state, input and measurement vectors, respectively. It was assumed that $p > m$. $\varphi \in C^0([-h_m, 0] \in \mathbb{R}^m$ was the initial condition of the system control inputs, and D the corresponding distribution matrix with full column rank. $\omega \in \mathbb{R}^r$ is an external excitation.

The system matrices A, B, B_h, C, D, E were assumed to be constant known matrices of appropriate dimensions. Without loss of generality, the matrix D was assumed to have the following structure:

$$D = \begin{bmatrix} 0 & D_2^T \end{bmatrix}^T \text{ where } D_2 \neq 0, D_2 \in \mathbb{R}^{(p-m) \times (p-m)} \text{ and}$$

was invertible.

Besides, the following assumptions necessary for the design of observers should be mentioned at this initial stage (Edwards & Spurgeon 1998):

A1. $Rank(CB_h) = rank(B_h) = q$

A2. $q < p \leq n$

A3. The invariant zeros of $([A,[B_h \mid E],C)$ lie in $\mathbb{C}^-$

A4. $Im(B) \cap Im(E) = \{\varnothing\}$ with $rank(CE) < rank(E)$

The goal of this research was to design a robust sliding mode observer to estimate the system's states (1). The robust stability and asymptotic approaching of the sliding mode should be satisfied.

Before the main results are presented, the following two lemmas should be noted:

Lemma 1 (Edwards & Spurgeon 1998) Consider System (1), there exists a solution $P = P^T > 0$ such that $B^T P = FC$ if and only if $rank(CB) = rank(B)$

Lemma 2 (Raoufi 2010) Given System (1) with rank $(CB_h) = rank(B_h)$, there exist nonsingular transformation matrices T and S such that:

$$TAT^{-1} = \begin{bmatrix} A_1 & A_2 \\ A_3 & A_4 \end{bmatrix}, TB = \begin{bmatrix} B_1 \\ B_2 \end{bmatrix}, TB_h = \begin{bmatrix} B_{h1} \\ 0 \end{bmatrix}$$

$$TE = \begin{bmatrix} 0 \\ E_1 \end{bmatrix}, SCT^{-1} = \begin{bmatrix} C_1 & 0 \\ 0 & C_4 \end{bmatrix}, SD = \begin{bmatrix} 0 \\ D_2 \end{bmatrix}$$

where $A_1 \in \mathbb{R}^{q \times q}$, $A_4 \in \mathbb{R}^{(n-q) \times (n-q)}$, $C_1 \in \mathbb{R}^{q \times q}$, $C_4 \in \mathbb{R}^{(n-q) \times (n-q)}$, $rank(E_1) = q$, and C_1 is invertible.

Remark 1 Due to Lemma (1), the inequality:

$$rank(CE) < rank(E)$$

implies that E did not satisfy the matching condition of the form $E^T P = F_E C$, which meant that the external excitation was categorized as an unmatched disturbance.

3 DESIGN OF ROBUST SLIDING MODE OBSERVER

According to Lemma (2), the nonsingular transformation was a key to deal of the observer design. System (1) can be transformed into the following equations in new coordinates $\tilde{x} = (x_1^T, x_2^T)^T = Tx$ and $\tilde{y} = (y_1^T, y_2^T)^T = Sy$:

$$\begin{cases} \dot{x}_1(t) = A_1 x_1(t) + A_2 x_2(t) + B_1 u(t) \\ \quad\quad + B_{h1} u(t - \tau(t)) + E_1 \omega(t) \\ y_1 = C_1 x_1 \\ \dot{x}_2(t) = A_3 x_1(t) + A_4 x_2(t) + B_2 u(t) + E_2 \omega(t) \\ y_2 = C_2 x_2 + D_2 \omega(t) \end{cases} \quad (2)$$

The transformation matrix S can be partitioned into:

$$S = \begin{bmatrix} \bar{S}_1 \\ \bar{S}_2 \end{bmatrix}, \bar{S}_1 \in \mathbb{R}^{q \times p}, \bar{S}_2 \in \mathbb{R}^{(p-q) \times p}$$

Thus the variable x_1 can be obtained by the measured output y by:

$$x_1 = C_1^{-1} \bar{S}_1 y(t) \quad (3)$$

The following observer structure will be used:

$$\begin{cases} \dot{\hat{x}}_1(t) = A_1 \hat{x}_1(t) + A_2 \hat{x}_2(t) + B_1 u(t) + B_{h1} u(t - h_m) \\ \quad\quad + L_1(y_1 - \hat{y}) + B_{h1} v(t) \\ y_1 = C_1 \hat{x}_1 \\ \dot{\hat{x}}_2(t) = A_4 x_2(t) + B_2 u(t) + L_3 y_1 + L_4(y_2 - \hat{y}_2) \\ y_2 = C_4 \hat{x}_2 \end{cases} \quad (4)$$

where the discontinuous switching part $v(t)$ and the observer gain $\tilde{L}$ are defined in the following form, respectively,

$$v(t) = \begin{cases} (\rho + \rho_0) \dfrac{B_{h1}^T P_1 S}{\| B_{h1}^T P_1 S \|}, & \text{if } S \neq 0 \\ 0, & \text{otherwise} \end{cases} \quad (5)$$

$$L = \begin{bmatrix} L_1 & L_2 \\ L_3 & L_4 \end{bmatrix} \begin{bmatrix} \bar{A}_1 C_1^{-1} & 0 \\ A_3 C_1^{-1} & P_2^{-1} K \end{bmatrix} \quad (6)$$

where $S = C_1^{-1} \bar{S}_1 y - \hat{x}_1$, ρ and ρ_0 are two constant positive scalars. P_1, P_2 and K will be determined during the proof of stability and $\bar{A}_1 = A_1 - A_1^s$ where A_1^s is a stable design matrix.

Furthermore, suppose that:

$$z(t) = H \begin{bmatrix} x_1 - \hat{x}_1 \\ x_2 - \hat{x}_2 \end{bmatrix} \quad (7)$$

is the controlled output for the error system where H is a full rank design matrix with the following structure:

$$H = \begin{bmatrix} H_1 & 0 \\ 0 & H_2 \end{bmatrix} \quad (8)$$

Consider the standard introduced $\mathcal{L}_2$ gain (or $\mathcal{H}_\infty$ gain) between z and the disturbance ω:

$$\| \mathcal{H} \|_\infty^2 = \gamma = \sup_{\|\omega\|_2 \neq 0} \frac{\| z \|_{\mathcal{L}_2}^2}{\| \omega \|_{\mathcal{L}_2}^2} \quad (9)$$

Now the main result of this section is given in Theorem (1) which established sufficient conditions for the existence of a sliding mode observer with $\mathcal{H}_\infty$ performance for System (1) and provided a constructive design procedure for a class of systems with time-varying input time delay.

Theorem 1 *Given System (1) with time-varying input time delay under those assumptions (2), considering the SMO structure (4). The observer error dynamics were asymptotically stable for the case where $\omega = 0$ with an $\mathcal{H}_\infty$ disturbance attenuation level $\sqrt{\gamma} > 0$ subject to $\| \mathcal{H}_\infty \| \leq \sqrt{\gamma}$, if there exist matrix K and positive symmetrical matrices P_1, P_2 such that the following LMI optimization problem is feasible:*

Minimize γ subject to

$$P_1 > 0, P_2 > 0$$

$$\begin{bmatrix} \Sigma_{11} & P_1 A_2 & P_1 \tilde{E}_1 \\ * & \Sigma_{22} & P_2 E_2 + KD_2 \\ * & * & -\gamma I \end{bmatrix} < 0 \qquad (10)$$

where matrices Σ_{11}, Σ_{22} are given by the following equations, respectively:

$$\Sigma_{11} = P_1 A_1^s + A_1^{sT} P_1 + H_1^T H_1$$
$$\Sigma_{22} = P_2 A_4 + A_4^T P_2 - KC_4 - C_4^T K^T + H_2^T H_2$$
$$\tilde{E}_1 = E_1 - L_1 D_1$$
$$\tilde{E}_2 = E_2 - L_4 D_1 - L_3 D_1$$

Proof: Using the system described by Eq.(2) and the sliding mode observers (4) along with the observer gain (6), the error dynamics in the new coordinate can be obtained:

$$\dot{\tilde{e}} = \tilde{A}_0 \tilde{e} + \begin{bmatrix} B_{h1} \\ 0 \end{bmatrix} f(u(\tau, \tau_m, t) - v) + \begin{bmatrix} \tilde{E}_1 \\ \tilde{E}_2 \end{bmatrix} \omega(t) \quad (11)$$

where

$$\tilde{e} = \begin{bmatrix} e_1 \\ e_2 \end{bmatrix} = \begin{bmatrix} x_1 - \hat{x}_1 \\ x_2 - \hat{x}_2 \end{bmatrix}, \tilde{A}_0 = \begin{bmatrix} A_1^s & A_2 \\ 0 & A_4 - L_4 C_4 \end{bmatrix}$$

$$f(u(\tau, \tau_m, t) = u(t - \tau(t)) - u(t - h_m)$$

Noticing the fact that $f(u(\tau, \tau_m, t) = \int_{t-\tau_m}^{t-\tau(t)} \dot{u}(s)ds$, it was reasonable to assume that the rate of the control input $\dot{u}(s)$ had an upper bound μ. Thus one can have $| f(u(\tau, \tau_m, t) | < 2h_m \mu = \rho_0$. And it followed that $\lambda(\tilde{A}_0) = \lambda(A_1^s) \cup \lambda(A_4 - L_4 C_4)$.

Considering the Lyapunov function $V(\tilde{e}) = \tilde{e}^T \tilde{P} \tilde{e}$, where $\tilde{P}$ had the following structure:

$$\tilde{P} = \begin{bmatrix} P_1 & 0 \\ 0 & P_2 \end{bmatrix}, P_1 \in \mathbb{R}^{q \times q}, P_2 \in \mathbb{R}^{(n-q) \times (n-q)} \quad (12)$$

where the symmetrical matrices P_1, P_2 are still to be determined. The derivative of $V(\tilde{e})$ along Eq. (11) is:

$$\dot{V} = \tilde{e}^T (\tilde{P}\tilde{A}_0 + \tilde{A}_0^T \tilde{P})\tilde{e} + 2\tilde{e}^T \tilde{P} \begin{bmatrix} B_{h1} \\ 0 \end{bmatrix} + 2\tilde{e}^T \tilde{P} \begin{bmatrix} E_1 \\ E_2 \end{bmatrix} \omega^T$$

$$= \tilde{e}^T \begin{bmatrix} A_1^{sT} P_1 + P_1 A_1^s & P_1 A_2 \\ A_2^T P_1 & P_2 \bar{A}_4 + \bar{A}_4^T P_2 \end{bmatrix} \tilde{e}$$

$$+ 2\tilde{e}^T \tilde{P} \begin{bmatrix} \tilde{E}_1 \\ \tilde{E}_2 \end{bmatrix} \omega^T + e_1^T P_1 B_{h1}(f(u(\tau, \tau_m, t) - v(t))$$

$$\qquad (13)$$

where $\bar{A}_4 = A_4 - L_4 C_4$.

To attain the robustness to the disturbance in then sense of $\mathcal{L}_2$ norm, the following constraint on the stability criteria should be added (Boyd 1994):

$$\dot{V} + z^T(t)z(t) - \gamma \omega^T(t)\omega(t) \leq 0 \qquad (14)$$

Noticing the expression of $v(t)$ and the upper bound of $f(u(\tau, \tau_m, t)$, one will have:

$$e_1^T P_1 B_{h1}(f(u(\tau, \tau_m, t) - v(t))$$
$$= e_1^T P_1 B_{h1} f(u(\tau, \tau_m, t) - (\rho + \rho_0) \| e_1^T P_1 B_{h1} \|$$
$$\leq \| f(u(\tau, \tau_m, t) \| \times \| e_1^T P_1 B_{h1} \| - (\rho + \rho_0) \| e_1^T P_1 B_{h1} \|$$
$$\leq -\rho \| e_1^T P_1 B_{h1} \| < 0$$

Thus substituting Eq. (8) and Eq. (13) into Eq. (4), and letting $K = P_2 L_4$, the LMI in Theorem (1) was obtained after simple algebraic manipulation; this was a sufficient condition to make Eq. (14) negative and thus completes the proof. ◆

Remark 2 *Theorem (1) is independent of the delay function $h(t)$, thus it was theoretically stable for all input delay. However from the definition of ρ_0, the scattering effect would be enormous which will make the SMO unrealizable when the delay effect becomes sufficiently large.*

4 SYNTHESIS OF SLIDING MOTION

It is well known that a sliding surface should be maintained by a switching component. The reachability condition of the sliding mode surface should be satisfied. The following sliding surface is defined here:

$$S_0 = \{e_1 \in \mathbb{R}^q : e_1 = 0\} \qquad (15)$$

Suppose that the external excitation $\omega(t)$ is bounded subject to $\omega(t) \leq \omega_0 < \infty$. The following was the main result of this section:

Theorem 2 *Under the observer structure of Eq.(4) and Theorem (1), an ideal sliding mode motion takes place on the hyper plane $S_0 = \{\bar{e}_1 = 0\}$ in finite time.*

Proof: Define the Lyapunov function:

$$V_2(t) = \bar{e}_1^T P_y \bar{e}_1(t) \tag{16}$$

The error dynamics were governed by the following equation:

$$\begin{aligned}
\dot{e}_1 &= A_1^s e_1 + A_2 e_2 + B_{h1} f(u(\tau, \tau_m, t)) + E_1 \omega(t) \\
\dot{e}_2 &= (A_4 - L_4 C_4) e_2 + E_2 \omega(t)
\end{aligned} \tag{17}$$

Differentiating the function along the system described by Eq.(17) gave:

$$\begin{aligned}
\dot{V}_2 &= e_1^T (A_1^{sT} P_y + P_y A_1^s) e_1 + 2 e_1^T P_y B_{h1} (f(t) - v(t)) \\
&\quad + 2 e_1^T P_y (E_1 + P_1 L_1) \omega(t)
\end{aligned} \tag{18}$$

Noting the fact that A_1^s is a Hurwitz matrix which means there is a positive definite matrix Q_y satisfying $G_l^T P_y + P_y G_l = -Q_y$. The following upper bound of $\dot{V}_2(t)$ may then be obtained:

$$\begin{aligned}
\dot{V}_2 &\leq 2 e_1^T P_y B_{h1} (f(\tau, h_m, t) - v(t)) + 2 e_1^T P_y E_1 \omega(t) \\
&\leq -2 \rho e_1^T P_y + 2 e_1^T P_y \| E_1 \omega_0 \|
\end{aligned} \tag{19}$$

Thus if $\rho = |(E_1 + P_1 L_1) \omega_0| + \delta$ is selected, where δ is a positive constant, this would lead to:

$$\begin{aligned}
\dot{V}_2 &\leq -2\delta \| P_y \bar{e}_1(t) \| \\
&\leq -2\delta \sqrt{\lambda_{min}(P_y)} \sqrt{V_2(t)}
\end{aligned} \tag{20}$$

where $\lambda_{min}(P_y)$ is the minimum eigenvalue of P_y. Thus from Eq.(20), the sliding motion defined will take place in finite time. This completes the proof. ◆

5 NUMERICAL SIMULATION

Consider the system described by Eq.(1) with time-varying delay and the following matrices:

$$A = \begin{bmatrix} 0 & 0 & -1 & 0 \\ 0 & 0 & 0 & 0.1 \\ 2 & 3 & -1 & 0 \\ 2 & -1 & 0 & -1 \end{bmatrix}, \quad B = \begin{bmatrix} 0 \\ 0 \\ 0 \\ 1 \end{bmatrix}, \quad B_h = \begin{bmatrix} 0 \\ 0 \\ 0 \\ 1 \end{bmatrix}$$

$$E = \begin{bmatrix} 0 \\ 0 \\ 0 \\ 1 \end{bmatrix}, \quad C = \begin{bmatrix} 1 & 0 & 0 & 0 \\ 0 & 1 & 0 & 0 \\ 0 & 0 & 0 & 1 \end{bmatrix}, \quad D = 0$$

The delay is chosen as $h(t) = h_m / 2(1 + \sin(\omega_1 t))$ with $h_m = 0.3s, \omega_1 = 0.5s^{-1}$. The control law was:

$$u(t) = u_0 \sin(\omega_2 t) \tag{21}$$

with $u_0 = 2, \omega_2 = 3s^{-1}$.

A constant $\rho_0 = 2$ was chosen to suppression the function $\alpha_2(t, \bar{x}_1, x_2, e_2, u)$ in the switch function for the open loop system to be both stable and bounded.

The initial conditions of the system and error dynamics were chosen to be $[2, 2, 2, 1]^T$ simultaneously. The simulation results are shown in Fig.(1). The estimated and real states of the system are shown in Fig.(2).

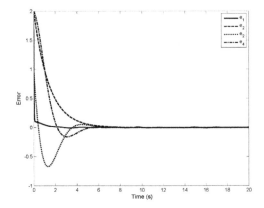

Figure 1: The errors of system states

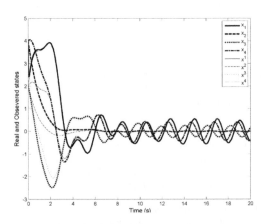

Figure 2: The real and observed system states

As one can see, the error of the system rapidly converged to zeros, especially e_1 for the sliding motion synthesis. And e_2 also converged to zero as the external excitations approached zero.

6 CONSLUSIONS

New sufficient conditions for a class of robust sliding mode observers with unknown time-varying time delays in the system's inputs was proposed and the reaching law and reaching property of the observer were formulated and proved, based on LMI condition, which had guaranteed the observer's asymptotic stability. Simulations example from a fourth order dynamic system verified the efficiency and feasibility of the observer.

ACKNOWLEDGMENTS

The authors thank the reviewers for their helpful comments and suggestions that have improved the presentation of this paper. The work of Faxiang XIE was supported by the Science Foundation of Jiangsu Province under Grant BK2012412 and the National Postdoctoral Fund support under Grant 2012M520986.

REFERENCES

Boyd, S. (1994). Linear Matrix Inequalities in System and Control Theory, Volume 15. Philadelphia.

Choi, H. H. & M. J. Chung (1996). Observer-based h controller design for state delayed linear systems. Automatica 32(7),1073–1075.

Choi, H. H. & M. J. Chung (1997). Robust observer-based h controller design for linear uncertain time-delay systems. Automatica 33(9), 1749–1752.

Edwards, C. & S. Spurgeon (1998). Mode Control: Theory and Application, Volume 7 of Systems and Control Book Series. Taylor Francis.

Fridman, E., U. Shaked, & L. Xie (2003). Robust h filtering of linear systems with time-varying delay. IEEE TRANSACTIONS ON AUTOMATIC CONTROL 48(1).

Liu, X. (2002). sliding mode control of uncertain time delay systems. Ph. D. thesis.

Raoufi, R. (2010). nonlinear robust observers for simultaneous state and fault estimation. Ph. D. thesis.

Richard, J.-P. (2003). Time-delay systems: an overview of some recent advances and open problems. Automatica 39(10), 1667–1694.

Seuret, A., T. Floquet, J. P. Richard, & S. K. Spurgeon. A sliding. mode observer for linear systems with unknown time varying delay. In American Control Conference, 2007. ACC '07, pp.4558–4563.

Spurgeon, S. K. (2008). Sliding mode observers: a survey. International Journal of Systems Science 39(8), 751–764.

Tan, C. P. & C. Edwards (2010). An LMI approach for designing sliding mode observers. International Journal of Control (931613092).

Wu, M. & Y. He (2008). Robust control of time-delayed systems-Free weighting matrix method (In Chinese). System and control series. Beijing: Science Press.

Wu, M., Y. He, J. She, & G. Liu (2004). Delay dependent criteria for robust stability of time-varying delay systems. Automatica 40(8), 1435–1439.

Energy, Environment and Green Building Materials – Sheng (ed.)
© *2015 Taylor & Francis Group, London, ISBN 978-1-138-02718-3*

Groundwater environmental impact assessment of Red Bull beverage due to mining water in Guangdong Province

L.S. Tang
School of Earth Sciences and Geological Engineering, Sun Yat-Sen University, Guangzhou, Guangdong, China

H.T. Sang
School of Engineering, Sun Yat-Sen University, Guangzhou, Guangdong, China
Guangdong Province Key Laboratory of Geological Processes and Mineral Resources, Guangzhou, Guangdong, China

Y.L. Sun & Z.G. Luo
School of Earth Sciences and Geological Engineering, Sun Yat-Sen University, Guangzhou, Guangdong, China

ABSTRACT: Environmental impact assessment of groundwater system is difficult due to its complexity and various influencing factors. MODFLOW Software can stimulate the flow and perform environmental impact assessment of groundwater. In this paper, the hydro-geological conditions of a water source in Guangdong Province are analyzed, and then the hydro-geological conceptual model and mathematical model are established. By inputting the hydrogeology parameters, which are calculated and obtained by the pumping test into the MODFLOW Software to simulate the environmental influence of groundwater under two kinds of working condition of mining water by two factories, the outputs of the modeling, such as the radius of influence, groundwater table, and groundwater balance figure, are analyzed. The results showed that water intake of Red Bull Beverage will affect the groundwater supply of Budweise, whereas it has no apparent effect on the nearby Coca-Cola. The modeling process and groundwater environment impact assessment method in this paper can provide reference for the environmental impact assessment of groundwater in Guangdong region, which also has the instruction function to the corresponding production practice.

KEYWORDS: MODFLOW; Groundwater; Environmental impact assessment

1 INTRODUCTION

Groundwater system is made up of groundwater aquifer system and groundwater flow system; it is influenced by natural factors (meteorological, hydrological, geological, etc.) and man-made factors (mining, irrigation, reservoir permeability, etc.). The influence factors are different, especially for the different types of groundwater. So the study of groundwater system is more difficult than surface water system. Because of the flexibility, efficiency, and convenience of numerical simulation, it becomes the important technical means in simulation of groundwater flow and groundwater environment impact assessment. MODFLOW (Modular Three-dimensional Finite Difference Groundwater Flow Model) is a three-dimensional groundwater flow numerical simulation software that was developed by the USGS in the

1980s. MODFLOW adopts a modular structure that is made up of a main program and several relatively independent packages. The packages include Modflow, Modpath, MT3D, and Zone Budget [1] and other features of the package are still being developed. Based on the finite difference method, its advantage is that the various steps are closely connected from start modeling, input, changing hydrogeological parameters and geometric parameters, operating model, and inverting correction parameters to real output. The whole process is systematic and normative from beginning to end. Currently, it is being used in the foundation pit precipitation simulation [2], groundwater resources evaluation [3], mine draining analysis, and mine water inflow forecast [4] in the domestic sphere.

In this paper, combined with the characteristics of MODFLOW, the exploitation and utilization of the groundwater resources in Foshan city are calculated and evaluated, which provides a reference for the groundwater environment impact assessment in Guangdong area.

2 HYDROGEOLOGY SITUATION OF THE STUDY AREA

The study area is located in the subtropical, belonging to Sanshui district, Foshan city. The geomorphic type is the Pearl river delta plain with a subtropical maritime monsoon climate and abundant rainfall. The average annual rainfall is 1687 mm, which is mostly concentrated in the spring and summer (from April to September).

There are two types of groundwater, loose rock pore water and carbonate rock fissure karst water, in the study area. Loose rocks pore water exists in the silt clay, silty clay, and sandy soil of quaternary strata. What mainly exists in the silt clay and silty clay is phreatic water, whose buried depth is 1.5m to 2.3m, the average is about 1.8m, and the thickness is about 30m. What mainly exists in the sand is confined water, whose buried depth is 28.5m to 32.3m, the average is about 30m, and the thickness is about 22m. The purpose water layer is the carbonate rock fissure karst water, which occurs in weathering limestone strata (rock). Some of the fractures and dissolved gaps of stratum are well developed, have good connectivity and the water content is relatively rich. The groundwater buried depth of this layer is 50.0m to 53.5m, the average is about 52.0m, and the thickness is more than 40m.

The main recharge sources of the groundwater in the study area are rainfall infiltration, lateral infiltration, and the leaking recharge of confined water, and the main excretion pathways are artificial mining, lateral outflow, and phreatic water. The dynamic level of underground water is restricted by artificial mining, meteorology, buried conditions, and hydrological factors. The groundwater in the study area belongs to the type of penetrating mining, and it is greatly influenced by artificial exploitation and seasonal change.

3 HYDROGEOLOGY MODEL OF THE STUDY AREA

3.1 Conceptual model

Hydrogeological model is derived from the hydrogeological conditions of studying area, but it is not completely equal to its actual hydrogeological conditions. The stratum of the simulating area is conceptualized into three layers, as shown in Table 1. The top boundary of the simulating area is the surface of phreatic water, which is the water exchange boundary with its level changing, such as rainfall infiltration and phreatic water evaporation. So the upper plane boundary is

Table 1. Formation conditions of the simulating area.

Number	Name of rock and soil layer	Thickness(m)
1–1	silt clay and silty clay	30
1–2	sandy soil	22
2	moderately weathered limestone	48

conceptualized as the second flow boundary. The bottom boundary of the simulating area is the sand layer, with confined water in the pores and fissures of loose rock. Due to the groundwater level and the confined water pressure level being equal, the transfluence can be negligible. The impact range of the study area is small based on the hydrogeological investigation and the fact that the area of the study area is some small. The range of simulation area is much more greater than the impact range; therefore, the arounds of the simulation area are conceptualized as the first kind of boundary of constant water level. As a summary, the study area can be conceptualized as an inhomogeneous and isotropic three-dimensional unsteady groundwater flow system.

3.2 Mathematic model

Based on the hydrogeologic conceptual model of the study area, the mathematical model is established as follows:

$$
\begin{cases}
\dfrac{\partial}{\partial x}\left(K_{xx}\dfrac{\partial H}{\partial x}\right)+\dfrac{\partial}{\partial y}\left(K_{yy}\dfrac{\partial H}{\partial y}\right)+\dfrac{\partial}{\partial x}\left(K_{zz}\dfrac{\partial H}{\partial z}\right) \\[2mm]
\qquad +w=\mu_s\dfrac{\partial H}{\partial t},\ (x,y,z)\in\Omega, t\geq 0 \\[2mm]
K_n\dfrac{\partial H}{\partial n}\Big|_{S_2}=q(x,y,z,t),\ (x,y,z)\in S_2, t\geq 0 \\[2mm]
H(x,y,z,t)\big|_{S_1}=H_1(x,y,z,t)\ (x,y,z)\in S_1, t\geq 0 \\[2mm]
H(x,y,z,t)\big|_{t=0}=H_0(x,y,z),\ (x,y,z)\in\Omega\cup S_2 \\[2mm]
H(x,y,z,t)\big|_{t=t_0}=H_t(x,y,z),\ (x,y,z)\in\Omega\cup S_2
\end{cases}
\tag{1}
$$

where K_{xx}, K_{yy}, and K_{zz} are, respectively, the component of the permeability coefficient in the direction x, y, and z, m/d; H is the hydraulic head, m; H_0 is the initial hydraulic head of the aquifer, m; H_t is the initial hydraulic head of the aquifer at t moment, m; w is source sink term, which is the flowing in to or out from aquifer in unit volume per unit time (flowing in is positive, whereas flowing out is negative), m/d; μ_s

is the elastic storativity of the confined aquifer, 1/m; t is time, d; Ω is the three-dimensional computational domain; K_n is the permeability coefficient in the normal direction; n is the outside normal direction of the boundary; S_1 is the first kind boundary; S_2 is the second kind boundary; and q is the flow rate per unit area of the second kind of boundary, m³/d.

4 HYDROGEOLOGY SIMULATION

4.1 Simulation scope

To evaluate the water mining impact on the surrounding environment, the simulation scope of the studying area is greater than the investigation area. Simulation calculation scope is 500m×1700m, as is shown in Fig. 1, which is divided into 30×34 finite difference grids, 50m per grid. The ground elevation is +4.4m; the vertical stratum is divided into three layers, whose thicknesses are, respectively, 48m, 22m, and 30m. For fine simulation, the water level change is better near the well; the grids of the ranges between 400m and 1300m on the x axis and between 400m and 1200m on the y axis are further subdivided, 25m per grid.

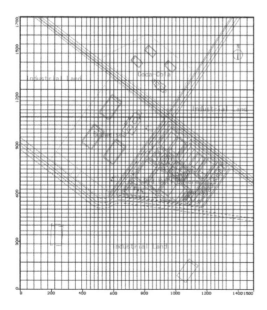

Figure 1. Grid graph of the simulation area.

4.2 Hydrogeological parameters

To proceed with the three-dimensional numerical simulation of the groundwater with MODFLOW, the hydrogeological parameters, including permeability coefficient, elastic storativity, and water storage

coefficient, are necessary, among which the most important is the permeability coefficient.

Due to the shallow depth of groundwater, the groundwater of silt clay and silty clay layer does not belong to the productive goal groundwater. The permeability coefficients of the muddy clay and silty clay layers are small, so the muddy clay and silty clay layers are generally regarded as water-resisting layers. Therefore, according to the existing experience, the value of the permeability coefficient is 5.0×10^{-7} m/s.

The permeability coefficient of sand layer is determined by a group of holes pumping tests. During the test, wells SJ1 and SJ2 are pumping wells; drill holes ZK1, ZK2, ZK3, ZK4, ZK5, GC1, and GC2 are observation wells. The confined water nonholonomic formula for calculating the permeability coefficient of steady flow pumping well is as follows:

$$K = \frac{0.366Q}{ls} \lg \frac{al}{r} \qquad (2)$$

where K is the permeability coefficient, m/d; Q is water discharge of the well, m³/d; l is the length of the filter, m; s is the drawdown, m; a=1.6; and r is the radius of the filter, m. The permeability coefficients of the sand layer in all the calculated directions are 2.0×10^{-5} to 2.6×10^{-5} m/s, the average of which is 2.3×10^{-5} m/s.

The permeability coefficient of intermediary weathered limestone layer is determined by a group of holes pumping tests. During the tests, well ZK6 is a pumping well; ZK1, ZK2, ZK3, ZK4, and ZK5 are observation wells. The confined water nonholonomic formula for calculating the permeability coefficient of steady flow pumping well is as follows:

$$K = \frac{0.16Q}{L(s_1 - s_2)} \left[arsh\frac{L}{r_1} - arsh\frac{L}{r_2} \right] \qquad (3)$$

where K is the permeability coefficient, m/d; Q is water discharge of the well, m³/d; L is the length of the filter, m; s_i is the drawdown, m; and r_i is the distance from observation well to pumping well, m. The permeability coefficients of intermediary weathered limestone layer in all the calculated directions are 2.6×10^{-5} to 3.5×10^{-5} m/s, the average of which is 3.0×10^{-5} m/s.

5 SIMULATION RESULTS

5.1 Groundwater level

The change of groundwater level in two kinds of pumping conditions is simulated. The first condition is pumping the two wells in Red Bull Beverage (as is shown in Fig. 2), and the extraction quantity per well is 160 m³/d. The second condition is pumping the four

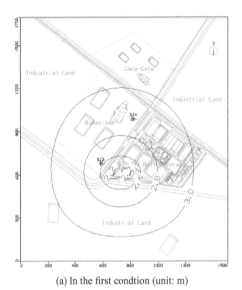

(a) In the first condtion (unit: m)

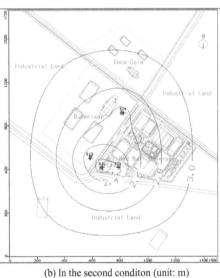

(b) In the second conditon (unit: m)

Figure 2. Groundwater isogram after one year pumping.

wells in both Budweiser and Red Bull Beverage (as is shown in Fig. 4), and the extraction quantity per well of Budweiser factory is 100 m³/d. Figure 2 shows that the influenced radius of pumping the two wells in Red Bull Beverage is about 500m and there is less influence on the nearby Coca-Cola. Figure 2 shows that the influenced radius of pumping the four wells in Red Bull Beverage and Budweiser is about 500m and there is more influence on the nearby Coca-Cola. Therefore, the coordination work of pumping water between Budweiser and Red Bull Beverage should be well done and the pumping water work should not be done in

the meantime to minimize the influence on the nearby Coca–Cola and the around underground water level.

5.2 *Groundwater resources amount*

The calculation of the increment, output, and balance of groundwater can be achieved with MODFLOW. By defining Budweiser as Zone 1 and Red Bull Beverage as Zone 2, their water balance is calculated. When the two wells in Red Bull Beverage are pumped, as observed in Fig. 3(a), the main groundwater supply of the Budweiser is the fixed water head boundary and the second supply is the rainfall infiltration, which shows that the pumping of Red Bull Beverage will impact the water supply of Budweiser. Figure 3(b) shows that the main groundwater supply of Red Bull Beverage is Budweiser factory, which indirectly proves that the water pumping of Budweiser will impact the groundwater supply of Budweiser.

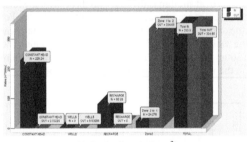

(a) Budweiser (unit: m³)

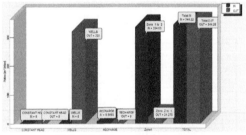

(b) Red Bull Beverage (unit: m3)

Figure 3. Groundwater equilibrium figures.

6 CONCLUSIONS

Based on a water source region in Foshan city, the groundwater environment impact assessment in Guangdong area via MODFLOW software is introduced; the conclusions are as follows:

(1) For groundwater environment impact assessment, the basic steps of MODFLOW software are as follows: analyzing the hydrogeological condition of evaluation district, establishing the hydrogeological conceptual model, establishing the hydrogeological mathematics model, confirming the

hydrogeological parameters, calculating, and visual outputting.

(2) With MODFLOW software, the groundwater environment impact assessment can be analyzed from different aspects, including the change of water level, the radius of influence, and groundwater equilibrium figures; these can prove, support, and recorrect each other.

(3) The water pumping of Budweiser will impact the groundwater supply of Budweiser, whereas it has no apparent effect on the nearby Coca-Cola.

ACKNOWLEDGMENTS

This research work was supported by the National Natural Science Foundation of China (Grant No. 40872205), the Specialized Research Fund for the Doctoral Program of Higher Education (No. 20120171110031), the Guangdong Natural Science Foundation (No. 07003738), and the Science and Technology Planning Project of Guangdong Province, China (No. 2008B030303009).

Contact E-mail: sanght@mail2.sysu.edu.cn

REFERENCES

[1] Q.Y. Wang, et al. (2007) MODFLOW and its application to groundwater simulation, Journal of Water Resources & Water Engineering, 18, 90–92.

[2] B. Wu, et al. (2010) Application of ModFlow to precipitation design of an engineering, Journal of Xinjiang Agricultural University, 33, 369–372.

[3] H.M. Zheng, et al. (2007) Application of visual ModFlow in groundwater numerical simulation in Tianjin City, Journal of North China Institute of Water Conservancy and Hydroelectric Power, 28, 8–11.

[4] M.X. Lei & Q. Xu. (2011) Application of visual MODFLOW in the research on water control in certain mine, Uranium Mining And Metallurgy, 30, 44–49.

Energy, Environment and Green Building Materials – Sheng (ed.)
© 2015 Taylor & Francis Group, London, ISBN 978-1-138-02718-3

Experimental study of Cl and F releasing of straw combustion

Y. Zhao, Y. Xiang, G.S. Cheng, Y. Wang, Y. Zhang, C.Q. Dong & Z.M. Zheng
National Engineering Laboratory for Biomass Power Generation Equipment, School of Renewable Engineering, North China Electric Power University, Beijing, China

H.L. Zhang
Electric Power Research Institute of Guangdong Power Grid Company, Guangzhou, China

ABSTRACT: The process of chlorine and fluorine releasing of straw combustion has an important influence on the biomass boiler corrosion. In this paper, factors, such as temperature, excess air ratio, residence time, and straw mass, affecting the releasing of chlorine and fluorine are taken into account. The method of measuring the content of chlorine and fluorine is high-temperature hydrolyze-ion chromatography. The results show that the precipitation rate of chlorine and fluorine increases with the increase of temperature. However, with regard to the effect of residence time and straw mass, the precipitation rate of fluorine has no change but chlorine increases with the increase of residence time and drops with the increase of straw mass. Furthermore, the precipitation rate shows no significant change as the excess air ratio varies.

KEYWORDS: Combustion; Precipitation rate; Chlorine; Fluorine; Straw

1 INTRODUCTION

As the energy crisis and environmental problems have become more and more serious, biomass as a renewable energy has obtained much attention and has been used in large-scale pyrolysis, gasification, and combustion using biomass, which are several effective methods used. However, using biomass brings many disadvantages, including contamination and corrosion problems. The problems are urgent and pressing to solve. The straw is rich in Cl and F compared with coal. Chlorine is easy to be released in HCl, NaCl, KCl, or other forms that will corrode the heating surface of the boiler as well as pollute the environment while being discharged into the atmosphere[1]. What is more, fluorine tends to be released in gaseous state (HF, SiF_4), resulting in significant damage to the environment and great harm to the human body after entering the food chain[2]. The damage and corrosion can be effectively reduced by controlling the release of F and Cl during combustion of straw; thus, it is quite necessary to study the chlorine and fluorine emission properties during straw combustion.

Currently, a number of domestic and foreign studies are being carried out on chlorine and fluorine emission properties[3,4,5]. Xu studied the migration of F and Cl in straw, rice husk and sawdust during pyrolysis and concluded that K, Na, and the increase of temperature and volatiles are conductive to the release of F and Cl[6]. Johansen drew a conclusion from the experiment in large-scale fluidized beds that the emission form of Cl becomes KCl from HCl with the increase of temperature; in addition, precipitation rate of chlorine at 900–1000°C has decreased[7]. Wen used the thermodynamic computer package FactSage to study the relationship between the two factors (temperature and atmosphere) and emission properties of Cl[8]. Lu investigated the correlation between the five factors (types of biomass, fuel characteristics, different kinds of additives, temperature, and ratio) and emission properties of Cl with tube furnaces[9]. Guo did the experiment on steam, temperature, and atmosphere[10]. These studies are mainly related to fluorine emission properties during co-combustion or pyrolysis, which use limited kinds of stock and factors. Furthermore, limited research is correlative to fluorine. Therefore, the main theme of this paper is to obtain the emission properties of chlorine and fluorine from straw during combustion, considering the factors of temperature, residence time, straw mass, and excess air ratio.

2 MATERIAL AND EXPERIMENTAL METHODS

2.1 Straw samples

Raw materials become straw samples after the crushing, drying, and screening process. Element compositions of straw sample are presented in Table 1.

2.2 Experimental system

The experimental system is shown in Figure1. The system developed by our team consists of six parts. The inner diameter of the quartz tube is 50mm, the outer diameter is 60mm, and the length is 700mm.

Table 1. Element composition of straw (air dried).

Element	N	C	H	S	Cl	F	Others
Content (wt.%)	0.83	43.68	5.42	0.99	0.74	0.023	48.31

2.3 Experimental process

When the furnace was heated to the given temperature, small crucibles (30 mL) with straw were (crushed sieving with 80 mesh) quickly placed in the quartz tube. Atmosphere in the tube was controlled by the flowmeter, and temperature was controlled by the temperature control system. The method of measuring the content of chlorine and fluorine was combustion hydrolysis at a high temperature-ion chromatography.

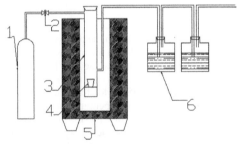

1.Cylinder 2.flowmeter 3.quartz tube 4.crucible
5.furnace 6.gas-washing bottle

Figure 1. Experimental system.

3 RESULTS AND DISCUSSION

3.1 Effect of temperature

Figure 2 shows the effect of combustion temperature on the precipitation rate of F and Cl of 2g straw in air with the excess air ratio of 1.5 for 30 min at 500°C–900°C. It could obviously be found that F increases as the temperature increases in the entire temperature range. The precipitation rate of F has already been more than 80% at 500°C and reaches 97%, indicating that most of the fluoride has been released. The precipitation rate of Cl is about 50% at 500°C–700°C, telling us that there are half of Cl in straw. However, it soars when it is more than 700°C and reaches 96% at 900°C. Thus, the release of Cl is at content temperature of combustion, because the volatiles separate out mainly at 200°C–400°C.

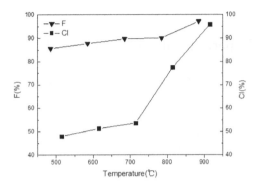

Figure 2. Effect of temperature on the precipitation rate of F and Cl.

3.2 Effect of excess air ratio

Figure 3 presents the effect of excess air ratio on the precipitation rate of F and Cl of 2g straw at 700°C for 30 min. The highest precipitation rate of Cl is at a ratio of 1.5, but F is on the contrast. The changes of precipitation rate of Cl and F with the increase of excess air ratio straw, however, are very small. Thus, the increase of ratio results in no change in precipitation rate.

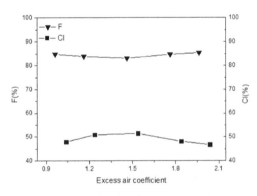

Figure 3. Effect of excess air ratio on the precipitation rate of F and Cl.

3.3 Effect of residence time

Figure 4 displays the effect of residence time on the precipitation rate of F and Cl of 2g straw in air with the excess air ratio of 1.5 at 700°C. As shown in Figure 4, the change of precipitation rate of fluorine with the increase in residence time is not very noticeable, which means the content of F remains marginal. The precipitation rate of chlorine increases as the residence time increases. At 700°C, it rises to 50% for 30min, and to 70% for 150 min; this interesting finding may contribute to the result that at 700°C chlorine spends more time on release.

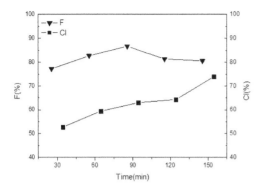

Figure 4. Effect of residence time on the precipitation rate of F and Cl.

3.4 Effect of straw mass

Figure 5 points out the effect of straw mass on the precipitation rate of F and Cl in air with the excess air ratio of 1.5 for 30 min at 700°C. It is found in Figure 5 that mass has little influence on the releasing of F. However, as the slope pf curve drops with the increase of mass, the precipitation rate of Cl is lower, which is due to the increase of thickness of the straw with the increase of mass.

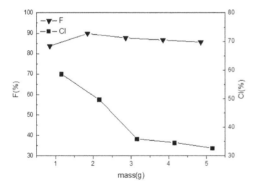

Figure 5. Effect of straw mass on the precipitation rate of F and Cl.

4 CONCLUSIONS

From the experimental results, the precipitation rate of F and Cl increases gradually with the increase of temperature and has little correlation with the excess air ratio. Besides, the residence time is attributable to the increase of precipitation rate of chlorine but has little effect on fluoride. High-temperature hydrolysis and ion chromatography analysis are used in processing of samples and measurement of content of F and Cl, respectively.

ACKNOWLEDGMENTS

The authors are grateful for the financial support for this work provided by the National Science and Technology Support Program (2012BAA09B01), the Beijing Higher Education Young Elite Teacher Project (YETP0712), the Fundamental Research Funds for the Central Universities (12MS42, 12QN12, and 2014ZP08), and the Sciences Foundation of Guangdong Power Grid Company of China (Grant No. K-GD2013-057).

REFERENCES

[1] Pindoria R. V, Lim J.Y., Hawkes J. E, Lazaro M.J., AHerod A., Kandiyoti R.. 1997. Structural character-ization of biomass pyrolysis tars/oils from eucalyptus wood waste: effect of H2 pressure and sample configu-ration. *Fuel* 76:1013–1023.
[2] Yang Z.F., etc. 1999. Modern Environmental Geochemistry. *geological publishing house* 6:181–185.
[3] Ren Q.Q., Zhao C.S., Wu X., Liang C., Chen X.P., Shen X.Z., Tang G.Y., Wang Z.. 2009. Formation behavior of sulfur and chlorine during co-pyrolysis of stalk with municipal solid waste. *Journal of Engineering Thermophysics* 08:1431–1433.
[4] Yu C.J., Luo Z.Y., Zhang W.N., Fang M.X., Zhou J.S., Cen K.F.. 2000. Norganic material emission during bio-mass pyrolysis. *Journal of fuel chemistry and technol-ogy* 05:3–425.
[5] Qian Z.K.. 2011. Release behavior of inorganic ele-ments during agricultural straw thermal conversion. *Huazhong University of Science & Technology.*
[6] Xu G.F.. 2009. Experimental study on pyrolysis and fluorine, chlorine migrate of biomass. *Huazhong University of Science & Technology.*
[7] Johansen J.M., Aho M., Paakkinen., Taipale R., Egsgaard H., Jakobsen J.G., Frandsen F.J., Raili T., Glarborg P.. 2013. Release of K, Cl, and S during com-bustion and co-combustion with wood of high-chlorine biomass in bench and pilot scale fuel beds. *Proceedings of the Combustion Institute,* 34 (2):2363–2372.
[8] Wen LH, Wu P. 2013. Discipline of chlorine release during biomass pyrolysis and combustion. Energy pro-jects 04:29–33.
[9] Lu C.. 2012. Study on characteristics of ash deposi-tion and release of potassium/chlorine during biomass co-fired with coal.*ShangDong University.*
[10] Guo X. J.. 2009. Study on Release and Control of Chlorine during Biomass Combustion. *Huazhong University of Science & Technology.*

Energy, Environment and Green Building Materials – Sheng (ed.)
© 2015 Taylor & Francis Group, London, ISBN 978-1-138-02718-3

Properties of gelatin and poly (3-hydroxybutyrate-co-3-hydroxy-valerate) blends

Ya Li, Jing-Kuan Duan, Ya-Juan Wang, Lan Jiang, Kai-Qi Shi & Shuang-Xi Shao
Institute of Materials Engineering, Ningbo University of Technology, Ningbo, China

ABSTRACT: Poly (3-hydroxybutyrate-*co*-3-hydroxyvalerate) (PHBV) and gelatin blending films were prepared via solution casting method. Thermal properties of the blends were measured by DSC, increment of T_m was observed comparing with pure PHBV. The spherulitic morphology and crystal structure of the blends were studied with POM and WAXD. It is found that the ring-banded structure of the spherulites became obscure with the addition of gelatin and the crystallinity decreased with increasing content of gelatin. Furthermore, contact angle of the blends with the fraction of 25/75 decreased to 56.8°.

KEYWORDS: gelatin/PHBV; blends; thermal properties; crystal morphology

1 INTRODUCTION

Biodegradable materials have received considerable attention for the growing interest about environmental impact of discarded plastics and resources conservation in recent years[Avella, M. and Errico, M. E. 2000, Qiu, Z. B.et al 2005]. In particular products from natural sources, such as starch, gelatin, cellulose or other natural polymers has been widely studied owing to their biological origin, biodegradability, biocompatibility, and commercial availability at relatively low cost, especially in the preparation of synthetic/polymer blends[Avella, M. & Errico, M. E. 2000, Chiellini, E.et al 2001, Gross, R. A. & Kalra, B. 2002, Carvalho, R. A. & Grosso, C. R. F. 2004, Mendieta-Taboada, O.et al 2008, Pawde, S. M. & Deshmukh, K. 2008, Pérez-Mateosa, M. 2007].

Poly(3-hydroxybutyrate-*co*-3-hydroxyvalerate) (PHBV) is one of bacterially derived polyesters. These polymers have recently attracted considerable attention by scientists from academic and industry mainly because they are biodegradable thermoplastics and elastomers that can be processed through conventional extrusion and moulding process[Avella, M. & Errico, M. E. 2000, Chiellini, E. 2001]. The drawbacks of PHBV are the high cost compared to that of petroleum-based commodities plastics. To reduce the cost or improve the performance properties of PHBV, studies on the modification of PHBV through copolymerization and blending with other polymers has been made in the area of industrial production, medicine and surgical[Avella, M.et al 2000, Chen, G. X.et al 2002, Ferreira, B. M. P.et al 2002, Fei, B.et al 2004, Liu, H. L.et al 2007, Meng, W.et al 2007].

Gelatin can be widely found in nature and is the major constituent of skin, bones and connective tissue, it can be obtained by a controlled hydrolysis of fibrous insoluble protein and collagen[Fan, L. H. et al 2005, Zhang, Y. Z.et al 2005, Pawde, S. M. and Deshmukh, K. 2008]. It is widely used in various applications as manufacturing of pharmaceutical products, X-ray and photographic films development and food processing et al.[Yannas, I. V. 1972]. Gelatin is characterized by having a high content of the amino acid glycine (33 mol%) and the presence of the amino acid hydroxyproline (10 mol%) and hydroxylysine (0.5 mol%). The properties of gelatin as a typical rigid-chain high molecular weight compound are in many respects similar to those of rigid-chain synthetic polymers, but it is different from the common biopolymers for the presence of both acidic and basic functional groups in the gelatin macromolecules[Kozlov, P. V. & Burdygina, G. I. 1983].

Aimed for the preparation and evaluation of natural biopolymers in the application as biodegradable materials, PHBV/gelatin blend films were developed. Attempts were made to characterize the thermal properties, crystal morphologies and the hydrophilicity of blend films.

2 EXPERIMENTAL

2.1 *Materials*

PHBV with a 12 mol % of HV was supplied by Ningbo Tianan Biologic Material Co., Ltd. The number-average molecular weight (M_n) and the molecular weight distribution index (M_w/M_n) was measured to be 182 000 g/mol and 1.52, respectively. PHBV was purified before use. The samples were dissolved in chloroform and then precipitated with methyl alcohol. Gelatin was purchased from Sinopharm Chemical Reagent Co., Ltd.

2.2 Preparation of PHBV/gelatin blends

PHBV and gelatin was dissolved separately in 2,2,2-trifluoro ethanol (TFE), at a concentration of 5% (w / v) and stirred for complete dissolution. The final compositions of the PHBV/gelatin blends were 100/0, 75/25, 67/33, 50/50, 33/67, 25/75 and 0/100, the mixture was poured onto a polytetrafluoroethylene mould. After evaporation of the solvent, films with a thickness of approximately 0.04 mm were obtained. Thermal properties, hydrophilicity and crystallization morphology of the samples were tested.

2.3 Characterization

Thermal transitions of the blending films were measured by differential scanning calorimetry (DSC) 200 F3 (Netzsch Instruments). All scans were made at a heating rate of $10°C$ min^{-1} under a dry nitrogen atmosphere. Samples were first heated from 20 °C to 200 °C and maintained at 200 °C for 5 min until the samples completely melted, thus eliminating the thermal history. Then they were cooled to 20 °C and heated again to 200 °C to study the behavior in the absence of previous thermal histories.

The mixed solutions of PHBV and gelatin with various compositions were cast on glass slides and films with the thickness of about 2 μm were obtained after the complete evaporation of solvents. The samples were first heated to 200 °C for 3 min and then cooled to 90 °C for crystallization. An Olympus BX51 polarized optical microscope (POM) equipped with a hot stage (Linkam THMS 600) was used in the morphology observation. Crystal structure of the samples were characterized by wide-angle X-ray diffraction using a D8 Advance system (Bruker Corporation) with Cu Kα (λ=0.15418 nm) operating at 40 kV and 40mA in the diffraction angle range between 3 to 40°.

Contact angles were measured at room temperature using a Dataphysics OCA20 Contact Angle system. The drop volume was about 6 μL, and dropped onto the films for five different positions of the same sample.

3 RESULTS AND DISCUSSION

3.1 DSC analysis

Figure 1 shows the DSC curves of PHBV/gelatin blends obtained from the second heating and the corresponding thermal parameters are summarized in Table 1. Endothermic peaks can be observed in Figure 1 which was ascribed to the melting peaks of PHBV. It can be known in Figure 1 that the melting

temperature (T_m) of pure PHBV is about 167.6 °C, T_m increased with the increasing content of gelatin in the blends and reached to the maximum for the 67/33 blend composites, and then it has a slightly decrease with the increasing content of gelatin, however there was little variation in the melt temperatures in the blends with the fractions of 50/50, 33/67 and 25/75. According to Thomson-Gibbs equation, T_m of the polymers was related with the crystal thickness l_c, increasing T_m of illustrated an increasing l_c of PHBV crystals formed in the blends.

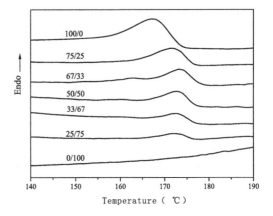

Figure 1. DSC heating thermograms of PHBV/gelatin blending films in different compositions.

Table 1. Melting temperatures (T_m), degree of crystallinity (x_c) and spherulites growth rate (G) of PHBV/gelatin blending composites.

PHBV/gelatin	$T_m/°C$	$x_c^*/\%$	$G^{**}/(\mu m/s)$
100/0	167.6	57.9	0.21
75/25	172.1	53.6	0.23
67/33	173.8	52.8	0.21
50/50	172.9	51.4	0.23
33/67	172.7	45.7	-
25/75	172.6		-

* Crystallinity calculated from WAXD patterns.
** Crystal growth rate (G) calculated from POM graph.

3.2 Crystallization morphology of PHBV/gelatin blends

Figure 2 shows POM images of pure PHBV and PHBV/gelatin blends crystallized at 90 °C. Ring-banded spherulites with Maltese cross of pure PHBV can be observed (Figure 2a) and formation of the bands is ascribed to lamellae twisting[Wang, Z. B. et al 2010]. With the addition of gelatin, radius of the spherulites increased. For the blends, the spherulites became ambiguous and the band structure disappeared for 67/33 and 50/50 blend (Figure 2c and 2d), and radius of the spherulites was larger than that of neat PHBV. With the addition of non-crystallizable gelatin, gelatin scattered in the blends and so the spherulites formed in PHBV phase can't be clearly seen which shows vague POM images (Figure 2b to 2d). Phase separation can be observed from Figure 2, indicating the immiscibility of PHBV and gelatin.

The spherulitic growth rate of neat PHBV and the blends at 90 °C was also measured by POM and the results were listed in Table 1. The results indicated that all the samples have a linearly growth of the spherulites. As can be concluded from Table 1 that the spherulitic growth rate of neat PHBV and the blends are similar (about 0.2 μm/s), blend fraction have little influence on the growth rate of PHBV spherulites.

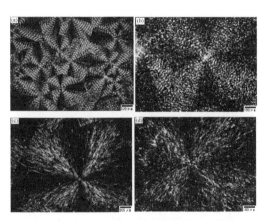

Figure 2. POM micrographs of neat PHBV and PHBV/gelatin blends crystallized at 90 °C. (a) 100/0; (b) 75/25; (c) 67/33; (d) 50/50; (e) 25/75.

Figure 3 shows the WAXD patterns of neat and blended PHBV crystallized at 90 °C. Main characteristic diffraction peaks of PHBV appears around 13.4°, 16.7°, 20.9°and 25.2°, which are corresponded to the (020), (110), (111) and (121) crystallographic planes, respectively[Wang, X.J. et al 2010]. It can be known in Figure 3 that both pure and blended PHBV exhibit peaks at the same locations, indicating that blending with amorphous gelatin does not modify the crystal structure of PHBV. Intensity of the diffraction peaks decreased with the increasing content of gelatin. The degree of crystallinity (x_c)

of the blend films were calculated by computer peak splitting program and the results were listed in Table 1. The results showed that the values of x_c decreases from 57.9% of pure PHBV to 45.7% of 33/67 blend. This may be attributed to the decreasing crystallization ability of PHBV in the blends.

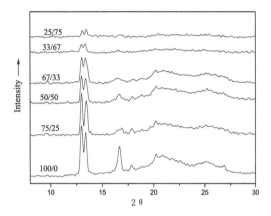

Figure 3. WAXD patterns of neat PHBV and PHBV/gelatin blends crystallized at 90 °C.

3.3 Contact angle of blend films

Gelatin can swell in water and dissolve in hot water, it has been reported that the average amount of bound water in helical gelatin was 0.37 g/g gelatin[Kozlov, P. V. & Burdygina, G. I. 1983], while PHBV is relatively hydrophilic compared with gelatin. Figure 4 gave the contact angles of PHBV and the blend films determined by deionized water via the static contact angle measurement. It is found that the contact angles of the 75/25 and 67/33 blend films are close with that of neat PHBV film, with the increasing content

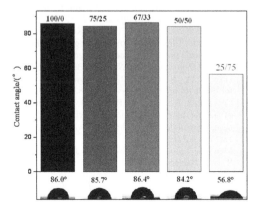

Figure 4. The contact angles of PHBV and the blending films with various PHBV/gelatin fractions.

of gelatin, the contact angle decreased to 84.2° for 50/50 blend and 56.8° for 25/75 blend. This can be ascribed to the hydrophilicity of gelatin, which has hydrophilic groups such as $-NH_2$, $-OH$, $-CONH$ and $-COOH$ et al. The increased hydrophilicity was an important factor to affect cell adhension and the biocompatibility of materials.

4 CONCLUSIONS

The above results revealed that PHBV/gelatin blends are immiscible. T_m of the blends increased as revealed by DSC. Both ring-banded spherulites and normal spherulites formed in the blend films and the crystallinity of PHBV in the blends decreased with increasing content of gelatin. Contact angle of the films decreased from 86.0° for PHBV to 56.8° for 25/75 blend films which implies the increase of hydrophilicity.

ACKNOWLEDGMENTS

Project (Y201326844) supported by the Department of Education of Zhejiang Province, China; Project (2009A610050, 2012A610089, 2013D10006) supported by the Natural Science Foundation of Ningbo City, China.

REFERENCES

[1] Avella M., Errico M. E. 2000. Preparation of PHBV/ starch blends by reactive blending and their characterization. *Journal of Applied Polymer Science* 77(1):232–236.
[2] Qiu Z. B., Yang W. T., Ikehara T., Nishi T. 2005. Miscibility and crystallization behavior of biodegradable blends of two aliphatic polyesters. Poly (3-hydroxybutyrate-co-hydroxyvalerate) and poly (epsilon-caprolactone). *Polymer* 46(25):11814–11819.
[3] Chiellini E., Cinelli P., Corti A., Kenawy E. 2001. Composite films based on waste gelatin: thermalmechanical properties and biodegradation testing. *Polymer Degradation and Stability* 73(3):549–555.
[4] Gross R. A., Kalra B. 2002. Biodegradable polymers for the environment. *Science* 297(5582):803–807.
[5] Carvalho R. A., Grosso C. R. F. 2004. Characterization of gelatin based films modified with transglutaminase, glyoxal and formaldehyde. *Food Hydrocolloids* 18(5):717–726.
[6] Mendieta-Taboada O., Sobral P. J. D., Carvalho R. A., Habitante A. M. 2008. Thermomechanical properties of biodegradable films based on blends of gelatin and poly(vinyl alcohol). *Food Hydrocolloids* 22(8):1485–1492.
[7] Pawde S. M., Deshmukh K. 2008. Characterization of polyvinyl alcohol/gelatin blend hydrogel films for biomedical applications. *Journal of Applied Polymer Science* 109(5):3431–3437.
[8] Pérez-Mateosa M. 2007. Formulation and stability of biodegradable films made from cod gelatin and sunflower oil blends. *Food Hydrocolloids* 23(1):53–60.
[9] Chiellini E. 2001. Biorelated polymers: Sustainable Polymer Science and Technology. New York: Kluwer Academic Plenum Publishers.
[10] Avella M., Martuscelli E., Raimo M. 2000. Review Properties of blends and composites based on poly (3-hydroxy)butyrate (PHB) and poly (3-hydroxybutyrate-hydroxyvalerate) (PHBV) copolymers. *Journal of Materials Science* 35(3):523–545.
[11] Chen G. X., Hao G. J., Guo T. Y., Song M. D., Zhang B. H. 2002. Structure and mechanical properties of poly (3-hydroxybutyrate-co-3-hydroxyvalerate) (PHBV)/ clay nanocomposites. *Journal of Materials Science Letters* 21(20):1587–1589.
[12] Ferreira B. M. P., Zavaglia C. A. C., Duek E. A. R. 2002. Films of PLLA/PHBV: Thermal, morphological, and mechanical characterization. *Journal of Applied Polymer Science* 86(11):2898–2906.
[13] Fei B., Chen C., Wu H., Peng S. W., Wang X. Y., Dong L. S. 2004. Comparative study of PHBV/ TBP and PHBV/BPA blends. *Polymer International* 53(7):903–910.
[14] Liu H. L., Dong L. M., Wang C., Zan Q. F., Tian J. M. 2007. Properties of the calcium phosphate cement/ PHBV microspheres composites. *Rare Metal Materials and Engineering* 36:40–42.
[15] Meng W., Kim S. Y., Yuan J., Kim J. C., Kwon O. H., Kawazoe N. 2007. Electrospun PHBV/collagen composite nanofibrous scaffolds for tissue engineering. *Journal of Biomaterials Science-Polymer Edition* 18(1):81–94.
[16] Fan L. H., Du Y. M., Huang R. H., Wang Q., Wang X. H., Zhang L. N. 2005. Preparation and characterization of alginate/gelatin blend fibers. *Journal of Applied Polymer Science* 96(5):1625–1629.
[17] Zhang Y. Z., Ouyang H. W., Lim C. T., Ramakrishna S., Huang Z. M. 2005. Electrospinning of gelatin fibers and gelatin/PCL composite fibrous scaffolds. *Journal of Biomedical Materials Research Part B-Applied Biomaterials* 72B(1):156–165.
[18] Yannas I. V. 1972. Collagen and gelatin in solid-state. *Journal of Macromolecular Science-Reviews in Macromolecular Chemistry and Physics* C 7(1):49-&.
[19] Kozlov P. V., Burdygina G. I. 1983. The structure and properties of solid gelatin and the principles of their modification. *Polymer* 24(6):651–666.
[20] Wang Z. B., Li Y., Yang J., Gou Q. T., Wu Y., Wu X. D. 2010. Twisting of lamellar crystals in poly(3-hydroxybutyrate-co-3-hydroxyvalerate) ring-banded spherulites. *Macromolecules* 43(10):4441–4444.
[21] Wang X. J., Chen Z.F., Chen X.Y., Pan J.Y., Xu K.T. 2010. Miscibility, crystallization kinetics, and mechanical properties of poly(3-hydroxybutyrate-co-3-hydroxyvalerate)(PHBV)/poly(3-hydroxybutyrate-co-4-hydroxybutyrate)(P3/4HB) blends. *Journal of Applied Polymer Science* 117(2):838–848.

Energy, Environment and Green Building Materials – Sheng (ed.)
© 2015 Taylor & Francis Group, London, ISBN 978-1-138-02718-3

Treatment of simulated acid mine drainage by air cathode microbial fuel cell

C.F. Cai, F.Z. Qi, J.P. Xu, H.C. Zhu & L. Jiang

School of Chemical and Biological Engineering, Anhui Polytechnic University, China

ABSTRACT: Simulated Acid Mine Drainage (AMD) was treated using air cathode Microbial Fuel Cell (MFC) and the results indicated air cathode MFC was feasible in treating AMD. 94.4% Chemical Oxygen Demand (COD) and 94.5% sulfate were successfully removed. The metal ions were 96% removed.

KEYWORDS: Air cathode; Acid mine drainage; Metal ions

1 INTRODUCTION

Acid mine drainage (AMD) is a harmful wastewater caused by the biological oxidization of metal sulfides to metal sulfates (Chang, Shin & Kim, 2000, Neculita, Zagury & Bussiere 2007). AMD with its low pH and the heavy metals such as lead, copper, and arsenic can do harm to aquatic life. Considering its adverse effects, a lot of measures are taken (Neculita & Zagury 2008, Sheoran, Sheoran, & Choudhary 2010). Passive treatment, including permeable reactive barrier(Akcil & Koldas 2006, Johnson & Hallberg 2005) and active treatment, including biological treatment(Costa & Duarte 2005, Oncel, Muhcu, Demirbas & Kobya 2013) were conducted and the benefits were obvious.

It is recently proposed that AMD can be treated by microbial fuel cell (Cheng, Dempsey & Logan 2007, Yang et al. 2012). In a MFC, the substrates are oxidized by the bacteria that are separated from the electron acceptor by proton exchange membrane. Electrons pass through the bacteria to the anode and then via a circuit to the cathode where water is formed through the combination of protons and oxygen (Sevda et al. 2013, Zhang et al. 2011). Air cathode MFC is one of the promising technologies and can be used to treat AMD.

In this study one single-chamber, air-cathode MFC was constructed to investigate the effects of COD/SO_4^{2-} (C/S) and the interval between the anode and the cathode on the performance of MFC. The other objectives are to select the best C/S and the interval in treating AMD using air cathode MFC.

2 MATERIALS AND METHODS

2.1 *Materials*

The simulated AMD used in the experiment was composed of (per liter): NH_4Cl 191mg, K_2HPO_4 75mg, Na_2SO_4 2215mg, $MgSO_4 \cdot 7H_2O$ 3844mg, $CuCl_2 \cdot 2H_2O$ 79.7mg, $FeSO_4 \cdot 7H_2O$ 149mg, $Pb(NO_3)_2$ 47.9mg, $ZnCl_2$ 62.8mg, $Cd(NO_3)_2 \cdot 4H_2O$ 82.5mg. The HCl solution was used to adjust the pH of AMD to 5.

The activated sludge and SRB were collected from Zhu Jiaqiao wastewater treatment plant, Wuhu, China and Anhui University of Technology, Hefei, China.

2.2 *Experimental setup*

One single-chamber, air-cathode MFC was constructed. The size of MFC was 16 cm in length, 16 cm in height and 17 cm in width. The anode electrode (projected surface area = 225cm²) was made of carbon cloth (without wet proofing) and the cathode was made of carbon cloth (without wet proofing) containing a Pt/C (0.5 mg/cm², water-facing side). Three anodes were fixed in the MFC at the interval of 3cm, 7cm and 11cm from the cathode. The proton exchange membrane (PEM) was place between the simulated AMD and the cathode. PEM was sequentially boiled in H_2O_2 (30%): deionzied water, 0.5 M H_2SO_4 and then deionized water (each time for 1h). PEM was placed between the anode carbon cloth and the reactor. Copper wire was used to connect the circuit containing a 1000Ω load unless stated otherwise.

500 ml sulfate reducing bacteria and 1000 ml activated sludge were added in the chamber and then 1500ml AMD was filled in the chamber. MFC was allowed to activate for 6 days before measurements were taken.

3 ANALYSES

Voltage was continuously measured every 30 min by a multimeter with a data acquisition system (PISO-80, ICP CO.Ltd). Chemical oxygen demand (COD) and sulfate measurements were carried out using the standard analysis methods. Metal ions were measured using TASS 990 atomic absorption spectrophotometer, purchased from Beijing Purkinje

General Instrument Co., Ltd. Current density was calculated as i = E/RA, where E (mV) is cell-measured voltage, R (Ω) the external resistance and A (cm2) the projected surface area of the anode. Power density in MFC was calculated according to P (mW/m2) = 10iE, where 10 is needed for the given units.

4 RESULTS AND DISCUSSION

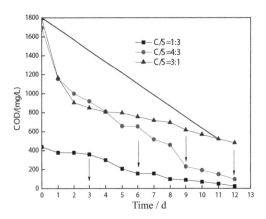

Figure 1.   The variation of COD with the increasing time.

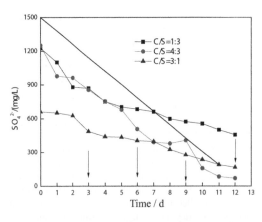

Figure 2.   The change in sulfate with the increasing time.

Fig.1 presents the effect of C/S on COD. The COD values decreased with the increasing time, varying from1783.23 mg/L to <10 mg/L with the maximum removal efficiency of some 94.4%. It can be seen that the removal efficiency of C/S=4:3 was high than that of C/S=1:3 and 3:1. It is considered that the C/S=4:3 was close to the theoretical C/S, 0.67 and achieve the best removal efficiency. Compared with the interval of 3cm, 7cm, 11cm and anode connecting in series, it

is found that at the interval of 3cm, the removal efficiency reached the maximum, 52.6%, higher than that of 7cm, 10.8%, of 11cm, 18.4% and anode connecting in series, 22.2%. The removal efficiency deceased with the increasing distance between the anode and cathode and connecting the anode in series didn't enhance the removal efficiency because that electrically connecting anode in series will produce voltage reversal. Voltage reversal is easily produced in MFC when connecting anode in series by having various COD concentrations in the MFC reactors.

The effect of C/S on sulfate is shown in Fig. 2. The sulfate concentrations decreased with the time, consistent with COD. The removal efficiency in descending order was C/S=4:3 94.5%, C/S=1:3 62.3% and C/S=3:1 74.5%. As mentioned above, C/S=4:3 is more close to the theoretic C/S and reached the best result. The removal efficiency of C/S=4:3 was lower than that of C/S=4:3, indicating that too much COD was not helpful in enhancing the removal efficiency of sulfate. At the interval of 3cm, the removal efficiency of sulfate at C/S =4:3, 31.6%, was higher than that at C/S=1:3 29.2% and 3:1 26.4%. Similarly at the interval of 7cm and 11cm, C/S=4:3 and C/S=3:1 reached the best results respectively, 40.7% and 31.1%.

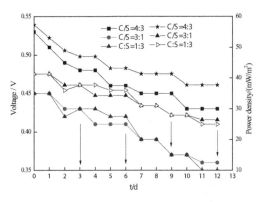

Figure 3.   Voltage and power density with the increasing time.

The effect of C/S on voltage is similar with that on power density (Fig. 3). The maximum voltage in descending order is C/S=4:3, C/S=1:3 and 3:1, reaching 57.32mV, 41.42mV, 39.56mV respectively. With the continuous reduction of the substrates (COD) the voltage deceased slowly. The decreasing rate of voltage was the same with COD reduction at the interval of 3cm, 7cm, 11cm and anode connecting in series. The variation of power density was consistent with the change in voltage. The maximum power density was obtained at C/S=4:3 55.18 mW/m2 at the interval of 3cm and the minimum power density was got at C/S=1:3 at the interval of 11cm 19.65 mW/m2. The power density obtained here was not high and it is considered that the style of the reactor has some influence on power density.

The removal efficiencies of metal ions, including Cu^{2+}, Zn^{2+}, Cd^{2+} and Fe^{2+} are all above 96% and the data are not shown here.

5 SUMMARY

Air cathode MFC was constructed to assess the effects of C/S and the interval of anode and cathode on the performance of MFC and the results indicated that air cathode MFC was feasible in treating AMD. It can be seen from the results that C/S=4:3 is optimal in the present experiment and the most suitable interval is 3cm. In the present experiment the maximum removal efficiency of COD and sulfate is 94.4% and 94.5% respectively. 96% metal ions are removed.

ACKNOWLEDGMENTS

This work was financially supported by the National Natural Science Foundation of China (51274001) and the Opening Project of State Key Laboratory of Coal Recourses and Safe Mining (SKLCRSM10KFA05).

REFERENCES

Akcil, Ata, & Koldas, Soner. 2006, Acid Mine Drainage (AMD): causes, treatment and case studies, Journal of Cleaner Production 14(12): 1139–1145.

Chang, In Seop, Shin, Pyong Kyun, & Kim, Byung Hong. 2000, Biological treatment of acid mine drainage under sulphate-reducing conditions with solid waste materials as substrate, Water research 34(4): 1269–1277.

Cheng, Shaoan, Dempsey, Brian A., & Logan, Bruce E. 2007, Electricity Generation from Synthetic Acid-Mine Drainage (AMD) Water using Fuel Cell Technologies, Environmental Science & Technology 41(23): 8149–8153.

Costa, MC, & Duarte, JC. 2005, Bioremediation of acid mine drainage using acidic soil and organic wastes for promoting sulphate-reducing bacteria activity on a column reactor, Water, air, and soil pollution 165(1–4): 325–345.

Johnson, D. B., & Hallberg, K. B. 2005, Acid mine drainage remediation options: a review, Sci Total Environ 338(1–2): 3–14.

Neculita, C. M., & Zagury, G. J. 2008, Biological treatment of highly contaminated acid mine drainage in batch reactors: Long-term treatment and reactive mixture characterization, J Hazard Mater 157(2–3): 358–366.

Neculita, C. M., Zagury, G. J., & Bussiere, B. 2007, Passive treatment of acid mine drainage in bioreactors using sulfate-reducing bacteria: critical review and research needs, J Environ Qual 36(1): 1–16.

Oncel, MS, Muhcu, A, Demirbas, E, & Kobya, M. 2013, A comparative study of chemical precipitation and electrocoagulation for treatment of coal acid drainage wastewater, Journal of Environmental Chemical Engineering 1(4): 989–995.

Sevda, Surajbhan, Dominguez-Benetton, Xochitl, Vanbroekhoven, Karolien, De Wever, Heleen, Sreekrishnan, TR, & Pant, Deepak. 2013, High strength wastewater treatment accompanied by power generation using air cathode microbial fuel cell, Applied Energy 105: 194–206.

Sheoran, A. S., Sheoran, V., & Choudhary, R. P. 2010, Bioremediation of acid-rock drainage by sulphate-reducing prokaryotes: A review, Minerals Engineering 23(14): 1073–1100.

Yang, Qiao, Wang, Xin, Feng, Yujie, Lee, He, Liu, Jia, Shi, Xinxin, Ren, Nanqi. 2012, Electricity generation using eight amino acids by air–cathode microbial fuel cells, Fuel 102: 478–482.

Zhang, X., Sun, H., Liang, P., Huang, X., Chen, X., & Logan, B. E. 2011, Air-cathode structure optimization in separator-coupled microbial fuel cells, Biosens Bioelectron 30(1): 267–271.

Energy, Environment and Green Building Materials – Sheng (ed.)
© *2015 Taylor & Francis Group, London, ISBN 978-1-138-02718-3*

Biochemicals of acid mine drainage using sulfate-reducing bacteria activity on a column reactor

C.F. Cai, L. Jiang, C.F. Wu & F.Z. Qi
School of Chemical and Biological Engineering, Anhui Polytechnic University, China

ABSTRACT: Acid Mine Drainage (AMD) is a prominent environment problem due to the high heavy metals present and sulfate content. Laboratory column was used to study the function and capacity of Sulfate-Reducing Bacteria (SRB). The results showed that the removal efficiencies of Cd, Zn, Fe, and Cu were above 90%. On account of no addition of carbon source, the growth of SRB was limited, and the reducing capacity was weakened more and more slowly with passing of time. So the addition of carbon source at regular intervals is very necessary.

KEYWORDS: Acid mine drainage; heavy metals; sulfate-reducing bacteria

1 INTRODUCTION

Heavy-metal pollution is a salient representation of environment problems (Costa, Martins, Jesus, & Duarte, 2008). Mining and metallurgical operations are considered the principal sources of heavy-metal pollution and sulfate in the environment. Acid mine drainage is a form of wastewater that contains a large quantity of heavy metals and sulfate. AMD has been traditionally treated by physical adsorption, chemical precipitation, and biological reduction (Jeen, Gillham, & Gui, 2009; Sahinkaya, Gunes, Ucar, & Kaksonen, 2011; Tsukamoto, Killion, & Miller, 2004). Sulfate-reducing bacteria (Costa & Duarte, 2005) have been considered an effective, economic approach for removing sulfate, because this method can use sulfate as the terminal electron acceptor during metabolism of organic matter, producing H2S that can be used as a metal precipitating agent (Gibert, De Pablo, Cortina, & Ayora, 2002; Neculita, Zagury, & Bussière, 2007; Rinck-Pfeiffer, Ragusa, Sztajnbok, & Vandevelde, 2000). The equation is expressed as follows:

$$2H^+ + SO_4^{2-} + 2CH_2O \xrightarrow{SRB} H_2S + 2H_2CO_3$$

$$M^{2+} + H_2S \rightarrow MS \downarrow + 2H^+$$

The corncob can release polysaccharide during the fermentation, which can provide organic matter for SRB (Elliott, Ragusa, & Catcheside, 1998; Ruhl, Weber, & Jekel, 2012).

So, in this research Cd, Zn, Cu, and Fe were the metals chosen to monitor the efficiency of the sulfate-reducing biological process under study (Jong & Parry, 2003).

2 MATERIALS AND METHODS

AMD water sample is a compound of nine different kinds of chemicals, and the formula is shown in Table 1.

Table 1. Formula of AMD(g/L).

NH$_4$cl	0.191	K$_2$HPO$_4$·3H$_2$O	0.075
Na$_2$SO$_4$	2.215	MgSO$_4$·7H$_2$O	3.844
Cucl$_2$·2H$_2$O	0.0797	FeSO$_4$·7H$_2$O	0.149
Pb(NO$_3$)$_2$	0.0479	Zncl$_2$	0.0638
Cd(NO$_3$)$_2$·4H$_2$O	0.0825		

The laboratory-scale investigation was performed with the purpose of developing a low-cost and simple biochemical process for the treatment of AMD water. The column reactor was carried out using polyvinyl chloride resin (PVC) filter columns of 70-cm height and 16-cm internal diameter. The bottom was closed with PVC plate. The column was packed with 67% ceramsite (2-4 mm in diameter) and 33% corncob (3–5 mm in diameter); 520-ml sulfate-reducing bacteria were added to the column. Thereafter, an acclimatization procedure was started, lasting 1 week, during which time the bacteria grew and adapted to the simulated conditions. The AMD water was constantly supplied at the bottom of the column with peristaltic pumps, the flow rate was 0.06 cm3s-1m and hydraulic residence time was 24 hours. The continuous flow lasted 8 days.

The metal concentrations of the AMD and the effluents of the system were determined by Atomic Absorption Spectroscopy (AAS). The values were accepted just when a reasonable standard deviation was obtained.

3 RESULTS AND DISCUSSION

The removal efficiency of Cd was tested with time, increasing it to ensure the removal of the system. The result is presented in Fig. 1. The initial removal of Cd was 76.07%, which indicates that SRB growth just occurred in the system. The removal efficiency rapidly increased in the first 5 days, and the greatest efficiency was 99.35% on the 5th day, which indicated SRB growth and the generation of CdS. As time passed, the organic matter was slowly worked out and the quantity of SRB was decreased, so the ability of reduction slowed down. Then, the tendency was slowly decreased.

It could be seen in Fig. 1 that the removal of Zn on the 4th day was abruptly increased from 91.18% to 94.92%. On the whole, the initial removal of Zn was much bigger than Cd, which showed that it was easier to reduce and precipitate Zn than Cd. The biomass reached a maximum, and the capacity of reduction was the strongest at this time. It was consistent that the removal of Cd began reducing after five days, but the average of the removal was above 90%. The nutrient medium began to be gradually replaced by AMD, so the removal decreased during the next few days.

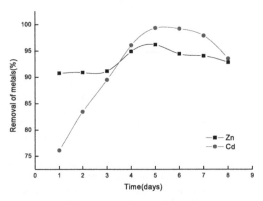

Figure 1. Variations of Cd, Zn removal with time.

The removal efficiencies of Cu and Fe are shown in Fig. 2. The tendencies of the two curves increased with the passing of time. The initial removal of Fe was lower than 40%; then, on the third day, the efficiency rose very quickly and reached 75.65%. On the

7th day, the value reached its highest and then began to decline.

The removal rate of Fe in the system was relatively rapid when compared with Cu. But the initial efficiency was bigger than Fe; it was 67.38%. The removal efficiencies of Cu maintained a rising trend. After the first two days, the values were smaller than Fe all through the next few days. On the last day, the removal efficiencies of the two heavy metals were nearly the same.

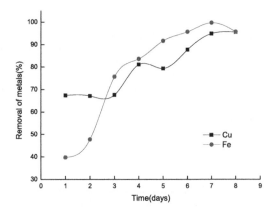

Figure 2. Variations of Cu, Fe removal with time.

On comparison with Fig. 2, we could draw the conclusion that Cu and Fe were reduced by SRB antecedent to Cd and Zn after five days; namely, Cu and Fe could be efficiently reduced with the decline of SRB biomass.

The decline in the removal rate of SO42- in the system was relatively rapid in the first three days: It was from 66.60% to 19.55%. On comparing Fig. 1 and Fig. 2, it could be illustrated that chemical precipitation played a leading role apart from biological precipitation. In the last 3 days, the trend was nearly gentle.

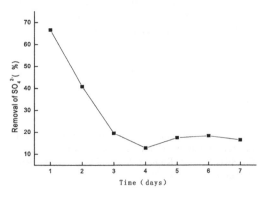

Figure 3. Removal efficiency of SO_4^{2-}.

4 CONCLUSIONS

This work demonstrated microbial sulfate reduction and precipitation of Cd, Zn, Cu, and Fe. Among the heavy metals, it was demonstrated that zinc was the first heavy metal to be removed, then cadmium, followed closely by copper, and lastly iron. The heavy metal removal was attributed to the insoluble metal sulfides produced by the biological activities of SRB.

Further studies are very necessary to extend the cycle of operation, as well as to select appropriate and economic carbon sources for SRB.

ACKNOWLEDGMENTS

This study was supported in part by the grants from the National Natural Science Foundation of China (51274001) and the Opening Project of State Key Laboratory of Coal Recourses and Safe Mining (SKLCRSM10KFA05). The authors express their sincere gratitude to Professor Changfeng Cai for help rendered.

REFERENCES

Costa, MC, & Duarte, JC. 2005. Bioremediation of acid mine drainage using acidic soil and organic wastes for promoting sulfate-reducing bacteria activity on a column reactor. Water, air, and soil pollution, 165(1–4), 325–345.

Costa, MC, Martins, M, Jesus, C, & Duarte, JC. 2008 Treatment of acid mine drainage by sulfate-reducing bacteria using low cost matrices. Water, air, and soil pollution, 189(1–4), 149–162.

Elliott, Phillip, Ragusa, Santo, & Catcheside, David. 1998. Growth of sulfate-reducing bacteria under acidic conditions in an upflow anaerobic bioreactor as a treatment system for acid mine drainage. Water Research, 32(12, 3724–3730.

Gibert, O, De Pablo, J, Cortina, JL, & Ayora, C. 2002. Treatment of acid mine drainage by sulfate-reducing bacteria using permeable reactive barriers: a review from laboratory to full-scale experiments. Reviews in Environmental Science and Biotechnology, 1(4), 327–333.

Jeen, Sung-Wook, Gillham, Robert W, & Gui, Lai. 2009 Effects of initial iron corrosion rate on long-term performance of iron permeable reactive barriers: Column experiments and numerical simulation. Journal of contaminant Hydrology, 103(3), 145–156.

Jong, Tony, & Parry, David L. 2003. Removal of sulfate and heavy metals by sulfate reducing bacteria in short-term bench scale upflow anaerobic packed bed reactor runs. Water Research, 37(14), 3379–3389.

Neculita, Carmen-Mihaela, Zagury, Gérald J, & Bussière, Bruno. 2007. Passive treatment of acid mine drainage in bioreactors using sulfate-reducing bacteria. Journal of Environmental Quality, 36(1), 1–16.

Rinck-Pfeiffer, Stéphanie, Ragusa, Santo, Sztajnbok, Pascale, & Vandevelde, Thierry. 2000. Interrelationships between biological, chemical, and physical processes as an analog to clogging in aquifer storage and recovery (ASR) wells. Water Research, 34(7), 2110–2118.

Ruhl, Aki S, Weber, Anne, & Jekel, Martin. 2012. Influence of dissolved inorganic carbon and calcium on gas formation and accumulation in iron permeable reactive barriers. Journal of contaminant hydrology, 142, 22–32.

Sahinkaya, Erkan, Gunes, Fatih M, Ucar, Deniz, & Kaksonen, Anna H. 2011. Sulfidogenic fluidized bed treatment of real acid mine drainage water. Bioresource technology, 1022, 683–689.

Tsukamoto, TK, Killion, HA, & Miller, GC. 2004. Column experiments for microbiological treatment of acid mine drainage: low-temperature, low-pH and matrix investigations. Water Research, 38(6), 1405–1418.

Energy, Environment and Green Building Materials – Sheng (ed.)
© 2015 Taylor & Francis Group, London, ISBN 978-1-138-02718-3

Intelligent monitoring system design of coal mine ventilation based on internet of things technology

W.C. Li, G.F. Hao & G.Q. Li
Hebei Engineering and Technical College, Cangzhou, Hebei, China

H.J. Wang
Chengjiao Coal Mine of Henan Yongcheng Coal Group, Yongcheng, Henan, China

ABSTRACT: In order to improve environmental conditions of coal mine and enhance the safety factor of downhole operation and the coal mine management level, an intelligent monitoring system of coal mine ventilation based on internet of things technology has been designed to prevent and reduce the occurrence of mine accidents. The system could be operated to analyze and judge the fan operating, the environmental parameters of the mine, and the video status that are collected by the wireless data collection terminal. When the environmental parameter exceeds the set value, the system increases the fan speed, and when the gas concentration exceeds the standard in the mine, the system starts the buzzer to give an alarm, which could gain precious time for operation personnel to evacuate safely.

KEYWORDS: Internet of things; Mine ventilation; Mesh router; Wireless data terminal

1   INTRODUCTION

In recent years, the major coal mine safety accidents have occurred now and then, and the central government attaches great importance to the safety production of coal mine industry. *Mine ventilation system* is an important equipment to realize the air exchange between the downhole operation and the ground, it is responsible for the delivery of fresh air to the downhole area, and it is to remove harmful, toxic gases, adjust the air volume, humidity and temperature inside the mine, and the purpose of it is to ensure safety in coal mine production[1]. Mine ventilation system is designed for online safety monitoring and real-time control of the downhole operation condition, which is applied to reduce the probability of accidents, and it also can improve the miners' survival chances in mine disaster. At present, most of China's coal mine ventilation system is a manual control and management, and it is lack of on-line monitoring and real-time control means in downhole operation. The introduction of intelligent monitoring system of coal mine ventilation can further enhance the coal mine production safety and its management level, prevent and reduce the accidents, and it is of great significance for improving the safety factor in underground works.

The Internet of things technology as an emerging technology provides a new idea for design of coal mine ventilation monitoring system[2, 3]. On the basis of the powerful function of ARM single chip computer technology and the Internet of things new technology, a scientific design of a wireless intelligent monitoring system for mine ventilation has been discussed in details in this thesis. This intelligent monitoring system collects the operating parameters of fans and fire alarm devices via wireless sensor network machine, and this system could achieve dynamic monitoring of coal mine air volume, temperature and harmful gas content, and based on the DSP microprocessor powerful signal processing ability, real-time data processing could be realized, and through the wireless Mesh network data can be transmitted to the main fiber network, and then the data can be uploaded to the ground command center. The system has the following functions, such as real-time condition monitoring, equipment state display, wireless remote control, real-time data acquisition, data storage and transmission, fault early warning device, intelligent control and etc.

2   THE WHOLE DESIGN OF THE SYSTEM

2.1   *Coal mine ventilation system state parameters*

Many parameters need to be monitored in the coal mine ventilation system, and they can be divided into fans parameters and external environmental parameters.

The fans parameters include the winding temperature, speed, voltage, current and air quantity. It is the basis and precondition for realizing the intelligent mine ventilation control to conduct the real-time and accurate acquisition of these parameters. External environmental parameters include environmental temperature, relative humidity, dust concentration, gas concentration and so on[4], and the real-time acquisition and processing of the data can be realized through all kinds of mines sensors, finally, it could provide an intuitive and reliable data for intelligent monitoring system to make the right judgments, as well as scheduling and coordination control.

2.2 The whole structure of the system

The mine ventilation intelligent monitoring system is composed of the wireless network, the roadway monitoring center, and the ground command center, integrated command center (figure 1). In the wireless network, wireless data terminals consist of a fans' own parameter acquisition device, environmental temperature and humidity parameters acquisition device, and wireless camera etc. Fans' own parameters are collected through the wireless sensor acquisition with specific functions, and external environmental parameters are collected through all kinds of special wireless sensors, and then all the data are processed by DSP, then they are sent to the roadway monitoring center and the downhole fiber, and lastly, the data are sent to ground command center through gigabit fiber ring network. Ground command center would conduct a comprehensive analysis and data process of all kinds of information received, when there is a parameter that exceeds the safety fixed value, a warning signal will be issued, and then a corresponding fan device will be started or adjusted. The roadway monitoring center could not only receive dispatching command and control from the ground command center but also independently carry out real-time downhole monitoring, so when the unexpected events occur there will be quickly judge and treatment.

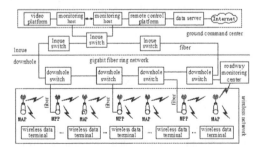

Figure 1. The Whole Structure of the Intelligent Monitoring System of Coal Mine Ventilation.

2.3 The wireless network structure

The system adopts wireless Mesh network to establish the topological structure of wireless data terminal. Wireless Mesh network is the communication system that has been built up on the basis of Mesh MEA (Mesh Enabled Architecture), and it has good expansibility and anti-interference feature, and it has been widely applied in the field of broadband wireless access[5]. Wireless Mesh is a multi hop, dynamic broadband network structure with free organization and self configuration, which is mainly composed of Mesh routers and data collecting terminals[6]. The Mesh router not only has the wireless routing and wireless data access function, but also has the function of gateway. The router node with a gateway function is also known as the Mesh entrance point (MPP), it is the entrance point that is from Mesh network to the cable network, which is responsible for the data forwarding between the cable network and wireless network and the wireless network management[7]. The router node with wireless routing and wireless data access function is called a Mesh access point (MAP), which is consisting of two wireless network cards. One supports 802.11g protocol that is for wireless coverage for the surrounding area, providing access for wireless data terminal. The other supports 802.11a protocol that is for data transmission among the Mesh nodes, achieving packets' fast forwarding. In wireless Mesh network, the backbone network is composed of MPP and MAP, which are applied to realize the downhole roadway wireless full coverage the realization of the mine roadway wireless full coverage, and the downhole Mesh router is used to realize the interconnection between wireless backbone network and the external network.

3 MESH ROUTER DESIGN

The Mesh router is the hinge part of the system, and it is mainly to complete two functions. One is responsible for receiving fans' real-time operation state, fans real-time parameters, environmental real-time parameters and downhole video data that are uploaded by wireless data terminals uploads the of the fan, and then the data will be directly sent to the roadway monitoring center, lastly the data will be sent to ground command center via fiber. The other is to receive the adjustment of the fan from the ground command center and roadway monitoring center to achieve fan's state regulation and parameters' setting.

System of MPP and MAP were operated with the help of MSR4000 type mesh Azalea Networks router. The router has 4 multi-function RF (radio frequency), and they can work in the 2.4GHz, 5GHz or 4.9GHz frequencies, each RF (radio frequency) can be used to realize 802.11a/b/g/n access point mode and 802.11a/b/g/n return mesh router mode, and they adopt 2x2 MIMO (Multiple-Input Multiple-Output, referred to as MIMO) with a radio frequency that could achieve two space division, and they can provide the data transmission speed that could get the highest RF speed of 300Mbps, and there is automatic channel selection and automatic detection and the elimination of radio frequency interference could also be realized, while it provides 10/100/1000BASE-T Ethernet (RJ45) and the USB console interface and eight N type antenna connectors. MSR4000 mesh router can completely meet the needs of coal mine data and video signal transmission requirements.

MSR4000 mesh router RJ45 interface is connected with the downhole switch, wireless data terminal sends data to the ground command center through fiber ring network, and is responsible for receiving and forwarding the fan state regulation order that is from the ground command center.

4 WIRELESS DATA TERMINAL DESIGN

Wireless data terminal consists of fan operation parameter acquisition device, real-time data acquisition device of environmental parameters and wireless video monitoring terminal. Terminal device is connected with mesh router wirelessly through the respective communication unit.

4.1 The design of the fan operation parameter acquisition device

Fan operating parameters acquisition device is composed of a data acquisition module, a data processor, a wireless communication module, a fan control module and the power supply, and the hardware structure is as shown in figure 2.

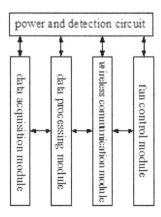

Figure 2. Fan operating parameter acquisition terminal hardware structure.

Data acquisition module is composed of voltage transformer and current transformer, YD62 magnetic sensitive speed sensor, WD4110 air flow sensor and level matching circuit. The collected data is subject to level matching and then it is converted to digital signal via A/D, and lastly it is transmitted to the I/O interface of data processor.

Data processor comprises a CC2430 microprocessor and its peripheral circuit, including crystal circuit, reset circuit, 3.3V circuit and 1.8V power supply. The data processor is responsible for the entire terminal data processing, data storage and data receiving and sending etc.

The wireless communication module is composed of CC2591. Although the CC2430 microprocessor has integrated with wireless transceiver, it is susceptible to outside interference, and the communication distance is limited, thus it is necessary to add a CC2591 wireless module between the antenna and the CC2430 microprocessor, which is used to amplify the receiving and transmitting signal power, ensuring the distance of data transmission.

Fan status control module is composed of a relay and a control circuit. And it is responsible for the real-time control of the motor working state according to the control command so as to improve the environmental parameters.

Power supply and detection circuit provides the power that is required for the terminal, and it is responsible for real-time detection of power supply, when the terminal electric quantity of power supply is lower than the set warning value, the signal is transmitted to roadway monitoring center through the wireless communication module to replace battery or adjust the working power supply.

4.2 Real time environmental parameter acquisition design

Environmental parameters include temperature, humidity, the gas concentration and dust concentration. In this system, the wireless temperature and humidity sensor, gas concentration sensor and dust concentration sensor that are distributed in coal mine roadway to realize real-time measurement of the environment, and the data will be transmitted automatically to the nearest MPP or MAP.

4.3 The video terminal design

Downhole video uses KT37-A mine portable wireless camera that is based on WiMAX to get the data, and the data can be uploaded to roadway monitoring center via MPP or MAP, so the fans' real-time running state, working surface condition and downhole roadway situation can be observed on the spot, and then it can be transmitted to the ground command center through the fiber ring network.

5 THE SYSTEM SOFTWARE DESIGN

The intelligent monitoring system of coal mine ventilation is developed by VB software, and the program flow diagram is as shown in figure 3. After the completion of the system power, it is to control circuit initialization, start the wireless collection terminals, collect fan operating parameters, environmental parameters and downhole video status, and the system's control state can be judged through the analysis and processing of the collected data. If it is in the manual mode, the program will automatically

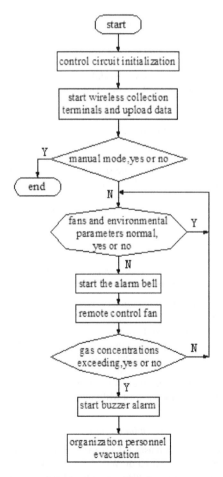

Figure 3. Monitoring program flow diagram.

end, and it is the Inoue and the downhole control on the spot.underground scene. If it is in automatic mode, if the wireless terminal acquisition data is abnormal, it starts alarm bell, and there is the remote control to increase the wind turbine, achieving the improvement of downhole roadway environment; if there is excessive gas concentrations in the environment, it starts buzzer alarm, on-site personal evacuation will be organized under the remote command of ground command center and fault handling center.

6 CONCLUSION

In this paper, the Internet of things technology is used in mine ventilation monitoring system, and the systematic wireless network has been constructed with wireless Mesh routers and wireless data acquisition terminals, and it realizes the real-time monitoring and control of coal mine fan operating conditions, and the real-time monitoring and transmission of downhole environmental parameters have been achieved, realizing accident forecast and early warning of disasters, and the remote intelligent control has been realized through the software platform that is developed by VB, improving the management level of coal mine and safety margin in the downhole operation. The experiments have showed that the system has high practicability.

REFERENCES

[1] Xu Hai. 2007. Design and application of Online monitoring system of main fan. *Mining Machinery* (11): 38–40.
[2] Zhao Wentao, Dong Jun. 2011. Application of networking technology in the coal mines. *Micro Computer Information* 27 (2): 121–122.
[3] Lin Shuguang, Zhong Jun, Wang Jiancheng. 2010. Application of the Internet of things in the coal mine safety production. *Mobile Communication*7 (24): 46–50.
[4] Xin Daxue.2013.Application of automatic control technology in the coal mine ventilation system. *Coal Technology* (10): 88–90.
[5] Li Zhijie, Fang Xuming.2012. In the wireless Mesh network, a QoS guaranteed cross layer scheduling method. *Journal of the China Railway Society*34 (10): 61–67.
[6] Li Qiwei, Huo Zhonggang, Wen Liang. 2012. Study and design of coal emergency rescue base on wireless Mesh network. *Coal Mining Automation* (6): 39–43.
[7] Wang Haidan, Shao Xiaotao. 2014. The design and implementation of downhole wireless Mesh network navigation system35 (1): 18–22.

Energy, Environment and Green Building Materials – Sheng (ed.)
© 2015 Taylor & Francis Group, London, ISBN 978-1-138-02718-3

Public opinion propagation model based on deffaunt model

Xin Zhou, Bin Chen, Zhi-Chao Song, Liang Ma & Xiao-Gang Qiu
College of Information and Management, National University of Defense Technology, China

ABSTRACT: Aiming at compensating for deficiencies of previous research conducted on public opinion propagation, this paper provides a new public opinion propagation model based on Deffaunt model. We extract an individual's character and construct an individual's behavior mechanism. According to Two-Step Flow Theory, we divide the behavior mechanism of an individual into two categories. By simulating a phenomenon that people who are unaware of the truth are deluded by rumormongers in XinJiang 7•5Incident, we model a possible process by which rumors spread among the crowd of Uyghur people. Through this simulation, we obtain a series of curves of opinion in the crowd. The simulation results show that the research work plays an important role in opinion monitoring and intervention.

KEYWORDS: Deffuant Model; Two Step Flow Theory; Rumor propagation; Simulation

1 INTRODUCTION

Public opinion refers to a certain tendency of comments, opinions, or attitudes to a particular event of most people (LIU, 2008). With the rapid development of Internet, the Internet is gradually becoming a new media of public opinion propagation. People can not only get information they need from the Internet but also express their opinions through it. On the one hand, Internet has become a platform of public opinion; on the other hand, it has become a mixed place of real information and misinformation. Although the Internet cannot represent the mainstream people completely, it is still a barometer of social emotion that cannot be ignored. Therefore, it is necessary to study the public opinion on the Internet.

2 RELATED WORKS

Currently, there are two research methods on public opinion. One method is analyzing the data crawled from the Internet. The other method is constructing a model of the individual based on multi-agent theory. Our work is based on the multi-agent theory.

There is a lot of research on the public opinion propagation based on the multi-agent model. Among them, some models excavate the substantive characteristics of public opinion to provide the theoretical basis of the evolution, such as Sznaid Model(Sznaid, 2000), Bounded Confidence Model(Deffuant,2000), and Galam Model (Galam, 2008). Some other models are put forward based on former achievements. Woloszyn(2006) studies a phenomenon of phase transition of opinion dynamics model based on Sznajd model. Based on the idea of Bounded Confidence Model, Kozma (2008) presents the influence of complex network on opinion evolution.

3 PROPAGATION MODEL BASED ON DEFFAUNT MODEL

3.1 *Principle of propagation model*

The macroscopic process of propagation in our model is based on Two-Step Flow Theory. When a hot spot occurs, it will be published in a website or somewhere else. In general, a few active netizens will focus on this hot spot. They will process the information and then spread it to other netizens. After that, the information will be spread and discussed in a substantial number of netizens.

The microscopic process of propagation in our model is based on Deffaunt Model. We choose two agents to meet at each time step. Agents re-adjust their opinion when their difference of opinion is in a threshold.

Therefore, in the process of public opinion propagation, it is necessary to model the information, individual, and the relationship among individuals. However, we focus on the modeling of information and individual in this paper.

3.2 *Modeling of information*

The attributes of information are index, opinion, and authority. Index presents the ID of information. Each information has only one ID, which is different from others. Opinion is the attitude toward information, which is divided into five categories: completely accept, slightly accept, neutrality, slightly opposite, and completely opposite. Authority presents the power of information that makes people believe it, which is determined by the source of information.

3.3 *Modeling of individual*

An individual is modeled as an agent, and the network society consists of a group of agents. Each agent has its own attributes and behaviors.

3.3.1 *Attributes of the agent*

The attributes of the agent are index, opinion, event role, stubbornness, and influence. Index presents the ID of agent. Each agent has only one ID, which is different from others. Opinion is the attitude toward information. Each agent plays a role in a certain event. We call it event role. A small group of agents easily attract others' attention. Their opinions always have a significant influence on others. Most agents find it hard to attract others' attention. Their opinions have a low influence on others. So we divide event roles into two categories: opinion leader and ordinary netizen. Influence is the ability by which an individual persuades others. However, in the process of interaction, it is not equal between two individuals. If someone has a higher qualification, is older, and possesses more knowledge, they may persuade others more easily and vice versa. Stubbornness is the attitude of rejecting others' opinions. Everyone would like their opinion to be accepted by other people. So, people will reject others' opinions to some degree.

3.3.2 *Behavior mechanism of agent*

We divide the behavior mechanism into two categories. One behavior is receiving the information from websites. The other behavior is communicating with other agents.

For the convenience of describing the model, we use agent i as the object of study.

A   Receiving the information from website

Figure 1 shows the process of receiving the information from website.

Figure 1.   Receiving the information from website.

- Step1: Authority Judging

When agent i receives information from the Internet, it can get the authority of information. Then, agent i will compare the authority of information with itself. If agent i accepted the information, it will change its opinion. If agent i does not accept the information, it will not change its opinion and will just know the information.

The algorithm is as follows: $If(A_e(t)>I_i(t))$. *Then,* enter Step2; *Else* exit, where $A_e(t)$ is currently the authority of information; $I_i(t)$ is currently the influence of agent i.

- Step2: Opinion Calculation

If agent i accepts the information, it will calculate its opinion the next moment. Agent i will take its

stubbornness into account. If its stubbornness is low, the agent may change its opinion easily and vice versa. The formula is as follows.

$$O_i(t+1)=O_i(t)+(1-S_i(t))[O_e(t)-O_i(t)],$$

where $O_i(t)$ and $O_i(t+1)$ are the opinion of agent i; $O_e(t)$ is the opinion of information; $S_i(t)$ is the stubbornness of agent i; and t represents the simulation time at the moment.

B   Communicating with the other agents

Figure 2 shows the interaction between two agents.

Figure 2.   Communicating with the other agents.

- Step 3: Stubborn Judging

When communicating with other agents, agent i will compare the opinion between the opposite and itself. If the divergence of opinion is acceptable, they will exchange ideas further. If the divergence of opinion is unacceptable, they will not exchange their opinions and will only know the information.

The algorithm is as follows: *If* $(|O_i(t)-O_j(t)|<(1-S_i(t)))$, *Then,* enter Step4; *Else,* exit, where $O_i(t)$ is the opinion of agent i; $O_j(t)$ is the opinion of agent j; $S_i(t)$ is the stubbornness of agent i; and t represents the simulation time at the moment.

- Step4: Opinion Change

If agent i accepts the information, it will calculate its opinion the next moment. The opinion of agent i is influenced by the opposite and itself. The formula is as follows:

$$O_i(t+1)=O_i(t)+I_j(t)(1-S_i(t))(O_j(t)-O_i(t)),$$

where $O_i(t)$ and $O_i(t+1)$ are the opinions of agent i; $I_i(t+1)$ and $I_i(t)$ are the influence of agent i; $I_j(t)$ presents the influence of agent j; $S_i(t)$ presents the stubbornness of agent i; and t presents the simulation time at the moment.

4   SIMULATION EXPERIMENT

In recent years, the number of terrorist attack incidents is persistently growing in China. The XinJiang 7•5 Incident is one of the most hazardous incidents. There is a phenomenon that many perpetrators are unaware of the truth, and they just follow what their friends do in the incident. So we simulate a case about how ordinary people are persuaded by rumormongers.

In this experiment, there are two types of people: One is ordinary people and the other is the rumormonger. Ordinary people are considered ordinary netizen, and rumormongers are considered opinion leaders. The aim of opinion leaders is to persuade ordinary netizen to accept their opinions. Based on the background of the case, we simulate that rumors spread in 10000 agents, whose relationship is a small-world network.

4.1 *Specifying the value range of parameters*

Section 3 has modeled the information and individual, but it does not specify the value range of attributes and the method of assignment. Based on the method of assignment of continuous interval in Deffaunt Model, we set the interval of attributes as follows:

4.1.1 *Attributes of information*
The opinion is denoted by O_e. We define $O_e \in [0,1]$ and $O_e \in R$. For the convenience of analyzing, we divide O_e into five categories: completely accept, slightly accept, neutrality, slightly opposite, and completely opposite. The opinion can be abbreviated as follows: a = complete accept, b = slightly accept, c = neutrality, d = slightly opposite, and e = complete opposite. Information belongs to one category only. Then, opinion can be presented as follows:

$$\text{Opinion} = \begin{cases} a, O_e \in [0.8, 1.0] \\ b, O_e \in [0.6, 0.8] \\ c, O_e \in [0.4, 0.6] \\ d, O_e \in [0.2, 0.4] \\ e, O_e \in [0.0, 0.2] \end{cases}$$

The authority is denoted by A_e. We define $A_e \in [0,1]$ and $A_e \in R$. The upper limit of A_e presents the highest authority, whereas the lower limit of A_e presents the lowest authority.

4.1.2 *Attributes of agent*
The model of opinion of agent is the same with the model of opinion of information, which is denoted by O_i.

The event role is denoted by E_i. It includes opinion leader and ordinary netizen.

The influence is denoted by I_i. We define $I_i \in [0,1]$ and $I_i \in R$. The upper limit of I_i presents the highest influence, whereas the lower limit of I_i presents the lowest influence.

The stubbornness is denoted by . We define $S_i \in [0,1]$ and $S_i \in R$. The upper limit of S_i presents the highest stubbornness, whereas the lower limit of S_i presents the lowest stubbornness.

4.2 *Relationship network*

Here, we suppose the relationship network is a small world network (D J Watts, 1998). The parameters of the network are as follows:

Table 1. Parameters of relationship network.

Attribute	Value
Average degree	20
Average clustering coefficient	0.691
Average path length	7.45

4.3 *Assignment of initial value*

4.3.1 *Assignment of information*
Based on the background of case study, the rumor is that "Uygur people become slaves of mainland boss." It is incredible to the people who know the truth. However, it will confuse people who do not know the truth. Some of these rumors come from the website of WUC, where opinion leaders believe the information is authoritative. The assignment of information is as follows:

Table 2. Assignment of information.

Attribute	Value
Opinion	1
Authority	1
Start time	0

4.3.2 *Assignment of agent*
First, we give the proportion of opinion leaders. We assume that the proportion of opinion leaders is 0.1 in this case, and they are randomly assigned to any node in the network.

For the sake of their profit, opinion leaders accept rumors completely. Opinion leaders possess background knowledge. It is easy for opinion leaders to persuade ordinary netizens who do not know the truth. Hence, their influence should be high. Because the aim of opinion leaders is to persuade others, it is hard for them to be persuaded by other people. On the other hand, ordinary netizens may not accept these rumors initially. However, most ordinary netizens will not investigate rumors and have little background knowledge about them. We assume ordinary netizens will easily accept the opinions that are different from their own. Based on this, the assignments of agents are listed as follows:

Table 3. Assignments of agents' attributes.

Event role	Opinion leader	Ordinary netizen
Number	999	9001
Opinion	U[0.8,1]	U[0,0.8]
Influence	U[0.8,1]	U[0,0.1]
Stubborn	U[0.4,0.6]	U[0.1,0.3]

where U[a,b] presents the uniform distribution whose minimum number is a and maximum number is b.

4.4 *Simulation*

Because no random factors are considered in our model, as long as the parameters are determined, all the simulation results are the same. Figure 3 shows the change of opinion of the crowd whose iteration number is ten. The abscissa represents simulation iterations, and simulation forward t = one unit each step. The vertical axis represents the number of agents.

We can find several interesting phenomena in the simulation. 1) If all the netizens are willing to spread information, the speed of propagation is very fast in a crowd. All the agents have known the information at

t = three. 2) The trend is that all the agents will accept the rumor. 3) At t = one, the number of agents who completely accept the rumor is around 4500; however, it is only 999 initially.

There are several reasons for these phenomena. 1) Because of the large number of agents who spread rumor initially and the average number of friends being 20, the propagation speed is fast enough throughout the entire crowd. The average path length is short, which is nearly 7.45. If only one agent spreads the rumor initially, all the netizens will know the rumor around t = seven. 2) From the perspective of initial number, the number of agents who accept the rumor is about 3000, and the scale is large enough. From the perspective of the agent's attributes, the influence and stubbornness of opinion leaders are higher than those of ordinary netizens. Hence, the change of opinion of opinion leaders is small, whereas that of the netizen is big. 3) The opinion leaders spread rumors at t = zero. Ordinary netizens can only interact with opinion leaders. We find that the number of agents who slightly accept the rumor and stand in the middle decreases quickly. It means that they are transferred into the crowd who completely accept the rumor.

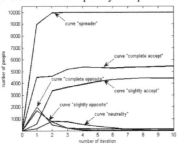

Figure 3. Varied trend of opinions of the crowd.

Consequently, if we intend to prevent the persistent propagation of rumors, several measures could be taken. 1) The influence of crowd should be improved. In other words, the ability of distinguishing the true from the false should be improved. However, it is a process spanning over a long period. 2) Some punitive measures toward rumor spreaders can be taken, such as cutting of the communication channels or controlling the rumor spreaders. 3) The truthful information of the incidence should be published to refute the rumor.

5 CONCLUSION

Aiming at compensating for the deficiencies of research conducted on public opinion propagation, this paper provides a new propagation model of public opinion based on Deffuant Model. The model aims at observing the emerging behavior of the microscopic group and at giving some explanations of the macroscopic phenomenon.

Compared with the previous classic model of opinion propagation, our model has the following advantages.

1 We draw on the experience of a part of the outstanding ideas of the classic model, which makes the model more verisimilar. First, under the guidance of the multi-agent system and the Two-Step Flow Theory, we establish the individual model. As for the interactions between two agents, we use one-to-one form of interaction based on Deffaunt Model.

2 The information is modeled, and the common attributes of information are extracted. The model of information is not taken into account in the classic model.

However, there are many disadvantages in this paper.

1 Although the model of the individual is more verisimilar than traditional studies, there is still a gap between our model and the real individual. Our model is suitable to simulate a certain situation of the real world, whereas it cannot simulate whole situations.

2 The assignments of attributes of the individual need to be validated by real data. In this paper, the assignments of initial attributes are carried out by experience. A large amount of work requires to be done to collect real data and to use these data.

Hence, we need to do more research on the individual, information, and network based on real data in future work.

ACKNOWLEDGMENTS

This work was supported by the National Nature and Science Foundation of China under Grant (Nos. 91024030 and 91324013).

REFERENCES

LIU Chang-yu, HU Xiao-feng, LUO Pi, SI Guang-ya. (2008). Study on Consensus Emergency Model Based on Asymmetric Personal Relationship Influence. Journal of System Simulation, 20(4): 990–996.

SznajdWeron K, Sznajd J. (2000). Opinion Evolution in Closed Community. International Journal of Modern Physics C, 11(6): 1157–1165.

Deffuant G, Neau D, Amblard F, et al. (2000). Mixing Beliefs Among Interacting Agents. Advances in Complex Systems, 3: 87–98.

Galam S. (2008). Sociophysics: A Review of Galam Models. International Journal of Modern Physics C, 19(3): 409–440.

Woloszyn M, Stauffer D, Kulakowski K. (2006). Phase Transition in Nowak-SznajdOpinion Dynamics. Physica A, 378(2): 453–458.

Kozma B, Barrat A. (2008).Consensus formation on adaptivenetworks. Physical Review E, 77(1): ID016102.

D J Watts, S H Strogatz. (1998). Collective dynamics of small world networks. Nature (S0028-0836), 393(4): 440–442.

Energy, Environment and Green Building Materials – Sheng (ed.)
© *2015 Taylor & Francis Group, London, ISBN 978-1-138-02718-3*

Effects of learners' motivation in second language acquisition

J.B. He & Z. Wang
Jiangxi Science & Technology Normal University, Nanchang, Jiangxi, China

ABSTRACT: Motivation is the inner power and the most important factor in promoting second language acquisition; it can also determine effects of foreign language learning. Based on some theories and research conducted in the past as well as the main factors of learning motivation, this thesis aims at putting forward the means and strategies for stimulating students' motivation for foreign language learning, as well as at cultivating students' strong motivation for language knowledge, turning passive learning into active learning.

KEYWORDS: Effect; Motivation; Learners; Second Language Acquisition; Strategy; Stimulate

1 INTRODUCTION

Motivation is an important factor for determining the success or failure of second language learning. It can directly affect the fluency of learning strategies for learners, accept input of the second language, degree of interaction with native speakers, the level of goal setting, the perseverance, and the persistence of developing second language skills. Back in the late 1950s, Lambert and Gardner started studying the motivation of the second language from social psychology. However, due to a narrow perspective, simple research methods, the second language motivation research is stagnating. In contrast, the motivation of studying a foreign language in China started a bit late. On the one hand, this is not unrelated to people's misconceptions. It seems to be unworthy of study, because it is a natural matter that motivation will affect scores, and motivation of achievement is a matter of course, which does not seem worth studying. On the other hand, a single perspective is the main cause for hindering its development. Due to the impact of the Gardner (1979) model of social psychology, foreign language learning motivation research is limited to interest in learning, self-confidence, and purpose for study such as a few motivation structures. It is proved that it is difficult to explain complex motivation in foreign language learning from a single view. To fully understand the impact of foreign language learning motivation, we should not ignore the findings of research in psychology.

Motivation is one of the significant factors that is used to determine the different learners with varying degrees of success factors in foreign language learning. The so-called motivation (or power) lies in the inner power and motivation to inspire people to act, which includes personal intentions, aspirations, pulses, trying to achieve the purpose, and so on. People who have a strong motivation and purpose in learning a foreign language and who learn actively and conscientiously can achieve good results. In contrast, people who lack a learning purpose and have no motivation are frustrated to meet difficulties that get unsatisfied sheets, let alone the possible success on the study. "Motive" as a psychological term refers to inner motivation caused by the need to inspire or encourage people to action to achieve a certain purpose. It plays a role in arousing adjustment, keeping and stopping action. Motivation generated on the basic need is determined by the cognition and the interact affects caused by encouragement. This paper illustrates the theory of English learning motivation, motivation and foreign language learning, and how to arouse the motivation of students to learn a foreign language.

2 LEARNING MOTIVATION AND SECOND LANGUAGE AQUISITION

H. Brown said that motivation is generally regarded as an inherent power, the role of an emotion, and a method to promote the desire of people to take action. Different linguists have studied the function of motivation in foreign language learning. For instance, Canadian linguists Lambern and R. Gardner studied much on motivation of second language acquisition. They proposed Instrumental Motivation and

Integrative Motivation by a ten-year follow-up survey of foreign language learners. Learners with instrumental motivation to learn a foreign language, passing the foreign language examination, aim at looking for a good career, understanding target material, and improving their social status. They regarded foreign language as a tool. However, learners with integrative motivation are mainly interested in association and culture of target language. For people who are learning a foreign language, it is better to use the target language to communicate within the community, enjoying the arts and culture of the target language.

Lambert and Gardner's theory on motivation, methods, and findings make a huge impact on foreign language learning. They investigated the motivation of English learners in Canada to learn French. Through the results, we know that integrative learners learn grammar better than instrumental learners. However, the results in North America are on the contrary. Instrumental learners learn better. Lambert and Gardner found that the motivation of integrative learners is stronger than that of instrumental ones.

Language learning motivation theory discusses the meaning and type of motivation, and the effects that the motivation causes in foreign language learning. It can help us understand the psychological learning process of foreign language learners, their social factors, and individual differences better, so as to provide a systematic and comprehensive guidance for us to correctly understand and grasp the students' foreign language learning motivation.

3 FACTORS INFLUENCING FOREIGN LANGUAGE LEARNING MOTIVATION

3.1 Cognition

The cognitive factor influences foreign language learning motivation, mainly because of self-awareness, including self-concept, self-efficacy, valence, and sense of agency. Self-concept consists of self-awareness on the advantages and disadvantages of foreign language learning, judges for success, and failure during foreign language learning and recognition for the rise and decline of self-respect in foreign language learning. Self-efficacy means the judging of one's ability in completing a specific assignment. Therefore, self-efficacy decided the task selection for foreign language learning activities. Only if the task is proper, the motivation could be stronger; if not, the motivation would disappear quite soon. Valence lays its meaning on awareness of purpose of foreign language learning activities. When learners believe the purpose is important for them, motivation would be stronger. Sense of agency means an awareness of

responsibility of the learners themselves. If they have a strong sense of responsibility, either their motivations would be stronger or their motivations would decline.

3.2 Emotion

The emotion factor that affects the foreign language learning is that the learning motives also change as the learners' emotions change. The emotion includes interests, attitudes, requirements, confidence, anxiety, and so on. Attitude is seen as a part of the motivation. Many researchers studied attitude with motivation, such as Garner's Attitude / Motivation Test Battery (AMTB) with no need for the original power of motivation. The relationship between self-confidence and motivation is positively correlated; the lack of confidence cannot devote strong motives; and the higher anxiety reduces motivation.

3.3 Social environment

Social environmental factor is not related to the learners but it still greatly influences them. These factors include community personnel's higher requirements on foreign language of job appliers, foreign language course curriculum requirements, the impact of foreign language teachers, the demands of parents, and peer influence.

4 STRATEGIES THAT STIMULATE AND CULTIVATE LEARNERS' MOTIVATION

4.1 Training meta-cognitive skills of learners

Cognitive strategies refer to the skills of learners that are exhibited while they take part in activities; these aim at inspiring learners' English learning motivation. Teachers must guide students to use scientific cognitive strategies, especially metal cognitive strategies, to make learners feel the joy of success so that they will achieve greater motivation of learning. Metal cognitive knowledge is composed of metal cognition and metal cognitive strategies. Metal cognitive knowledge can be divided into the following: personal knowledge, task knowledge, and strategic knowledge. Meta-cognitive strategies include planning, monitoring, and evaluation. Only by develop the meta-cognition can the learners realize for themselves that their current level has an objective evaluation, make the right plan to reduce blindness, and improve foreign language proficiency. The progress and success is the reward for the efforts of the learners and will inspire them to become more motivated for learning.

4.2 Strengthening emotional teaching

Strengthening of the new behaviorism theory suggests that in the learning process, the students' skills are enhanced, their learning motivation will also be strengthened, and otherwise it will be weakened. Foreign language learning motivation of students mainly comes from their teachers and their attitudes. Even a commendable job or a little praise will enable students' hearts to be full of sunshine; great encouragement will stimulate greater interest in foreign language learning. Therefore, teachers should always encourage their students, try not to criticize them, give them confidence in language learning, and provide them with an incentive to overcome all difficulties in completing learning tasks. Do not project, in front of students, a vice-serious face, always pick fish bones in front of students, or make a large job up the "wrong," as a great tendency to turn iron into steel. This will seriously dampen the learner's self-esteem and self-confidence and students will develop a fear of foreign language learning. In short, reward leads to positive emotions, whereas punishment leads to negative emotions. Modern American educator Bloom said: "A positive emotional learning with students, should be better than those who lack enthusiasm, student interest or pleasure, or learning materials than those who felt anxiety and fear boring students, learning by easier and faster." As foreign language teachers, we should keep this in mind.

4.3 Optimizing social environment

In a sense, personal motivation is the result of social and cultural factors. In the present situation, compound talents for both knowing a foreign language and being professional in a specific area are much needed. Universities are getting into action, preparing teaching materials, and teaching methods of reform to strengthen the teaching equipment and buildings. More and more people have realized the importance of foreign language learning, but in the whole social environment, the attention on foreign language education is still not enough. To fundamentally solve this problem of foreign language learning motivation, advantages of parents, teachers, and role models should be taken together, in the context of the whole society, to form a wonderful environment of foreign language learning. Specifically, each of the parents should make an effort to inculcate in their children the importance of foreign language learning, to encourage and push their foreign language learning. To study the foreign language learning motivation, it is important to explore ways in which to stimulate students' language learning motivation and strategy. Intrinsic motivation is the power you devote to learning a foreign language and achieving good teaching purposes.

4.4 Motivation and objectives

With motivation, you would have enough energy, be well prepared to overcome various setbacks and difficulties, and be perseverant. We will adopt a positive attitude toward foreign language learning, target language, its sociocultural connotations, and so on. Thus, when fresh students enter school, emphasizing the establishment of learning goals is necessary. As long as we have a new goal, we can provoke a new motivation. Motivation and objectives are closely linked; if there is no motivation, goals do not matter. On the contrary, if there is no goal, motivation would also no longer exist. Successful learners have a clear purpose and plan that is good at making long-term and periodic plans for their learning activities. Take the initiative to set goals for yourself, take the initiative to be involved in foreign language learning. Unsuccessful learners are passive, lack purpose, rely too much on the teacher, and adopt a negative attitude to foreign language learning. One can imagine that this negative and indifferent attitude is not ideal for language learning. Therefore, when studying, not only students should learn from textbooks but it is also important to have clear learning objectives, correct attitudes toward learning. This is the only way to obtain a multiplier effect in foreign language learning.

4.5 Foreign language teaching

During foreign language teaching, language teachers should create a relaxed learning environment, enhance student interest in learning and learning initiative, and use flexible and diverse teaching methods. To inspire students' motivation to learn the language, the key to improving the effectiveness of language teaching is to create for students a pleasant, easy environment; this would attract students to being interested in the language environment. Modern American educator Bloom said: "A student learns with a positive emotional, he should be better than those who lack of enthusiasm, interest or pleasure, or easier and faster than those students who felt anxiety and boring for learning materials." English teaching, because of its particularity to stimulate and develop student's interest in learning, is especially important. Interest stimulation in student learning motivation plays the regulatory role of teaching, and it is the primary condition that improves English teaching. Many students learn a language to achieve a certain level, mainly because there is no interest in their language learning, and thus there is no such motivation. Teachers should try to enable students to know more about the target language culture, history, habits, and customs. The teachers must be amiable toward the students; they should use novel and unique teaching methods and create a flexible and good classroom atmosphere as well as a language environment to encourage students to actively participate in language practice. A good classroom teaching and learning environment can enhance the students' memory, creative thinking, enable them to be active, develop full mental potential, and achieve the best mental state to exhibit the best intelligence potential. Teaching methods can be used in the humanistic approach, the communicative and accessible approach, and so on. The humanistic approach as a teaching method emphasizes integrity; lays emphasis on a human's respect and attention in the learning process; pays attention to the development of the inner world of learners; and advocates emphasizing the process of teaching and the cultivation of learning autonomy. The humanistic approach to human concerns of the inner world is in line with the requirements of quality education in China these days, with the new trends of modern teaching being consistent. Therefore, teachers should not just teach but also help students understand their inner self and the development of a healthy personality. If students get a sound personality development, they will have a learning drive and will double to study hard. In addition, the communicative approach can be used with the multimedia English teaching method. Flexible teaching methods will give students a funny and lively studying environment, and it will stimulate students' learning motivation as well.

The teachers should possess good knowledge about students' learning conditions and state of mind. Therefore, there should be more communications between teachers and students, not only exchange of learning but also exchange of emotion. Teachers' encouragement to the students is the motivation and driving force of their progress and it acts as a "catalyst" to stimulate students' interest in language learning. "Pro-his teacher, and believes in the Road" is a vivid description of this reason. A good work ethic, a wealth of teaching experience, and energetic, optimistic, and understanding classroom teachers will, no doubt, result in an easy, harmonious, and pleasant atmosphere in which students are encouraged to have a strong thirst for knowledge, thus promoting teaching.

4.6 Second class for foreign language learning

Thus, in teaching, teachers' feedback to students by checking through their work, quizzes, and exams in a timely manner with regard to their foreign language learning will greatly influence stimulating students' learning motivation and will make them more confident in foreign language learning. Teachers need to be sure of students' achievements and should offer them more encouragement. Teachers should try to make each student realize that they have harvested more or less from their paid work and learnt something. Every point in the small progress of students should

be given recognition and praise, in other words, to make them more demanding and encourage them to strive for greater progress. Usually, the students should be absorbed in suggestions and opinions from teaching and further improve the foreign language teaching methods based on these suggestions and comments. These results show that the terms from psychology, "teachers should not only pay attention to cultivating motivation for learning a foreign language learner, but also has to stimulate the motivation they have owned. It is necessary to by using some of the incentives change motivation from the latent state into active state, to make it a internal motivation of promoting learning in fact ,and to mobilize the enthusiasm to learn, to solve the current learning task. At the same time, the formed motivation should have be continued to be consolidated, deepen and improved."

In addition, we should organize a variety of activities in students' second class. Second-class activities will help students not only increase their knowledge but also broaden their horizons, and, most importantly, stimulate the interest of students in English learning as well as increase more opportunities for language practice. For instance, organizing English speech contest, English party, English horn, learning to sing English songs, talking with foreign guests, showing English movies, video, holding the English newspapers, radio broadcasting, and creating a wonderful English atmosphere. Such activities can not only stimulate students' interests in learning English but also improve their language skills as well as practical communication skills.

5 SUMMARY

In summary, the motivation on effects of foreign language learning is enormous. As we all know, language learning is a difficult task, apart from complex system engineering. In foreign language learning, the impact of motivation on learning must not be ignored, which is the key to success in learning a foreign language. Teachers should try their best to discover ways to stimulate students' strong thirst for language and culture knowledge, turning passive learning into active learning. It is the only way for teaching to be effective while foreign language learning is successful.

REFERENCES

Cook, V. Second Language Learning and Language Teaching, Beijing: FLTRP and Edward Arnold Limited, 2000.
Ames, C. Classrooms, goals, structures and student motivation. Journal of Educational Psychology, 1992,(84):267–271.

Ashton, P.T. & Webb, R.B. Making a difference: Teachers' sense of efficacy and student achievement. New York: Longman, 1986.

Platt, C. W. Effects of casual attributions for success on first-term college Performance: A covarinance structure model. Journal of Educational Psychology,1988,(3):80.

Williams, M. & R. Burden. Psychology for Language Teachers. Cambridge: Cambridge University Press, 1997.

Dornyei, Z. Motivation and motivating in the foreign language classroom. The Modern Language Journal, 1994.

Martin V Covington. The Will to Learn: A Guide for Motivating Young People. Cambridge: Cambridge University press,1998.

Energy, Environment and Green Building Materials – Sheng (ed.)
© 2015 Taylor & Francis Group, London, ISBN 978-1-138-02718-3

Large section roadway excavation rapid technology

Z.F. Yao, W.B. Wei, B. Shi & Z. Wei

Faculty of Resources and Safety Engineering, China University of Mining and Technology, Beijing, China

ABSTRACT: Based on the characteristics of the deep large cross-section rock tunnel, inclined empty wedgy cutting technique, mechanized working line composing of full-hydraulic excavating drill-camel and loader buckets, secondary bolt/shotcreting support technique are proposed. These techniques effectively solve the problems of small footage in the rock tunnel, slow transferring rate and the unstable support in the deep tunnel. The shaping quality is improved; the time of working procedure is shortened and the driving speed is boosted.

KEYWORDS: Medium deep hole blasting; Mechanized working line; Secondary support technique

1 INTRODUCTION

Main text Underground mining dominates the coal production in China. Rock tunnel construction is significant procedure during period of the new mine construction and the tunnel developing in producing mines. Recently, with the development of the fully mechanized equipments and techniques, the production rate has been improved remarkably. The working face that product million or more than ten million tons of coal is not uncommon, which increases the work of tunneling. To meet the requirement of full-rate production, improving the shaping quality and accelerating the tunnel driving are of great importance for the coal industry.

Tunnel excavation is a complicated working procedure which is composed of blasting, loading and transporting and supporting [1–2]. Based on these three key techniques, domestic scholars have conducted a large number of researches [3–5]. From the blasting prospective, empty wedgy cutting technique is proposed in the medium deep hole, which achieve one charge per blasting in the large cross-section rock tunnel. The time of transporting and supporting is significantly shortened by the mechanized driving equipment. With the increasing of mining depth, the conditions of the surrounding rock are changing. According to the flow characteristic, new support techniques which could enforce the stability of tunnels are proposed.

2 MEDIUM DEEP HOLE BLASTING TECHNIQUE

Blasting is the major methodology of tunnel driving in China. With the development of heading machine, high-power tunneling equipments appear, which facilitates the application and development of medium deep hole blasting technique. By applying this technique, excavation footage cycle is increased. On the other hand, the burst rock quantity per blasting is increased and the time aided working procedure is decreased, which boost the speed and efficiency of driving.

2.1 Cutting blasting

The function of cutting is to cave the rock in the working face and create a free surface for the following blasting. Therefore, underhole blasting is the key point of tunnel excavating.

2.1.1 Cutting method

Usually, the ration of crater radius r in minimum burden w is defined as the blasting index.

$$n = r/w \qquad (1)$$

where :r = crater radius; w = minimum burden and n = lasting index.

When the *n* = 1, the blasting funnel is called normal cast blasting crater; when 1<n<3, it is called intensive cast blasting crater; when 0.75<n<1 it is called subdued blasting crater and when n<0.75 the blasting could not emerge. The blasting index can be 0.8~1.0 during the tunnel excavating and it can be larger during cutting.

The cutting blasting is approximately equivalent to underpart free surface crater blasting under the condition of concentrated charge at intervals. The space between holes is confirmed by the size of plastic zone. The comparison between straight cutting and inclined cutting is shown in the following Figure.1.

As can be seen from Figure 1, inclined cutting blasting aggravate the rock crush, especially in the large cross-section rock tunnel and it is almost not limited by the width of the section. Therefore, inclined cutting has better application than the straight cutting in the large cross-section rock tunnel driving.

(a) Inclined cutting   (b) Straight cutting

Figure 1.   Blasting effect.

2.1.2 *The mechanism and effect of central empty hole*

The cutting of central empty hole could induce cracks in the surrounding rock, which will alter the direction of the minimum burden and the direction of blasting effect and create attached free surface for the cut working. According to the free surface hypothesis of rock busting mechanism and stress wave interference hypothesis, the stress wave triggered by underhole blasting could radiate at the free surface and induce tensile failure zone heading to the blasting cartridge. Because extrusion crush is generated by columnar charge blasting and it need little energy to cast the cracked rock in the holes. And the central empty holes can enhance the rock casting and prepare for the following blasting.

Taking advantage of the hole space and wedgy cutting technique, the central empty hole technique that facilitates the casting rock and deepen the effective depth of the underholes improves blasting efficiency. At the same time, the central empty holes could orient for the blasting and promote the cracked rock.

2.1.3 *Calculation of the charge length*

The cutting blasting is approximately equivalent to underpart free surface crater blasting under the condition of concentrated charge at intervals. The quantity of concentrated charge can be calculated according to the following M. M. Bopeckob formula:

$$Q = KW^3 \left(0.4 + 0.6n^3\right) \qquad (2)$$

Wher, Q = quantity of concentrated charge(kg); K = quality index of charge(kg·m–3); N = blasting index and W = concentrated charge burden distance (m)

2.2 *Assistant and peripheral holes*

Holes in the roof, floor and the later wall are included by peripheral holes and they are the blastholes for the construction of tunnel section design. They are often fixed right on the outline. The holes in the roof and the later wall are designed to be parallel and the holes in the floor are in the same level. Assistant holes are located between the cutting holes and the peripheral holes and evenly arranged at the intervals of 600mm. The depth of assistant and peripheral holes is almost the same and the intervals should be controlled relatively small. The low density and low detonation velocity explosives are selected to control the shaping quality and decrease the surrounding rock failure.

2.3 *Charge structure and detonation*

Charge structure and detonation method are the key points to the blasting result. To guarantee the blasting effect, continuous, decoupling and reverse structure charging are applied in the tunnel excavating. Before charging process, the rock power and water should be cleaned. Millisecond delay detonators are applied for the disposable whole section blasting.

3 MECHANIZED WORKING LINE COMPOSING OF FULL-HYDRAULIC EXCAVATING DRILL-CAMEL AND LOADER BUCKETS

By now, borehole-blasting method dominates the tunnel excavation in China. Boom-type roadheader is still rare in home. There are two types of mechanized working lines based on borehole-blasting method: ①air-leg rock drill with bucket mechanical mucker or rake mucker; ②all-hydraulic rock-drilling jumbo with side discharge loader.

The former is mainly used in small section tunnel of small coal mine and its heading speed is relatively. The latter is common in the large section rock tunnel in China. Because the side discharge loader needs to move to and from the reversed loader or conveyors, the distance and time could not be controlled very much and it is not so efficient. Electric full hydraulic pressure drive is applied in the crawler-type bucket loader whose conveyor could load the cracked rock into the loader, which can make the working procedure more efficiently.

4 SECONDARY BOLT AND SHOTCRETING SUPPORT TECHNIQUE

Based on the bearing capacity of the surrounding rock, the non-deformability could be enhanced by the

270

shotcreting support technique which makes the surrounding rock become part of the supporting system and control the deformation after excavation. It has several obvious advantages: rapid construction, high mechanization and low cost and is applied widely in the tunnel excavation.

Traditional disposable bolt and shotcreting support technique mainly enhance the compressive strength of the tunnel surrounding rock. With the increasing of the mining depth, the properties of the surrounding rock will be changed and the rheological behavior will be increasingly obvious. The disposable support focusing on the compressive strength is not effective for the tunnel deformation control any more. Therefore, according to the rheological behavior in the deep and the supporting thesis, the secondary bolt and shotcreting support technique is proposed.

4.4 The mechanism of secondary bolt and shotcreting support technique

In the mechanical equilibrium system of "support-surrounding rock", the roof separation, plastic zone developing and continuous deformation can be partly controlled. It could only undertake a small part of the compressive deformation of the supporting depends on the motion state of upper rock layer, After the excavation procedure, there emerge obvious displacements towards the tunnel space, especially in the deep surrounding rock. The disposal supporting limits the deformation of the surrounding rock to some degree. The secondary support is conducted after the deformation energy is released.

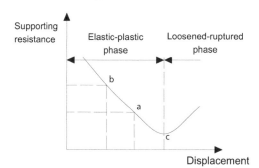

Figure 2. Surrounding rock displacement characteristic.

4.5 Optimal supporting time

The secondary support should be conducted at a prosper time that the surrounding rock deformation tends towards stable. According to the soft rock tunnel supporting theory, the optimal supporting time should be controlled at the time when the supporting force of the surrounding rock and the engineering force research the summit.

In the engineering project, the shrinkage distortions of the tunnels are often monitored. The curve of displacements with time can be plotted according to the monitoring data. The inflection point of the curve is often seen as the right time for supporting.

Figure 3. The optimal supporting time.

5 CONCLUSIONS

1 The slow footage driving cycle can be effectively controlled by the medium deep hole blasting technique and the layout of the central empty holes can facilitate the rock loading and transporting.
2 Mechanized working line composing of full-hydraulic excavating drill-camel and loader buckets is applied in the tunnel excavating and promote the heading speed.
3 Deep rock tunnel deformation could be effectively controlled by the secondary bolt and shotcreting support technique and the supporting time is of significant importance for the tunnel stability.

REFERENCES

[1] Huo X L, Chen S G, Zhang X M, "Study on Construction Characteristic and Temporal-spatial Effect of Tangjiashan Tunnel", CONSTRUCTION TECHNOLOGY, pp 79–83, 2012.
[2] Zuo Q J, "Study on Mechanical Effect of Surrounding Rock for Super-large Cross Section Soft Slate Tunnel during Construction Period", China University of Geosciences, 2013.
[3] Sun X M, Cai F, Yang J, Hong W S, "Numerical simulation of the effect of coupling support of bolt-mesh-anchor in deep tunnel", Mining Science and Technology, pp 352–359, 2009.
[4] HENNING J. G, MITRI H. S, "Examination of hanging-wall stability in a weak rock mass", Metallurgy and Petroleum, Montréal, pp 40–44, 1999.
[5] He M C, Duan Q W, Sun X M, "Computer numerical simulation of soft rock in China" Computer Applications in the Minerals Industries, pp 697–700, 2001.

Energy, Environment and Green Building Materials – Sheng (ed.)
© *2015 Taylor & Francis Group, London, ISBN 978-1-138-02718-3*

Structural design of cyclonic microbubble flotation column system

B.Q. Dai
College of Mechanical and electronic Engineering, Shandong University of Science and Technology, Taian, China

J.J. Yuan
Department of Mechanical and electronic Engineering, Shandong University of Science and Technology, Taian, China

X.Y. Liu
Department of Information Engineering, Shandong University of Science and Technology, Taian, China

J.X. Ge
Department of Mechanical and electronic Engineering, Shandong University of Science and Technology, Taian, China

ABSTRACT: With the introduction of flotation theory and flotation process, this paper discusses working principle of Microbubble Flotation Column, designs the structure of Flotation Column system, and analyzes Operating Features of the main parts of flotation column, bubble generator, tailings box, microbubble generator.

KEYWORDS: flotation, cyclonic microbubble flotation column, concentrate, tailing

Mineral resources are important material basis and guarantee of the human survival and social development. Limited quantities and one-off development are the main reasons for declining tendency of the mineral resources, so the recycling of mineral resources becomes a topic with general concern in the modern society.

1 THE INTRODUCTION OF FLOTATION

Flotation is a technology of selective separation of useful minerals from vapor-liquid-solid boundary, according to the differences of physical and chemical properties of mineral surface,

The theory basis for different flotation technique is consistent, that is, the hydrophobic property is from itself or external condition. Flotation process includes grinding, classification, size mixing, flotation, scavenging, and selection.

The specific implementation way of the technique is crushing ore firstly, and then grinding ore in order to get mineral particles till the size of particle meets standard of flotation process. Then mineral particles and water are proportionally mixed, and flotation reagent is added and stirred in this stirring process in order to mix uniformly. After a certain period of time, pulp is transported to flotator by circulating pump, rubbed together, collided with mineral particles which will be sent to the top of flotator and formed foam layer, and concentrates are piped by overflow canal. Mineral particles with poor hydrophobic which are called tailings cannot adhere to bubble, and cannot rise with bubble, so they will subside in the bottom of flotator(Y. G. Song. Z & X. Zhu. 2010).

Cyclone-static micro-bubble flotation column is widely used and has got a good result in application among series of flotator. It is a new and high efficient equipment which has good separation indexes, good processing capability, high energy-efficient, high flotation rate, little floor space needed, high efficiency (G. Cheng et al .2011).

2 THE WORKING PRINCIPLE OF CYCLONE-STATIC MICRO-BUBBLE FLOTATION COLUMN

Cyclone-static micro-bubble flotation column is a new style which includes three parts of separation section, cyclone separation section, and pip tray section. Cyclone separation section is actually magnified Cyclone overflow pipe. At the top of the cylinder separation zone, overflow weir is placed in order to collect concentrate and sprinkler water system set is for foam layer washing, in order to achieve better separation feeding equipment is placed at the top of the cyclone; meanwhile, the final tailings discharges from outlet at the bottom of the sweeping through tailings tank. Bubble generator is connected with flotation device, which makes sure the discharge flow into flotation column at a tangent.

When flotation column is working, the mineral pulp, including gas, liquid and solid phase, is transported into the barrel at the top of the cylinder via circulating pump, jetted into the flotation column through the radial nozzle. This process not only realizes separation, but also provides advantages for the later flotation, and then the pulp slowly declines within the column because of the gravity. Pressurized mineral pulp is transported into Bubble generators which suction amount air by suction tube at the same time. Bubble and amount gas generate strong turbulence mineralization (X.C.Zhang.2003). The two materials flow into flotation column at a high speed at a tangent and form rotational flow field. So enough energy is provided for flotation process in pip tray section maintaining flotation during flotation process. A large number of bubbles are separated continuously from cyclone separation part to cyclone separation part, and count flow collision action with pulp occurs. In this process, hydrophobic particles and bubbles collide and adhere with each other because of gravity and spray water. In this case, part of the floating mineral will cause desorption event and form mineralized bubbles named concentrate overflowing top flotation column. Another part of the not floating mineral particles cannot be attached to the bubbles, which are discharged from the tailing lifting device or go to the next job processing, namely tailings.

3 THE DESIGN OF THE STRUCTURE OF CYCLONE-STATIC MICRO-BUBBLE FLOTATION COLUMN

Its structure (S.B.WANG.2008.) mainly includes seven parts of the upper column, the middle column, the under column, tailings box, bubble generator, sprinkler system, tailing box and control system(figure 1). The cylinder can be divided into three sections, selecting, roughing and sweeping (Z. J. Zhao & Wang Fan 1998).

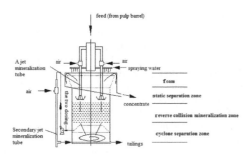

Figure 1. cyclone-static micro-bubble flotation column structure diagram.

3.1 The upper column

The upper column is the selected section of microbubble flotation column, and composed of upper cylindrical tube, spray water system, inverted cone, and concentrate collection and so on. Concentrate overflow structure is designed into spiral tank bottom according to performance characteristics of flotation minerals and concentrate. Inverted cone and cruciform plates are installed in the mid-column, and prevent concentrate foam layer from connecting and reducing concentrate recovery. So this section can increase overflow speed, save flotation time, reduce flotation cost.

3.2 The middle column

The middle column is rougher section of cyclone-static micro-bubble flotation column, and composed of middle cylindrical tube, support beam, and sieve plate. Sieve plays an important role in forming bubbles, and sieve plates are usually made of wear-resistant and good corrosion-resistant steel plate. The hole spacing must be uniformity, and the plate porosity is more than 60%. This sieve plate can assure Uniformity of bubbles, benefit the quality of concentrate.

3.3 The lower column

The lower column is the end section of cyclone-static micro-bubble flotation column, and composed of under cylindrical tube, funnel, water inlet, and tailing pipe. Water inlet connects to cyclone-static separation part at a tangent. A large number of bubbles and hydrophobic particles are jetted into filtrate section together. Strong vortex agitation can not only accelerate mineralized bubble separation, but suspend ore grain fully. This section improves separation efficiency, and tailings can be discharged of tailing pipe directly even working.

3.4 Tailing box

Tailing box is one of the essential parts of the flotation column, which is composed of box, support frame, and level regulator and so on. Tailings box connects to flotation column by steel pipe and valve. After ore pulps are separated, the final tailings are discharged though tailings pipe. Level regulator is used for adjusting liquid level of tailings box, and is used with flotation reagent dosages to adjust ash of concentrates and tailings.

3.5 Feeding and water spray system

Feeding and water spray system is mainly composed of ore storage, feeding pipe and spray pipe. Ore storage is installed on the top of the flotation column. Mineral particles are added to the flotation column through feeding pipe, which ensures that one flotation column can conduct separation several times, and also can enlarge the range of particle size and improve recovery rate of concentrate. Spry system is used for washing dust bubbles and improving concentrate quality.

3.6 *Bubble generator*

Bubble generator is the key part for the flotation effect, and the generator plays an important role on flotation result. The main role of the generator is generating enough micro-bubbles and exciting bubble mineralization.

4 CONCLUSION

Cyclonic microbubble flotation column not only has good separation indexes, strong processing ability, high efficiency, high speed of flotation, small occupation area, energy-efficient and high efficiency, but also is easy to repair and operate. The equipment selection has the extremely widespread application prospect not in metallic ore and nonmetal ore.

REFERENCES

Y. G. Song. Z & X. Zhu. 2010. Development and Practical Application of FWX Series Flotation Column, Coal Mining Machinery 10:185–186.
S.B. WANG. 2008. Features and Application of FWX Series Cyclonic Micro- bubble Flotation Column 36(7):1–4.
G. Cheng et al .2011, Development of Technology and Equipment of Flotation Column, Coal Preparation Tecnology1:66–69.
Z. J. Zhao & Wang Fan. 1998, Study on Bubble Producer of Flotation Column, Coal Mining Machinery 4(3):22–23.
X.C. Zhang. 2003, Working principle and application of CPI flotation column, Nonferrous Metals 2:22–24.

Energy, Environment and Green Building Materials – Sheng (ed.)
© 2015 Taylor & Francis Group, London, ISBN 978-1-138-02718-3

Effects of inorganic ions on phosphorus removal with Fe/Mn oxide formed *in situ* by $KMnO_4$-Fe^{2+} process

K. Liu, J.H. Sun & T. Zhou

College of Aerospace Engineering, Nanjing University of Aeronautics and Astronautics, Nanjing, China

ABSTRACT: Effects of inorganic ions on the removal of phosphorus with Fe/Mn oxide formed *in situ* by $KMnO_4$-Fe^{2+} process were investigated. The following results were discovered. 1mmol/L or 10mmol/L nitrate had no significant effect on the removal of phosphorus with Fe/Mn oxide formed *in situ* by $KMnO_4$-Fe^{2+} process. In the presence of 1mmol/L bicarbonate, the removal of phosphorus had a certain effect, but it was negligible. 1mmol/L chloride or 1mmol/L sulfate had a negligible effect on the removal of phosphorus. Under the same experimental conditions, the effect of chloride was weaker than the effect of sulfate. Addition of calcium increased the removal of phosphorus with Fe/Mn oxide formed *in situ* by $KMnO_4$-Fe^{2+} process. In the presence of 0.5mmol/L calcium and at pH 7-9, the removal efficiency of phosphorus was significantly improved by 9.9% to 28.7%. Silicate could promote the phosphorus removal in acidic condition, and significantly inhibited phosphorus removal in alkaline condition.

KEYWORDS: Fe/Mn oxide formed *in situ*; coagulation; inorganic ion; calcium; silicate

1 INTRODUCTION

Phosphorus plays an important role to the development of plants, animals and the industrial manufacture (Choi et al. 2012, Nguyen et al. 2014). However, its excess supply can cause eutrophication, leading to the deterioration of water quality and threatening the life of aquatic creatures (Ismail 2012, Jyothi et al. 2012). Therefore, the removal of phosphorus from water and wastewater becomes a necessary approach to fight eutrophication.

Various treatment technologies have been developed for phosphorus removal from water and waste-water, including chemical precipitation, adsorption, reverse osmosis and biological methods (Thörneby & Persson 1999, Mohamed 2002, Zhu et al. 2011). Most of these approaches are generally more suitable for the control of high phosphorus concentration. The control of low phosphorus concentration is relatively difficult. Chemical precipitation is recommended as one of the most effective removal processes for the low concentrations of phosphorus (Moharami & Jalali 2013). The most important characteristics of the chemical precipitation are high efficiency and simple operation conditions. In this study, Fe/Mn oxide formed *in situ* by $KMnO_4$-Fe^{2+} process worked as coagulant to remove phosphorus.

It is well known that inorganic ions and organic compounds are ubiquitously present in natural water bodies including surface water, groundwater and seawater (Zhao et al. 2006). Because the complex reaction occurs in the solid-liquid surface, coexisting ions have great impacts on removal of phosphorus with Fe/Mn oxide formed *in situ* by $KMnO_4$-Fe^{2+} process. In this study, coexisting ions such as NO_3^-, HCO_3^-, Cl^-, SO_4^{2-}, Ca^{2+} and SiO_3^{2-} were added to the reaction system in order to evaluate the removal efficiency of phosphorus with Fe/Mn oxide formed *in situ* by $KMnO_4$-Fe^{2+} process. It is worth pointing out that ionic strength and carbonate alkalinity are ubiquitous in water and wastewater. Thus, additions of 1mmol/L $NaHCO_3$ and 10mmol/L $NaNO_3$ were respectively used to regulate alkalinity and ionic strength in part of experiments.

2 MATERIALS AND METHODS

2.1 *Materials*

The solutions of phosphorus and MnO_4^- were obtained from KH_2PO_4 and $KMnO_4$, respectively. The solution of Fe^{2+} supplied from ferrous sulfate as coagulant was freshly prepared for each set of experiments. $NaNO_3$, $NaHCO_3$, $NaCl$, Na_2SO_4, $CaCl_2$ and Na_2SiO_3 as coexisting inorganic ions were prepared for experiments. Background electrolyte solutions were prepared from the salts of $NaNO_3$ and $NaHCO_3$. The pH of the solutions was adjusted by introducing appropriate amount of acid (HCl) or base (NaOH) solutions. All chemicals were reagent-grade and were used without any purification. All the experimental solutions were prepared with distilled water.

2.2 Methods

The experiments of coagulation were carried out on a laboratory scale. Jar tests were carried out using a program controlled two paddle stirrers (TA2-I, Wuhan, China). In each test, the Fe/Mn oxide formed *in situ* with a Mn:Fe molar ratio of 1:3 was prepared. $KMnO_4$ was added into the jar of 1L synthetic water containing phosphorus and inorganic ion, subsequently, ferric sulfate was dosed. And then, rapid mix took place for 1 min at a speed of 120 rpm, followed by slow mix for 30 min at 40 rpm. The settling time lasted for 30 min. Then the supernatant were immediately filtered with cellulose acetate membranes of 0.45μm-pore size for the determination of residual total phosphorus concentration. If not otherwise specified, the pH was adjusted to keep constant level throughout the jar tests with the addition of 0.1 mol/L solutions of HCl or NaOH. All jar tests were carried out in a controlled temperature at 18±1°C.

2.3 Analytical methods

The total phosphorus concentration was analyzed by spectrophotometric measurement at 700 nm (UV–visible T6-spectrophotometer, Pgeneral, China) after reaction with molybdate and ascorbic acid in accordance with the standard methods (Water and Wastewater Monitoring and Analysis Method, the Fourth Edition, China, 2002).

3 RESULTS AND DISCUSSION

3.1 Effect of NO_3^- on the removal of phosphorus by $KMnO_4$-Fe^{2+} process

The experiment was performed at initial phosphorus concentration of 1mg/L (calculated as P) and ferrous sulfate dosage of 2.5mg/L (calculated as Fe). Different NO_3^- concentrations were applied to the solution to investigate the effects of NO_3^- on phosphorus removal with Fe/Mn oxide formed *in situ* by $KMnO_4$-Fe^{2+} process, as shown in Fig. 1. When the pH was increased from 4 to 9, the removal of phosphorus decreased from 89.1% to 19.8%. When the NO_3^- concentration was 1mmol/L or 10mmol/L, the removal of phosphorus by $KMnO_4$-Fe^{2+} process was no significant change. According to the result, 10mmol/L NO_3^- was used to regulate ionic strength as the background electrolyte in subsequent experiments.

3.2 Effect of HCO_3^- on the removal of phosphorus by $KMnO_4$-Fe^{2+} process

Bicarbonate and carbonate are widely present in natural water bodies. There is a balance between HCO_3^- and CO_3^{2-}, and HCO_3^- is considered to be the main

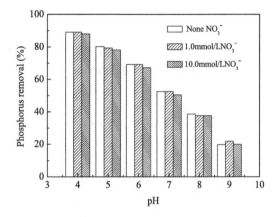

Figure 1. Effect of NO_3^- on the removal of phosphorus with Fe/Mn oxide formed *in situ* by $KMnO_4$-Fe^{2+} process. P dosage 1.0mg/L, Fe dosage 2.5mg/L, Mn/Fe molar ratio=0.33.

existing form at neutral pH value (Zhao et al. 2006). The effect of HCO_3^- on the removal of phosphorus with Fe/Mn oxide formed *in situ* by $KMnO_4$- Fe^{2+} process at different pH was illustrated in Fig. 2.

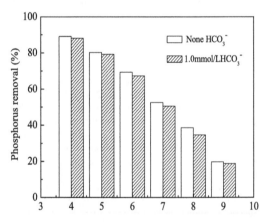

Figure 2. Effect of HCO_3^- on the removal of phosphorus with Fe/Mn oxide formed *in situ* by $KMnO_4$-Fe^{2+} process. P dosage 1.0mg/L, Fe dosage 2.5mg/L, Mn/Fe molar ratio=0.33.

As shown in Fig. 2, when 1mmol/L HCO_3^- was present, the removal of phosphorus decreased from 88.1% to 18.6% at pH 4-9. It was found that HCO_3^- species had little negative effect on phosphorus removal. It was explained that HCO_3^- could compete with phosphorus for the binding sites of Fe/Mn oxide formed *in situ*. Therefore, the present of HCO_3^- led to a little decrease of phosphorus removal by $KMnO_4$-Fe^{2+} process. Compared with the removal of phosphorus by $KMnO_4$-Fe^{2+} process, the influence of 1mmol/L HCO_3^- could be neglected. Therefore, considering the presence of alkalinity in actual water, 1mmol/L HCO_3^- was used to adjust in follow-up experiments.

3.3 Effect of Cl⁻ and SO₄²⁻on the removal of phosphorus by KMnO₄-Fe²⁺ process

When 1mmol/L Cl^- or 1mmol/L SO_4^{2-} was present respectively, the removal of 1mg/L phosphorus in the solution by $KMnO_4$-Fe^{2+} process at different pH was illustrated in Fig. 3.

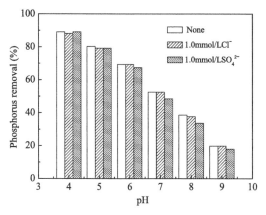

Figure 3. Effect of Cl^- and SO_4^{2-} on the removal of phosphorus with Fe/Mn oxide formed *in situ* by $KMnO_4$-Fe^{2+} process. P dosage 1.0mg/L, Fe dosage 2.5mg/L, Mn/Fe molar ratio=0.33.

It showed that 1mmol/L Cl^- or 1mmol/L SO_4^{2-} had a negligible effect on the removal of phosphorus. When the pH was increased from 4 to 9, the removal of phosphorus decreased from 89.1% to 19.8%. In the presence of 1mmol/L SO_4^{2-}, the removal of phosphorus by $KMnO_4$-Fe^{2+} process decreased from 89.1% to 17.9%. In the presence of 1mmol/L Cl^-, the removal of phosphorus by $KMnO_4$-Fe^{2+} process decreased from 88.1% to 19.8%. Compared with SO_4^{2-}, the influence of Cl^- was weaker on phosphorus removal by $KMnO_4$-Fe^{2+} process.

3.4 Effect of Ca²⁺on the removal of phosphorus by KMnO₄-Fe²⁺ process

Ca^{2+} is widely present in the water which may affect the surface characteristics of iron species and removal of anions phosphorus (Anazawa & Ohmori 2001, Genz et al. 2004). The effect of 0.5mmol/L Ca^{2+} on phosphorus removal by $KMnO_4$-Fe^{2+} process at different pH was examined in Fig. 4. When 0.5mmol/L Ca^{2+} was present, the removal of phosphorus at pH 4-6 was enhanced by 2.0%-4.0%. Phosphorus removal at pH 7-9 was obviously improved by 9.9%–28.7%. It can be seen that the removal of phosphorus by

$KMnO_4$-Fe^{2+} process was increased in the pH ranging from 4 to 9 and enhancement was more significant at higher pH. The result can be explained that the presence of Ca^{2+} could effectively compensate the surface negative charge generated by reaction of phosphorus and Fe/Mn oxide formed *in situ*, which favored the phosphorus anions removal. A similar result was obtained from Liu's study which investigated the effect of calcium in enhancing arsenic removal by ferric chloride in the presence of silicate and found that calcium increased the arsenic removal through the introduction of calcium increased the surface charge of ferric hydroxide. Additionally, increasing the amount of precipitated solids might be one reason of enhancement phosphorus removal when Ca^{2+} was dosed (Liu et al. 2007).

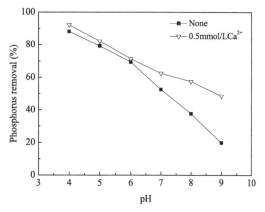

Figure 4. Effect of Ca^{2+} on the removal of phosphorus with Fe/Mn oxide formed *in situ* by $KMnO_4$-Fe^{2+} process. P dosage 1.0mg/L, Fe dosage 2.5mg/L, Mn/Fe molar ratio=0.33.

3.5 Effect of SiO₃²⁻on the removal of phosphorus by KMnO₄-Fe²⁺ process

SiO_3^{2-} exists widely in natural water bodies. The study found that the silicate can effectively occupy the surface of solid metal oxide activity, and have a great impact on metal oxide surface (Meng et al. 2000). The effect of SiO_3^{2-} on the removal of phosphorus by $KMnO_4$-Fe^{2+} process was shown in Fig. 5. When the pH was less than 7, the removal of phosphorus increased a little in the presence of 0.5mmol/L silicate. Phosphorus removal was not affected by silicate when the pH was 7. The removal of phosphorus had significantly inhibited at pH 8-9. When the pH changed from 8 to 9, phosphorus removal was obviously decreased by 32.6% and 16.4%, respectively. It was explained that when the pH was less than 7, silicate existed in non-ionized species, which were difficult to adsorb on the surface of Fe/Mn oxide. Non-ionized species of silicate would produce coagulation effect, thus it could promote the removal of phosphorus by $KMnO_4$-Fe^{2+} process. When the pH was greater than 7, silicate could compete with phosphorus for the binding sites of Fe/Mn oxide formed *in situ* (Blaney et al. 2007). Therefore, the present of silicate led to a significantly decrease of phosphorus removal by $KMnO_4$-Fe^{2+} process in alkaline solution.

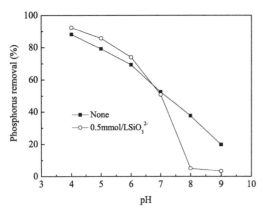

Figure 5. Effect of SiO_3^{2-} on the removal of phosphorus with Fe/Mn oxide formed *in situ* by $KMnO_4$-Fe^{2+} process. P dosage 1.0mg/L, Fe dosage 2.5mg/L, Mn/Fe molar ratio=0.33.

4 CONCLUSIONS

Effects of inorganic ions including NO_3^-, HCO_3^-, Cl^-, SO_4^{2-}, Ca^{2+} and SiO_3^2 on phosphorus removal by $KMnO_4$-Fe^{2+} process were examined. The results supported the following conclusions. With the present of 1mmol/L or 10mmol/L NO_3^-, the removal of phosphorus by $KMnO_4$-Fe^{2+} process was no significant change. In the presence of 1mmol/L HCO_3^-, the removal rate of phosphorus had a certain effect by $KMnO_4$-Fe^{2+} process, but the removal efficiency was negligible. 1mmol/L Cl^- and 1mmol/L SO_4^{2-} had a negligible effect on the removal of phosphorus. Under the same experimental conditions, the effect of Cl^- was weaker than the effect of SO_4^{2-}. 0.5mmol/L Ca^{2+} could promote the removal of phosphorus by $KMnO_4$-Fe^{2+} process in the pH ranging from 4 to 9 and enhancement was more significant at pH 7-9. The presence of 0.5mmol/L SiO_3^{2-} could promote the phosphorus removal at lower pH, and significantly inhibited phosphorus removal at higher pH.

ACKNOWLEDGMENTS

The authors are grateful for the financial support provided by the National Natural Science Foundation of China (No.21407077), the Postdoctoral Science Foundation of Jiangsu Province (No. 1002013C), the Fundamental Research Funds for the Central Universities (No.308201NS2012079) and A Project Funded by the Priority Academic Program Development of Jiangsu Higher Education Institutions.

REFERENCES

Anazawa, K., Ohmori, H. 2001. Chemistry of surface water at a volcanic summit Area, Norikura, central Japan: multivariate statistical approach. *Chemosphere* 45(6): 807–816.

Blaney, L.M., Cunar, S., Sengupta, A.K. 2007. Hybrid anion exchange for trace phosphate removal from water and wastewater. *Water Research* 41: 1603–1613.

Choi, J., Lee, S., Kim, J., Park, K., Kim, D., Hong, S. 2012. Comparison of surface modified adsorbents for phosphate removal in water. *Water Air and Soil Pollution* 223: 2881–2890.

Genz, A., Kornmuller, A., Jekel, M. 2004. Advanced phosphorus removal from membrane filtrates by adsorption on activated aluminium oxide and granulated Ferric hydroxide. *Water Research* 38(16): 3523–3530.

Ismail, Z.Z. 2012. Kinetic study for phosphate removal from water by recycled datepalm wastes as agricultural by-products. *International Journal of Environmental Studies* 69: 135–149.

Jyothi, M.D., Kiran, K.R., Ravindhranath, K. 2012. Phosphate pollution control in waste waters using new biosorbents. *International Journal of Water Resources and Environmental Engineering* 4: 73–85.

Liu, R.P., Li, X., Xia, S.J., Yang, Y.L., Wu, R.C., Li, G.B. 2007. Calcium-enhanced ferric hydroxide co-precipitation of arsenic in the presence of silicate. *Water Environment Research* 79: 2260–2264.

Meng, X.G., Bang, S., Korfiatis, G.P. 2000. Effects of silicate, sulfate, and carbonate on arsenic removal by Ferric chloride. *Water Research* 34(4): 1255–1261.

Mohamed, A.M.O. 2002. Development of a novel electro-dialysis based technique for lead removal from silty clay polluted soil. *Journal of Hazardous Materials* 90: 297–310.

Moharami, S., Jalali, M. 2013. Removal of phosphorus from aqueous solution by Iranian natural adsorbents. *Chemical Engineering Journal* 223: 328–339.

Nguyen, T.A.H., Ngo, H.H., Guo, W.S., Zhang, J., Liang, S., Lee, D.J., Nguyen, P.D., Bui, X.T. 2014. Modification of agricultural waste/by-products for enhanced phosphate removal and recovery: Potential and obstacles. *Bioresource Technology* 169: 750–762.

Thörneby, L., Persson, K. 1999. Treatment of liquid effluents from dairy cattle and pigsusing reverse osmosis. *Journal of Agricultural Engineering Research* 74:159–170.

Zhao, L., Ma, J., Sun, Z.Z. 2006. Effect of inorganic ions degradation of trace nitrobenzene in aqueous solution by catalytic ozonation. *Environmental Science* 27: 924–929.

Zhu, R., Wu, M., Zhu, H.G., Wang, Y.Y., Yang, J. 2011. Enhanced phosphorus removal by a humus soil cooperated sequencing batch reactor using acetate as carbon source. *Chemical Engineering Journal* 166: 687–692.

Energy, Environment and Green Building Materials – Sheng (ed.)
© *2015 Taylor & Francis Group, London, ISBN 978-1-138-02718-3*

Scale fingerprint clustering-based gait tracking algorithm under high-frequency gait mode switching environment

Y. Liu, S.L. Wang, L. Wang & L.L. Li
Institute of Optical Communication Technology, Chongqing University of Posts and Telecommunications, China

ABSTRACT: The MEMS inertial accelerometer has the ability to sense the acceleration information of the human body. By making use of this ability, plus the simple mode recognition algorithm, human gait could be identified from several dynamic locomotion gait modes; however, this algorithm is only effective in a low-frequency gait mode switching situation. Once the gait mode switching frequency comes to 0.05Hz or higher, its tracking accuracy drops down to 80% or less. This paper presents a gait tracking algorithm based on clustered Reference Point (RP) fingerprint, to improve the tracking accuracy under high-frequency gait mode switching situation. By running on an Android mobile phone, the new algorithm is proved by at least 13%.

KEYWORDS: Clustered RP Fingerprint; High-frequency Gait Mode Switching; MEMS Inertial Accelerometer

1   INTRODUCTION

In recent years, various kinds of handheld devices (eg: Android phones) integrate MEMS inertial sensors, which makes gait information tracking of pedestrian based on Android phones possible. The conventional gait tracking algorithm can track the gait information only when switching the gait mode (includes "walk," "run," "jump," "upstairs," and "downstairs") under the situation of low frequency. However, gait information that can be tracked based on switching the gait mode under the situation of high frequency is still unresolved, and this gets more and more extensive attention.

Many gait tracking algorithms are developed to determine the kind of gait of the pedestrian by analyzing the MEMS sensors data. In literature[1-2], Jonghee Han and Hyo Sun Jeon et al make use of an MEMS accelerometer to monitor Parkinsonian Patient's gait. The sensors are mounted on the ankles; peak value and interactive relationship between left/right ankle accelerations are taken into account for gait tracking. In literature[3], Rueterbories, Jan et al use the MEMS inertial accelerometer for tracking gait information of the pedestrian only under the situation of the activity of walking. In literature[4], Herman KYChan and Huiru Zheng et al start the gait research by using an iPhone in which MEMS accelerometers are integrated. However, traditional gait algorithms are only effective in a stable locomotion mode. They lose the accuracy when used in a high-frequency gait switching situation.

2   SYSTEM ARCHITECTURE

This paper designs and implements a gait tracking algorithm system, and the algorithm is based on clustered scale fingerprint. The system is based on the recognition algorithm of modes, then tracking gait information when switching gait mode under the situation of high frequency. In this paper, before the phase of online tracking, a clustering processing for all gait modes' RP is added before online tracking; this can reduce the complexity of the algorithm and also maintain its precision when switching gait mode under the situation of high frequency. The detail function modules of the system are shown in Figure 1:

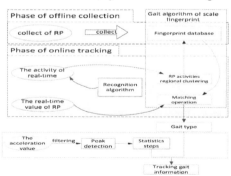

Figure 1.   Function block diagram of the system.

2.1   *Recognition algorithm module of activities*

This paper is a continuous work to the previous research on the recognition model of human locomotion gait modes[5]. The model uses BP artificial neural network and secondary classification algorithm as classifiers. It reaches 95% recognition accuracy for the gait mode in low-frequency gait mode switching, and the accuracy of gait tracking can reach 99% if we introduce further simple processing. But the recognition accuracy for the gait mode will drop down greatly during high-frequency gait mode switching (high-frequency=0.05Hz), so the gait tracking accuracy will get bad. The interesting part is regardless of the low-frequency or high-frequency gait mode switching, the classification algorithm always makes a positive recognition between gait mode combination I ("walk," "upstairs," and "downstairs") and gait mode combination II ("jump" and "run"). This feature makes it possible for the RP clustering, and we take this condition as the standard of clustered RP fingerprint. In other words, recognition of activities is the cornerstone of tracking gait information when switching gait mode under the situation of high frequency.

2.2 Number of step detection

To detect the number of steps, this paper uses the number of peaks of the Y-axis acceleration data as the number of steps. This paper detects the number of steps every 4s. Due to the acceleration, raw data contain a lot of glitches. This paper uses a smoothing filter to filter the raw acceleration data, and the result of the filtering is shown in Figure 2.

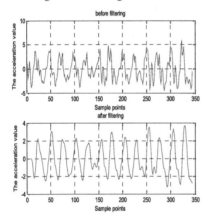

Figure 2. Result of filtering.

2.3 Scale fingerprint algorithm

Scale fingerprint technology is the collected RP of each mode to compose the scale fingerprint database. Then, the real-time measure of scale features are taken in the scale fingerprint database for matching to speculate current gait information. The scale fingerprint algorithm proposed in this paper has two stages: the phase of offline collection and online tracking. In the phase of offline collection, the RP of five gait modes is collected, and each RP will be pasted along with its corresponding label. In the phase of online tracking, real-time measure of scale features is taken into the scale fingerprint database for matching to speculate current gait information. The scale fingerprint algorithm proposed in this paper has

the advantage of high precision, but it brings an overloaded calculation to the system. To resolve the problem, the clustering method is applied to all RP based on some characteristics. Consequently, measured RP is sent to one cluster at first, and then the best matched RP is found out in this cluster.

Due to the recognition algorithm can distinguish mode combination I and gait combination II effectively. Therefore, the recognition algorithm not only provides modes, but also as the processing module of clustered RP fingerprint. The RP of gait mode combination I as cluster II, and the RP of gait mode combination as cluster II. The three scale features of Skew[6], per_h3 and quar_v[7] of the three modes in mode combination I change stable relatively, no matter the situation is high-frequency or low-frequency gait mode switching, therefore, this paper take Skew, per_h3 and quar_v as RP of modes in mode combination I. In the same way, take Skew, per_h3 and quar_h[8] as RP of modes in mode combination II similarly. In the real-time gait tracking stage, the classification algorithm adaptively select three corresponding features by deciding which combination the current gait belongs to, then matching the fingerprint with database. The four features mentioned above are derivated from acceleration data after filtering. The detailed formulas are shown as following:

$$per\_h3 = \sqrt{x^2{}_h[\frac{3}{4}N] + y^2{}_h[\frac{3}{4}N] + z^2{}_h[\frac{3}{4}N]} \quad (1)$$

Formula (1) represents the module value which is the 3/4 point of accelerometer values of 3 axis mapping in horizontal direction within 4s, where x_h is the mapping value of X-axis in the horizontal direction, y_h is the mapping values of Y-axis in the horizontal direction, z_h is the mapping value of Z-axis in the horizontal direction;

$$quar\_v = \sqrt{x^2{}_v[\frac{3}{4}N] + y^2{}_v[\frac{3}{4}N] + z^2{}_v[\frac{3}{4}N]} - \sqrt{x^2{}_v[\frac{1}{4}N] + y^2{}_v[\frac{1}{4}N] + z^2{}_v[\frac{1}{4}N]} \quad (2)$$

Formula (2) represents the spacing module value which is calculated by the 4s accelerometer values at 3/4 and 1/4 point in the vertical direction, where x_v is the mapping value of X-axis in the vertical direction, y_v is the mapping values of Y-axis in the vertical direction, z_v is the mapping value of Z-axis in the vertical direction;

$$quar\_h = \sqrt{x^2{}_h[\frac{3}{4}N] + y^2{}_h[\frac{3}{4}N] + z^2{}_h[\frac{3}{4}N]} - \sqrt{x^2{}_h[\frac{1}{4}N] + y^2{}_h[\frac{1}{4}N] + z^2{}_h[\frac{1}{4}N]} \quad (3)$$

Formula (3) represents the spacing module value that is calculated by the 4s accelerometer values at 3/4 and 1/4 point in the horizontal direction, where x_h is the mapping value of X-axis in the horizontal direction, y_h is the mapping values of Y-axis in the horizontal direction, and z_h is the mapping value of Z-axis in the horizontal direction:

$$Skew = \frac{1}{N}\sum_{i=1}^{N}(\frac{a_i - a}{Std}) \qquad (4)$$

Formula (4) represents the skewness of the acceleration value of 3 axes within 4s, where a_i represents the module value of 3 axes; Std represents the average value of ai within 4s.

2.4 Matching model

The gait tracking algorithm mainly follows four rules: 1) k locomotion clusters (k=2); 2) m locomotion models (m=5); 3) n Reference Points in each cluster (n=3); 4) in the paper, using Euclidean distance to evaluate the similarity between measured RP and database. The Euclidean distance [9] is defined in formula (5):

$$E_p = \sqrt{p_i - r_i} \qquad (5)$$

where E_p represents dissimilarity for the real-time measure of scale features and RP, where r_i represents the value of real-time measure of scale features, and p_i represents the value of RP. The value of E_p is larger, indicating that dissimilarity is larger for the real-time measure of scale features and RP, and dissimilarity is smaller in the contrary.

The three features in mode combination I have unequal weights on the locomotion matching, and their weights are 1/12, 2/3, and 1/4. Similarly, the weights of features in mode combination II are 1/12, 2/3, and 1/4 for the best. Figures 3 and 4 show the features distribution in each cluster.

In the real-time tracking stage, the classification algorithm finds out which mode combination belongs to, and then calculates three features and their dissimilarity corresponding to this cluster; finally, a comprehensive dissimilarity is obtained by multiplying the weights. The smallest comprehensive dissimilarity is the best matched gait mode. The comprehensive dissimilarity is shown in Formula (6):

$$E_z = \sum_{i=1}^{n} x_i E_{p,i} \qquad (6)$$

where x_i represents weights for corresponding scale features, and $E_{p,i}$ represents the dissimilarity for corresponding scale features.

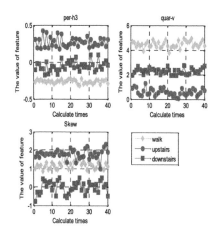

Figure 3. Distribution of three scale features in cluster I.

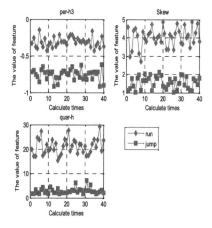

Figure 4. Distribution of three scale features in cluster II.

3 RESULTS AND DISCUSSION

Ten testers with Huawei Mate mobile phones in their pockets participated in the test to investigate the tracking accuracy of the scale fingerprint algorithm. Statistical results of the test are shown in Figures 5 and 6.

The statistical results show that the tracking accuracy of the traditional gait tracking algorithm and that of the SFA (scale fingerprint algorithm) are almost consistent when switching gait mode under the situation of low frequency, but the tracking accuracy of the traditional gait tracking algorithm is significantly lower than that of the SFA when switching gait mode under the situation of high frequency.

Each tester has to test 500 steps for each gait mode and switch the mode every 120 steps when switching the gait mode under the situation of low frequency (switching frequency < 0.0083Hz). On the contrary, they need to switch the mode every 20 steps when switching the gait mode under the situation of high frequency (switching frequency > 0.05Hz). According to the formula of accuracy, we realize that the tracking accuracy of the traditional gait tracking algorithm and the SFA is more than 99% under the situation of low frequency, as depicted in Figure 5,

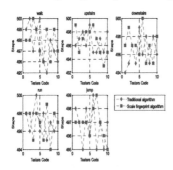

Figure 5. Tracking accuracy of two algorithms under low-frequency switching.

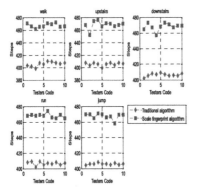

Figure 6. Tracking accuracy of two algorithms under high-frequency switching.

which shows that gait information is stable and easy to track under the situation of low frequency. From Figure 6, we realize that the tracking accuracy of the traditional gait tracking algorithm is less than 80%, but the tracking accuracy of the SFA is more than 93.3% when switching the gait mode under the situation of high frequency.

4 CONCLUSIONS

This paper designs and implements a gait tracking algorithm based on clustered scale fingerprint. Massive testing is performed in a real working environment, and the results show that the algorithm proposed in this paper has at least 13% improvement with regard to the tracking accuracy in the high-frequency gait mode switching situation. The following study will focus on using the model to improve the classification accuracy of the gait mode in the high-frequency gait mode switching situation and then to further promote the gait tracking accuracy.

ACKNOWLEDGMENTS

This work was financially supported by the Natural Science Foundation of Chongqing (No.CSTC-2012jjB40003); the National Natural Science Foundation of China (No.51175535); and the Open Research Program from Key Laboratory of Optoelectronic Devices and Systems (Education Department / Guangdong Province) of Shenzhen University.

REFERENCES

[1] Han J, Jeon H S, Jeon B S. Gait detection from three dimensional acceleration signals of ankles for the patients with Parkinson's disease[C]//Proceedings of the International Special Topic Conference on Information Technology in Biomedicine. 2006.
[2] Niazmand K, Tonn K, Zhao Y. Freezing of gait detection in parkinson's disease using accelerometer based smart clothes[C]//Biomedical Circuits and Systems Conference (BioCAS), 2011 IEEE. IEEE, 2011: 201–204.
[3] Rueterbories J, Spaich E G, Larsen B, et al. Methods for gait event detection and analysis in ambulatory systems[J]. Medical engineering & physics, 2010, 32(6):545–552.
[4] Chan H K Y, Zheng H, Wang H, et al. Feasibility study on iPhone accelerometer for gait detection[C]//Pervasive Computing Technologies for Healthcare (Pervasive Health), 2011 5th International Conference on. IEEE, 2011:184–187.
[5] Yu Liu, Le Wang, Cao Yang, et al. The improvement of behavior recognition accuracy of Micro inertial accelerometer by secondary recognition algorithm[J]. Sensors & Transducers, 2014.
[6] Gupta P, Dallas T. Feature Selection and Activity Recognition System using a Single Tri-axial Accelerometer[J]. Biomedical Engineering, IEEE Transactions on, 2014.
[7] Preece S J, Goulermas J Y, Kenney L P J, et al. A comparison of feature extraction methods for the classification of dynamic activities from accelerometer data[J]. Biomedical Engineering, IEEE Transactions on, 2009, 56(3):871–879.
[8] XUE Y. Human motion Patterns Recognition based on Signal Triaxial Accelerometer [D]. South China University of Technology Guangzhou, 2011.
[9] Schouten T E, Van den Broek E L. Fast Exact Euclidean Distance (FEED): A New Class of Adaptable Distance Transforms[J]. IEEE transactions on pattern analysis and machine intelligence, 2014.

Energy, Environment and Green Building Materials – Sheng (ed.)
© *2015 Taylor & Francis Group, London, ISBN 978-1-138-02718-3*

Positive scheme numerical study for gas jet in the direction pipe

J.L. Zhong & G.G. Le
School of Mechanical Engineering, NUST, Nanjing, China

ABSTRACT: The gas jet in the direction pipe is researched. The mathematical expression of positive scheme axis-symmetric Euler equations is derived. Flow parameters in the direction pipe are provided and treated with the non-dimension. The gas jet in the composite material direction pipe is simulated, and the pressure and temperature responses of the inner direction pipe are obtained. The results show that a significant gradient change exists near the peak values of the pressure curve of the central axis, which means the positive scheme method has a high resolution in the discontinuity.

KEYWORDS: gas jet, positive scheme method, direction pipe, high resolution

1 INTRODUCTION

The aero-defense rocket that is used to realize the multi-object end defense for the cities has the characteristics of great shooting accuracy, low price, convenient operation, and so on. To improve the shooting dispersion of the aero-defense rocket, the research on the gas jet problem of the rockets is of great importance. Bernardo C etal[1,2] proposed a Runge-Kutta discontinuous Galerkin method, which has the same high-order accuracy and high resolution in the spatial and time dimension. Chang I S etal [3,4] carried out the numerical calculation for the unsteady-state underexpanded jet flow with the CE/SE method; the complex wave structure such as shear layer, expansion waves, incident shock, mach shock disk, and so on is captured. Liu etal[5-7] proposed a positive scheme method for computational fluid mechanics that can be used to solve hyperbolic conservation laws. So far, there has been no research literature published about the numerical calculation for gas jet with a positive scheme.

The positive scheme method is developed to solve the axis-symmetric Euler equations, and the relevant calculation code is programmed. With the developed positive scheme method used in the numerical simulation for the gas jet in the composite material direction pipe, the unsteady change process of the complex wave structure and flow field parameters of the flow field are revealed. The research method established a theoretical method to analyze the combustion gas jet load of the aero-defense rocket, which is of great theoretical significance and practical value.

2 DEVELOPMENT OF POSITIVE SCHEME METHOD

2.1 *Mathematical model*

On the assumption that the stickiness and the heat conduction effect are ignored, the conservation forms of the axis-symmetric flow Euler equations are

$$\frac{\partial U}{\partial t} + \frac{\partial F(U)}{\partial x} + \frac{\partial G(U)}{\partial y} = S(U), \quad \Omega \times (0,T) \tag{1}$$

where $U = [\rho, \rho u, \rho v, E]^T$, $F(U) = [\rho u, \rho u^2 + p, \rho uv, (E + p)u]^T$, $G(U) = [\rho v, \rho uv, \rho v^2 + p, (E + p)v]^T$, and $S(U) = (-\rho v / y)[1, u, v, (E + p) / \rho]^T$. T is the time variable, ρ is the density, u and v are the velocity components in the x and y direction, p is the pressure, γ is the specific heat capacity, e is the internal energy in unit mass, $E = \rho e + \rho(u^2 + v^2) / 2$, represents the total energy in unit volume, and the state equation is $p = (\gamma - 1)\rho e$.

2.2 *Spatial discretization*

Ensuring the control equation (1) is discrete with the positive scheme method, we get the semi-discrete finite-difference equation of the axis-symmetric Euler equations:

$$\left(\frac{\partial U}{\partial t}\right)_{i,j} = -\left[\frac{1}{\Delta x}(F_{i+1/2,j} - F_{i-1/2,j}) + \frac{1}{\Delta y}(G_{i,j+1/2} - G_{i,j-1/2})\right] + S_{i,j} \tag{2}$$

where Δx and Δy are the mesh size in x and y direction.

The numerical flux in equation (2) consists of a central difference flux $F^c_{i+1/2,j}$ and an upwind flux $F^{up}_{i+1/2,j}$. Adopting the limiter $L^0 = Rdiag(\phi^0(\theta^k))R^{-1}$, the flux limiter $\phi^0(\theta)$ satisfies

$$0 \leq \phi^0(\theta), \frac{\phi^0(\theta)}{\theta} \leq 2, \phi^0(1) = 1 \tag{3}$$

The numerical flux could be structured as follows:
$$F^0_{i+1/2,j} = F^{0,up}_{i+1/2,j} + L^0(F^{0,c}_{i+1/2,j} - F^{0,up}_{i+1/2,j}) \tag{4}$$

where $F^{0,c}_{i+1/2,j} = \frac{1}{2}[F(U_{i,j}) + F(U_{i+1,j})]$, $F^{0,up}_{i+1/2,j} = \frac{1}{2}[F(U_{i,j}) + F(U_{i+1,j})] - \frac{1}{2}|\overset{o}{A}|(U_{i+1,j} - U_{i,j})|\overset{o}{A}|$ is the module of $A = \nabla F$; it equals $R|\Lambda|R^{-1}$, where $|\Lambda| = diag(|\lambda^k|)$. Therefore, the weak dissipation scheme is:

$$F^0_{i+1/2,j} = \frac{1}{2}[F(U_{i,j}) + F(U_{i+1,j})] - \frac{1}{2}|\overset{o}{A}|(U_{i+1,j} - U_{i,j})$$
$$+ \frac{1}{2}L^0|\overset{o}{A}|(U_{i+1,j} - U_{i,j}) \tag{5}$$

If the module of $A = \nabla F$ is $|A| = Rdiag(\overset{|}{\mu})R^{-1}$, the diagonal matrix $diag(\overset{|}{\mu}) \geq |\Lambda|$. Adopting the limiter $L^1 = Rdiag(\phi^1(\theta^k))R^{-1}$, $\phi^1(\theta)$ is the minmod flux limiter that satisfies

$$\phi^1(\theta) = 0, (\theta \leq 0)\theta, 0(< \theta \leq 1)1, (\theta > 1) \tag{6}$$

The numerical flux can be structured as follows:

$$F^1_{i+1/2,j} = F^{1,up}_{i+1/2,j} + L^1(F^{1,c}_{i+1/2,j} - F^{1,up}_{i+1/2,j}) \tag{7}$$

where $F^{1,c}_{i+1/2,j} = \frac{1}{2}[F(U_{i,j}) + F(U_{i+1,j})]$, $F^{1,up}_{i+1/2,j} = \frac{1}{2}[F(U_{i,j}) + F(U_{i+1,j})] - \frac{1}{2}|A|(U_{i+1,j} - U_{i,j})$. The strong dissipation scheme of the numerical flux can be obtained further:

$$F^1_{i+1/2,j} = \frac{1}{2}[F(U_{i,j}) + F(U_{i+1,j})] - \frac{1}{2}|A|(U_{i+1,j} - U_{i,j})$$
$$+ \frac{1}{2}L^1|A|(U_{i+1,j} - U_{i,j}) \tag{8}$$

Combining the weak dissipation scheme (5) with the strong dissipation scheme (8), the positive scheme numerical flux can be structured as follows:

$$F^{\alpha,\beta}_{i+1/2,j} = \frac{1}{2}[F(U_{i,j}) + F(U_{i+1,j})] - \frac{1}{2}[\alpha|\overset{o}{A}|(I - L^0)$$
$$+ \beta|A|(I - L^1)](U_{i+1,j} - U_{i,j}) \tag{9}$$

The positive scheme numerical flux satisfies the CFL condition:

$\frac{\Delta t}{\Delta x}(\alpha \max_{1 \leq k \leq n, U} |\lambda^k| + \beta \max_{1 \leq k \leq n, U} \overset{|}{\mu}) \leq \frac{1}{2}$, Δt is the time steps, and the constant α and β satisfy $0 \leq \alpha \leq 1$, $\alpha + \beta \geq 1$.

2.3 Time discretization

To match the spatial accuracy, the time discretization uses the Runge-Kutta method, which is the second-order accurate:

$$U^*_{i,j} = U^m_{i,j} - [\frac{\Delta t}{\Delta x}(F^{\alpha_x,\beta_x}_{i+1/2,j} - F^{\alpha_x,\beta_x}_{i-1/2,j}) + \frac{\Delta t}{\Delta y}(G^{\alpha_y,\beta_y}_{i,j+1/2} - G^{\alpha_y,\beta_y}_{i,j-1/2})] + \Delta t S_{i,j} \tag{10a}$$

$$U^{**}_{i,j} = U^*_{i,j} - [\frac{\Delta t}{\Delta x}(F^{\alpha_x,\beta_x*}_{i+1/2,j} - F^{\alpha_x,\beta_x*}_{i-1/2,j}) + \frac{\Delta t}{\Delta y}(G^{\alpha_y,\beta_y*}_{i,j+1/2} - G^{\alpha_y,\beta_y*}_{i,j-1/2})] + \Delta t S^*_{i,j} \tag{10b}$$

$$U^{m+1}_{i,j} = \frac{1}{2}U^m_{i,j} + \frac{1}{2}U^{**}_{i,j} \tag{10c}$$

3 CALCULATION PARAMETERS OF THE DIRECTION PIPE

3.1 Flow parameters

The specific heat at constant volume is

$$C_v = C_p/\gamma = 1906.43/1.229 = 1551.204 \text{ J/(kg·K)} \tag{11}$$

The gas constant is

$$R = C_p - C_v = 355.226 \text{ J/(kg·K)} \tag{12}$$

On the assumption that the flow is a one-dimensional steady isentropic one, we ignore the flow loss. The area ratio of the nozzle can be obtained according to its structure size.

$$\zeta^2_e = A_e/A_t = 30.65^2/15.65^2 \approx 3.8356 \tag{13}$$

where ζ_e is the nozzle area expansion ratio; A_e and A_t are the cross-sectional areas of the exit and the throat of the nozzle.

According to the area ratio formula of the one-dimensional steady isentropic flow, we get

$$\frac{A_e}{A_t} = \frac{1}{Ma_e}\left[\frac{2}{\gamma+1}\left(1 + \frac{\gamma-1}{2}Ma^2_e\right)\right]^{\frac{\gamma+1}{2(\gamma-1)}} \tag{14}$$

where the Ma_e is the mach number of the exit nozzle. Substituting the parameters in Table 2 and equation (13) into equation (14), we get $Ma_e = 2.63$.

Since the flow of the nozzle is isentropic, the stagnation parameter in the whole nozzle is the same; the flow parameters of the section of the nozzle exit cone can be calculated by the following formula:

$$p_e = \left(1 + \frac{\gamma - 1}{2} Ma_e^2\right)^{-\frac{\gamma}{\gamma - 1}} p_0 = 5.058 \, (atm) \qquad (15)$$

$$T_e = \left(1 + \frac{\gamma - 1}{2} Ma_e^2\right)^{-1} T_0 = 1780.134 \, (K) \qquad (16)$$

$$\rho_e = p_e / RT_e = 0.81 \, (kg/m3) \qquad (17)$$

$$a_e = \sqrt{\gamma RT_e} = 881.57 \, (m/s) \qquad (18)$$

$$U_e = Ma_e \cdot a_e = 2318.53 \, (m/s) \qquad (19)$$

where p_e, T_e, ρ_e, a_e, and $U_e = 0.5$ are the pressure, temperature, density, sound velocity, and velocity, respectively, of the gas in the nozzle exit cone.

3.2 Treatment of non-dimension

The inside diameter of the direction pipe r = 35mm, the inside diameter of the nozzle r_e = 30.65mm, and the length of the direction pipe L =1650mm.

Making the three parameters characteristic length: $L_0 = r_e = 30.65$ mm, characteristic density $\rho_0 = \rho_e = 0.81$ kg/m3, characteristic velocity $U_0 = a_e = 881.57$ m/s as the independent reference quantity, the treatment of non-dimension is conducted and we get the following:

Characteristic pressure: $p_0 = \rho_0 U_0^2 = 629504.1886$ Pa

Characteristic time: $t_0 = 0.34767 \times 10^{-4}$ s

Therefore, the non-dimensional flow parameters and geometric parameters are as follows:
(1) Flow parameters of the nozzle

$$\bar{\rho}_e = 1.0, \bar{u}_e = Ma_e = 2.63, \bar{v}_e = 0, \bar{p}_e = 0.8141 \qquad (20)$$

(2) Parameters of atmosphere at rest

$$\bar{\rho}_\infty = 1.5963, \bar{u}_\infty = 0, \bar{v}_\infty = 0, \bar{p}_\infty = 0.16096 \qquad (21)$$

(3) Geometric parameters

Radius of the nozzle exit cone: $\bar{r}_e = 1.0$ \qquad (22)

Radius of the direction pipe:

$$\bar{r} = 0.035/0.03065 = 1.14 \qquad (23)$$

Length of the direction pipe: $\bar{L} = 53.8$ \qquad (24)

4 POSITIVE SCHEME NUMERICAL STUDY FOR THE GAS JET

Half of the calculation domain cut by any meridian plane through the nozzle jet is shown in

Fig 1; the non-dimensional calculation domain is $[0:53.8] \times [0:1.14]$; and the calculation domain is made discrete with $\Delta x = \Delta y = 1/20$ uniform grid. The central axis of the direction pipe AB, the inner surface of the direction pipe CD, and the surface of the nozzle solid wall EF use the mirror reflection boundary condition; the exit cone boundary of the nozzle AF uses the cone inflow condition; the lower boundary BC uses the extrapolation boundary condition; the upper boundary DE uses the static atmosphere condition and assigns the static atmosphere condition to the initial flow field.

Figure 1. Calculation domain.

The pressure curves of the inner surface and central axis of the direction pipe as well as the temperature curves of the inner surface of the direction pipe when t=200.0s are shown in Fig 2; all the parameters are non-dimensional.

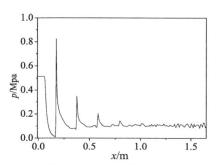

(a) Pressure of the central axis of the direction pipe

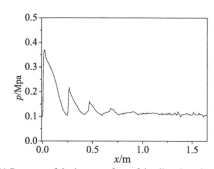

(b) Pressure of the inner surface of the direction pipe

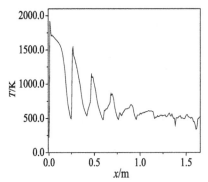

(c) Temperature of the inner surface of the direction pipe

Figure 2. Flow distribution parameters of the inner surface and central axis of the direction pipe.

As shown in fig 6, if the direction pipe is not very long, the impinging shock is intersected and reflected periodically inside the direction pipe; the distance of two pressure spikes that are caused by the impinging shock is approximately the same; in certain lengths of the direction pipe, the structure of the impinging shock has nothing to do with the length of the direction pipe; and a significant gradient change exists near the peak values of the pressure curve of the central axis, which means the positive scheme method has a high resolution in the discontinuity. From the pressure curves in fig 6(a) and fig 6(b), it can be seen that the strength of the impinging shock decreases along the direction pipe, and the variation of pressure curves matches well with the literature[8].

5 SUMMARY

The positive scheme method is developed to solve the axis-symmetric Euler equations. The corresponding calculation codes are programmed, and they are applied to the study of the supersonic flow field problem. On comparing the numerical results obtained by the developed positive scheme method with the experimental results and the numerical results obtained by the already existing high accuracy scheme method, the correctness of the developed positive scheme method is verified.

REFERENCES

[1] Bernardo C, Chi-Wang S. Runge-Kutta Discontinuous Galerkin Methods for Convection-Dominated Problems[J]. Journal of scientific computing, 2011, 16(3).

[2] Demkowicz L, Gopalakrishnan J, Niemi A H. A class of discontinuous Petrov–Galerkin methods. Part III: adaptivity[J]. Applied numerical mathematics, 2012, 62(4): 396–427.

[3] Chang I S, Chang C L, Chang S C. Unsteady Navier-Stokes Rocket Nozzle Flows[J]. AIAA Paper, 2005, 4353: 2005.

[4] Bianchi D, Turchi A, Nasuti F, et al. Coupled CFD Analysis of Thermochemical Erosion and Unsteady Heat Conduction in Solid Rocket Nozzles[J]. AIAA Paper, 2012, 4318: 2012.

[5] Liu X D, Lax P D. Positive schemes for solving multi-dimensional hyperbolic systems of conservation laws II[J]. Journal of Computation Physics, 2003, 187(2): 428–440.

[6] Nevskii Y A, Osiptsov A N. Slow gravitational convection of disperse systems in domains with inclined boundaries[J]. Fluid Dynamics, 2011, 46(2): 225–239.

[7] Fazio R, Jannelli A. Second Order Positive Schemes by means of Flux Limiters for the Advection Equation[J]. International Journal of Applied Mathematics, 2009, 39(1).

[8] Xu Qiang. The experimental research and testing technology for the gas jet dynamic effect of the aero-defense rocket.[D] Nanjing university of science and technology, 1994.

Energy, Environment and Green Building Materials – Sheng (ed.)
© 2015 Taylor & Francis Group, London, ISBN 978-1-138-02718-3

Risk identification and evaluation for decision-making in integrated urban-rural water supply system constructions

Y.H. Mao, B. Yang & H.Y. Li
College of Civil Engineering and Architecture, Zhejiang University, Hangzhou, Zhejiang, China

ABSTRACT: The integrated water supply system is proposed to improve the water supply service in China. As a crucial stage in the system construction, decision-making is faced with many risks. To assist the construction of integrated urban-rural water supply system, the paper identified and evaluated the risks occurring in the decision procedures. It proved that the overrun of construction cost, the misprediction of water demand, the low water pressure and the low reliability of water supply are the main risks to be addressed.

KEYWORDS: Integrated water supply system; decision-making; risk identification; risk evaluation

1 INTRODUCTION

As one of the most important basic facilities in modern society, water supply systems, composed of water sources, treatment plants, transmission pipes and distribution networks, are closely related to every aspects of our lives.

The rapid urbanization process in China has driven the fast development of urban water supply systems. By the end of 2010, the daily water supply capacity of urban water supply systems reached 276,000,000 cubic meters, supplying 139,000,000 cubic meters water in average every day. As a result, urban areas in China are enjoying a relatively high quality of water supply service. Comparatively speaking, rural areas fall behind in the water supply system constructions, mainly due to its undeveloped economy. Centralized water supply systems haven't been widely adopted in rural areas, and private small water supply facilities, which are in short of water treatment technology and whose pipeline materials may be harmful, are the basic form in many villages, leading to the bad water quality in rural families.

In face of the imbalanced development in water supply systems in urban and rural areas, the integrated urban-rural water supply system is proposed, aimed at taking advantage of the water supply facilities in urban areas and its experience in engineering and management, promoting water supply system constructions in rural areas, improving water quality in rural families, and realizing the optimal allocation of water resources. The main idea of the integrated urban-rural water supply system is to satisfy the water users in rural and urban areas with one water supply system by integrating water resources and related water supply facilities such as treatment plants, transmission pipes and distribution networks. Just as other urban water supply systems, the integrated water supply system is a huge project faced with many uncertainties. These uncertainties, also known as risks, influence the whole life circle of the system. The decision-making stage, which is concerning core issues like the water resources, treatment technology selection and pipelines arrangement, is risk sensitive particularly.

2 LITERATURE REVIEW

2.1 *Literature on decision-making of water supply systems construction*

The researches focusing on the decision-making of water supply systems construction are named with planning or preparing stages. Mo et al. (2013) presented their findings that the lack of systematic planning resulted in three main problems, namely, water supply systems cannot meet the needs of modern cities, the emergency reaction ability of water supply systems is weak, and water resources and water supply facilities are not well allocated. They suggested that the preparedness stage should pay more attention to these issues to avoid similar problems. By studying a construction project of water supply system, Jia et al. (2009) demonstrated the regular methods taken in water resources analysis and the arrangement of transmission pipes and distribution networks.

2.2 *Literature on risk management of water supply systems construction*

The developments of theories on risk management and related practices in other field lay a good foundation for risk management of urban water supply systems construction. Given the fact that a systematic risk management method hadn't been established, Zhou et al. (2013) explored the risks occurred from source to tap, including the risks in water sources, treatment plants, transmission pipes and distribution networks, and built up a risk evaluation system. Lu (2010) built a model to analyze risk communications in water supply systems, and concluded the basic risk communication mode, providing a foundation for further studies. Besides, Lu et al. (2010) also conducted a literature review, summarizing the most frequently used approaches in risk evaluation in water supply systems.

2.3 *Literature on integrated urban-rural water supply systems*

The integrated urban-rural water supply systems is in its initial phase, related practices can be found only in several areas in China, and many existing articles are related to these practices. Hu (2005) proposed that the integrated urban-rural water supply system is a result of the urbanization, and taking Wenzhou as an example, he analyzed the water demand in rural and urban areas and the marginal value of integrated urban-rural water supply systems. Yang & Cao (2006) summarized the main forms of integrated urban-rural water supply systems, suggesting that the planning work should take water demand, geographic conditions and geomorphic conditions into account to choose the appropriate supply form. Zhen & Shu (2011) surveyed the integrated urban-rural water supply system in Shanghai, analyzed the main problems existed, and suggested further research directions to solve these problems.

The related researches suggest that the construction of integrated urban-rural water supply systems is a complex project, and the decision-making in the preparedness stage will have a crucial influence on the system. Although much can be learned from the urban water supply system, the integrated urban-rural water supply systems planning is quite challenging for its exclusive features and seldom practice experience.

3 DECISION-MAKING OF INTEGRATED URBAN-RURAL WATER SUPPLY SYSTEMS

3.1 *The issues to be addressed in decision-making*

The goal of water supply systems is to provide water with good quality and suitable pressure to water users economically and reliably. From the life circle of the system, as is shown in Figure 1, we know that the decision-making concerns all life stages. The issues which the decision-making has to address are as follows. (1) which region will the system serves and what is the water demand;(2) where the water resources

locate and whether they can be used;(3) what water treatment technology should be taken;(4) how the transmission pipes and distribution networks should be arranged;(5) what measures can be taken to make the system reliable;(6) what measures should be taken to make the system sustainable.

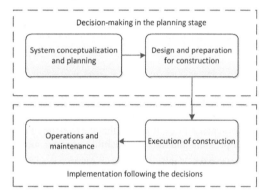

Figure 1. The life circle of water supply systems.

3.2 *The interactions between issues to be addressed*

To integrated urban-rural water supply systems, the water supply area is vast, the water demand varies a lot from time to time especially in rural areas, and the pipeline winds a long distance. These features influence many aspects of the system, making the issues to be addressed in the planning interact with each. The issues, like which region to serve and whether water resources are available, are influenced by the vast demand area and the demand variation. What's more, the selection of water treatment technology and the arrangement of transmission pipes and distribution networks are designed to promote the quality, pressure and reliability of the system, while they are all affected by the long distance of the pipeline.

Besides, the construction of the system is subject to the cost. All issues above can be perfectly solved if cost is not considered. To meet the criteria of cost control, a balance must be reached between the issues, which strengthens the interrelated relationships between the issues. All the interactions between the issues make the decision-making a dynamic adjustment process.

3.3 *The decision-making process in the construction of integrated urban-rural water supply systems*

To address the issues mentioned above, the decision-making follows a procedure, as Figure 2 shows, and the interactions between the issues are also shown in it.

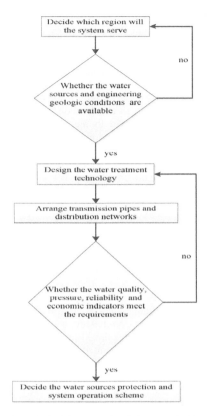

Figure 2. The decision procedure in integrated urban-rural water supply systems planning.

4 RISK IDENTIFICATION IN THE CONSTRUCTION OF INTEGRATED URBAN-RURAL WATER SUPPLY SYSTEMS

4.1 *Approaches taken to identify risks*

To identify the risks involved in the decision-making of integrated urban-rural water supply systems, document retrieval and expert consultation methods are taken. Urban water supply systems and integrated urban-rural water supply systems have much in common, and literature on risk identification of urban water supply systems can be found easily, which made related articles our priority. However, integrated urban-rural water supply systems have some unique features, as have been mentioned before, and these features may bring out risks that haven't appeared in urban water supply systems. Therefore, the experts, both from the limited practices in China and related scholars in universities, are consulted about what they think are risks in the construction of integrated urban-rural water supply systems.

4.2 *The results of risk identification*

By systemizing and analyzing the risks from document retrieval and expert consultation, we found that the risks encountered in the decision-making can be attributed to two categories, namely, risks concerning the functions of the system, and risks from constraints to the system. To meet the demands of the system functions, like water quality, pressure and reliability, a series of decisions will be made, and risks concerning the functions of the system come along in the decision process. These risks, occurring when the technological means adopted lose efficacy, can also be called technical risks. Besides, the decision-making is also subject to cost and sustainable development of the system, and these risks can be called economic risks and sustainable risks. The results of risk identification are presented in Table 1 and Table 2.

Table 1. Risks concerning functions of the system.

Occurring stages	Risks
Deciding supply region	The misjudgment on geolocal conditions The misprediction of water demand
Water sources selection	Insufficient water resources Deterioration of raw water
Water treatment technology design	The failure of Water treatment technology design
Pipelines arrangement	Low reliability of water supply Low water pressure High leak rate water contamination in pipelines
Deciding operation scheme	Equipment failure Slow responses in emergencies

Table 2. Risks from constraints to the system.

Categories of risks	Risks
Economic risks	The overrun of construction cost High operating cost
Sustainable risks	Deterioration of water sources The insufficient of supply ability

5 RISK EVALUATION IN THE CONSTRUCTION OF INTEGRATED URBAN-RURAL WATER SUPPLY SYSTEMS.

5.1 *Approaches taken to evaluate risks*

By analyzing the probability and consequence of the risks and judging the severity of them, risk evaluation aims at supporting the decision-making. The development of the risks in integrated urban-rural water supply systems hasn't been revealed, and the practices of the system construction are limited, and as a result, the data for quantitative evaluation is unavailable. To evaluation the risks, analytic hierarchy process is taken. Analytic hierarchy process has been discussed in many articles, and more detailed information about it can be found in the literature. To compare the relative importance of different risks, experts from the limited practices in China and related scholars in universities are consulted. To comply with the idea of risk evaluation, we suggest that the experts should take both the probabilities and consequences into consideration when they are judging the risks. And the results are presented in Table 3.

Table 3. The results of risk evaluation.

Categories	risks	Weight
Technical risks	The misjudgment on geological conditions	0.036
	The misprediction of water demand	0.121
	Insufficient water resources	0.072
	Deterioration of raw water	0.081
	The failure of Water treatment technology design	0.015
	Low reliability of water supply	0.084
	Low water pressure	0.086
	High leak rate	0.069
	water contamination in pipelines	0.035
	Equipment failure	0.024
	Slow responses in emergencies	0.035
Economic risks	The overrun of construction cost	0.163
	High operating cost	0.080
Sustainable risks	Deterioration of water sources	0.040
	The insufficient of supply ability	0.060

5.2 Results of risk evaluation

It can be learned that technical risks have the most risk factors and become a major concern of integrated urban-rural water supply systems. As with every risk factor, the probability and consequence varies a lot, which is revealed by the weight of the risk. Of all the risks, the overrun of construction cost, the misprediction of water demand, the low water pressure and the low reliability of water supply are the four risks with the maximum weight, and the results are consistent with the researches and practices. With the development of techniques in construction, more and more attention has been paid to the economic analysis, as it is with the integrated urban-rural water supply systems. Besides, the decision-making of the integrated urban-rural water supply systems followed the main idea of the urban water supply system constructions, emphasizing the prediction of water demand, the promotion of water pressure and system reliability.

6 CONCLUSIONS

To improve the water services in rural areas, the integrated urban-rural water supply system is proposed in China. The decision-making, as a crucial stage in the life circle of the system, is faced with many risks. To support the decision-making, a series of risks in the decision procedures are identified. For decision-makers, resources are limited and how to allocate them to address the risks depends on the severity of the risks. Therefore, the analytic hierarchy process is used to evaluate the risks. The results are consistent with the findings in related research and practices.

REFERENCES

Jia H.F. et al. 2009, The arrangement of the city water supply system in Shunyi, Water & Wastewater Engineering 35(1):23–26.

Lu R.Q. et al. 2010,Study on risk transmission mechanism model of urban water supply systems, Journal of Natural Disasters19(6):119–123.

Lu R.Q. Niu Z.G. & Zhang H.W.2010, Research of risk assessment of urban water supply system, Water & Wastewater Engineering 36:4–8.

Mo L. et al. 2013, Research on regulation and control techniques in urban water supply system, Water & Wastewater Engineering 39(7):13–18.

Hu Z.H. 2005, Economic analysis of integrated water supply system and its mechanism innovation, Journal of Wenzhou University 18(1):28–35.

Zhou Y.Z. et al.2013, Risk evaluation and safety management in urban water supply systems, Water & Wastewater Engineering 39(12):13–16.

Energy, Environment and Green Building Materials – Sheng (ed.)
© 2015 Taylor & Francis Group, London, ISBN 978-1-138-02718-3

Research and simulation on the integration of marine SCR and exhaust muffler

Y. Chen & L. Lv

Key Laboratory of High Performance Ship of Ministry of Education, Wuhan University of Technology, Wuhan, China

ABSTRACT: This study introduced a new approach that integrated Selective Catalytic Reduction (SCR) and reactive muffler for marine diesel engine. The optimization procedure used in the present investigation is based on the finite element method and focuses on the reduction of exhaust noise and NO_x simultaneously while maintaining a reasonable installation size and pressure loss. In the analysis of acoustics, it was shown that the TL of Case E that added an intubation pipe and a perforation plate into the expanding tube before the catalyst was significantly improved, even more than the original muffler. Simultaneously, pressure loss of Integrated SCR Muffler can meet the design requirement. For the rated condition of engine, the blocking degrees have a negligible impact on TL at the frequency below 800Hz, but it is influential for the high frequency. The simulated results propose a new idea to monitor the degradation degree of the SCR through the engine exhaust noise.

KEYWORDS: Diesel; muffler; SCR; Transmission Loss; NO_x emissions

1 INTRODUCTION

The nitrogen oxides that are harmful emissions generated by marine diesel engines can lead to a series of environmental problems and cause actual harm to the health of people. International Maritime Organization (IMO) has formulated MARPOL 73/78 Annex VI to restrict marine engine NO_X emissions. Selective Catalytic Reduction (SCR) is recognized as the most effective technology for the control of marine NO_X emission, both in China and abroad.

Environmental Protection Agency, U.S. 2009, Herdzik, J. 2011 & Johnson, D. R. 2009. In order to control and reduce exhaust noise in operate ship, the general method is to install the exhaust muffler. However, having an SCR after muffler will cause technical issues, such as the limitation of space on board, total cost, and exhaust back pressure. Thus, integrating the SCR with a muffler is a good choice. The integrated system can not only be used to reduce the NO_X emission and exhaust noise but also has the advantages of compact structure, small volume, and easy installation.

Dokumaci, E. 1995 & Peat, K. 2000. In recent years, the problem of transmission of sound in catalyst and perforated structure has been investigated by many researchers. Kirchhoff, G. 1968. Kirchhoff presented a complete transmission equation of acoustic wave for circular tubes having large or small cross-sections, and an accurate analytic solution was obtained. Tijdeman, H. 1975. Tijdeman extended the Kirchhoff equation to obtain numerical solutions for different wave number and helmholtz number. Astley, R. J. 1995. Astley investigated the fluid-dynamical equations governing wave propagation in catalytic converter elements containing isothermal mean flow and acting as the basis for a finite element solution scheme involving time harmonic variation.

In this paper, different integration solutions between SCR and muffler are compared to achieve excellent performance of reducing exhaust noise and NO_X emissions. With the simulation results, the role of Integrated SCR Muffler and the significance of methodology are illustrated.

2 MATERIALS AND METHODS

2.1 Design schemes of integrated SCR muffler

Figure 1 shows the structural layout of the original muffler. Obviously, the structure of the exhaust system is too compact, which is the major restrictive factor to the development of SCR. Thus, by maintaining the exhaust pipe unchanged, the overall dimensions of the Integrated SCR Muffler must remain unchanged.

The original muffler contains two expansion cavities: Cavity I and II are connected by four intubations. There is a tapered plate between the intubations, as shown in Figure 2a. Munjal, M.L. 1987. As an expansion chamber muffler, the original muffler can reflect waves by introducing a sudden change in the cross-sectional area.

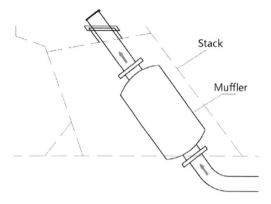

Figure 1. Structural layout of original muffler.

Generally, the marine SCR is composed of an expanding tube, a catalyst, and a shrinkable tube. In order to improve the acoustic performance of SCR, the muffler elements can be installed in the expanding tube where the exhaust diffusion is, such as tapered plate, perforated pipe, and perforated plate. According to the basis of the plane-wave theory, it can eliminate odd times of the basic frequency when the intubation length is a half of the expansion chamber. On the other hand, the muffler elements mentioned earlier make for the break-up of spray droplets, urea decomposition, and mixing of the spray with exhaust. Therefore, this paper presents a comparative study of different Integrated SCR Mufflers that added tapered plate, perforated pipe, or perforation plate into the expanding tube before the catalyst, as shown in Figure 2b–e. The exhaust

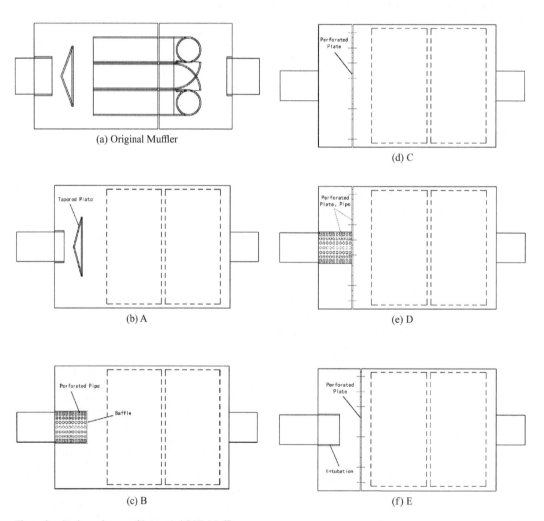

Figure 2. Design schemes of Integrated SCR-Muffler.

passes through Integrated SCR Muffler to over muffler elements to the cavity, and through the catalyst to the outlet. As we consider that no matter what changes take place in the structure of the shrinkable tube, there can always be a sharp increase of back pressure, so the shrinkable tube remains the same.

2.2 Numerical computation method of monolith

The transmission loss may be more precise than insertion loss in simulation, which is only related with the structure of the muffler and flow character-istics but not the acoustic source or the environment. For this purpose, the numerical computation method of monolith is proposed and finite element simulation is presented and analyzed.

The equation for wave propagation in the tube bun-dles of the catalyst is, however, rather involved, and requires a few simplifying assumptions. The capillary tube walls are assumed to be impervious in the present investigation. On a macroscopic scale, the problem of wave propagation in the catalyst is analogous to the problem of wave propagation in an equivalent fluid that has an effective density and an effective sound speed, both of which can be complex.

The linearized version of the continuity, momen-tum, and the energy equations for acoustic wave prop-agation are given, in the absence of flow, by

$$\frac{\partial \rho_m}{\partial t} + \rho_0 \nabla \cdot u_m = 0 \tag{1}$$

$$\nabla p_m + \rho_0 \frac{\partial u_m}{\partial t} - \mu[\nabla^2 u_m + \frac{1}{3}\nabla(\nabla \cdot u_m)] = 0 \tag{2}$$

$$\frac{\partial T_m}{\partial t} - \frac{1}{\rho_0 c_p}\frac{\partial p_m}{\partial t} = \frac{\kappa}{\rho_0 c_p}\nabla^2 T_m \tag{3}$$

Dokumaci, E. 1995 & Easwaran, S. V. 1988. At low frequencies, or when the cross-sectional dimen-sions of the capillary are very small compared with the wave-length, acoustic laminar conditions exist in the capillaries, and the flow in the capillary can be assumed to be fully developed. Under such con-ditions, the equivalent sound speed and equiva-lent density can be deduced by using the following momentum equation:

$$\nabla p_m + \rho_m \frac{\overline{\partial u_m}}{\partial t} = 0 \tag{4}$$

$$\rho_m = \rho_0 + \frac{Rf}{i\omega}G_s(s) \tag{5}$$

where ρ_m is the equivalent density; ρ_o is the air density; R is the flow resistance; and f is the open-area ratio.

Viscous effects play a role in determining the effective density, whereas thermal effects become important in obtaining the equivalent sound speed as described in the following equation:

$$c_m = c_0 / \left\{ \left[1 + G_s(s)Rf/i\omega\rho_0\right]\left[\gamma - \frac{\gamma-1}{1+(\Pr Rf/i\omega\rho_0)G_s(s\sqrt{\Pr})}\right]\right\}^{1/2} \tag{6}$$

The $G_s(s)$ is hence characterized by Equation 7.

$$G_s(s) = \frac{-\frac{s}{4}\sqrt{-i}\,\frac{J_1(s\sqrt{-i})}{J_0(s\sqrt{-i})}}{1 - \frac{2}{s\sqrt{-i}}\frac{J_1(s\sqrt{-i})}{J_0(s\sqrt{-i})}} \tag{7}$$

$$s = \alpha\sqrt{\frac{8\omega\rho_0}{Rf}} \tag{8}$$

where J_0 and J_1 are the Bessel functions of order zero and unity, respectively; α is a factor that depends on the shape of the cross-section of the capillary; and Pr is the Prandtl number (the ratio of momentum to thermal).

2.3 Evaluating indicators

2.3.1 Acoustic performance

Acoustic attenuation performance of a muffler is measured by the capacities of noise elimination in the necessary frequency range. By applying different test-ing methods, the performance evaluating indicators can be divided into transmission loss (TL), insertion loss (IL), and noise reduction (NR). The TL of muf-fler is researched in the present paper.

Equation 9 shows how, for example, the TL of the muffler is calculated.

$$TL = 10 * \log\left|\frac{S_n *(p_n + \rho * c * v_n)^2}{4\sum_i p_i^2 * S_i}\right| \tag{9}$$

where S_n and S_i are the cross-sectional area of the inlet and outlet, respectively; p_n and p_i are the pressure of the inlet and outlet, respectively; ρ is the density of flow; c is the sound velocity; and v_n is the particle vibration velocity.

2.3.2 Pressure loss

When airflow passes through the Integrated SCR Muffler, the pressure loss can reach several kPa from either muffler elements or catalyst. An opti-mized type of Integrated SCR Muffler is required to match the requirement of aerodynamic performance.

Anyway, the pressure loss can never be too high, or else it will reduce engine power and increase fuel consumption. Research shows that the exhaust temperature decreases with the increase of exhaust back pressure, because of the poor combustion quality. Nevertheless, the catalyst used under low temperature for a long time would result in sulfur poisoning. In this paper, pressure loss of the exhaust system should be limited to a 5kPa maximum.

2.4 Modeling of muffler

In order to consider the complex acoustic fields inside the Integrated SCR Muffler, the TL of the exhaust system was calculated from the three-dimensional finite element method. By dividing the Integrated SCR Muffler into catalyst and cavity, and using different sound pressure and speed as the boundary conditions, the finite element model of Integrated SCR Muffler can be obtained, as shown in Figure 3. It should be noted that the equivalent fluid should be used to describe the problem of wave propagation in the catalyst; the general expression for effective density and effective sound speed is given in Chapter 1.3.

LMS Virtual.Lab, a finite element program, is selected to carry on the numerical simulation of TL with meshes generated by Hypermesh. For the linear acoustic model, the biggest grid cell must be less than one sixth of the shortest wavelength, which can guarantee precision. To coordinate the calculation of the cost and accuracy, a compromised mesh size and mesh number are used in this paper, as shown in Figure 4.

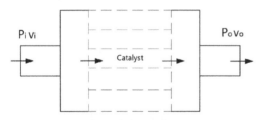

Figure 3.   Simplified model.

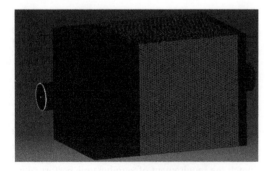

Figure 4.   FEM model of integrated SCR muffler.

3   RESULTS AND DISCUSSION

3.1 Numerical study of TL

In this chapter, the transmission loss was analyzed and assessed. Figure 5 illustrates that the one-third octave of TL comes from different design solutions. It was found that the effect of noise reduction of the original muffler is good for the range of the whole frequency, especially at the middle and low frequency. When taking a look at the result of other cases, the TL of Case E at the center frequency 500Hz and 1250Hz were significantly improved, even more than the original muffler. It proves that the muffler elements, including intubation pipe and perforated plate, play active roles in improving the performance of noise attenuation. On the other hand, Case D is the optimal solution from A-D design solutions. Weltens, H. 1993 & Bhattacharjee, S. 2011. As the exhaust passes through Case D, it has to cross a perforated pipe and a perforated plate, which can promote a mixture of urea droplets and gas. Thus, it is capable of increasing homogeneity in front of the catalyst and improving NO_X conversion efficiency.

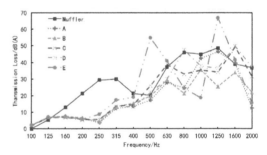

Figure 5.   One-third octave of TL.

By integrating the interval for one-third octave of TL, the computed transmission losses of Case D and Case E were 53 and 66 dB(A), respectively (Fig. 6). Both of them are greater than the original muffler (52 dB(A)), and it is necessary to pick them out for further investigation.

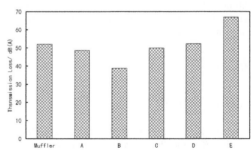

Figure 6.   Computed TL over the entire frequency range.

3.2 Effects of silencing element on the pressure loss

In view of the compressibility, complexity, and turbulence of flow field inside the Integrated SCR Muffler, the three-dimensional numerical methods should be used to predict the pressure loss. Hirata, K. 2009. In this paper, a series of numerical simulations were performed by using the v2013 version of the AVL FIRE code. The k-zeta-f turbulent model for high Reynolds numbers is applied to simulate the internal flow field of the exhaust system. The developed code is based on the pressure correction method and uses a PISO algorithm, which is an efficient method to solve the Navier–Stokes equations in unsteady problems.

The rated condition of the marine diesel was selected to assess the potential of the system at high load, taking into account further research. The cross-section of the flow field was obtained from the numerical simulations (Fig. 7), which showed that the pressure loss of the original muffler and Case A was mainly concentrated in the tapered plate, Case C and D in the perforated pipe, and Case B and E in front of the perforated plate. The maximum pressure loss

of the Case E is 4.34 kPa, which is less than the limit value of 5 kPa (Tab. 1). By contrast, the pressure loss of the Case D is too high. This can be used to explain, in part, that Case E was the best compromise.

Table 1. Comparison of pressure loss.

Load	Speed	Pressure Loss (kPa)					
		Muffler	A	B	C	D	E
% 100	Rpm 1500	6.23	4.58	5.31	4.09	5.43	4.34

3.3 Acoustic performance of blocked catalyst

During the operating of marine SCR, clogging of the catalyst can easily occur due to a build-up of soot, SOF, ash and other matter. It will reduce engine power and increase fuel consumption when the pressure loss becomes too high. On the other hand, it can be supposed that the blocked catalyst may change the acoustic performance of the SCR. In that case, we can monitor the degradation degree of the SCR through the engine exhaust noise.

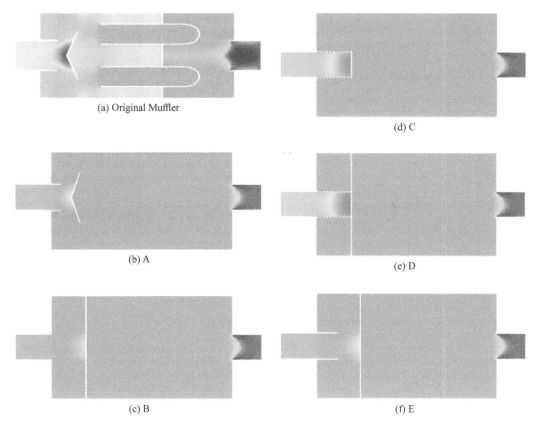

(a) Original Muffler

(b) A

(c) B

(d) C

(e) D

(f) E

Figure 7. Sectional view of pressure distribution.

The Honeycomb-type catalytic converter consists of hundreds (thousands) of individual channels. The exhaust gas flows through these channels and catalytically reacts. The catalytic reactions take place at active sites that are spread within the so-called washcoat of the monolith. This washcoat is a porous solid layer that covers the solid substrate, as shown in Figure 8. This article focused on the simulation of catalyst blocking in a simplified manner, which set different values of δ_{wall}. As shown in Table 2, the pressure loss increases along with the increase of δ_{wall} and decrease of open-area ratio. What is more, the pressure loss rises to 6.76 kPa when the blocking degree is 50%.

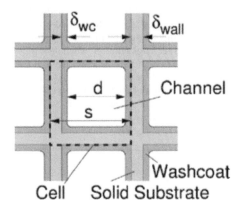

Figure 8.   Structure of a squared cell monolith.

Table 2.   Comparison of different blocking degrees.

Case		Pressure Loss (kPa)	δ_{wall} (mm)	Open-Area Ratio (-)
Blocking Degree	0%	0.23	0.450	0.741
	10%	0.35	0.623	0.667
	20%	0.44	0.798	0.593
	30%	0.78	0.984	0.519
	40%	1.11	1.185	0.445
	50%	6.76	1.402	0.371

Figure 9 shows the effects of blocking degrees on the TL based on Case A. It can be seen that the blocking degrees have a negligible impact on TL at the frequency below 800Hz, but it is influential for the high frequency. With the degree of blocking becoming more and more serious, the TL at the center frequency 1000Hz and 1250Hz first increased and then decreased. On the whole, the locations of peak in the TL move to a high frequency when catalyst blocking becomes severe. However, there are some general trends between blocking degree and TL, though it

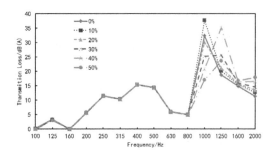

Figure 9.   Effects of blocking degrees on the TL.

looks complicated. Further research is needed to monitor the degradation degree of the SCR through the engine exhaust noise.

4   CONCLUSIONS

This study introduced a new approach that integrated SCR and muffler for the marine diesel engine. The simulation of Integrated SCR Muffler illustrated the excellent performance of reducing exhaust noise, particularly when considering the installation size and pressure loss.

In the analysis of acoustics, it was shown that the TL of Case E at the center frequency 500Hz and 1250Hz was significantly improved, even more than was the original muffler. It proves that the muffler elements, including intubation pipe and perforated plate, play active roles in improving performance of noise attenuation.

The maximum pressure loss of the Case E is 4.34 kPa, which is less than the limit value of 5 kPa. By contrast, the pressure loss of the Case D is very high. This can be used to explain, in part, that Case E was the best compromise.

For the rated condition of the engine, the blocking degrees have a negligible impact on TL at the frequency below 800Hz, but it is influential for the high frequency. With the degree of blocking becoming more and more serious, the TL at the center frequency 1000Hz and 1250Hz first increased and then decreased.

ACKNOWLEDGMENTS

This work was supported by the National Natural Science Foundation of China (No. 51379165) and the Natural Science Foundation of Educational Department of Hubei Government (No. 20520005). The authors would like to thank Prof. Lv Lin for his analytical support, and those who have not been mentioned here for their valuable services rendered in their individual technical disciplines, which formed the basis for preparing this article.

REFERENCES

Astley, R. J. & Cummings, A. 1995. Wave Propagation in Catalytic Converter: Formulation of the Problem and Finite Element Scheme. Journal of Sound and Vibration, 188(5): 635–657.

Bhattacharjee, S. & Haworth, D. C. 2011. CFD Modeling of Processes Upstream of the Catalyst for Urea SCR NOX Reduction Systems in Heavy-Duty Diesel Applications. SAE paper 2011-01-1322.

Dokumaci, E. 1995. Sound Transmission in Narrow Pipes With Superimposed Uniform Mean Flow and Modeling the Automobile Catalytic Converters. Journal of Sound and Vibration, 182(5): 799–808.

Dokumaci. E. 1998. On Transmission of Sound in Circular and Rectangular Narrow Pipes With Superimposed Mean Flow. Journal of Sound and Vibration, 210(3): 375–389.

Easwaran, S. V. 1988. Wave Attenuation In Catalytic Conveners: Reactive Versus Dissipative Effects. J. Acoustic. Soc. Am, 102(2): 935–943.

Environmental Protection Agency, U.S. 2009. Regulatory Impact Analysis: Control of Emissions of Air Pollution from Category 3 Marine Diesel Engines. EPA-420-R-09-019.

Hirata, K., Niki, Y. & Kawada, M., et al. 2009. Development of Marine SCR System and Field Test on Ship. International Symposium on Marine Engineering, BEXCO, Busan.

Herdzik, J. 2011. Emissions from Marine Engines versus IMO Certification and Requirements of Tier 3. Journal of KONES Powertrain an d Transport, Vol. 18, No. 2 .

Johnson, D. R., Bedick, C. R. & Clark, N. N., et al. 2009. Design and Testing of an Independently Controlled Urea SCR Retrofit System for the Reduction of NOX Emissions from Marine Diesels. Environmental Science and Technology, 43 (10): 3959–3963.

Kirchhoff, G. 1968. Ueber Den Einflufs Der Warmeleitung In Einem Gase Aufdie Schallbewegung. J.Ann. Phys. Chem, 134: 177–193.

Munjal, M.L. 1987. Acoustics of Ducts and Mufflers. Wiley, New York.

Peat, K. & R. Kirby. 2000. Acoustic Wave Motion Along a Narrow Cylindrical Duct in the Presence of an Axial Mean Flow and Temperature Gradient. The Journal of the Acoustical society of America, 107(4): 1859–1867.

Tijdeman, H. 1975. On the Propagation of Sound Waves in Cylindrical Tubes. Journal of Sound and Vibration, 39: L-33.

Weltens, H., Bressler, H. & Terres, F., et al. 1993. Optimization of Catalytic Converter Gas Flow Distribution by CFD Prediction. SAE Paper 930780.

Energy, Environment and Green Building Materials – Sheng (ed.)
© 2015 Taylor & Francis Group, London, ISBN 978-1-138-02718-3

Pedestrian seamless positioning algorithm based on low-cost GPS / IMU

Z.S. Tian, H.M. Huang, Z.W. Yuan & M. Zhou
Chongqing Key Laboratory of Mobile Communications Technology, Chongqing University of Posts and Telecommunications, china

ABSTRACT: This paper proposes a pedestrian seamless positioning algorithm based on the low-cost GPS / IMU to enlarge the application of Location-based Services (LBSs) in GPS-denied indoor and underground environments. First of all, we convert the latitude and longitude coordinates collected by GPS into the coordinates in ENU and then achieve the GPS-based positioning bay using the ENU; Second, the data from the inertial accelerometer, gyroscope, and magnetometer are fused by Extended Kalman Filter (EKF), to conduct Pedestrian Dead Reckoning (PDR) positioning based on IMU. Finally, we rely on the coordinate transformation in ENU, and the fusion of heading and velocity data by EKF to realize the seamless positioning based on the low-cost GPS/IMU. Experimental results demonstrate that the proposed algorithm can guarantee high-enough positioning accuracy with errors lower than 8%.

KEYWORDS: Pedestrian seamless positioning; GPS/IMU; Low-cost; EKF; PDR

1 INTRODUCTION

In most GPS/IMU-based seamless positioning scenarios, the modules of GPS and IMU are very large, and meanwhile the hardware is generally expensive and possesses a high complexity of algorithm. On this basis, the technology of seamless positioning has still not been widely used. With the rapid development of MEMS and integration technologies, the future MEMS inertial sensors and GPS positioning modules are expected to exhibit miniaturization and low price. In this case, the modules of GPS and IMU could be well integrated in many hand-held devices, such as the smartphones. Therefore, the pedestrian seamless positioning based on low-cost GPS / IMU begins to catch much more attention in recent years.

In [1], Xiufeng He and Yongqi Chen proposed the concept of seamless positioning based on GPS / IMU. In their system, the attitude information is obtained based on the navigation algorithm of AHRS, the latitude and longitude information is provided by GPS, and the seamless positioning is achieved by using the data fusion. In [2–3], the authors proposed some new seamless positioning algorithms that are applied to the vehicle environment and with a high complexity of algorithm. In [4–7], the authors conducted the seamless positioning of aircrafts with the help of high-accurate modules of IMU and GPS; the GPS / IMU modules are very expensive; and the complexity of

the algorithm is very high. In [8], Chen et al proposed to use the PDR algorithm based on the high-accurate module of IMU integrated with the GPS receiver to achieve seamless positioning, but the experimental platform of their system is very complex.

This paper proposes a novel seamless positioning algorithm based on low-cost GPS / IMU, in which the heading and velocity information are provided by IMU and the east and north coordinates are obtained from GPS.

2 IMU POSITIONING

The IMU, which consists of the MEMS inertia accelerometer, gyroscope, and magnetometer, is generally integrated into state-of-the-art smart-phone. In this paper, we rely on the IMU to calculate the attitude of the pedestrian and then use the PDR algorithm to speculate the locations and directions of the pedestrian. The PDR algorithm involves three steps: step detection, step length estimation, and reckoning.

2.1 Attitude estimation

Using the Extended Kalman Filter (EKF), the attitude of the pedestrian is calculated by data fusion based on the MEMS inertia sensors [9–12]. The flow chart of attitude estimation is shown in Figure 1.

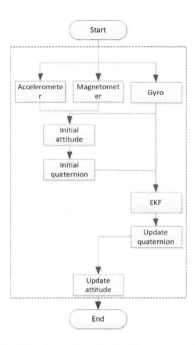

Figure 1. Flowchart of attitude algorithm.

2.2 Step detection

The combined acceleration of triaxial inertial accelerometer is used to step detection of the pedestrian, and it calculates the formula as follows:

$$Acc\_norm = \sqrt{a_x^2 + a_y^2 + a_z^2} \tag{1}$$

The raw acceleration data generally contain a large number of glitches, which can be eliminated by using Hamming window filtering. The acceleration data before and after Hamming window filtering are compared in Figure 2.

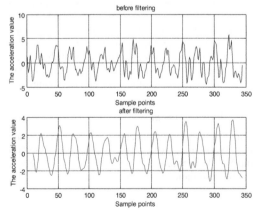

Figure 2. Effect of filtering.

In this paper, using Acc_norm, threshold and interval of step can detect the step effectively. The result of step detection is shown in Figure 3.

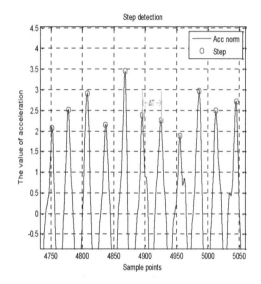

Figure 3. Result of step detection.

2.3 Step length and velocity estimate

The models used for step length estimation are generally classified into four categories: constant / pseudo-constant model, linear step length model, nonlinear step length model, and artificial intelligence model. Since the difference of these four models is slight, we use the nonlinear step length model [10] to estimate the step lengths of pedestrian, S_L, as described in (2).

$$S\_L = K \times \sqrt[4]{A_{max} - A_{min}} \tag{2}$$

where K represents the step coefficient; Amax and Amin stand for the values of the positive and negative peaks, where K represents the step coefficient; and Amax and Amin stand for the values of the positive and negative peaks.

$$V = \frac{S\_L}{\Delta T} \tag{3}$$

where ΔT represents the time interval between every two positive peaks.

2.4 Dead reckoning

To conduct dead reckoning, we assume that at the k-th moment, the coordinates in the east and north are E(k) and N(k), heading is (k), and step length is S_L(k). The relationship between the locations of the pedestrian at the k-1-th and k-th moment is as follows:

$$\begin{cases} E(k) = E(k-1) + S\_L(k-1).\cos\sigma(k-1) \\ N(k) = N(k-1) + S\_L(k-1).\sin\sigma(k-1) \end{cases} \quad (4)$$

3   GPS POSITIONING BASED ON ENU

To achieve GPS positioning based on ENU, the coordinate transformation in (5) should be considered:

$$\begin{cases} N(k) = a / \sqrt{1 - e^2 \sin^2 \Phi(k)} \\ X(k) = (N + h(k))\cos\lambda(k)\cos\Phi(k) \\ Y(k) = (N + h(k))\sin\lambda(k)\cos\Phi(k) \\ Z(k) = (N(1 - e^2) + h(k))\sin\Phi(k) \end{cases} \quad (5)$$

where λ, Φ, and h represent longitude, latitude, and height by GPS; X, Y, and Z are the coordinates of λ, Φ, and h under the Geocentric Coordinate System (GCS), λ_0, Φ_0, and h_0.

$$\begin{bmatrix} x \\ y \\ z \end{bmatrix} = \begin{bmatrix} -\sin\lambda_0 & \cos\lambda & 0 \\ -\sin\Phi_0\cos\lambda_0 & -\sin\Phi_0\sin\lambda_0 & \cos\Phi_0 \\ \cos\Phi_0\cos\lambda_0 & -\cos\Phi_0\sin\lambda_0 & \sin\Phi_0 \end{bmatrix} \begin{bmatrix} X - X_0 \\ Y - Y_0 \\ Z - Z_0 \end{bmatrix} \quad (6)$$

where x, y, and z represent the coordinates of λ, Φ, and h under ENU.

4   SEAMLESS POSITIONING ALGORITHM

After the coordinate transformation, the heading and velocity data are fused by using the EKF to achieve the seamless positioning based on low-cost GPS/IMU. The EKF model is described as follows:
In the EKF model, the state vector is defined as

$$X = [E \quad N \quad V \quad \sigma]^T \quad (7)$$

Then, the state equation is

$$\begin{cases} E(k) = E(k-1) + V(k-1)\sin\sigma(k-1) + w_E \\ N(k) = N(k-1) + V(k-1)\cos\sigma(k-1) + w_N \\ V(k) = V(k-1) + w_V \\ \sigma(k) = \sigma(k-1) + w_\sigma \end{cases} \quad (8)$$

The measurement vector should be

$$Z = [E_{GPS} \quad N_{GPS} \quad V_{PDR} \quad \sigma_{PDR}]^T \quad (9)$$

Thus, we obtain the observation equation in (10).

$$\begin{cases} E(K)_{GPS} = E(k) + w_{EGPS} \\ N(k)_{GPS} = N(k) + w_{NGPS} \\ V(k)_{PDR} = V(K) + w_{VPDR} \\ \sigma(k)_{PDR} = \sigma(k) + w_{\sigma PDR} \end{cases} \quad (10)$$

where $w_{EGPS} \sim N(0, \delta_E^2)$, $w_{NGPS} \sim N(0, \delta_N^2)$, $w_{VPDR} \sim N(0, \delta_V^2)$ and $w_{\sigma PDR} \sim N(0, \delta_\sigma^2)$.

Since the equation of state is a nonlinear equation, based on the EKF, the state transition matrix equals to

$$F = \frac{\partial f_k}{\partial X}\Big|_{X = X_k^-} = \begin{bmatrix} 1 & 0 & \sin\sigma & V\cos\sigma \\ 0 & 1 & \cos\sigma & -V\sin\sigma \\ 0 & 0 & 1 & 0 \\ 0 & 0 & 0 & 1 \end{bmatrix}\Big|_{X = X_k^-} \quad (11)$$

The observation matrix, system noise matrix, and measurement noise matrix are shown in (12)-(14).

$$H = \begin{pmatrix} 1 & 0 & 0 & 0 \\ 0 & 1 & 0 & 0 \\ 0 & 0 & 1 & 0 \\ 0 & 0 & 0 & 1 \end{pmatrix} \quad (12)$$

$$Q = \begin{bmatrix} (3m)^2 & 0 & 0 & 0 \\ 0 & (3m)^2 & 0 & 0 \\ 0 & 0 & (0.1m/s)^2 & 0 \\ 0 & 0 & 0 & (5^o)^2 \end{bmatrix} \quad (13)$$

$$R = \begin{bmatrix} \delta_E^2 & 0 & 0 & 0 \\ 0 & \delta_N^2 & 0 & 0 \\ 0 & 0 & \delta_V^2 & 0 \\ 0 & 0 & 0 & \delta_\sigma^2 \end{bmatrix} \quad (14)$$

where δ_E^2 and δ_N^2 represent the errors in the east and north by GPS; δ_V^2 and δ_σ^2 represent the errors with respect to V and σ by PDR.

According to changes in the measurement noise matrix of observation, the equation adjusts weights of GPS and PDR positioning in the seamless positioning. The adjustment of measurement noise matrix is based on the following two criteria: First of all, when the quality of signal of GPS is well, the positioning accuracy of GPS is higher compared with the PDR. In this case, we reduce the error coefficient of GPS positioning in the measurement noise matrix to enlarge the weight of GPS positioning.

Otherwise, we increase the error coefficient of GPS positioning in the measurement noise matrix to decrease the weight of GPS positioning. Second, since the performance of MEMS inertial sensors generally suffers from the accumulated error, we modify the measurement noise matrix based on the time duration of positioning. The process of measurement noise matrix modification is as follows: We set the error coefficient of PDR equal to 2m when the time duration of positioning is smaller than 15 minutes, and increase the error coefficient of PDR as the time duration of positioning increases.

The quality of satellite signal of GPS is generally characterized by using Horizontal Dilution of Precision (HDOP). The small value of HDOP indicates the better quality of the satellite signal of GPS. The variation of HDOP during the pedestrian positioning is shown in Figure 4.

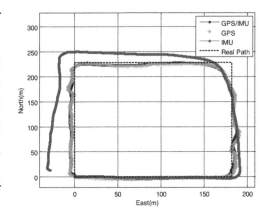

Figure 5. Contrast of positioning results under open environment.

Table 1. Error Analysis of different methods under open environment.

Positioning method	Minimum error	Maximum error	Average error
GPS	0.75m	5.32m	3m
IMU	0.96m	7.4m	5.2m
GPS / IMU	0.84m	4.17m	2.6m

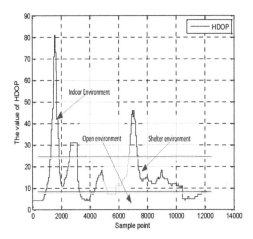

Figure 4. Change of HDOP.

5 EXPERIMENTAL RESULTS

Both the open and shelter environments are selected for the testing to verify the effectiveness of the proposed seamless positioning algorithm based on low-cost GPS / IMU. The comparison of the positioning results by GPS / IMU, IMU, and GPS in an open environment is shown in Figure 5.

As illustrated in Table 1, the average positioning errors by using the GPS, IMU, and the proposed GPS / IMU are 3m, 5.2m, and 2.6m. Compared with the GPS and IMU, the average positioning error achieved by GPS / IMU is decreased by about 13 and 50 percentages. The positioning error of GPS / IMU is still less than 1%.

The positioning results in a shelter environment are shown in Figure 6. Different from the open environment, the existence of shelters, such as the trees on the road, deteriorates the quality of satellite signal of GPS. Based on the results in Figure 6, it is proved that the proposed GPS / IMU still performs well when the satellite signal of GPS is weak.

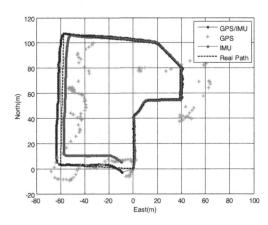

Figure 6. Contrast of positioning results under shelter environment.

As shown in Table 2, the positioning result of GPS is worse when the quality of satellite signal of GPS is bad, but the positioning error of GPS / IMU is still less than 8%.

Table 2. Error Analysis of different methods under shelter environment.

Positioning method	Minimum error	Maximum error	Average error
GPS	62m	9m	52m
IMU	0.7m	8.7m	5.6m
GPS / IMU	0.5m	4.15m	3.15m

6 CONCLUSIONS

In this paper, we proposed a novel pedestrian seamless positioning algorithm based on low-cost GPS / IMU. Using the EKF, we conduct coordinate transformation in ENU, fuse the heading and velocity data from IMU, and, consequently, achieve the seamless positioning. Furthermore, we modify the measurement noise matrix of observation equation to improve positioning accuracy further. The extensive experimental results demonstrate that the proposed algorithm can not only maintain the positioning accuracy when the quality of satellite signal of GPS is good but also significantly improve the accuracy performance when the satellite signal of GPS is weak.

ACKNOWLEDGMENTS

This work was supported in part by the Program for Changjiang Scholars and Innovative Research Team in University (IRT1299), the National Natural Science Foundation of China (51175535 and 61301126), the Special Fund of Chongqing Key Laboratory (CSTC), and the Natural Science Foundation of Chongqing (cstc2012-jjB40003, cstc2013jcyjA40032, and cstc2013jcyjA40041).

REFERENCES

[1] He X, Chen Y, Liu J. Development of a low-cost integrated GPS/IMU system [J]. Aerospace and Electronic Systems Magazine, IEEE, 1998, 13(12): 7–10.
[2] Toledo-Moreo R, Zamora-Izquierdo M A. Collision avoidance support in roads with lateral and longitudinal maneuver prediction by fusing GPS/IMU and digital maps[J]. Transportation research part C: emerging technologies, 2010, 18(4): 611–625.
[3] ZHU Yan-hua, CAI Ti-jing, YANG Zhuo-peng. MEMS-IMU/GPS integrated navigation system [J]. Journal of Chinese Inertial Technology, 2009, 17(5): 552–556.
[4] Rios J A, White E. Fusion filter algorithm enhancements for a MEMS GPS/IMU[C]//ION NTM. 2002: 28–30.
[5] Brown A, Lu Y. Performance test results of an integrated GPS/MEMS inertial navigation package[C]// Proceedings of ION GNSS. 2004: 21–24.
[6] Qi H, Moore J B. Direct Kalman filtering approach for GPS/INS integration [J]. Aerospace and Electronic Systems, IEEE Transactions on, 2002, 38(2): 687–693.
[7] Kim K H, Lee J G, Park C G. Adaptive two-stage extended Kalman filter for a fault-tolerant INS-GPS loosely coupled system [J]. Aerospace and Electronic Systems, IEEE Transactions on, 2009, 45(1): 125–137.
[8] Wei Chen. Research on GPS/Self-Contained Sensors Based Seamless Outdoor/Indoor Pedestrian Positioning Algorithm [D]. Hefei: University of Science and Technology of China, 2010.
[9] Zengshan Tian, Yuan Zhan, Mu Zhou, et al. Pedestrian dead reckoning for MARG navigation using a Smartphone [J]. Journal on Advances in Signal Processing, 2014.
[10] LIU Y, XIONG W J. An Indoor Positioning Method with MEMS Inertial Sensors[J]. Sensors and transducers, 2013, 11(118): 9–14.
[11] Abdulrahim K. Understanding the Performance of Zero Velocity Updates in MEMS-based Pedestrian Navigation [J]. International Journal of Advancements in Technology, 2014, 5(1): 53–60.
[12] F, Hoppe J, Zhang R, et al. Acoustic indoor-localization system for smart phones [C]//Multi-Conference on Systems, Signals & Devices (SSD), 2014 11th International. IEEE, 2014: 1–4.
[13] Lan K C, Shih W Y. On Calibrating the Sensor Errors of a PDR-Based Indoor Localization System[J]. Sensors, 2013, 13(4): 4781–4.

Energy, Environment and Green Building Materials – Sheng (ed.)
© 2015 Taylor & Francis Group, London, ISBN 978-1-138-02718-3

Nitrogen oxide emission comparison of biodiesel and diesel in diesel engine

Q.L. Zhang

Mechatronics Department, Handan Polytechnic College, Handan, China

ABSTRACT: NO_x, NO, NO_2, and N_2O emissions for a direct injection diesel engine were studied using diesel and biodiesel. NO_x emission curves of biodiesel and diesel are quite identical for the engine, especially for the low engine load. NO emission increases accompanied by the increasing engine load. NO_2 emission is low when the engine is in low or high load, and it is the highest when the engine is in medium load. To be in contrast, N_2O emission is almost zero when the engine is in medium or high load. For a diesel engine, NO_x, NO, and NO_2 emissions are higher for the engine burning biodiesel compared with the engine burning diesel, whereas N_2O emission is decreased in the same scenario.

KEYWORDS: Diesel engine; Biodiesel; Nitrogen oxides; Emission characteristics; Diesel

1 INTRODUCTION

NO_x is one of the main harmful pollutants of diesel engine endangering both public health and the environment (Leskinen et al. 2007). Compared with regular diesel, biodiesel burned by a diesel engine will increase NO_x emission, which arouses public concerns. Emission research on biodiesel engine mainly focuses on NO_x emission in total, but rarely on the specific emission of NO, NO_2, N_2O, and their relevant emission trends. A direct injection diesel engine was used as a sample engine to carry out emission characteristics check on NO_x, NO, NO_2, and N_2O using biodiesel and regular diesel.

2 TEST EQUIPMENT AND SCHEME

The engine applied in the test is a 4-cylinder, intercool turbo-charged direct injection diesel engine. Its main technical parameters are listed in table 1.

Table 1. Main technical parameters of the engine used in the test.

Parameters	Values
Displacement / L	1.896
Compression	19:1
Cylinder diameter / mm	78.5
Stroke / mm	95.5
Rated power / (kW / rpm)	96 / 4000
Maximum torque / (N·m / rpm)	285 / 1900

The main equipment used in the test include AVL-PUMA automatic test frame and AVL-PEUS multi ingredient gas analysis device. It applies FTIR (Fourier Transform Infrared) method, which can measure 30 different gas contents, including NO, NO_2, N_2O, and some other contents. According to its working principle, the measured NO_x concentration is the total concentration of NO, NO_2, and N_2O.

The FTIR method involves the incoming ray emitted from the infrared source reaching the beam splitter after passing a simulated radiation optical frequency amplifier. The two split beams can incur an optical path difference after different routes. Interference phenomenon can occur and refocus on MCT (Mercury Cadmium Telluride) infrared receiver, so it can be tested. If the sample were placed on the path of the interference beam, the intensity energy in the interferogram will change because the sample absorbs a certain frequency energy. The interferogram can be collected and fast Fourier transmission can be performed. Infrared spectrogram of absorbency or transmittance changes accompanied by the wavelength or wave number can be obtained. Different matters in the exhaust can be qualified and quantified through comparing the infrared spectrogram with the standard spectrum.

Biodiesel tested in the experiment is an engine replacement fuel consisting of fatty acid methyl ester obtained through a reaction between the waste animal and vegetable oil and alcohols under the influence of a catalyst (Demirbas 2003). Biodiesel can also be called "waste cooking oil manufactured biodiesel." The physical and chemical properties of biodiesel and diesel are listed in Table 2.

Experiment fuels used in the research are regular No. 0 diesel and waste cooking oil biodiesel. Load-bearing characteristic experiments of two special working conditions, that is, the rotate speed at

Table 2. Properties of biodiesel and diesel.

Properties	Biodiesel	Diesel
Density (20°C) / kg·m⁻³	876	834
Kinetic viscosity (40°C) / mm²·s⁻¹	4.412	3.232
Flash point (close-cup method) / °C	140	70
Cetane number	57.8	48
Low heat value / MJ·L⁻¹	32	35
Mass fraction of sulfur / %	150	446

4000 rpm (the rated power rotate speed) and 1900 rpm (the maximum torque rotate speed), were carried out on the diesel engine burning biodiesel and diesel.

3 EXPERIMENT RESULTS AND ANALYSIS

Engine power is reduced by 5.9% at 4000 rpm when the diesel engine burns biodiesel compared with burning diesel from the experiment. Engine power is reduced by 9.4% at 1900 rpm when the diesel engine burns biodiesel compared with burning diesel from the experiment. Engine fuel consumption is increased by 13.5% and 13.8%, respectively, for the two working conditions mentioned earlier.

The main reasons for the power reduction and fuel consumption increase are that without any adjustment to the engine fuel supply system, volume fuel supply quantity is maintained constant. Biodiesel volume low heat value is lower than diesel volume low heat value, leading to its lower engine power and decreased kinetics. Biodiesel mass is greater than diesel mass for the same volume, because biodiesel density is greater than diesel density Considering biodiesel mass low heat value is lower than diesel mass low heat value, fuel consumption will be increased in order to achieve the same amount output power.

3.1 NO_x emission

Figures 1 and 2 show NO_x emission changing with engine load when burning biodiesel and diesel at 4000 rpm and 1900 rpm, respectively.

Figures 1 and 2 also show that NO_x emission concentrations are low when the engine is under load, and their values do not exhibit difference regardless of whether burning biodiesel or diesel, because the cylinder temperature is low when the engine is under load.

NO_x emission increases slowly with increasing engine load. The reasons are that along with the increasing engine load, circulating fuel delivery increases, maximum combustion temperature increases, and excess air coefficient is slowly reduced to the amount that can produce the maximum NO_x (Leenus et al. 2011).

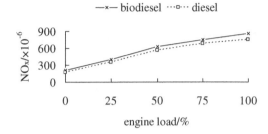

Figure 1. NO_x emission of biodiesel and diesel at 4000 rpm.

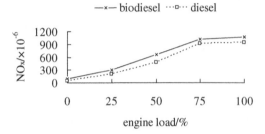

Figure 2. NO_x emission of biodiesel and diesel at 1900 rpm.

NO_x is usually formed while the combustion temperature is above 1500°C (Hoekman & Robbins 2012).

Figures 1 and 2 show that NO_x emission increases when burning biodiesel compared with burning diesel. The reason is that oxygen in biodiesel may provide additional oxygen supply, leading to a complete combustion process. This, in turn, results in high combustion temperature, thus facilitating NO_x generation.

Figures 1 and 2 also show that NO_x emission is reduced with increasing rotate speed, because the increase in rotate speed may reduce the high temperature lasting time (Johnson 2010).

3.2 NO emission

NO is one of the most important ingredients of NO_x emission. Figures 3 and 4 show NO emission changing with engine load when burning biodiesel and diesel at 4000 rpm and 1900 rpm.

Figures 3 and 4 also show that NO emission is low when the engine is under load regardless of burning biodiesel or diesel. NO emission increases with the increasing engine load, showing high temperature contributing to NO formation (Wang et al. 2000). Figures 3 and 4 show that NO emission increases when the engine burns biodiesel compared with diesel under the same engine load. The NO emission curve is similar to the NO_x emission curve because of the high NO emission percentage in the total NO_x emission.

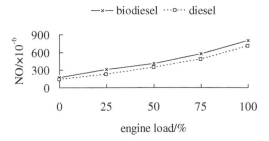

Figure 3. NO emission of biodiesel and diesel at 4000 rpm.

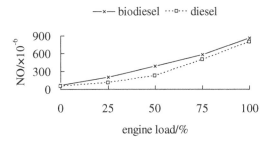

Figure 4. NO emission of biodiesel and diesel at 1900 rpm.

With the increase of rotate speed, NO emission decreases because the increasing rotate speed may reduce the high temperature lasting time. NO emission displays a peak value at 1400 rpm, showing that the peak value is caused by the high temperature inside the cylinder.

3.3 NO_2 emission

NO_2 is another important ingredient of NO_x emission. NO_2 may react with HC to form photochemical smog under the sunlight, which can damage the environment. It is necessary to conduct thorough research on NO_2. NO_2 is the further reaction result of NO inside the cylinder. The reaction process is

$$NO+HO_2 \rightarrow NO_2+OH$$

After this reaction, NO is regenerated from NO_2.

$$NO_2+O \rightarrow NO+O_2$$

Figures 5 and 6 show NO_2 emission changing with engine load when burning biodiesel and diesel at 4000 rpm and 1900 rpm.

Figures 5 and 6 also show that the NO_2 emission concentration change trend of biodiesel is basically identical with the change trend of diesel. NO_2

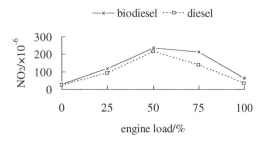

Figure 5. NO_2 emission of biodiesel and diesel at 4000 rpm.

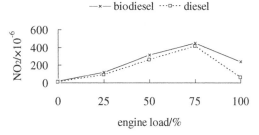

Figure 6. NO_2 emission of biodiesel and diesel at 1900 rpm.

emission is low, as the engine is under load regardless of the rotate speed being 4000 rpm or 1900 rpm.

Figures 5 and 6 show that NO_2 emission increases as the engine load increases. NO_2 emission increases slightly for the biodiesel compared with the diesel for the same engine load. NO_2 emission is relatively low even for the 100% engine load showing heavy engine load, and high temperature goes against NO_2 formation (Torkian et al. 2011).

According to the NO_2 formation mechanism, the most suitable reaction condition helps form NO_2 (Graboski & McCormick 1998). NO_2 emission reaches its peak value as temperature and air fuel are coefficient in favor of NO_2 generation. NO_2 can be transferred into NO as the engine load continues to increase, air fuel coefficient decreases, and temperature inside the cylinder increases.

NO_2 emission curves of biodiesel and diesel display that NO_2 emission increases first, then decreases as the rotate speed increases, and reaches its peak value at the rotate speed of 2500 rpm. As the rotate speed increases, the air inhalation increases; full combustion can be achieved; and temperature inside the cylinder increases. The result is that NO_2 emission increases. If the rotate speed carries on increasing, the high temperature duration time will be reduced and it will result in a decrease in NO_2 emission. Compared with NO formation, NO_2 formation is relatively small.

3.4 N_2O emission

N_2O is one of the six greenhouse gases defined in Kyoto Protocol, although its emission is extremely low in the experiment.

Figures 7 and 8 show N_2O emission changing with engine load when it burns biodiesel and diesel at 4000 rpm and 1900 rpm. Figures 7 and 8 also show that N_2O emission concentration change trend of biodiesel and diesel is basically identical. The N_2O emission is very low for biodiesel regardless of the rotate speed being 4000rpm or 1900rpm. The maximum value of N_2O emission for the diesel does not exceed 3.5×10^{-6} for the low engine load, and N_2O emission is almost zero for 50% of the engine load and above.

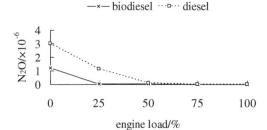

Figure 7. N_2O emission of biodiesel and diesel at 4000 rpm.

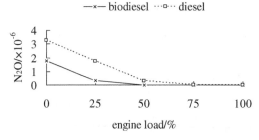

Figure 8. N_2O emission of biodiesel and diesel at 1900 rpm.

4 CONCLUSIONS

The NO_x emission curve for the diesel engine burning biodiesel is very similar to the diesel engine burning diesel. NO_x emission is low for the low engine load and increases as the engine load increases.

The NO emission increases with the engine load increasing, because the high temperature is in favor of NO formation.

NO_2 emission achieves its peak value when the diesel engine is under the moderate engine load, and NO_2 emission decreases when the engine load is either up or down. Heavy engine load and high temperature are not in favor of NO_2 formation.

N_2O emission is extremely low. There are trivial amounts of N_2O for the low engine load, and the amount is almost zero for moderate to high engine load.

The change in trend of NO_x, NO, NO_2, and N_2O is identical for the diesel engine burning biodiesel or burning diesel; that is, burning biodiesel can only change the emission quantity but not alter the emission change law.

REFERENCES

Demirbas, A. 2003. Biodiesel fuels from vegetable oils via catalytic and noncatalytic supercritical alcohol transesterifications and other methods: a survey. *Energy Conversion and Management.* 44: 293–109.

Graboski, M.S. & McCormick, R.L. 1998. Combustion of fat and vegetable oil derived fuels in diesel engines. *Progress in Energy and Combustion Science.* 24: 125–164.

Hoekman, S.K. & Robbins, C. 2012. Review of the effects of biodiesel on NOx emissions. *Fuel Processing Technology.* 96: (237–249).

Johnson, T. 2010. Review of diesel emissions and control. *SAE International Journal of Fuels and Lubricants.* 3(1): 16–29.

Leenus, J.M.M. Edwin, G.V. & Prithviraj D. 2011. Effect of diesel addition on the performance of cottonseed oil fuelled DI diesel engine. *International Journal of Energy and Environment.* 2(2): 321–330.

Leskinen, A.P. Jokiniemi, J.K. & Lehtinen, K.E.J. 2007. Transformation of diesel engine exhaust in an environmental chamber. *Atmospheric Environment.* 41: 8865–8873.

Torkian, B. Reza E. Kamran, K. & Ali M.B. 2011. Effect of some BED blends on the equivalence ratio, exhaust oxygen fraction and water and oil temperature of a diesel engine. *Biomass & Bioenergy.* 35: 4099–4106.

Wang, W.G. Lyons, D.W. Clark, N.N. Gautam, M. & Norton, P.M. 2000. Emissions from nine heavy trucks fueled by diesel and biodiesel blend without engine modification. *Environmental Science & Technology.* 34: 933–939.

Energy, Environment and Green Building Materials – Sheng (ed.)
© 2015 Taylor & Francis Group, London, ISBN 978-1-138-02718-3

Research on electrolysis factors of galvanizing slag's soluble anode effect on current efficiency

X.F. He, Y.G. Li, J. Chen & Z.Y. Cai
Hebei United University, TangShan, HeBei, China

ABSTRACT: Using single-factor experimentation, technique conditions related to the current efficiency of dissoluble anode's electrolysis of zinc slag were studied. The results showed that optimum conditions for the electrolysis were determined as follows: concentration of Zn^{2+} 55 g·L^{-1}, temperature of 30°C~40°C , current density of 300 A·m^{-2}~500 A·m^{-2}, and current efficiency more than 98%.

KEYWORDS: galvanizing slag; soluble anode; current efficiency

INTRODUCTION

Direct utilization of zinc is 60% (only 40% in the malleable iron body) in the process of hot dip galvanized, with about 20% forming cadmia, which mainly comes from zinc-iron alloy that is formed by zinc liquid reacting with plating pieces, cell body iron of spelter pot, and not rinsing clean iron salt residues on the surface of plated part. When steel wire galvanizes zinc, the iron of the zinc-iron reaction comes from the iron pot wall, iron compression roller, steel wire, and other sundries containing iron. Besides, the iron tool inserts zinc liquid in the operation process and the steel that is not processed clean before galvanized zinc will drag iron salt into the zinc liquid[1,2]. Galvanized zinc slag is not liquated by the hydrometallurgy process[3] in this passage, and it is researched that factors of soluble anode electrolysis have effects on current efficiency.

1 EXPERIMENTAL PORTION

1.1 Experimental raw material

Galvanized zinc slag in the experiment comes from a certain domestic steel tube hot-dip galvanized factory, and mainly its chemical composition is listed in table1. Phase analysis results indicate that the forms of zinc are Fe-Zn alloy and metallic state zinc.

In the experiment, the water is deionized, and the chemical reagents such as zinc sulfate and others are analytically pure.

1.2 Experimental method

The cathode (aluminum) and anode are inserted in the electrolyte, and then, the direct current is switched on in addition to record time and current. After a 1h power dump, the cathode and anode are taken out, and then the dry cathode is weighed. Besides, current efficiency is successively calculated. Electrolytic beaker is placed in a constant temperature water bath to control temperature.

Depositional zinc is stripped from the cathode, and EDS is used to analyze zinc content.

2 EXPERIMENTAL RESULT AND DISCUSSION

2.1 Galvanized zinc slag liquation

At the temperature of 820°C (zinc boiling point 907°C), the hot galvanizing slag is heated, and after melting, this temperature is maintained at 30 minutes

Table 1. Main chemical composition of galvanized zinc slag.

Element	Zn	Si	Fe	Cl	S	Ni	P	Ca	Ti
Content (mass percent)/%	85.93	2.30	8.81	1.66	0.45	0.48	0.20	0.10	0.06

to separate zinc liquid and slag and to obtain metal zinc and hot galvanizing slag. Melting galvanized zinc slag is cast into the anode plate that is in reserve.

Chemical composition of zinc and slag after liquation is listed in tables 2 and 3.

Table 2. Chemical composition of supernatant after liquation

Element	Zn	Si	Fe
Content(mass percent)/%	98.13	1.50	0.37

Table 3. Chemical composition of galvanized zinc slag after liquation

Element	Zn	Si	Fe	Cl	S	Ni	P	Ca	Ti
Content (mass percent)/%	70.77	5.01	15.90	3.22	1.17	1.67	1.07	1.04	0.15

In liquated supernatant, zinc content reaches 98% and iron content is lower. On the contrary, in liquated galvanized zinc, slag's zinc content is lower and iron content is very high.

2.2 Soluble anode electrolysis

Considering that the anode is zinc containing impurities, in this study, the electrolyte is zinc sulfate aqueous solution without sulfuric acid.

Current efficiency refers to the ratio between the quality of actual depositional zinc and the theory quality of theory depositional zinc using the same electric quantity. The influence factors of current efficiency contain current, temperature of electrolyte, concentration of electrolyte, and so on.

2.2.1 Influence of zinc-iron concentration on the current efficiency

The size of the anode that is made up of galvanized zinc slag is 20mm×30mm, and the size of the aluminum cathode is 35mm×35mm. In this condition, current is 325mA, electrolytic time is 60 minutes, temperature is 40°C , current efficiency is measured, the concentration of Zn^{2+} range is 40~60g·L^{-1}, and the step width is set at 5.

In the electrolyte, zinc sulfate is ‖–‖ type strong electrolyte; thus, the corresponding activity coefficient is small and an effective concentration of zinc ions is much smaller than the apparent concentration. It can be seen from figure 1 that the influence of Zn^{2+} content in electrolyte on current efficiency is little in the research range of Zn^{2+} content.

2.2.2 Influence of electrolytic temperature on current efficiency

In the condition of current for 325mA, electrolytic time for 60 minutes, the current efficiency is measured, the electrolytic temperature range is 30~50°C, and the step width is set at 5. It can be seen from figure 2 that current efficiency is slightly reduced along with an increase in the electrolytic temperature[7]. The solution does not contain sulfuric acid, so it can avoid producing hydrogen, thus making overvoltage decrease and electric fluid acidity increase. Thus, the depressed trend of current efficiency is not obvious. But the reasons of electrolytic resistance, electrochemical reaction, short circuit, burning board, and so on can make electrolytic temperature increase; therefore, electrolytic must be cooled. Generally, electrolytic temperature is controlled in 30~40°C.

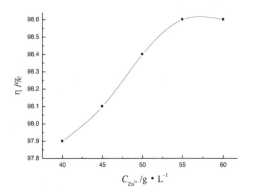

Figure1. Influence of Zn^{2+} concentration on current efficiency.

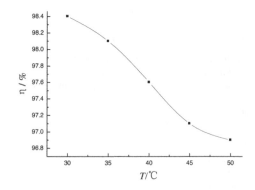

Figure 2. Influence of electrolytic temperature on current efficiency.

2.2.3 Influence of current density on current efficiency

In the condition of Zn^{2+} concentration for $50g \cdot L^{-1}$, temperature is 40°C, electrolytic time is 60 minutes, current efficiency is measured, and current densities are 135 $A \cdot m^{-2}$, 188 $A \cdot m^{-2}$, 208 $A \cdot m^{-2}$, 265 $A \cdot m^{-2}$, and 314$A \cdot m^{-2}$.

It can be seen from figure 3 that current efficiency can increase with current density increasing in the same condition. Due to high current density, there is a higher requirement for equipment and operation; a decrease in current density can reduce voltage drop in the electrolyte. However, low current density is bad for production. With the increase of current density, hydrogen overvoltage increasing current efficiency can increase and can obtain the crystalline pyknotic cathode zinc. But electrolytic resistance and temperature will simultaneously increase and there will be impurities deposition. So, the current density generally selected is 300 $A \cdot m^{-2}$~500 $A \cdot m^{-2}$.

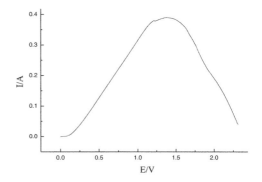

Figure 4.   Cathodic polarization curve.

3 CONCLUSIONS

Using wet processing to recycle and reuse zinc slag technology is simple and feasible, and zinc recovery rate is high. It can determine reasonable and economic conditions as well as lay a solid foundation for the industrialization of galvanized zinc slag using wet processing by means of the research, that is, the influence of electrolysis process conditions of galvanized zinc slag soluble anode on current efficiency.

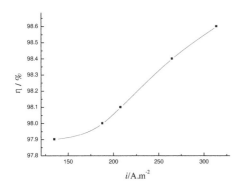

Figure 3.   Influence of current density on current efficiency.

2.2.4 Polarization curve

In the condition of Zn^{2+} concentration for $50g \cdot L^{-1}$ and room temperature, the change in current along with changing cathode potential is measured. It can be seen from figure 4 that with the increase of cathode voltage, the current curve appeared to peak; in other words, with the increase of cathode voltage, the current will decline when it increases to a certain value. In order to improve the current efficiency, reduce the wastage of energy, and reduce the occurrence of the secondary reaction, cathode potential should be controlled under 1.5V to make cathodic polarization decrease.

REFERENCES

[1] Shen Juncong. 2006: the zinc market extraordinary one year[J]. China Nonferrous Metals, 2006,(3):62–63.
[2] Zeng Zhiyu. Control of steel wire pot galvanizations dregs[J]. Hunan Metallurgy, 2003, 31(6):37–39.
[3] He Xiaofeng, Li Yungang, Chenjin. Comments on hot galvanizing zinc dross and residue recovery process[J]. China Nonferrous Metallurgy, 2008,4(2):55–58.
[4] Peng Genfang. Analyses and optimization exploration of D. C. consumption in zinc electrolytic winning[J]. Energy Saving of Non- ferrous Metallurgy, 2003,20(2):17–20.
[5] Tang Shouceng. Comprehensive Production of Reduced Direct Current Consumption with Zinc Electrolysis Process[J]. Hunan Nonferrous Metals, 2004,20(3):22–24.
[6] Wang Yanjun, Xie Gang, Yang Dajin. A review on energy saving in zinc electrowinning[J]. Southern Metals, 2006,(2):13–15.
[7] Liu Xianbin. Production Practise on Increasing Current Efficiency of Zine Electrowinning[J]. Nonferrous Mining and Metallurgy, 2006.22(1):23–25.

Energy, Environment and Green Building Materials – Sheng (ed.)
© 2015 Taylor & Francis Group, London, ISBN 978-1-138-02718-3

Treatment of phenol wastewater by titanium catalyst

R. Wang, Y.X. Xiao & J.C. Gu
School of Energy and Environment, Xihua University, China

ABSTRACT: For the wet catalytic oxidation method, we use ilmenite concentrate as a catalyst and simulate phenol wastewater as the subject to carry out the experiment. We explore the catalyst and oxidant dosing quantity, pH, and reaction time on the impact of the removal rate of phenol. The experimental results show that for 10°C, catalyst and oxidant dosing quantity are 5 g/L and 0.5 ml/L, pH 2.92, after 90 minutes of the reaction; the removal rate of 10 mg/L phenol wastewater can reach 88%.

KEYWORDS: Phenol, Titanium1 concentrate, CWAO wastewater treatment, Catalyst

With the acceleration of the industrialization process, environmental pollution problems have become increasingly prominent. Phenol has a high toxicity and poor biodegradability; wastewater discharge oil is widely used in pharmaceutical, printing, and dyeing and other enterprises. Environmental hazards are very large. Currently, the treatment of phenol containing wastewater is carried out by mainly using microwave, extraction, electrolysis, catalytic wet air oxidation, and other advanced oxidation methods[1-4].

The method of potential catalytic wet air oxidation has caused wide attention. The current research on the catalytic wet oxidation is mainly focused on the catalyst research, with most of the porous material loading of the active components of the prepared catalyst [5-12]. This kind of catalyst for some organic wastewater removal shows a good effect. But the load-type catalyst on active component load or smaller catalyst loss is inevitable; this, to a certain extent, has restricted the application of such a catalyst.

China is a big country of ilmenite resources and abundant reserves. The titanium iron ore ilmenite concentrate has a selected high iron content, and its compounds account for more than 40% of the total, which results in the advantage in terms of the repeated use of the catalyst.

This study aims at attempting after treatment to prepare a catalyst using titanium concentrate and to explore the effects and influence factors of this catalyst on removal of phenol wastewater.

1 PART OF THE EXPERIMENT

1.1 *Reagents and instruments*

PHS-3B type acidity meter; gas bath thermostats oscillator; TU-1901 ultraviolet spectrophotometer.

Phenol; 30% hydrogen peroxide; 4- amino phenol; Antje Billing; potassium ferricyanide; and other reagents were analytically pure.

1.2 *Catalyst preparation method*

The separation and enrichment was conducted for ilmenite concentrate. The washing and drying of titanium concentrate took place in the roasting furnace at 450°C for 2 hours after cooling, roasting, removing, and grinding to the required size.

1.3 *Study on the catalytic performance*

The catalyst prepared by 250ml 1g was added into the conical flask; phenol was added, configuring 200ml of simulated wastewater (10mg/L). A certain amount of commercially available 30% hydrogen peroxide was added. Using PHS-3B type acidity meter, the solution was regulated at about pH=2.9; afterward, gas was added in a bath thermostatic oscillator at 130r/min, respectively, at 10 degrees Celsius for 90 minutes under reaction.

1.4 *Methods of analysis*

(1) The residual phenol concentration test:
The "determination of 4- amino antipyrine spectrophotometric method water quality of volatile phenols" (HJ503-2009) was used as the standard for the determination.
(2) Calculate the catalyst effect:
The performance of the catalyst for phenol removal rate is expressed.
The calculation formula of the phenol removal rate is as follows:

The phenol removal rate $= \dfrac{C_{AO} - C_{Ae}}{C_{AO}} \times 100\%$, where CWAO is the initial phenol concentration; afterward, C_{Ae} is used for phenol by a concentration reaction.

2 RESULTS AND DISCUSSION

2.1 Effect of catalyst dosage

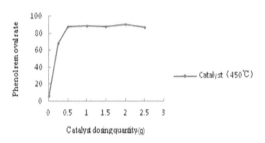

Figure 1. Effect of additive amounts of different catalysts on phenol removal rate.

Reaction conditions are as follows: Reaction temperature is 10°C, 30% hydrogen peroxide 0.5ml/L, pH 2.92, 90min of the reaction time, and catalyst dosage is 0–2.5g/200ml.

We can see from Figure 1 that the removal rate of catalyst investment eventually adds more effects of phenol to some extent. When the catalyst in the reaction system of the dosage is 0.5g/200ml or above, finally the phenol removal rate will no longer exist because the amount of catalyst increases. When the catalyst amount is insufficient, the amount of catalyst will immediately become a restriction of the reaction system, which results in the phenol removal rate and the amount of catalyst present in a certain interval positive correlation. In the solid–liquid reaction system, a solid catalyst material will be formed on the surface of a solid ionic state (here is mainly Fe^{2+} and Fe^{3+} as the core of the whole reaction catalyzed. If the catalyst dosage is insufficient, it will result in the solid catalyst and the reactant reduces the contact area, so that with regard to the catalytic material in the ionic state the whole reaction system required is insufficient. This will indirectly affect the efficiency of the whole catalytic material; eventually, at the macro level, the decrease in the overall phenol removal rate is reflected. When the catalyst amount reaches a certain time, and then by adding excess of solid catalyst, the final removal rate will not increase. The reason may be that the reaction system of the catalytic oxidation reaction reached saturation, and this excess of catalyst was not involved in the reaction system. Experimental results show that the amount of catalyst in the 2.5g/L has been able to achieve the best removal effect.

2.2 Effects of hydrogen peroxide content

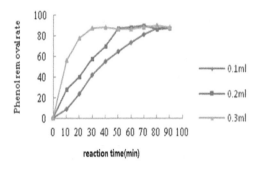

Figure 2. Effect of additive amounts of different antioxidants on phenol removal rate.

The reaction conditions were as follows: reaction temperature 10°C, the dosage of catalyst 5g/L, pH 2.92, reaction time 90min, and 30% hydrogen peroxide dosage is 0.1–0.3ml/200ml.

We can see from Figure 2 that different hydrogen peroxide dosages are required to maintain a margin in hydrogen peroxide (can be calculated by carbon conservation in the phenol) circumstances. To have much of an effect on phenol removal rate is not final, but this has a great influence on the removal rate of phenol. Within the 10–40min group, the amount of H_2O_2 reaction compared with the investment response group plus a small amount on the phenol removal rate is higher. Hydrogen peroxide dosage is more and it results in the concentration of hydrogen peroxide in the reaction system becoming larger as a whole. According to the related interpretation of chemical equilibrium, an increased concentration of reactants is advantageous for the reaction in the positive direction, to enable the reaction rate of the larger group of a high concentration of hydrogen peroxide on removal of phenol. The results show that the oxidant is sufficient, the phenol removal rate does not increase with the amount of oxidant investment increase, and its effect is only on the removal rate of phenol.

2.3 Effect of reaction conditions

2.3.1 Effect of pH

Reaction conditions are as follows: Reaction temperature is 10°C, the dosage of catalyst is 5g/L, 30% hydrogen peroxide solution is 0.5ml/L, and reaction time is 90mi.

The study demonstrated that the [1,13-15] catalyst under mild reaction conditions of catalytic activity and pH had a very big relationship. From the relation graph on the conversion of phenol and pH value (as shown in Figure 3), it can be seen that with the value of pH in 2.2–3.6, the catalyst (450°C) showed

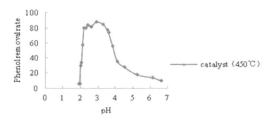

Figure 3.　Effect of pH on phenol removal rate.

higher catalytic activity (removal rate reached more than 75%), pH to achieve an optimum point near 2.92. Although the pH value is less than 2.2 or greater than 3.8, the catalytic efficiency of dramatic decline is exhibited. The main active components of the solid catalyst are divided into Fe_2O_3 and Fe_3O_4. The following reaction should occur in acidic condition:

$$Fe_2O_3+H^+\rightarrow Fe^{3+}+H_2O \qquad (1)$$
$$Fe_3O_4+H^+\rightarrow Fe^{3+}+Fe^{2+}+H_2O \qquad (2)$$

According to the Habei–Weiss theory, the H_2O_2 reaction of Fenton decomposition experiences the following process[16]:

$$H_2O_2+Fe^{2+}\rightarrow Fe^{3+}+HO\cdot+HO\cdot \qquad (3)$$
$$RH+HO\cdot\rightarrow R\cdot+H2O \qquad (4)$$
$$R\cdot+Fe^{3+}\rightarrow Fe^{2+}+product \qquad (5)$$
$$H_2O_2+HO\cdot\rightarrow HO_2\cdot+H_2O \qquad (6)$$
$$Fe^{2+}+HO\cdot\rightarrow Fe^{3+}+HO\cdot \qquad (7)$$
$$Fe^{3+}+H_2O_2\rightarrow Fe^{2+}+H^++HO_2\cdot \qquad (8)$$
$$Fe^{3+}+HO_2\cdot\rightarrow Fe^{2+}+H^++O_2 \qquad (9)$$

Through a series of reactions, it can be seen that if the reaction is carried out in an alkaline environment, the oxidation state of iron is not effective in the ionic states of iron for the catalysis of the reaction. Iron oxide that is in a strong acidic environment is dissolved, and the general pH<2 can produce more dissolved. As you can see from Figure 3, the pH that is in the range of 2.2–3.6 between iron and titanium oxide in the ionic state condition is good. According to the Habei–Weiss theory, the H^+ concentration that is very high is not conducive to the reactions (8) and (9) of Fe^{3+}; the effect of the transition to Fe^{2+}, an indirect influence on the generation of hydroxyl radicals, is not conducive to the oxidation of organic compounds. If the pH is very high, it not only affects the reaction (3) that are carried out but also may make the iron ions appear in hydroxide form. Under the joint effects of several major factors, we found that pH reached about 2.92, to achieve the best combination of several major factors.

2.3.2　Influence of reaction time

Reaction conditions are as follows: Reaction temperature is 10°C; the dosage of catalyst and hydrogen peroxide is 5g/L, 0.5ml/L 30%, pH 2.92.

Through the experiment, we can see that the removal rate of phenol increased with time and reached the maximum value in response to the final 80 minutes (as shown in Fig. 4). Using the formula,

$$K_n = \frac{R_n - R_{n-1}}{R}$$

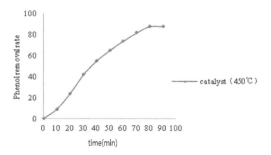

Figure 4.　Effect of reaction time on phenol removal rate.

Phenol type in Kn -- the N time period of removal rate and the removal rate of phenol ratio. eventually, every 10min for a period of time; the n value range is 19, followed by a similar

Phenol, Rn -- the N time period of removal rate.

The calculated K values, in turn, were 0.102, 0.170, 0.205, 0.148, 0.114, and 0.102,0 (corresponding to 0.091, 0.068, K1 to K9).

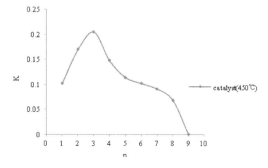

Figure 5.　Different periods of phenol removal rate.

Making a chart of the relationship between K and n (Fig. 5), we can see that the conversion rate of the whole catalytic wet oxidation system of phenol showed a rapid increase at first; then sharply decreased; slowly reduced; and finally tended to stop, thus displaying a kind of development trend. We can see that the phenol oxidation rate in the whole reaction after 10 to 40 minutes in this period is relatively large. The reaction of phenol drop in this period accounted for 59.4% of the total reaction. Through the display of a sampling of 80 minutes and 90 minutes after the reaction of sampling results, there is no difference between their phenol concentration; that is, the whole reaction system in response to 80 minutes has reached a balance, and the reaction time is no longer an effective factor influencing the whole reaction process.

3 CONCLUSION

Based on the phenol removal rate influence inquiry conditions, the main conclusion is as follows:

1 Catalyst dosage inadequacy can have an impact on the removal rate of phenol under certain conditions. When reaching an equilibrium point, the amount of catalyst increases and will not improve the removal effect of the whole system. At the temperature of 10 DEG C, catalyst dosage is 5g/L, pH is 2.92, reaction time is 90min, and phenol wastewater on the removal rate of 10mg/L can reach 88%.
2 The oxidant (here 30% hydrogen peroxide) maintains margin, the rate of different concentrations will only affect the overall reaction, and the removal rate has little effect on the final concentration.
3 The reaction time of pH will have a great influence on the catalytic wet oxidation of phenol, acidic or insufficient, and this will have an impact on the phenol removal. In the use of processing titanium concentrate low magic city phenol waste water experiment, pH is about 2.92 more ideal.
4 Through research on reaction time, we found that the whole catalytic system after a reaction of 10–40min has higher efficiency, and response to 80min basically reached the end point of the reaction.

ACKNOWLEDGMENTS

The authors are grateful to The Ministry of education "chunhui plan" (NO. Z2011099), the Educational Commission of Sichuan Province of China (NO.11ZA008), and Innovation Fund of Postgraduate, Xihua University (NO. ycjj201383).

REFERENCES

[1] J.J.Zhao, Y.F.Wu, L.F.Jia,et al.application of chemical effects.[J]. removal of phenol wastewater COD on Fe /Al2O3 catalyst reaction conditions, 2012,41(11): 1915–1918.

[2] P.Gao, N.Li, A.Q.Wang,et al. Perovskite LaMnO3 hollow nanospheres:The synthesis and the application in catalytic wet air oxidation of phenol[J]. Materials Letters, 2013, 92:173–176.

[3] S.Nousir, S.Keav, J.Barbier Jr., et al. Deactivation phenomena during catalytic wet air oxidation (CWAO) of phenol over platinum catalysts supported on ceria and ceria-zirconia mixed oxides[J]. Applied Catalysis B: Environmental, 2008, 84:723–731.

[4] X.X.Chen and F.Zhang. Treatment of wastewater containing phenol[J]. chemistry teaching, 2012, 8: 75–77.

[5] P.Luo and Y.Q.Fan. Journal and catalytic wet oxidation properties of [J]. environmental engineering, preparation of CuO / R -Al2O3, 2009,3 (5):782–786.(in Chinese)

[6] I-Pin Chen, Shiow-Shyung Lin, Ching-Huei Wang, et al. CWAO of phenol using CeO2/γ-Al2O3 with promoter—Effectiveness of promoter addition and catalyst regeneration[J]. Chemosphere, 2007, 66:172–178.

[7] S.L.Zhao, H.D.Liang and X.M. Ju.Fe2O3 / CNT catalytic wet oxidation of phenol with [J]. H2O2 of Applied Chemistry, 2010, 27 (2):197–200.

[8] G.J.Li, L.X.Zhu and Y.C.He. Treatment of catalytic wet oxidation of phenol wastewater of cleaning of [J]. industrial water treatment, 2010, 30 (5): 38–41.

[9] J.C.Zhang, X.F.Zhang and W.L.Cao.et al.Study on[J]. molecular supported iron-based composite oxides for hydroxylation of phenol catalyzed by the catalyst, 2003,17(1): 40–45.(in Chinese)

[10] Mirjana Bistan,Tatjana Tišler,Albin Pintar.Ru/TiO2 catalyst for efficient removal of estrogens from aqueous samples by means of wet-air oxidation[J]. Catalysis Communications, 2012, 22:74–78.

[11] A.Quintanilla, A.F.Fraile, J.A.Casas,et al. Phenol oxidation by a sequential CWPO–CWAO treatment with a Fe/AC catalyst[J]. Journal of Hazardous Materials, 2007, 146:582–588.

[12] C.He, H.X.Xi and J.Zhang.at al. Comparison of [J]. ion for catalytic oxidation of Fe3+ and Cu2+ type catalyst for phenol carrier zeolite and activated carbon adsorption and, 2003, L9 (4):289–296.(in Chinese)

[13] F.C.Ban, C.B.Li and L.Chen,at al. Treatment of phenol wastewater by Fenton reagent [J]. chemical technologyresearch and development, 2009,38 (4):47–49.(in Chinese)

[14] M.B.Kasiri, H.Aleboyeh, A.Aleboyeh. Degradation of Acid Blue 74 using Fe-ZSM5 zeolite as a heterogeneous photo-Fenton catalyst [J]. Applied Catalysis B: Environmental, 2008, 84:9–15.

[15] M.L. Luo, Derek Bowden, Peter Brimblecombe. Catalytic property of Fe-Al pillared clay for Fenton oxidation of phenol by H2O2[J]. Applied Catalysis B:Environmental, 2009, 85:201–206.

[16] H.W. Hu, X.Y.Li. The reaction mechanism of Fenton catalytic oxidation and the influence factors of the progress of [J]. Bulletin of science and technology, 2012, 28 (4):220–222.

Energy, Environment and Green Building Materials – Sheng (ed.)
© *2015 Taylor & Francis Group, London, ISBN 978-1-138-02718-3*

Study on the beneficiation process of a fine ilmenite in YunNan province

Q.R. Yang
Kunming University of science and technology, Kunming, China
Kunming metallurgy college, Kunming, China

G.F. Zhang & P. Yan
Kunming University of science and technology, Kunming, China

ABSTRACT: A fine ilmenite in YunNan province with TiO_2 4.16% and TFe 18.15% mainly exists in the type of ilmenite, which consists of a little of rutile and perovskite. Based on the conditional experiment results, the optimistic experimental conditions were confirmed. A laboratory beneficiation study with the combined flowsheet of low-intensity magnetic separation, high-intensity magnetic separation, and different-sized fraction shaking table gravity separation showed that a rough titanium concentrate with a TiO_2 grade of 41.26% and a recovery of 45.82% was obtained.

KEYWORDS: fine ilmenite; mineral processing technology; magnetic separation; gravity separation

1 INTRODUCTION

Titanium and titanium alloys that have many good characteristics of high strength, corrosion resistance, and so on are widely used in many areas, such as aerospace, marine development, chemicals, and medicine (Wang, Y.L., Qu, H.L., 2011). The materials producing titanium sponge, titanium dioxide, and titanium alloys come from ilmenite ($FeTiO_3$) and rutile (TiO_2). The resource reserve of our primary titanic magnetite and menaccanite is rich, but its distribution is uneven. Meanwhile, the explored reserve of rutile and easy mining ilmenite is small, and the most is primary ore with a low grade (Long, Y.B., Zhang, Y.S., 2007). So, the beneficiation experiments about the fine low-grade ilmenite in Yunnan province are significant in solving the contradiction between supply and demand of titanium resources in China (Zhang, G.Q., Ostrovski, O., 2002).

Nowadays, these methods such as gravity concentration, magnetic separation, flotation, electric separation, joint separation method, and so on have been reported in the recovery of ilmenite (Wang, Z. et al., 2010). The joint separation method, which has united the advantage of the single separation method, is suitable for treating difficult separated fine ilmenite (Xiao, J.H. et al., 2007) (Li, L., Yang, C., 2011). This paper has adopted the joint separation method to conduct beneficiation on the ilmenite in Yunnan province.

2 NATURE OF SAMPLE

2.1 Main chemical constituents of sample

The sample of this experiment is derived from an ilmenite ore in Yunnan province. The raw ore contains 4.16% of TiO_2 and 18.15% of TFe. The spectrum analysis results, the multi-elementary analysis results, and the phase results of titanium are shown in tables 1–3.

Table 1. Results of spectrum analysis of sample / %.

element	Ag	Al	As	B	Ba	Be
content	0.002	3	0.006	>0.1	0.05	<0.001
element	Bi	Ca	Cd	Cr	Cu	Fe
content	0.01	0.2	0.007	0.005	0.07	1
element	Ga	Ge	Mg	Mn	Mo	Ni
content	0.005	<0.001	0.3	0.008	0.001	0.003
element	P	Pb	Sb	Si	Sn	Ti
content	0.05	0.09	0.01	>>10	0.2	1.2
element	V	W	Zn	In	Ta	Nb
content	0.01	0.01	0.03	<0.01	<0.005	<0.01

Table 2. Multi-elementary analysis results of sample / %.

element	Fe	TiO_2	P	S	As
content	18.15	4.16	0.201	0.066	0.0177
element	SiO_2	CaO	MgO	Al_2O_3	
content	28.20	1.05	2.65	18.11	

Table 3. Phase results of titanium of sample / %.

phase	Total TiO$_2$	Rutile	Ilmenite and titano-magnetite	Perovskite	Sphene, silicate
Content	4.16	0.25	2.702	0.198	1.01
Rate	100.00	6.01	64.95	4.76	24.28

As shown in table 1–3, the elements possessing recovery value are titanium and iron; contents of other valuable elements such as silver, copper, cobalt, and lead are too low to be recovered; and harmful elements content of sulfur, phosphorus, and arsenic also do not exceed the standard. The titanium grade of sample is not high; the valuable titanium minerals are rutile, ilmenite, and perovskite; and the titanium in gangue of sphene and silicate are unable to recover.

2.2 Granularity characteristics and distribution situation of titanium

The analysis results about particle size and metal distribution can ascertain the distribution situation of titanium dioxide in different-sized fractions, so the control of grinding fineness is particularly important during beneficiation. The size analysis result is shown in table 4.

Table 4. Size composition of sample and distribution of titanium in all size fractions.

Size / mm	Yield / %	TiO$_2$ grade / %	TiO$_2$ distribution ratio / %
−3 + 1	14.46	3.65	12.58
−1 + 0.154	23.26	4.01	22.27
−0.154 + 0.074	13.96	4.83	16.05
−0.074 + 0.038	27.25	5.25	34.15
−0.038	21.06	2.97	14.94
Total	100	4.16	100

The sample was broken to -3mm after being dried; particle size analysis results show that the content of $0.038 \sim 0.074$mm is the maximum, and the content of $0 \sim 0.038$mm is also more. But the grade of TiO$_2$ in different-sized fractions is not very concentrated, and the metal distribution of TiO$_2$ is also not too concentrated. The TiO$_2$ distribution of the sample is very scattered, and TiO$_2$ does not have an obvious concentration in a certain size fraction.

3 STUDY RESULTS AND DISCUSSION

Due to the sample, ore contains more mud and titanium is disseminated in fine ilmenite and titanomagnetite ore. It is hard to get the ore with a low grade of titanium-qualified concentrate by conventional beneficiation methods. So, this study has processed grinding fineness and field intensity of strong magnetic separation experiments, as well as fineness of shaking table experiments. An optimum beneficiation process has been confirmed by a series of probe experiments.

3.1 Grinding fineness experiments of weak magnetic-strong magnetic separation

The grinding fineness is an important indicator of responded dissociation degree and also a key influence factor of beneficiation (Zhang, Q., 2009). The dissociation degree of a single mineral will improve with the increase of grinding fineness. The overtop fineness not only increases cost but also influences the beneficiation results by ore argillization. The condition experiments about grinding fineness whose strength of weak magnetic separation is 0.1T and high-intensity separation is 0.8T were conducted. The results are shown in Figure 1.

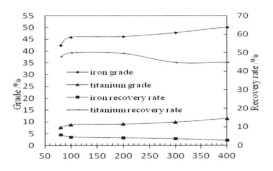

Figure 1. Results of low-intensity magnetic separation–high-intensity magnetic separation experiment with different grinding fineness.

It is shown that the grade of iron and titanium increases with the increase of grinding fineness, the increased range is little, and titanium recovery rises first and then decreases (Fig.1). In a comprehensive consideration, the grinding fineness is determined to be -200 mesh accounting for 90%, and the rough concentrates with the TiO$_2$ and Fe grades are, respectively, 9.10% and 31.86%; a titanium recovery rate of 61.65% was obtained.

3.2 Magnetic field strength experiments of strong magnetic separation

The magnetic field strength is an important influencing factor of beneficiation index (Zhang, Y.M., 2007). The condition experiments of magnetic strength during separation of titanium by strong magnetic separation were conducted to improve the grade of titanium concentrate, and the magnetic matrix model is 1/2 tooth plate. The experiment results are shown in Figure 2.

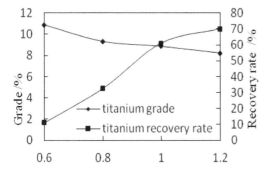

Figure 2. Results of magnetic field intensity experiment.

As shown in Figure 2, when the magnetic field intensity of strong magnetic separation was determined to be 1.2T under the premise of guarantee recovery rate, the rough concentrate with a TiO_2 grade of 8.47% and a recovery of 69.37% was obtained. Meanwhile, the results have proved that a single magnetic separation could not get qualified titanium concentrate.

3.3 Feed density experiments of shaking table

The feed density is an important influencing factor of shaking table processing capacity, and it also has a significant impact on the separation process. The higher the feed density is, the more serious the inclusion mixes between mineral during the process of separation. When the feed density was very low, the stability pulp flow could not be formed on the surface of the shaking table and its processing capacity would be lowered. So the detailed condition experiments about the feed density of the shaking table were conducted, and the results were shown in Figure 3.

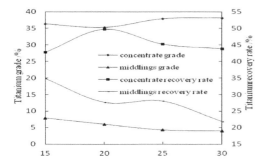

Figure 3. Results of shaking table experiment with different feeding concentrations.

Figure 3 has shown that the rough concentrate with a high content of TiO_2 could be obtained by a shaking table. Under the comprehensive consideration of grade and recovery, a titanium rough concentrate with a grade of 35.54% and a recovery of 50.75% was obtained when the feeding density was 20%. But the single gravity separation technology of shaking table still could not get qualified titanium concentrate, and further studies should be done on this basis.

3.4 Joint process experiments of weak magnetic–strong magnetic different graded shaking table

It is shown that the useful minerals could be enriched to a certain degree by magnetic separation and gravity separation through the front trial experiments. In order to get a better beneficiation index, the joint process of weak magnetic–strong magnetic different graded shaking table was proposed, and its feasibility has been investigated. The weak magnetic separation of field strength 0.1 T and the strong magnetic separation of field strength 1.2 T technology were adopted in the front. This technology could separate strong magnetic minerals, slough part of slurry and obtain rough titanium concentrate. During the two-stage process, the rough titanium concentrate was divided into three sized fractions that contained 100–200 mesh, 200–300 mesh, and less-than 300 mesh and the three size fractions were handpicked by a shaking table. respectively. The test results were shown in table 5.

Table 5. Results of the complete process of the closed circuit of low-intensity magnetic separation–high-intensity magnetic separation-shaking table.

Product	Yield /%	Grade /%	Recovery rate /%
Titanium concentrate	4.62	41.26	45.82
tails	95.38	2.36	54.18
Raw ore	100	4.16	100

The closed-circuit result of the joint process has shown that, compared with rough concentrate, the TiO_2 grade of titanium concentrate has significantly improved. The titanium concentrate with a grade of 41.26% and a recovery of 45.82% has been obtained by the joint process of the weak magnetic–strong magnetic different graded shaking table. This has created a good foundation for further wet processing (Tab.5).

4 CONCLUSIONS

1 Beneficiation experiments about a fine ilmenite in Yunnan province have been conducted. The process mineralogy study indicates that the TiO_2 content of the sample in sphene and silicate reaches

24.28%. This disadvantage determines that it is difficult to obtain a high titanium recovery rate.

2 Because of the valuable element in this ore showing the characteristics of scattered distribution and fine granularity, the single gravity separation and magnetic separation could not obtain a qualified product. In a comprehensive consideration of the advantages of each single process, a joint process of weak magnetic–strong magnetic different graded shaking table was proposed in this paper, and a rough titanium concentrate with a grade of 41.26% and a recovery rate of 45.82% was obtained.

REFERENCES

Wang, Y.L. & Qu, H.L. 2011. Experimental research on magnetic separation of certain ilmenite in ShanDong, *Non-ferrous metals (concentration)*. 2010, 1(1): 29–32.

Long, Y.B. & Zhang, Y.S. 2007. Experimental research on comprehensively recovering titanium from iron tailings of low grade vanadic titanomagnetite, *Express information of mining instustry*. 2007, 23(7): 22–24.

Zhang, G.Q. & Ostrovski, O. 2002. Effect of preoxidation and sintering on properties of ilmenite concentrates, *Mineral Processing*. 2002, 5(64): 201–218.

Wang, Z. & Sun, T.C., Ji, J. 2010. Research on beneficiation tests of a fine ilmenite ore, *Conservation and utilization of mineral resources*, 2010, 10(5): 25–27.

Xiao, J.H. & Zhang, Z.H., Zhang, Y. 2007. The mineral separation research on weathering granule ilmenite ore which associated rutile ore. *Non-ferrous metals (concentration)*, 2007, 8(3): 12–14.

Li, L. & Yang, C. 2011. Basic study on vanadic titanomagnetite milling in Panzhihua, *Iron steel vanadium titanium*, 2011, 32(1): 29–33.

Zhang, Q. 2009. Beneficiation overview. Beijing: Metallurgical industry press, 2009: 91–98.

Zhang, Y.M. 2007. Soild material separation theory and technology, Beijing: Metallurgical industry press, 2007: 136–145.

Energy, Environment and Green Building Materials – Sheng (ed.)
© *2015 Taylor & Francis Group, London, ISBN 978-1-138-02718-3*

Utilization status and expectation of phosphogypsum

Q.R. Yang
Kunming University of science and technology, Kunming, China
Kunming metallurgy college, Kunming, China

P. Yan & G.F. Zhang
Kunming University of science and technology, Kunming, China

ABSTRACT: This paper has fully introduced the phosphogypsum, including its nature, utilization status, insufficiency, and suggestions.

KEYWORDS: Phosphogypsum, Nature, Utilization status, Insufficient, Suggestion

1 INTRODUCTION

Phosphogypsum is the by-product of Phosphoric acid and the Phosphate fertilizer industry, and its main component is $CaSO_4 \cdot 2H_2O$ (Ye, X.D., 2009). According to incomplete statistics, the phosphogypsum production of the world reached 150 million tons a year ago, and China's production was 30 million tons with a growth rate of 15% (Cao, G.X. etal., 2005). The increase of phosphogypsum emissions with the years has taken up a lot of land, and the contained detrimental impurity would impact the environment (Hu, Z.P., 2006). The huge stacking cost and increasing environmental pressure have seriously restricted the sustainable development of enterprises. Thus, the use rate of world phosphogypsum production could not reach 4.5%, and China did not reach 10% (Dong, Y., 2010).

2 PROPERTY OF PHOSPHOGYPSUM

The main component of phosphgypsum is $CaSO_4 \cdot 2H_2O$, and its content is about 70%. Except calcium sulfate, it contains many impurities, such as not washed phosphoric acid, fluorid, iron-aluminum oxides, acid-insoluble substance, organics, and so on (Wu, D.L., 2008). Light gray indicates some impurities. It is soluble in acid, ammonium salt, and glycerin and slightly soluble in water. Phosphgypsum is turned into hemihydrates gypsum at 128°C and into gypsum at 160°C (Ye, H.Z., Fan, Y.C., 1997). The content of phosphgypsum bears on the grade and production technology. Currently, the two water processes are the wettest processes in phosphoric acid technology; the reaction equation is depicted as follows (Equation 1):

$$Ca_5(PO_4)_3F + 5\,H_2SO_4 + 10H_2O \rightarrow$$
$$3H_3PO_4 + 5CaSO_4 \cdot 2H_2O + HF \qquad (1)$$

3 UTILIZATION STATUS OF PHOSPHOGYPSUM

According to the characters of phosphogypsum, the scale comprehensive utilization mainly focuses on three things: building materials, chemical products from calcium sulfate, and improving soil in agriculture.

3.1 Utilization in building materials

3.1.1 Cement-sustained release
China is the first industry country of natural gypsum, and 85% of natural gypsum is used for cement-sustained release. It has provided market support for phosphogypsum as the substitution of natural gypsum used in cement-sustained release, whose Chinese cement output is huge. These enterprises have conducted similar experiments, such as HangZhou cement plant, NanJing Gangnam cement plant, Ma'anshan cement plant, ShangHai cement plant, and so on. The results of tests showed that they can increase the long-term strength of cement and drop the cost of production without a reduction of the cement's strength, which used phosphogypsum as the substitute of natural gypsum (Peng, J.H. etal., 2007).

3.1.2 Steam-pressing brisk
Guizhou Kailin Group has successfully developed a production technology that directly uses dihydrate

calcium sulfate, making high-strength water-resistant phosphorus gypsum bricks by industry-university-research cooperation. Meanwhile, it has built the largest production line of high-strength water-resistant phosphorus gypsum bricks.

Yuntianhua Group has developed a production line of building brick and block by using low-temperature ceramic-modified phosphogypsum, with the cooperation of Kunming university of science and technology (Li, G.M. *etal.*, 2012).

Nowadays, the technology of making steam-pressing bricks by phosphogypsum is in the experimental research stage, and the reliability of this technology needs further verification.

3.1.3 *Gypsum board, gypsum block, or gypsum powder*

Gypsum board has many advantages, such as its lightweight, hot insulation, sound insulation, waterproof and fireproof, simple construction, low cost, and so on. In the early 1970s, China began developing gypsum hollow sticks and gypsum blocks that were used for thermally insulated walls of high-rise buildings. Nowadays, the production technology of gypsum blocks is developing very fast in China. Shandong province has built 4~6 production lines, and the Chinese output of gypsum blocks would exceed 20 million m³.

3.1.4 *Mine-filling agent*

Guizhou Kailin group developed "cementing filling mining technology," which used phosphorus gypsum and mining waste gangue as the main filling aggregate. After 4 years of field tests, this technology has completed engineering design and application; it has achieved obvious economic, social, and environmental benefits. This technology could utilize 2 million tons of phosphogypsum, 0.73 million tons of waste gangue, and decrease 500 mu of land stockpiling of phosphogypsum and waste bottles.

3.1.5 *Road materials*

The use of phosphogypsum to enhance and modify a two-dust pavement base has resulted in the development of high-performance lime-ash stabilized soil dry base materials. This technology could avoid the weakness that the early strength of the two-dust pavement base is low, improving crack resistance of the two-dust pavement. This technology could consume a lot of phosphogypsum.

3.2 *Chemical products from calcium sulfate*

3.2.1 *Production of sulfuric acid and cement*

Sulfur resource shortage has become the bottleneck that restricts the development of Chinese sulfuric acid

industry. The inventor of making sulfuric acid from phosphogypsum is Austria Linz chemical company. Currently, China is building 7 sets of "sixth-four" engineering plants; LuBei group in Shangdong province has built a 400-thousand ton production line that made sulfuric acid and cement from phosphogypsum in 2000. But it was difficult to control the technological conditions for the component waving of phosphogypsum. So these production lines could not be generalized in China.

3.2.2 *Production of potassium sulfate*

Production of potassium sulfate from phosphogypsum involves a one-step method and a two-step method. The by-product of the one-step method is calcium chloride, which is difficult to handle, so its prospect is not good. The by-products of the two-step method are calcium chloride and calcium carbonate, the first of which was used for fertilizer and the latter was used for cement. This technology conforms to circular economy and the demand of reducing CO_2 emissions, but it has some disadvantages, such as cost of production is too high and economy is poor[10].

3.3 *Application in agriculture*

3.3.1 *Soil improvement*

The pH of phosphogypsum is generally 1–4.5, and it has important nutrients requiring plant growth, such as phosphorus, sulfur, calcium, silica, magnesium, iron, and so on. So phosphogypsum can be used instead of natural gypsum to improve the soil physiochemical properties and conditions of microbial activity. The experiments showed that the soil physicochemical properties could make substantial improvement after one year's usage of phosphogypsum to improve saline-alkali soil.

3.3.2 *Production of calcium superphosphate*

Phosphogypsum can be used instead of low-grade phosphate ore to produce eligible superphosphate, which is mixed with high-grade phosphate ore, and the output of superphosphate could increase by more than 20 % (Duan, L., Huang, Y., 2012).

4 PROBLEMS EXISTING IN THE UTILIZATION OF PHOSPHOGYPSUM

1 The differences between production technologies and processing methods caused by different components of phosphate ore cause a lot of fluctuation in phosphogypsum's physicochemical properties and impurity content. So, the disadvantages have

brought a series of problems for resource utilization and resulted in a lot of quality fluctuation of target products.

2 The pretreatment of phosphogypsum has increased some cost to enterprise, so most comprehensive utilization projects of phosphogypsum have disadvantages for which the contributed capital is big, the benefit of economy is bad, and the initiative of enterprise is low.

3 The gypsum that is produced in flue gas desulfurization of the thermal power industry has high purity and less moisture content, so it has an obvious location and price advantage. This has brought a big impact on utilization of phosphogypsum, so the market competitiveness of phosphogypsum is less without the support of national policy.

4 The key technology of phosphogypsum utilization is not perfect. For instance, the technology of making sulfuric acid and cement has disadvantages for which production control is difficult, energy consumption is high, investment is big, and the produced building materials are substandard.

5 SUGGESTIONS ABOUT UTILIZATION OF PHOSPHOGYPSUM

5.1 *Strengthen propaganda, raise awareness*

From the view of resource, phosphogypsum is not only a kind of solid waste but also a kind of resource. Because our country has a large population and there is a shortage in per capita of resources, so any resource is precious, should be protected, and would enable us to get comprehensive utilization in a long-term view. So, we should strengthen propaganda and increase the awareness of importance that utilizes phosphogypsum as a natural resource as well as give them enough attention.

5.2 *Improve industrial policy*

The utilization of phosphogypsum has formed industrialization and achieved better development. It has played a positive role in the development of the national economy and protection of the environment. So it is important that the related department should publish related policy to encourage rapid growth of

phosphogypsum resource utilization and exhibit a huge difference of preferential policy between phosphogypsum and natural gypsum. This will provide legal protection and support to the healthy development of phosphogypsum resource utilization.

5.3 *Increase the research and development of key technology*

The government should encourage related research departments studying of new ways and channels of phosphogypsum utilization by setting a special fund.

REFERENCES

Ye, X.D. 2009. Quality status and five noted points during comprehensive utilization of phosphogypsum, *Phosphate and Compound Fertilizer*, 2009, (6):60–61.

Cao, G.X. & Cao, G.L., Ma, W.L. 2005. Study on one step method of potassium sulfate from phosphogysum, *Application of Chemical Engineering*, 2005, 34(7): 421–423.

Hu, Z.P. 2006. Study on preparation of calcium sulfide by reducing phosphogypsum, *Enviromental protection of chemical industry*, 2006, (11):24–25.

Dong, Y. 2010. Study on the process of phosphogypsum package prepared slow-release urea, *Enviromental protection of chemical industry*, 2010, (3):14–15.

Wu, D.L. 2008. Study on Application of phosphogypsum as cement retarder, *Scientific guide*, 2008, 27(6):76–77.

Ye, H.Z. & Fan, Y.C. 1997. Effect of phosphogypsum on improvement of red soil, *Plant nutrition and Fertilizer Science*, 1996, 2(2):181–185.

Hu, L.B. 2010. Utilization and Recovery of phosphogypsum, *Technology and development of chemical industry*, 2010, 4(2):35–36.

Peng, J.H. & Hu, C.Q. 2007. Study on the disposal of phosphogypsum non washing pretreatment, *Gypsum and Cement for Building*, 2007, 5(5):4–9.

Li, G.M. & Li, X., Jia, L. 2012. General situation of treatment and disposal of phosphogysum, *Inorganic chemicals industry*, 2012, 10(10):11–13.

Yuan, W. & Tan, K.F., He, C.L. 2009. Properties of cementations binder consisting of phosphogypsum, *Journal of Wuhan University of Technology*, 2009, 10(31):39–42.

Duan, L. & Huang, Y. 2012. Experimental study on the mechanical properties of phosphogypsum used in the plaster mold-concrete composite hollow floor, *Journal of Guizhou University (Natural Sciences)*, 2012, 10(29):112–117.

Energy, Environment and Green Building Materials – Sheng (ed.)
© 2015 Taylor & Francis Group, London, ISBN 978-1-138-02718-3

Life prediction of the subsea tunnel based on chloride migration model

X.H. Lv

College of Civil Engineering, Southwest Jiao Tong University, Chengdu City, Sichuan Province, China

ABSTRACT: In marine corrosion environment, chloride seeps into the inner of concrete through the diffusion effects, and has gathered on the surface of steel bar. This paper comprehensively analyzes chloride migration based on subsea tunnels' form and environmental condition. Based on chloride migration model, we can predict the service life of subsea tunnel.

KEYWORDS: life prediction; subsea tunnel; chloride migration model; chloride ion concentration

1   RESEARCH BACKGROUND

In recent decades, ground transport has failed to meet the demand of the modernization process because of the economic development and population growth. Therefore, at the end of the twentieth century, ITA (International Tunnellling Association) put forward the proposal, "developing underground space, begins man's Cave days". Construction of the subsea tunnel has become the inevitable trend. Subsea tunnel is a very safe passage of the straits, and it has the following advantages. First, subsea tunnel doesn't occupy the land. Second, it doesn't hinder navigation. What's more, it does no harm to the ecological environment.

Based on the rapid development of subsea tunnel and complex service environments, the structure of high importance and high construction costs, and poor serviceability and high maintenance costs, the life prediction of the subsea tunnel has become inevitable.

2   LIFE PREDICTION METHOD FOR SUBSEA TUNNEL

2.1   *The pressure permeability model of chloride ion in concrete*

The migration of chloride ion in concrete under hydrostatic pressure are coupled pressure permeability and chloride ion diffusion of the two basic physical and chemical process. Pore fluid in concrete is in process of permeability from outside to inside because of the action of pressure gradient, along with the infiltration of external chloride ions into concrete. At the same time, chloride ion occurs directional migration in pore fluid into concrete, under the

effects of the concentration gradients. Since the size of the chloride ion is less than the water molecules, the rate of chloride ion penetration in pore fluid is much smaller than speed of diffusion migration, in the same pore structure of concrete. Therefore, the combined effect of the penetration and diffusion is more likely to occur in the shallow layer of concrete (the position close to the surface). When it comes to the deep layer of concrete, penetrating effect is significantly decreased and the diffusion effect will dominate.

The conservation law of chloride ion showed that:

$$\frac{\partial C}{\partial t} = D\frac{\partial^2 C}{\partial x^2} - u\frac{\partial C}{\partial x} \tag{1}$$

Initial conditions: $C(x,0) = 0 \quad 0 \le x \le \infty$

Boundary conditions: $C(0,t) = C_s; C(\infty,t) = 0 \quad t > 0$

$C(x,t)$ for solute concentration (%); D for chloride diffusion coefficient (mm^2/s); u for the average pore velocity (mm/s).

In 1961, Ogata & Banks draw on analytical solution of the formula (1):

$$C(x,t) = \frac{C_s}{2}\left[erfc(\frac{x-ut}{\sqrt{4Dt}}) + \exp(\frac{ux}{D})erfc(\frac{x+ut}{\sqrt{4Dt}}) \right] \tag{2}$$

The water flow rate in unsaturated concrete is:

$$u(x,t) = \frac{kp_0}{\mu\sqrt{\pi}}e^{-\frac{\mu x^2}{4kEt}}\sqrt{\frac{\mu}{kEt}} \tag{3}$$

Combining formula (2) with formula (3), is used to calculate the chloride ion concentrations associated with time and space.

2.2 The chloride migration model in concrete

When transporting in the pore solution in concrete, chloride ions will have chemical and physical functions with hardened cement paste, so that part of the chloride ions form a hardened cement paste. Among them, the chemical function mainly means that chloride ion combined with calcium aluminate forms Friedel salt. The physical function mainly means chloride ions are adsorbed on the surface of hydrated calcium silicate, and is closed in C-S-H layer and difficult to escape. Therefore, it is necessary to describe the relationship between the adsorption amount of chloride ion and the free chloride ion concentration of the pore solution quantitatively. This paper considers the binding capacity of chloride ion with the method of linear combination.

The total flow for the transport of chloride ion in concrete:

$$J_{cl} = J_{d,cl} + J_{c,cl} \tag{4}$$

The mass conservation of total chloride ion can be expressed as:

$$\frac{\partial C_t}{\partial t} = -\nabla J_{cl} \tag{5}$$

From the above equations, chloride ion transport equations can be obtained based on the C variable:

$$\frac{\partial C}{\partial t} = \frac{1}{1 + R\beta} \nabla \left[D_{cl}(h) \frac{\partial C}{\partial x} + C \cdot D_m \frac{\partial h}{\partial x} \right] \tag{6}$$

Initial conditions: $C\big|_{x>0,t=0} = C_0$

Boundary conditions: $C\big|_{x=0,t>0} = C_s, C\big|_{x=l,t>0} = C_0$

On the basis of the above analysis, this paper predicts the service life of subsea tunnel by using mathematical model method, based on the chloride threshold value criterion. The lifetime (T) of subsea tunnel is divided into two parts: inside and outside, that is Ti and To. As long as chloride concentration of outer row at steel-concrete reaches the critical concentration (Ccr), no matter the inside or outside, the service life of subsea tunnel is considered to reach its end. The lifetime (T) means the smaller value between Ti and T_0.

$$T = \min\{T_0, T_i\} \tag{7}$$

Therefore, chloride ion concentration (C) of inside and outside is a function of position (x) and time (t). When chloride concentration of outer row at steel-concrete reaches the critical concentration (Ccr), the time means the service life of that side.

3 ENGINEERING EXAMPLE

3.1 Engineering background

Shenjiamen Harbour Subsea Tunnel is located in Putuo District of Zhoushan City, Zhejiang Province. The tunnel is only used for pedestrians, rather than traffic. North of the tunnel is the ferry terminal of Shenjiamen Bingang Road, and south is the passenger terminal building of Lujiazhi Island. The whole length of the tunnel is 266.70 m. The engineering environment is coastal or marine, and the tunnel is consisted of pipeline sinking section, north shore section, south shore section and dry dock.

Figure 1.   Landform situation and tunnel site.

High performance concrete is used for this engineering, which has characteristics of high strength, high anti permeability, high waterproof and corrosion resistance. In consideration of durability, mix composition by adding active mineral admixture is the inevitable choice. Mineral powder uses S95 above granulated blast-furnace slag, and high range polycarboxylate water reducer is used as water reducing agent. The concrete strength grade is C50, and the impermeability grade is greater than S12. Double-mixture technology is used for the preparation of concrete, and the minimum amount of cementitious materials is 360 kg/m³. The concrete is made of Portland cement, the strength grade of which is 52.5 and P.II. The content of C_3A in cement is not greater than 5%, and water-binder ratio is 0.34. The chloride diffusion coefficient of concrete is not greater than 4×10^{-12} m²/s.

3.2 Life prediction of engineering

According to requirements of the concrete strength grade, water impermeability grade, buried depth, water cement ratio and chloride ion diffusion coefficient, the values of the parameters for the subsea tunnel can be obtained as follows: water pressure $P_0 = 0.15$ MPa, absolute permeability coefficient $Ks = 1.47 \times 10^{-6}$ mm^2, cover thickness is 60 mm, and chloride ion diffusion coefficient $D = 4 \times 10^{-6}$ mm^2/s.

As the literature shows that, when the critical concentration of chloride ion reaches 0.3%, steel bar is likely to begin corrosion, so the critical concentration $C_{cr} = 0.36\%$. With reference to the value of surface chloride concentration, which is taken out from the concrete underwater part of a bridge in Norway, concrete chloride ion accounts for 0.42% of the total quality of concrete.

1 Lifetime of the outside
Substituting the above parameters values into Formula (2), we obtain the time for 106 years, and this is when the concentration of chloride ion at x = 60 mm reaches the critical chloride ion concentration. As is shown in figure 2. That is, lifetime of the outside tunnel is 106 years.

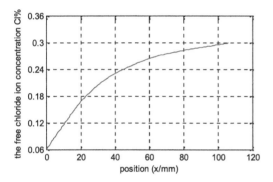

Figure 2.   The changes of chloride ion concentration with time at x = 60 mm.

2 Lifetime of the inside
With reference to the results of the investigation of Jiaozhou Bay Subsea Tunnel, the environmental

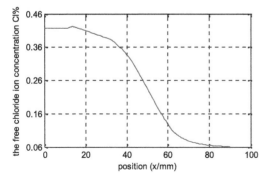

Figure 3.   The chloride ion concentration of the outside and inside in 100th year.

humidity is 60%. Based on the experiment measured, initial relative humidity (h_0) of concrete is 10%. Other parameters values are the same as engineering conditions. With the Matlab programming and analysis, the concentration distribution at various locations can be obtained in 100 years, as shown in figure 3. From the figure we can see, the concentration of chlorine ion is only 0.14% in 60 mm of cover thickness in 100th year, and that concentration hasn't exceeded the critical concentration (Ccr).

Figure 4 shows changes in the concentration of chloride ion with time of the inside tunnel at x = 60 mm. When the chloride ion concentration reaches 0.3% (the critical chloride ion concentration), the time (t) is 106 years. That is, lifetime of the inside tunnel is 160 years.

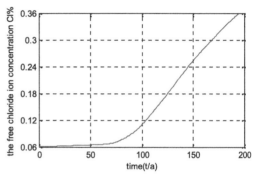

Figure 4.   The changes of chloride ion concentration with time at x = 60 mm.

The lifetime of subsea tunnel can be expressed as: $T = \min\{T_0, T_i\}$. By the above, To = 106 years, Ti = 160 years, so T = 106 years. To sum up, the lifetime of Shenjiamen Harbour Subsea Tunnel is 106 years in the chloride erosion.

4 BRIEF SUMMARY

According to the chloride threshold value criterion, the formula for chloride ion content of the inside and outside tunnel is used as a calculation model for the lifetime of the inside and outside respectively. The lifetime of subsea tunnel is defined as the smaller value between the inside and outside. At the same time, this paper introduces the background and general situation of the engineering of Shenjiamen Harbour Subsea Tunnel in Zhoushan City. According to service environment, various parameters of subsea tunnel can be obtained. The chloride ion migration model is used to analyze the service time, and the time is when the chloride ion concentration of the outer row of steel bars reaches the critical concentration respectively. The final conclusion is this subsea tunnel meets the design requirements of service life for 100 years.

REFERENCES

[1] Chen Cong, Research in chloride environment underground concrete structure durability [D]. Shanghai: Tongji University, 2009.

[2] Jin Weiliang, Zhaoyu. 2002. Durability of concrete structures [m]. Beijing: Science Press.

[3] Kim Jin, Keun Lee, Chit-Sung. Moisture diffusion of concrete considering self-disiceation at early ages [J]. CCR, 1999, 29: 1921–1927.

[4] Vladimir. Corrosion of reinforcement induced by environment containing chloride and carbon dioxide [J]. Bulletin of Materials Science, 2003,26: 605–608.

[5] Luo Gan, Durable life under the chloride ion penetration environmental prediction of reinforced concrete members [D]. Xiamen: Huaqiao University, 2003.

Energy, Environment and Green Building Materials – Sheng (ed.)
© 2015 Taylor & Francis Group, London, ISBN 978-1-138-02718-3

Optimization of control parameters based on an improved detailed urea-SCR model

T. Feng & L. Lu
Wuhan University of Technology, Wuhan, China

ABSTRACT: This paper introduced an improved detailed SCR system. This model took into account the pore diffusion process in the catalyst, which can accurately describe the system-out NO_X concentration under transient engine operating conditions. Based on this model, the parameters of the Adblue control strategy can be optimized offline, which reduced the cost and time. In this paper, an ideal ETC test results were obtained within a short time with the help of this detailed model.

KEYWORDS: SCR; Pore diffusion; Parameters optimization

1 INTRODUCTION

Selective catalytic reduction (SCR) of NO_X with Adblue has been proven to be the promising solution for meeting future diesel NO_X emissions standards. The SCR system is complicated. Willems, 2007, Johnson, 2013 & Chi, 2005. Therefore, to reduce the cost and development efforts, modeling and simulation approaches represent an essential step in the development process of the controller. Hsieh, 2011. Hsieh presented a continuous stirred tank reactor model to describe the characteristic SCR system. Lu, 2013. Lu used a full chemistry SCR model in the controller development. However, the existing models are not accurate enough to fully describe the transient response of NO_X concentration with the change of Adblue dosage due to the neglecting of the pore diffusion process in the catalyst.

This paper presents an improved detailed SCR model which takes into account the pore diffusion process in the catalyst. Following that, the Adblue dosage controller was introduced. Finally, the control parameters were optimized based on the detailed model and verified by the ETC test.

2 DETAILED SCR MODEL

2.1 Modeling

The Honeycomb-type catalytic converter consist of hundreds of individual channels. The reactions including NH_3 ad/desorption, SCR reaction, NH_3 oxidation were take place at active sites within the washcoat that covers the solid substrate (Figure 1).

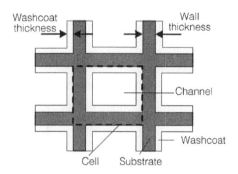

Figure 1. Structure of the catalyst.

A 12L V_2O_5-WO_3/TiO_2 catalyst was used in this paper. Considering of the structural symmetry of catalyst, the entire converter was represented by one single channel. The convection, diffusion, conduction in the gas phase, conduction in the solid phase, and the energy transfer between solid and gas were modeled (Figure 2).

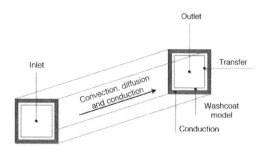

Figure 2. Modeling of the catalyst channel.

2.2 Washcoat model

In the surface of the catalyst, the catalytic reaction can be divided into four main process: mass transfer, pore diffusion, ad/desorption and catalytic reaction (Figure 3). The pore diffusion was modeled as shown in the Figure 4. The gas concentration in the washcoat was solved by the partial differential equations.

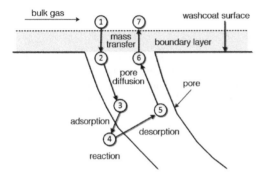

Figure 3.   Steps of a catalytic reaction.

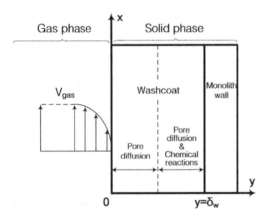

Figure 4.   Pore diffusion model.

The balance equation for the gas species k is described by Eq. (1).

$$\varepsilon_w \cdot \frac{\partial \left(\rho^L \cdot w_k^L \right)}{\partial t} = \frac{\partial}{\partial y} \left(D_{k,eff} \cdot \rho^L \cdot \frac{\partial w_k^L}{\partial y} \right) + MG_k \cdot \sum_i^I v_{i,k} \cdot \dot{r}_i \left(c_k^L, T_s \right) \tag{1}$$

Where, ε_w is the porosity of the considered washcoat layer, ρ_L is the density of the gas mixture (kg/m³), w_k is the mass fraction of the gas species k, $D_{k,eff}$ is the effective diffusion coefficient (m²/s), MG_k is the molecular weight of gas species k (kg/mol), $v_{i,k}$ is the

stoichiometric coefficient of gas species k in reaction i. r_i is the reaction rate (mol/m³s).

The boundary condition at solid surface (y=0) is determined by the balance of diffusive flux and mass transfer, which is described by Eq. (2).

$$D_{k,eff} \cdot \rho^L \cdot \frac{\partial w_k^L}{\partial y} = k_{k,m} \cdot \left(\rho^L \cdot w_k^L - \rho^B \cdot w_k^B \right) \tag{2}$$

Where, $k_{k,m}$ is the mass transfer coefficient of the gas species k (m/s), ρ_B is the density of bulk gas (kg/m³).

Assuming that no diffusive flux out of the last washcoat layer, the boundary condition at the total washcoat layer thickness (y=δw) is described by Eq. (3).

$$\frac{\partial w_k^L}{\partial y} = 0 \tag{3}$$

Wakao, 1962. In the paper, the effective diffusion coefficient in pore diffusion process is given by the random pore model (RPM). Assuming that the washcoat pores can be divided into macro-pores and micro-pores and the proportion is decided by the geometrical characteristics of the catalyst. In the RPM model, the effective diffusion coefficient of macro and micro pore is described by Eq. (4)–(6).

$$D_{Kn,M/\mu} = \frac{1}{3} \cdot d_{por,M/\mu} \cdot \sqrt{\frac{8 \cdot R \cdot T_s}{\pi \cdot M}} \cdot \tau \tag{4}$$

$$\frac{1}{D_{M/\mu}} = \frac{1}{D_{k,g}} + \frac{1}{D_{Kn,M/\mu}} \tag{5}$$

$$D_{k,eff} = \varepsilon_{w,M}^2 \cdot D_M + \frac{\varepsilon_{w,\mu}^2 \cdot \left(1 + 3 \cdot \varepsilon_{w,M} \right)}{1 - \varepsilon_{w,M}} \cdot D_\mu \tag{6}$$

Where, $D_{Kn,M/\mu}$ is the Kndsen diffusion coefficient (m²/s), $d_{por,M/\mu}$ is the size of macro and micro pore (m), $D_{M/\mu}$ is the diffusion coefficient of macro and micro pore (m²/s), τ is the correction factor.

2.3 Chemical reaction

The paper focuses on the following SCR chemical reaction including NH_3 ad/desorption, standard SCR reaction and NH_3 oxidation. The reaction rates are presented below:

$$NH_3 + S \rightarrow NH_3(s) \tag{7}$$

$$r_{NH_3,ads} = K_{ads} \cdot \exp\left(-\frac{E_{ads}}{RT} \right) \cdot c_{NH_3} \cdot \left(1 - \theta_{NH_3} \right) \tag{8}$$

$$NH_3 + S \rightarrow NH_3(s) \tag{9}$$

$$r_{NH_3,des} = K_{des} \cdot \exp\left(-\frac{E_{des} \cdot (1 - \varepsilon \cdot \theta_{NH_3})}{RT}\right) \cdot \theta_{NH_3} \quad (10)$$

$$4NH_3(s) + 4NO + O_2 \rightarrow 4N_2 + 6H_2O \quad (11)$$

$$r_{NO} =$$
$$K_{NO} \cdot \exp\left(-\frac{E_{NO}}{RT}\right) \cdot c_{NO} \cdot \theta_{NH_3}^* \cdot \left(1 - \exp(-\frac{\theta_{NH_3}}{\theta_{NH_3}^*})\right) \quad (12)$$

$$4NH_3(s) + 3O_2 \rightarrow 2N_2 + 6H_2O \quad (13)$$

$$r_{NH_3,O_2} = K_{O_2} \cdot \exp\left(-\frac{E_{O_2}}{RT}\right) \cdot \theta_{NH_3} \quad (14)$$

Where, r is the reaction rate (mol/m³s), T is the temperature (K), c is the mole concentration (mol/m³), E is the activation energy (K), K is the frequency factor (1/s, mol/m³s), θ_{NH_3} is the ammonia surface coverage, $\theta_{NH_3}^*$ is the critical surface coverage, ε is the coverage dependency.

2.4 Parameters estimation

The synthetic gas reactor test (SGB test) was conducted to estimate the reaction kinetics parameters. In the SGB test, the catalyst was fully milled into powder in order to eliminate the effect of diffusion. The space velocity was 30000h⁻¹ and the temperature was varied from 200°C to 450°C. The inlet gas concentration is shown in the Table 1.

Table 1. Inlet gas concentration.

Gas	Mole fraction
NO	7.50×10^{-4}
NH$_3$	8.25×10^{-4}
H$_2$O	1.00×10^{-1}
O$_2$	1.00×10^{-1}
SO$_2$	1.00×10^{-4}
N$_2$	7.98×10^{-1}

Based on the results of SGB test, the optimized kinetics parameters are show in the Table 2.

After the estimation of the reaction kinetics parameters, the diffusion correction factor τ was also calibrated according to the bench test results. Finally, as shown in the Figure 5, the simulation results well agreed with the test after the estimation.

3 BENCH TEST

3.1 Adblue dosage control strategy

Todd, 2014. Given the key role of ammonia storage in SCR control, a reasonable control strategy should be

Table 2. Reaction kinetics parameters.

Parameter	Value	Unit
K$_{ads}$	5.22E+09	m/s
E$_{ads}$	6.99E+04	J/mol
K$_{des}$	6.28E+04	kmol/(m²•s)
E$_{des}$	9.55E+04	J/mol
ε	3.70E-01	-
θ^*_{NH3}	6.00E-02	-
K$_{NO}$	1.69E+06	m/s
E$_{NO}$	6.04E+04	J/mol
K$_{O2}$	3.74E+03	kmol/(m²•s)
E$_{O2}$	1.66E+05	J/mol

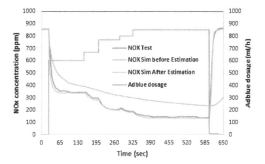

Figure 5. Comparison between test and simulation results.

able to adjust the quantity of ammonia stored in the catalyst by precisely control of NSR, thus to increase the conversion efficiency of NO$_X$ and avoid NH$_3$ slip. Therefore, a fuzzy ammonia storage control strategy was presented in this paper. The scheme of the Adblue dosage control strategy is shown in the Figure 6.

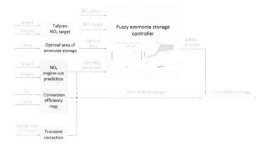

Figure 6. Block scheme of model-based control strategy.

As shown in the Fig.6, a fuzzy ammonia storage strategy was used in the controller. In the controller, the potential for NH$_3$ slip is evaluated by the fuzzy evaluation system. Based on the evaluation results, an emendation of Adblue dosage is proceed so the ammonia storage changed along the chosen process to the optimal area. The parameters of the evaluation system and optimal area definition need a delicate optimization in order to meet both NO_x emission requirements and NH_3 slip targets.

3.2 ETC test results

The ammonia storage control strategy was evaluated by the ETC test. The baseline engine out NO_x of the ETC test was 11 g/kWh and the catalyst outlet NO_x target was 2.0 g/kWh because of the NO_x reduction capability of the catalyst.

With the help of the detailed SCR model, the control parameters can be optimized offline, which would reduce the cost and time. As shown in the Table 3, the calibration process of the control parameters was finished within four ETC tests.

Table 3. ETC test results.

NO.	Test results		
	$BSNO_x$ (g/kwh)	Avg. NH_3 slip (ppm)	Max. NH_3 slip (ppm)
1	1.78	2.82	46.16
2	1.82	1.12	20.51
3	1.73	2.51	32.48
4	1.65	1.42	20.64

The test result of the 4th ETC test is shown in the Figure 7. The test results indicates that, the detailed model presented in this paper can accurately describe the system-out NO_x concentration.

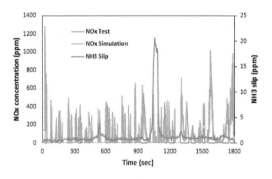

Figure 7. Test results of the 4th ETC test.

4 CONCLUSIONS

This paper presents an improved detailed SCR model which considers the pore diffusion process in the catalyst. This model can accurately describe the system-out NO_x concentration in the transient engine operating conditions. With the help of this detailed SCR model, the control parameters can be optimized offline, which would reduce the cost and time.

ACKNOWLEDGMENTS

This work was supported by the National Natural Science Foundation of China (No. 51379165) and the Natural Science Foundation of Educational Department of Hubei Government (No. 20520005). The authors would like to thank Prof. Lu Lin for his analytical support, and those who haven't been mentioned here, for their valuable services rendered in their individual technical disciplines, which formed the basis for preparing this article.

REFERENCES

Frank Willems, Robert Cloudt, Edwin van den Eijnden, et al. 2007. Is closed-loop SCR control required to meet future emission target? SAE paper 2007-01-1574.
Tim Johnson. 2013. Vehicular Emissions in Review. SAE paper 2013-01-0538.
John N. Chi, Herbert F. M. DaCosta. 2005. Modeling and Control of a Urea-SCR Aftertreatment Sytem. SAE paper 2005-01-0966.
Ming-Feng Hsieh, Junmin Wang. 2011. Development and experimental studies of a control-oriented SCR model for a two-catalyst urea-SCR system. Control Engineering Practice, 219(4):409–422.
L. Lu, L. Wang. 2013. Model-based optimization of parameters for a diesel engine SCR system. International Journal of Automotive Technology, 14(1):13–18.
Wakao W., Smith J.M. 1962. Diffusion in catalyst pellets. Chemical Engineering Science, 17(1962):825–834.
Todd J. Toops, Josh A. Pihl, William P. Partridge. 2014. Fe-Zeolite Functionality, Durability, and De-activation Mechanisms in the Selective Catalytic Reduction (SCR) of NOX with Ammonia. Funda-mental and Applied Catalysis, 15(3):97–121.

Energy, Environment and Green Building Materials – Sheng (ed.)
© 2015 Taylor & Francis Group, London, ISBN 978-1-138-02718-3

Research for the influence of light transmissivity of air humidity in fixed distance

Y.N. Wu, Z.J. You & J.N. Ma
North China Electric Power University, Beijjing, China

L. Chen
Mathematics and Physics Department, North China Electric Power University, Beijjing, China

ABSTRACT: Experiments have been done to measure the influence of light penetration of pure fog with fixed thickness between two layers of glass. Using a natural light and a laser light source for experiments, the result is obtained from two different sources. Then, analysis and comparison of the experiment result can imply that the curve of light transmissivity is changing with humidity. And, it demonstrates the influence of humidity on light transmissivity.

KEYWORDS: atomization; humidity; luminous power of transmission; optical transmittance

1 INTRODUCTION

Water spray is a water vapor condensation phenomenon that consists of a large number of small droplets suspended in a low altitude atmosphere, making air turbidity and visibility drop[1]. Experiments show that the humidity level can intuitively reflect the mist concentration[2], so this experiment is based on the atomizing chamber where mist concentration simulation is in [3]. We measure the optical power by changing the fixed thickness between the glass thickness of air humidity and that of the fixed light source. After a lot of summary and analysis of experimental data, it is concluded that net humidity has an influence on the luminous power of transmission with the fixed light and is the main reason of visibility to be reduced in the air. Due to the limitation of experimental conditions, the experiment is only a preliminary exploration. In the future, this can be developed for transparency controlled in double-deck glass and other related projects. The economic benefit is in the process design, privacy protection, and artificial landscape design, and other fields have broad prospects for development.

2 ANALYSIS OF THE NECESSITY OF MAKING ARTIFICIAL FOG AND FOG METHOD

2.1 The necessity:

2.1.1 The experiment need is a relatively closed environment. The natural environment cannot meet the demand of the experiment, because it is complicated and sealing ability is poor.

2.1.2 The fog concentration of artificial fog is easier to be controlled. It enables the fog on the surface of the glass space to produce a stable concentration gradient, which makes it easy to measure the luminous power of transmission between two layers of glass of a corresponding different humidity.

2.1.3 The experiment needed the fog has a certain stability. We should reduce as far as possible the air flow between the experimental process and the environment.

2.1.4 We should measure relatively few impurities such as dust, particles, and so on when we study the effect of humidity on the luminous power of transmission. The four points show that the experiment cannot use the fog in the natural environment and it must be conducted in the laboratory.

2.2 Method of making fog:

"Mist" in this experiment, in essence, is formed by saturation with water vapor in the air that condenses into water droplets because of the temperature drop. This experiment adopts the crystals piece atomizer for atomized water vapor to simulate the fog. The right amount of water in the double deck glass is used in the sink. In the experiment, a small atomizer is placed in the water and the crystals get close to the surface to produce artificial fog. But some water drop pops up when fog is produced and it influences the measurement of humidity and luminous power of transmission; so, we put the atomizer in the corner between the double glazing. We get the datum on the other side after the fog is produced and stabilized.

3 PROCESS OF EXPERIMENTS

3.1 *Experimental apparatus as following table 1:*

Table 1. Experimental apparatus.

Apparatus	Specification & Range	Quantity
Glass container without bottom and cap	25*35*45 (cm)	1
Transparent glass (with hydrophobic membrane)	35*45 (cm)	1
Optical power meter	20mW	1
Incandescent lamp	40W	1
Laserlight source	3mW (635nm)	1
Water channel	50*60*8 (cm)	1
Plank	40*50*0.3 (cm)	1

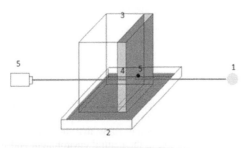

Figure 1. Diagram of experimental set-up:
1. Incandescent lamp or Laser light source 2. Water channel 3. Transparent glass (with hydrophobic membrane) 4. Fog-filled glass layer 5. Optical power meter.

3.2 *Experimental principle*

Optical power is the work done by light in a unit time. The reflection and refraction of light through different media enable energy dissipation and light power to decrease. Water fog is composed by small water droplets floating in the air. Diffuse reflection and refraction occurs between the water droplets when the light penetrates the fog. This experiment uses the characteristic to study the blocking effect of air humidity on optical power. To prevent water droplets condensed by moisture on the glass surface from affecting the experimental results, a layer of hydrophobic membrane should be present on the inner wall of the container that is going to be measured.

3.3 *Experimental operating steps*

Assembling the experiment device as shown inFigure1.

3.3.1 *Measuring the influence of humidity on natural light transmittance:*

1 As shown in Figure 1, position of the light source away from the side of the glass device is 50 cm (this experiment adopts a 25W incandescent lamp as light source to simulate daily indoor light). On the other side, position of the optical power meter opposite the light source away from the glass is 20 cm. Measure the optical power meter and hygrometer before the fog is produced.
2 Turn on the atomizer power supply, and open the atomizer at the bottom of the glass container. We can begin measuring the data till the fog tends to be stable and has an obvious mist concentration gradient from bottom to top.
3 Data measurement: Start at the point that is away from the top of the container by 2cm. Keep the light source and optical power meter in the same horizontal plane. Put the hygrometer into the fog slowly, then fix it into the container wall by keeping it at the same height with the light source and optical power meter. Wait for 2 minutes after the set-up is in place and make sure the fog is stable. Simultaneously, observe the hygrometer readings and optical power meter readings; then, record them.
4 Move the light source, optical power meter, and hygrometer, respectively, down by 2 cm. Repeat the earlier steps until the measurement point is away from the glass at the top by 37 cm.
5 Light transmittance = through optical power/bottom light power. Input test data into MATLAB software and use the MATLAB software for data tracing point. According to the dot, make the optical power of the general curve change with

the humidity and it can intuitively reflect the air humidity effects on light. Chart 2 is one group of curves.

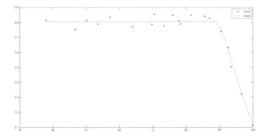

Figure 2.   Experimental curve in incandescent light.

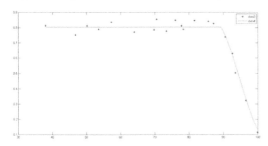

Figure 3.   Experimental curve in laser light.

6   Repeat the experiment content to get more groups of experimental data. It is advisable to get more than 5 groups in order to reduce error.

3.3.2   *Measuring the influence of humidity on laser light transmittance:*

Using the red laser (635 nm, 3 mw) as a light source[4], study the influence of air humidity on light transmittance under the condition of monochromatic light. The experimental steps are similar to those in step 2, but the incandescent lamp is replaced with a laser light source.

Figure 3 is the curve of the optical power that changes with humidity for a set of laser light source experiments.

4   DATA ANALYSIS

After multiple sets of experimental measurements of average, we, respectively, obtained the experimental data under the laser light and the incandescent light. After we analyzed the data, in a single second, three functions and exponential functions were carried out on the experimental data fitting to get a set of reasonable descriptions of the experimental curve. But we did not get an ideal result. Finally, we decided to use

forms of piecewise function fitting and we found that the curve could be better described.

The experimental data and the fitting function are as follows:

Table 2.   Experimental data.

X	Y(natural light)	Y(laserlight)
38.0	0.8032	0.8134
46.8	0.7521	0.7536
50.2	0.7822	0.8175
53.6	0.8190	0.7875
57.3	0.8060	0.8337
64.0	0.7513	0.7773
69.8	0.7760	0.7884
70.5	0.8247	0.8537
73.4	0.7567	0.7728
76.0	0.8173	0.8457
77.8	0.7820	0.8141
78.3	0.7587	0.7837
81.6	0.8267	0.8419
85.6	0.8073	0.8462
87.1	0.7967	0.8263
90.5	0.6987	0.7334
92.6	0.5907	0.6353
93.5	0.4933	0.5036
96.5	0.2820	0.3262
99.9	0.0747	0.1169

Fitting functions:
Under the incandescent light:

$$\begin{cases} y = 0.8023 & 38.0 \leq x \leq 87.6 \\ y = 3.5 \times 10^{-4}x^3 - 0.10x^2 + 9.46x - 294.7 & 87.6 \leq x \leq 100 \end{cases}$$

Under the laser light source:

$$\begin{cases} y = 0.8182 & 38.0 \leq x \leq 89.4 \\ y = 3.8 \times 10^{-4}x^3 - 0.10x^2 + 10.14x - 315.11 & 89.4 \leq x \leq 100 \end{cases}$$

5   CONCLUSION OF THE EXPERIMENT

The experimental results show that regardless of incandescent lights or laser, the light transmittance between the glass layers that have a fixed thickness is reduced with the increase of air humidity. The light transmittance changes slightly with the increase of humidity when it is less than 90%. But when the humidity increases to 90%, the light transmittance quickly decreases with the increase of air humidity, showing that the fog is clearly blocking the spread of light. According to the experimental data, the light transmittance of incandescent lights is lower than that of laser when the humidity is less than 90% and the light transmittance of incandescent lights falls faster than that of laser. This is because laser has high convergence degree and strong penetrability[5]. During the experiment, we find that when we cannot see through the glass layers there is at least 99% humidity. That is to say, if we want to block the sight with fog, humidity must be more than 99%.

6   INTERPRETATION AND APPLICATION

According to the result, if we only think about the influence of humidity and ignore the other factors, we can find a phenomenon. The light transmittance nearly stays still when the humidity is less than 90%, and it falls fast when the humidity is more than 90%. In our normal life, humidity is between 40% and 70%, and the light transmittance is not clearly seen under the influence of humidity. So if we feel visibility in the air, it is not good. It has nothing to do with the humidity, and the particles and smoke are actually the real cause. We can place the smoke and particles in a container to research the real reason that makes the visibility decrease.

The experiment intuitively shows the general rule of the light transmittance with the change of humidity, but it is only under the condition of pure water vapor. It can be used as an experimental basis for more complex related experiments; for instance, to explore the relationship between light transmittance and steam concentration in other gas or liquid vapor or to explore the influence of mist concentration on the light transmittance when it is under the condition of suspended particulate matter. In our life, we also can further develop applications, such as transparency controllable double glass door. We can install the fog generator and humidity detector between double glass layers; under the condition of controllable humidity, we can control the transparency of the glass door. In this way, we implement the traditional use of the door and blend the process design into it. This can be considered a new train of thought to design furniture. Besides, the experimental results can be applied to packaging design and even can be used as an experimental basis to make a new type of wall that can change transparency by changing humidity. Then, we can use the wall to build a new type of building. In the military aspect, we can use it as a basis to design a barrier that is light and easy to carry, of course, made of other material. In art, we can use it for designing decorative items to create a new art and visual perception. Due to limitations, we could not further apply it to these aspects and only provide the reference for related researchers and developers, hoping it would be applied more usefully and widely for various aspects.

REFERENCES

Zheng Lv,Sophie,etc. The experiment to determine the best wavelength offog light penetration. [J]. applied optics, 2008, 29 (4).

ZhonghengYuan,JingZhang,etc. Indoor simulation experiment researchfortransfer characteristic of dual wavelength laser. [J]. laser technology, 2010, (4):478–481.

Xiang Xiao, XiaoluZhao, LianchengLu, etc. Numerical simulation study for The mist concentration distribution in fog turns room. [J]. Journal of Engineering Thermophysics, 2005, 26 (5):776–778.

Linqian Xia. Single color transformation of through the fog system based on the near infrared spectrum. [J]. Infrared and Laser Engineering, 2007, 4 (1):234–237.

KewenXia,Jianping Song. Technology oflaser optical fiber nuclear logging. [J]. Petroleum Instruments., 2001, (2):8–10.

Energy, Environment and Green Building Materials – Sheng (ed.)
© *2015 Taylor & Francis Group, London, ISBN 978-1-138-02718-3*

Evaluation of rural territorial functions: A case study of Henan Province, China

Chao Fu

Institute of Geographic Sciences and Natural Resources Research, CAS, Beijing, China
University of Chinese Academy of Sciences, Beijing, China
International Cooperation Center, National Development and Reform Commission, Beijing, China

ABSTRACT: The new urbanization process requires a coordinative development of cities and rural areas. Territorial functions of rural areas are defined as the advantageous effects on nature and human society that, in particular, rural systems perform by using their own property and interaction with other systems in certain social development stages. This paper establishes an index system for evaluating rural territorial functions, including agricultural function, social function, economic function, and ecological function. By establishing a model based on General Regression Neural Network (GRNN) with county level as the basic unit, this paper makes a comprehensive evaluation on the rural territorial functions of 109 counties and/or cities in Henan Province during the years 2000, 2005, and 2010. Results show that compared with that in 2000, each function in 2010 has obviously improved, with the spatial heterogeneity of economic function being the most evident, social service function being comparatively balanced, and spatial distribution of agricultural production function changing insignificantly. Cluster analysis is adopted to study the major functions of rural regions. Thus, Henan Province is divided into six major function zones, so as to enhance administrative management and developmental policy.

KEYWORDS: Rural territorial function, General Regression Neural Network, Henan Province, Rural development model

1 INTRODUCTION

China is under rapid urbanization, with urbanization rate reaching 52.27% in 2012 (Bai et al., 2014). Rural area owns irreplaceable functions compared with cities in terms of the economic, social, ecological, and cultural aspects (Liu et al., 2011). Rural area also owns evident spatial and chronological heterogeneity (Li et al., 2011; Li et al., 2014; Long et al., 2011). Along with the rapid expansion of construction land, population urbanization, and new countryside construction, many problems such as conflicts between people and land, widening gap between urban and rural areas, homogeneous appearance of the countryside, and imbalanced function of rural area become more and more prominent (Liu et al., 2011). Analysis of the functional distinctions of rural area and its evolution process integrates and differentiates the value contributions and function judgments of a particular rural area in supporting regional development and satisfying social needs; some research also spatially

divides rural area into different functional areas (Li et al., 2011; Liu et al., 2013). Some papers disclose problems of the current rural area, certify their future direction, and promote a commercial, professional, and modern development path for rural areas (Liu et al., 2009; Zhang and Liu, 2008). Different operative models of new rural construction area are also proposed (Cui et al., 2006; Chen et al., 2010).

Research by domestic scholars is carried out in the background of overall urban–rural development, new countryside construction, and other national strategies (Liu, 2007). Therefore, most of the research results are related to the land usage, economic structure, social problems, and plan implementation. Liu et al (2011) start from the perspective of land use and believes that four major functions of land include agricultural production function, economic leading function, ecological conservation function, and social security function. Long et al (2009) define four rural development types, namely agriculture-dominant type, industry-dominant type, commercial and tourist service type, and balanced development type, from an industrial perspective. Long et al (2009) also make rural evaluations based on indicators, including the change rate of cultivated lands, rural population change rate, proportion of agricultural employee, and so on. From the perspective of rural development differentiation, Liu (2002) selects 30 indicators out of 5 indicators, including development level, industrial structure, and development potential; classifies China into six regional types with provinces as the basic unit; and attempts to analyze the reasons behind the regional differences of China's rural development. By adopting the map superposition method, Luo et al (2009) divide China into four function regions, including the East, West, Middle, and Qinghai-Tibet, according to the spatial differentiation of five basic functions such as the agricultural supply, employment, living guarantee, cultural inheritance and leisure, and ecological adjustments. They also plan and locate leading functions and supplementary functions of agricultural industries in different regions.

Through a literature review, the existing findings are mainly evaluated based on economic functions with a lack of study on multi-functional comprehensive zoning (Long et al., 2009; Weng et al., 2002; Yao and Guo, 1992). Previous literature focused on analyzing the current circumstances of functions but rarely revealed the multifunctional evolution patterns (Liu et al., 2010). In most cases, these study set counties act as the basic research unit, while leaving the research system of multiscale comprehensive zoning unconstructed. As for zoning methods, most literature adopts cluster analysis, with the minority using self-organization feature map (SOFM). By establishing a territorial function evaluation system of the rural based on General Regression Neural Network (GRNN) with county level in Henan Province as the basic unit, this paper discloses the spatially distinctive pattern of regional multifunction in rural areas in 2000, 2005, and 2010. This paper also constructs the predominant function zoning of rural regions in Henan.

2 EVALUATION INDEX SYSTEM AND METHODS

2.1 Constructing the evaluation index system

Considering the comparative advantage of Henan Province, functional characteristics of rural region, and the need for function zoning, this paper defines the leading functions of rural regions in Henan into four major functions: economic development function, social service function, agricultural production function, and ecological civilization function. The

paper builds an index system based on altogether 16 indicators, including territoriality, comprehensive natural elements, socioeconomic conditions, relative integration and stability, and the negotiation of scientific nature and operability (Table 1).

2.2 General regression neural network

Due to the interrelations and interactions of different territorial functions of the rural, the operations of each sub-function create a complicated system; thus, the widely used quantitative evaluation methods and linear modeling are not easily applicable to the sophisticated reality (Feng, 2003). The general regression neural network under rapid development recently provided a new approach in solving complicated problems of non-linear modeling. The general regression neural network has many advantages that traditional methods do not have: Network, for instance, can "imitate" and "memorize" any sophisticated "function" relation between the input and output variables, and it can process all kinds of vague and non-linear data (Li, 2004).

3 EMPIRICAL ANALYSIS

3.1 Regional background of Henan Province

Henan Province lies in the hinterland of major grain-producing areas in Huang-Huai-Hai Plain. With fertile soil and a warm climate, it is the birthplace of Chinese people, Chinese civilization as well as the agricultural civilization of China. Henan Province has long been an important strategic resource area of China's agricultural development, due to its distinct geographical location and its agriculturally inclined natural environment. In 2011, the total population of Henan Province was 93.88 million, of whom 55.79 million were rural population, which was 59.5% of the total provincial population.

Henan is a typical agricultural province of huge grain production; its arable land amounts to 7926.4 thousand hectares, constituting 6.51% of the nation's total. As a typical agricultural zone, Henan plays an important role in the spatial patterns of China's urbanization. Studying the territorial functions in Henan's rural area poses huge significance to optimizing the adjustments to locate spatial development and functions of rural areas and to guarantee food security and the development of urban and rural areas as a whole.

3.2 Data sources and method

This paper uses the annual statistics (during 2000, 2005, and 2010) of Henan Province and 109 counties

Table 1. Evaluation index system of rural territorial function.

Function layer	Sub-function layer	Indicator layer	Factor layer
Economic development function	Industry development	Rural GDP per capita	Output value of primary industry/rural population
		GDP per capita	Investment in fixed assets/total population
	Capital input and output	Financial income per capita	General budgetary financial revenue/ total population
		Investment in fixed assets	GDP/total population
Social service function	Employment and distribution	Income level of rural residents	Net income per capita of farmers
		Employment level in rural areas	Rural non-farm employment/ population proportion in rural area
	Public Services	Social Welfare	Number of beds in social welfare institutes/total population* ten thousand people
		Sanitary and health development level	Number of beds in hospitals and clinics/ total population* ten thousand people
Agricultural production function	Cultivated land resources	Land use degree	Area of arable land/regional area
		Grain yield per unit area	Area of arable land/rural population
		Area of cultivated land per capita	Total grain yield/ area of arable land
	Agricultural production supplies	Food Supply per capita	Total grain yield/total population
		Meat production per capita	Total meat production/ total population
Ecological conservation function	Ecological conservation	Ecological land area per capita	Ecological area/ total population
	Ecological service	Ecological land	Ecological land/ regional area
		Ecological system service value	Ecological value/total population

Note: This paper resorts to the land use type in 2010 to study ecological function pattern of rural regions, due to the effects of land use data.

(cities) in Henan spatially as the basic unit for a detailed analysis. All the indicator statistics in this research paper are from *Statistical Yearbook of Henan Province (2000, 2005, and 2010), Statistical Yearbook of China's regions (2000, 2005, and 2010),* and *Statistical Yearbook of Socio-Economy of China's Counties and Cities (2000, 2005, and 2010).* The Spatial statistics come from Prefecture-level City GIS Base Map of Henan Province (Figure 1).

Table 2 presents the descriptive statistics of territorial function evaluation in 2000, 2005, and 2010. As we can see, the overall economic indicators and social indicators of Henan Province have made significant elevation, with rapid economic development and social service development. However, with the widening differentiation of socioeconomic development indicators, the gap between the rich and poor is also increasing. Henan's agricultural production function remained stable, which is in accordance with its status of agricultural production in China.

Figure 1. Spatial distribution of prefecture-level cities in Henan Province.

Table 2.   Descriptive statistics of rural territorial function evaluation index.

Function Layer	Factor Layer	2000		2005		2010	
		Average	Standard deviation	Average	Standard deviation	Average	Standard deviation
Economic development function	Value of primary industry per capita (yuan)	1464	459	2629	835	4029	1157
	Investment of fixed assests per capita (yuan)	426	404	5592	4157	16430	10749
	Financial income per capita (yuan)	150	84	397	421	771	777
	GDP per capita (yuan)	4649	2222	11872	7240	21574	13958
Social service function	Per capita income of farmers (yuan)	2072	438	3416	840	5749	1435
	Numbers of rural employees (ten thousand people)	39.56	16.67	39.75	17.42	40.75	17.73
	Numbers of non-farm employees (ten thousand people)	9.50	5.55	14.44	8.04	18.21	9.32
	Number of beds in welfare houses (ten thousand people)	5.7	3.4	8.9	7	21.8	10.62
	Number of beds in hospitals (ten thousand people)	15.0	9.5	17.4	10.8	21.85	9.9
Agricultural production function	Arable Land Use Degree	0.089	0.038	0.049	0.019	0.10	0.04
	Area of arable land per capita (ha)	0.166	0.0425	0.091	0.023	0.175	0.047
	Grain yield per unit (kg)	3318	970	7385	2012	4139	1010
	Food supply per capita (kg)	475.6	149.2	573.5	191.1	602.4	207.1
	Meat production per capita (kg)	59.0	26.3	83.0	47.2	70.0	43.1
Ecological conservation function	Ecological area per capita (ha)					0.42	0.42
	Ecological land use proportion					0.20	0.03
	Ecological value per capita (yuan)					3672	5086

3.3   Regional multi-function evaluation

3.3.1   Statistical analysis of the regional multi-function evaluation

Statistical analysis based on the four types of territorial functions in Henan rural counties leads to the following findings (and Table 3): (1) From 2000 to 2010, the frequency of EFE changes from concentrated to discrete distribution, whereas the spatial pattern varies from homogeneity to differentiation and then to a more balanced status. (2) The frequency distribution of ECFE is significantly skewed, obviously lacking symmetry. This indicates that the contribution from the ecological conservation function of the rural areas to the counties varies to a relatively large extent from region to region. (3) Except for ECFE, kurtosis of the evaluation on the other three functions is less than zero, indicating a great degree of variation in ecological conservation function of a small number of counties, whereas the majority of the county units are distributed in a centralized way. (4) The coefficients of skewness of both SFE and AFE are less than zero, with their average less than their median, indicating that the distribution of high-value data is more concentrated and that of low-value data is more discrete. (5) The coefficients of skewness of both ECFE and

EFE are more than zero, being distributed in a positively skewed way. The coefficients of skewness of both SFE and AFE are less than zero, which shows a negatively skewed distribution. The deviation coefficient of ECFE is the largest among the four, indicating the most obvious spatial difference.

Table 3. Statistical characteristics comparison of rural territorial functions in Henan Province (2010).

	EFE	SFE	AFE	ECFE
Max.	92.05	85.54	78.23	87.20
Min.	36.65	36.92	22.10	33.01
Median	64.23	66.34	54.17	36.67
Average	67.68	65.16	51.51	44.00
Standard deviation	16.66	10.40	11.79	15.99
Kurtosis	−1.39	−0.04	−0.23	1.83
Coefficient of Skewness	0.14	−0.53	−0.49	1.83

3.3.2 Changes in rural regional multifunctional pattern

Research on the rural multifunctional evaluation in Henan Province shows that the multifunctional index of the rural areas is unevenly distributed, which leads to different patterns of regional multifunction in rural areas:

1 From 2000 to 2010, the rural areas in Henan Province maintained a rapid growth in terms of economic development. In 2010, the EFE of Henan rural areas doubled in general compared with that in 2000, indicating a strengthening of economic development function. However, the pattern of economic development function varies significantly between different regions due to various reasons such as economic base, geographic location, transportation factors, and effects of policies. A shift from a balanced development pattern to a regional uneven development pattern widens the gap between the rich and the poor. For example, rural areas that are surrounding the City Clusters in China's Central Plain Area not only enjoy considerable economic growth but also have the potential for rapid economic development; the economic base is relatively weak in Eastern Henan and Southern Henan, which leads to a slow economic development. In other words, this spatial pattern is characterized by a rapid development at the core and a slow development at the peripherals, with the City Clusters in China's Central Plain Area in the center.

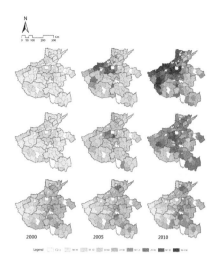

Figure 2. Spatial pattern change of EFE (top), SFE (middle), and AFE (bottom) in Henan Province.

2 Social-service Function Evaluation (SFE) in 2000 generally fell between 29 and 40; the overall social services were at a low level. In 2005, in Henan county-level regions, the average value of SFE rose to 43.66, showing some improvement of social services. The spatial distribution pattern was similar to that in 2000. In 2010, the average SPE increased to 65.15 as social services significantly improved. Seen as a whole, the spatial pattern of SPE distribution is relatively balanced in county-level areas of Henan Province. Spatial differentiation is not obvious.

3 From 2000 to 2010, agricultural production function of Henan county-level regions fluctuated, but the overall performance was on the rise. Compared with 2000, in 2010, the AFE increased in 95 counties, indicating the enhancement of agricultural production function in most of the counties, although AFE in 14 counties declined with the fastest decline in counties in Zhengzhou city. Overall, as a major agricultural province, Henan remained stable in agricultural production function only with an insignificantly slight fluctuation. Regions strong in agricultural production function mainly lay in Eastern Henan or Southern Henan, which were weak in economic development. This phenomenon showed a complementary relationship between spatial pattern and economic development pattern.

3.3.3 Evolution of the rural territorial functions

The spatial pattern of distribution (Figure 3) is a result of a Comprehensive Function Evaluation (CFE) in the county-level regions in Henan Province, by using the General Regression Neural Network to integrate multifunctional indicators of economic development, social services, and agricultural production. In this manner, we obtained the comprehensive functional assessment of spatial pattern distribution. In 2000, the average value of CFE in counties in Henan was 33.71. The counties that scored less than 40 in CFE accounted for 89.00% of the total area of Henan. The overall rating was not high but had a balanced spatial distribution. In 2005, the average CFE increased to 48.59; counties with CFE less than 40 were 24.77% in total, which was mainly in Eastern and Western Henan. Counties with CFE between 40 and 60 amounted to 57.80% of Henan's total area, most of which were located in the surroundings of Zhengzhou City or in Northern and South Henan. Overall, 17.40% had a CFE more than 60, mainly in Zhengzhou City, Jiaozuo City, and Jiyuan City. The coefficient of skewness was relatively big, showing an obvious spatial agglomeration surrounding the City of Zhengzhou. By 2010, the rural comprehensive function in Henan significantly improved, with the average CFE rising to 71.51 (the highest 89.17 in Xingyang City; the lowest 45.95 in Nanzhao country). Overall, 87.16% of Henan's total area scored more than 60 in CFE, with a skew from the north to the south. However, the coefficient of skewness was relatively small, indicating a balanced spatial development pattern.

To sum up, back in 2000, the county-level economic development function and social services function in Henan Province as a whole were well balanced. However, the heterogeneity in agricultural production function was so obvious that the comprehensive function was significantly affected by AFE. By 2005, Henan county-level economic development was balanced not only with an increase in regional differences but also with an enhancement in social services as well as agricultural production function. There seemed no prominent changes in spatial pattern compared with that in the year 2000. So we can see that the comprehensive function of the rural was more affected by the dual impact of agriculture and economy, which led to the formation of a CFE central-high, peripheral-low pattern with Zhengzhou City at the core. In 2010, the regional deference in

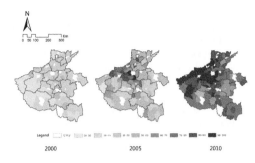

Figure 3. CFE spatial patterns of county-level areas in Henan in 2000, 2005, and 2010.

county-level economic development became notable. Meanwhile, social services function was strengthened. Agricultural production function fluctuated at a certain degree. Three evaluation indicators together exerted an influence on the Comprehensive Function Evaluation of the rural areas, forming a balanced distribution of the rural territorial functions.

3.4 Pattern of zoning based on predominant territorial functions

3.4.1 Predominant rural territorial functions recognition method

We first use "R-type cluster analysis" to process the data from the Rural Regional Multifunction Evaluation of Henan Province in 2010. Then, we use "Q-type cluster analysis" to process the data from 109 counties (cities) in Henan Province, during which the following methods are applied: Average Linkage method, Within-groups Linkage method, Nearest Neighbor (Single Linkage) method, and Furthest Neighbor (Complete Linkage) method. We finally choose the Furthest Neighbor (Complete Linkage) method to take the stepwise cluster analysis and use squared Euclidean distance to measure the inter-cluster distance of every pair. The results of cluster analysis indicate that the counties in Henan can be divided into six types of functional areas based on the rural regional multifunction (Figure 4).

The overall features concluded from results of the clustering analysis are as follows: (1) A small proportion of county-level units in Henan Province show large heterogeneity whereas the majority are

Table 4. Descriptive statistics of EFE, SFE, and AFE in Henan Province.

	2000		2005		2010	
	Average	Standard Deviation	Average	Standard Deviation	Average	Standard Deviation
EFE	29.76	2.22	47.07	15.72	67.68	16.66
SFE	34.81	4.31	43.66	9.57	65.16	10.40
AFE	45.98	9.48	47.20	10.07	51.51	11.79

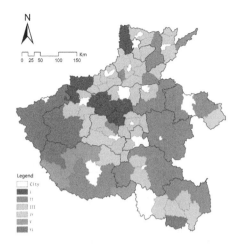

Figure 4. Spatial distribution of clustering at county scale in Henan Province.

concentrated, showing limited difference. In terms of six types of functional areas, Sanmenxia City, Luoyang City, Nanyang City, Xinyang City, and other 15 counties constitute three of them; the remaining 90 counties constitute the other three. (2) There exists a divergence between county-level units within the same city. Even though these counties are within the same administrative city, there are certain differences in terms of functional clustering. As can be seen from Figure 4, except for Sanmenxia and Zhoukou, all other cities have more than one rural function. (3) There is consistency to a certain extent between

the rank in the assessment of comprehensive function of rural areas and the type of cluster; for example, all the 10 counties in Type I are ranked in top 30; Type II counties are ranked from 60 to 108; Type III counties are mainly ranked from 20 to 50; and no counties in Type IV and Type V are ranked in top 50.

3.4.2 *Predominant rural territorial functions recognition*

According to the results of the cluster analysis, 109 counties of Henan Province are divided into six typical types. Four basic indexes and comprehensive index averages in Table 6 show that the listed six types of counties are significantly different from other types, which is the basis for analyzing functional features of all types of counties.

Type I: Core economic development zone. This zone includes some county-level units in and around Zhengzhou City, as well as Linjiyuan City and Anyang City. It ranks at the top not only in Comprehensive Function

Index column but also in Economic Function and Social Function index. It is a key area not only in Index column but also in Economic Function and Social Function index. It is a key area with strong economic functions and comprehensive livelihood construction.

Type II: Agricultural production safeguarding zone. This area is mainly located in Eastern Southwestern Henan. It has a well-developed performance in agricultural function whereas it lags behind the rest in

Table 5. ANOVA results.

		Sum of squares	df	Mean square	F	Statistical significance
Agricultural Function	Between groups	9423.357	5	1884.671	34.806	.000
	Within group	5577.252	103	54.148		
	Sum	15000.610	108			
Economic Function	Between groups	21357.996	5	4271.599	51.122	.000
	Within group	8606.310	103	83.556		
	Sum	29964.306	108			
Social Function	Between groups	5544.973	5	1108.995	18.626	.000
	Within group	6132.726	103	59.541		
	Sum	11677.699	108			
Ecological Function	Between groups	24484.498	5	4896.900	161.474	.000
	Within group	3123.608	103	30.326		
	Sum	27608.106	108			

Table 6. Average of territorial function Index for 6 Types of County-level Areas.

Types	Economic Function Index	Social Function Index	Agricultural Function Index	Ecological Function Index	Comprehensive Function Index
I	89.46	74.23	35.40	39.62	85.44
II	51.52	58.66	57.02	36.29	61.78
III	76.57	70.18	57.05	38.32	78.70
IV	58.27	70.90	40.63	56.69	63.73
V	70.73	44.40	39.43	80.88	61.18
VI	83.52	68.34	32.97	83.71	78.94

economic function. Thus, it ranks low in comprehensive function. It acts as an important agriculture security function area for Henan Province.

Type III: function-improving zone for rural areas. This zone is located in the north and central of Henan, lying in the buffer area of the core economic development and agriculture-safeguarding zone. Seen from the aspect of rural territorial functions, it is the best-developed zone in Henan with a satisfying overall economic development, a higher degree of social insurance, and a strong agriculture production function.

Type IV: Model zone of livelihood and social services. This zone includes Lushan County of Pingdingshan City, Ruyang County and Luoshan County in the surroundings, and Guangshan County and Shangcheng County of Xinyang City. These county-level units are located between the mountains and plains, with the latter strong in economic development. Although their economic development is lagging behind, their social services are well developed. What is more, this zone is endowed with a certain degree of conservation function.

Type V: Economic restructuring and development zone. This zone lies within Nanyang City, including Nanzhao County, Neixiang County, Xichuan County, and Tongbai County. The terrain is relatively flat in Nanyang City, so agricultural production function in Nanyang is relatively lagging behind. But its conservation function is at a moderate level; its economic development function is at an average level; and its social security function is relatively low. Seen from a general rural development trend, this zone is in a special transition period of rural economic development. To conserve its ecological environment, we should pay attention to environment protection when seeking economic development. Simultaneously, we should also focus on strengthening social and livelihood functions.

Type VI: Nature conservation areas. The area is mainly mountainous county, mainly including western Henan Sanmenxia, Luoyang City, part of the county, and the new county south of Henan Xinyang.

With mountains and rivers within its scope, rich in biodiversity, it is an important ecological barrier in Henan Province. Meanwhile and also in the forefront of economic development function, social services are more forward, and the terrain is limited, being the most backward agricultural function. Therefore, to ensure the economic development function is strictly controlled, we should prevent the destruction of the ecological environment. In order to protect the ecological function of the priority primary goal, we should restrict access to the high energy consumption and high pollution industries and projects in order to build a "society" as the focus.

4 CONCLUSION AND DISCUSSION

This paper establishes an index system for evaluating rural territorial functions, including agricultural function, social function, economic function, and ecological function. By establishing a model based on the General Regression Neural Network (GRNN), this paper also makes a comprehensive evaluation and analyzes the chronological evolution process in terms of the rural territorial functions of 109 counties and/or cities in Henan Province during the years 2000, 2005, and 2010. The results show that in 2000 economic development and social service function of counties in Henan were more or less balanced, and the characteristics of spatial heterogeneity of agricultural production function are distinctive. In 2005, there was a widening economic development difference in various counties; social service function and agricultural production function improved with no evident change in spatial distribution. In 2010, there were evident spatial differences of economic development function in counties in Henan, with social service function being strengthened and agricultural production function fluctuating. All these functions contribute to a complement with economic distribution.

Furthermore, this study makes a systematic clustering of territorial multifunction evaluation in the year 2010 in rural areas. Analysis shows that county units in Henan Province are characterized by a strong heterogeneity of 19 counties, including Sanmenxia, Luoyang, and Xinyang, which are less concentrated distinctions of major counties and different major functions of county units within the same administrative region of a city. Cluster analysis divides major function districts of rural areas in Henan province into six categories, namely, economic development core area, agriculture production support area, rural function improvement area, people's livelihood model area, economic transformation development area, and natural ecological protection area. This paper suggests that different development goals and development strategies should be regarded according to different major function areas, so as to further achieve a coordinated development of urban and rural areas.

This paper provides a preliminary study on the identification, evaluation, and evolution of rural territorial function in Henan Province of China. However, due to the restriction of data, it is rather hard for us to establish an evaluation framework at the township level that can better depict the disparities of territorial function of rural areas. Moreover, the mechanism of function forming and value realizing should be further discussed in the follow-up studies.

REFERENCES

Bai Xuemei, Shi Peijun, Liu Yansui. Realizing China's urban dream. Nature, 2014, 509(1799):158–160.

Cao Shuyan, Xie Gaodi. The evolution of the perspective of functional zoning driven by development issues in China. Resources Science, 2009, 31(4):539–543. (in Chinese)

Chen Wenke, Liu Tianxi, Chen Hanhua, et al. Typical models and corresponding paths of new countryside construction in central China: a case study of Hunan Province. Chinese Rural Economy, 2010(5):4–14.

Cui Ming, Qin Zhihao, Tang Chong, et al. Categorization and mode study of new village construction in China. City Planning Review, 2006, 30(12):27–32.

Feng Zhipeng. General regression neural network based prediction of time series. Journal of Vibration, Measurement & Diagnosis, 2003.6. 23(2):45–48. (in Chinese)

Li Jie, Wang Ke, Wang Hang. Highway freight prediction method based on GRNN. Computer and Communications, 2004.6.30 (24) 112–117. (in Chinese)

Li Yurui, Liu Yansui, Long Hualou. Study on the pattern and types of rural developmentin the Huang-Huai-Hai region. Geographical Research, 2011, 30(9), 1637–1647. (in Chinese)

Li Yurui, Wang Jing, Liu Yansui, et al. Problem regions and regional problems of socioeconomic development in China: A perspective from the coordinated development of industrialization, informatization, urbanization and agricultural modernization. Journal of Geographical Sciences, 2014. 24(6):1115–1130.

Liu Hui. Study on the regional disparities of rural development in China. Geography and Territorial Research, 2002, 18(4):71–75. (in Chinese)

Liu Yansui, Liu Yu, Chen Yufu. Territorial multi-functionality evaluation and decision-making mechanism at county scale in China. Acta Geographica Sinica, 2011, 66(10):1379–1389. (in Chinese)

Liu Yansui, Zhang Fugang, Zhang Yingwen. Appraisal of typical rural development models during rapid urbanization in the eastern coastal region of China. Journal of Geographical Sciences, 2009, 19(5):557–567.

Liu Yansui. Rural transformation development and new countryside construction in eastern coastal area of China. Acta Geographica Sinica, 2007, 62(6):563–570. (in Chinese)

Liu Yu, Liu Yansui, Guo Liying. Comprehensive evaluation and optimization strategy of the territorial function for grain production: A case of the area along Bohai Rim in China. Progress In Geography, 2010, 29(8):920–926. (in Chinese)

Liu Yu, Liu Yansui, Guo Liying. SOFM-based functional subareas of rural area along the Bohai Rim in China. Human Geography, 2013, (3):114–120. (in Chinese)

Long Hualou, Liu Yansui, Zou Jian. Assessment of rural development types and their rurality in eastern coastal China. Acta Geographica Sinica, 2009, 64(4):426–434. (in Chinese)

Long Hualou, Zou Jian, Pykett J et al. Analysis of rural transformation development in China since the turn of the new millennium. Applied Geography, 2011, 31(3): 1094–1105.

Luo Qiyou, Tang Huajun, Tao Tao, et al. Research on regional differentiation and orientation of agricultural multifunction in China. Research of Agricultural Modernization, 2009, 30(5):519–523.

Weng Lili, Li Yongshi, Wang Xiaowen, et al. The demarcating of the rural economy type of Fujian Province. Journal of Fujian Teachers University (Philosophy and Social Sciences Edtion), 2002(3):48–53. (in Chinese)

Yao Jianqu, Guo Huancheng. A rural functional classification of the Huang-Huai-Hai Region and their areal development modes. Geographical Research, 1992,11(4):11–19. (in Chinese)

Zhang Fugang, Liu Yansui. Dynamic mechanism and models of regional rural development in China. 2008, 63(2):115–122.

Energy, Environment and Green Building Materials – Sheng (ed.)
© 2015 Taylor & Francis Group, London, ISBN 978-1-138-02718-3

Thermal environment simulation analysis of a new regenerated glass pumice external wall insulation building in hot summer and cold winter zone

S. Shu, Q. Gu, X. Zhou & B. Li

Wuhan University of Technology, School of Civil Engineering and Architecture, Wuhan, China

ABSTRACT: A certain office building in Wuhan city was selected in this paper as an example of the hot summer and cold winter zone. The thermal environment simulation of the building was calculated by energy-saving analysis software Ecotect Analysis when the building external walls were, respectively, used by glass pumice external insulation system, glass pumice concrete block self-insulation system, and specified by limit requirements. The results show that building energy consumption, heat losses of envelope, and total discomfort value of two glass regenerated pumice external insulation systems can meet the requirements in specification. And compared with the pumice concrete block self-insulation external wall system, the glass pumice external insulation external wall building can be more effective for energy consumption, decrease envelope heat losses proportion, and reduce discomfort value caused by cold. The results will provide a reference to promote this environmentally material used for the wall insulation system in the energy conservation building in the region.

KEYWORDS: regenerated waste glass pumice; pumice board external insulation exterior wall system; pumice concrete block self-insulation exterior wall system; Thermal environment simulation analysis.

1 INTRODUCTION

Regenerated glass pumice is a kind of artificial porous lightweight material made from waste glass after a series process of crushing, combustion, and foaming. The manufacturing patent of the product was authorized in June 2003. By controlling the production process, we can obtain different performance pumice materials that have a wide range of uses in many fields. Using new glass pumice as insulation material, we can not only reuse resource but also improve the performance of wall insulation.

First, this article experimentally measured the thermal conductivity of glass pumice and glass pumice concrete block. Then, from a thermal performance analysis perspective we can design pumice external insulation and pumice concrete block self-insulation, two systems to be compared with limit requirements. Citing a certain office building in Wuhan as an example, conducting a thermal environment simulation analysis for three situations by Ecotect Analysis, the results will provide a reference basis for practical application of this new environmentally regenerated material to the wall insulation system in energy conservation building.

2 PROJECT SUMMARY

The office is located in Wuhan city, toward north, with a total construction area of 5245m² that is subjected to an eight-layer frame structure; the building height is 36m.

Plan layout diagram of office second floor is shown in figure 1.

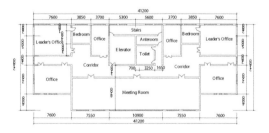

Figure 1. Plan layout diagram of office second floor.

3 MATERIAL PERFORMANCE OF GLASS REGENERATED PUMICE EXTERIOR WALL INSULATION SYSTEM

According to Technical Regulation of Lightweight Aggregate Concrete, both coarse and fine grain in the mixture ratio design of the regenerated pumice concrete test cube adopt glass pumice, and the design strength of concrete is 5MPa to ensure that the self-insulation glass pumice concrete block can also meet the basic mechanics performance of the filler wall. The thermal conductivity of glass pumice and glass pumice concrete block was measured by the thermal conductivity detector, as shown in figure 2; the results are shown in table 1. In order to enable the heat preservation performance of optimal glass pumice, this paper takes the category of pumices with the smallest thermal conductivity in table 1 as the external wall thermal insulation material.

Regarding EPS board external insulation system, a thickness of 30mm was applied to the regenerated

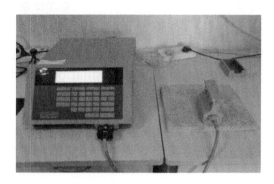

Figure 2.   Measurement process of pumice.

glass pumice board exterior insulation system. The width of the glass regenerative pumice concrete insulation block is given by 200 mm, referring to the standard size of the air concrete block in Autoclaved Aerated Concrete Block. Some other material parameters of the insulation system are shown in Table 2, with reference to table 2 of Analysis and Explanation of Green Building Example Used by Autodesk Ecotect Analysis 2011. Surface layer, leveling layer, and other tectonics that have a small influence on the exterior wall system's thermal performance are not listed here, aiming at simplifying the modeling.

Table 1.   Measured result of pumice sample thermal conductivity.

Sample Name	Thermal Conductivity
	W/m•K
Glass Pumice	0.104-0.133
Glass Pumice Concrete block	0.246

4 ANALYSIS OF BUILDING THERMAL ENVIRONMENT

The heat transfer coefficient of building roof, windows, and external wall in specification limits are

Table 2.   Material parameters of each layer in different external wall insulation systems.

Constructional System	Materials	Thickness	Thermal Conductivity	Specific Heat	Density
		mm	W/m•K	J/(kg) •K	kg/m³
External Wall External Insulation System	Cement Mortar	20	0.93	1050	1800
	Air Concrete Block	200	0.22	1050	700
	Cement Mortar	10	0.93	1050	1800
	Glass Pumice Board	30	0.104	837	690
	Mixed Mortar	10	0.87	1050	1700
External Wall Concrete Block Self-Insulation System	Cement Mortar	20	0.93	1050	1800
	Pumice Concrete Block	200	0.246	920	890
	Mixed Mortar	10	0.87	1050	1700

Table 3. Heat transfer coefficient of building envelope structure.

| Constructional System | Heat Transfer Coefficient of Building Envelope Structure | | |
	Roof	External Wall	Window
	W/m²•K	W/m²•K	W/m²•K
Glass Pumice External Insulation	0.7	0.7	2.5
Glass Pumice Concrete Block Self-Insulation	0.7	0.98	2.5
Specification Limits	0.7	1.0	2.5

given by Public Building Energy Efficiency Design Standards, whereas heat transfer coefficient of external wall with pumice external insulation and pumice concrete block self-insulation system are calculated by Ecotect Analysis, as shown in table 3.

According to the calculation conditions and parameters in Public Building Energy Efficiency Design Standards, thermal environment simulation was carried out for the whole office building by Ecotect Analysis. Calculation model of the office building, as shown in figure 3. Meteorological data in the process of thermal environment simulation adopts CSWD format, which is the domestic measured data coming from Tsinghua University and China meteorological administration.

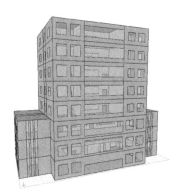

Figure 3. Calculation model of the office.

4.1 Simulation analysis of energy consumption

Building energy consumption in three situations was, respectively, analyzed by Ecotect Analysis, and calculated data are shown in figure 3.

Figure 3 depicts the annual total energy consumption of a building with glass regenerated pumice external insulation and glass regenerated pumice concrete block self-insulation, which are, respectively, 174712928Wh and 179724224Wh: Both meet

the specification limited building energy consumption value of 180000128Wh. Due to the smaller heat transfer coefficient, the heating energy consumption value of glass regenerated pumice external insulation is smaller, which embodies the glass pumice external insulation system and has obvious advantages for heat preservation performance in winter. The cooling energy consumption values of two insulation constructional systems are similar, so the glass pumice external thermal insulation system can effectively save energy consumption for the building.

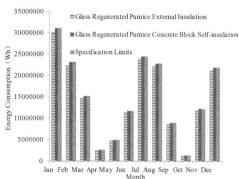

(a) Heating, cooling, and total energy consumption

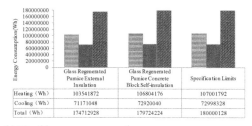

(b) Monthly energy consumption

Figure 4. Energy consumption under different thermal insulation structures.

4.2 Analysis of discomfort value

Ecotect Analysis provides three calculation methods of discomfort. Flat Comfort Bands method was used in this paper, which evaluates the strength value of discomfort through degree hours by calculating the degree hours of the regenerated glass pumice external insulation system, the glass regenerated pumice concrete block self-insulation system, and the specification limits, respectively. The results of annual discomfort hours are shown in figure 4.

Figure 4 depicts that building discomfort values caused by too cold are more than those caused by too hot. The discomfort values of a building with glass pumice external insulation and glass pumice concrete block self-insulation external wall are, respectively, 13206.9 and 13251.2 degree hours, which can meet the specification limited value of 13254.4 degree hours. Discomfort caused by extreme cold of glass pumice external insulation is better than glass pumice concrete block self-insulation. Meanwhile, discomfort values caused by extreme heat of glass pumice external insulation external wall building is 3593.1 degree hours, which is a little higher than the specification limited value of 3560.5 degree hours. Due to

the heat transfer coefficient of glass pumice external insulation system being smaller, when the outdoor temperature is fitting, it is not suitable for indoor heat to be transferred, which leads to the indoor and outdoor heat not being effectively exchanged.

4.3 Analysis of passive gains breakdown

Followed by the glass regenerated pumice external insulation system, passive gains breakdown of glass regenerated pumice concrete block self-insulation system and the specification limits constructional system have been analyzed; the data throughout the year are shown in Table 4.

The conclusions from Table 4 show that building heat losses of ventilation under two exterior wall insulation systems are, respectively, 62.8% and 61%. It is apparent that ventilation is the main way leading to heat losses of building, factors influencing the proportion related to the ratio of window to wall and the air tightness of windows. There are many windows in this office building, so the proportion of ventilation heat losses is very large. The heat conduction of building envelope is another main route of heat loss; 35.5% and 37.3% are the conduction heat losses under two exterior wall insulation systems, whose heat losses are mainly influenced by heat transfer coefficient of external envelope. Due to the smaller heat transfer coefficient, the envelope heat losses proportion of glass pumice external insulation building is less than glass pumice concrete block self-insulation building.

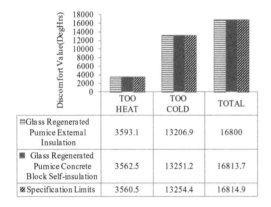

	TOO HEAT	TOO COLD	TOTAL
Glass Regenerated Pumice External Insulation	3593.1	13206.9	16800
Glass Regenerated Pumice Concrete Block Self-insulation	3562.5	13251.2	16813.7
Specification Limits	3560.5	13254.4	16814.9

Figure 5. Discomfort value of different constructional systems.

5 CONCLUSION

Except for the annual discomfort value caused by extreme heat of pumice concrete block self-insulation building being slightly higher than the standard limits, the energy consumption, heat losses of envelope, and total discomfort value of two glass regenerated pumice external insulation systems can meet the requirements in specification.

Table 4. Passive gains breakdown of different constructional systems (%).

Constructional System / Project	Glass Pumice External Insulation		Glass Pumice Concrete Block Self-Insulation		Specification Limits	
	Heat Losses	Heat Gains	Heat Losses	Heat Gains	Heat Losses	Heat Gains
Conduction	35.5	14.5	37.3	15.1	37.4	15.1
Solar	0	28.6	0	30.0	0	30.0
Ventilation	62.8	27.2	61.0	26.3	60.9	26.2
Internal	0	26.3	0	25.4	0	25.4
Inter-Zonal	1.7	3.3	1.7	3.2	7.9	3.2

Compared with the pumice concrete block self-insulation external wall system, although discomfort value caused by extreme heat of glass pumice external insulation external wall building is a little higher, it is better for energy consumption, decreases envelope heat losses proportion, and reduces discomfort value caused by cold for the building.

ACKNOWLEDGMENTS

This research was supported by the Urban Construction Project of Wuhan City (Grant No. 2013CDA153) and the Graduate Student Innovation Funds Project of Wuhan University of Technology (Grant No.2014-zy-083), whose assistance is highly appreciated here.

REFERENCES

Ren S., Gu Q., Wang Y.& Luo H., 2013. Analysis of thermal performance of a new regenerated glass pumice external wall insulation system in hot summer and cold winter zone. *Advanced Materials Research*, Proceeding of 3rd International Conference on Structures and Building Materials, Guiyang, March, 2013.

Gu T.S., Xie L.Y.& Chen G., 2006. Energy saving of building and heat preservation of wall. *Engineering Mechanics*: 167–184.

Autodesk Inc, Bomu China. 2012. *Analysis and explanation of green building example used by Autodesk Ecotect Analysis 2011*. Beijing: Electronic Industry PRESS: 123–168.

Technical regulation of lightweight aggregate concrete. 2002. Beijing: China Architecture & Building PRESS.

Autoclaved aerated concrete block. 2006. Beijing: China Architecture & Building PRESS.

Public building energy efficiency design standards. 2005. Beijing: China Architecture & Building PRESS.

Zhang Q.Y. & Yang H.X., 2011. *Typical meteorological database handbook for buildings*. Beijing: China Architecture & Building PRESS.

Gao Y. & Zhang J., 2010. Thermal insulation performance of aerated concrete self-insulation compared with polystyrene board exterior insulation wall system. *New Building Materials*: 37(3):48–52.

Chen S.& Di H.F., 2001. Thermal environment analysis of several buildings in Beijing city. *Xi An University of Architecture and Technology*: 33 (3):255–260.

Energy, Environment and Green Building Materials – Sheng (ed.)
© *2015 Taylor & Francis Group, London, ISBN 978-1-138-02718-3*

Analysis on the space–time difference characteristics of inbound tourism in Gansu province

J.P. Yan & X.Z. Wang

College of Geographic and Environmental Science, Northwest Normal University, Lanzhou, Gansu, China

ABSTRACT: Based on the data of inbound tourism travelers of 14 cities in Gansu province from 2001 to 2012, the evolution of space–time difference characteristics of inbound tourism was analyzed by using market share, year border concentration index, geography concentration index, and concentration ratio. The results show that in recent years, the inbound tourism of Gansu province developed slowly and is influenced by the international environment; inbound tourism spatial distribution is unbalanced with a high geographical concentration and mainly concentrated in 8 major tourist cities located in Hexi area along the Silk Road, which has rich ethnic customs, local culture, and exhibits political, economic, and cultural development. However, inbound tourists of Tianshui and Zhangye have relatively large annual fluctuations and Gannan presents recession signs; With the increase of overall economic development, strength, and implementation of related tourism industry policy, the spatial distribution of inbound tourists will gradually spread, move toward stability, and, ultimately, be balanced in some degree of aggregation levels.

KEYWORDS: Gansu province; inbound tourism; spatial agglomeration; geography concentration index; year border concentration index

1 INTRODUCTION

Inbound tourism is an important component of the tourism industry, an important indicator to measure a country and the local tourism comprehensive development level, and a good way to improve the tourism economic benefits and increase tourism income. Simultaneously, it is also expanding to the outside world, which is an important means to promote international cooperation and exchange[1]. But because of differences in resource endowments, tourism infrastructure, and the socioeconomic development level, the tourism development in each area shows differences as well as an imbalance in space with the space–time dynamic change[2]. Strengthening research differences in spatiotemporal distribution of inbound tourism at the national and regional level, developing the right industrial policy, determining the regional development priorities, and improving inter-district coordination mechanisms has important theoretical and practical significance. Therefore, scholars have extensively studied the spatiotemporal distribution of inbound tourism and its evolution from different aspects. Based on research, the yardstick can be divided into national[3], regional[4], provincial[5], and municipal[6] areas. From the research point of view, the main aspects are discussed from the source country of the origin market[7], the factors[8] of inbound tourism and inbound tourism consumption characteristics[9], and so on. From the quantitative evaluation indicators, the evaluations are Preference Scale[10], geographic concentration index, and Theil index. Visible, relevant scholars try to conduct a multifaceted in-depth study on inbound tourism from different research perspectives, but much of China's economy is relatively developed for the central and eastern regions. The study of the economy in the relatively backward Northwest is relatively small, especially for the northwest–southeast direction of the narrow strip landform in Gansu Province.

Located in the northwest of China's Gansu Province, economic backwardness, 2012 GDP was 565.02 billion yuan, is located in the first 27, but with the beautiful natural scenery, a wide variety of ancient cultural relics, and unique contemporary folk customs, large-scale development has presented a good opportunity. Gansu Province inbound tourism has experienced rapid development. In 2007, with 331 500 inbound tourists trips, tourism foreign exchange earnings amounted to $ 70,210,500, respectively, in 2002 and 1.57 times 1.49 times, in 2003 and 2008 affected "SARS" epidemic and the financial crisis, subject to greater impact inbound tourism followed by gradual recovery. In 2012, the number of inbound tourists reached 102,000. In this paper, spatiotemporal differences in income distribution and the consolidation of the 14 cities (state) with regard to tourist arrivals in Gansu Province in 2006–2012 try to depict market share, quantitative evaluation, interannual concentration index, and geographic concentration index inbound tourist flow in Gansu. Their evolution is discussed in order for the relevant government departments and enterprises to develop tourism-targeted, effective regional tourism development policy and planning as well as to provide a useful reference.

2 DATA SOURCES AND METHODOLOGY

2.1 Data sources

The number of inbound tourists is an important indicator of the level of regional and international tourism development. This paper collected data pertaining to 14 cities (prefectures) of tourist arrivals in Gansu Province in 2001–2012, as well as aimed at understanding the geographical distribution and variation of inbound tourist traffic. Relevant data are collected from the years 2001 to 2012, "Gansu Yearbook," Gansu Tourism Bureau network, and related literature.

2.2 Research methods

Based on the time and spatial dimensions, the use of market share, annual concentration index, geographic concentration index, the concentration ratio of the 2001–2012 annual distribution pattern of inbound tourist flow in Gansu Province, and dynamic spatiotemporal evolution of the empirical analysis are depicted.

Market share is the basic indicator of market structure analysis, which reflects the market segmentation of the market share of each sub. In this paper, inbound tourists during a year of municipal (state) sectional data analysis of market structure, the use of inbound tourists cities (prefectures), and changes in the structure of time series data analysis are indicated.

Annual concentration index refers to the number of visitors in a given period each year for centralized, discrete, or uniform distribution metrics.

$$V = \sqrt{\dfrac{\sum\limits_{i=1}^{N}(T_i - \overline{T})^2}{N}}$$

(1)

where V is the number of inbound tourists travel interannual concentration index, V, a smaller value, indicates the number of inbound tourists whose interannual variability is smaller and more stable over time, whereas the number of inbound tourists is instable, and the magnitude of interannual variability is bigger. T_i is a city (state) i-year period, where the number of visitors in the city's accounting study is the percentage of total visitors (state) molecule value; $\overline{T}$ stands for the study period in a city (state), where the annual average number of visitors accounted for within the study period city (state) is the percentage of total visitors molecular value; and N is the number of years included in the study period.

Table 1. Inbound tourist flow in every city and state of Gansu province.

	2006	2007	2008	2009	2010	2011	2012
Lanzhou	5.82	5.95	1.33	1.26	2.01	1.22	2.24
Jiayuguan	2.53	1.99	1.29	1.20	0.61	0.74	0.76
Jinchang	0.04	0.00	0.02	0.02	0.02	0.05	0.05
Baiying	0.06	0.04	0.01	0.01	0.01	0.01	0.01
Tianshui	1.78	2.16	0.07	0.03	0.02	0.41	0.23
Wuwei	1.87	0.72	0.31	0.58	0.77	0.79	0.55
Zhangye	0.01	0.78	0.12	0.02	0.06	0.53	0.55
Pingliang	0.17	0.09	0.03	0.04	0.00	0.07	0.12
Jiuquan	9.33	10.46	4.52	2.65	3.38	4.39	5.19
Qingyang	0.03	0	0	0.02	0	0	0.01
Dingxi	0.04	0.04	0.08	0.01	0.02	0.02	0.04
Longnan	0.07	0.02	0	0	0	0	0
Linxia	0.23	0.21	0.06	0.07	0.05	0.24	0.08
Gannan	8.34	10.63	0.47	0.16	0.05	0.62	0.38
Total	30.33	33.09	8.32	6.07	7.02	9.09	10.20

Geographic concentration index is used to measure the concentration of geographic phenomena in space or time distribution, and the distribution may reflect the time variation of inbound tourists and city[11]. The geographic concentration index value is closer to 100; the more concentrated the distribution of tourists, the stronger the geographic clustering. The formula is

$$G = 100 \times \sqrt{\sum_{i=1}^{n} (\frac{X_i}{X})^2} \qquad (2)$$

where G is the geographic concentration index that reflects the inbound tourist flow in Gansu province cities (prefectures) of clustering, X_i for each i, Gansu Province, the number of tourists (state) entry, X is the total tourist arrivals in Gansu Province, and n is the city (state) number.

Concentration ratio is an important indicator of market monopoly. Usually, ordering the front of several major market shares reflects their monopoly[13]. The general indicators used are CR1, CR4, and CR8. In this paper, CR1 and CR8 are used to characterize the first degree, before the eight cities (prefectures) tourist reception status inbound tourists destination.

3 SPATIOTEMPORAL DISTRIBUTION OF INBOUND TOURIST FLOW DIFFERENCES AND EVOLUTION

3.1 Overall change in the size of inbound tourist flow

Overall, inbound tourism in Gansu Province can be divided into two phases: 2001–2007 continued to rise (except for 2003), and the slow recovery after 2008 plummeted. During the "Fifth" period, Gansu Province takes advantage of good opportunities brought by western development; tourism serves as a new economic growth point to be nurtured. Enforcement points, areas, and groups, with the development of strategies and X skeleton key development strategy, improve infrastructure and services;

are rich forms of tourism, tourism product structure adjustment, and spatial layout; and form the center of Lanzhou, West, Central, East Third District linkage, seven scenic groups' overall development, six tours' radial development pattern, forming a tourism economic zone along the Silk Road, the Yellow River Basin distribution. In 2005, inbound tourists were 28.8484 million passengers, but in 2003 due to the "SARS" epidemic affecting the province, inbound tourism suffered a temporary decline in 2004, which has basically been restored. Because of inbound tourism during the infancy of Gansu, Lanzhou inbound tourism focus, Jiuquan, Jiayuguan, Gansu, Linxia, and other cities' (states) share of a smaller market share was observed in 2003 despite a temporary decline in inbound tourism. However, Tianshui, Lanzhou, Wuwei, and Zhangye were four outstanding tourist cities that attracted a large number of international tourists in 2007, amounting to 33.0878 million passengers inbound tourists. The impact of the international financial crisis in 2008, the domestic "3.14" incident, and the impact of the earthquake "5.12" resulted in the province's inbound tourism slump, further developing the tourism market and aggressive marketing campaign to make inbound tourism increase. However, due to the overall economic strength of the impact of Gansu, the further development and investment in the construction of service facilities for outstanding tourism resources was lacking, and the inbound tourism recovery was slow.

3.2 Differences in the spatial distribution of inbound tourism and Cause Analysis

As can be seen from Figure 2, Gansu exhibited an unbalanced spatial distribution of inbound tourists that was quite different. Lanzhou, Jiayuguan, Jiuquan, and Gansu inbound traffic was in a leading position; Tianshui, Wuwei, and Zhangye were in a mid-size passenger Linxia immigration; Jinchang, silver, Pingliang, Qingyang, Dingxi, and Gansu immigration traffic continued to be small. In addition, to

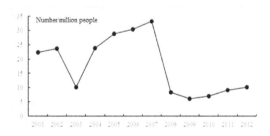

Figure 1. Total number of changes in inbound tourism in Gansu province.

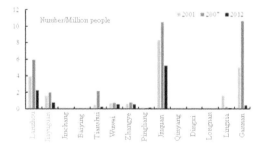

Figure 2. Inbound travelers received in every city and state of Gansu province.

357

Linxia, but in more than a medium-sized city (state) of inbound tourists in 2007 compared with 2001 increased to varying degrees, but in 2012 compared with 2007 they were significantly attenuated.

Northwest Gansu Province experienced a lack of a variety of factors and regional economic development to Southeast narrow banded, tourist attractions were scattered, accompanied by effective regional cooperation in tourism development mechanism, tourism resource endowments, traffic location, and so on together to depict uneven spatial distribution in various regions of the province with regard to inbound tourism. Jiuquan, Gansu Province is the largest city, accounting for 42% of the area of Gansu Province. Dunhuang is an international tourist attraction; it is characterized by a prosperous economy, convenient transportation, and tourism service facilities that are relatively complete. The number of inbound tourists ranked first in Gansu Province. Lanzhou is the capital city of Gansu Province, northwest regional transport hub, and a tourist center, with a 4A level scenic spots, 3A-class attractions 2, and 2A-class attractions 2. In 2004, it successfully created for China an excellent Tourism City, and the market share of inbound traffic rate was about 20%. Jiayuguan West Ming Dynasty Great Wall is the starting point of the world's cultural heritage; China's excellent Tourism City has a higher grade of tourism resources, 5A level scenic spots 1, 4A level scenic spots 3, 3A-class attractions, and an entry traffic market share of around 8%. Tianshui is located in Shaanxi, the midpoint of Gansu, Sichuan border region, and two major cities of Xi'an. Lanzhou is in the southeastern economic and cultural logistics center, the largest tourist attraction, 5A-class attractions 1, 4A level scenic spots 4 Office, 3A-class attractions 3, and 2A-class attractions 7. But market share fluctuations should improve the ability of travel services to form a relatively stable source and strengthen improvement of tourism facilities to further improve quality and attractiveness of tourism resources. Wuwei and Zhangye were created in 2005 as outstanding tourist cities, although there were relatively few high-quality tourist attractions in Wuwei City. But excellent location, calendar entry traffic market share in recent years are higher than in Zhangye. Gannan Tibetan Autonomous Prefecture, strong national characteristics, strange and mysterious Tibetan culture, and beautiful prairie scenery have attracted a large number of inbound tourists who indulge in sightseeing. In 2007, the market share of inbound traffic ranking first in the province reached 32.11 percent. In 2008, due to the impact of the financial crisis and the earthquakes, snowstorms, competitive effects of natural disasters, as well as the six tourist cities in Gansu Gannan, the market share in 2012 fell by 3.77 percent. Linxia Hui Autonomous Prefecture has a unique style and Hui

Yellow River Three Gorges scenery, but poor traffic conditions, lack of tourism infrastructure have restricted its inbound tourism economy. Jinchang, silver, Pingliang, Qingyang, Dingxi, and Gansu experienced a widespread lack of tourism resources; thus, due to the low level of economic development issues, market share reached a maximum of only 2% in 2012.

3.3 Time changes and causes analysis of spatial distribution of inbound tourists

Figure 3 depicts Gansu cities (prefectures) concentration index inbound tourism interannual differences, that is, the number of inbound tourists in the measure are unevenly distributed among the annual period of time. Inbound tourism in Gansu Province "leading" position in Jiuquan had an annual minimum 3.67% of concentration index, tourist arrivals., aAlthough thesere are greatly reduced in recent years, but the development of tourism and the overall similarity of Gansu Province are observed.; Lanzhou, Jiayuguan, Wuwei, in and Pingliang are 5% or less, interannual changes in are small, and there is a relatively stable destination.; Gansu, Linxia, and Qingyang interannual concentration index is close to 10%, greater inter-annual variation.; oOther cities (prefectures) annual concentration index in the middle level are relatively stable destination areas. Among them, Lanzhou, Jiayuguan annual fluctuation is small, indicating a long-term tourism development, has formed a certain steady flow of tourists., bBut the inter-annual fluctuations in the newly created outstanding tourist city of Tianshui, Zhangye is are relatively high, and this should further improve the quality of tourism services, and transportation, and tourism infrastructure., tThe formation of long-term international tourism influences the formation of their relatively stable source of inbound tourism.; Gannan inbound tourists experienced rapid growth before 2007, followed by the province's tourist traffic restrictions and other high-grade tourist attractions. competition, has a The market share of competition was reduced to 2012 of 3.77% in 2012, and this should increase tourist traffic infrastructure

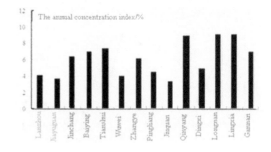

Figure 3. Year border concentration index of every city and state of Gansu province.

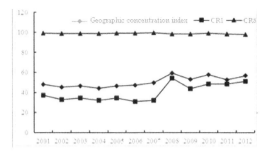

Figure 4. Variation of geographic concentration index and concentration ratio.

investment and prevent inbound tourism market recession.

As can be seen from Figure 4, 2001–2007, Gansu and Jiuquan compete with each other to make the first index decline after 2007. Gannan, due to tourist traffic restrictions and other cities (states) tourism competition, experiences a significant decline in the amount of inbound tourists, but Jiuquan with relatively mature Dunhuang tourism keeps the first place. Inbound tourists are almost concentrated in eight major destinations, such as Lanzhou, Jiuquan, Jiayuguan, Gannan Prefecture, and so on, as well as in other cities (prefectures) occupy the sum of the market share of less than 2%. Overall, the inbound tourism geographic concentration of Gansu Province is high, the distribution of inbound traffic area is uneven, and its development changes can be divided into three parts: a small decrease in 2001–2004, an upward trend in 2004–2008, and a fluctuating downward trend in 2008–2012. In 2003 and 2004, NWS Tianshui and Lanzhou, two outstanding tourist cities, shunted some tourists and made geographic concentration decrease. In 2004–2008, total passenger traffic increased immigration, but Gansu Jiuquan and increased traffic volume higher than Tianshui, Lanzhou, and mighty Zhangye, so that geographic concentration index rose. In 2012, geographic concentration index was 56.63, compared with 59.10 in 2008, where there was a small decline, indicating that inbound tourism in Gansu Province has reduced spatial concentration trend. Overall, immigration in Gansu Province Tourism geographic concentration is high, there is uneven regional distribution of inbound traffic, immigration passengers in Gansu, and the tendency is not only obviously destination but also relatively fixed. Mainly for Lanzhou, Jiuquan, Jiayuguan, Gannan Prefecture, and so on, eight major destinations are concentrated in the west along the Silk Road region and are characterized by rich ethnic culture and customs, local culture, and prominent political, economic, and cultural relatively developed cities (states).

4 CONCLUSIONS AND COUNTERMEASURES

This paper selects 14 cities (state) of tourist arrivals in Gansu Province during the years 2001–2012 years for the study. It selects the appropriate measurable differences in spatiotemporal distribution of tourism index formula for quantitative analysis. The results showed that during the past five years in Gansu Province, inbound tourism and "Ten development of five" period seemed to be slow and influenced by changes in the international environment. The 2008 financial crisis almost stumbled with regard to inbound tourism. Gansu experienced an uneven spatial distribution of inbound tourism as well as a high geographic concentration, concentrated in Hexi silk region. The road along the Jiuquan, mighty Zhangye, and Jiayuguan exhibits a rich ethnic culture and customs; the local cultural highlights of Gannan and Linxia are political, economic, and cultural relatively developed in Lanzhou, Tianshui. Tianshui, Zhangye of inbound tourists exhibit interannual fluctuations that are relatively large in Gannan and there are signs of recession, but with gradual implementation of the economic development and improvement, the overall strength of Gansu Province, and the associated tourism industry policies, the spatial distribution of inbound tourists will gradually spread, gradually move toward stability, and, ultimately, achieve a certain level of aggregation balance.

Gansu Province inbound tourism has made some achievements, but the development is slow, geographic concentration is high, and a part of the city (state) interannual fluctuations, recession, and the long-term trends continue to affect the healthy and sustainable inbound tourism in Gansu development. Therefore, it is necessary to put forward corresponding countermeasures and suggestions for related questions: First, analyze inbound tourist market, hierarchical development, and stable major source markets. Then, actively explore the market potential, expand customer base, and form a relatively stable flow. Reduce fluctuations in the intensity of inbound tourism during the international environmental change; improve tourism management system; strengthen tourism personnel training; standardize management of the tourism market; improve service levels; strengthen regional cooperation; enrich the connotation of inbound tourism products; improve inbound tourism industry competitiveness; and promote tourism resources and the neighboring provinces of Gansu complementary, thus promoting the rapid development of tourism in the Yellow River culture and cultural crossroads of the Silk Road from the X-type architecture, West, Central, three districts linkage in the East, and seven scenic groups coordinated development. This was the basis of six tours "wood" shaped spatial pattern on the basis of consolidating tourism development, optimizing the spatial layout, nurturing growing tourist routes, building a modern tourist destination, and promoting industrial upgrading and transformation, which, in Jiuquan, Lanzhou, and Jiayuguan should strengthen the tourism product design as well as resource development to maintain tourism advantages. Tianshui and mighty Zhangye should strengthen the tourism infrastructure construction, as well as improve the tourism environment, with an excellent tourist destination. Jiuquan, Lanzhou, and Jiayuguan form a regional linkage, gradually form a stable source, and are driven by relatively backward areas. In Jinchang, silver, Dingxi, Pingliang, and Qingyang immigration tourism development, the Gannan and Linxia experience an urgent need to increase tourist traffic to improve tourism infrastructure and service capabilities, to prevent customer loss. Overall, economic strength is weakened in Gansu Province to a certain extent, the development of inbound tourism is restricted, and there is a need to develop the local economy to steadily improve, support, and promote the economy on tourism.

REFERENCES

[1] Zhong Jing.Chinese inbound tourism to foreign markets spatial and temporal characteristics Research [J]. Economic Issues, 2008, (9):122–126.

[2] Sun Gengnian. On regional development and regional joint development of tourism [J] Geography, 2001, 16 (4):1–5.

[3] Wang Kai, YI Jing, Li Hao. Spatial and temporal differences in the evolution of Chinese Inbound Tourism Development Analysis: 1991–2010 [J] Geography, 2014, 01:134–140.

[4] Li Chuangxin, Ma Yaofeng, Zhang Ying. From 1993 to 2008 the dominance of space-time region inbound tourist flow dynamic evolution model - based on empirical research Geographic Entropy Method [J] improvements, 2012, 31 (02):257–268.

[5] Ji Xiaomei, Chen Jinhua spatiotemporal Fujian international tourism market dynamics change and expand countermeasures [J] Economic Geography, 2013, 33 (05):158–164.

[6] Ma Xiaolong Xi'an tourism flow of inbound tourism spatial and temporal evolution and regulation systems [J] Geography, 2006, 04:88–93.

[7] Tu Jianjun. Study on the dynamic model of [J]. inbound tourism distribution of touristresources and environment in Sichuan province in the Yangtze Basin, 2004, 13 (04):338–342.

[8] Wan Xucai, Wang Houting, Fu Zhaoxia, Ma Hong zhuan. Factors of China inbound tourism development of urban diversity and its influence - Taking key tourist city as study cases of [J]. in geography, 2013, 32 (02): 337–346.

Energy, Environment and Green Building Materials – Sheng (ed.)
© *2015 Taylor & Francis Group, London, ISBN 978-1-138-02718-3*

Research on multi-level water bloom decision-making method based on information entropy of vague set

Qing-Wei Zhu, Xiao-Yi Wang, Li Wang, Ji-Ping Xu & Yu-Ting Bai

School of Computer and Information Engineering, Beijing Technology and Business University, Beijing, China

ABSTRACT: This paper proposes an information entropy calculation based on the multilevel decision-making model in Vague set for the characters of water bloom decision-making in lakes and reservoirs with multi-targets and multi-attributes. First, the target-solution-attribute decision-making model is built by strictly distinguishing the decision targets and solution attributes. Second, this paper puts forward the calculation method of information entropy in Vague set to obtain the relative weights; then, it uses the TOPSIS (Technique for Order Preference by Similarity to Ideal Solution) method to get the ideal solution. Finally, it uses the similarity measurement algorithm to evaluate the alternatives and determines the decision-making results according to the sorting of alternatives. The method was applied to the decision-making problem for water bloom emergency governance in city lakes and reservoirs. The results prove that the method is reasonable and effective, which can improve the degree of differentiation and the pertinence of decision-making.

KEYWORDS: Water bloom; Vague Set; Information Entropy; Similarity Measure; Multi-level Decision-making

1 INTRODUCTION

The concept of Vague set was proposed by Gau and Buehrer in 1993[1]. Vague set has a stronger representation capability than traditional fuzzy set in dealing with uncertain information[2]; it expresses the decision-makers' support, opposed and neutral information to a certain thing simultaneously. Song Huang proposed a multi-target decision-making method based on the entropy weight and vague set[3]; Xiaoguang Zhou introduced the entropy weight into fuzzy matter-element decision according to the theory of Vague set and the matter-element theory, combined with the concept of information entropy[4]. Jianqiang Zhang solved a decision-making problem in which the attributes were expressed by Vague set under the condition that the weights were unknown[5]. On the basis of previous studies, this paper combined Vague set with information entropy and then proposed the entropy computing method of Vague set. The decision theory and methods have been widely studied and applied since Nicholas Bernoulli proposed the concept of utility values in the first half of the 18th century[6]. Aiming at the imperfection of the existing decision-making methods, this paper builds up the multi-level decision-making model, according to the TOPSIS method to get the ideal solution using Vague set theory, setting targets as a basis for the selection of ideal solutions and letting impact attributes act as an evaluation index for alternatives. The information entropy calculation method of Vague set is mainly used, and the similarity measurement algorithm is used to evaluate and sort the alternatives. This decision-making method has been applied to the emergency governance of water bloom in lakes and reservoirs, and the result shows the feasibility of the method.

2 ENTROPY METHOD OF VAGUE VALUES

2.1 *Vague set*

Assuming that U is a domain of discourse, V is a set consisting of all Vague sets in U and $A \in V$, $B \in V$. $M(A, B)$ can be named similarity measurement between A and B, if $M(A, B)$ meets the following properties[7]:

1 $0 \leq M(A,B) \leq 1$;

2 if $A = B$, $M(A,B) = 1$;

3 $M(A,B) = M(B,A)$.

Assume that $x = [t_x, 1 - f_x]$, $y = [t_y, 1 - f_y]$ are two Vague values in U, when discussing Vague value similarity measurement. t is true membership function of Vague set A, and it represents the lower bound of membership degree for the evidence supporting $x \in A$. f is the false membership function of Vague set A, and it represents the lower bound of membership degree for the evidence against $x \in A$; moreover, $0 \leq t_x + f_x \leq 1$, $0 \leq t_y + f_y \leq 1$.

Literature [8] added the calculation factor and improved the existing similarity measurement:

$$M_Z(x,y) = 1 - |t_x - t_y - (f_x - f_y)|/8 -$$
$$\left(|t_x - t_y| + |f_x - f_y|\right)/8 - |t_x - t_y + f_x - f_y|/4. \quad (1)$$

2.2 Entropy calculation of vague set

Information entropy method is a kind of important objective weighting method[9] that uses the information existing in data itself to determine the weights. The classic information entropy method is only applicable to real numbers. Therefore, the information entropy method was put forward under the condition of Vague set.

The advantage function $S(x) = t_x - f_x$ is used to estimate the fitness of row elements to column elements in the matrix of the Vague set, in order to convert Vague variables into the classic value of relative membership degrees. Assuming that x_{ij} stands for fitness of row element i to column element j, then

$$x_{ij} = t_{ij} - f_{ij}. \quad (2)$$

The fitness matrix $M = [x_{ij}]_{m \times n}$ can be built, which synthesizes the supportive and opposed information of Vague value. Normalizing the classic value converted from Vague value, the evaluation of row element i to column element j is defined as follows:

$$X_{ij} = \frac{x_{ij}}{\sum_{i=1}^{m} x_{ij}} \forall i,j \quad (3)$$

The entropy E_j of row element i to column element j is[10]

$$E_j = -k \sum_{i=1}^{m} X_{ij} \ln X_{ij} \forall j, \quad (4)$$

k is a constant, $k = 1/\ln m$; this ensures that $0 \le E_j \le 1$.

Information deviation degree d_j:

$$d_j = 1 - E_j. \quad (5)$$

If decision makers have no preference between attributes, the weight ω_j is defined as follows:

$$\omega_j = \frac{d_j}{\sum_{j=1}^{n} d_j} \forall j. \quad (6)$$

3 MULTI-LEVEL DECISION MODEL BASED ON VAGUE SET

3.1 Build decision-making level

The core of multi-target decision is layering the decision problem, that is, forming the target layer, solution layer, and attribute layer, respectively.

For the target layer and solution layer, the fitness matrix is formed by an expert's advice, whose element means the fitness of every solution to multiple targets. For the solution layer and attribute layer, the monitoring and survey results are used to construct the influence matrix. Optimal relative degree between levels (can be specificied to the fitness and the influence degree) can be achieved by language variables (linguistic term set). A suitable language variable can be chosen to represent the preference information according to the research question. Literature [11] adopts an 11-level linguistic term set of Vague values to represent the subjective evaluation information.

3.2 Method of obtaining ideal solutions

(1) Obtain decision matrix

A stands for the solution-target correlation matrix. According to the linguistic variables and vague value transformation rules, the solution character under the target is represented by Vague set:

$$A_{u \times v} = [t_{ij}^A, 1 - f_{ij}^A], i = 1,2,\cdots u; j = 1,2,\cdots v. \quad (7)$$

t_{ij}^A stands for the degree that alternative solutions a_i meet target c_j, f_{ij}^A stands for the degree that alternative solutions a_i do not meet binding target c_j, and $0 \le t_{ij}^A + f_{ij}^A \le 1$. To simplify the mark, set $1 - f_{ij}^A = t_{ij}^{A*}$.

Similarly, B stands for the solution-attribute correlation matrix; the influence of attribute under the solution is represented by Vague set:

$$B_{u \times w} = [t_{ij}^B, 1 - f_{ij}^B], i = 1,2,\cdots u; j = 1,2,\cdots w. \quad (8)$$

(2) The determination of ideal solution VPIS and negative ideal solution VNIS

In the fitness matrix M, we assume that $r_j^* = \max_{1 \le i \le m} x_{ij}$, $r_j^- = \min_{1 \le i \le m} x_{ij}, 1 \le j \le n$. When the target is efficiency type, the attribute of solution a_i corresponding to r_j^* is the optimal attribute of c_j, and the attribute of solution a_i, corresponding to r_j^- is the worst attribute of c_j. Similarly, when the target is cost type, the attribute of solution a_i, corresponding to r_j^- is the optimal attribute of c_j, the attribute of solution a_i, corresponding to r_j^* is the worst attribute of c_j. If r_j is the same, it can be considered that all solutions are the optimal or the worst, and the optimal or worst attribute is $\left(\frac{1}{n} \cdot \sum_i^n t_{ij}^B, 1 - \frac{1}{n} \cdot \sum_i^n f_{ij}^B \right)$.

The optimal attribute of the target matrix has been obtained:

$$C_{v \times w}^* = [t_{ij}^{C*}, 1 - f_{ij}^{C*}](i = 1,2,\cdots v; j = 1,2,\cdots w.), \quad (9)$$

And the worst attribute of the target matrix is

$$C_{v \times w}^- = [t_{ij}^{C-}, 1 - f_{ij}^{C-}](i = 1,2,\cdots v; j = 1,2,\cdots w). \quad (10)$$

According to the entropy weight calculation method of the Vague set, calculating the weight of the target, respectively ω_j^* and ω_j^-, is based on the two matrices given earlier.

The solutions can be ordered using positive-ideal solution (PIS) and negative-ideal solution (NIS) of multi-objective problem, and then the attribute value of the ideal solution will be got. Use VPIS and VNIS to represent PIS and NIS based on Vague set, then

$$VPIS = \omega_j^* \cdot C_{v\times w}^{*\ \prime}, \qquad (11)$$

$$VNIS = \omega_j^- \cdot C_{v\times w}^{-\ \prime}. \qquad (12)$$

3.3 Alternative sorting

According to the similarity measurement method of Vague values, we calculate the distance of alternative, VPIS and VNIS:

$$d_i^* = \sum_{j=1}^{n} \omega_j M_Z([t_{ij}, t_{ij}^*], VPIS), i = 1, 2, \cdots, m \qquad (13)$$

$$d_i^- = \sum_{j=1}^{n} \omega_j M_Z([t_{ij}, t_{ij}^*], VNIS), j = 1, 2, \cdots, m \qquad (14)$$

ω_j is caculated using the entropy weight method in the solution-attribute matrix.

Relatively closeness coefficient of each alternative is

$$\sigma_i = \frac{d_i^-}{d_i^* + d_i^-}, i = 1, 2, \cdots, m \qquad (15)$$

The larger σ_i is, the closer is the solution to the ideal solution and the further away it is from the negative ideal solution. The solutions can be sorted by the value of σ_i, according to the posted schedule principle.

4 APPLICATION OF WATER BLOOM EMERGENCY GOVERNANCE IN LAKES AND RESERVOIRS

Lake algae blooms are a typical manifestation of eutrophication, whose harmfulness not only lies in the serious pollution of scarce water resources but also is a direct threat to human health through the food chain[12].

The lake's governance methods for water blooms are mainly divided into four categories: biological, physical, chemical, and ecological methods, including lake endogenous nutrient, biological control, water diversion flushing, artificial aeration, and so on. The research on the water bloom emergency management decision-making is rare, though the emergency decision-making methods have been researched and applied in many fields. Therefore, the method proposed in this paper can be used in this field.

Step 1 Build a decision matrix

According to the research purpose, choose governance solutions such as endogenous nutrient method as alternatives and set the decision targets, including "influence on surroundings" and so on. Among the targets, the "appeared time of effect" is efficiency type, and other targets are cost type. The decision targets, solutions, and attributes are shown in Figure 1.

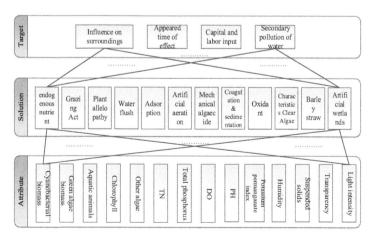

Figure 1.   Three-layer decision model.

The evaluation value of experts for the alternatives to targets and attributes is collected by way of questionnaire investigation. The decision matrix is composited based on the linguistic term set (Table 1).

Table 1. Solution-target evaluation matrix, solution-attribute evaluation matrix.

	A1	A2	A3	A4	...	A9	A10	A11	A12
C1	MP	FP	MP	MP	...	FG	M	M	MG
C2	MP	FP	MP	FP	...	MG	MP	MP	P
C3	M	MP	MP	MP	...	P	M	M	G
C4	M	MP	M	P	...	MG	MG	MP	FG

The linguistic term set can be transformed into a Vague set, and the Vague set matrix can be got by corresponding the grade to the typical Vague values.

Step 2 Determine VPIS and VNIS

According to Formula (2), the Vague set solution-target matrix is transformed into a fitness matrix. We can get the optimal (worst) attribute-target matrix according to Section 3.2(2).

On the basis of the earlier two matrices, we get the weights of targets, as shown in Table 2, according to Formulae (3)-(6). According to Formulae (11) and (12), we get the ideal solution VPIS and negative ideal solution VNIS.

Table 2. Target of entropy weight.

	C_1	C_2	C_3	C_4
ω_j^*	0.1940	0.1685	0.2402	0.3973
ω_j^-	0.1755	0.3185	0.2200	0.2860

Step 3 Sort alternatives

According to Formula (1), we calculate the similarity measurement of solutions with VPIS and VNIS. Calculate the distance of the various solutions to the VPIS and VNIS using Formulae (13) and (14). Calculate relative degree of each solution with Formula (15), and the calculation results are shown in Table 3 and Figure 2.

The decision result in this paper shows that the relative degree of artificial wetland management method (solution 12) is much larger than other solutions; namely the solution 12 is best, followed by artificial aeration (solution 8) and coagulation precipitation (solution 6). However, the endogenous nutrient method (solution 1) is relatively the worst. The difference among each solution is large and it can be concluded that the decision result can provide effective guidance for emergency management of blooms in city's lakes and reservoirs.

Table 3. Distance and relative degree of the various solutions to the VPIS and VNIS.

	A_1	A_2	A_3	A_4		A_9	A_{10}	A_{11}	A_{12}
d_i^*	0.86	0.85	0.88	0.92	...	0.92	0.83	0.86	0.87
d_i^-	0.81	0.82	0.86	0.88	...	0.90	0.82	0.84	0.94
$\sigma(A_i)$	0.48	0.49	0.49	0.48	...	0.49	0.49	0.49	0.51

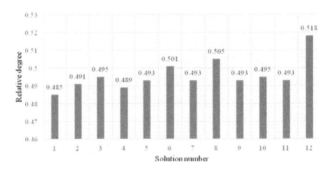

Figure 2. Relative degree of various solutions.

5 CONCLUSIONS

This paper proposes the calculation method of information entropy weight on the basis of Vague set and constructs' multilevel decision model. The method is applied in emergency management decisions in water blooms of lakes and reservoirs in cities. The decision result shows the following:

1 Multilevel decision model is put forward innovatively. Attributes and targets are strictly distinguished in decision problems, which improves the richness of decision factors. Simultaneously, this model also gives scientific and reasonable division and weight distribution for all kinds of factors.

2 The information entropy calculation method of Vague set reflects the characteristics of the data itself, mining and reflects fuzzy information through Vague set.

3 The result of emergence governance of water bloom in lakes and reservoirs corresponds to expert evaluation opinions and practical application. The decision-making result has a great degree of differentiation, and its guiding role is obvious.

The method on information entropy calculation of Vague set combined with the multilevel decision model improves the pertinence and effectiveness of the decision-making. The method fills the blanks of the related fields and can play a role of reference in case of a similar problem.

ACKNOWLEDGMENTS

This work was financially supported by The Innovation Ability Promotion Project of Beijing Municipal Commission of Education College (PXM2014_014213_000033) and The Importation and Development of High-Caliber Talents Project of Beijing Municipal Institutions (CIT&TCD201404031). The support rendered by these institutions is gratefully acknowledged.

REFERENCES

[1] Gau W L, Buehrer D J. Vague sets [J]. IEEE Transactions on Systems, Man, and Cybernetics, 1993, 23:610–614.

[2] Zadeh L A. Fuzzy sets [J]. Information and control, 1965(8):338–356.

[3] HUANG Song, HUANG Weilai, Approach to Multiple Objectives Decision Making Based on Entropy Coefficient and Vague Set [J]. Chinese Journal of Management, 2005:120–123.

[4] ZHOU Xiao-guang. Research on Method of Vague Matter-element Decision Making Based on Entropy Weight [J]. Journal of Systems & Management, 2009, 18(4):454–458.

[5] ZHANG Jian-qiang, HANG De-cai. Fuzzy TOPSIS multi-attribute decision method based on the entropy weight of Vague set [J]. Journal of Zhejiang University of Technology, 2012, (5):524–527.

[6] FANG Zhigeng, LIU Sifeng, ZHU Jianjun, et al. Decision-making Theory and Methods [M]. Beijing: Science Press. 2009:4.

[7] Dengfeng L, Chuntian C. New similarity measurement of intuitionistic fuzzy sets and application to pattern recognitions [J]. Pattern Recognition Letters, 2002, 23(1):221–225.

[8] ZHOU Xiao-guang, ZHANG Qiang. Comparision and improvement on similarity measurement between vague sets and between elements [J]. Journal of Systems Engineering. 2005, 20(6):613–619.

[9] Nedeltchev S, Shaikh A. A New Method for Identification of the Main Transition Velocities in Multiphase Reactors Based On Information Entropy Theory [J]. Chemical Engineering Science, 2013, 100:2–14.

[10] Chen C J, Chen S M. Fuzzy risk analysis based on similarity measurement of generalized fuzzy numbers [J]. IEEE Transactions on Fuzzy System, 2003, 11(1): 45–46.

[11] ZHOU Xiaoguang, ZHANG Qiang. Decision-making Theory and Methods Based on Vague Set [M]. Beijing: Science Press. 2009.

[12] LIU Zaiwen, WU Qiaomei, WANG Xiaoyi, et al. Algae growth modeling based on optimization theory and application to water-bloom prediction[J]. Journal of Chemical Industry and Engineering. 2008, 59(7): 1869–1873.

Energy, Environment and Green Building Materials – Sheng (ed.)
© *2015 Taylor & Francis Group, London, ISBN 978-1-138-02718-3*

Development and review of application of low-noise asphalt concrete in urban roads

T.F. Nian & P. Li
College of Civil Engineering. Lanzhou University of Technology, China

L. Yang
Shandong Zhonghong Road & Bridge Construction Co. Ltd, China

Y.L. Zhang & Z.C. Liu
College of Civil Engineering. Lanzhou University of Technology, China

ABSTRACT: The noise caused by urban road traffic has already seriously influenced the environment of life and work. The low-noise asphalt pavement could be widely used in the construction of environmental protection in the future. Dense asphalt concrete is the main asphalt concrete pavement in the domestic realm, and porous low-noise asphalt pavement is a new kind of functional surface. Due to the traffic composition, material properties, and the difference of the natural environment, the mixture design method of foreign porous low-noise asphalt, raw material selection, and construction techniques may not be applied in our country, so it is necessary to explore the technology of the porous low-noise asphalt roadbed applied in our country. Combined with the practical application of engineering analysis of low-noise asphalt concrete in the urban road, we should discuss the design, production, and transportation of low-noise asphalt mixture; comprehensively analyze the application process; put forward improvement measures on the shortcoming; and promote the application of low-noise asphalt concrete.

KEYWORDS: Urban road; Porous asphalt concrete; Low noise pavement; Reduction noise

1   INTRODUCTION

With the development of the economy and an increase in the living standard, urban transportation is rapidly expending, motor vehicles are significantly increasing, and the interference degree and scope of road traffic noise in peoples' normal life, study, and work environment besides roads are growing and expending. In 2005, some research showed that people living in the near-road buildings in Beijing constituted about 16 percent of the urban population, nearly 100 million, and they had to suffer from serious traffic noise pollution[1]. The choice of reasonable measures to reduce road traffic noise is very important; the porous low-noise asphalt concrete pavement may be considered an ideal and radical reduction measure.

2   POROUS ASPHALT PAVEMENT

Porous asphalt pavement (Referred PA) is paved with high porosity power asphalt mix material on a cement concrete pavement or other pavement structure layers [2].

2.1   *Mechanism of low-noise road surface noise reduction*

The effect of sound reduction is related to the size of asphalt surface mix aggregate, thickness, porosity binder, and porosity. The thickness and porosity are the most important factors for reducing the noise. Porous asphalt pavement, which is included in connected pores and closed pores, is generally called "whole porosity." Connected pores will play an effective role in draining off water and absorbing noise, so it is effective porosity [3]. Porous asphalt pavement has a good macro structure; a developed and connected void (Interconnected Pores) is formed on the road surface and internal pavement, and it should be an innovation in the asphalt pavement structure [2]. Compared with the ordinary asphalt concrete pavement, the fundamental difference is its porosity of approximately 15% ~ 20%, or even more than 20%. However, the porosity of the common asphalt pavement is only 3% to 6%. The porosity is a key factor in reducing noise, under the same thickness. Mechanisms of the low-noise pavement can be summarized as the following four points:

a) Reducing attachment noise.
Compared with dense roads, the contact between the tire and the road is decreasing. It is helpful to reduce attachment noise.

b) Reducing pump noise.
As the inter-working gap between surface layers, the air of surface pattern slot will overflow through the space when the tire comes into contact with the road surface, reducing the noise generated by compressed air blasting and changing the high frequency of pump noise into low frequency. This is shown in Figure 1. The air between common asphalt pavement tires and pavement layer cannot be discharged, and produces air compression and expansion sound, whereas the porous asphalt pavement allows air to escape from the gap, reducing the road noise of the tire.

c) Good flatness of road reduces the impact of noise.
The pavement can make the vehicle run steadily, effectively reduce the vehicle shock on the road, and then lead to a reduction in the impact of noise.

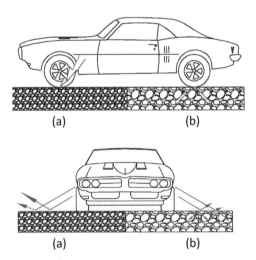

Figure 1. Contrast of common asphalt pavement and porous asphalt to reduce noise contrast.
Note: The part (a) is the common asphalt pavement, and the part (b) is the porous asphalt pavement.

d) Sound absorption effect of gap on the surface.
It can absorb not only the noise from the engine but also the noise radiation by the transmission parts to the pavement outside, and it can still absorb the tire noise caused by car chassis' action that is reflected back to the road and the noise caused by other interface reflections to the road. The sound absorption mechanism is similar to the absorption mechanism of the porous sound-absorbing material. The common asphalt pavement cannot absorb radiation to the pavement noise, and this is shown in Figure 1.

3 ENGINEERING APPLICATION ANALYSIS

3.1 *Case analysis*

In a certain urban road renewal project, the municipal group mostly adopted the porous asphalt concrete pavement and paved it about 1230 meters long[4]. The porosity of the pavement design is 20%, and the optimum proportion of the mixture is 4.6%.

3.2 *Pavement materials and requirements* [5]

Porous asphalt mixture and common asphalt mixture are the same, using asphalt binder as binder, which contains coarse aggregate, fine aggregate and filler. The technical requirements of the coarse aggregate and asphalt tamer hoof fat macadam mixture (SMA) are basically the same, and require hard stone and good particle shape. But for the low noise road surface, the grinding value PSV of coarse aggregate is demanded more, that is the road of gravel applied in noise and anti-sliding resistance require mill instead of sliding. Fine aggregate can be stone chips, artificial sand or natural sand and aggregate should be dry, clean and free of impurities, no proper mix of weathered, angular and strong requirements. Packing appropriate USES limestone or magmata rocks in strong alkaline rocks such as hydrophobic stone fine grinding of mineral powder, mineral powder to dry and clean. Table 1 is shown the parameters of aggregate.

3.3 *Analysis on the mixture of porous asphalt concrete*

Low-noise porous asphalt concrete mixture ratio designs mainly focus on the high porosity, ensuring it has certain stability, strength, and durability, and select the gradation and determination of optimum asphalt content. According to the low-noise asphalt mixture gradation range currently recommended by the premise, in the case of 20% void fraction, asphalt content is around 4.5%, flying loss is around 13%, and durability is poor. Reference NCAT REPORT NO.99-3 proposed a grading range [4] and in combination with the practical situation of this project it was low-noise asphalt mixture proportioning design; the grading of the reference range is shown in figure 2.

Table 1. Used in the surface of the coarse aggregate quality requirements.

Aggregate	Target	Units	Technical requirement	Target	Units	Technical requirement
Coarse aggregate	Crushed stone value	%	≤16	Polished stone value	BPN	≥42
	Apparent density	g/ cm³	≤2.60	Content of soft rock	%	≤2
	Abrasion loss	%	≤25	water absorption	%	≤2
Fine aggregate	Apparent density	g/ cm³	≥2.50	Silt content	%	≤1
	Sand equivalent	%	≥70	Firmness (>0.3mm)	%	≤12
Mineral powder	Particle size<0.6mm		100	Apparent density	g/ cm³	≥2.50
	Particle size<0.15mm	%	90-100	Plasticity index	/	≤4
	Particle size<0.075mm		70-100	Coefficient of hydrophilic	/	≤1

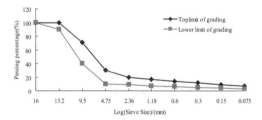

Figure 2. Gradation reference range of low-noise porous asphalt mixture.

3.4 Porous asphalt concrete pavement construction technology

During the construction of porous asphalt concrete, temperature should be controlled to avoid the phenomenon of wrinkled skin. Before paving, the heating temperature of the paver screed should be above 100°C, and asphalt mixture temperature should be controlled at 155~170°C; after the appropriate paving, the loose spread coefficient is recommended to be about 1.18 and the specific adjustments are according to the test situation of the road. Before construction, cleanliness and drying of waterproof layer should be ensured. Blowing and drying process should be done, if necessary. When determining the pavement thickness to accurately position lateral baffles a reasonable control of pavement flatness. Asphalt mixing of materials should be used immediately; when it needs to be stored in the storage bin, there should not be drainage, and it should not be stored in use for the next day. After stirring the mixture, equipment should be cleaned regularly to ensure the cleanliness of the equipment, and the release agent should be brushed when used as production equipment in the production process.

After the porous asphalt paving, the road should be rolled under high temperature conditions as soon as possible. Besides, in addition to the necessary water and other short rest, the roller compaction process at each stage should be uninterrupted, while at low temperatures not repeatedly rolling asphalt to prevent worn stone or crushed stone edges, thereby undermining an aggregate embedded crowd [6].

3.5 Transportation of porous asphalt concrete

Porous asphalt concrete requires constantly stirring and heating it during transport to ensure the quality of the product. When the material transport vehicles are transporting the mixed process, tarpaulin should be covered to prevent the mixture from cooling and forming a surface crust, and there is no stop during transport. During transport to the construction site, the temperature of the asphalt mixture should be checked, not less than paving temperature requirements. Lorry discharge should be clean, and if remains are found, they should be cleaned up in time.

3.6 Test analysis of porous asphalt pavement

Noise monitoring is designed to compare and draw a recommended low-noise pavement structure by measuring the noise level in the test section of different pavement structures and tire when the cars pass. Noise monitoring generally uses mobile method or fixed-point method. Mobile method involves the noise test equipment installed near the location of car tires, fixed to the vehicle body, and measured in the whole process of moving car tire noise. Fixed-point method involves the sound-level meter fixed at the measuring point, and then the test vehicle is passed through at a certain distance from the measuring point by measuring the noise level [7].

a) Noise measurement and analysis.
Noise test method took fixed-point method. Figure 3 shows the test results of the analysis. Compared with the conventional asphalt pavement, the noise of porous asphalt in the dry state can be maximumly decreased above 4dB(A), and in a humid environment it can be maximumly decreased above 7dB(A). Porous asphalt has a significant noise reduction.

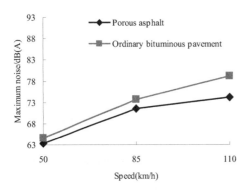

Figure 3-a. Test results in dry environment.

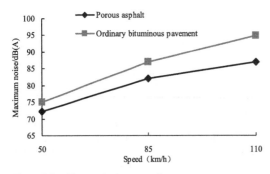

Figure 3-b. Test results in wet environment.

Figure 3. Results of the road noise test in dry and wet environment outside the car.

b) Test analysis of anti-slide.
Along with pendulum coefficient of friction tester and test artificial sanding method, the results are shown in Figure 4. The value of the coefficient of friction pendulum and texture depth of porous asphalt pavement was significantly greater than that of the ordinary asphalt pavement. Especially in wet conditions, the value of the coefficient of friction pendulum declined

a little compared with dry conditions. The general decline in the asphalt pavement is higher. This fully demonstrates the porous asphalt pavement skid resistance in the safety of the rain.

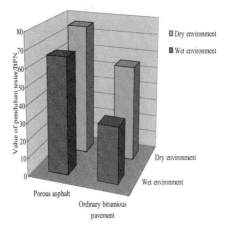

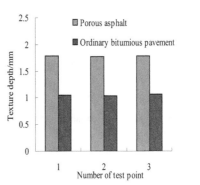

Figure 4. Test results of texture depth under dry conditions and the value of pendulum tester.

3.7   Inadequate and improvements existing in porous asphalt concrete pavement

a) The void clogging.
Excellent road noise porous pavement performance originates mainly from its large porosity. With the role of transportation and the natural environment, its porosity will decrease year by year, or the gap is blocked. Its proposed improvement involves using high-pressure water to rinse porous pavement noise on a regular basis.

b) Early loose.
The main reason of loose is considered inadequate asphalt membrane thickness, excessive binder aging, and the loss of asphalt aggregate adhesion under freeze-thaw state. Its improvement measure is to adopt modified asphalt or high viscosity asphalt, plus fiber, increase the asphalt film thickness, and

accurately control with the heating temperature of material and aggregate.

c) Separating the lower lying layers.

Porous asphalt pavement is usually a paved surface, whereas the lower part of the structure still consists of commonly used materials. Construction cycle that dragged too long and insufficient consumption of sticky layer oil will result in the separation of permeable asphalt surface layer and the underlying layer. Therefore, we need to determine the amount of waterproof adhesive layer oil depending on the surface texture sufficient of the underlying layer; increase design to accelerate the necessary subsidiary discharge pavement moisture; and arrange the construction schedule reasonably.

d) Short life.

Large porosity increases the contact area of the sun, binder, and air. Porous asphalt pavement is characterized by the fast binder aging, relatively low durability, and shorter life expectancy. In response to this shortage, its proposed improvements that is modified asphalt, mixed with a certain amount of fiber to increase the amount of asphalt and increase the aggregate surface coating of asphalt film thickness wrapped.

4 SUMMARY

(a). Low-noise performance of porous asphalt concrete has some advantages that other types of concrete cannot match, especially in urban road noise, and has a strong suitability.

(b) Porous low-noise asphalt pavement skid resistance is not only better than ordinary asphalt pavement but can also reduce road traffic noise. Because of differences in climate, materials, and environment, we should choose according to qualification.

(c) Porous low-noise asphalt pavement is a new kind of functional pavement; the structural design, construction technology, and a large number of promoters in different regions still need further research and analysis.

REFERENCES

[1] C.G. LIU et al. Study on application of low noise asphalt concrete pavement. Urban Roads Bridges & Flood Control, 2010, 04(4):157–157, 159.

[2] W.D. CAO et al. International and domestic status quo of research on low noise asphalt pavement. Petroleum Asphalt, 2005, 19(01):50–54.

[3] J.D. XU et al. Low noise asphalt pavement permeability applications. Highway, 1998(01):14–18.

[4] B. ZHANG. Research on the application of low-noise porous asphalt pavement: [dissertation]. Shandong Normal Univ., 2005.

[5] Code for construction and acceptance of asphalt pavements (GB 50092-96). Beijing: China Communications Press, 1996.

[6] Y.S. YIN. Research on low-noise porous asphalt pavement: [dissertation]. Chang'an Univ., 2005.

[7] L.S. QIN. Study on Low noise asphalt pavement structure. In: Proc. of the first national highway technological innovation top BBS. Beijing, 2002, 206–212.

Energy, Environment and Green Building Materials – Sheng (ed.)
© *2015 Taylor & Francis Group, London, ISBN 978-1-138-02718-3*

Research of spatiotemporal schedule model of agents' behavior in artificial society

Liang Ma, Bin Chen, Lao-Bing Zhang, Peng Zhang, Xin Zhou & Xiao-Gang Qiu
College of Information and Management, National University of Defense Technology, China.

ABSTRACT: In this article, a spatiotemporal schedule model of agents' behavior is established. With the main concern for the behavior of agent at the individual level, a hierarchical model is designed, in which activities are classified as mandatory activity, traveling activity, and flexible activity. Naturally, activities linked by time order constitute the one-day behavior pattern, which may differ on weekdays and at weekends. The one-week behavior pattern is composed of weekday behavior pattern and weekend behavior pattern. It is significant to make the model data driven when more data sources are available in the era of big data, so a hierarchical model configuration mechanism is proposed in this article. Then, an algorithm of activity chain generation is put forward. As a case study, taking student agent as an example, the activities chain generating results proved to be temporally heterogeneous at the individual level. We conclude that the proposed model may be effective in modeling agents' behavior in an artificial society.

KEYWORDS: Artificial Society; Agents' Behavior; Hierarchical Model

1 INTRODUCTION

1.1 *Artificial society*

The key of computational social science is to establish an artificial society (Wang Fei-yue, 2004), in which a person in the real world is modeled as an agent with certain attributes such as age, gender, role, and so on. Besides, the relationship among human beings and the relationship between human beings and the environment are also considered in artificial society.

Modeling the behavior of agents is a meaningful and hard-pressed part in artificial society, especially for social phenomena such as epidemic control, transportation planning, and emergent evacuation driven by agents' behavior (ZHOUTao, 2013). This article focuses on modeling city-level agents' behavior.

1.2 *Related works*

Studying the behavior pattern of human beings had been a very tough and expensive task. The data collected in Uppsala and Mobidrive project were recorded by questionnaires or other paper material, thus limiting the quantity and accuracy of the data (KW Axhausen, 2002). Deficiencies exist in the general modeling paradigm of agent behavior. This is because it was not easy to collect the original data of human behaviors earlier. But the era of big data is just around the corner, in which wireless sensor devices (GPS, mobile phone, etc.) are used to collect contact data on the time-resolved face-to-face proximity among individuals. Marta C. González et al. studied a

trajectory of 100000 anonymous mobile phone users whose position is tracked for a six-month period and found that human trajectories show a high degree of spatiotemporal regularity, indicating that despite the diversity of their travel history, humans follow simple reproducible patterns (Marta C. González, 2008).

In this article, a spatiotemporal schedule model of agents' behavior is discussed in section 2, an extensive frame, which can fill up-to-date data into the model to bridge the gaps between an individual agent's behavior and a person's behavior in the real world. Then, in section 3, the generation of activities chain is discussed. In the next section, a case study is presented to validate the model. Finally, the work is summarized and the future work is discussed in section 5.

2 MODEL ESTABLISHMENT

2.1 *Agent*

Agent, a "random realization" of the census data for the region being simulated, means a virtual or synthetic person in artificial society who maintains the demographic "structure" of the actual population (GeYuanzheng, 2014). In other words, a census taken from the synthetic population would, within statistical limits, return the original census. An agent is composed of attributes such as gender, age, role, and so on. In the case of some specific situation, the domain attributes should be added. For example, disease-related attributes are added in the modeling of the epidemic.

2.2 Hierarchical spatiotemporal schedule model

Each individual is characterized by a time-independent characteristic travel distance and a significant probability to return to a few highly frequented locations (ZHOUTao, 2013). This suggests that the behavior pattern is of great regularity.

In the case of a student agent, table 1 could be used to describe the agent's one-day behavior pattern.

Table 1. One-day activities of student agent.*

Start time	Duration	End time	Activity Set	Type
[00:00,00:00]	/	[06:30,07:30]	1 sleep (100%)	MA
/	[00:10,00:20]	/	1 breakfast (100%)	FA
[07:50,08:05]	/	[11:55,12:20]	1 class (60%)	MA
			2 experiment(20%)	
			3 library(20%)	
/	[00:10,00:30]	/	1 lunch(100%)	FA
/	/	[14:00,14:30]	1 rest(100%)	MA
[15:10,15:30]	/	[17:50,18:00]	1 class (50%)	MA
			2 experiment(20%)	
			3 library(30%)	
/	[00:30,01:00]	/	1 recreation(50%)	FA
			2 do exercise(50%)	
/	[00:10,00:30]	/	1 dinner(100%)	FA
/	[01:30,02:00]	/	1 class(70%)	MA
			2 recreation(30%)	
/	/	[24:00,24:00]	1 sleep(100%)	MA

*(1) blank area in the first three columns means there are no constraints on the corresponding element (including start time, duration, end time). a:b in the column of start time and end time means clock time, whereas in the column of duration it means the activity will last for (a*60+b)minutes.
(2) Activity Set column gives a rather rough description of activity branches in a certain period, which may be configured more accurately with more statistical data.
(3) MA and FA act as labels to define the types of activities that will be discussed later.

One-day behavior pattern (DP) is defined as given next:

$$\begin{cases} DP_p =< Popu, ActivityFrame > \\ ActivityFrame = \{ActivitySet\} \\ ActivitySet = \{Activity\} \end{cases}$$

Popu means the role of the agent, and DP_p defines the behavior framework of the agent whose role is p. *ActivitySet* defines the activity branches. *Activity* is the basic element that is used to describe the behavior of the agent, which is defined next:

$$\begin{cases} Activiy =< Time, ActionType, Location, Probability > \\ Time =< Start, Duration, End, StartDist, DurationDist, EndDist > \\ \sum_{i=1}^{n} probability_i = 1, \text{n is the size of } ActivitySet. \end{cases}$$

Time is constrained in three aspects: start time, duration, and end time, which are in line with a certain statistical distribution. In each period defined by *Time*, the agent can only select one activity and the probability value to be selected is determined by *probability*. The type of selected activity is described by *ActionType*, and where to execute the activity is determined by *Location*.

Activities are different from each other regarding their importance throughout the whole day. Taking students as an example, study is a necessity for a student every day and its time constraints are stricter than sleep and exercise. We define activities such as study with strict time constraints as mandatory activities (MA), whereas other activities with less strict time constraints are called flexible activities (FA). In addition, traveling activities (TA) are used to transfer agents from one location to another. The categories of activities are described by table 2.

Table 2. Activity types.

Type	Characteristic	Example
MA	must-do activity	study, work
FA	Influenced by MA	eat, recreation
TA	transport agent	car, bicycle

Though one-week behavior pattern (WP) of an agent is stable, difference exists between behavior patterns on weekdays and at weekends. WP comprises behavior patterns on weekdays and at weekends, so it is natural to define WP as given next:

$$WP = \sum_{i=1}^{5} DP_p^{weekday} + \sum_{i=6}^{7} DP_p^{weekend}$$

2.3 Hierarchical configuration mechanism

In artificial society, agents' behavior is concerned to a great extent with individual level. But collecting original data related to the behavior of human beings is always a tough work. Thus, it leads to a gap between the modeling behavior and human beings' behavior in the real world. As more and more research would be done in the near future, there will be plenty of data and experiences regarding the human beings' behavior. So we design a mechanism facilitating absorbing up-to-date data into the suggested model.

DP is classified by agents' roles. On the first level, we categorize agents into several types: child, student, worker, retired, and unemployed. Different DP is then assigned to corresponding types of agents. So every type of agent gets a fundamental time framework for executing activities. The type can further be divided into sub-types in detail. The sub-types of students include primary student, middle school student, and college student. The *ActivityFrame* in corresponding DP in the same period varies among sub-types.

In order to configure the *ActivityFrame* among sub-types on the basis of the first-level time framework, we define *ActivityDetail* to make the sub-types and DPs correspond with each other. This is the second-level configuration.

$$\begin{cases} ActivityDetail = < SubPopu, DPType, ActivityDetailSet > \\ ActivityDetailSet = \{ActivityIndex, ActivitySet\} \\ ActivitySet = \{ActionType, Probability, Location\} \end{cases}$$

Obviously, *ActivityDetail* and corresponding DP are matched by *SubPopu* and *DPType*. *ActivityDetail* corresponds to *ActivityFrame* in DP with the help of *ActivityIndex*, and *ActivityIndex* determines which *ActivitySet* in *ActivityDetail* should be matched with that in *ActivityFrame*.

We can classify agents on the basis of the first-level categorization, which could be also divided further. The model could be built in a higher-resolution mode without any change if original data of agents' behavior were summarized to be consistent with the form *ActivityDetail* defined. Therefore, it suggests that the model is data driven.

3 GENERATION OF ACTIVITY CHAINS

The hierarchical configuration mechanism determines the fundamental time framework and activity branches in the corresponding period for every agent in artificial society. Then, discrete event lists generated according to the assigned DP are needed to drive artificial society. Two problems remain to be solved. First, selecting one activity from every *ActivitySet* in every period according to the DP and creating a TA when activity is executed at a different location from the former one, thus formulating an activity pattern called "primitive activities chain" (PAC). PAC cannot be scheduled by simulation engine for the activity lacking the timestamp. Therefore, a second step is urged to convert PAC into an executive activity chain (EAC). In step 2, the start time, duration, and end time get an explicit hour, obeying corresponding statistical distribution. Third, in the process of generating EAC, the time constraint of TA is given by traveling activity generator (TAG) and the location is allocated by location allocator (LA); both of them will be discussed later.

3.1 Traveling Activity Generator (TAG)

In TAG, all the vehicles taken into account are listed in the set *Vehicle*.

$$Vehicle = \{Walk, Bicycle, Personal\_Car, Bus, Metro, Taxi\}$$

Here, two main problems are addressed. The first is to decide which vehicle should be used. In order to solve the problem, we consider two main factors: One is the distance between start location and target location, and another is whether the agent has a personal car.

The second problem to be tackled is the calculation of the time cost by the transferring process.

$$TravelTime = \frac{Distance}{AverageSpeed_{mode}} + ExtraTravelTime_{mode}, mode \in Vehicle$$

ExtraTravelTime is used to measure time cost in waiting-for-bus, traffic jam, and other events delaying vehicles.

3.2 Location Allocator (LA)

According to the location type, the activities are divided into two categories: location-specified activity (LSA) and location-random activity (LRA). The location of LSA is fixed in the whole simulation cycle, and the fixed location is assigned during the generation of synthetic population in artificial society (GeYuanzheng, 2014). Compared with LSA, the location of LRA is more likely to be chosen randomly regarding the factors of proximity and crowdedness.

3.3 Executive Activity Chain Generator (EACG)

The main task of EACG is to assign timestamps to every activity in PAC as well as take the characteristic of activity types (depicted in table 2) into account.

3.3.1 Unit behavior pattern splitting

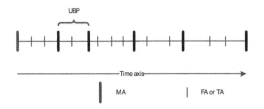

Figure 1. Process of splitting PAC.

We split PAC by MA into several unit behavior patterns (UBP), which are illustrated in figure 1. By planning each UBP, an EAC is generated by combining all the planned UBPs.

3.3.2 Unit behavior pattern planning

Planning the UBP involves two relevant procedures: One is to assign timestamps to every activity according to corresponding statistical distribution, and another is to guarantee the time constraint of MA. The former process is a simple random-number generating problem, and the second process acts as an observer to the former process. The observer will ensure that the start time of an activity is behind the end time of the predecessor. If the observer detects any confrontation, it will try to adjust the PAC from current activity to previous activity to meet the time constraint of MA. In the process of adjusting, the duration of FA limited by a minimum duration is cut down.

4 CASE STUDY

Middle school students from Hepingli Street, Chaoyang District in Beijing, are chosen to testify the spatiotemporal schedule model. There are 2601 middle school students in Hepingli Street. Their one-day behavior pattern is described in table 1. The element in "type" column defines the types of activities while referring to table 2.

Table 3. One of the executive activity's chains.

start time /s	end time /s	action type*	activity type
0	23745	sleep	MA
23745	24647	breakfast	FA
24647	24805	Bicycle	TA
24805	39449	experiment	MA
39449	40992	lunch	FA
40992	50604	rest	MA
50604	59970	study	MA
59970	63552	do exercise	FA
63552	63710	Bicycle	TA
63710	64868	dinner	FA
64868	70440	recreation	MA
70440	86400	sleep	MA

*This column describes the action type except for the FA, for which the vehicle is filled in the column.

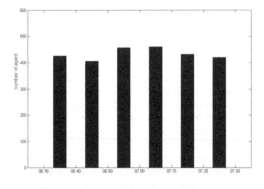

Figure 2. Number of agents per five minutes.

All the selected agents' executive activities chains have been generated, and one of them is listed in table 3.

The executive time of mandatory activities among agents is analyzed. In this case, we take "sleep" as an example. Figure 2 shows the number of agents when executing mandatory activities per five minutes. The results indicate that the numbers of agents are evenly distributed in the limited period. Other mandatory activities follow the same rule: that the activity chain generated for every agent is different from each other.

5 CONCLUSION AND OUTLOOK

In this article, a modeling method for human spatiotemporal behavior is proposed. The method can be used in some scenarios that are driven by agents' behavior, including epidemics, transportation planning, emergent evacuation, and so on. For a special design of hierarchical configuration mechanism, the proposed model is convenient and flexible in the loading of up-to-date statistical data of human behaviors.

The modeling of the dynamics of the activities choices is usually severely limited by the lack of information about habits and previous choices that are not fully involved in the proposed model. We will do further research to take these psychological factors into account in the process of making choices of activities.

ACKNOWLEDGEMENTS

This work was supported by the National Nature and Science Foundation of China under Grant (Nos. 91024030 and 91324013).

REFERENCES

Wang Fei-yue.(2004).Artificial Societies, Computational Experiments,and Parallel Systems: A Discussion on Computational Theory of Complex Social-Economic Systems. Complex Systems and Complexity Science, 1(4):25–35.

MartaC. González, CésarA. Hidalgo & Albert-LászlóBarabási. (2008) Understanding individual human mobility patterns.Nature, 453(5):779–78.

ZHOUTao, HanXiao-pu, YAN Xiao-yong, et al. (2013). Statistical Mechanics on Temporal and Spatial Activities of Human. Journal of University of Electronic Science and Technology of China, 42(4):481–540.

K.W Axhausen, AZimmermann, SSchönfelder, et al. (2002). Observing the rhythms of daily life: A six-week travel diary. Transportation, 29(2):95–124.

GeYuanzheng, MengRongqing, CaoZhidong, et al. (2014). Virtual city: An individual-based digital environment for human mobility and interactive behavior. Simulation, DOI:0037549714531061.

Energy, Environment and Green Building Materials – Sheng (ed.)
© *2015 Taylor & Francis Group, London, ISBN 978-1-138-02718-3*

Research status of magnetic plasma mass separator

H.L. Zhao
Harbin Institute of Technology, China
Northeast Petroleum University, China

B.H. Jiang, C.S. Wang, H.C. Liu & Z.L. Zhang
Harbin Institute of Technology, China

ABSTRACT: The developmental level of nuclide separation technology is directly related to the production and application of nuclide. The magnetic plasma mass separation, which inherits the advantages of traditional electromagnetic nuclide separation and fundamentally resolves its shortcomings, is an important innovation in the area of nuclide separation. This article compares and summarizes various kinds of magnetic plasma mass separation devices, lists their important parameters, summarizes the advantages and disadvantages of various methods of isotope separation on the basis of energy consumption, presents the calculation method of productivity index, and provides a basis for reasonable selection and intensive study of nuclide separation.

KEYWORDS: Nuclide separation; Magnetic plasma separator; Productivity; Energy consumption

1 INTRODUCTION

All countries in the early 1970s began studying about the industrial magnetic plasma separator (MPS) under ionization state, which is more efficient and where there is less energy consumption compared with the traditional electromagnetic isotope separation. Its essence method is the plasma method, including ion cyclotron resonance method by means of plasma cyclotron heating, plasma centrifugation method, and plasma optical mass separation method.

In 1975, Аскаръян first introduced the ion cyclotron resonance separation method[1]. By the end of 1976, the U.S. TRW company had completed and published experimental results of potassium K nuclide separation[2]. Ni nuclide accumulation data are given in[3, 4]. France ERIC device had performed radionuclide separation of Ca, Cr, Zn, Ba, and Yb[5-8]; the device configuration is shown in Fig 1. Russian technology center Курчатовский институт СИРЕНЬ devices had performed Li isotope separation experiment[9]; the device configuration is shown in Fig 2. In 1997, Japan had done Li and k nuclide experiments [10,11]. Systems developed by Kharikov National University (KNU) Department of physics[12]and the international scientific Center Kharikov Institute of physics and technology (ННЦХФТИ) [13,,14] had performed ⁶Li and ⁷Li experiment.

In a series of papers, not only the possibility of constructing MPS is discussed[12,15-20] but also the possibility of applying it to the nuclear waste (RAW)

and reprocessing spent fuel (SF)[21-26] as well as the proposal of some new devices simultaneously is discussed[27,28].

RAW and SF retreatment thought lies in the fact that working substance(RAW or SF) is transformed from solid (liquid) to gas when feeding before the entrance of ionization area. Those ions located in the magnetic field used to create plasma are heated selectively; this urges the path change of ions in the magnetic field. The cold ion and thermion separate and descend in the ion collection panel, performing subsequent removal of sedimentary element from collection panel. Thus, RAW and SFmass ion separation is realized. This is called "separation" according to the constituent, and it is also called "the whole

Figure 1. France ERIC device.

separation." But "the local separation" is the solution to reducing the radioactive granule proportion in RAW and does not require overall retreatment.

Figure 2. Russia СИРЕНЬ device.

The main research work of plasma isotope separation methods currently includes the following: comparative analysis of effective magnetic plasma separation devices of RAW and SF separation in experimental or production period; basic process of current and isotope; plasma speed change formed through linear or non-linear process determined by mathematical means; and so on. This article focuses on the first content, namely the comparison of various plasma separation devices in the main operating parameters and crucial key indicators for assessment; it is also convenient in quickly finding a reasonable and effective isotope separator.

2   PRACTICAL MPS AND COMPARISON

The following two cases to apply MPS to RAW and SF reprocessing are possible: 1) Use MPS as the first level of SF reprocessing; 2) use RAW reprocessing after SF chemical process in chemical radiation factory.

Currently, about six countries are studying and realizing the magnetic plasma separator. Devices can be divided into two categories:

1 Plasma creation, heating, and selective isotope separation in theory (see Table 1).
2 In terms of means of production, a different level of productive forces in RAW and SF reprocessing (see Table 2).

In the early 1990s, the U.S. Department of Energy (DoE) decided to resolve research problems related to RAW and SF reprocessing production devices based on three former production devices. These 3 units had accumulated about 379,000 M3 of waste. According to DoE assessment, with the help of existing technology,

RAW reprocessing would cost about 200 billion dollars, and the plan cannot be achieved until 2028. But in accordance with other assessments, RAW reprocessing based on existing process needs 44 years, that is, this work could not have been accomplished within the time limit mentioned earlier.

The U.S. company ATG use MPS during liquid RAW processing. This technology can significantly reduce crystal quantity (each bucket of M3 volume of the crystallization RAW worth 0.5 ~1 million US dollars), save processing and preservation resources. RAW reservation demands are more than those related to preservation of natural or light radioactive wastes.

Figure 3. America PMF (DEMO) device.

The ATG company plans to create a demo unit - Plasma Quality Filter (PMF). A corresponding experiment had been performed and simultaneously established a reprocessing plant containing two PMF RAW; the project cost 16 billion dollars. This PMF unit is a single hole unit with a cylindrical-shaped magnetic field and a radial electric field along the system axis[23,28,29]; the unit profile photo is shown in Fig 3. RAW enters the ejector in the form of gas or steam, ionizing forms at about 5000 C of hot plasma, and then the thermal plasma is sprayed into the center area. Ions proceed with mass separation in perpendicular magnetic field and electric field, and at last, they are collected in the receiver. The light ion is collected in a vacuum tank face on the PADS, whereas the heavy ion is collected on the side of the center area in the vacuum tank.

In the separator ПС-1, the working medium is ionized into plasma after entering into plasma source in the form of gas, vapor, or microparticle. The plasma depart in accordance with ion cyclotron resonance principle after entering the uniform magnetic field[24].

The existing MPS characteristic carries on ion heating and separation in the orthogonal electromagnetic field, in which electric field foundation may be realized in the following ways:

1 Separator ПС - 1 is inductive. An electric field is induced that can accelerate ions through the weak

Table 1. Separation device for research.

Separation device, country, and year	TRW.Inc[2] USA 1976	ERIC[5-8] France 1989	СИРЕНЬ[9] Russia 1993
Significance of device	Separation	Enrichment、Separation	Separation
Overall dimensions: L- length, R- Device radius, r_p- Plasma radius	L-1m $2r_p$-5.7cm	L-3.4m, R-0.15m $r_p\approx$6cm	L-1.6m, R-0.3m r_p-3cm
Magnetic system type Magnetic field strength/T	Thermomagnetism system≤0.4	Superconducting System3	Thermomagnetism system 0.1~0.28
Magnetic field uniformity ΔH/H	-	5×10^{-3}	10^{-2}
Creation method and mechanisms of plasma	Thermal ionization, high-frequency discharge	Thermal evaporation, ultra high-frequency discharge	Thermal evaporation, arc, high-frequency discharge
Working frequency of external heater/kHz	$(0.8~1)\times10^5$	$(1~3.75)\times10^{10}$	$(3~6.6)\times10^5$
n_p/cm^{-3} Plasma density n_p/cm^{-3} T_e, T_i/eV Plasma temperature T_e, T_i/eV	$10^9~10^{11}$ 0.2, 0.2	$10^{11}~10^{12}$ 1.4~6, 0.2~0.6	10^{12} 4~7, 7~10
working fluid	Ne, Ar, Xe, ^{39}K, ^{41}K	Ca, Cr, ^{132}Ba, ^{179}Yb	^{6}Li, ^{7}Li

Table 2. Separation device for production.

Separation device, country, and year	PMF(DEMO) [23,28,29] USA 2001	ПС-1[24] Russia 2004	ОПН-1[21,22,27] Ukraine 2007	ДИС-1[36,37] Ukraine 2009
Significance of device	RAW and SF Reprocessing	RAW and SF Reprocessing	RAW and SF Reprocessing, Isotope enrichment	SF Reprocessing Display analog
Status of device	Comprehensive Experiment	Create	Design	Effective
Geometric dimensions: L- length, R- Device radius, r_p- Plasma radius	L-3.89m R-0.4, r_p-0.375m	- r_p-0.5m	L-4m R-0.75, r_p-0.5m	L-1.75m R-0.19m
Type of magnetic system magnetic field strength/T	Thermal-magnetic system 0.16	-	Superconducting System ≤3	Thermal-magnetic system 0.35
Plasma density n_p/cm^{-3} Plasma temperatureT_e, T_i/eV	$>10^{13}$ 1.5~2, -	$>10^{13}$	$\geq10^{12}~10^{13}$ 50, 20	$\approx10^{11}$ $\approx$5, 5
Ionization consumption /(eV/ion)	500	500	-	<500
Plasma ionization and heating energy consumption/(MW)	5	0.6	1~4	0.01
productivity of reprocessing device	0.63~0.99t/circadian U	150t/yer U	30t/yer U	280g/yer Ar

horizontal constant magnetic field superposition method.

2 Separator PMF is electrostatic. The plasma potential difference is established through contact of the ring electrode and the power.

3 Beam plasma discharge. During the discharge process, electron and ion heating is conducted in accordance with the beam interaction mechanism. Due to the self-excited ion cyclotron oscillation, plasma swing and plasma heating occurred

during the discharge process[30], rather than an external high-frequency generator and high-frequency antenna. Due to the abandoning of external high-frequency generator and high-frequency antenna, ОПН-1 (Table 2) structure is simplified, and a similar description is provided in [23,24,28,29].

3   PRODUCTIVITY COMPUTATION

MPS productivity (total number of particles) for separating substances can be determined by the following formula[38]:

$$N = Q_n \cdot t \tag{1}$$

where N is particle quantity, Qn is plasma flow, and t is the element accumulating time.

Constant density and temperature of the plasma in the condition of unchanged magnetic field dimensions and the separating device productivity N

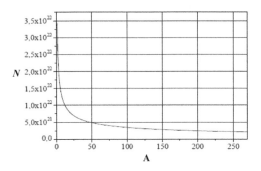

Figure 4. Relation curve of separation device productivity and relative atomic mass.

calculated value is listed in figure 4 depending on the relative atomic mass A.

For element separation, MPS productivity can be described in the next equation:

$$\dot{m} = M \cdot \Delta\mu \cdot n_p(r) \cdot v_p \cdot S_p(r) \cdot K_c \tag{2}$$

Table 2 productivity evaluation is based on this standard, where M - relative Atomic mass g; $\Delta\mu$ -percentage composition of material elements; $n_p(r)$ - the unit radius R of the plasma density cm^{-3}, v_p-plasma speed cm/s; $S_p(r)$ - plasma beam cross-sectional area cm^2; and Kc - separator efficiency factor.

Coefficient K_c describes the gaseous state material transfer efficiency, the plasma ion separation efficiency according to the quality, and separated ion

collection efficiency in the separation area on collection panel during the ionization process. These are determined according to the equation given next:

$$K_c = K_1 K_2 K_3 K_4 \tag{3}$$

where K_1-ion factor; K_2- working material separation factor; K_3 - receive factor, apart from ion landed on the inside surface of vacuum tube, statistical quantity of deposition ion on the receiving surface; and K_4 – separator internal ion loss coefficient due to the scattering and charge exchange in neutral atoms.

4   ENERGY CONSUMPTION EVALUATION

Energy consumption is an important characteristic index that is used to evaluate MPS. ОПН-1 unit energy consumption mainly includes the following processes: 1) solid material into a gas; 2) gas working medium ionization and plasma heating; 3) the establishment of a magnetic field; and 4) maintainence of MPS vacuum tank working pressure.

The literature[31] completed the evaluation of minimum energy consumption from solid into gas. To ensure that uranium plasma beam is maintained at 2.2×10^{21}(cm^{-3}) level, the lossless gas-ionization-demanded power infused into the vacuum tank is about 10kW[32].

In accordance with the rp/rb≈5~10cm (rp-plasma beam radius; rb - electron beam radius) size relationships in[33], we know that when rp=50cm, rb should be not less than 10~5cm. This requires that the cross-section of the electron beam with the launch-pad is approximately 200~400cm^2. When electron density is 1~2A/cm^2, consumption of power ranges from 1 MW to 4 MW. In essence, this power is used for ionized and plasma heating.

To create 3.14m^3 in volume with 0.15T field strength and water-cooled electrodes of a uniform magnetic field, demanded power is 0.28 MW. In order to make the unit change in a larger scope of the magnetic field strength, energy consumption used to create a magnetic field is not constant. For example, for ERIC and ОПН-1, power consumption is dependent on the purpose of this experiment, the unit work mode, and tasks. The RAW "knockout" energy of the magnetic field is not more than 2~3% of total power consumption, and the elements of the overall separation" indicate an achievement of 10%. Separate methods of enrichment of heavy isotopes achieve 50%. For the sake of efficiency and unit large volume (>0.5m^3), plasma heat needs a high magnetic field strength value of H>0.05~0.1T. MPS vacuum tank maintains the work pressure, and the energy consumption of low-temperature vacuum pumping speed of approximately 18,000 is about 16.4 kW[34].

In accordance with plasma creation and heating process of ion cyclotron resonance method, we can determine the unit structure, configuration and components, and working and maintenance program. In these conditions, the plasma internal and external geometrical cross-cutting dimension, the vacuum tank and the magnet system size have an average of 1:3:4.5:5.75. All of these determine the magnetic field homogeneity, necessary structures, and these strictly require the magnetic field coil to produce accuracy with regard to the installation site and the magnetic property of applied materials ($\mu \leq 1.05$)[35].

In addition, the magnetic plasma separator ДИС-1 through the magnetic mirror and the external injection method[36,37] mainly studies the heavy ion plasma source working condition. The unit can perform separation of heavy ion component from ionized gas mixture.

5 CONCLUSIONS

This article summarize the MPS status in the world's major nuclear countries and its advantages and disadvantages in terms of separation device geometry dimension; the type of magnetic system and magnetic field strength; and plasma creation method and mechanism. We should summarize and list the productivity calculation method; calculate separation device assessment method and process in energy consumption for a comparative evaluation of various separation devices; and provide a basis for reasonable selection and intensive study of isotope separation devices.

Fund project:
The authors would like to acknowledge the support of the National Fundamental Science Research Grant (No. 51207033) and International scientific and technological cooperation projects of China (No. 2011DFR60130).

REFERENCES

[1] Г.А. Аскаръян, В.А. Намиот, А.А. Рухадзе. Изменение массового состава плазмы в плазменных ловушках при ионном циклотронном нагреве[J]. Письма в ЖТФ. 1975, 1(18):820–823.

[2] J.M. Dawson, H.C. Kim, D. Arnush. Isotope Separation in Plasmas by Use of Ion Cyclotron Resonance[J]. Phys. Rev. Lett. 1976, 37:1547–1550.

[3] M. Mussetto, T.E. Romesser, D. Dixon et al. IEEE Int. Conf. on Plasma Science[C], San Diego, Calif. 1983. IEEE Conf. Records Abstracts. 1983:70.

[4] M. Mussetto, H. Bull. Isotope separation using selection ion cyclotron resonance heating[C]. American Phys. Soc. 1983, 28:1029.

[5] P. Louvet. II In Proc: 2nd Workshop on Separation Phenomena in Liquids and Gases[C], Versailles. 1989, I: 71.

[6] A.C. La Fontaine, P. Louvet. Compte rendu des[C]. Journees sur les Isotopes Stables. Saclay. France. 24–25 Novembre. 1993:332.

[7] P. Louvet, A.C. La Fontaine, B. Larousse, M. Patris. The 4th Int[C]. Workshop on Separation Phenomena in Liquids and Gases. Beijing. P.R. China. 19–23. August 1994:83.

[8] P. Louvet, A.C. La Fontaine. Proc. of Int. Conf. on Chem. Exchange and Uranium Enrichment[C]. Tokyo. Japan. Bulletin of the Research Lab. for Nucl. Reactors. 1990:289.

[9] А.И. Карчевский, В.С. Лазько, Ю.А. Муромкин. Исследование разделения изотопов лития в плазме при изотопически селективном ИЦР-нагреве[J]. Физика плазмы. 1993, 19:411–419.

[10] T. Suzuki, Sh. Kugai, N. Fujita et al[C]. Fall Meeting of the Atomic Energy Society of Japan. 1997:L67.

[11] Y. Kawai, T. Suzuki, M. Nomura, N. Fujita[C]. Fall Meeting of the Atomic Energy Society of Japan. 1997: L66.

[12] А.Н. Довбня, А.М. Егоров, В.Б. Юферов. Сравнительный анализ проектов плазменных сепараторов изотопов с колебаниями на циклотронных частотах[J]. Вопросы атомной науки и техники. 2004(4):51–57.

[13] Б.С. Акшанов, Н.А. Хижняк. Новый эффективный метод разделения изотопов[J]. Письма в ЖТФ. 1991, 17(6):13.

[14] Б.С.Акшанов, В.Ф.Зеленский, Н.А.Хижняк. Метод разделения изотопов в системе встречных, аксиально-симметричных магнитных полей[J]. Вопросы атомной науки и техники. 2000(4):198.

[15] О.М. Швец, В.Б. Юферов, Е.И. Скибенко и др. Труды Украинского вакуумного общества[M]. Киев. 1995, 1:195.

[16] В.Б. Юферов, Д.В. Винников, О.С. Друй, В.О.Ильичеваи д.р. Вестник Национального технического университета ХПИ. Тематический выпуск[D]. Электроэнергетика и преобразовательная техника. 2004, 35:169–179.

[17] В.И. Волосов, И.А. Котельников, И.Н. Чуркин, С.Г. Кузьмин. Установка для разделения изотопов методом ИЦР-нагрева[J]. Атомная энергия. 2000, 88(5):370–378.

[18] А.Г. Беликов, В.Г. Папкович. Некоторые возможности получения изотопов в системе с остроугольной геометрией магнитного поля[J]. Вопросы атомной науки и техники. 2004(4):58–63.

[19] И.Н. Онищенко. Параметрический магнитный сепаратор[J]. Вопросы атомной науки и техники. Серия: Плазменная электроника и новые методы ускорения. 2004(4):64–66.

[20] Ю.А. Кирочкин, А.Ю. Кирочкин. Теоретическое исследование возможности разделения изотопов при движении заряженных частиц в постоянном электромагнитном поле цилиндрического

конденсатора и линейного тока, протекающего вдоль его оси[J]. ЖТФ. 2007, 77(10):89–96.

[21] Е.И. Скибенко, В.Б. Юферов, Ю.В. Ковтун. Концептуальный проект плазменного источника на основе пучково плазменного разряда для сепарационных технологий[J]. Сборник докладов ОТТОМ-8. Харьков, 2007, 1:232–238.

[22] Е.И. Скибенко, В.Б. Юферов, Ю.В. Ковтун. Фор инжектор разделяемого вещества на основе пучково плазменного разряда для ионно атомных сепарационных технологий[J]. Вестник Национального технического университета ХПИ. Тематический выпуск. Техника и электрофизика высоких напря- жений. 2007, 20:180–199.

[23] A. Litvak, S. Agnev, F. Anderegg et al. 30th EPS Conference on Contr[C]. Fusion and Plasma Phys., St. Petersburg, 7–11 July 2003 ECA, v. 27A, O-1.6A.

[24] V.A.Zhil'tsov, V.M. Kulygin, N.N. Semashko et al. Plasma separation of the elements applied to nuclear materials handling[J]. Atomic Energy. 2006, 101(4): 755–759.

[25] В. Б. Юферов, С.В. Шарый, В.А. Сероштанов, О.С. Друй. Криогенная система откачки плазменного сепаратора элементов ДИС[J]. Харьковская нанотехнологическая ассамблея. Вакуумные нанотехнологии и оборудование, Харьков, 2006, 1: 58–61.

[26] В.Б.Юферов, С.В.Шарый, В.А.Сероштанов, О.С.Друй. О некоторых проблемах при выборе плазменного источника для сепаратора элементов[J]. Сб.докладов ОТТОМ-8. Харьков, 2007, 2:76–79.

[27] Патент.№24729 Україна. Пристрій для розділення речовини на елементи[P]. Є.І. Скибенко, Ю.В. Ковтун, В.Б. Юферов.Бюллетень 2007, 10:3.

[28] Tihiro Ohkawa. Plasma mass filter[P]. US Patent №6.096.220 /T. Ohkawa. Aug. 1. 2000:54.

[29] J. Gilleland, T. Ohkawa, S. Agnew et al. WM'02 Conference[C]. February 24–28, 2002, Tucson, AZ.

[30] Я.Б. Файнберг. Взаимодействие пучков заряженных частиц с плазмой[J]. Атомная энергия. 1961, 11:313.

[31] Ю. В. Ковтун, Е. И. Скибенко, В. Б. Юферов. Тепловые характеристики блока фазовых превращений форинжектора металлической плазмы для резонансного сепаратора элементов[J]. Вопросы атомной науки и техники. Серия: Вакуум, чистые материалы и сверхпроводники. 2007,16(4): 179–183.

[32] И.С. Григорьев. Разделение изотопов с помощью света[M]. Препринт ИАЭ 6246/12, РНЦ «Курчатовский институт», 2002.

[33] М.Ю. Бредихин, А.И. Маслов. Исследование нагрева электронов плотного пучково-плазменного разряда в сильных магнитных полях[J]. Украинский физический журнал. 1973, 2:315–316.

[34] Ю.В. Холод, Б.В. Гласов, В.И. Курносов, Е.И. Скибенко. Вопросы атомной науки и техники[J]. Серия: Общая и ядерная физика. 1984(1):20–25.

[35] М.Ю. Бредихин, Б.В. Гласов, Р.В. Пайл. Магнитная система для исследования лоренцевой ионизации высоковозбужденных атомов водорода[J]. Сб. статей Физика плазмы и проблемы управляемого термоядерного синтеза. Киев, 1971(2):251–260.

[36] В.А. Сероштанов, С.В. Шарый, В.Б. Юферов и др. Двухступенчатый плазменный источник со сжатым вакуумно-дуговым разрядом сепаратора[J]. Вісник Харківського університету. Серія фізична: Ядра, частинки, поля. 2008, №794:111–114.

[37] С.В.Шарый, В.А.Сероштанов, В.Б. Юферов и д.р. Стационарный газовый плазменный источник тяжелых ионов с дрейфом электронов[J]. Вісник Харківського університету. Серія фізична: Ядра, частинки, поля. 2008, №794, 1(37):121–124.

[38] Ю.В. Ковтун, Е.И. Скибенко, В.Б. Юферов. Оценка производительности магнито- плазменных (электромагнитных) сепараторов при получении изотопно чистых веществ[J]. Вопросы атомной науки и техники. 2008, 1(17):35–41.

Energy, Environment and Green Building Materials – Sheng (ed.)
© 2015 Taylor & Francis Group, London, ISBN 978-1-138-02718-3

Antimicrobial effect mechanism of polyphenols against campylobacter jejuni

Qi Tian, X.J. Du, Rui Xue, J.J. Gen, X.F. Xie, Bin Liang & J.P. Wang
Tianjin University of Science and Technology, Tianjin, China

ABSTRACT: Eugenol is obtained by distillation after extraction from the dry flower bud of the volatile oil, which belongs to the myrtle family plant clove, using a certain amount of sodium hydroxide solution to separate the phenol oil. The tea is impregnated in polar solvent, and the column separation method is used so that the purity of the tea polyphenols can reach 98%. This is the first time that these two kinds of materials are used to dispose the jejunum campylobacter (it can cause human fever, acute enteritis, and guillain-barre syndrome of the food source pathogenic bacteria) for the research of the antibacterial mechanism. The possible targets of the EG and TP in bacteria might be cell wall, cytoplasmic membrane protein, and DNA as indicated by Scanning electron microscope, Confocal laser scanning microscope analysis, cytoplasmic membrane permeability, DNA binding, and Proteomics experiments, respectively.

KEYWORDS: eugenol; tea; polyphenols; jejunum; campylobacter; antibacterial mechanism

1 INTRODUCTION

Eugenol (EG) is a major component of clove essential oil. It has demonstrated several biological activities such as anti-inflammatory properties, anti-oxidation, and so on. Tea polyphenols (TP) is a kind of natural compound extracted from green tea; it can be used as a natural food preservation agent (Almajano et al. 2008; Shumin et al. 2011; Perumalla and Hettiarachchy2011) and functional food (Wu and Wei2002; Kler et al. (2009). Some studies also show that TP has a broad-spectrum antibacterial activity, and it can inhibit the growth of most food-borne pathogens such as Serratia marcescens, Pseudomonas aeruginosa, Escherichia coli, Salmonella typhimurium, Listeria monocytogenes, and Staphylococcus aureus (Almajano et al.2008; Cooper et al. 2005; Hamilton-Miller 1995; Carson et al. 2002; Anna et al. 2003). TP has great potential in food preservation by antimicrobial activity.

Campylobacteriosis exhibits a worldwide zoonosis. In industrialized countries, Cumpylobacter is one of the most frequently identified causes of intestinal disease. Campylobacter jejuni is found as a commensal in the intestinal tract of a wide range of warm-blooded animals, both domestic and wild.

2 MATERIALS AND METHODS

2.1 Materials

The materials used were as follows: Eugenol (content is 99%, the batch number 200306091), Tea polyphenol (content is 95%, the batch number 84650602), and jejunum campylobacter (campylobacter jejuni) ATCC F38011.

2.2 Minimal inhibitory concentrations of the EG and TP

The MIC was determined by the microbroth dilution method (Branen and Davidson, 2004, H. Wang, Y. Lu, 2009) using the sterile emulsifier diluted eugenol mother liquor in different concentration gradients of diluent. The serial twofold dilutions of the EG were 0.005, 0.0025, 0.00125, and 0.00062(mg/ml). Using the MHB diluted Tea Polyphenols liquor into different concentration gradients of diluent, The serial twofold dilutions of the TP 80, 60, 40, and 20(mg/ml), with joining DMSO (1%) and blank as a control. Campylobacter Jejunum in MH medium, 37°C, 13000r/min, three gas 24 h (5% O_2 and 10% CO_2), measured by the double-broth dilution method to counting the Colony in 0 h, 5 h, 10 h, and 24 h, respectively. The minimal inhibitory concentrations (mic) of Eugenol and Tea Polyphenols were determined.

2.3 Scanning Electron Microscopy (SEM) was performed on campylobacter jejuni by the EG and TP

While dealing with the C.jejuni with antimicrobial substances, no treatment group was found for comparison. 37°C, training for 24 h with three gas. First, 1ml of microbial is taken and washed thrice with PBS (pH=7.2). Then, the 2.5% glutaraldehyde is used to fix. In turn, 30%, 50%, 75%, and 95% ethanol is used to dehydrate, once per 10min. The bacteria till critical-point drying after dehydration after metal spraying coating, bacteria morphology, and the change in the structure were observed under scanning electron microscopy (SEM).

2.4 Confocal laser scanning microscope analysis of campylobacter jejuni treated by the EG and TP

While dealing with the C.jejuni with antimicrobial substances, no treatment group was observed for a comparison. 37°C,training for 24 h with three gas. First, 1ml is taken and centrifuged to get a clear liquid in 13000r/min. Double fluorescent dye with FITC-ConA and PI is used first with FITC-ConA dark staining for 10 min and then with PI dark staining for 10 min. Then, the sample is added onto a clean glass slide, observing under the Laser scanning confocal microscope and collecting the images.

2.5 DNA binding assay

It has been reported that antimicrobial peptides could reach the inner structure of cells through the membrane and bind to bacterial DNA with or without the disrupting membrane. (Brogden, 2005; Tang, Shi, Zhao, Hao, & Le, 2009; Du et al. 2005) reported that tea polysaccharide had a remarkable degradation effect for plasmid DNA by agarose gel electrophoresis, which indicated that DNA was a possible target of the antibacterial drug.

Plasmid (pCAMBIA 1303) DNA (0.2lg) was placed in 15 of µl TE buffer consisting of 10 mM Tris and 1 mM EDTA (pH 8.0) after it had been mixed with EG (0.01mg/ml and 0.005mg/ml) and TP (100mg/ml and 60mg/ml).

Table 2. The MIC of Tea polyphenols of jejunum campylobacter.

| Groups (mg/ml) | The number of colony(lgCFU/ml) | | | |
| | Time(h) | | | |
	0	5	10	24
control sample	6.51	6.43	7.62	9.25
DMSO	6.41	6.48	7.84	9.23
TP80	6.32	0	0	0
TP60	6.36	6.00	5.78	5.32
TP40	6.30	6.20	6.10	5.84

The reaction mixtures were incubated at 30°C for 1 h before being loaded on a 1.2% agarose gel, and the electrophoresis was performed for 30 min at 30°C under 100 V. Starch was used as a control.

3 RESULTS

3.1 Minimal inhibitory concentrations of the EG and TP

From **Tables 1** and **2**, we can see that both Eugenol and Tea polyphenols have significant inhibitory effects on jejunal campylobacter. According to the treatment group in the colony growth, lgCFU within a 24-h value is lower than the control group 3-log value of the minimum inhibitory concentrations MIC value. The MIC values were 0.00125 mg/ml and 40 mg/ml. Specific results are shown in table.

Table 1. The MIC of Eugenol of jejunum campylobacter.

| Groups (mg/ml) | The number of colony(lgCFU/ml) | | | |
| | Time(h) | | | |
	0	5	10	24
control sample	6.51	6.43	7.62	9.25
DMSO	6.41	6.48	7.84	9.23
EG0.005	6.36	6.52	5.08	0
EG0.0025	6.60	6.40	5.30	4.85
EG0.00125	6.46	6.30	5.40	5.94

3.2 Electron microscope analysis of campylobacter jejuni treated by the EG and TP

1 Scanning electron microscopy (SEM) was performed on campylobacter jejuni by the EG and TP.

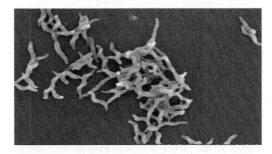

Figure 6. Normal C.jejunum.

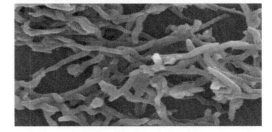

Figure 7. Treated C.jejunum with EG.

386

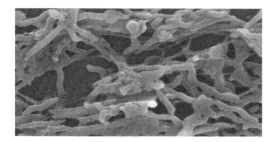

Figure 8.   Treated C.jejunum with TP.

From **figure 6**, we can see that the normal jejunum campylobacter is spiral, and the surface is smooth full. From **figure 7**, we can see that after the eugenol role in which the S shape of the Jejunum campylobacter bacteria was stretched and rashined, the surface of the bacteria was damaged. Some bacterial somatic cells ruptured, and content overflowed. From **figure 8**, we can also clearly see that bacterial cells are broken up by the role of tea polyphenols. Obvious things occur between the bacteria and the bacterial adhesion phenomenon.

3.3   Confocal laser scanning microscope analysis of campylobacter jejuni treated by the EG and TP

Figure 9.   C.jejuni contrast.

Figure 9 is the negative control of jejunum campylobacter: From the figure, we can see that there are no dying cells in the negative control. Membrane intact cells constitute the vast majority in the vision.

Figure 10.   EG processing.

The C.jejuni were treated with 0.0025 mg/ml eugenol jejunum after 3h. In significant reductions observed in the whole cells, the dead cells compared with the control group exhibited a large increase. Thus, we can see that the Eugenol has a strong inhibitory effect on C.jejunum.

In the 40 mg/ml of tea polyphenol treatment related to the C.jejunum after 3h, the complete cells started being drastically reduced. Dead cells compared with the control group exhibited a vast increase; thus, we can see that the tea polyphenol has a strong inhibitory effect on C.jejunum.

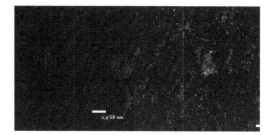

Figure 11.   TP processing.

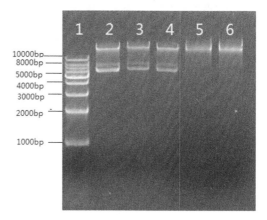

Figure 12.   Interaction of the EG and TP with plasmid DNA. Lane 1,0.2 µg Mark, Lane 2,0.2µg plasmid DNA alone; lanes 3, 0.2µg plasmid DNA and 0.2µg EG; lanes 4, 0.2µg plasmid DNA and 0.2µg EG,lanes 5, 0.2µg plasmid DNA and 0.2µg TP, lanes 46,0.2µg plasmid DNA and 0.2µg TP u.

3.4   DNA binding assay

Lane 1 is 1Kb Mark; Lane 2 is the comparison of the Plasmid DNA3300; and lanes 3 and 4 are a result of the role of eugenol(0.01mg/ml and 0.005mg/ml) with reference to no obvious change. Lanes 5 and 6 are a result of the role of tea polyphenol (100mg/ml and 60mg/ml) with reference to a different obvious change. In lanes 1 and 2, the DNA might be broken down, and the ring is cut to form a linear structure.

4 CONCLUSION

The eugenol and tea polyphenols of C.jejunum have significant bacteriostatic effects. Ultra-structural analysis by SEM provided evidence that EG and TP cause alterations in the outer membrane's integrity as well as disruption of cell walls. During CLSM analysis, we can see that the bacterial membrane is damaged. In the study, the plasmid DNA was bound to the EG and TP and was then decomposed into small pieces. Thus, DNA might be considered another target for the TP, and the impact of the EG is seldom.

ACKNOWLEDGMENTS

This study was supported by Tianjin university of science and technology of China. The authors are deeply grateful to professor Shuo Wang.

REFERENCES

Branen, J.K., Davidson, P.M., 2004.Enhancement of nisin, lysozyme, and monolaurin antimicrobial activities by ethylenediaminetetraacetic acid and lactoferrin. International Journal of Food Microbiology 90, 63–74.

Almajano MP, Rosa CJ, Angel LJZ, Michael HG (2008) Antioxidant and antimicrobial activities of tea infusions. Food Chem 108:55–63.

Mamoru O, Takuo S, Kazuhiko H, Reiichiro S, Mizuki H, Eiji H, Chizuko Y, Hajime K (2011) Antimicrobial susceptibilities and bacteriological characteristics of bovinePseudomonas aeruginosa and Serratia marcescens isolates from Mastitis. Vet Microbiol 154:202–207.

Shumin Y, Jian-rong L, Jun-li Z, Yi L, Linglin F, Wei C, Xuepeng L (2011) Effect of tea polyphenols on microbiological and biochemical quality ofCollichthysfish ball. J Sci Food Agric 91(9):1591–1597.

Wu CD, Wei GX (2002) Tea as a fun evaluation of antimicrobial efficacy of herbal alternatives (Triphala and Green Tea Polyphenols), MTAD, and 5% sodium hypochlorite againstEnterococcus faecalis biofilm formed on tooth ctional food for oral health. Nutrition 18(5):443–444.

Kler A, Zenger R, Dimpfel W (Jun 2009) Green tea extract, especially for use as a functional food item, food supplement or corresponding ingredient, the use thereof and method for producing said green tea extract, European patent application, Patno:Ep2066186.

Backert S, Kwok T, Schmid M, Selbach M, Moese S, Peek RM, Konig W, Meyer TF, Jungblut PR (2005) Subproteomes of soluble and structure-bound Helicobacter pyloriproteins analyzed by two-dimensional gel electrophoresis and mass spectrometry. Proteomics 5:1331–1345.

Anna RB, La Simona TM, Gioia B, Spiridione G, Vincenzo E, Dario R (2003) (-)Epigallocatechin-3-gallate inhibits gelatinase activity of some bacterial isolates from ocular infection, and limits their invasion through gelatine. Biochim Biophys Acta 1620:273–281.

Energy, Environment and Green Building Materials – Sheng (ed.)
© 2015 Taylor & Francis Group, London, ISBN 978-1-138-02718-3

Improving the mechanical properties of two aluminum alloys by thermal cycling treatment

C.Y. Chen

Handan Polytechnic College, Handan, China

ABSTRACT: In order to improve the mechanical properties and the rigidity of the materials after heat treatment, we should use 7075 aluminum alloy and 7A04 aluminum alloy for test with the thermal cycling treatment. The influence of different number of cycles and cycle alternating time on the two kinds of materials should be tested, and the test results and microstructure should be analyzed. The result shows that the increase of the thermal cycling time has an obvious effect within a certain range, and the alternating cycle time influence on the two kinds of materials is very different. So the hardness can be improved only by choosing a reasonable arrangement method based on the material characteristics.

KEYWORDS: Thermal Cycling Treatment; 7075 aluminum alloy; 7A04 aluminum; hardness

1 INTRODUCTION

It is common to eliminate the residual structure stress of aluminum alloy by mechanical stretching, but it cannot work on some complex shape of blank pieces. In this case, we usually use the aluminum alloy's characteristic promotion at low temperature to improve the dimensional stability.

7075 and 7A09 belong to AL-Zn-Mg-Cu series heat treatment hardening alloy and are widely used in aviation. The former is mainly used for the aircraft's high stress structure parts and mold, whereas the latter is mainly used for the aircraft bolts and landing gears. With the development of aviation, structure performance is more important, in order to improve the structural strength of the 7075 and 7A09. We attempt to improve their mechanical properties by using the thermal cycling treatment.

2 CONCEPT OF THE THERMAL CYCLING TREATMENT

The thermal cycling treatment is one form of cryogenic treatment. The cryogenic treatment process originated in Switzerland; it drew universal attention in the 1960s and then rapid development. In the 1990s, the research on supersteel, carbide, and other materials started in China. Aluminum alloy, whose metallographic structure has different expansion coefficients, can use the thermal cycling treatment to improve the mechanical properties. According to some research on LY12 and ZL102 alloy, the thermal cycling treatment can improve the mechanical

properties significantly and eliminate the effect of stress relaxation.

However, the use of thermal cycling treatment has a time requirement; for some alloys, a slow alternating treatment can reduce residual stress, but for Steel35 and other materials, only a rapid alternating treatment can be effective. There is little research on the 7075 and 7A09 thermal cycling strategy and the alternating time effects on mechanical properties, so these are our key points in this article.

3 TEST AND RESULTS ANALYSIS

3.1 Materials and methods

The test samples are $\Phi18mm*20mm$, 7075, and 7A09 Aluminum alloy bars. The chemical composition of the test samples is shown in table 1.

Table 1. Chemical composition of the test samples (element mass fraction %).

Material	AL	Zn	Mg	Cu	Mn	Cr	Si	Fe	Ti
7075	rest	6.0	2.21	1.77	0.30	0.20	0.40	0.50	0.2
7A09	rest	5.59	2.27	1.67	0.15	0.18	0.50	0.50	0.10

Liquid nitrogen is used for the cold treatment; the test method and test group are shown in table 2. The two aluminum alloys used solid solution treatment and were heated for 2 hours (480 degC). Then, the cryogenic treatment was used; the cryogenic temperature for 7075 was -196 degC, 0.5h. Then, there was holding at a high temperature of 175 degC in the medium for a certain time, respectively, such as 0.125 hours, 0.25 hours, and 0.75 hours. 7A09's cryogenic temperature was -140 degC; the time was 0.5h and 0.8h; then, there was holding in the medium at a high temperature of 120 degC for 1h, 2h, and 3h. Finally the two materials went through 120 degC for 4.5h and 175 degC for 10h. Cryogenic treatment of the thermal cycling process flow chart is shown in Figure 1, and 2 cycles are selected as an example (corresponding to test for A2 and B2).

Table 2. Test method.

Material	Test number	Heat treatment process +120°×4.5h + 175°×10h
7075	A1	480°×2h + (-196°× 0.5h + 175°× 0.5h) ×1
	A2	480°×2h + (-196°× 0.5h + 175°× 0.5h) ×2
	A2 add	480°×2h + (-196°× 0.5h + 175°× 0.5h) ×3
	A3	480°×2h + (-196°× 0. 5h + 175°× 0.125h) ×2
	A4	480°×2h + (-196°× 0. 5h + 175°× 0.25h) ×2
	A5	480°×2h + (-196°× 0. 5h + 175°× 0.75h) ×2
7A09	B1	480°×2h + (-140°× 0.5h + 120°× 2 h) ×1
	B2	480°×2h + (-140°× 0.5h + 120°× 2 h) ×2
	B2 add	480°×2h + (-140°× 0.5h + 120°× 2 h) ×3
	B3	480°×2h + (-140°× 0.5h + 120°× 3 h) ×2
	B4	480°×2h + (-140°× 0.5h + 120°× 1 h) ×2
	B5	480°×2h + (-140°× 0.8h + 120°× 2 h) ×2

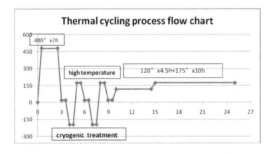

Figure 1. Cryogenic treatment of thermal cycling process flow chart.

3.2 Test result

HB-3000 type Brinell hardness measurement was used for hardness test; the comparison results were shown in table 3.

Table 3. Test result.

Test number	Hardness /HB	Test number	Hardness /HB
A1	102	B1	141
A2	139	B2	177
A2 add	141	B2 add	179
A3	154	B3	180
A4	149	B4	166
A5	147	B5	190

3.3 Analysis

Analyzing the earlier results, the following conclusions are obtained:

a. Effect of the thermal cycling times

More cycling times can improve the mechanical properties of aluminum alloy, and this was reflected in the 7075 and 7A09 aluminum alloy samples. During the first test, we did not consider the impact of cycle times above two, but because of the great improvement during the second thermal cycle, we added another comparison test to find the effect above two cycles, and the result shows there was little effect on the hardness when we used a third cycle.

b. Effects of cycle alternating time

For the 7075 aluminum alloy, after solid solution treatment, the second phase particle on the original α(Al) was integrated into the α(Al), strengthening the solid solution effect, so the mechanical hardness and strength is improved, but the matrix lattice deformation became brittle. After cryogenic treatment at -196 degC for the past 0.5 hours, the second phase that dissolved in the matrix separated out in the form of a small patch, the characteristics of the original structure that were filled with bulk of the second phase were changed; and the small second phase can enhance the mechanical properties of materials as well as get higher hardness and higher plasticity. Then, heating at a temperature of 175 degC, the small second-phase particle distribution and intensity was changed, so as to obtain higher and more stable mechanical performance. The number of cycles determines the size and quantity of the second-phase precipitation, whereas the smaller the second phase the better the quality of performance. As the result shows, the preservation time in the high temperature medium should not be too long: Under the condition of 0.125 hours, we get the highest hardness of 154 HB. As the preservation time grew, hardness decreased, eventually become stable, and the stable value should be slightly less than 147HB.

There is the same treatment progress for 7A09, after solid solution treatment: Chunks of second phase melted into the α(Al) and separated out after deep cryogenic treatment of -140 degrees. Different from 7075, when heating at 120 degC, a long heating time can make the mechanical properties of 7A09 better, from 177 HB at 2 hours to ascend to 180 HB at 3 hours. The reason may be that longer heating time makes the boundary between the second phase and α(Al) matrix more ambiguous and longer soaking time makes the second phase crystal much more finer, but it should not be an upward trend with the prolonging of. When we made a contrast test B4 to evidence the speculation, the result is similar to 7075, where the improved effect obtained from longer heating time would become weaker as the time kept on being prolonged. The heating time grew from 1 hour to 2 hours with the temperature staying at 120 degC; the hardness increased by 11HB, but the number was 3 HB when the time was prolonged from 2 hours to 3 hours. Another situation appeared in the test of B5: When we extend the cooling time, the hardness will be obviously improved. Extending the cooling time from 0.5 hours to 0.8 hours at -140 degC, then warming for 2 hours at 120 degC, the hardness would reach 190 HB. As the metallographic aspect indicated, cryogenic treatment increased soluble stripping driving force; $MgZn_2$ phase appeared in and out of the grain crystal; strengthening phase appeared in the grain crystal, whereas lengthening cryogenic time increased the number of the second phase; and distribution was more uniform as well as produced a significant strengthening effect.

4 ANALYSIS OF THE EFFECT OF TEMPERATURE DIFFERENCE DURING THERMAL CYCLING TREATMENT

4.1 Additional test method

According to the earlier test results, we face a new problem of how the temperature difference during the thermal cycling treatment influences the mechanical properties. So, we added additional tests and

Table 4. Additional test method.

Material	Test number	Heat treatment process +120°×4.5h + 175°×10h
7075	A3	480°×2h + (-196°× 0. 5h + 175°× 0.125h) ×2
	A3a	480°×2h + (-196°× 0.5h + 100°× 0.125h) ×2
	A3b	480°×2h + (-196°× 0.5h + 125°× 0.125h) ×2
7A09	B5	480°×2h + (-140°× 0.8h + 120°× 2 h) ×2
	B5a	480°×2h + (-140°× 0.8h + 100°× 2 h) ×2
	B5b	480°×2h + (-140°× 0.8h + 140°× 2 h) ×2

materials, and the testing process was maintained, as shown in Table 4. The focus is the medium temperature whose difference may have an effect on the material's hardness. 7075 selected medium temperature at 100 degC, 125 degC, and 175 degC; 7A09 selected 100 degC, 120 degC, and 140 degC for comparison.

4.2 Additional test result

We also use the HB-3000 type Brinell hardness measurement for hardness test; the comparison results are shown in table 5.

Table 5. Additional test result.

Test number	Hardness /HB	Test number	Hardness /HB
A3/175	154	B5/120	190
A3a/100	160	B5a/100	150
A3b/125	165	B5b/140	178

4.3 Additional test analysis

As with the two aluminum alloy materials, there is no similar trend between temperature difference at thermal cycle and hardness. So the optimal temperature difference exists.

As with the 7075 aluminum alloy, the temperature difference at (-190~175) did not have the highest hardness; the highest hardness appeared at (-190~125); and the increase or decrease of temperature would decrease the hardness. In this test, 125 degC is chosen as the medium temperature and it results in better mechanical properties.

As with the 7A09 aluminum alloy, the temperature difference at (-140~120) had the highest hardness; the increase or decrease of temperature difference would decrease the hardness. But for the same temperature difference at 20 degC, the hardness change rate is not the same and the hardness at a temperature difference (-140~140) is higher than (-140~100). So the higher the temperature difference, the slower the hardness decreased.

Different materials have different sensitivities in terms of temperature difference, and the sensitivity of 7075 is much lower than that of 7A09.

5 CONCLUSION

The following are the conclusions from the earlier test: (a) Thermal cycling makes the second phase crystal smaller and finer, whereas it produces a lot of dislocation and plastic deformation so as to prevent the deformation of the materials, and this can improve the hardness of this material. (b) Reference test results found in other articles such as a test on

7050 and 7A04 aluminum alloys and 7075 and 7A09 used in this article indicated that thermal cycling time can be controlled at two, and more time will bring no more revenue. (c) The best temperature difference between the heating and cooling circulation is due to the materials; a large temperature difference may not result in performance improvement. (d) The effect of alternating cycle time is obvious; especially for 7A09, a long enough cooling time can greatly improve the material hardness; for 7075, it is necessary to shorten the time of heat preservation, so for the different material properties, the alternating time's function is not the same. However, an appropriate strategy can make the mechanical properties more remarkable. (e) Cooling temperature difference has an optimal value; the reactions of different materials on the temperature difference are different, and there is no linear relationship between hardness and temperature difference even with the same material.

REFERENCES

[1] Kou Shengzhong, Zhao Yanchun, Xue Shuwei. Effect of Supercooling Treatment on Thermal Stability and Mechanical Properteis of $Cu_{45}Zr_{42}Al_8Ag_5$ Bulk Metallic Glasses[J]. Hot Working Techonology, 2011, 40(2):58–62.

[2] Huang Shuhai, Zhao Zude, Xiao Yuanlun. Influence of Thermal-cooling Cycle on Both Quenching-induced Residual Stress and Machining-induced Distortion of Aluminum Cone-shaped Part[J]. Chinese Journal of Mechanical Engineering, 2010, 7(14):73–78.

[3] Shao Hua, Liu Zhaojing, Liu jun. Effects of interrupted ageing treatment and thermal cycling on the 2024 aluminum alloy micro plastic deformation resistance [J]. The Chinese Journal of Nonferrous Metals, 2000, 10(3):330–333.

[4] Wang Qiucheng. The research on the elimination and evaluation technology about residual stresses in aluminum alloys for aircraft[D]. Hangzhou:Zhejiang Univercity, 2003.

[5] Wu Haihong, Zhang Yuping, Zhang Shixing. Experimental research and strength prediction on aged strengthening of 7A04 aluminium alloy[J]. Light Alloy Fabrication Technology, 2009(37):37–41.

[6] Podgornik B, Leskovsek V, Vizintin J. Influence of deep-cryogenic treatment on tribological properties of P/M high-speed steel[J]. Materials and Manufacturing Processes, 2009, 24(7–8):734–738.

[7] Wang P, Lu W, Wang Y H, et al. Effects of cryogenic treatment on the thermal physical properties of Cu76.12Al23.88 alloy [J]. Rare Metals, 2011, 30(6):644–649.

Energy, Environment and Green Building Materials – Sheng (ed.)
© *2015 Taylor & Francis Group, London, ISBN 978-1-138-02718-3*

Exploring the relationship between buildings' temperature and eco-environment components around them in Futian District, Shenzhen, China

X.Q. Sun & M.M. Xie
School of Land Science and Technology, China University of Geosciences (Beijing), Beijing, China

ABSTRACT: Urban heat island effect exacerbated by urbanization has recently attracted people's attention all over the world. Futian District with a high density urban development is the center of Shenzhen, also the ideal study area. Exploration of the relationship between the sample building's temperature and the eco-environment components, including percent impervious surface area, vegetation fraction within the buffer around the sample buildings, and shape index of them in Futian District was made. Our explorations were based on the land surface temperature image, the percent impervious surface imagery and the vegetation fraction image derived from the Landsat-5 Thematic Mapper images, supplemented by the records of the survey about the building's characteristics in 2005. Our analysis indicates that the percent of impervious surface, vegetation fraction within the buffer and the shape index indeed has effects on building's temperature, further findings suggest there is an obvious negative correlation between buildings' temperature and percent impervious surface, meanwhile, buildings' temperature increases with the increments of vegetation fraction and buildings' shape index, especially in low rise buildings (4–9), this trend is apparent. Our results contribute to a deeper understanding of the effect of eco-environment components on building's temperature.

KEYWORDS: buildings' temperature; percent impervious surface; vegetation fraction; shape index

1 INTRODUCTION

Urban heat island (UHI) is a phenomenon that higher surface temperature occurring in the urban areas than in the rural areas (Voogt et al. 2003), meanwhile, an important part of urban eco-environment effect. UHI not only disturbs the social environment people living in but also threats people's health. UHI is exacerbated by the increasing impervious surface caused by urbanization, especially the extension and the expansion of buildings (Zhou et al. 2008). Studies about the UHI have become the focus of the world since factors affecting the UHI were explored domestic and overseas. Positive correlation between land surface temperature (LSI) and percent impervious surface was verified (Xu 2009, Lin et al. 2007, Lin et al. 2013, Xu et al. 2013), and negative correlation between LSI and vegetation fraction (Chen 2010, Luan et al. 2014, Cheng et al. 2004, Jia et al. 2009, Chen et al. 2014). Only within a certain radius can eco-environment components around buildings affect building's temperature, nevertheless, studies about buildings' temperature and eco-environment components are barely reported. Parklands' cooling effect determined by their characteristics was found (Jia et al. 2009, Chen 2014, Chang et al. 2007), Wiens et al suggested

that the shape index was a critical parameter of patch shape, the higher the value of shape index was, the easier for the inner part of patch to interact with the outer environment, at the same time the even more complicated the patch shape was.

As for studies about the relationship between buildings' temperature and eco-environment components are relatively few, our research considers the percent impervious surface, the vegetation fraction within the buffers of the sample buildings referring to the threshold of effect of garden on temperature in Chang's paper (Chang et al. 2007), and the shape index of buildings as the eco-environment components, in addition, we make a correlation analysis and a regression analysis of eco-environment components and buildings' temperature to explore their relationships.

2 STUDY AREA AND METHODS

2.1 *Study area*

Shenzhen, located in the south of Guangdong of China, has experienced a full range of urbanization since Reform and opening up. Shenzhen with a great economic development is no longer a fishing village, but a modern metropolis. Futian district, located in the center of the special economic zone in Shenzhen, starts from Red hill road to the east next to the Luohu district and reaches Overseas Chinese Town next to Nanshan district, bordering to Beacon Hill, Lotus hill and Longhua Town in the north and Shenzhen river and Shenzhen Bay in the south. Futian district where the Shenzhen government locates is the center of development and construction with a high density urban development. Considering the characteristics mentioned above on Futian district, there is no doubt to choose Futian district as our study area. Following is the location of Futian District in Shenzhen (Fig. 1).

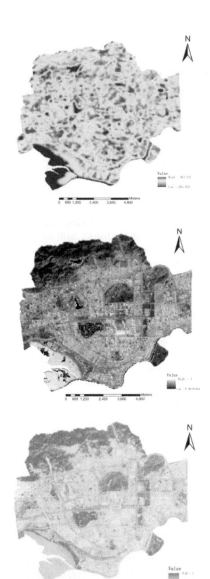

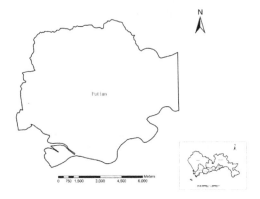

Figure 1.   Futian District, Shenzhen.

2.2 *Datasource and image pre-processing*

Construction census about the sample buildings in Futian District, Shenzhen, China was obtained in 2005, and 15418 objects were included. LST was derived from the Landsat-5TM image acquired in 2005 based on Qin's mono window algorithm (Xie et al. 2013a). A linear spectral mixture analysis (LSMA) was utilized to calculate the percent impervious surface and vegetation fraction according to the Ridd's vegetation-impervious surface-soil (VIS) model (Xie et al. 2014b; Fig. 2).

To simplify our later calculation and analysis, the resolution of LSI image, percent impervious surface image and vegetation fraction image were rescaled to 0.5 meter. All the resolution procedures were done in Arcgis10.0 and SPSS 20.0.

2.3 *Statistical analysis*

We classified the samples by the number of floors as following classes: (1) 0–3(4942 samples), (2) 4–6(4742 samples), (3) 7–9(5209 samples), (4) 10 (525 samples).

Buffers of the sample buildings were determined by the fourth root of building area as radius. Percent impervious surface data and vegetation fraction data within buffers were obtained by respectively

Figure 2.   LSI, ISA and vegetation fraction derived from Landsat-5 TM imagery.

overlapping the buffer vector data with impervious surface data and vegetation fraction data. We made a correlation analysis between buildings' temperature and their eco-environment components in all four classes.

To further investigate these relationships, a zonal analysis and a regression analysis were utilized to respectively evaluate the mean buildings' temperature at each 5% increment of percent ISA and vegetation fraction in the second class and the third class.

3 RESULTS AND DISCUSSIONS

3.1 Buildings' temperature patterns and statistics

Statistical analysis of building's temperature derived from overlapping the building vector data with the LSI data suggests that 213 samples with their temperature less than 296K mainly locate in the southwest of Futain District, 164 residential buildings are included. Meanwhile, Shenzhen bay located in the southwest of Futian District may cause cooling effect. 50 buildings' temperature is more than 303K including 22 energy-intensive industry buildings, this kind of buildings consume and absorb large amounts of heat thus increasing their own temperature (Xie et al. 2008).

3.2 Buildings' temperature relationships of eco-environment components

The negative correlation between buildings' temperature and percent impervious surface within buffers of constructions is significant (Table 1), building's temperature decreases with the increment of percent impervious surface. This trend is especially remarkable in the second class, whereas not in the fourth class resulted from the cooling effect of shadow of huge buildings and dense vegetation around them.

Table 1. Pearson correlation coefficient of Buildings' Temperature (BT) and eco-environment components of sample buildings in Futian.

	BT in the second class (4–6)	BT in the third class (7–9)	BT in the fourth class (10-)	BT
ISA	−0.277**	−0.348**	−0.219**	−0.103*	−0.347**
Vegetation fraction	0.034*	0.112**	0.184**	–	0.137**
Shape index	0.073**	0.159**	0.218**	–	0.099**

*− indicates the confidence interval is 95%, **− indicates the confidence interval is 90%

Positive correlation can be found between buildings' temperature and vegetation fraction within the buffer, meanwhile, obvious in the third class. Many previous studies proved the cooling effect of Greenspace thus causing the faint correlation in the first class. Nevertheless, the increment of Greenspace may lead to the decrease of impervious surface area which means reduce of shadow giving rise to the uniform variation trend of vegetation fraction and buildings' temperature. In the case of the fourth class, the cooling effect of Greenspace is weaker accompany with less shadow of huge buildings.

Buildings' temperature with complicated shape is relatively high. This effect of shape index of building's temperature can be found especially in the third class. Some earlier research found that the inner energy and materials of the Greenspace patch with complicated shape could easily communicate with the outer environment (Wiens et al. 1993, Wang 2006). Luan et al suggested that there was a positive correlation between the mean temperature and the shape index of the Greenspace patch (Luan et al. 2014), there is no doubt that the same trend can be applied to constructions. So without considering other factors, no matter construction or Greenspace, complicated shape contributes to the mean temperature.

Overall, the eco-environment components, including percent impervious surface, vegetation fraction and shape index do have effects on building's temperature, and especially in the second class and the third class, this relationship is obvious.

The random sampling method was utilized to select the sample buildings, other methods may lead to different results. In addition, we classified the sample buildings, according to the numbers of floor. Nevertheless, different classifications may affect the accuracy of the results. Correlation analysis on building's temperature and eco-environment components didn't take structures and materials of buildings into account however (Liu et al. 2011).

3.3 Regression analysis

From Figure 3 (a) and Figure 4 (a), we could see that from 10%-50%, the LSI decrease decreased slightly, however, the decrease began to grow steeply after that. The gradients of the rate capability from 10%–50% decreased, over 50% the gradients of the rate capability increased steeply. Figure 3 (b) indicated that there is a nonlinear correlation between LST and vegetation fraction of the second class, and the building's temperature stopped increasing when the vegetation fraction reached 10%, however after that it began to increase again until 45% where it reached the maximum. In the third class, the linear correlation which could be seen in Figure 4 (b) was apparent between buildings' temperature and vegetation fraction, namely, building's temperature increased steeply with the increment of vegetation fraction. Figure 3 (c) showed that the building's temperature increased slowly before the shape index reached 2.5 after that it increased more rapidly. Although we could find an obvious relevant relationship between buildings' temperature and shape index in the second class, we couldn't find the similar correlation in the third class.

Though increasing percent impervious surface, keeping vegetation fraction at 0.1 and at the same time simplifying the shape of the building can contribute to the decrease of building's temperature, Urban Heat

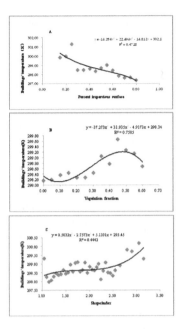

Figure 3. Relationship between buildings' temperature and shape index in the second class (4–6).

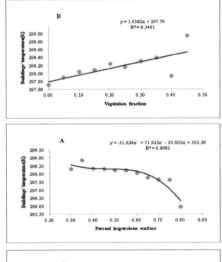

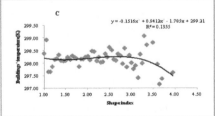

Figure 4. Relationship between buildings' temperature and shape index in the third class (7–9).

Island Effect would be exacerbated by the increment of percent impervious surface overall, further studies are needed to balance the two.

4 CONCLUSIONS

Correlation analysis and regression analysis of the 15418 building's temperature and eco-environment components, including percent impervious surface, vegetation fraction within the buffers of constructions and shape index of sample buildings were used, coupled with the records of construction census made in 2005 in Futian District. The followings are our findings after our analyses,

Percent impervious surface within buffers does have an effect on building's temperature and a negative correlation can be found between the two, especially in the second class and the third class the relationship is apparent. We consider that the cooling effect of shadow of huge buildings and dense vegetation around the construction may account for this phenomenon.

Vegetation fraction contributes to the increment of building's temperature, especially in the third class. The cooling effect of Greenspace patch is not obvious beyond its impact threshold. The decrease of shadow resulted from the increase of vegetation fraction leads to the increment of building's temperature in Futian District.

Positive correlation can be found between buildings' temperature and shape index of building, the inner part of construction with complicated shape can easily communicate with the outer part, namely, the inner environment is determined by the outer parts.

In conclusion, our results suggest that eco-environment components do have effects on building's temperature, further investigation shows a positive correlation between buildings' temperature and vegetation fraction within buffers and shape index respectively, however an inverse relationship between buildings' temperature and percent impervious surface. Meanwhile, buildings in the second class are more easily affected by the outer environment around them.

Liang Boying Mountain, Top Dragon Mountain and Da Naoke Mountain, located in the north of Futian District, and the Shenzhen Bay, located in the southwest of Shenzhen, just like two cold islands, have cooling effect on all the buildings in Futian District. Our findings focused on single building may not be suitable for other regions, different from Futian District as for the specific limitations, so they can't be applied widely. Further, studies of additional metropolitan areas and correlation analysis on multiple temporal and spatial scales are recommended to provide more information about the relationship between buildings' temperature and eco-environment components.

ACKNOWLEDGMENTS

This research was sponsored by the Natural Science Foundation of China (NSFC41101175). We thank Prof. Wang Yanglin and Prof. Wu Jiansheng for supplying parts of data. We are grateful to the editors and anonymous reviewers for their helpful comments on the original manuscript.

Corresponding author: Miaomiao Xie, E-mail: xiemiaomiao@cugb.edu.cn kenanxiaoqing@sina.cn.

REFERENCES

[1] Chang, C.R., Li, M.H. & Chang, S.D. 2007. A preliminary study on the local cool-island intensity of Taipei city parks. Landscape and Urban Planning 80(4):386–395.

[2] Chen, F.M. 2010. Effects of Urban Park on Urban Heat Island: A Multi-scale Study in Shanghai. East China Normal University, Department of Environmental Science.

[3] Chen, T. & Zhang, X.Y. & Li, P.Y. et al. 2014. Effect of Greenland on Thermal Environment in Wuhan. Geospatial Information 12(2):53–55.

[4] Cheng, C.Q., Wu, N. & Guo, S.D. 2004. A Study on the Interaction Between Urban Heat Island and Vegetation Theory, Methodology, and Case Study. Research of Soil and Water Conservation (3):172–174.

[5] Jia, L.Q. & Qiu, J. 2009. Study of Urban Green Patch's Thermal Environment Effect with Remote Sensing: A Case Study of Chengdu City.LA Investigation and Research (12):97–101.

[6] Lin, Y.B., Xu, H.Q. & Zhou, R. 2007. A Study on Urban Impervious Surface Area and Its Relation with Urban Heat Island: Quanzhou City, China. Remote Sensing Technology and Application 22(1):14–19.

[7] Lin, D. & Xu, H.Q. 2013. Urban Impervious Surface and its Effects on Thermal Environments in Xiamen: An Analysis Based on Remote Sensing Technology. Journal of Subtropical Resources and Environment 8(3):78–84.

[8] Liu, W.Y., Gong, A.D. & Zhou, J. 2011.Investigation on Relationships between Urban Building Materials and Land Surface Temperature through a Multi-resource Remote Sensing Approach. Remote Sensing Information (4).

[9] Luan, Q.Z., Ye, C.H. & Liu, Y.H. et al. 2014. Effect of urban green land on thermal environment of surroundings based on remote sensing: A case study in Beijing, China. Ecology and Environmental Sciences (2).

[10] Voogt J. A. & Oke, T. R. 2003.Thermal remote sensing of urban climates. Remote sensing of environment 86(3):370–384.

[11] Wang, X. 2006. A Remote Sensing Study of Urban Green Space Distribution and Its Thermal Environment Effect. Beijing Forestry University, Specialty of Forest Management.

[12] Wiens, J. A., Stenseth, N.C. & Van Horne, B. et al. 1993.Ecological mechanisms and landscape ecology. Oikos:369–380.

[13] Xie, M.M., Zhou, W. & Wang, Y.L. et al. 2008. Thermal Environment Effect of Land Use in Urban Area: A Case Study in Ningbo Urban Area. Acta Scientiarum Naturalium Universitatis Pekinensis (5):815–821.

[14] Xie, M.M., Wang, Y.L. & Fu, M.C. et al. 2013a. Pattern dynamics of thermal-environment effect during urbanization: A case study in Shenzhen City, China. Chinese Geographical Science 23(1):101–112.

[15] Xie, M.M., Wang, Y.L. & Chang, Q. et al. 2013b. Assessment of landscape patterns affecting land surface temperature in different biophysical gradients in Shenzhen, China. Urban Ecosystems 16(4):871–886.

[16] Xu, H. 2009. Quantitative analysis on the relationship of urban impervious surface with other components of the urban ecosystem. Acta Ecologica Sinica 29(5):2456–2462.

[17] Xu, Y. & Liu, Y. 2013. Study on the thermal environment and its relationship with impervious surface in Beijing city using TM image. Ecology and Environmental Sciences (4).

[18] Zhou, H., Gao, Y. & Ge, W. et al. 2008.The Research on the Relationship Between the Urban Expansion and the Change of the Urban Heat Island Distribution in Shanghai Area. Ecology and Environment (1):163–168.

Energy, Environment and Green Building Materials – Sheng (ed.)
© 2015 Taylor & Francis Group, London, ISBN 978-1-138-02718-3

Research on damage identification of grid structures based on frequency response function

M.H. Wang, S.Y. Dong, C.M. Ji & X.T. Yang

School of Civil and Transportation Engineering, Beijing University of Civil Engineering and Architecture, Beijing, China

ABSTRACT: Due to extracting modal parameters from fitting frequency response function of structure could lose some useful original information, one direct kind of structural damage identification method is presented by using the frequency response function in this paper. Also, owing to more members in grid structure and more sampling points generally, the order of the frequency response function matrix formed is larger. Hence, considering a structural damage identification method that frequency response function is combined with principal component analysis is feasible. Take a scaled model of the grid structure to numerical analysis, the results show that the method is effective, damage identification method based on structural frequency response function could detect structural damage location accurately in a single damage state.

KEYWORDS: Damage Identification; Frequency Response Function; Principal Component Analysis

1 INTRODUCTION

A larger number of hanging equipment and moving loads exist in spatial grid structures include clinker library and pre-homogenization library in cement industry and A380 hangar, long-term operation of these equipments can cause structural damages, such as loose bolts, bending and rupturing of members in the structure of local, lead to catastrophic and sudden accidents easily. It is difficult for workers to observe and detect damages due to the spatial grid structure has hundreds of thousands members. Therefore, research on damage identification of grid structure during the period of service has important practical significance.

Currently, most damage identification methods which are put forward by the scholars at home and abroad are based on the structural modal analysis, use the modal parameters as basic variables of damage identification. Pandey & Biswas (1995) found that modal curvature change can be a good characterization of structural damage from calculating finite modal of the beam. Luo & Wang et al (2013) combined modal curvature and wavelet transform to research damage identification of the grid structure. But damage indexes based on dynamic behavior of structures are extracted from fitting frequency response function, it may be lost some useful original information. Therefore, compared to other damage indexes, structural frequency response function is more direct, the error is smaller, and include all

information about structural modal parameters. In addition, grid structure has a plenty of members and lots of sampling points are needed to analyze the structural dynamic response, it makes structural frequency response function matrix relatively huge. As a frequency response function matrix which has hundreds or even thousand orders, it must bring great inconvenience in later data processing, however principal component analysis can be able to translate multiple indexes into several comprehensive indexes at little lose information and play the role of dimensionality reduction. Consequently, in this paper, apply numerical simulation to undamaged and damaged states of grid structure, frequency response function matrix is compressed and dimensionality reduction by principal component analysis, use damage detection method based on frequency response functions to identify damage to grid structure.

2 BASIC THEORY AND SELECTION OF DAMAGE INDEX

2.1 *Basic theory of frequency response function (Li et al. 2006) (Yang et al. 2007)*

Assuming that system coupled is harmonic excitation $f(t) = Fe^{jwt}$, where F = excitation amplitude; ω = exciting frequency, then the steady response of harmonic vibration, $x = Xe^{jwt}$, where X = steady amplitude of displacement response; ω = steady frequency of displacement response.

The vibration differential equation:

$$m\ddot{x} + c\dot{x} + kx = f(t) \tag{1}$$

Where m = quality; c = viscosity damping coefficient; k = stiffness, and x = displacement; $\dot{x}$ = velocity; $\ddot{x}$ = acceleration; and $f(t)$ = exciting force.

In the harmonic excitation, Eq. (1) can be transformed into:

$$(k - m\omega^2 + jwc) = F \tag{2}$$

Define displacement frequency response function of the system as the ratio of steady displacement response and excitation amplitude, the ratio, $H(\omega)$ is given as follows:

$$H(\omega) = \frac{X}{F} = \frac{1}{k - m\omega^2 + j\omega c} \tag{3}$$

Then:

$$X = H(\omega)F \tag{4}$$

Frequency response function can be defined as the ratio of steady state response and Fourier Transform of excitation if the system is subject to any force, the equation is as follows:

$$H(\omega) = \frac{X(\omega)}{F(\omega)} \tag{5}$$

2.2 Algorithm of principal component analysis

The principal component analysis method takes advantage of dimensionality reduction mainly and translates multiple indexes into several comprehensive indexes at least lose information, is also known as a main component analysis method. The comprehensive indexes of transformation are referred to as principal component generally, each of principal components is a linear combination of primitive variables and not correlative, moreover, principal component owns some better performance than primitive variable.

Assuming that the number of samples is n, m indexes are observed by each sample, $X = (X_1, X_2, ..., X_n)'$, the original data matrix X is given as follows:

$$\begin{bmatrix} x_{11} & x_{12} & \cdots & x_{1i} & \cdots & x_{1m} \\ x_{21} & x_{22} & \cdots & x_{2i} & \cdots & x_{2m} \\ \vdots & \vdots & \vdots & \vdots & \vdots & \vdots \\ x_{j1} & x_{j2} & \cdots & x_{ji} & \cdots & x_{jm} \\ \vdots & \vdots & \vdots & \vdots & \vdots & \vdots \\ x_{n1} & x_{n2} & \cdots & x_{ni} & \cdots & x_{nm} \end{bmatrix}$$

Then the original data matrix is standardized, the new matrix Y is given as follows:

$$\begin{bmatrix} Y_{11} & Y_{12} & \cdots & Y_{1m} \\ Y_{21} & Y_{22} & \cdots & Y_{2m} \\ \vdots & \vdots & \vdots & \vdots \\ Y_{n1} & Y_{n2} & \cdots & Y_{nm} \end{bmatrix}$$

For n samples is given, seek covariance matrix R after data matrices of standardization, each element in covariance matrix can be expressed by corresponding covariance, the covariance matrix R is given as follows:

$$R = cov(Y_i, Y_j) = \begin{bmatrix} r_{11} & r_{12} & \cdots & r_{1m} \\ r_{21} & r_{22} & \cdots & r_{2m} \\ \vdots & \vdots & \vdots & \vdots \\ r_{n1} & r_{n2} & \cdots & r_{mm} \end{bmatrix}$$

Where $r_{ij} = cov(Y_i, Y_j) = \sum_{k=1}^{n}(x_{ki} - \overline{x_i})(x_{kj} - \overline{x_j})$.

The covariance matrix R is a real symmetric matrix of $m \times m$, thereby solving eigenvalues ($\lambda_1 \geq \lambda_2 \geq ... \geq \lambda_m$) and corresponding eigenvectors ($u_1, u_2, ..., u_m$) of covariance matrix. We define $a_k = \frac{\lambda_k}{\lambda_1 + \lambda_2 + \cdots + \lambda_m}$ $(k = 1, 2, \cdots, m)$ as variance contribution rate of any principal component Y_k and $\frac{\sum_{i=1}^{p}\lambda_i}{\sum_{i=1}^{m}\lambda_i}$ as cumulative contribution rate of variance of principal component, it is appropriate that the number of principal components p makes cumulative variance contribution more than 85%(He. 2006). Then p principal components are given as follows:

$$\begin{cases} Z_1 = u_{11}R_1 + u_{12}R_2 + \cdots + u_{1P}R_P \\ Z_2 = u_{21}R_1 + u_{22}R_2 + \cdots + u_{2P}R_P \\ \vdots \\ Z_p = u_{P1}R_1 + u_{P2}R_2 + \cdots + u_{PP}R_P \end{cases}$$

Wherefore obtaining p-order vector after principal component analysis replaces original m-order vector, reduces the dimension of the matrix greatly and brings the convenience of calculation.

2.3 The selection of damage index

After using the principal component analysis method to reduce dimensions of primitive frequency response function matrix, the new matrix can retain structural primitive information. We define the vector angle of damage corresponding point as damage index (Jin. 2011), the index θ is given as follows:

$$\theta = cos^{-1}\left(\frac{x_D \cdot x_U}{|x_D||x_U|}\right) \tag{6}$$

Where x_D = vector of frequency response functions at damaged state; x_U = vector of frequency response functions at undamaged state.

Obviously, if the structure does not have injures, the vectors of undamaged and damaged states are unchanged substantially, the damage index θ, should be close to zero infinitely; if the structure has injures, the vectors of undamaged and damaged states are changed, the damage index θ, should be changed correspondingly.

3 NUMERICAL SIMULATION EXAMPLE

In this paper, a scale model of grid structure is used for numerical simulation. The situation is as follows: the size is 3m×7m, 80 nodes and 237 members in total, members are Q235 steel pipes and the cross-sectional size is Φ42mm×2.5mm, elastic modulus was 200GPa. Damaged state is to narrow the cross-sectional size of member 34 to Φ40mm×1.5mm. The model and damage location are shown in Fig 1.

First, the structural dynamic response under external loads is obtained by SAP2000. In this paper, the acceleration responses of each node are used as original signals. The sampling frequency is 100Hz, and the sampling time is 2s. Next, transforms the original signals from time domain to the frequency domain by Fourier Transformation, and then get the acceleration frequency response functions, $x(\omega)$. Finally, the frequency response function matrix $X = (X_1, X_2, ..., X_{80})'$ is composed by the acceleration frequency response function of each node. The principal component analysis method is implemented by SPSS. The calculation results of accumulative contribution rate are shown in Table 1 under the damaged state, and Table 2 showed the results that under the undamaged state.

The above results shows that the accumulative contribution rate of variance of the fourth principal component reached 91.477% and 92.653%, so the first four principal components have been able to retain most of the information of the original signal. The 80×201 original frequency response function matrix is transformed into 80×4 matrix, the dimension reduction is more convenient to calculate.

Input the matrix to MATLAB, and calculate the damage index of each node (vector angle). The results are shown in Fig 2.

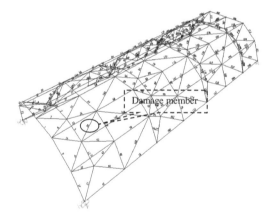

Figure 1. The model and damage location.

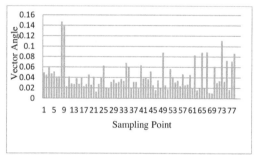

Figure 2. The results of damage identification.

Table 1. Accumulative contribution rates to variance under the damaged condition.

Principal Component	Initial eigenvalue			Extracting square sum and input		
	Total	Variance rate (%)	Accumulative rate (%)	Total	Variance rate (%)	Accumulative rate (%)
1	36.341	46.591	46.591	36.341	46.591	46.591
2	20.781	26.643	73.234	20.781	26.643	73.234
3	7.934	10.171	83.405	7.934	10.171	83.405
4	6.296	8.071	91.477	6.296	8.071	91.477
5	2.272	2.913	94.390	2.272	2.913	94.390
6	1.339	1.716	96.106	1.339	1.716	96.106

Table 2. Accumulative contribution rate to variance under the undamaged condition.

Principal Component	Initial eigenvalue			Extracting square sum and input		
	Total	Variance rate (%)	Accumulative rate (%)	Total	Variance rate (%)	Accumulative rate (%)
1	37.046	47.495	47.495	37.046	47.495	47.495
2	20.735	26.583	74.078	20.735	26.583	74.078
3	8.295	10.635	84.713	8.295	10.635	84.713
4	6.194	7.941	92.653	6.194	7.941	92.653
5	2.196	2.815	95.468	2.196	2.815	95.468
6	1.278	1.638	97.106	1.278	1.638	97.106

From figure 2 we can see that the vector angles at point 8 and 9 are comparatively large, while member 34 is between point 8 and 9. It means that the changes are comparatively obvious after member 34 damaged. Thus proving that the damage identification method based on the frequency response function is effective.

4 CONCLUSIONS

Structural response signal in time domain is extracted from SAP2000 in this paper, the frequency response function matrix is achieved by Fourier transform of a signal, the structural damage matrix is reduced dimension by principal component analysis method by SPSS, finally, according to the changes of damage index, the effect of damage identification is achieved, the results shows that the method is ideal.

1 Through theoretical derivation, the damage identification method is established by combining frequency response function with the principal component analysis method.
2 The principal component analysis method can decrease dimension of matrix effectively, reduce the workload in later data processing, make damage detection of large and complex structures more feasible.
3 The damage index, θ is able to judge the structural damage location accurately and provide a reference for specific engineering examples.

ACKNOWLEDGMENTS

This research was financially sponsored by the Beijing National Science Foundation (8132023), Engineering Research Center of Scientific and Technological Achievements Transforming - Beijing. Higher Institution Engineering Research Center of Structural Engineering and New Material (02058214001) and BUCEA Urban Rural Construction and Management Industry Research Development Collaboration Post Graduate Training Centre.

REFERENCES

He, X.Q. 2006. Multivariate Statistical Analysis (The second edition). Beijing, China Renmin University Press.
Jin, R.R. 2011. Structural damage detection based on principal components analysis. Changsha. Hunan University.
Li, X.P. & Yu, Z.W. 2006. Structural damage identification based on frequency response function. Journal of China & Foreign Highway 26(1).
Luo, S.S. & Wang, M.H. et al. 2013. Research based on modal curvature and wavelet transform for identifying damage of reticulated shell structures. Building Technique Development 11:8–10.
Pandey, A. K. & Biswas, M. 1995. Damage diagnosis of truss structures by estimation of flexibility change. Modal Analysis-the International Journal of Analytical and Experimental Modal Analysis 10(2):104–117.
Wang, J. & Hu, X. 2006. Application of MATLAB in Vibration Signal Processing. Beijing, China Water Power Press.
Yang, H.F. & Wu, Z.Y. and Wu, D. 2007. Structural damage detection method based on acceleration frequency response function. Journal of Vibration and Shock 26(2).

Energy, Environment and Green Building Materials – Sheng (ed.)
© 2015 Taylor & Francis Group, London, ISBN 978-1-138-02718-3

Damage detection of cable force relaxation based on HHT

M.H. Wang, Y.S. Liu, S.S. Shi & X.T. Yang

School of Civil and Transportation Engineering, Beijing University of Civil Engineering and Architecture, Beijing, China

ABSTRACT: As the outside enclosure structure of buildings, glass curtain wall favored to win the majority of architects with its good transparence and beautiful structure. Because the cable cross-section area of monolayer cable net glass wall is small, the transparence is more advantageous than other form of curtain wall and the application prospect is widespread. But the cable net will occur to pre-stressed relaxation under the long-time load, which will bring losses for people. In this article we simulate the cable net relaxation through numerical simulation. Also, we have researched how to identify the site of injury using the Hilbert marginal spectrum and intuitively grasped the law of development of structural damage through the time-varying characteristics of the instantaneous frequency vibrations of the modal response.

KEYWORDS: Dynamic detection; Numerical simulation; Hilbert-Huang transform; Anomaly index

1 INTRODUCTION

The cable net curtain wall structure is a complex system which relies on large tonnage tension to achieve plane stiffness and the internal force worked between cable net and the main body structure. The large cable net curtain wall transforms thousands of tons tension to the structure. So great pull not only exists for a long time, but also constantly changes dramatically under the wind load and temperature load, which should not be neglected to the main structure. The pre-stressed relaxation of cable has obvious effect on the cable net wall which supported by flexible cable and brittle glass. The accumulation of pre-stressed relaxation causes hundreds of tons glass deformation greatly and cause the durability and safety of the cable net curtain wall.

The traditional damage identification is based on Fourier transform. The Fourier transform is a statistical average of the signal in time domain, and cannot reflect the local characteristics of the signal, and require the analyzed signal is in the linear-steady state. But in the field of civil engineering, most of the vibration response of the measurements are unsteady and non-linear. Therefore the time-frequency analysis method based on wavelet transform has received a widespread attention (Li, 2003). However, the wavelet analysis method is still on the basis of Fourier transform, and still have limitations to identify the nonlinear and transient characteristics contained by the vibration signal (Huang, 1998).

Hilbert-Huang Transform (HHT) is a new data processing method proposed by the Chinese-American Dr. Huang E. This method is applied to non-linear, non-stationary signals, and is a new self-adaptive data processing method (Huang, 1998). Comparing to Fourier transform and wavelet analysis, HHT method has obvious advantages in terms of objectivity and resolution (Hu, 1997). This paper has analyzed the detecting method for damage identification of cable net curtain through numerical simulation experiment under the foundation of HHT.

2 THE CONSTRUCTION OF DAMAGE FEATURE BASED OD HHT

2.1 Hilbert-Huang transform

We define the intrinsic mode function(IMF) is $c_j(t)$. Through HHT, we can get $c_j'(t)$,

$$c_j'(t) = H[c_j(t)] = \int_{-\infty}^{+\infty} \frac{c_j(u)}{\pi(t-u)} d \qquad (1)$$

The analytical signal of $c_j(t)$ is $z_j(t)$.

$$z_j(t) = c_j(t) + jc_j'(t) = a_j(t)e^{i\theta_j(t)} \qquad (2)$$

So we can get the instantaneous frequency is:

$$\omega_j(t) = \frac{d\theta_j(t)}{dt} \qquad (3)$$

2.2 The analysis of Hilbert spectrum

We can get the Hilbert amplitude spectrum after HHT for the each IMF.

$$H(\omega,t) = \text{Re} \sum A_i(t) e^{i\int \omega_i(t)\,dt} \qquad (4)$$

$\omega_i(t)$ is the instantaneous frequency.

The Hilbert amplitude spectrum described the distribution of amplitude signals along with the time and frequency. We can get the marginal spectrum through integration for the time.

$$h(\omega) = \int_0^T H(\omega,t)\,dt \qquad (5)$$

And the marginal spectrum of energy is:

$$E(\omega) = \int_0^T H^2(\omega,t)\,dt \qquad (6)$$

Comparing to Fourier spectrum, the Hilbert marginal spectrum describe the distribution of the signal components in different frequency values. In practical application and numerical calculation, the signal frequency must be divided into a number of bands $\Delta\omega_i (i=1,n)$. The bigger of the n, the greater of the time-frequency resolution. And the maximum value depends on the length and sampling intervals of the signal. The spectrum that signal amplitude value or energy in each frequency is $h(\omega_i)$ or $E(\omega_i)$, i=1, n.

2.3 The component vector of signal modal

When the structural damage occurred, the different parts of the vibration response of their energy distribution in each mode will change in ambient excitation, and the amount of change have a relationship with the damage location. Therefore, the overall slip of each modal vibration frequencies and their relative composition changes will cause the characteristic quantity change. In order to make the information contained by modal component vector under different excitation be comparable, we normalize it and define the modal component vector as follows:

$$\varepsilon_i = \frac{E(\omega_i)}{\sum_{i=1,n} E(\omega_i)}, i = 1,...,n \qquad (7)$$

In addition, the ratio of modal components vary with position and the amount of changing is different. So

it can reflect the position information of structural damage through the modal component proportional relationship.

2.4 The anomaly index of modal composition

ε_i reflect the relative proportion of each signal frequency component. We can determine whether the structure is abnormal by comparing to the modal component vector in unknown condition and good condition at the same location. ε_i^u represent the undamaged modal component vector in undamaged condition and ε_i^d represent the modal component vector that may contain damage. We diagnose whether the structural modal composition distribution have changed by analyzing the matching degree. We define the abnormal index is:

$$D_f = 1 - \frac{[\sum_i^n \varepsilon_i^d]^2}{\sum_i^n (\varepsilon_i^u)^2 \sum_i^n (\varepsilon_i^d)^2}, D_f \in [0,1] \qquad (8)$$

$D_f = 0$ show that the modal component of the response signal have not changed and no damage occurred. The greater of the numerical value of D_f the lower of the matching degree of modal component of the present structure and lossless structure in dynamic response and the greater of the chance that structural damage occurring.

3 NUMERICAL SIMULATION EXPERIMENT

3.1 The model of numerical simulation experiment

For Figure 1. the experimental model is a monolayer cable net glass curtain wall structure and its size is 3m×3m. The ring beam and four support columns use 32#A joist steel and the pre-stressed cable use stainless steel wire with a diameter of 10mm. By using the SAP2000 finite element to simulate the model of cable net curtain wall under impact load which simulated by Gaussian white noise whose peak is 500N, acting on the 16th node. To simulate damage occurring by the method of reducing the elastic modulus of the structure. As shown in Figure 2, the injury unites are between the 9th and 15th node. And the injure can be divided into the following condition:

The first condition: E reduction 30%; The second condition: E reduction 60%; The third condition: E reduction 90%.

We select the nodes of 6,10,17,4 as the measuring points, and the sampling frequency is 1000HZ. According to the dynamic response data of the each measuring point and the analysis results of model eigenvalue, it can be seen that even for the simplest monolayer cable net system the loss of natural frequency caused by the structural damage is not very obvious, and make it not feasible to identify damage happened only from the modal frequency.

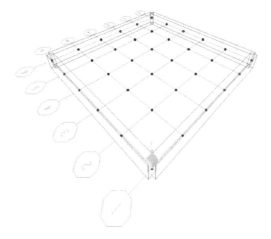

Figure 1. The stereogram.

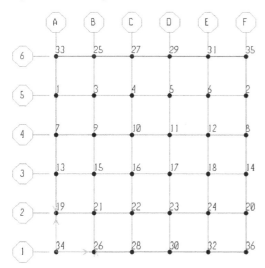

Figure 2. The planar graph.

3.2 Abnormal index calculation and damage identification

3.2.1 The Empirical Mode Decomposition (EMD) of the signal and choosing of Intrinsic Mode Function (IMF)

We have made the EMD for the response signal of time-acceleration in different measuring points and conditions. The EMD result of the 10th point is shown in Figure 3. And then we selected the previous four IMF to make a Hilbert spectrum analysis. Because they have focused on the most of the signal energy.

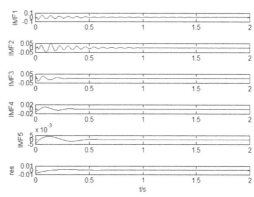

Figure 3. The EMD of intact structure.

3.2.2 The forming of modal component vector

We made Hilbert transform for the selected IMF and then obtained the frequency distribution ($H(\omega,t)$) of the signal. $\omega = [0{\sim}100]$, $t = [1,2]$. For Figure 4.

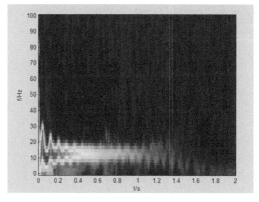

Figure 4. The Hilbert spectrum of intact structure.

Then we can get the marginal spectrum of the signal energy $E(\omega_i)$ through the empirical mode decomposition (i=1, 100) and the frequency resolution is 50HZ. At this point the signal energy distribution within 0~100hz is described. Then we can get the modal component vector through normalization. Figure 5 shows the Hilbert marginal spectrum of the modal component vector of the 11th point. Figure 6. show the normalized result of the damage sensitive vector of the first 20.

3.3.3 The calculation of the anomaly index

Through calculation we can get the anomaly index of each measuring point as shown in Table 1.

From Table 1, we can get that with the increasing of the degree of structural damage the abnormal index becomes larger which indicates that using abnormal index to identify abnormal structural damage is feasible. For the same damage case, the closer the distance between the measured position and the injury the larger their correlation.

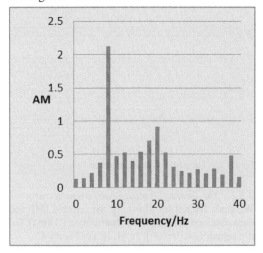

Figure 5.   The Hilbert marginal spectrum.

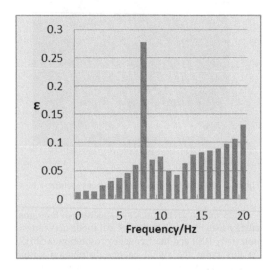

Figure 6.   The normalized figure.

4 CONCLUSION

① The Hilbert marginal spectrum got by the HHT reflected the distribution of signals on different frequency components and it would change with the damage occurring. The abnormal index characterized the change of signal distribution in different periods. So it can be more sensitive to predict the occurrence of damage.

② Analysis shows that the HHT is very good to deal with the nonlinear and non-stationary signal

Table 1.   The anomaly index of each measuring point in different condition.

Condition	6th	10th	17th	24th
Unit9,15damage 30%	0.2658	0.8973	0.4551	0.2374
Unit9,15damage 60%	0.2743	0.9124	0.4621	0.2352
Unit9,15damage 90%	0.2812	0.9425	0.5274	0.2641

produced by the cable net structures, and it also has a good anti-noise performance. The signal frequency components reflected by Hilbert marginal spectrum have statistical properties and the damage feature information extracting from it can be better adapted to the dynamic response of the modal analysis of the actual project.

③The abnormal index can effectively confirm whether the injury occurred and qualitatively characterize the severity of injury. It does not require the modal parameters and incentive information, just need to detect the response at any location. So it is more suitable for the dynamic and real-time requirements of the structural health monitoring.

ACKNOWLEGMENTS

This work was financially supported by Beijing National Science Foundation(8132023), Engineering Research Center of Scientific and Technological Achievements Transforming - Beijing Higher Institution Engineering Research Center of Structural Engineering and New Material (02058214001), BUCEA Urban Rural Construction and Management Industry Research Development Collaboration Post Graduate Training Centre.

REFERENCES

Li, H.Q. et al. 2003. Structural damage detection and analysis based on wavelet transform. *Journal of civil engineering*, 36(5):52–57.
Huang, N.E. et al.1998. The empirical mode decomposition and Hilbert spectrum for nonlinear and nonstationary time series analysis. *Proceedings of the Royal Society of London*, A454:903–995.
Huang, N.E. et al.1998. The empirical mode decomposition and the Hilbert spectrum for non-linear and non-stationary time series analysis. *A-Mathematical Physical and Engineering Sciences*, (454):928–933.
Hu, G.S. 1997. *Digital signal processing theory, algorithms and implementation*. Beijing: Tsinghua University Press.

Energy, Environment and Green Building Materials – Sheng (ed.)
© 2015 Taylor & Francis Group, London, ISBN 978-1-138-02718-3

State-feedback control for stochastic high-order nonlinear systems with time-varying delays

Z.G. Zhong

College of Science and Technology, Ningbo University, Ningbo, Zhejiang, China

ABSTRACT: In this paper, state-feedback stabilization for stochastic high-order nonlinear systems with time-varying delays is considered. By adding one power integrator, a state-feedback based controller is proposed for robust stabilization of a class of general uncertain nonlinear systems. Globally asymptotical stability is achieved in probability by choosing an appropriate Lyapunov Krasoviskii function.

KEYWORDS: stochastic high-order system; time-varying delay; state-feedback control; globally asymptotically stable

1 INTRODUCTION

Stochastic uncertain nonlinear systems exist widely in engineering applications. The investigation on stochastic nonlinear systems has attracted a large amount of attention in past decades. Various types of stochastic systems have been studied in literature [1, 2]. Meanwhile, numerous studies have been proposed for the control of high-order nonlinear systems [3, 4]. Time-delay is very common in real systems, which causes system instability and deterioration of system performance. Therefore, the stochastic time-delay systems have been concerns of [5, 6]. To overcome this problem, [7] introduces a promising technique by adding a power integrator. In this paper, this technique is extended to a class of more general stochastic high-order nonlinear systems with delays.

2 PROBLEM FORMULATION

Consider a system as follows

$$
\begin{cases}
d\eta_i(t) = h_i \eta_{i+1}^{p_i}(t)dt + f_i(t, \bar{\eta}_i(t), \bar{\eta}_i(t-d(t)))dt \\
+ g_i^T(t, \bar{\eta}_i(t), \bar{\eta}_i(t-d(t)))d\omega, i=1,\cdots,n-1, \\
d\eta_n(t) = h_n v(t)dt + f_n(t, \eta(t), \eta(t-d(t)))dt \\
+ g_n^T(t, \eta(t), \eta(t-d(t)))d\omega, \\
y(t) = \eta_1(t)
\end{cases}
\tag{1}
$$

where $v(t) \in R$ is the system input, $y(t) \in R$ is the system output, $\eta(t) = (\eta_1(t),\cdots,\eta_n(t))^T \in R^n$ is the system state vector, $\eta_2(t),\cdots,\eta_n(t)$ are unmeasurable, and $\bar{\eta}_i(t) = (\eta_1(t),\cdots,\eta_i(t))^T, i=1,\cdots,n.$ $d(t): R_+ \to [0,d]$

is the time-varying delay. $p_i \geq 1$ $(i=1,\cdots,n)$ is the order of the system. $h_i, i=1,\cdots n$ are unknown constants denoting the control coefficient. ω is an m-dimensional standard Wiener process defined on a complete space $(\Omega, \mathbb{F}, \{\mathbb{F}_t\}_{t\geq 0}, P)$ (Ω is a sample space, $\mathbb{F}$ is a σ field, $\{\mathbb{F}_t\}_{t\geq 0}$ is a filtration, P is a probability measure). The mapping $f_i : R_+ \times R^i \times R^i \to R$, $g_i : R_+ \times R^i \times R^i \to R^m$ $i=1,\cdots,n$ is locally Lipschitz, such that $f_i(t,0,\cdots,0)=0$, $g_i(t,0,\cdots,0)=0$.

Given a system

$$
\begin{aligned}
dx(t) = f(x(t), x(t-d(t)),t)dt + g(x(t), \\
x(t-d(t)),t)d\omega
\end{aligned}
\tag{2}
$$

with initial condition $\{x(\theta): -h \leq \theta \leq 0\}$ $= \xi \in C_{\mathbb{F}_0}^b([-h,0]; R^n)$, ω denotes the standard Wiener process, time-varying delay $d(t): R_+ \to [0,h]$ is a Borel measurable function. We have following lemmas.

Lemma 1: For system (2), if there exists a function $V(x(t),t) \in C^{2,1}(R^n \times [-d,\infty); R_+)$ and two K_∞ class functions α_1, α_2, a K class function α_3 such that $\alpha_1(|x(t)|) \leq V(x(t),t) \leq \alpha_2(\sup_{-d\leq s\leq 0}|x(t+s)|)$, and $V(x(t),t) \leq -\alpha_3(|x(t)|)$, then the solution $x(t)=0$ of system (2) is globally stable in probability, and moreover $P\{\lim_{t\to\infty}|x(t)|=0\}=1$ [8].

Lemma 2: For any $x \in R, y \in R$, constant $p \geq 1$, the following inequalities hold: $|x+y|^p \leq 2^{p-1}|x^p+y^p|$, $(|x|+|y|)^{\frac{1}{p}} \leq |x|^{\frac{1}{p}} + |y|^{\frac{1}{p}}$. If $p \in R^* = \{q \in R: q = \frac{n}{m} \geq 1\}$

(m, n are odd integers), then $|x-y|^p \leq 2^{p-1}|x^p-y^p|$, $\left|x^{\frac{1}{p}}-y^{\frac{1}{p}}\right| \leq 2^{1-\frac{1}{p}}|x-y|^{\frac{1}{p}}$ [9].

Lemma 3: For any $x \in R, y \in R$ and any constants $c > 0, d > 0, \varepsilon > 0$, the following inequality holds: $|x|^c|y|^d \leq \frac{c}{c+d}\varepsilon|x|^{c+d} + \frac{d}{c+d}\varepsilon^{\frac{-c}{d}}|y|^{c+d}$ [3].

Lemma 4: For any $c_1, \cdots, c_n$ and constant $p > 0$, the following inequality holds: $(c_1 + \cdots + c_n)^p \leq \max\{n^{p-1}, 1\}(c_1^p + \cdots + c_n^p)$ [1].

Lemma 5: If $p \in R^* = \left\{q \in R : q = \frac{n}{m} \geq 1\right\}$ (m, n are odd integers), x, y are real functions, then it exists a constant $c > 0$ such that $|x^p - y^p| \leq p|x-y|$ $(x^{p-1} + y^{p-1}) \leq c|x-y|\left|(x-y)^{p-1} + y^{p-1}\right|$ [9].

3   STATE-FEEDBACK CONTROLLER DESIGN

Defining positive constants $h_i, i = 1, \cdots, n$, introduce a rescaling transformation

$$\eta_1(t) = x_1(t), \eta_i(t) = \frac{M^{q_i} x_i(t)}{\tilde{h}_i}, i = 2, \cdots, n$$

$$v(t) = M^{q_{n+1}} u(t)$$
(3)

where $\tilde{h}_i = h_1^{\frac{1}{p_1 \cdots p_{i-1}}} \cdots h_{i-1}^{\frac{1}{p_{i-1}}}$, $q_i = \frac{q_{i-1}+1}{p_{i-1}}$, $i = 2, \cdots, n$. $q_1 = p_n = 1$, $M \geq 1$. The original system can be expressed as

$$\begin{cases} dx_i(t) = Mx_{i+1}^{p_i}(t)dt + \overline{f}_i(t, \overline{x}_i(t), \overline{x}_i(t-d(t)))dt \\ + \overline{g}_i^T(t, \overline{x}_i(t), \overline{x}_i(t-d(t)))d\omega, i = 1, \cdots, n-1 \\ dx_n(t) = M\tilde{h}_{n+1}u(t)dt + \overline{f}_n(t, x(t), x(t-d(t)))dt \quad (4) \\ + \overline{g}_n^T(t, x(t), x(t-d(t)))d\omega \\ y(t) = x_1(t) \end{cases}$$

where $\overline{f}_1 = f_1, \overline{g}_1 = g_1$, $\overline{f}_i = \frac{\tilde{h}_i f_i}{M^{q_i}}$, $\overline{g}_i = \frac{\tilde{h}_i g_i}{M^{q_i}}$, $i = 2, \cdots, n$

Assumption 1: For any $i = 1, \cdots, n$, it exists constants $a_1 \geq 0$ and $a_2 \geq 0$, such that

$$\left|\overline{f}_i(t, \overline{x}_i(t), \overline{x}_i(t-d(t)))\right|$$

$$\leq a_1(|\overline{x}_1|^{\frac{1}{p_1 \cdots p_{i-1}}} + \cdots + |\overline{x}_{i-1}|^{\frac{1}{p_{i-1}}} + |\overline{x}_i| + |\overline{x}_1(t-d(t))|^{\frac{1}{p_1 \cdots p_{i-1}}}$$

$$+ \cdots + |\overline{x}_{i-1}(t-d(t))|^{\frac{1}{p_{i-1}}} + |\overline{x}_i(t-d(t))|)$$

$$\left|\overline{g}_i(t, \overline{x}_i(t), \overline{x}_i(t-d(t)))\right|$$

$$\leq a_2(|\overline{x}_1|^{\frac{1}{p_1 \cdots p_{i-1}}} + \cdots + |\overline{x}_{i-1}|^{\frac{1}{p_{i-1}}} + |\overline{x}_i| + |\overline{x}_1(t-d(t))|^{\frac{1}{p_1 \cdots p_{i-1}}}$$

$$+ \cdots + |\overline{x}_{i-1}(t-d(t))|^{\frac{1}{p_{i-1}}} + |\overline{x}_i(t-d(t))|)$$
(5)

Assumption 2: It exists a constant $\chi < 1$, such that the time-varying delay $d(t)$ satisfies $d(t) \leq \chi$.

Assumption 3: Constants $h_i(i = 1, \cdots, n)$ are all positive, and it exists positive numbers γ and $\overline{\gamma}$, such that $\gamma \leq |h_i| \leq \overline{\gamma}$, $i = 1, \cdots, n$.

Introduce a new variable as

$$\xi_1(t) = x_1(t), \xi_i = x_i^{p_1 \cdots p_{i-1}} - \alpha_{i-1}^{p_1 \cdots p_{i-1}}(x_{[i-1]}), i = 2, \cdots n \ (6)$$

and the control law is $u = \alpha_n(x)$, where $\alpha_i : R^i \to R$ is continuous and satisfies $\alpha_i(0) = 0$.

First, choose a Lyapunov function $V_1(\xi_1(t)) = \xi_1^4(t)/4$. From lemma 3 and (4), (5):

$$V_1 = M\xi_1^3 x_2^{p_1} + \xi_1^3 \overline{f}_1(t, x_1, x_1(t-d(t)))$$

$$+ \frac{3}{2}\xi_1^2 \left|\overline{g}_1(t, x_1, x_1(t-d(t)))\right|^2$$

$$\leq M(\xi_1^3 x_2^{p_1} + (a_1 + \frac{2}{3}a_2^2 + \frac{3}{4}l_1 + \frac{1}{2}l_2)\xi_1^4$$
(7)

$$+ (\frac{1}{4}l_1^{-3} + \frac{1}{2}l_2^{-1})\xi_1^4(t-d(t)))$$

where $l_1 > 0, l_2 > 0$ are constants. The virtual control law is

$$\alpha_1^{p_1}(t) = -m_1\xi_1(t),$$

$$m_1 = c_1 + \frac{3}{4}l_1 + \frac{1}{2}l_2 + a_1 + \frac{2}{3}a_2^2, c_1 > 0$$
(8)

such that $V_1 \leq M(-c_1\xi_1^4(t) + \xi_1^3(t)(x_2^{p_1}(t) - \alpha_1^{p_1}(t))$

$$+ \mu_1\xi_1^4(t-d(t))), \mu_1 = \frac{1}{4}l_1^{-3} + \frac{1}{2}l_2^{-1}$$.

In step n, assuming that the controller design in steps $(k-1)(k = 2, \cdots, n)$ has been completed, then:

$$V_{k-1} \leq M(-c_{k-1}\sum_{i=1}^{k-1}\xi_i^4 + \xi_{k-1}^{\frac{4p_1 \cdots p_{k-2}-1}{p_1 \cdots p_{k-2}}} \times (x_k^{p_{k-1}} - \alpha_{k-1}^{p_{k-1}})$$

$$+ \sum_{i=1}^{k-1}\mu_i\xi_i^4(t-d(t)))$$
(9)

Now we prove the kth Lyapunov function V_k.

$$V_k = V_{k-1}(x_{[k-1]}) + \int_{\alpha_{k-1}}^{x_k} (s^{p_1 \cdots p_{i-1}} - \alpha^{p_1 \cdots p_{i-1}})^{\frac{4p_1 \cdots p_{i-1}-1}{p_1 \cdots p_{i-1}}} ds$$

$$=: V_{k-1} + W_k \tag{10}$$

Follow the trace of system (4), one can conclude that

$$V_k = -Mc_{k-1} \sum_{i=1}^{k-1} \xi_i^4 + M\xi_{k-1}^{\frac{4p_1 \cdots p_{k-2}-1}{p_1 \cdots p_{k-2}}} \times (x_k^{p_{k-1}} - \alpha_{k-1}^{p_{k-1}})$$

$$+ M\sum_{i=1}^k \mu_i \xi_i^4(t-d(t)) + M\xi_k^{\frac{4p_1 \cdots p_{k-1}-1}{p_1 \cdots p_{k-1}}} (x_{k+1}^{p_k} - \alpha_k^{p_k})$$

$$+ M\xi_k^{\frac{4p_1 \cdots p_{k-1}-1}{p_1 \cdots p_{k-1}}} \alpha_k^{p_k} + \frac{1}{2} \sum_{i=1}^k \sum_{j=1}^k \frac{\partial^2 W_k}{\partial x_i \partial x_j} g_i^T g_j$$

$$+ \sum_{i=1}^{k-1} \frac{\partial W_k}{\partial x_i} (Mx_{i+1}^{p_i} + f_i) + \xi_k^{\frac{4p_1 \cdots p_{k-1}-1}{p_1 \cdots p_{k-1}}} f_k \tag{11}$$

To estimate the second term and the last three terms of (11) with lemma 1, 2, 3, 5 and assumption 1, (10), the following statements hold

$$\xi_{k-1}^{\frac{4p_1 \cdots p_{k-2}-1}{p_1 \cdots p_{k-2}}} (x_k^{p_{k-1}} - \alpha_{k-1}^{p_{k-1}}) \le c\xi_{k-1}^{\frac{4p_1 \cdots p_{k-2}-1}{p_1 \cdots p_{k-2}}} |\xi_k|$$

$$\left| (x_k - \alpha_{k-1})^{p_{k-1}-1} + \alpha_{k-1}^{p_{k-1}-1} \right|$$

$$\le c_{k,1} \xi_{k-1}^4 + \rho_{k,1} \xi_k^4 \tag{12}$$

$$\frac{1}{2} \sum_{i=1}^k \sum_{j=1}^k \frac{\partial^2 W_k}{\partial x_i \partial x_j} g_i^T g_j$$

$$\le 2 \sum_{i=1}^k \sum_{j=1}^k \left| \int_{\alpha_{k-1}}^{x_k} (s^{p_1 \cdots p_{k-1}} - \alpha^{p_1 \cdots p_{k-1}})^{\frac{3p_1 \cdots p_{k-1}-1}{p_1 \cdots p_{k-1}}} ds \right|$$

$$\cdot \left| \frac{\partial^2 \alpha_{k-1}^{p_1 \cdots p_{k-1}}}{\partial x_i \partial x_j} g_i^T g_j \right|$$

$$+ 6 \sum_{i=1}^k \sum_{j=1}^k \left| \int_{\alpha_{k-1}}^{x_k} (s^{p_1 \cdots p_{k-1}} - \alpha^{p_1 \cdots p_{k-1}})^{\frac{2p_1 \cdots p_{k-1}-1}{p_1 \cdots p_{k-1}}} ds \right|$$

$$\cdot \left\| \frac{\partial \alpha_{k-1}^{p_1 \cdots p_{k-1}}}{\partial x_i} g_i \right\|$$

$$\cdot \left\| \frac{\partial \alpha_{k-1}^{p_1 \cdots p_{k-1}}}{\partial x_j} g_j \right\| + 4 |\xi_k|^{\frac{3p_1 \cdots p_{k-1}-1}{p_1 \cdots p_{k-1}}} \sum_{i=1}^{k-1} \left\| \frac{\partial \alpha_{k-1}^{p_1 \cdots p_{k-1}}}{\partial x_i} g_i \right\| \cdot \|g_k\|$$

$$+ 2p_1 \cdots p_{k-1} |\xi_k|^{\frac{3p_1 \cdots p_{k-1}-1}{p_1 \cdots p_{k-1}}} |x_k|^{p_1 \cdots p_{k-1}-1} \|g_k\|^2$$

$$\le M(\sum_{i=1}^{k-1} c_{k,3} \xi_i^4 + \sum_{i=1}^{k-1} \mu_{k,3} \xi_i^4(t-d(t)) + \rho_{k,3} \xi_k^4) \tag{13}$$

$$\sum_{i=1}^{k-1} \frac{\partial W_k}{\partial x_i} (Mx_{i+1}^{p_i} + f_i)$$

$$\le 4 \sum_{i=1}^{k-1} \left| \int_{\alpha_{k-1}}^{x_k} (s^{p_1 \cdots p_{k-1}} - \alpha^{p_1 \cdots p_{k-1}})^{\frac{3p_1 \cdots p_{k-1}-1}{p_1 \cdots p_{k-1}}} ds \right|$$

$$\cdot \left| \frac{\partial \alpha_{k-1}^{p_1 \cdots p_{k-1}}}{\partial x_i} (Mx_{i+1}^{p_i} + f_i) \right|$$

$$\le M(\sum_{i=1}^{k-1} c_{k,2} \xi_i^4 + \sum_{i=1}^{k-1} \mu_{k,2} \xi_i^4(t-d(t)) + \rho_{k,2} \xi_k^4) \tag{14}$$

$$\xi_k^{\frac{4p_1 \cdots p_{k-1}-1}{p_1 \cdots p_{k-1}}} f_k \le \xi_k^{\frac{4p_1 \cdots p_{k-1}-1}{p_1 \cdots p_{k-1}}} a_1(|\bar{x}_1|^{\frac{1}{p_1 \cdots p_{k-1}}} + \cdots + |\bar{x}_{k-1}|^{\frac{1}{p_{k-1}}}$$

$$+ |\bar{x}_k| + |\bar{x}_1(t-d(t))|^{\frac{1}{p_1 \cdots p_{k-1}}}$$

$$+ \cdots + |\bar{x}_{k-1}(t-d(t))|^{\frac{1}{p_{k-1}}} + |\bar{x}_k(t-d(t))|)$$

$$\le M(\sum_{i=1}^{k-1} c_{k,4} \xi_i^4 + \sum_{i=1}^{k-1} \mu_{k,4} \xi_i^4(t-d(t))$$

$$+ \rho_{k,4} \xi_k^4) \tag{15}$$

where $c_{k,i}, \mu_{k,i}, \rho_{k,i}, i = 1,2,3,4$ are constants and positive numbers.

Substitute (12)-(15) to (11), then exists a constant $\rho_k > 0$ such that

$$M\xi_{k-1}^{\frac{4p_1 \cdots p_{k-2}-1}{p_1 \cdots p_{k-2}}} \times (x_k^{p_{k-1}} - \alpha_{k-1}^{p_{k-1}}) + \frac{1}{2} \sum_{i=1}^k \sum_{j=1}^k \frac{\partial^2 W_k}{\partial x_i \partial x_j} g_i^T g_j$$

$$+ \sum_{i=1}^{k-1} \frac{\partial W_k}{\partial x_i} (Mx_{i+1}^{p_i} + f_i) + \xi_k^{\frac{4p_1 \cdots p_{k-1}-1}{p_1 \cdots p_{k-1}}} f$$

$$\le M(\sum_{i=1}^{k-1} c_k \xi_i^4 + \sum_{i=1}^{k-1} \mu_k \xi_i^4(t-d(t)) + \rho_k \xi_k^4) \tag{16}$$

Substitute (16) into (8) gets

$$V_k \le -M(c_k - 1) \sum_{i=1}^{k-1} \xi_i^4 + 2M \sum_{i=1}^k \mu_i \xi_i^4(t-d(t))$$

$$+ M\xi_k^{\frac{4p_1 \cdots p_{k-1}-1}{p_1 \cdots p_{k-1}}} \alpha_k^{p_k} + M\xi_k^{\frac{4p_1 \cdots p_{k-1}-1}{p_1 \cdots p_{k-1}}} (x_{k+1}^{p_k} - \alpha_k^{p_k}) \tag{17}$$

$$+ (c_k - 1 + \rho_k) M\xi_k^4$$

then the kth virtual control law is

$$\alpha_k^{p_1 \cdots p_k} = -(c_k - 1 + \rho_k)\xi_k =: -m_k \xi_k \tag{18}$$

(17) can be expressed as

$$V_k \leq -M(c_k-1)\sum_{i=1}^{k-1}\xi_i^4 + M\sum_{i=1}^{k}\mu_i\xi_i^4(\text{t}-d(\text{t}))$$
$$+M\xi_k^{\frac{4p_1\cdots p_{k-1}-1}{p_1\cdots p_{k-1}}}(x_{k+1}^{p_k}-\alpha_k^{p_k}) \tag{19}$$

At step $k = n$, one can choose a virtual control law $\alpha_n : \text{R}^n \to \text{R}$ as follows

$$\alpha_n^{p_1\cdots p_n}(x) = -m_n\xi_n \tag{20}$$

where $p_n = 1$, m_n are known positive constants. For (18), when $k = n$

$$V_n \leq -M\sum_{i=1}^{n}c_i\xi_i^4 + M\sum_{i=1}^{n}\mu_i\xi_i^4(\text{t}-d(\text{t}))$$
$$+M\tilde{h}_{n+1}\xi_k^{\frac{4p_1\cdots p_{k-1}-1}{p_1\cdots p_{k-1}}}(u-\alpha_n(x)) \tag{21}$$

then the continuous state-feedback controller $\alpha_n(x)$

$$\alpha_n(x) = -(b_1x_1 + b_2x_2^{p_1} + \cdots + +b_nx_n^{p_1\cdots p_{n-1}})^{\frac{1}{p_1\cdots p_n}} \tag{22}$$

where $b_1,\cdots,b_n$ are known constants and depend on $m_1,\cdots,m_n$, but independent on M.

Theorem 1: If system (1) satisfies assumption 1, 2, 3, and the state-feedback controller (20),

The following statements hold:

(I) The closed system consists of (1), (6), (8), (18) and (20) have only one solution on $[0,\infty)$.

(II) For the closed system consists of (1), (6), (8), (18) and (20), the equilibrium is globally stable in probability. The system state can be tuned to zeros, that is $P\left\{\lim_{t\to\infty}\sum_{i=1}^{n}|x_i(\text{t})| = 0\right\} = 1$.

It can be readily obtained by (6), (21) and lemma 1.

4 CONCLUSION

This paper studies the design of state-feedback controller for a class of stochastic high-order nonlinear time-delay systems by adding a power integrator. It is theoretically proved that the proposed design can obtain globally asymptotical stability in probability and all states can be maneuvered to zero.

REFERENCES

[1] X.J.Xie & W.Q.Li, Output feedback control of a class of high-order stochastic systems. International Journal of Control, 82(9):1692–1705, 2009.
[2] R. Has'minskii, Stochastic Stability of Differential Equations, Alphen aan den Rijn: Sijthoff & Noordhoff, 1980.
[3] C.J. Qian & W. Lin, Non-Lipschitz continuous stabilizers for nonlinear systems with uncontrollable unstable linearization. Sys. and Control Letters, 42(3):185–200, 2001.
[4] W. Lin & C. Qian, A continuous feedback approach to global strong stabilization of nonlinear systems, IEEE Trans. on Automatic Control, 46(7):1061–1079, 2001.
[5] W.S. Chen, L.C. Jiao & J. Li, Adaptive NN back-stepping output feedback control for stochastic nonlinear strict feedback systems with time-varying delays. IEEE Trans. on Systems, Man and Cyber., 40:939–950, 2010.
[6] S.Y. Xu & T.W. Chen, Robust H-infinite control for uncertain stochastic systems with state delay. IEEE Trans. on Automatic Control, 47:2089–2094, 2002.
[7] L. Liu & X.J. Xie, Output-feedback stabilization for stochastic high-order nonlinear systems with time-varying delay. Automatica, 47(12):2772–2779, 2011.
[8] S.J. Liu & S.Z. Ge, Adaptive output-feedback control for a class of uncertain stochastic nonlinear systems with time delays. International Journal of Control, 81:1210–1220, 2008.
[9] J. Polendo & C.J. Qian, A generalized framework for global output feedback stabilization of nonlinear systems. Proc. of the 44th IEEE Conf. on Decision and Control. 2646–2651, 2005.

Energy, Environment and Green Building Materials – Sheng (ed.)
© 2015 Taylor & Francis Group, London, ISBN 978-1-138-02718-3

Numerical simulation on structure and anti-explosion performance of new ASA building plates

H.J. Wang, W.Z. Liao, M. Li & S.J. Shen
College of Civil Engineering and Traffic Engineering, Beijing University of Civil Engineering and Architecture, Beijing, China

ABSTRACT: The new type of lightweight ASA (Acrylonitrile Styrene Acrylate copolymer) compound wall plates with insulation properties are widely used in light steel structure residential architecture[1]–[3]. Because of their main raw material are fly ash and industrial solid waste, the plates not only conform to the economic and social benefits, but also meet the sustainable development strategy of China in this contemporary society which advocates green buildings, energy conservation and emissions reduction. This paper simulated the flexural bearing capacity of the wall panels test process using three dimensional finite element software ANSYS, and compared the data tested by China Academy of Building Research[3], then verified the correctness of the analytic model. Meanwhile, the simulation analysis increased the load on plates to make up for the deficiencies of the experiment. In addition, in order to adapt to the special environment of anti-explosion requirements, a performance improvement was conducted on the ASA plates. The research results using the explicit finite element software LS-DYNA indicated that the displacement of general ASA wallboard was 195.5 times of the new type of lightweight ASA wallboard at 6ms in the explosion process. The new plates effectively controlled the deformation. The foam slurry in ASA plates played a good role in energy absorption. The effective stress increased 23.5 times, and the components maintained a good integrity.

KEYWORDS: Numerical Simulation; Anti-explosion; ASA wallboards; green material

1 INTRODUCTION

With the rapid development of tall and super-tall buildings in China, and the deepening reform of wall, the demand of new building materials for light weight, high strength, environmental protection and energy conservation continuously increases fast. This series board of ASA (Acrylonitrile styrene acrylate copolymer) belonging to new architecture board is the key part of the light steel residential systems. The new type of lightweight ASA compound wall plates with insulation properties used double combinations: the one layer embedded in the steel framework, which works with the framework, not only constitute the lateral force resisting system architecture, but also significantly reduce the amount of steel structure of architecture; the other layer that mounted on the outside of the steel frame, effectively prevents the "cold bridge" problem appearing in the steel frame of exterior wall[1]. And the ASA plates with industrial solid wastes as their main raw material, made by the physical foaming process, are the typical low carbon environmental protection material in today's advocacy of green building, energy conservation and emissions reduction[2]. This paper referred to the data on bearing capacity of ASA plates tested by

China Academy of Building Research. And this study simulated the bending capacity of the testing process, also analyzed the bearing capacity using three-dimensional finite element software ANSYS to expand the scope of application of this system. Meanwhile, for the anti-explosion protection engineering field as well as civil defense engineering, etc., this paper developed a new type of lightweight anti-explosion wallboards, and created models to analyze anti-explosion properties by LS-DYNA.

2 EXPERIMENTS

The testing and data are carried out by China Academy of Building Research and provided by Huali joint-tech co., LTD in Beijing[3]. Specimen cross-section is shown in Fig.1. The test used hierarchical load. Fig.2 shows the form of fixed manner. And the field test loading process is shown in Fig.3.

3 ANALYSIS OF NUMERICAL SIMULATION

3.1 *Theoretical basis*

Considering the plasticity of the structure, the explosion tests were modeled using the explicit analysis

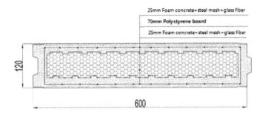

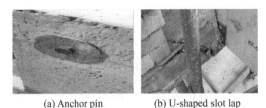

Figure 1. Sample cross section.

$$f_c = f_{ck}\left(2\frac{\varepsilon}{\varepsilon_0} - \left(\frac{\varepsilon}{\varepsilon_0}\right)^2\right) \qquad \varepsilon \leq \varepsilon_0$$

$$f_c = f_{ck}\left(1 - 0.15\left(\frac{\varepsilon - \varepsilon_0}{\varepsilon - \varepsilon_c}\right)\right) \qquad \varepsilon > \varepsilon_c$$

3.2 Unit selection

The material model of concrete chose unit SOLID65 which applied to three-dimensional entity with or without steel reinforcement. This type of unit possesses nonlinear properties such as cracking, crushing, plastic deformation and creep (in three orthogonal directions). The unit defined by eight nodes, each node has three degrees of freedom which are x, y, z direction [5].

(a) Anchor pin   (b) U-shaped slot lap

Figure 2. Fixation.

3.3 Numerical simulation and corresponding calculation

Based on the tests, the constrains used U-groove tied steadily (constrained on xyz directions), the board was fastened at one end (constrained on xyz directions) and lapped on the other end (constrained on xz directions), the loading chose plane loads, the model, constraints and loads are showed in Fig.5:

Figure 3. Load test on site.

Figure 5. Numerical simulation.

software, LS-DYNA. The numerical simulation was simplified according to the special conformation of the specimen. The density of polystyrene board is 15 ~ 20kg / m³; the density of the component is 650kg / m³. Due to the simplification, polystyrene board material was ignored, the definition of U-shaped steel tank material was modeled based on "Steel Design Code". The elastic modulus of steel is 2.06 × 10¹¹Pa, poison's ratio is 0.3, and the yield strength is the average yield strength obtained from the tests. The numerical simulation chosen bilinear kinematic hardening criterion and used the ideal elastic-plastic constitutive relationship models showed in the following:

The numerical simulation was based on the real experimental loads (as shown in Fig.6:

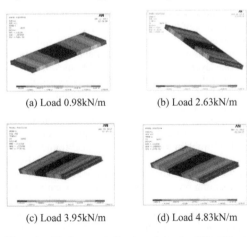

(a) Load 0.98kN/m   (b) Load 2.63kN/m

$$\sigma = E \cdot \varepsilon \qquad 0 \leq \varepsilon \leq \varepsilon_s$$
$$\sigma = \sigma_s \qquad \varepsilon \geq \varepsilon_s$$

On the basis of the experiments, the elastic modulus of ASA board is 2700MPa, and poison's ratio is 0.2. The constitutive relationship referred to the constitutive relationship model of concrete. The ascending section of the model is second-degree parabola, and the descending section is oblique straight line [4]:

(c) Load 3.95kN/m   (d) Load 4.83kN/m

Figure 6. Displacement of contour plot under different loads.

Numerical simulation were set up based on 10 sets of data obtained from experiments, the results were shown in Table 1.

Due to the lapping on one end of the board, the tests lead to the disability of bearing capacity of U-shaped clamps which triggered the disability of continuous

Table 1. The comparison between the results of numerical simulations and the experiments.

Loads (kN/m)	0.57	0.98	1.39	1.8	2.21	2.63	2.93	3.24	3.95	4.67	4.83
Experimental data Deflection (mm)	0.00	0.64	1.29	1.95	2.62	3.24	3.67	4.22	5.43	6.54	----
Numerical simulations (mm)	0.70	1.21	1.71	2.22	2.74	3.3	3.72	4.13	5.35	6.64	6.94

loading. The sizes of the clamps lead to the fracture on the center of the board before the end crash. The method of design resulted in instability of U-shaped clamps and crushing damage on the end of the wall board when the loads were not huge enough. Therefore, in order to compensate for the defects of experiments and obtain the deflection of the wall board under harder loads, the numerical simulation were builted to execute analysis under harder loads, whose values were 15kN/m, 10kN/m, 8kN/m. The results are shown in Table 2.

Table 2. The calculated deflection of the wall board under stimulated hard loads.

Loads (KN/m)	15	10	8
Simulated deflection (mm)	153.73	36.77	15.37

4 NEW ASA BLAST-RESISTANT BOARDS

Considering the disadvantages of construction convenience, the energy saving and environmental protection, and the good anti-seismic performance, the ASA board system strengthened with steel are applied to more and more engineering [1], however, in the field of anti-explosion such as protection and civil defense shelter engineering, it is essential to modify the ASA wall board appropriately. The best way is to develop a new lightweight blast-resistant wall board for such special anti-explosion environment.

4.1 Material conditions

Materials: common steel plate, studs, profiled steel sheet, foam concrete. The design configuration is shown in Fig. 7.

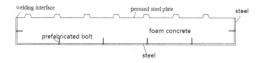

Figure 7. The design configuration of new lightweight blast-resistance wall board.

5 THE ANTI-EXPLOSION MODEL

5.1 The establishment of the model

The ASA-SX60 solid wall board which is retrieved from "Steel mosaic ASA board energy-saving building construction" (08CJ13 Atlas) was analyzed as the study objects, the size of the longitudinal direction is 600mm, the size of the board is 600mm×600mm×60mm (l×b×t). The wall board was tied at four sides as shown in Fig.8. The dynamite placed 0.1m above the center of the board, doses were 0.5kg. The specific sizes of the new blast-resistance wall board were presented in the following texts: the intermediate section formed by foam concrete was 600 × 600 × 53mm; the thickness of the profiled steel sheet faced explosives was 1mm. When established the models, the composite blast-resistance wall board was built by three parts, the profiled steel sheet was simplified to steel plate with the thickness of 1mm, steel plate faced away explosives was 6mm thick. The explosive tests were modeled using the explicit analysis software, LS-DYNA. The models were composited by explosives, ASA solid wall board and steel plate, these sections were all discrete as three-dimensional solid unit SOLID164, the minimum size of the unit was 1mm, the blast load was defined as *Load_Blast.

5.2 Material model

Material model of foam concrete choice *Mat_Crushable_Foam model, and steel plate use *MAT_PLASTIC_ KINEMATIC to simulate. Because of the explosion effect time is very short, it is usually assumed that the bond between steel and concrete is intact, and the bond slip between foam concrete and steel is defined as common node. At the same time, large deformation of concrete unit would lead to grid distortion. The *MAT_ADD_ EROSION is defined to simulate material failure [7]. When the concrete was damaged the unit could be out of work. This algorithm automatically delete concrete unit, and avoid reducing of calculation accuracy and smaller calculation steps and so on. The basic properties of the material are shown in Table 3.

Table 3. Basic material properties of foam concrete and steel.

Materials Category	p/ MPa	E / MPa	Poisson's ratio	Density / kg/m³
Foam concrete	6.5	3.0×103	0.2	650
Steel	235	2.1×105	0.2	7850

Annotate: p- compressive strength, E- Elastic modulus

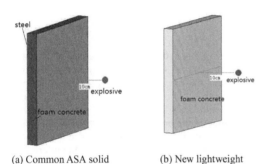

(a) Common ASA solid wallboard   (b) New lightweight anti-explosion ASA wallboard

Figure 8. ASA anti-explosion model.

5.3 Results of the analysis

Numerical simulation results show that (Figure 9, figure 10), the common wallboard's displacement in the Z direction is 4.2 times of the new under the action of the same distance and explosives equivalent at 2ms. At 6ms the common wallboard is 195.5 times of the new lightweight anti-explosion wall. The maximum effective stress loops on the wallboard. The effective stress of new anti-explosion wall panels increases 23.5 times comparing with ordinary solid wallboard. The common ASA solid wall has a larger displacement and lower effective stress under blast loading. And its anti- explosion performance is relatively poor. Considering engineering applications, even if the explosion did not cause serious structural damage to the body, the ASA wall as the maintenance structure would appear explosion fragments, which could cause casualties and equipment damage of house. This paper proposes a new type of lightweight anti-explosion ASA wallboard, which has both sides of steel plate, taking advantage of the characteristics of various materials: Holes inside foam concrete have formed numerous internal free faces, increasing the

Common ASA solid wallboard   New lightweight anti-explosion ASA wallboard

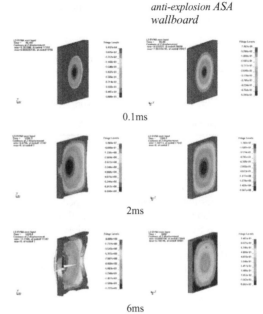

0.1ms

2ms

6ms

Figure 9. Comparison chart of wall displacement.

Common ASA solid wallboard   New lightweight anti-explosion ASA wallboard

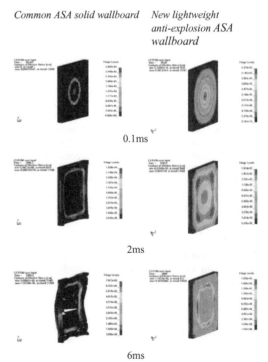

0.1ms

2ms

6ms

Figure 10. Effective stress comparison chart of wallboard.

shock wave reflection and refraction. explosion, so the explosion effects decay [8].While free faces have a good ability to absorb the impact energy[9]. The effective stress comparison chart of wallboard shows that surface steel of new lightweight anti-explosion ASA wallboard effectively increase the stiffness and bearing capacity of member, it not only effectively reduces the overall amount of deformation in order to control the structural deformation, but also maintain the structure integrity commendable.

6 CONCLUSION

In summary, compared with the actual load test, the model values are larger than the experimental values. The main reason for this situation is due to the design of the computer model cannot fully take into account the complexity of the situation in practice [10], while the simplified conditions for the model results also have some certain bias. the simulation results are compared with existing experimental data, and the paper verifies the correctness of the model, and provides the basis for structural design. The new lightweight Anti-explosion ASA wall proposed by this paper have a significantly lower effect on explosion shock wave overpressure. And its processing practices play a role in the reflection and refraction of shock waves of foam concrete, Steel plates have enhanced the integrity of the members, so that the amount of deformation under explosion reduced nearly 200 times, and effective stress increased nearly 24 times. The new wallboards have better anti-explosion properties. The new energy-absorbing effect of anti-explosion ASA wallboard can reduce the extent of damage to the structure under shock wave. The research may

be served as a reference in special environment engineering applications about anti-explosion, as well as has great practical value for protection works and civil defense engineering.

REFERENCES

[1] Yu Chongming. Light steel plate mosaic ASA energy-efficient buildings [J] building energy, 2012 (02):1–2.
[2] Zhu Hengjie. Innovation ASA board mosaic integrated energy-saving building [J] Urban Housing, 2009 (01): 54–55.
[3] Zhu Hengjie, YU Chongming. Development and application of lightweight foam concrete precast panels ASA, 2013 and Cement Products Symposium.
[4] Xu Gang. Mosaic infilled steel frame side force Experimental Research.[D] Tianjin: Tianjin University, 2007.
[5] Lu Xinzheng, Jiang Jianqiong. Analysis with ANSYS Solid 65 unit complex stress concrete composite members.[M] structures, 2003, 33 (6):22–24.
[6] 08CJ13 ASA board inlaid steel energy efficient building construction portfolio.
[7] Shi Shaoqing, Liu Renhui, etc. Steel - aluminum foam - Research blast performance of new composite structure to reduce steel [J]. Vibration and Shock, 2008,24 (4): 143–146.
[8] Zhang Jingfei, Fen Mingde, etc. Experimental study on anti-knock properties of foam concrete [J] Concrete, 2010 (10):10–12.
[9] Zhao Rong, Guo Weiguo, etc. An experimental study of the behavior of the new lightweight foam concrete squeeze [J] Experimental Mechanics, 2006 (26): 562–566.
[10] Zhang Feifei. Experimental and theoretical thermal analysis of composite panels [D] Hebei: Hebei University of Technology, 2011.

Energy, Environment and Green Building Materials – Sheng (ed.)
© *2015 Taylor & Francis Group, London, ISBN 978-1-138-02718-3*

Analyses of the exergetic efficiency of solar energy heat pump system

Y.Q. Di, W. Zhao & H.Y. Di

China Academy of Building Research, China Building Technique Group Co., Ltd. Beijing, China

ABSTRACT: The paper established exergy models of the parts in the solar energy heat pump system based on the second law of thermodynamics and the exergy efficiency and cost exergy efficiency are proposed on the basis of the evaluation index system of exergy analysis. And on this basis, according to the Beijing area, outdoor meteorological parameters and system parameters, verifying the accuracy of the established mathematical model, and it can be obtained that energy consumption in solar heat pump system from large to small in the order are: solar collectors, compressors, storage tanks, condensers, evaporators and throttles. The result provides a theoretical basis for design and other aspects of the system.

KEYWORDS: Solar heat pump system; Exergy models; Exergy efficiency; Energy consumption

1 INTRODUCTION

Solar energy and heat pump technology is the most effective way in saving conventional energy. Energy-saving effect is much better when solar energy and heat pump technology are applied together, but how to produce good energy saving effect using this method is the key of the research for relevant scholars and researchers. In the foreign literature, Torres-Reyes et al. made exergy analysis of solar-assisted heat pump used for air heating in theoretical and experimental fields, and the ways to optimize evaporation temperature and condensation temperature of R-22 are proposed. The experiment of solar heat pump system is performed by Cervantes de Gortari and Torres-Reye, and the maximum exergy efficiency is determined. They pointed that the exergy loss of solar collectors/evaporator is the biggest in solar heat pump system. There are also some researches in this field in China, Kuang Yuhui and Zhang Kaili put forward energy balance and exergy balance equations of the main equipments in solar energy heat pump system from the theory of heat and analyzed the performance of the system and its energy saving effect. Wang Haiying working in Qingdao Technological University analyzed equipment energy utilization of regular operation of a solar heat pump system based on the first and second law of thermodynamics, and gave calculation formula of exergy loss about each link of the system. Through the actual system, energy consumption in solar heat pump system from large to small in the

order are: compressors, condensers and evaporators and system improvement measures are put forward.

This paper set up exergy models of every part of solar energy heat pump, through the experiment to verify the accuracy of the model, and draws the system exergy efficiency and the exergy loss of various components, to provide certain reference basis for the design of solar energy heat pump system and future research.

2 MATHEMATICAL MODELS OF SOLAR ENERGY HEAT PUMP

Energy has the dual nature of quantity and quality, the energy analysis of the first law of thermodynamics, although did not distinguish between energy quality, only pay attention to the amount of energy, the direction of energy flow of the system or process was indicated. The second law of thermodynamics taken the useful energy exergy as research index, conducted focusing both on the quantity and the quality and indicated the link of exergy loss in the process or system to provide guidance for optimization of system or process. Therefore, this article established the exergy models of a solar heat pump system, providing a theoretical basis for the design and operation optimization of the system. The solar heat pump system can realize three ways: solar heating alone, solar heat pump combined heating, heat pump heating alone. Figure.1 is the schematic diagram.

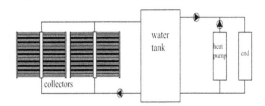

Figure 1. The principle diagram of the solar energy heat pump.

2.1 Load model

The useful energy provided by solar heat pump system can be shown by the supply and return water temperature, namely, the exergy of indoor benefits (taken outdoor environment condition (Pa, Ta) as the ground state):

$$Exq = m_g c_p (T_g - T_h - T_a \ln \frac{T_g}{T_h})$$ (1)

In the formula, Exq: indoor exergy; Ex_g: exergy of Heating in the water supply; Ex_h: exergy of heating in the water return; m_g: the mass flow of heating medium; T_g, T_h: temperature of heating supply and return water; T_a: outdoor temperature; C_p: the specific heat at constant pressure of the water, KJ / Kg.

2.2 Solar thermal collector

Solar collector as one of the core component of solar heat pump system, absorb solar radiation to provide heat source for the whole system, on the other hand, it is also one of the radiating device, and at the same time, inevitably there is a part of the heat loss when the heat of absorption. Taking the outdoor environment condition (Pa, Ta) as the ground state, according to the general form of exergy equation: exergy loss= input exergy-output exergy-system exergy variable. If the system is in a steady state flow conditions, system exergy variable is 0, the exergy equation for solar collector is:

$$Ex_{Lc} = \partial A_c I_c \left(1 - \frac{T_a}{T_{sun}}\right) - m_c c_p \left(T_{co} - T_{ci} - T_a \ln \frac{T_{co}}{T_{ci}}\right)$$ (2)

In the formula, Ex_{LC}: exergy of solar collector; Ex_{co}, Ex_{ci}: the exergy of outlet and inlet of solar collector; ∂: solar heat collector absorption rate; Ac: the solar collector area; Ic: solar radiation intensity; Ta: outdoor temperature; T_{sun}: the solar core temperature; m_c: solar collector heat transfer medium flow; T_{co}, T_{ci}: solar heat collector outlet and inlet temperature.

Solar collector exergy loss coefficient is the ratio of solar heat collector exergy loss and total system input exergy:

$$\chi_c = \frac{Ex_{Lc}}{F_c \left[\partial A_c I_c \left(1 - \frac{T_a}{T_{sun}}\right) + W_{P1} \right] + F_r (W + W_{P3}) + F_{P2} W_{P2} + F_D W_D}$$

$$= \frac{\partial A_c I_c \left(1 - \frac{T_a}{T_{sun}}\right) - m_c c_p \left(T_{co} - T_{ci} - T_a \ln \frac{T_{co}}{T_{ci}}\right)}{F_c \left[\partial A_c I_c \left(1 - \frac{T_a}{T_{sun}}\right) + W_{P1} \right] + F_r (W + W_{P3}) + F_{P2} W_{P2} + F_D W_D}$$ (3)

In the formula, x_c: solar collector exergy loss coefficient; Fc: solar heat collector operation control function, when solar heat collector operation, Fc = 1, conversely, Fr = 0 ; W: heat pump power consumption; W_{P3}: power consumption of circulating pump between the water tank and the heat pump; F_{P2}: control functions intermittent heating, when Heating, $F_{P2} = 1$, conversely, $F_{P2} = 0$; W_{P2}: heating circulating pump power consumption; F_D: electric heating control function, when electric heating operates, $F_D = 1$, conversely, $F_D = 0$; W_D: electric heating power consumption.

2.3 Water storage tanks

The water storage tank is a heat storage device, but also thermal equipment, plays a vital role in the whole system. The water tank temperature not only affects the efficiency of heat collection of solar collector, but also directly affects the evaporation temperature of heat pump and heat pump efficiency as a heat source heat pump. Therefore, the water storage tank must be reasonable design, good heat preservation measures to reduce the heat loss. Taking the outdoor environment condition (Pa, Ta) as the ground state, according to the general form of exergy equation, water tank input exergy comprises exergy of solar heat collector water, the exergy of load side return water and electric heating input exergy; Output exergy including exergy of solar collector, exergy that output to the load side; The internal exergy change of water storage tank is caused by its higher water temperatures, which can be said by the useful energy. Therefore, the exergy equation of water storage tank can be expressed as:

$$Ex_{Ls} = F_c m_c c_p \left(T_{co} - T_{ci} - T_a \ln \frac{T_{si}}{T_{so}}\right) - Q_s \left(1 - \frac{T_a}{T_s}\right) -$$

$$F_r m_g c_p \left(T_s - T_{eo} - T_a \ln \frac{T_s}{T_{eo}}\right) - F_{P2} \bar{F}_r m_g c_p \left(T_{sm} - T_h - T_a \ln \frac{T_{sm}}{T_h}\right)$$ (4)

In the formula, Ex_{LS} : the exergy loss of water storage tank; Qs : tank storage rate; Tsm : the average temperature in the water storage tank; e_{xi}, e_{xo} : the ratio of outlet exergy and inlet exergy of The heat pump evaporator; T_{eo}: the evaporator outlet temperature.

The water storage tank exergy loss coefficient is the ratio of water storage tank exergy and total system input exergy.

$$\chi_s = \frac{Ex_{Ls}}{F_c\left[\partial A_c I_c\left(1-\frac{T_a}{T_{sun}}\right)+W_{P1}\right]+F_r(W+W_{P3})+F_{P2}W_{P2}+F_D W_D} \quad (5)$$

2.4 Heat pump

The heat pump is mainly composed of evaporators, compressors, condensers and expansions. Taking the outdoor environment condition (Pa, Ta) as the ground state, four part exergy equation of heat pump as follows.

2.4.1 Evaporators

The heat loss in the evaporator heat transfer process does not exist, heat released of the evaporator water side was completely absorbed by the refrigerant side, but because of the two fluids of evaporator are different, there is a temperature difference in heat transfer process. Therefore, the exergy loss (Ex_{Lev}) is existed in the evaporator heat transfer process. According to the general form of the exergy equilibrium equation, evaporator input exergy including entrance exergy of water side (Ex_{ei}) and entrance exergy of refrigerant side (Ex_{eri}); Output exergy including outlet exergy of water side (Ex_{eo}) and outlet exergy of refrigerant side (Ex_{ero}). The evaporator flow process can be regarded as a steady state flow process, the system does not exist exergy variable. Therefore, the evaporator exergy loss equation can be expressed as:

$$Ex_{Lev} = m_g c_p\left(T_s - T_{eo} - T_a \ln\frac{T_s}{T_{eo}}\right) - $$
$$m_r\left[(h_{ero} - h_{eri}) - T_a(s_{ero} - s_{eri})\right] \quad (6)$$

Se_{ro}, Se_{ri}: outlet entropy and entrance entropy of the refrigerant.

The evaporator exergy loss coefficient is the ratio of the evaporator exergy loss and the total of the whole system input exergy:

$$\chi_{ev} = \frac{Ex_{Lev}}{F_c\left[\partial A_c I_c\left(1-\frac{T_a}{T_{sun}}\right)+W_{P1}\right]+F_r(W+W_{P3})+F_{P2}W_{P2}+F_D W_D}$$
$$= \frac{m_{w1}c_p\left(T_s - T_{eo} - T_a \ln\frac{T_s}{T_{eo}}\right) - m_r\left[(h_{ero} - h_{eri}) - T_a(s_{ero} - s_{eri})\right]}{F_c\left[\partial A_c I_c\left(1-\frac{T_a}{T_{sun}}\right)+W_{P1}\right]+F_r(W+W_{P3})+F_{P2}W_{P2}+F_D W_D} \quad (7)$$

2.4.2 Compressors

If the compression process is adiabatic process, compressor input power is completely converted into heat absorbed by the refrigerant, electrical energy can be completely converted into useful energy, that belongs to the category of exergy. But only a part of heat energy can be converted into useful energy. In the process of compression, electrical energy converted into heat energy, in other words, part of the useful energy is transformed into unavailable energy, that is anergy, That is to say, energy loss Ex_{Lom} is existed in the process. Taking the compressor as a control volume, its Input exergy includes input power w and entrance refrigerant exergy Ex_{comi}, that can be seen as a process of steady state flow, no system exergy change, therefore, the compressor exergy loss equation is:

$$Ex_{Lcom} = W - (Ex_{como} - Ex_{comi}) \quad (8)$$

The compressor exergy loss coefficient is the ratio of the compressor exergy loss and total system input exergy:

$$\chi_{com} = \frac{Ex_{Lcom}}{F_c\left[\partial A_c I_c\left(1-\frac{T_a}{T_{sun}}\right)+W_{P1}\right]+F_r(W+W_{P3})+F_{P2}W_{P2}+F_D W_D}$$
$$= \frac{W - m_r(ex_{como} - ex_{comi})}{F_c\left[\partial A_c I_c\left(1-\frac{T_a}{T_{sun}}\right)+W_{P1}\right]+F_r(W+W_{P3})+F_{P2}W_{P2}+F_D W_D} \quad (9)$$

2.4.3 Condensers

Assuming that there is no heat loss in the heat transfer process of the condenser, namely the refrigerant condenser heat release was completely absorbed by the water side, The heat transfer process in the condenser is the temperature difference of heat transfer for heat transfer, so there is exergy loss Ex_{Lcon}. According to the general form of exergy equations, condenser input exergy includes the entrance of refrigerant side exergy Ex_{conri} and entrance exergy of water side Ex_h. Output exergy includes refrigerant outlet exergy Ex_{conro} and the water outlet exergy Ex_g. The condenser flow process can be regarded as a steady state flow process, no system exergy change. Therefore, in the exergy loss equation of condenser is:

$$Ex_{Lcon} = m_r\left[(h_{conri} - h_{conro}) - T_a(s_{conri} - s_{conro})\right] - $$
$$m_{w2}c_p\left(T_g - T_h - T_a \ln\frac{T_g}{T_h}\right) \quad (10)$$

Se$_{ro}$, Se$_{ri}$: outlet entropy and entrance entropy of the condensers.

The condenser exergy loss coefficient is the ratio of the condenser exergy loss and the total of the whole system input exergy:

$$\chi_{con} = \frac{Ex_{Lcon}}{F_c\left[\partial A_c I_c\left(1-\frac{T_a}{T_{sun}}\right)+W_{P1}\right]+F_r\left(W+W_{P3}\right)+F_{P2}W_{P2}+F_DW_D}$$

$$=\frac{m_r\left[\left(h_{conri}-h_{conro}\right)-T_a\left(s_{conri}-s_{conro}\right)\right]-m_{w2}c_p\left(T_g-T_h-T_a\ln\frac{T_g}{T_h}\right)}{F_c\left[\partial A_c I_c\left(1-\frac{T_a}{T_{sun}}\right)+W_{P1}\right]+F_r\left(W+W_{P3}\right)+F_{P2}W_{P2}+F_DW_D}\quad(11)$$

2.4.4 Expansion values

The process of refrigerant pass the expansion valve is isenthalpic process, there is no heat loss, but the entropy is different before and after expansion, therefore, there is exergy loss Ex_{Lf} in the process. Its input exergy Ex_{fi} is the entrance exergy of expansion valve, output exergy Ex_{fo} is the outlet exergy of the expansion valve. The exergy loss equation of expansion valve is:

$$Ex_{Lf}=Ex_{fi}-Ex_{fo}=m_rT_a\left(s_{rfi}-s_{rfo}\right)\quad(12)$$

Se$_{ro}$, Se$_{ri}$: outlet entropy and entrance entropy of the expansion valve.

The expansion valve exergy loss coefficient is the ratio of the condenser exergy loss and the total of the whole system input exergy:

$$\chi_f=\frac{Ex_{Lf}}{F_c\left[\partial A_c I_c\left(1-\frac{T_a}{T_{sun}}\right)+W_{P1}\right]+F_r\left(W+W_{P3}\right)+F_{P2}W_{P2}+F_DW_D}$$

$$=\frac{m_rT_0\left(s_{rfi}-s_{rfo}\right)}{F_c\left[\partial A_c I_c\left(1-\frac{T_a}{T_{sun}}\right)+W_{P1}\right]+F_r\left(W+W_{P3}\right)+F_{P2}W_{P2}+F_DW_D}\quad(13)$$

3 EVALUATION INDEX SYSTEM BASED ON THE EXERGY ANALYSIS

3.1 Solar energy heat pump system exergy efficiency

The solar heat pump system has three kinds of operating conditions: solar heating alone, a combination of solar energy and heat pump heating and heat pump heating alone. Exergy efficiency can be expressed as:

$$\eta_{ex}=1-\sum\chi=1-F_c\chi_c-\chi_s-F_r\left(\chi_{ev}-\chi_{com}-\chi_{con}-\chi_f\right)\quad(14)$$

3.2 Cost energy efficiency

In the solar energy heat pump system, the system input exergy comprises exergy obtained by solar heat collectors, namely free exergy and exergy obtained by the conventional energy (electricity) input, that is cost exergy. In this system, the cost exergy includes all cost exergy of circulating pump, they were recorded as Exd$_{p1}$, Exd$_{p2}$, Exd$_{p3}$, the cost exergy consumed by the heat pumps- Exd$_r$ and the cost exergy consumed by auxiliary heating devices- Exd$_D$. Therefore, the system cost energy efficiency is:

$$\eta'_{ex}=\frac{Exq}{Exd_{P1}+Exd_{P2}+Exd_{P3}+Exd_r+Exd_D}\quad(15)$$

4 EXPERIMENTAL STUDY OF SOLAR ENERGY HEAT PUMP

Experimental study is a way to explore the relationship between independent variable and dependent variable, is the inevitable stage of theory applied in practice. In this paper, taking Beijing area as the experimental object, by the solar heat pump heating system experiment, data processing, verify the correctness of the mathematical model.

Solar heat pump experimental system consists of solar collector systems, water source heat pump system, control system and solar energy testing system. According to the solar radiation in Beijing area, heat collecting efficiency of the vacuum tube solar collector in the winter is 58%, designing 12 configurations, single block area of solar collector is 1.62m2, a total of 19.44 m². The water storage tank is stainless steel water tank of 1.1 tons, in order to ensure the system can run continuously all day, there is an electric auxiliary heating device with 3kW in it. The experimental platform using EWS60H as water source heat pump made by Wright air-conditioning manufacturers, refrigerant R22, heat 6000W, rated heat power is 1750W.

The experimental schemes during the heating season as follows:

1 Collecting meteorological parameters in the Beijing area during the heating season in 2013~2014;
2 Setting the control parameters of the system and testing;
3 Changing experimental conditions: changing the start temperature of electric heating, changing the cycle of solar heat collector, changing the heating temperature, changing the start temperature of water source heat pump, changing water yield of the water tank.

4.1 Validate the data processing and model experiment

Solar energy heat pump system experimental platform runs under the same operational control parameters, that is the solar energy heat pump cyclic temperature is 8 °C, heating pump cyclic temperature is 10 °C, the start temperature of water source heat pump is 30 °C, the start temperature of electric heating is 15 °C. The experimental bench structure parameters maintain the same. Taking the average outdoor meteorological parameters in the experimental period, the outdoor temperature is -1.1 °C, solar radiation intensity is 826W/m2.

Taking the average values of the test parameters when the system is in working condition that it is combined heat with solar energy and heat pump. The experimental results are shown in Table 1.

Table 1. The average parameters of experimental conditions.

Program	Inlet temperature / °C	outlet temperature /°C	Flow /kg/s
Solar thermal Collector	16.51	21.89	0.42
Evaporator Water side	16.94	15.16	0.46
Condenser Water side	27.83	34.58	0.14

Pressure enthalpy diagram of water source heat pump system is shown in Fig. 2, by the parameters and properties of R22 provided by manufacturers, the specific enthalpy and entropy of R22 in water source heat pump unit can be obtained, as shown in Table 2.

Table 2. The specific enthalpy and entropy of R22.

Number	h (kJ/kg·K)	s (kJ/kg)
1	415.2099	4.6898
2	440.20993	4.7031
3	240.9152	4.0678
4	240.9152	4.076

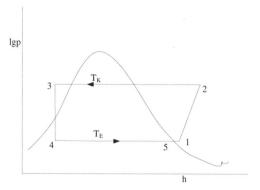

Figure 2. Water source heat pump pressure-enthalpy diagram.

The exergy loss and exergy loss coefficient of the solar collector, water tank, evaporator, condenser, compressor and expansion valve can be calculated by the exergy models of the components in the solar energy heat pump system. The results are shown in Tables 3 and 4.

Table 3. Input exergy and earning exergy of system.

Free exergy (W)	Cost exergy (W)	Total exergy (W)	Indoor earnings Exergy (W)
9395.58	1501.2	10896.28	421.13

Table 4. Exergy and exergy loss coefficient of system.

Programs	Power (W)	Heat loss (W)	Exergy Loss (W)	Exergy loss coefficient (%)
Collectors	9875.33	385.01	8736.85	80.19
Storage Water tanks	9490.32	54.05	112.91	1.04
Evaporators	3471.11		104.67	0.96
Condensers	3969		105.85	0.97
Compressors	1501.2		1002.12	9.2
Throttle valves	0		0.16	0.0015
Total	11376.53	439.05	10062.55	92.35

According to the above results, the average exergy efficiency of solar heat pump system in test period is 1-0.9235=0.0765, system cost exergy efficiency is 421.13/1501.2=0.28.

The Exergy loss rate of each component is the ratio of the exergy loss of components and total exergy loss which is shown in Fig. 3. It can be seen from the graph, exergy loss from large to small in the order are: the solar collectors of solar energy heat pump system, compressors, condensers, evaporators, water storage tanks and throttle valves.

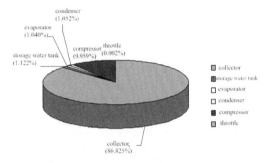

Figure 3.   Exergy loss rate of each component.

The solar core temperature is 5600K, solar radiation energy converts to cycle fluid from the solar heat collector and circulation medium temperature at 300 K to 330 K, resulting in most of the exergy loss of solar collectors. This part together with exergy loss caused by the structures of the solar heat collector play a significant role in all exergy loss of the system.

The exergy loss of heat pump compressors is generated by the process that useful mechanical power is converted to thermal energy which cannot be completely converted into useful power, this part together with the exergy loss caused by the friction between its structures will be higher than the exergy loss caused by the temperature difference of evaporators and condensers.

5   CONCLUSIONS

This paper describes the exergy models of components of a solar heat pump system, and it can be got that exergy loss from large to small in the order are: the solar collectors of solar energy heat pump system, compressors, condensers, evaporators, water storage tanks and throttle valves. At the same time, experimental exergy loss is the same with the qualitative trend analysis, which demonstrates the correctness of the thermodynamic model and provides a theoretical basis for the design and operation of solar energy heat pump system.

ACKNOWLEDGMENTS

A project supported by Natural Science Foundation of China (51108438); Issue supported by National "Twelfth Five-Year" Plan for Science & Technology Support (2012BAJ06B06); Low Carbon Development Strategy of China (201214).

REFERENCES

Torres-Reyes E, Picon Nunez M, Cervantes de Gortari J. Exergy *analysis and optimization of a solar-assisted heat pump, Energy, 23:337–44, 1998.*

Cervantes J G, Torres-Reyes E. *Experiments on a solar-assisted heat pump and an exergy analysis of the system, 22:1289–1297, Appl Therm Eng, 2002.*

Yuhui Kuang, Kaili Zhang, Liqiang Yu, etc. *Journal of Qingdao Institute of Architecture and Engineering. 2001, 22(4):80-83.*

Haiying Wang. *Shandong HVAC. 007, (2):348–353.*

Yanqiang *Di. Building efficient energy supply system integration technology and engineering practice. Beijing: China Architecture & Building Press, 2011.*

Tsatsaronis G. *Progress of Energy combustion 3, (13):199–227, Science, 1993.*

Jianing Sun, Qinglin Chen, Qinghua Yin. *Gas and heat, 2003, 23 (9):523–528.*

Ozer K, Koray U, Arif H. *Exergetic assessment of direct-expansion solar-assisted heat pump systems: review and modeling, (12) 1383–1401, Renewable and Sustainable Energy Reviews, 2008.*

Energy, Environment and Green Building Materials – Sheng (ed.)
© 2015 Taylor & Francis Group, London, ISBN 978-1-138-02718-3

Heat recovery ventilator technology development and application in the public buildings

L. Yuan

Design and Arts College, Beijing Institute of Technology, China

ABSTRACT: This article is a contribution and comparative studies to assess the proportion of the central air conditioning for fresh air heat recovery technology, will commence in the contrast between residents and public buildings. And domestic different stage in Germany's policies and regulations are more concerned about energy efficiency of residential buildings. But in fact, the domestic fresh air heat recovery technology in the non-residential building-level application has a vast market.

KEYWORDS: Energy-saving; Heat recovery; ventilator; Public buildings

1 BACKGROUND

As we all know, heat recovery ventilator technology have been widely used in foreign civil construction and public buildings; but for our present stage, there is no centralized ventilation in civil, so in this area to achieve a heat recovery ventilator technology is more difficult. The domestic public buildings with centralized multi-ventilated, and public buildings for energy conservation have been a concern, so the heat recovery ventilator technology used in the public construction sector has good prospects.

2 DEFINITION OF HEAT RECOVERY TECHNOLOGY

Heat recovery technology is called by energy, recycling process collectively. It is by heat conduction of thermal energy to achieve the purpose of re-use with a mass flow between the two different temperatures. The purpose is to reduce outdoor air heated to a preset for the indoor air temperature generated by the primary energy demand. Exhaust heat recovery technology and the use of outdoor air enthalpy difference through a separate system or with other systems in parallel to achieve the recovery of heat or cold purposes.

Overall, the heat recovery technology can be used for residential buildings can also be used for non-residential buildings. Mechanical ventilation with heat recovery equipment in residential and non-residential buildings can effectively save heat. In summer and winter in northern China and southern regions, the ventilation heat loss during the 40% to 48% of total heat loss of housing.

In recent years Germany adopted a series of energy initiatives, in 2009 promulgated the commencement of the Energy Saving Ordinance (EnEV). In order to effectively reduce primary energy consumption, heat recovery ventilation system in the field of non-residents and residents of the buildings have been considerably developed, particularly in the field of civil construction has been widely used.

3 THE PROPORTION OF RESIDENTIAL BUILDINGS AND NON-RESIDENTIAL BUILDINGS

In accordance with the use of functional architecture can be divided into residential buildings and public buildings. A public building is the architectural units of non-residents. Such as offices, administrative buildings, shops and shopping centers, hospitals and first aid centers, schools and commercial buildings and other residential areas of the buildings do not belong to. Unlike China, high-rise buildings are generally considered public buildings for the purpose of establishment in Germany. Other areas of the buildings do not belong to residential buildings also include the relevant universities and colleges, the stadiums, theaters, churches, and meeting places.

When more than half of the total area of the buildings used for a residential purpose, the buildings were known as residential buildings. When a building is surrounded by walls not closed, but only independent support structure, then he does not belong to the building, naturally does not belong to the non-residential building neither. The number of existing residents in the building for 18.2 million in Germany is much larger than just 1.7 million of non-residential buildings. However, this figure did not reflect the truth. By Figure 2 reflects another scene when people focus on the total construction area.

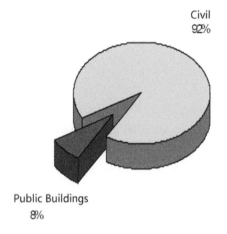

Figure 1. The existing building in Germany in 2011 (in 2011 the National Bureau of Statistics).

This is a German survey of 1991. Single-family homes (EFH), townhouse (RH) and multi-family housing (MFH) is 50.5% of the total building area. Rather than residential buildings is 49.5% of the total building area. In particular, the factory Industrie (10%), warehouse Lager (12%) and other non-residential buildings sonstige NWG (11%) with a high percentage.

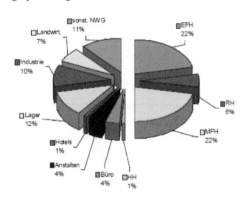

Figure 2. Federal Republic of Germany proportion of the volume of each type of housing in 1992.

The area of assigned residential buildings and non-residential buildings are basically balanced; in Germany, nearly 75 percent of the existing buildings were built before 1975. In order to impact on energy consumption, it is best to use as a basis to judge the volume of housing. Because the above data do not take into account the room height, but for different building categories, room height is different. Divided

by the building floor area ratio of the case, rather than residential buildings occupying 36.5%, which is up to 63.5% of non-residential buildings. The volume of the buildings' energy consumption evaluation as a reference, in the case of this premise consumption can be approximated proportion of different types of housing.

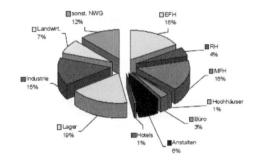

Figure 3. Federal Republic of Germany proportion of the volume of each type of housing in 2010 provided by Umwelt-Campus Birkenfeld.

The figure can be found through the energy consumption of non-residential buildings nearly two-thirds of the total building energy consumption. Indoor ventilation is concerned; we can guess the assumption by the facts, although there is no credible study to illustrate the use of the ratio of different types of ventilation technology in a house. Indoor ventilation with heat recovery technology application rate in the non-residential buildings is higher than in residential buildings today, because in residential buildings through ventilation technology can achieve energy savings are not applied in a non-resident building is so obvious.

In order to share both better estimate, the use of existing market research and studies were analyzed. According to market research reports building air conditioning professional associations, centralized ventilation technology showed the following developments in the residential construction sector in Germany.

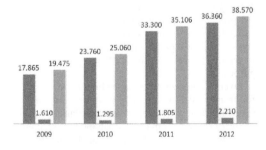

Figure 4. Number of supply (building air conditioning professional associations 2013 report) central air conditioning parts used in residential buildings.

424

This market research report covers 80% of the German market, and from which we can make the following energy-related inferred.

4 CENTRALIZED VENTILATION TECHNOLOGY IN NON-RESIDENTIAL BUILDINGS

According to UCB (Umwelt-Campus Birkenfeld) research reports and market research reports by air conditioning Equipment Manufacturers Association show that heat recovery technology trends in the field of non-residential central air conditioning architecture.

Back to the heat transfer coefficient of the heat recovery system diagram showing the positive trends, the average returnntralized ventilation technology in non-residential buildings reached 69.5% thermal coefficient in 2011.

The same heat recovery system in the utilization of non-residential buildings also had significant development in 2012, and the average of 83.2% of the viable device equipped with a heat recovery system.

Manufactured by the German Association of the comparative survey covers 75% of the German market, we can find, heat recovery in residential construction sector accounted for 165,429 megawatt-hours per year than 6.7% and contrasts with the non-residential construction sector was 2,301,568 as the ratio of 93.3% of total megawatts per year. These data are for centralized ventilation and air conditioning field.

5 THE POTENTIAL FOR ENERGY SAVING INDOOR VENTILATION TECHNOLOGY

July 2012, the European Commission published a survey report on energy demand and energy-saving potential in the field of ventilation equipment.

The report laid the foundation for ventilation equipment to green design criteria for continued development. The study includes not only residential building ventilation, but also contains the non-residential building's ventilation. According to the European Commission's report shows that 24% of European residential buildings equipped with mechanical ventilation. But only 1.5% of the residential buildings are equipped with heat recovery technology currently. Although this value is very low, but now there are 256 annual Coke beat the heat by heat recovery technology in the field of residential construction savings. According to this relative primary energy coefficient of 2.5 per year will have 168 shot coke electrical energy for mechanical ventilation

run into. Therefore, the annual net primary energy savings are 88PJ. (Note: PJ is Petajoule acronym for 10 of 15 times joules, called shot coke.) These are equivalent to seven million tons of carbon dioxide equivalent emissions per year. Thus, the EU concluded will have 60 percent more energy savings in the field of residential building ventilation in 2025. According to market forecast report, then there will be 500 shot coke energy savings for each year. Deduct the required electricity demand, net primary energy savings of around 360 shot coke. These are equivalent to twenty million tons of reducing carbon emissions.

Energy saving potential in the public construction sector was significantly greater. 70% of public buildings are now equipped with mechanical ventilation, seven percent of the devices with heat recovery systems. 2010 These devices saved 2210 shot coke heat, which is corresponding to the power loss of 532 shot coke and the net energy savings for 1678 shot coke. These are equivalent to an annual reduction of 100 Mt (million tonnes) of carbon dioxide emissions. UP To 2025, due to vigorously promote the application of heat recovery technology in the field of non-residential building, energy-saving are expected to take at least 3200 shot coke potentially. These are equivalent to more than 950PJ in 2010 to shoot the focus of primary energy savings potential, also, or carbon dioxide emissions by 50 million tons. According to forecasts to 2025, primary energy savings potential in the residential construction sector will reach about 448 PJ, while 14.6% of the total value; for the non-residential construction sector is about 2630PJ, accounting for 85% of the total. For the domestic market, according to these results can also be clearly drawn, building energy saving used in public buildings field is very important in the country.

6 CONCLUSIONS

Due to the development of mechanical ventilation technology is relatively backward, residential buildings rarely equipped centralized mechanical ventilation. For this stage, with the rapid development of the domestic economy, the rapid development of public buildings, public buildings have mostly built in recent years; these buildings for public buildings to develop a heat recovery ventilator technology can provide a good platform.

Public construction sector with respect to residential construction sector not only has a greater potential for energy savings, and because of its use of large equipment, the cost will be significantly less than the ordinary household equipment, in terms of economics and more with a clear competitive advantage.

Energy, Environment and Green Building Materials – Sheng (ed.)
© 2015 Taylor & Francis Group, London, ISBN 978-1-138-02718-3

Active distribution network application practice for grid

H.G. Zhao
Beijing Join Bright Digital Power Technology Co., Ltd., Beijing, China
School of Electrical Engineering, Tianjin Polytechnic University, Tianjin, China

X.L. Yang
Beijing Join Bright Digital Power Technology Co., Ltd., Beijing, China

M. Li, M. Chen & Z. Tang
Wuxi Power Supply Company, Wuxi, China

ABSTRACT: The active distribution network is not only a kind of distribution network, which is able to precept and adjust actively, but also the more advanced form of the power grid. However, the planning idea of the grid's "classification, hierarchy, system" is more suitable for the management of the active distribution network. The focus of the construction of the active distribution network is on the collection and application of information. Currently, the research focuses are mainly on the nine major aspects, such as active control over a wide area, advanced sensing and measurement technology and so on. Beijing Join Bright Digital Power Technology Co., Ltd. have done a lot of research work in this direction, and take the Wuxi power company as a pilot. Based on the regional grid planning idea, the construction project of the city cable internet supervision platform is proposed, the equipment perception layer has been constructed, the functional application layer construction for the directions of planning, energy saving, failure prediction and equipment management and so on has also achieved initial effective results.

KEYWORDS: active distribution network; grid; information; internet of things

1 INTRODUCTION

With the sustained growth of Power demand, the lacking of fossil energy and the global warming, needing to introduce new sources of energy to improve. The distribution network is the main way of consumptive distributed generation capacity, however, the status, distribution network consists of a lot of problems, including the weakness of the primary network, the low-level of automation, the low degree of information integration, which becomes a great problem of the incoming high-leveled permeability of renewable energy, and also a bad factor in the structural optimization of energy. In order to cope with the new challenges brought by large-scale distributed clean energy generation synchronization, traditional distribution network requires transforming from a passive distribution network of one-way power transmission (passive grid) to an active distribution network of two-way power transmission (active grid) in the major turning point.

In addition, the rapid growth of information, and the constant update of the new technology of electric power, is bound to promote the development of grid to a higher form. The combination of new communication technologies and new energy systems will occur a major revolution. The rapid development of the power information and communication network will comprehensively promote the construction of smart grid [1].

The several aspects, which are around the basic concept, feasibility techniques and research focus of the active distribution network, are analyzed and given a brief description in their applications. Contrasting with the current research focus of the active distribution network, Beijing Join Bright Digital Power Technology Co., Ltd. tries to carry out related research, and conducted a pilot work in Wuxi Power Supply Company. Based on the idea of gridded regional planning, and merging the internet of things technology, wireless communications technology and data mining technology, the company proposes the construction project of city cable internet of things supervision platform, and has achieved initial effective results.

2 THE INTRODUCTION OF GRID CONCEPT AND FEASIBILITY TECHNOLOGY

2.1 *The concepts of grid and active distribution network*

1 Grid

The so-called "grid" distribution network planning, refers to a kind of planning method, which put the electricity demand on the land blocks as a guide. According to the regional regulatory detailed planning, the different nature of land and development depth are classified, which are divided into a number of "small grids". With the typical load forecasting model, each "grid" is carried out a systematic load forecasting. While according to the difference of the planning standards, the layout of the primary network in 10kV is carried out subregionally, overall distribution automation, communication, protection configuration and so on.

2 Active distribution network

Active distribution network (ADN) is a distribution network with the high-leveled permeability of distributed power and two-way flow of power, which is an important part of the future smart distribution network. The active distribution network can delay investment, improve the response speed, network visibility and network flexibility, has a higher power quality and power reliability and higher level of automation, access DER more easily, reduce network loss, use assets better, improve load power factor, has a higher efficiency of the distribution network and availability of sensitive customer.

The main differences between active distribution network and passive distribution network are shown in Table 1 [2].

Table1. difference between the active distribution network and the passive distribution network.

Subject	The passive distribution network	The AND
Technical standard	Hard	Flexible
Management style	Focused	Dispersed
Network structure	Changeless	Flexible
Simulation	Average	Accurate
Controlling and protecting mode	Passive	Active

2.2 *The introduction of feasibility technology*

Existing ADN project mainly involves three aspects of technology, including hardware, monitoring and controlling and network operation. Considering the future development needs of developing ADN, viable technology needed to solve are:

1 The new distribution system

Passive distribution network adopts elimination on the spot of the intermittent energy model. When intermittent energy generates electricity in excess, because the distribution network itself didn't have the ability of regulation, and excess electricity cannot be into transmission and distribution grid, the only method is to reduce its output operation; while to solve the regulatory problem of elimination on the spot of intermittent energy, active distribution network should put forward a new distribution system, which adapts to multi-energy accessibility.

2 Monitoring and prediction

Network parameters are measured. The actual operating conditions are estimated. The required load, power generation and the forecast electricity price need to ensure security. Advanced sensing and measurement technology is applied to expand the monitoring range of the distribution network.

3 Operation and control

Distribution management system (DMS), the distributed control of DER, demand side management (DSM), microgrid/Freeder (island operation); power flow management, automatic voltage control (AVC), dynamic line rating (DLR) and so on.

4 Planning and design

Abandoning the traditional methods of operating independence, a variety of methods of uncertainty conditions may encounter to consider in the planning stage. To deal with DER access, which has high permeability, the new form of accessing to energy and energy storage devices simply are not being as ADN planning [3].

3 THE RESEARCH OF ADN

3.1 *The feature and research focus of ADN*

The basic definition of active distribution network (ADN) is: the distribution network, in which there is distributed or decentralized energy, possesses the ability of control and operation. ADN has four features. Firstly, there is a certain distributed controllable resource; secondly, there is a more perfectly impressive and manageable level; thirdly, there is a control center to achieve coordinated and optimized management; fourthly, there is the network topology, which can be flexibly adjusted. ADN can realize a lot of functions, such as the monitoring of the distribution network, load and energy regulation, the distribution network energy saving, the distribution network fault active prevention and so on.

The ADN can be monitored by sensing technology, collecting grid signal automatically, analyzing data and acquiring the relevant information, to provide the optimal solution to users. Meanwhile, likely dangers are predicted, and some coping strategies are developed. Rather than measures are taken only after a failure occurs like traditional distribution network, ADN achieves the "observability" and "controllability" of the distribution network [4-5].

In order to achieve operation and control of the active distribution network flexibly and intelligently, it is necessary to carry out extensive research. The research focus is mainly in the following respects.

1 The planning and Optimization of the active distribution network;
2 WAN active control;
3 Adaptive protection and control;
4 Real-time network simulation;
5 Advanced sensing and measurement technologies;
6 Distributed universal communication;
7 The knowledge extraction based on intelligent methods;
8 The management equipment of the distribution network;
9 The asset management for the distribution network with high permeability of distributed energy.

3.2 *Domestic and foreign research situation*

At home and abroad, technology about ADN has been researched, and there has been a small amount of research results. The main aspects are as follows.

1 ADN optimal planning method
   Currently, the research for ADN planning methods pays mainly attention to the mutual coordination problems between DREG and grid operation. And the significant value of integrated planning ideas begins to be noticed.

   For ADN planning, those block factors existed in every aspect of the production run, which impact the utilization efficiency of the renewable energy, need to be fully analyzed at first; on this basis, through researching the mechanism of the relevant factors, the corresponding planning strategy is proposed.

2 Active control
   Active control theory is the theoretical basis of the active distribution network. From the control theory, the traditional passive control is a process of appearing a problem and solving the problem, which lacks of predictability and has no concept of global optimization. In contrast with the passive control, active control pre-analyzes the possibility of deviation from the target, and formulates and adopts a series of preventive control measures. Through early perception and system control, the coordination optimization purpose is achieved, and program objective is achieved finally. In the distribution network run-time control process, control center must grasp the actual operating information on the power grid in real time, and understand the

load of each line in the power grid. When the grid appears blocked, through active control network structure and controllable devices in the network, and a variety of controllable power generation and demand response load, the load on the line on the power grid is eased, to ensure both the safety of the grid operation, and the purpose of the stability of the grid operation.

3 Active management
   In the process of the distribution network running, "combined cooling heating and power", distributed energy storage, voltage regulator load, demand response load, biomass stabilize PV randomness and volatility make full use. Through closely interacting with the grid run trend, the purpose of consumptive renewable energy generation in full is reached.

4 Overall perception
   In order to achieve active control and active management of the distribution network, the wider comprehensive perception of the distribution network must be achieved [6]. Adopting a more economical, reliable and advanced sensing, communication and control terminal technology, the real-time and comprehensive monitoring of the running status of the distribution network, distributed generation device status, assets device status and the power reliable is achieved, so the observability of the distribution network is achieved. In the traditional distribution network, measuring monitors range is from 110kV substation to 10kV network, without regard for the network below 10kV basically. For active distribution network, because the large number of distributed energies are connected to 10kV and below bus, and demand response resources are mostly connected to the low voltage bus, these low voltage power supply network must be monitored comprehensively; in order to achieve the full control of the distribution network resources, the operational status of the superior power grid must be understood, thereby requiring to expand the measurement rang down to 400V bus, and up to 220kV. The construction of the smart grid and the large-scale installation and application of the smart meters have laid a solid foundation for expanding the measurement range.

4 DIRECTION OF APPLICATION OF ADN TECHNOLOGY

4.1 *Application in the grid distribution planning*

The traditional planning method has no concern about the impact of DG connecting, nor does it concern of the flexible controlling characteristic of ADN, which makes the planed network has no safety insurance after the multi-energy connection.

At the same time, medium voltage distribution network facing characteristics of grid planning overall meticulous work currently, the traditional planning methods lack of forward-looking, and has been in a project driven passive state.

In the case of distributed power access, the ADN integrated planning technology is a unit with the minimum grid planning area. It aims at a low energy loss, high power supply reliability and high utilization of green energy. Taking into account the demand side response and other management tools, the problems of grid planning after DG accesses need to be solved, to make the grid planning and more comprehensive management and systematic.

4.2 Application in status monitoring and accident prevention of distribution network

Currently, distributed network devices can't be monitored anytime, particularly on the aspect of trouble shooting of the distribution network. Current technology can't achieve the prevention of failure, and the repair of lines in time, which brings hidden trouble to the safe operation of the distribution network.

The sensing, communicating technology in ADN can expand the monitoring range of distribution network status, realizing timely monitoring the running status of the distribution network through transducers, like smart sensor and RFID tabs and so on.

According to state data on distribution network from time to time, it is to extract knowledge of data mining techniques. Through setting up the intelligent network forecasting model, accident of the distribution network is prevented actively, to reduce the accident rate and improve operational safety and reliability of the distribution network.

4.3 Application in energy saving of power distribution

Current world has entered into a modern society, which is an information technology and network as features. Dissemination of knowledge and information becomes more convenient and quick. Information has reduced the cost of all kinds of trades significantly, and decrease energy costs rapidly. In addition, the increase of information investment can reduce energy consumption, and make production activities transforming from energy-intensive to information-intensive, and from extensive high-power in the development direction of intensive low-power line energy intensive, with unlimited information resources instead of the limited energy resources. With knowledge and information to replace the energy input, another way for saving will become true.

From the view of economics, energy saving should be the resources with lower values to replace the value of higher energy consumption. The value of other resources used in the energy saving should not be more than the value of energy itself, not at the cost of energy saving.

Therefore, strengthening the construction of the information, and realizing the energy alternative, are the only way to reduce energy loss and accelerate the construction of the intelligence distribution network.

4.4 Application in promoting the reliability of the distribution network

Because of the large scale of the distribution network and many equipment, branch lines and the user side of the equipment operation situation are not being known. Because of the lack of data collection methods, which leads to the data sources of the distribution network are diverse, basic statistical information is not accurate and comprehensive.

By sensing technology, the data based on distribution network are collected, which include the equipment foundation data, running status data, asset data, etc. By data mining technology, according to the characteristic of equipment to classify, the comprehensive analysis of the distribution network equipment is realized by knowledge extraction. Equipment defects are found and equipment utilization is improved, thereby improving system reliability.

5 THE CURRENT WORK

According to the requirements of the ADN, a lot of theoretical research and practical work have been done currently.

Main content:

1 Set up the evaluation system of energy efficiency of distribution network;
2 According to function of sensor, we select wireless temperature and humidity sensor, water sensor, fault current sensors to complete the construction of the perception layer in the monitoring system of distribution network perception;
3 According to a lot of r distribution network monitoring data, correlation analysis, clustering data mining methods are used to identify the distribution network data;
4 The application of distribution network fault monitoring data, using neural network, gray theory and other intelligent forecast method to build the fault forecasting model, realize the distribution network accident prediction, provide the operation reliability.

Wuxi city cable internet supervision platform is as an example. The first step is platform construction, in which the comprehensive perception of the power distribution equipment is achieved in the physical layer. The second step is to apply data mining technology, in which facing the vast monitoring data of the distribution network, through the knowledge extraction, generation, setting up an intelligent prediction model, active fault prevention and economic operation of equipment is implemented to reflect the initiative of the distribution network.

5.1 The overview of Wuxi grid region

Wuxi power grid covers the administrative area of 4627.47 km2, with a total power supply population of 6.4322 million (resident population). Distribution network planning should implement the principle of differentiation. Based on the level of economic and social development, user properties and environmental requirements in different regions, the differentiated planning and construction standards are used to reasonably meet the electricity needs of various users.

According the regional grid division standard is "classification, fragmentation", the 6 administrative areas of Wuxi is divided into 33 regional power grids. This project is in the New Area and Nanchang Area as a pilot to construct system platform.

5.2 The architecture of the city cable internet supervision platform

The system is mainly composed of perception layer equipment, transmission equipment, storage and application server, Internet access to information show the parts such as platform and business application.

1 Intelligence layer
  The equipment of the intelligence layer of Cable Internet of things, mainly includes the field of all kinds of wireless sensor nodes, Use of intelligent sensors and RFID tags from the Internet of things perception technology, in the face of mass sensor application, used the unified coding scheme, implementation of intelligence of cable trunk and branch line running status data, and the unity of the whole life asset information.

2 Data communication layer
  Transport layer devices are included gathering node and the convergence of the gateway. Gathered node of each sensor node is responsible for the net management and formatting data transmission task, Convergence of the gateway is responsible for the gathering node access and format conversion, and completes the sensor information model to IEC61850 information model transformation

and redefine and modeling. Assemble gateway will receive the sensory information of into the IEC61850 information model will this information encrypted before sent to the application layer storage server.

3 Function of the application layer
  Cable Internet supervision platform of the Function of the application layer includes equipment management, loss analysis, forecast warning, economic operation, and other functions achieve power grid equipment defects, failure rate, operation cost, and management to further improve the service life. System frame is as follows:

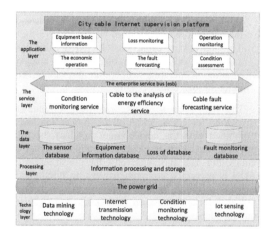

Figure 1. The system frame.

By the basis of the perception layer data, the data communication layer completing the data transmission and acquisition, in the application layer using pattern recognition, the intelligence of knowledge extraction and prediction technology to implement the following functions:

1 The management of cable basic data: Included cable type, length, the safety current, maximum load current, laying method, maximum allowable temperature of the cable conductor, equipment manufacturers, operating life, contractor, cable head production personnel, connecting device information;

2 The monitoring of cable fault: The sensors are installed on the cable trunk and branch line, to collect the data in the fault condition. Including actual operating voltage, current, load rate, cable body temperature, time, running time, which is under the cable fault state;

3 The anticipation of cable fault type: Combining with fault data of monitoring, accumulated a certain data, using data mining technology, to analysis

431

the correlation of data. According to the operation standard of cable, formulated fault early warning standard. Such as temperature can be controlled with temperature monitoring, when the temperature exceeds the standard, the system is alarmed. Forward, when accumulated to a certain degree of the data, through mining the particular relationship, such as temperature, humidity and the cable fault parameters. It is to summarize the potential law of data, classify the fault types, forecast the type cable fault which is according to the actual operation;

4 The analysis of cable properties: Included the basic information about cable equipment, operation information, fault information, contractor, vendor, date of commissioning, equipment loss of the different fixed number of years;

5 The analysis of cable loss: Included cable loss analysis of different operating life, loss analysis of different operating temperature, loss analysis of different load rate.

The next step of work, we will continue to strengthen the construction of the application layer. Combined with the development trend of the four information, which is cloud computing, internet of things, big data and smart city, the construction of ADN is promoted comprehensively by intelligence analysis methods, such as applying the internet of things, information fusion technology and big data mining and so on.

6 CONCLUSION

The popularization and application of ADN could greatly enhance the compatibility of the power grid for green energy and the efficiency of the power grid for the existing asset utilization, which is the future development trend of smart distribution network.

The analysis is carried around the several aspects of the basic concept of the ADN, feasibility technology and research focus. A contrast to the nine major research focuses, such as the current WAN active control, intelligent knowledge extraction, advanced sensor and measurement and so on, Beijing Join Bright Digital Power Technology Co., Ltd. has conducted a lot of research and practical work. In Wuxi city cable

internet supervision platform as an example, based on regional grid planning ideas, perception layer of the system, data communication layer and the functional application layer is structured. In future work, the hot technology of the ADN, which will expand to the construction of the application layer, should be given more attention. The construction of the ADN will be promoted comprehensively.

REFERENCES

[1] Celli G, Ghiani E, Mocci S, et al. From passive to active distribution networks: methods and models for planning network transition and development[C]. 42nd International Conference on Large High Voltage Electric Systems 2008, CIGRE 2008, Paris, France, 2008:1–11.

[2] You Yi, Liu Dong, Yu Wenpeng, et al. Technology and its trends of active distribution network[J]. Automation of Electric Power Systems, 2012, 36(18) :10–16.

[3] Fan Mingtian, Zhang Zuping, Su Aoxue, et al. Enabling technologies for active distribution systems[J]. Proceedings of the CSEE, 2013, 33(22):12–18.

[4] CIGRE Task Force C6.11, Development and operation of active distribution networks[R], 2011.

[5] Zhou Q, Guan W, Sun W. Impact of demand response contracts on load forecasting in a smart grid environment[C]. IEEE Power & Energy Society General Meeting, San Diego, USA, 2012:1–4.

[6] Miao Yuancheng, Cheng Haozhong, Gong Xiaoxue, et al. Evaluation of a distribution network connection mode considering micro-grid[J]. Proceedings of the CSEE, 2012, 32(1):17–23.

[7] Wang Zhaoyu, Ai Qian. Multi-objective allocation of microgrid in smart distribution network[J]. Power System Technology, 2012, 36(8):199–203.

[8] Arias J L C, Westermann D. Load forecasting scheme based on energy efficiency for planning the expansion of electrical systems[C]. IEEE Bucharest Power Tech, Bucharest, Romania, 2009:1–8.

ABOUT THE AUTHORS

H.G. Zhao(1977—), male, bachelor degree, The research directions are the power system and its automation, and power system planning. Email:zhbr_zhg@163.com;

X.L. Yang(1985—), female, master degree, The research directions are the power system and its automation. Email:yxlallok@126.com.

Energy, Environment and Green Building Materials – Sheng (ed.)
© *2015 Taylor & Francis Group, London, ISBN 978-1-138-02718-3*

Commercial complex advantages and development in Nanning City

Da-Yao Li
College of Civil Engineering and Architecture, Guangxi University, China

Yan-Qing Li
Qinzhou College, Qinzhou, Guangxi, China

Qi-Jin Zeng
Construction & Investment Co., Ltd. of Wuxiang New District, Nanning, Guangxi, China

Jia-Chen Yu
Construction Planning Information Center, Qinzhou, Guangxi, China

ABSTRACT: With the rapid development of social economy, making the urbanization process, the urban populations grow fast, leading to the sharp contradiction between human and land. People's life style and consumption concept also have been changed accordingly. Different from traditional commercial buildings (e.g., department store), a new type of commercial building form named "experiential consumption" which collects of consumption and leisure entertainment – the Commercial Complex in the City emerges as the times require. Through typical examples, this paper analyzes the development status of the Commercial Complex in Nanning City, and finds out the Nanning Commercial Complex's advantages in land, traffic, the pace of life, parking, and weather etc. Combining with Nanning Commercial Complex's development in different regions to make a preliminary study, and the study concluded that the development of Nanning Commercial Complex should adjust measures to local conditions and give full consideration to consumption object, market environment, traffic environment and other factors. This article will provide some theoretical basis and research direction for the theoretical research and practical application in the future, and at the same time, it also will provide theoretical basis and suggestions for the development of the City Commercial Complex in other regions.

KEYWORDS: Commercial Complex; Advantages; Development; Nanning

1 BACKGROUND

With the rapid development of social economy, making the urbanization process and the urban populations grow fast. Land resources are increasingly scarce, urban functions tend to diversify and commercial activities become more and more frequent. People's lifestyle and consumption concept also have been changed accordingly, The traditional commercial buildings (such as a department store) have not been able to adapt the transformation from the traditional single shopping mode to a new commercial building form named "experiential consumption "which collects of consumption and leisure entertainment. Therefore, the "Commercial Complex in City", a new type of commercial building form which can meet the need of a variety of activities emerges as the times require, and plays a more and more important role in the city.

2 THE CONCEPTION AND DEVELOPMENT OF THE COMMERCIAL COMPLEX

The concept of "commercial complex", is originated from the concept of "city complex", but there are obvious differences between them. "City Complex" is the "city within a city"(the urban economic aggregates of functional polymerization and city land intensive). It is based on building group, fusing five core functions of commercial retail, commercial office, hotels and restaurants(or catering), apartments and comprehensive entertainment into one. But the "Commercial Complex" is combined more than three kinds of urban living space functions, such as the city commercial, office, residential, hotel, exhibition, catering, conference, entertainment and so on, and establishes an active relation of interdependence and mutual benefit between the parts. Then it becomes the complex includes multi-function, high efficiency, complex and unified, the complex building based on function of commercial shopping.

As an important part of city construction, commercial construction is a city element of public, opening and full diversity. Also, it is one of the original motives of the regional development. With the rapid development of national economy in our country, urbanization process brings problems of the city land and traffic. The urban population need for high quality commercial facilities better than before in solving these problems. So the commercial building needs to have comprehensiveness of high efficiency and multiple functions. In this context, the development of commercial complex, which integrates shopping, hotel, office, clubs, apartments, and leisure entertainment has become the inevitable trend of. The development of commercial buildings in China is roughly divided into the following three stages.

The embryonic stage: From 1993 to 1995, there are about 10 commercial complexes in our country, most of their functions are hotel and business office and their service object mostly tourists and occupation persons. China-plaza of Guangzhou is a typical representative of them;

The early stage: From 1996 to 2001, with the continuous development of national economy, national economic policy is increasingly open, the investment in property is continuously growing, and its pattern is also increasingly diversified. The numbers of China's commercial complex run to 100, most of them are the function of the retail industry as the leading factor to integrate other consumer services, such as Chongqing metro plaza;

The development stage: From 2002 to 2011, China's urbanization process was accelerating, social economic developed at top speed. Large and super large cities gradually established, resulting in severe problems of land and traffic. At the same time, urban population requirements for the commercial facilities of high quality is also rising, therefore, there are about 1200 ~ 1500 commercial complexes have sprung up in this period. The famous commercial complexes in this period like Shenzhen Huarun center, Hongkong Langham Place etc.

The commercial complex has many advantages such as optimal land utilization, shorten the transportation distance, improve work efficiency, investment benefit and the others. It will be the development direction and emphasis of the commercial building at the present stage and even in the future.

3 COMMERCIAL COMPLEXES ADVANTAGES IN NANNING CITY

Nanning is in the south of China and located in the palace, which combines both Southwest Asia economic circle and Southeast Asia economic circle. Nanning has also played an important role in China's regional international city in ASEAN countries, it is the economic center of the Beibu Gulf Rim coast. The annual China-ASEAN Exposition, making an ASEAN business district, which initial positioning of the exhibition business circle into an urban international frontier trade area. As the permanent place of China-ASEAN Exposition and a central city of development strategy in the Beibu Gulf Economic Zone, plays a more and more important role in our country even in the world. Therefore, we should think seriously about that how to further define about the development direction and position of Nanning, how to make a good choice of its development mode.

In Nanning, there exist some kinds of commercial concentrate areas, such as the commercial complex, class commercial complex and the pedestrian street and others. But the developmental pattern of commercial complex and class commercial complex is better than the pedestrian street in general. We have investigated the reasons, and found that there are mainly some following points.

3.1 Land parcel

With the rapid development of social economy, making the urbanization process, the urban populations grow fast, leading to the sharp contradiction between human and land. The capacity of land resources in Nanning city becomes saturated gradually, resulting rising land prices, then the land development becomes a kind of status – land is getting expensive and expensive. While the construction of the commercial center, which is similar to the walking street needs a relatively wide space. That is a big difficulty under the condition of the limited land resources at present. However, the rapid development of the urban population and the improvement of residents' consumption demand need the supporting development of the business concentration area.

Figure 1. Traffic.

3.2 Traffic

The development of the social economy, the improvement of residents' consumption and the saturated state of the land use intensity, will inevitably lead to the traffic problems. Nanning's pedestrian street is usually located in Nanning urban center, but the urban center is usually traffic congestion. Commercial complex in Nanning although mostly located in suburb different regions of Nanning city, but the city bus routes covers almost any business district. People can take advantage of the convenient traffic conditions, according to the distance of different areas of commercial complex to choose the better one, which in any case increases the flow and consumption, too.

3.3 Parking

With the development of national economy, there are more and more private cars, whether the parking place is convenient becomes an important factorforo customers. The pedestrian street is mostly located in the urban center, its spatial arrangement is relatively scattered and the administration are not the same, so it's almost impossible to find an area under the normal management for parking. At the same time, we worry about the issues of illegal parking, vehicle scratches and so on. But the administration of Nanning city commercial complex is usually mature and unified; the parking space of the regular parking lot also can basically meet the needs of consumers. And at present, most of the commercial complex of Nanning provide consumers some personalized service, such as free parking with Consumption receipts.

Figure 2. Parking.

3.4 Pace of life

The pace of modern man's life rhythm is faster and faster. Many office workers disappear in the morning and reappear at night, and their working pressure is big, so the majority of office workers chooses the way of online shopping which is fast and convenient. However, there are some shortcomings of online shopping, such as that there exists a long delivery time and big gap of real sensory and so on. So the commercial complex is the better choice, because they are convenient transportation and have the functions of dining, leisure and shopping activities and so on. Most of the commercial complex in Nanning can achieve a complete supporting operating system of work and life, but most of the pedestrian streets to put particular emphasis on dining and shopping. Therefore, the commercial complex has more advantages than the pedestrian streets.

3.5 Weather factor

Commercial center as Pedestrian Street, which is enslaved to bigger influence of outdoor environment (such as weather factor), unable to maintain a steady passenger flow, and cannot achieve an all-weather operation. But the commercial complex is affected by the outdoor environment much smaller. In bad weather conditions, such as rainy and cloudy or hot summer weather, consumer's subjective emotion and consumer enthusiasm are affected by the objective conditions of Pedestrian Street bigger. On the contrary, consumers tend to choose the commercial complex, which has a good indoor environment, thereby, brings more stream of people and consumption to the commercial complex. And they can maintain steady passenger flow, keeping an all-weather operation in 24 hours.

4 COMMERCIAL COMPLEXES DEVELOPMENT IN EACH REGION OF NANNING

For a long time, the commercial zone of Nanning mainly concentrated in Chaoyang Road area. Chaoyang commercial district is the core commercial district of earliest-formation in Nanning. After decades of development, with commercial activities occur frequently in Nanning, Seven Star Shopping District, Feng Ling district and a series of commercial districts emerge as the times require.

Commercial complex in different districts have different development pattern, we should according to local conditions in the practical application.

Figure 3. Weather factor.

Development of commercial complex with different commercial district needs to take full account of the population quantity, population structure, economic development level and other factors of regional.

Development of commercial complex needs considered the combination of traffic, positioning, development pattern, parking consumer experience and other factors. The following two examples are an analysis which is the best suitable commercial development pattern for each area of Nanning.

Hangyang international city with a forward-looking international design concept, aiming at the goal of "to be the first city in the southwest of China" to build it into the large-scale international building complex of the central image area. Its collection of commercial retail, residential, office, hotel and catering, entertainment, urban transport and other functions in one, so it is a "city within a city" which has the optimum combination with multifunction. And it is the first building complex in southwest of China, and the fifth building complex of China.

But in reality, the development of Hangyang international city is not plain sailing as originally envisaged. With the opening of Huarun • wanxiang City, Hangyang international city is obviously affected. When I investigated and surveyed in this place in 2014, the scene of Hangyang international city is different from my impression. Many shops were closed for preparing to decorate them in order to reopen. There are many shops for clearance sale, even the Broadway Cinema was closed, and consumers are very few.

Hangyang international city is located in a superior position. Its international trade atmosphere is nothing comparable to this, but in the challenge of market development, it could not reach the expected goals, there even exists a big gap. Personally, I think one of the reasons is the surrounding commercial environment. The new dream island is not far away, Crystal City is nearer and Wanxiang City is also at the next intersection. Because the similar types of commercial buildings are relatively concentrated in the same commercial area, so we need to analyze and think fully. We should think about that how to break through the present situation to project its own characteristics to attract people and consumption. As we know, in order to cope with the social development and challenge better, Hangyang international city is making new arrangement and decoration. I hope the Hangyang international city can change the recent status in the fiercely competitive market.

A successful example is the new world of the Nanning department store. New world of Nanning department store broke up consumers' inherent impression of the Nanning department store. Its operating position is the youth, vogue, fashion, high-quality goods, the general planning for the department stores combines with department store, household electrical appliances, supermarket, restaurant and catering, cinema, entertainment and other business format in one, to provide for consumers a one-stop "Shopping Mall" which is a collection of "eating, drinking, playing and purchasing". At the same time, it is also the first large-scale comprehensive department store in the west area of Nanning, filling the gap in the retail areas of industry. New world of Nanning department store is nearby universities and business, adjacent to Guangxi University, the transportation is convenient, the geographical position is richly endowed by nature. It will work with Chaoyang store and Jinhu store on the Nanning Department store, connecting the dots, then connecting the planes, and gradually form a network layout of the Nanning Department store of "three points and one line, one axis with two wings" to provide more convenient service for citizens.

The reason for the new world of Nanning department store's success, not only a superior geographical position, but also its fuller consideration of the service object of surrounding regions, the market environment, traffic environment and other factors. So it can adjust measures to local conditions, then make its positioning clear, and further broaden the customer group coverage of the Nanning Department Store to bring the company into a new stage of rapid development. Even it has a shortcoming of insufficient parking space, but its consumption object is mainly students, and its transportation is convenient, so the new world of Nanning department store can still remain stable and good development.

5 CONCLUSION

With the rapid development of social economy, making the urbanization process, the urban populations grow fast, leading to the sharp contradiction between human and land. People's lifestyle and consumption concept also have been changed accordingly, so the city commercial complex, which can meet various requires of activities gradually become the general trend in today's development of social. Through typical examples, this paper analyzes the development status of the Commercial Complex in Nanning City, and finds out the Nanning Commercial Complex's advantages in land, traffic, the pace of life, parking, and weather etc. Combining with Nanning Commercial Complex's development in different regions to make a preliminary study, and the study concluded that the development of Nanning Commercial Complex should adjust measures to local conditions and give full consideration to consumption object, market environment, traffic environment and other factors. This article will provide some theoretical basis and research direction for the theoretical research and practical application in the future, and at the same time, it also will provide theoretical basis and suggestions for the development of the City Commercial Complex in other regions. Hope that the development of Nanning commercial complex much better than before, and the modernization of the city construction can develop stably and prosperously.

REFERENCE

[1] Guoli Cao, Xin Yang and Yong Lan. The pursuit of a modern business center: the creation of Nanning Wanda Commercial Plaza. Architectural Creation, 2004, (09): 120~125 In Chinese.

Energy, Environment and Green Building Materials – Sheng (ed.)
© 2015 Taylor & Francis Group, London, ISBN 978-1-138-02718-3

HPLC method for determination of concentration of Oxiracetam in human plasma

X. Xie
Henan Polytechnic College, Zhengzhou, Henan, China

ABSTRACT: To establish an HPLC method for determining the concentrations of Oxiracetam in human plasma and to evaluate its pharmacokinetic characteristics. A ZORBAX NH2 (200 mm × 4.6 mm, 5 µm) column was used to separate Oxiracetam in plasma with a mobile phase of a mixture of acetonitrile – water (95 : 5, V / V) at a flow rate of 1 mL/min. Oxiracetam was detected at 210 nm. The linear range of the standard curve of Oxiracetam was 2 ~ 300 µg/mL, and the determination limit was 2 µg/mL. The extraction recoveries were more than 80%, intra-day and inter-day RSD were less than 3.46%. The Oxiracetam plasma concentrations were determined after intravenous infusion and its pharmacokinetic parameters were calculated. The method is sensitive, fast and accurate. It is suitable for therapeutic Oxiracetam monitoring and its pharmacokinetic studies.

KEYWORDS: Oxiracetam; HPLC; Pharmacokinetics

1 MATERIALS AND METHODS

1.1 *Instruments and reagents*

The HPLC Waters515 system with automatic sampler (Waters717) was used to separate and detect Oxiracetam in human plasma, METTLER TOLEDO AX-205 electronic balance, XW-80A eddy mixer, PROINO high speed centrifuge, the PK514BP ultrasonic cleaner was supplied by the Amerain WATERS Company, Mettler-Toledo Instrument (Shanghai) Co. Ltd, Shanghai Jingke Company, American Kendro Laboratory Products and Germany Bandel Company, respectively.

Oxiracetam injection and reference substance were offered by the Shijizhuang Oyi Pharmaceutical Factory. Acetonitrile were chromatographies pure grade. Pure water was supplied by Wahaha.

1.2 *Conditions for chromatogram*

The separation was carried out with a mobile phase of a mixture of acetonitrile – water (95 : 5, V / V) at a flow rate of 1 mL/min and A ZORBAX NH2 (200 mm × 4.6 mm, 5 µm) column. 20µL of the purified sample was injected. Oxiracetam was detected at 210 nm.

1.3 *Method of pretreatment*

The stock solutions of Oxiracetam at the concentration of 1 mg/mL were all dissolved under mobile phase and kept at 4°C. A liquor of 0.2 mL plasma of sample plus 0.4 mL of ethyl acetate, the aim was to remove the protein. Then it was vortex-mixed 2 min, centrifuged at 5000 r/min for 5 min. The water phase was discarded and the organic phase was moved to a clean glass tube and dried under Nitrogen in a 40°C water bath. The residue was reconstituted with 0.1 mL of mobile phase, centrifuged for 3 min, and 20 µL of it was injected for analysis.

1.4 *Corroboration of methods*

Under above conditions, the retention time of Oxiracetam were 8.343 min. The reference substance of Oxiracetam, The blank plasma, blank plasma plus with Oxiracetam, volunteer samples are show in Figure 1.

0.5 mL blank plasma was added in each glass tube which contained Oxiracetam at concentrations of 2, 5, 10, 20, 50, 100 and 300 µg/mL. After drying with Nitrogen, purified, injected and analyzed the regression equation was as follows: $y = 15081 + 93483\ x$, r = 0.9999, the limit of quantity (LOQ) is 2 µg/mL. The linear relationship of Oxiracetam from 2 µg/mL to 300 µg/mL is good.

5, 50 and 250 µg/mL of Oxiracetam were spiked in blank plasma and analyzed at above conditions. The recovery rate, intra-day and inter-day RSD were calculated (Table 1).

1.5 *Subject and design*

Ten healthy volunteers were participated in this study ± physical examination and laboratory screening.

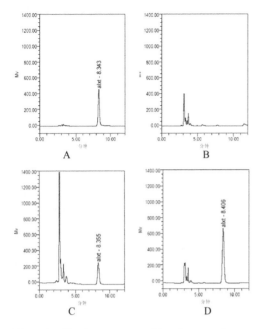

Figure 1. The chromatograms of the reference substance of Oxiracetam (A), blank plasma (B), blank plasma spiked with Oxiracetam (C), volunteer plasma (D).

Table 1. Recovery rate, intra-day and inter-day relative standard deviation (RSD) of Oxiracetam (n=5).

Concentration (ng/mL)	Recovery rate(%)	Intra-day RSD(%)	Inter-day RSD(%)
5	96.458	0.671	1.228
50	98.357	0.473	1.116
250	100.062	1.326	3.384

They were asked to avoid all prescription for at least 10 days before the study. Those who had a history of drug or alcohol abuse or allergy to the components of Oxiracetam and those who had concomitant drug therapy were excluded. All subjects gave their written informed consent at the beginning of the study and being explained the nature of the drug and purpose of this study.

Ten healthy volunteers were randomly divided into 2 groups, male-female, fasting overnight (at least 10 hours) until 7 o'clock in the morning the next day. Under the condition of continuous electrocardiogram monitoring, 6 g of Oxiracetam dissolved in 250 ml of 0.9% sodium chloride injection was intravenous dripped within 30 minutes. 4 mL venous blood samples were obtained before and 10, 20, 30, 45 min and 1, 1.5, 2, 3, 4, 6, 8, 12, 24h after the administration of Oxiracetam injection preparations. The blood samples were centrifuged and plasma was collected and

stored at -20°C for analysis. After the dosing volunteers were accepted electrocardiogram monitoring for 2 hours.

The second week, each subject was intravenous dripped 6 g Oxiracetam dissolved in 250 ml of 0.9% sodium chloride injection at am 7:00 for 7 days. 4 mL blood sample was collected from the 5th to 7th day before administration, and 10, 20, 30, 45 min t and 1, 1.5, 2, 3, 4, 6, 8, 12, 24h after the administration at 7th day after administration. The blood samples were centrifuged and plasma was collected and stored at -80 °C for analysis. After the dosing volunteers were accepted electrocardiogram monitoring for 2 hours.

2 RESULTS

2.1 Plasma concentrations of Oxiracetam in each group

The average plasma concentrations of Oxiracetam before and after single and multiple dose intravenous infusion injection of 6 g within 30 min are shown in Table 2. The average plasma concentrations of Oxiracetam after intravenous infusion injection of 0.4 mg/kg within 5 min, then intravenous dripping of 0.4 mg/kg/h for 6 h are shown in Table 3. The time-concentrations curves are shown in Figures 2 and 3.

Table 2. The Oxiracetam time-plasma concentrations of single and multiple dose after intravenous infusion injection. ($x \pm$ SD, n = 10).

Time (h)	Concentration (ng/mL)	
	Single dose	Multiple dose
-48	-	5.031 ± 0.854
-24	-	4.965 ± 0.834
0	-	5.002 ± 0.795
0.17	123.186 ± 37.892	154.645 ± 58.355
0.33	189.236 ± 26.235	205.179 ± 49.835
0.5	210.685 ± 20.508	211.831 ± 36.118
0.75	199.878 ± 20.750	176.821 ± 22.960
1	149.230 ± 17.10	137.152 ± 18.20
1.5	120.713 ± 16.189	98.828 ± 18.047
2	94.230 ± 18.654	82.674 ± 14.586
3	69.240 ± 15.341	53.191 ± 17.514
4	46.947 ± 12.607	38.834 ± 13.841
6	28.542 ± 7.682	24.105 ± 10.023
8	17.571 ± 5.062	16.225 ± 6.012
12	9.781 ± 3.024	8.201 ± 3.001
24	5.036 ± 0.895	4.644 ± 1.054

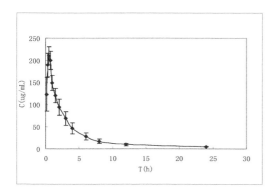

Figure 2. The Oxiracetam time-plasma concentration curves after single dose intravenous infusion of 6 g($\bar{x}$ ± SD, n = 10).

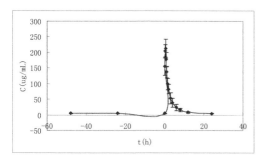

Figure 3. The Oxiracetam time-plasma concentration curve after multiple dose after intravenous infusion injection 6 g ($\bar{x}$ ± SD, n = 10).

2.2 *Pharmacokinetic parameters of Oxiracetam in each group*

The mean pharmacokinetic parameters of Oxiracetam in each group after were shown in Table 3.

Table 3. The Oxiracetam pharmacokinetic parameters after 5 min intravenous infusion of 0.3 mg/kg and 0.4 mg/kg and 6 hours intravenous dripping of 0.4 mg/kg ($\bar{x}$ ± SD, n = 10).

Parameters (unit)	Single dose	Multiple dose
$t_{1/2}$ (h)	4.754 ± 0.14	4.351 ± 2.23
T_{max} (h)	0.586 ± 0.16	0.375 ± 0.028
C_{max} (ng/mL)	230.89 ± 24.60	244.370 ± 38.12
AUC_{0-24} (ng/mL/h)	693.53 ±125.19	609.52 ± 148.20
$AUC_{0-\infty}$ (ng/mL/h)	687.90±148.12	617.03± 139.89
$MRT0-24$ (h)	4.18 ± 0.45	4.566 ± 0.501
CL (L/Kg/h)	0.150 ± 0.35	0.128 ± 0.026
V (L/Kg)	1.40 ± 0.44	1.61 ± 0.35

3 DISCUSSION

The serum concentration of Oxiracetam reached the maximum after 30 min of single and multiple intravenous infuse and the declined rapidly within 3 – 4 h. The mean residence time, an indication of the average persistence time of drug molecules in the human body, differ significantly between single and multiple dose (p < 0.05). Oxiracetan did not accumulate in serum after multiple intravenous infusions and the pharmacokinetic characteristics shows no significant difference between male and female volunteers.

An HPLC method for determining the blood concentration of Oxiracetam was developed and report. It is sensitivity, specialty and precision, and suitable for Oxiracetam therapy drug monitoring and pharmacokinetic studies. The whole experiment process was smooth, no adverse reaction occurred.

REFERENCES

Wei Chunmin, Wang Benjie, Guo Ruichen. Pharmacokinetics of Oxiracetam in Healthy Volunteers. *J Chin Pharmaceutical Sci; 2005; 14(1):2*

Lecaillon JB, et al. Determination of oxiracetam in plasma and urine by column- switching high performance liquid chromatography. *J Chromatogr 1989;497–223.*

Perucca E, Albrici A, Gatti G, et al. Pharmacokinetics of oxiracetam following intravenous and oral administration in healthy volunteers. Eur. *J. Drug Metab. Pharmacokinet. 1984; 9(3):267.*

Kondoh, K. Haohimoto, H. Nishiyama,H. et al. 1994. Effects of MS-551, a new class III antiarrhythmic drug, on programmed stimulation-induced ventricular arrhythmias, electrophysiology, and hemodynamics in a canine myocardial infarction model. *J Cardiovasc Pharmacol. 23(4) 1994. 675–679.*

Trovarellia G, et al. Biochemical studtes on the nootropic drug,oxiracetam,in brain.*Clin Neuropharmacol 1986; (Suppl 3):S 56.*

Kamiya,J. Ishii,M. Yoshihara,K. et al. MS-551: Pharmacokinetical profile of a novel class III antiarrhythmic agent. *Drug Development Research. 30 1993. 37–44.*

TM Soldatos C, et al. A new nootropic (a phase I safety and CNS-B efficacy study with quantitative pharmaco-EEG and pharmaco psychology study. *Curr Ther Res 1979; 26; 525.*

Donald,K. et al. Inhibition of ATP-sensitive potassium channels in cardiac myocytes by the novel class III antiarrhythmic agent MS-551. *Pharmacology & Toxicology. 77 1995. 65–70.*

Kamiya, J. Ishii, M. Katakami, T. et al. Antiarrhythmic effects of MS-551, a new class III antiarrhythmic agent, on canine models of ventricular arrhythmia. *Jpn J Pharmacol. 58(2) 1992. 107–115.*

Energy, Environment and Green Building Materials – Sheng (ed.)
© *2015 Taylor & Francis Group, London, ISBN 978-1-138-02718-3*

A domain-expert oriented modeling framework for unconventional emergency

Z.C. Fan, X.G. Qiu, L. Liu & P. Zhang
College of Information System and Management, National University of Defense Technology, China

X.Y. Zhao
Naval Aeronautical Engineering Institute, China

ABSTRACT: Recently unconventional emergencies have drawn wide attention in the field of social management. Constructing a computing scenario of a specific emergency is considered to be an effective way to facilitate decision-making in emergency response. The advent of artificial society method makes it possible to quantitate the complex social problem. Though some methodologies have been proposed, the modeling of unconventional emergencies is still facing many challenges from the view of modelers. Lack of a methodology to guide the normative modeling has puzzled the domain experts and modelers for a long time. In the paper, a domain-expert oriented methodology is introduced to guide the whole modeling process. Then we propose that the modeling framework for unconventional emergency is divided into three parts including agent, environment and event. Based on the idea of agent modeling language, meta-models of each part are designed in detail to specify the evolution of unconventional emergency. The modeling framework can offer a legible picture to describe an unconventional emergency scenario, and it can also become the basis of developing a suit of modeling tools in the future.

KEYWORDS: Artificial Society; Modeling Framework; Unconventional Emergency; Simulation

1 INTRODUCTION

Unconventional emergencies can impose a major threat on the societal stability and security [Wang 2007, 2011]. Recently unconventional emergencies have drawn widely attention in the field of social management. Constructing a computing scenario of a specific emergency in the computer world is considered to be an effective way to facilitate decision-making in emergency response. The complex evolution of unconventional emergencies isn't easy to capture, which involves many heterogeneous objects and their interaction behaviors.

The advent of artificial society (AS) method [Epstein 1997] makes it possible to quantitate the complex social problem. The main idea of AS is building a replication of the real social system using the method of agent-based modeling and simulation (ABMS). It has been widely acknowledged that AS and ABMS may be the appropriate way to model the evolution of unconventional emergency [Bandini 2009]. The achievements in the field of AS have provided more opportunity for modeling and understanding the social problem in the past years. The existing works of AS almost focus on a specific domain, such as public health, transportation, and economic.

Their modeling frameworks are different from each other, and the architectures have different merits depending on the purpose of the simulation. This makes it impossible to adapt their frameworks for the modeling of unconventional emergency. Moreover, it takes on greater indeterminism as a result of intangible factors such as emotion, psychology and opinion. So there is a great demand for a domain-expert oriented methodology [Jennings 2000; Wooldridge 2000] and a common modeling framework.

From the view of normative modeling, based on model-driven architecture (MDA) in the soft engineering [OMG 2003; Gareia 2004], people have proposed the concept of agent modeling language to allow the description of the complex systems at a more abstract level and meta-models are designed to describe the domain concepts from different views. The most representative products include Gaia, AUML, Tropos, AML, MESSAGE, etc. [Wooldridge 2000]. However, these methodologies neither aim at any special domain, nor serve any simulation platform. It's hard to use them for depicting the objects of unconventional emergency. To overcome the lack, we design a series of meta-models based on a modeling framework to guide the normative development.

2 DOMAIN-EXPERT ORIENTED METHODOLOGY

2.1 Development process

To build the model and run the simulation, making simulations of large, complex systems requires common standardized procedures. A domain-expert oriented methodology can support modeling and simulation of complex social systems [Garro 2010]. Besides, methodology for agent-based modeling and simulation should follow the standard development cycle. It can guide the whole development process from the system analysis to the set-up of virtual scenario. And the model development process can be divided into four stages: system analysis, conceptual modeling, simulation design, and code generation. The virtual scenario can be set up based on the code.

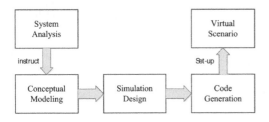

Figure 1. Development process.

- **System Analysis:** the purposes and requirements of the model development, the understanding of the system, and the design rules and guidance are described in details;
- **Conceptual Modeling:** firstly, the main components and elements of the system are extracted and summarized, and then a number of core concepts are proposed; secondly, the conceptual model framework of the system is designed as a whole.
- **Simulation Design:** based on the conceptual model framework, the simulation models are designed in a concrete modeling environment;
- **Code Generation:** the automatic generation from the simulation models to the code is implemented in the target simulation environment in the modeling environment. The most important thing is to ensure that the code could be deployed and run on the simulation platform accurately and effectively.

2.2 Transition based on model-driven paradigm

Model Driven Architecture (MDA) is a soft development methodology which is proposed by Object Manage Group (OMG) [OMG 2003]. It can separate system functions from implementations, and admit the modelers focus on modeling. In MDA, models must be described by a concise modeling language. To fill the gap, the concept of meta-model is

proposed, which can be considered as a set of concepts and relations highlighting the common properties of a class of models. It is usually used to define the syntax of a modeling language at an abstract level. As shown in Figure 2, meta-model can be directly transformed into a programming language. Some templates need to be defined to offer some mechanisms which can be used to support the transformation from meta-model to programming language.

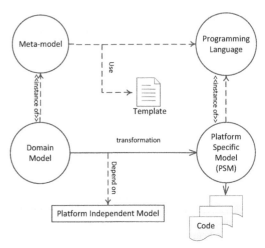

Figure 2. Development process.

For domain modeling, domain model (DM) describes the domain objects, concepts, relations and processes. At the conceptual level, DM can be viewed as the requirements of the software system. As said above, the meta-model is designed to address which concepts should be used in the simulation independent of a simulation platform. So, DM can be considered as an instance of the meta-model to some extent.

In MDA [Gareia 2004], the concept of platform specific model (PSM) is to put forward to build a bridge DM and Code for a specific simulation platform, such as Netlogo, Repast, Mason, *etc.* The transformation from DM to PSM requires mapping modules that explain what each concept is translated into. The mapping modules can be defined in a platform independent model (PIM), which contains a series of technological details independent of some implementation or development technologies.

3 MODELING FRAMEWORK FOR UNCONVENTIONAL EMERGENCY

3.1 Requirement analysis

With the complexity of the systems grows, a comprehensive, flexible and efficient modeling framework can help to express the objects and their relationships, and interpret the emergence and formation of some social phenomenon or processes. Moreover, the evolution of unconventional emergency is a very complex process, which involves some uncertain factors and intangible phenomenon [Wang 2011]. The complexity of unconventional emergency makes heavy demands on the design of the modeling framework. It should meet some requirements which can be listed as follows.

- **Integrality**: in a social context, the modeling framework should capture the main evolution mechanism of unconventional emergency, and reflect its key process reasonably. So the modeling framework can include all the related elements, phases, relations and processes. The system should be abstracted in an appropriate level.
- **Easy-to-understand:** as said above, our goal to develop a suit of modeling tools which can facilitate understanding and modeling of unconventional emergency. It requires the concepts in the modeling framework should be closed or similar to the domain concepts, which can be easily understood by the domain experts. It's very important to support them in modeling and simulation of the systems quickly and effectively.
- **Robustness**: as the basis for the design of visual modeling tools, the meta-modeling framework must have the combined, inheritable, extensible, and reusable characteristics. To meet these demands, the meta-models should be refined in a reasonable level.

Additionally, some auxiliary concepts need to be defined in order to maximize their reuse and understanding in the meta-modeling framework. That is to say, these meta-models can define a whole or part of each entity, which not necessarily reflect the whole entity. All the entities can be inherited from the meta-models. It's necessary to emphasize that the meta-models are abstract, and they can't be instantiated in the implementation of the systems.

3.2 System elements

The modeling framework for unconventional emergency can be divided into agent, event, and the environment (See Fig.2). The agent can be considered as a map of an individual or organization in the computer world (on behalf of individuals/companies). It's a concrete entity which has the social, cognitive, adaptive and autonomous ability within a particular environment.

An agent with a set of characteristics and rules can behave in a fashion on behalf of an individual or organization in the real world. The environment provides the space where the population exist

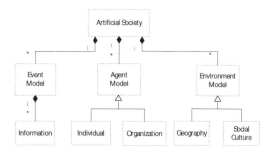

Figure 3.   System elements.

and function. It may represent the physical or non-physical (such as logical, cultural, institutional, etc.) Surroundings which can be heterogeneous, dynamic, open, and distributed. An event can be defined as a set of attributes and behavioral rules which have some effects on social phenomena. An event usually contains some pieces of information.

3.3   Meta-models

As said above, the meta-models represent a set of concepts and relations highlighting the common properties of a class of models. As shown in Figure 4, eight core concepts (*Agent, Individual, Organization, Event, Information, Environment, Geography, and socialCulture*) are defined, and besides the relations defined in UML specification, any other relations (*Apperceive, Situate, Act on*) are also defined to assist in describing the relationships among the system elements.

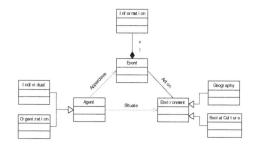

Figure 4.   Meta-models.

An agent with a set of characteristics and rules can represent an individual or organization in the real world [3]. *The individual* represents a person in the real society which has its own attributes and behaviors. *The organization* is composed of a series of individuals with the common goal, and can also act as an entity in the computer world. It provides a framework where the individuals, resources, applications, and roles coexist.

An individual can be defined by some aspects (See Figure 5). Attributes (*Attributes*) describe its own structure, including internal and external attributes. Mental State (*Mental state*) means that an individual has some psychological states, such as belief (*Belief*), goal (*Goal*), and emotion (*Emotion*). A set of roles (*RoleSet*) represents the status of an individual, which mean an individual can play several roles at the same time. Social networks (*Social Networks*) represent an individual's social relationships, such as family, acquaintanceship, friendship, colleague, etc. Behavior (*Behavior*) means that the individual has the ability to interact with others or environments. It can be constituted by a set of actions, which include action identity (*Action_ID*), preconditions (*Precondition*), process (*Process*), content (*Content*) and result (*Result*).

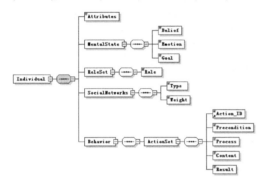

Figure 5. The individual meta-model.

Organization (*Organization*) is composed of a series of individuals with the common goal. It provides a framework where the individuals (*IndividualList*), resources (*Resource*), and rules (*Rule*) coexist (See Figure 4). It has its own type (*Type*) and structure (*Structure*).

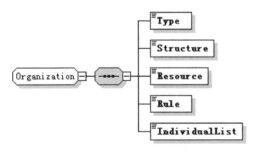

Figure 6. The organization meta-model.

The event meta-model mainly describes the related objects, process, and state of a real event in the society. Event contains pieces of information; the environment can be divided into two parts: geography and social culture. Geography (*Geography*) represents physical environment, such as building, roads, communication networks, which usually has its own resources and can provide some services; Social culture represents (*SocialCulture*) non-physical environment which can reflect the habits and development level of a society.

4 CONCLUSION

Lack of a methodology to guide the normative modeling has puzzled the domain experts and modelers for a long time. In the paper, a domain-expert oriented methodology is introduced to guide the whole modeling process and a modeling framework is proposed for unconventional emergency. Based on agent modeling language, meta-models of each part are designed in detail to specify the evolution of unconventional emergency.

The modeling framework can offer a legible picture to describe an unconventional emergency scenario, and it can also become the basis of developing a suit of modeling tools. In the future, we will continue our research and through a case study validate the effectiveness of the proposed modeling framework.

ACKNOWLEDGMENTS

This work was supported by National Nature and Science Foundation of China under Grant (Nos. 91024030 and 91324013).

REFERENCES

Wang, F.Y. (2007). Toward a Paradigm Shift in Social Computing: The ACP Approach. *IEEE Intelligent Systems*, 22(5):65–67.

Wang, F.Y., Zeng, D.J. and Cao, Z.D. (2011). Social Computing Methods for Non-Traditional Security Chanlleges Enabled by the Social Media in Cyberspace, *Science & Technology Review*, 29:15–22.

Epstein, J M R Axtell (1997). Artificial Societies and Generative Social Science, *Artificial Life and Robotics* (S1433-5298), 1(1):33–34.

Bandini, S., Manzoni, S. and Vizzari G. (2009), Agent Based Modeling and Simulation: An Informatics Perspective, *Journal of Artificial Societies and Social Simulation*, 12(4):4.

Jennings, N.R. (2000). On agent-based software engineering, *Artificial Intelligence*, 117:277–296.

Wooldridge, M., Jennings, N.R. and Kinny, D. (2000). The Gaia Methodology for Agent-Oriented Analysis and Design. *Journal of Autonomous Agents and Multi-Agent Systems*, 3(3):285–312.

Gareia, E., Lorenzo, P. and Gareia, L.R. (2004) A MDA-based framework to achieve high productivity in software development, In: *Proceedings of the Eighth IASTED International Conference Software Engineering and Applications*.

Object Management Group (OMG 2003), MDA guide version 1.0.1, OMG/03-06-01.

Garro A. and Russo W. (2010). easyABMS: a domain-expert oriented methodology for agent-based modeling and simulation. *Simulation Modelling Practice and Theory*, 18(10):1453–1467.

Energy, Environment and Green Building Materials – Sheng (ed.)
© *2015 Taylor & Francis Group, London, ISBN 978-1-138-02718-3*

The influence of David Pepper's ecological theory of socialism on China's era

Hong-Rui Chen
Harbin Institute of Technology, Harbin, China

ABSTRACT: David Pepper is an important figure in the 1990s on behalf of Ecological Marxism. Although he has a utopian eco-socialist theory, he has inherited the critical spirit of Marxism, the critique of the capitalist mode of production, and raised the values of harmony between man and nature, so that we can strengthen the belief in building socialism with Chinese characteristics, understanding the essence of contemporary capitalism, Pepper advances the socialism to the ecology to combine Marxism with ecology movement, and warns the developing countries not to follow eco-imperialist policies. His idea of eco-socialist society, not only has an important reference for the construction of the construction of ecological civilization, but also has the important value of the global ecological environment.

KEYWORDS: David Pepper; eco-socialism; ecological civilization.

1 INTRODUCTION

Pepper's eco-socialist theory implies not only theoretical significance, but has fresh significance. Pepper puts forward many ideas in building a future socialist model, which provide meaningful references for the construction of building a harmonious socialist society, we first discuss the economic dimensions of Pepper ecological socialist ideas enlightenment to China's economic construction.

2 THE DEVELOPMENT OF THE ECO ECONOMY

Pepper has opposed to the steady economic view that the ecological socialists advocated before the 1990s. He believes the most scientific, economic model is a moderate growth, economic model, he points out the problems of existence in developing countries, the implementation of this kind of economic growth is unrealistic, he believes the growth rate to be controlled within the range of moderate and rational. Pepper said: "Economic growth in eco-socialism must be rational, based on personal interests and the basis of equality, equality and harmony between man and nature is the goal of the planned growth can be coordinated." [1] Pepper advocated the moderate growth economy, indicating it has been recognized that negative economic growth and high economic growth is not conducive to social and economic development of the society. Pepper proposed modest economic growth is not only reasonable demands

of society, but people's rational choice, which has a significant reference for China's economic development.

Moderate economic growth means not only to meet people's needs, but without causing the destruction to the environment by pursuiting economic growth. As the primary stage of socialism of our country, we should base on our population, resources and environment, the development of the situation combined with our national conditions, we must not pursue economic growth at the expense of destroying the environment. China's traditional economic growth model is a single linear process that resources-products-process emissions, which is characterized by high consumption, high emission and low efficiency. While this economic model made rapid growth in China's economy, it does not only damage to the environment, but affects the survival and development of human beings. At the same time it does not calculate the environmental costs into economic costs, thus ignoring the limited resources. So we have to change the past traditional economic growth to develop eco-economy. Ecological economy is from the principles of ecology with people, man and nature living in harmony, considering both the needs of human survival and sustainable development without destroying the environment, the economy and the natural environment in order to achieve the dialectical unity of mode and more focus on sustainable development and harmonious development between man and nature and the economy in the process of economic development with both current and long-term interests of mankind to ensure sustainable human development model.

Ecological economy is the basic way of developing circular economy, namely the so-called circular economy is in the process of economic development, following the rules of ecology with the rational use of resources and economic development, using the modern science and technology to establish circulation mode of production. That is to say to reduce the resource consumption without polluting the environment to use the persistent products to reduce exhaust emissions, and consumption of resources. The resource circulation and the reuse of saving as the center, the process of recycled economy is resources-products-waste-renewable resources, characterized by low consumption, low emission, high efficiency based on principles of recycling and recycled economic model, realization of circular economy, which has great significance. [2] First of all, the development of circular economy and realize the high consumption of utilization of resources and recycling is to alleviate the economic development and the limited ways of resources. Promoting clean production, changing waste into use to the demand for natural resources and economic and social activities to minimize the influence of the ecological environment, to fundamentally solve the economic development and the contradiction between the shortage of resources and environmental protection. Second, the development of circular economy can not only improve the utilization rate of resources, reduce production cost, improve economic efficiency, and can make the products meet international environmental standards, enhance international competitiveness. Again, the circular economy is the premise of health and safety, adheres to the people-centered principle, to social benefit, economic benefit, environmental benefit all-round development as the goal. This not only conforms our country to the concept of sustainable development, and to building a harmonious socialist society in China.

So the circular economy is the best choice to solve the problem of economic development and environmental pattern, this pattern can ease pressure on scarce resources and economic growth in the solution to protect the ecological environment in our country and promoting the sustainable utilization of the resources and so on all play an important role, this economic model is also stressed on consumption and production cycle. For example: to limit the production and consumption of disposable supplies, so that it is conducive to the development of circular economy. Through ecological industrial park, our country should be the development of the western region, revitalize the northeast old industrial base, focus on the pearl river, the Yangtze river delta and the Bohai economic circle, the idea of circular economy, the development of regional circular economy as a whole the area development, strengthen the international

economic cooperation, the development of circular economy in the world, maintain the harmony between human and nature, save resources, protect the environment, realize the harmonious development of man and nature. This is not only the main content of the construction of ecological civilization, also is the inherent requirement of promoting comprehensive, balanced and sustainable development. We will adhere to the scientific concept of development to meet the needs of human beings, protect the ecological environment, promoting the coordinated development of economy and resources. On the basis of the promotion of advanced technology, and constantly optimize the industrial structure, the development of primary industry at the same time, also want to the second and third industry development, adjusting the unreasonable internal structure, change the previous high input, high consumption, low efficiency, improve the economic efficiency and the quality. Out of a low consumption, low investment, high output of economic model to realize the sustainable development of China's economy.

3 TO PROMOTE SOCIAL INJUSTICE

Pepper justice thoughts on ecological socialism theory to our country the construction of a harmonious socialist society has an important significance. Pepper believes the lack of fairness and justice is one of the biggest ecological problems, and China's largest ecological problem is the lack of fairness and justice, due to the weakness of democracy at the grassroots level cannot ensure fair distribution of interests, appear many unfair phenomenon, China should insist on democracy at the grassroots level construction and to strengthen the construction of the Party's democracy, the establishment of a social security system of justice.

First of all, Pepper ecological socialism theory advocates a democracy at the grassroots level, give power at the grass-roots level, and make the citizens directly involved in the decision-making and management of public practice, in Pepper view only in this way can sufficiently guarantee citizens' rights, it also has important implications for the political construction of our country. Since the third plenary session of the 11th CPC Central Committee on developing the socialist democracy construction in our country, arouse the enthusiasm of the masses of the people, such as the reform of the electoral system in our country, the election by township to expand the scope of the county, it's not only mobilized the political enthusiasm of the people, also make the election of the people's congress is more extensive. The 15th CPC National Congress emphasis on building socialism with Chinese characteristics, political, under the leadership of the Communist Party of China, it puts forward the foundation of the people are masters of the country to develop a socialist democratic politics. The first of the 17th CPC National Congress of the Autonomous System into the characteristic socialism democracy at the grassroots level range. Although emphasizing the democracy at the grassroots level in our country, but because our country's democracy at the grassroots level system weak slowly led to appear many disharmonious phenomena, so we should reform the political system innovation, and constantly improve the socialist system. Should expand democracy at the grassroots level because it can not only arouse the citizen's political enthusiasm, also can cultivate citizens living habits, make better citizens to participate in the country's economic and cultural management.

Development of democracy at the grassroots level is an important choice. To promote China's democratic political construction and social development, the urgent request of developing socialist democracy at the grassroots level is advantageous to the maintenance and realization of the broad masses of the people's fundamental interests, which is conducive to mobilize the positive factors to serve socialist modernization, and carry out the perfect democratic centralism, make citizens better serve the socialist construction.

Second, since its inception the Chinese-made Party, our country the realization and safeguard social fairness and justice as a value goal, although after the reform and opening up, China's sustained economic development, cultural growing prosperity, social stability, but along the way there are many unfair problems. Such as the regional development gap between urban and rural areas and residents income distribution gap is still large, social conflicts increased obviously. Jobs such as insufficient equality, the education opportunity equality, opportunities for health care is not equal, unreasonable income distribution and the judicial fair enough influence and restriction of the current the realization of a harmonious society. Some unfair phenomena also affected the social order, such as money worship, the relationship between the wind corruption phenomenon such as the existence of these phenomena contribute to the spread of unhealthy tendencies, harm the social justice, judicial field also corruption phenomenon in recent years, intensified social injustice.

These problems have affected national security and the important factors which restrict the development of the social harmony. We must take effective measures to adhere to safeguard social fairness and justice. In the primary stage of socialism in our country, we should be in accordance with the spirit of the eighteenth big, safeguard social fairness and justice, helping citizens in economic and social development, equality, competition, equal participation

to the maintenance and realization of fairness and justice, and promote social harmony, we need to do the following:

1 To build systems of social justice safeguard, social fairness and justice. The 16th CPC National Congress pointed out, "Our country should strengthen the construction of guarantee system of justice, the report also proposed that the 18th CPC National Congress must adhere to safeguard social fairness and justice. Justice is the inherent requirement of socialism with Chinese characteristics. Our country should establish the social justice guarantee system, fair chance and fair regulation of power as the main content." Power fair means any of the citizen's legal rights can be protected, everyone is equal before the law system, and equal right to survival and development of basic rights, not because of sex, birth, occupation, or difference, such as a property right to fair is the most basic requirements of social equity and justice. Opportunity fairness refers to the survival and development of every social members enjoy equal opportunities, equal participation in social life, and share the achievements of the resulting. Everyone is equal before the fair regulation referring to the rules, everyone is bound by the rules. So China should further improve the system of democratic rights, display the function of the judicial safeguard fairness and justice. Only by putting the rights and interests of social justice system into practice, can we truly achieve and maintain social fairness and justice, enable the people to share political, economic and cultural achievements of social development, and ultimately to promote social harmony.

2 The correct understanding and dealing with the relationship between fairness and efficiency, attaching great importance to the economic construction in our country. The efficiency has been raised, but does not very well with fairness and has produced some social problems. Building a harmonious socialist society to correctly handle the relationship between fairness and efficiency. At the same time, taking economic construction as the center, we also want to solve the unfair problems from the rights of the citizens, law enforcement and judicial, take measures of income distribution, and strive to create a fair social environment, to ensure people's right to equal participation and development.

3 We should solve the people concerning problems, which is often the issue of most concentrated social contradictions, and the social equity and justice are the core of these problems. Such as a more prominent problem of income distribution in recent years, corruption, demolition, industry

monopoly, education, medical problems, etc. Solving this kind of problem is related to social harmony and stability. These are the most realistic beneficial questions, which is both a priority for building a harmonious society, and the most appropriate breakthrough.

Fairness has laid a solid material foundation in building a socialist harmonious society, the government should establish decision-making mechanisms to put the policy of the people's fundamental interests as the starting point to strengthen the governance of justice, and formulate specific solutions. Such as speeding up the development of education, social security, medical and health care, housing and other social undertakings adjust the relations of income distribution, raise the proportion of household income in the national income distribution, increase the intensity of tax adjustment for high earners. Gaining perfect democracy, guarantee citizens' rights, we strengthen social management laws and regulations and in the process of law enforcement, we shall uphold the principle of fair justice, the exercise of power, in strict accordance with legal procedures to meet citizens of heart of fairness and justice, maximum in protecting the interests of the people. Although we have the implementation and safeguard social fairness and justice, building a harmonious socialist society is a long-term and arduous strategic task, as long as we keep governing idea for the people, conscientiously put the masses of the people's fundamental interests first to solve the difficulties and problems in the production. On the basis of the development to meet the people's growing material and cultural needs, we guarantee the people sharing the achievements of the development, fully implement of the socialist fairness and justice value. Finally, we will be able to reach the goal of building a harmonious socialist society. Realizing social fairness and justice to continuously enhance the centripetal force, the cohesion of socialism with Chinese characteristics, charisma, is of great significance. To realize social equity and justice is the inherent requirement of social harmony, reflecting the core value orientation of the construction of socialist harmonious society. Only by maintaining and safeguard social equity and justice, can we truly achieve social harmony.

REFERENCES

[1] David Pepper Ecological socialism: from the deep ecology to the social justice [M]. Becky, eds. Jinan: Shandong University Press, 2005.

[2] Guan Yanchun. Pepper ecological socialism China enlightenment [J]. Journal of learning and exploration, 2011 (4).

Energy, Environment and Green Building Materials – Sheng (ed.)
© 2015 Taylor & Francis Group, London, ISBN 978-1-138-02718-3

The theory of David Pepper about the blueprint of the construction of the ecological socialism theory

Hong-Rui Chen, Gui-Fen Liu & Shao-Feng Chen
Harbin Institute of Technology, Harbin, China
Middle School of Daqing, China
The Iron Man Memorial Hall of Daqing, China

ABSTRACT: In the face of the increasingly serious ecological crisis, western scholars from different angles have found the original cause of the ecological crisis and the basic way of solving the problems. Since the 1990s, David Pepper is one of the most representative persons in British ecological socialist theorists. He thinks that the capitalist mode of production is the root of the ecological crisis, and the pursuit of profit and cost externalization of capitalism, resulting in the degradation of the environment, which led to the ecological crisis. David Pepper points out that the capitalist system cannot solve the ecological crisis. In his opinion, to solve the ecological crisis of social change is a must. So he puts forward a conception of the future society, the construction of ecological socialism, the socialist values for the center with human, human and nature harmonious development of society and social justice of ecological socialism.

KEYWORDS: David Pepper; ecological crisis; the ecological socialism

1 INTRODUCTION

Pepper puts forward many ideas in building a future socialist model, which provide meaningful references for the construction of building a harmonious socialist society, in this passage, we will discuss the Ecological Socialism Theory of Blueprint on the Construction.

In Pepper's view, the ecological contradictions of capitalism are unable to find a way out of environmental problems, he thinks that the change of the social system is the only way to solve ecological problems. Pepper, therefore, constructs the ecological socialism to build the society based on anthropocentrism values, social justice as the value demands, value orientation of the harmonious development of man and nature, which is his three building ecological basic features of socialist society.

1.1 *Anthropocentric value position*

Since the 1990s, the anthropocentrism proposed loud slogans about ecology Marxism, first raised by the British ecological Marxist theorist Reny, Glen DE man, he advocated human in opposition to the ecological crisis, reviewing their own attitudes towards nature, and adhered to the human scale. Pepper made a positive response to this, and became the main advocate for the thought and elucidates. Pepper is a core category of the ecological socialism for Anthropocentrism, standing on the ground of anthropocentrism to build ecological socialist thoughts.

Pepper of anthropocentrism is carrying forward the humanitarian, loving natural things of survival anthropocentrism, which takes the common development of human and nature as the goal, but no matter when, it'll never let go of human subjectivity. Its core idea is advocated in the treatment of the relationship between man and nature. The human should be the value measure of human interests as the fundamental. In fact, anthropocentrism is a constant pursuit of freedom. Pepper pointed out: "This kind of anthropocentrism is not a technology center in the sense of anthropocentrism. Not only the non-human world as a means to achieve the goal of "strong" anthropocentrism, but for the purposes of eco-centrism is a beneficial natural "weak" anthropocentrism. [1] At the same time, as he puts it, "The anthropocentrism of ecological socialism is a kind of long-term collective of anthropocentrism, and humanism, rather than of new classical economics short-term anthropocentrism of individualism." The anthropocentrism which emphasizes the integrity and insufficiency of human interests, and attaches great importance to the natural human value and significance. Because it is committed to achieve the sustainable development of human beings. Sustainable development and the core is not only to consider the needs of the current, but also to consider the needs of the development of the future, not at the expense of later generations to satisfy modern people's interests. From the connotation of sustainable development, the first is the development of the people, the interests of the people, especially people's ultimate interests and long-term goals, which is the focus of development. As Pepper can imagine, this is a richer society in art, where people eat more diversified and delicate food, use more art construction technology, receive better education, have a diversity of leisure and tourism, and achieve sexual relations and so on.

Pepper points out that the human can only observe from the Angle of human nature, in a position to talk about the nature of anthropocentrism. Especially when human interests conflict with the interests of the nature, the human always needs to be considered over the interests of the nature. He opposed to talk about the intrinsic value of the nature without people, and also opposed to leave the rights of people to talk about the rights of nature. Pepper insists on ecological problems and he made further explanation of the cause of the anthropocentrism, he thinks that if it is not the people, but the nature that placed in a central position, which will reverse the relationship between human and nature, to the relationship between human and nature mystification, resulting in various human systems. In his view, the root cause of ecological problem does not lie in anthropocentrism, but the result of the current economic system, because human nature is rational, currently in treating nature of greed and craziness by the current social and economic system. As long as changing the current social and economic system, people can overcome their greed and terminate, and reasonably in a rational manner, the use of natural resources in a planned way. When we emphasize people are the main body, nature is the object at the same time, Pepper thinks the relationship between subject and object is the unity of opposites. Human as the main body moves through practice, at the same time, the activities of human beings must conform to nature, human beings from the overall interests to "dominate" the nature, the "dominant" contains the meaning of planning, management, governance. It means that the human according to the rational manner, in a planned way, reasonably use natural resources development production from social interests so as to satisfy the needs of the development of the human comprehensive and freely, which can realize the sustainable development of society to establish a new type of person and nature coordination development of society. In this society, the person in a leading position, and shall be responsible for the ecological environment to realize the harmonious coexistence between man and nature, thus they will not only satisfy the material needs of the people and form a new natural harmonious relation, but also truly achieve a high degree of unity naturalism and humanism. From the standpoint of the anthropocentrism, human beings should be responsible for the ecological crisis, also only anthropocentrism to put the interests of humanity and nature to unify, for human nature to establish reasonable limits.

In Pepper's view, people's reasonable needs and interests as the starting point of anthropocentrist, which requires ideas of humanity, in accordance with the persistent existence and development as a whole in the nature of the target to deal with the relationship between human and nature, against the arrogant attitude towards nature and wanton looting. The anthropocentrism emphasizes the integrity and insufficiency of human interests, and attaches great importance to the nature of the human value and significance. Pepper believes that if people really adhere to the interests of the human at the center, really stand in the Angle of the lasting existence and development for humanity as a whole to recognize and deal with the relationship between human and nature, they would not only plunder nature with subjective will, but would respect the ecological law, scientific and careful attitude to nature. This will make the current ecological crisis gradually resolved. Pepper believes position in dealing with the relationship between man and nature. They should take the attitude of human, which is not technology or ecological center, but only sticks to anthropocentrism to treat biological nature and actively play a role of man's subjective, and pave the way for human progress. He thinks that only by insisting on anthropocentrism, the humanist, the construction of ecological socialism society, can we alleviate the contradiction between man and nature.

Therefore, Pepper builds the ecological harmony of "weak" anthropocentrism ecological socialism is not only feasible, but also is a reality, accord with the reality of contemporary living conditions. The human saves itself, and fundamentally solve the ecological crisis to establish a kind of ecological principle and the combination of socialism, beyond the contemporary capitalism and the traditional mode of extensive socialist new society. Pepper's anthropocentrism is the position both on previous sublation of the anthropocentrism and reflection to the contemporary human social practice activities.

In short, Pepper offered human's overall interests and long-term interests as the basis, advocating anthropocentrism as the essence of ecological socialism, filled with humanistic concerns, which fully shows Pepper has more general ecological center theories and broader vision than other environmentalists. His anthropocentrism advocated to put people in the central position of the natural relationship between the human's overall interests and long-term interests as the fundamental starting point and the ultimate value measurement of human behavior, which provides a more realistic way of thinking for us to solve the current ecological crisis.

1.2 *Values of ecological harmony*

Pepper puts forward his own ideas on the future of social economy, politics. It seems that ecological socialist society should be harmonious development between man and nature of society. In the economic model, Pepper advocated the development of the principle of moderation. The so-called moderate development refers to the creation of meeting human needs without destroying the ecological environment of the harmonious development of man and nature, economy mode, which is not a rejection of all forms of growth, but the opposition blind, harmful to the environment and human economic growth should be willing to give up economic growth and resource consumption to maintain stable economic development of mankind. Pepper said: "Capitalism is an unreasonable error development of productive forces. Pepper believes in eco-socialism, not based on profits to the production, development and distribution of resources, but according to the need of production." He also said: "Production is based on the resources of labor and not based on slavery, most people will perform their talents to get along with others, so personal desire is largely consistent with the community." Human can better understand and dominate nature, according to this principle of resource allocation and development to establish a sustainable, rational, planned development of eco-socialist equality for each person, it is not only to meet people's needs, to the overall development and long-term interests of humanity, but also conducive to the protection of the ecological environment.

On the economic system, the traditional ecological socialists think seriously undermine the larger scale of human activities on the natural cause, they emphasized the implementation of Schumacher doctrine and small-scale, decentralized operating mechanical viewpoint of low pollution. Pepper was against this view, he said: This economic model is unrealistic, he advocated we should establish to meet the needs of people for the purpose of a moderate economic growth system, and such growth will not conflict with the natural environment, he believes that the economy is an equal need to have planned development for everyone rather than highly centralized. He does not oppose the use and development of technology, noting that technology can enhance and adapt to the transformation of nature and natural ability. Pepper emphasized the use of technology should be within reasonable limits, which can help the harmonious development between man and nature.

About the way of economic running, Pepper thinks the market of the capitalist system is characterized with blindness, disorder and spontaneity, which will lead to the destruction of the environment, Pepper does not reject the industrial and manufacturing but refuses to be alienated technology and production. Pepper advocates Frankel proposed economic operation that combines planning and market mixed with the economy, he believes that a mixed economy is to restrict competition in the market in a modest field, it will not be due to market the blind expansion of the environment caused by excessive pressure and excessive consumption of resources. He stressed the need to establish social and ecological economic model to replace the market economy to establish a way to protect the natural and beneficial economic system for the sake of future generations, starting from the long-term interests and the overall development of the human person, the "less production, but better" as the main lever of economic development. He believes that the evaluation of an economic system is reasonable and effective, not only to see whether it brings economic benefits, but also the environmental costs and social benefits.

In terms of the economic system, Pepper advocates common ownership of the means of production, he believes we should not overpass the limits of nature's laws, it should overcome alienation through common ownership of the means of production, he points out that we should be in the interest of the collective, there are plans to control the human relationship with nature and should not dominate or exploit nature. In other words, the implementation of man and nature, the collective conscious control, which is not only a fundamental guarantee of a green economy, it is also consistent with the long-term interests of human and ecological interests.

In politics, Pepper advocates economic, political, ecological should be combined with a decentralized overall combination of new democratic political system regionalization and internationalization of combining. First, he points out that modern democracy can not completely deny, but to transform the modern system by democratic participation, democratic self-service comprehensive development of man. Secondly, he emphasizes international democracy. 90 years ago, the idea of ecological Marxist political decentralization and non-bureaucratic, as well as workers 'autonomy and biological characteristics of grass-roots democracy centrism, and the implementation of workers' social life and production management, they think ecologically rational grass-roots democracy, political structure, which can not only mobilize the enthusiasm of the citizens, but also improve the quality of political decision-making. Pepper criticizes this view that it is unrealistic and cannot rely on grassroots organizations to address the ecological crisis, pointing out that social labor management should be organized democratization, while also criticizing the biological center, he thinks it separates the links between regions, and is not to solve ecological problems. He believes that ecological problems have become a global problem, it must rise to the international level area.

In addition, he believes that capitalist industrialization leads to mental decline and spiritual poverty, so he advocates focusing on spiritual civilization. We can establish and popularize scientific world outlook and reform traditional education to support cultural and other grassroots cultural movement. He also advocates the role of the State in the future society, Pepper does not agree to repeal anarchist viewpoint countries, Pepper agrees with the view of ecological socialist theorists Glenn Friedman's: A State is necessary to ensure that civil rights democratic participation, coordination of economic resources and ensure that the management plan to achieve functional departments. Pepper believes decentralized local ecological problems cannot be solved, so the eco-socialism must still rely on the power of the state, the state can play a positive role in the social life and solving environmental problems, that is, to achieve production and distribution, equality and social justice requires the State, so Pepper advocates the establishment of a similar system of the country or countries, he acknowledges the need for national or similar organizations in the economic society.

REFERENCES

[1] The ecology Marxism study [M]. Beijing: people's publishing house, 2011.
[2] David pepper. Ecological socialism: from the deep ecology to the social justice [M]. Becky, eds. Jinan: Shandong University Press, 2005.

Energy, Environment and Green Building Materials – Sheng (ed.)
© 2015 Taylor & Francis Group, London, ISBN 978-1-138-02718-3

Application of a 3D modeling software—Cityengine in urban planning

Y.W. Luo, J. He & H.J. Liu

Department of Architecture and Urban Planning, College of Civil Engineering and Architecture, Guangxi University, Guangxi, China

ABSTRACT: As a frontier three-dimensional modeling software, Cityengine is low cost and high-usage, and has great potential of secondary development. These advantages will make it more widely used in urban planning: it will not only assist urban planners to analyze urban space, but also urban geography and physics environment. Cityengine is beneficial to realize the integration of planning and practice, quantitative analysis of urban planning and promotion of public participation.

KEYWORDS: Cityengine; Rule-based modeling; Urban planning; Quantitative analysis

1 BACKGROUND

The city is a vivid three-dimensional (3D) space and urban planning is rational allocation of resources in this space. With the continuous development of the city, the information about this 3D space becomes more and more diversified. Therefore, it has to rely on computer technology to realize visual display of urban space by 3D models. Since China imported 3D modeling technology in the late 1980s, it has developed nearly twenty years. Many associated concepts such as digital city, green city and smart city have been proposed in urban planning circle. Based on its geometric properties and related information, 3D modeling technology is good at effectively allocating urban resource, reasonably choosing the direction of urban development and correctly managing cities (Yang et al. 2003).

Compared with conventional software, rules-based modeling has a huge advantage and would be a development trend of future 3D City Modeling. This paper describes the modeling principles, steps and characteristics of Cityengine, and discusses how to use in urban planning and what significance it would bring to urban planning.

2 DESCRIPTION OF CITYENGINE

CityEngine is a 3D modeling software that was initially invented by the Swiss Federal Institute of Technology in Zurich in 2011. After several upgrades and authorized, it is a product of Esri company (this promotes the realization of seamless combination of Cityengine and GIS). CityEngine can be applied to urban planning, rail transportation, electric power, pipelines, construction and other fields.

2.1 Modeling principles

The modeling principle of Cityengine is based on modeling rules written by Computer Generated Architecture (CGA) which is a unique edit language of Cityengine. There is a great difference between conventional 3D modeling software and Cityengine in modeling. Rule is a group of CGA languages that define and describe a series of geometric and texture features to determine changes of objects and assign the changes to other objects (Barroso et al. 2013). The idea of modeling rule is to define the rules and repeatedly optimize the design to create more details (Zhao & Coors 2012, Lv & Li 2013). It may spend more time to write rules at the beginning, but once this work was completed the rules can be created plenty of different design options and much faster than the conventional manual modeling (Figure 1).

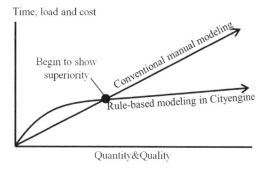

Figure 1. Cost advantages of Cityengine.

2.2 *Modeling steps*

The modeling on Cityengine can be divided into three steps. The first step is to obtain basic data, such as terrain, roads, buildings and other 2D data. As Cityengine supports a variety of model format, there are many ways to obtain the basic data, for example the.shp format data from GIS and .osm data from Open Street Map. The second step is to generate models by CGA, and it is a key step in modeling. Because of the limited space, this paper only presents the basic principles (Figure 2). Through writing the four basic function commands: extrude, comp, color and split, a 2D plane can be rapidly molded into a 3D model. Then one can save the rule files in .cga format and assign the rules to other 2D data. The third step is to use and output the models for another purpose, such as import it to 3D Web scene and ArcGIS online to realize a linkage editor, scheme comparison and model sharing.

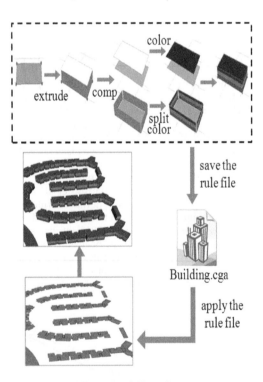

Figure 2. Modeling steps of Cityengine.

2.3 *Comparison with conventional software*

Compared with conventional 3D modeling software such as Sketch Up, 3D Max, the greatest advantages for Cityengine are the following three points. One is that the rapid and mass rule model has greatly reduced the load, time and cost. The second is that

Cityengine can be combined with a variety of software (especially the seamless combination of GIS) and increase the value of 3D models. Last is that this software has CGA language which has great potential secondary development.

3 CITYENGINE IN URBAN PLANNING

Modeled by Cityengine, the 3D models can intuitively display urban space, keep up with the urban construction changes timely and simulate urban physics and geographical environment (Laycock 2011). These applications are summed up in two aspects: For the first use of the model, it acts as an assistant tool of urban spatial analysis, reflecting in urban macro planning and micro planning. For the second use of the model, it is imported into other software to realize simulation analysis of geographical and physics environment (Figure 3).

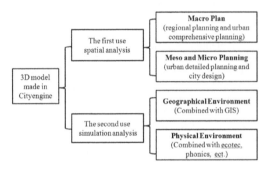

Figure 3. Application of Cityengine in urban planning.

3.1 *Applications in macro planning*

It assists urban planners to determine the direction of city development and land layout of the regional planning and urban comprehensive planning. Some scholars have proposed that the 3D modeling should not be blindly pursued exquisite, but placed an emphasis on practical application and left room for expansion (Gao & Chen 2013). So is Cityengine, unlike the conventional modeling software, it reflects a rough 3D space of a city within a short time, making it ideal to meet the needs of the urban macro planning.

Firstly, it has a guiding significance for the construction of new districts. According to the geographic information of urban 3D model, planners can grasp direction and new location of site selection; strengthen links of transport and infrastructure between old and new districts.

Secondly, it helps planners decide the land layout. Take green space layout as an example: although green ecosystem is a complex and continuous 3D system, the conventional green planning ignores the changes of green 3D structure, like plant species, plant size, vertical greening, etc. The model of Cityengine can present the 3D magnitude of vegetation occupied by the green leaves, and better to explain differences in the spatial structure than the ratio of green space (Zhang & Wang 2001). What is more, due to the function of secondary development of Cityengine, it is expected to calculate the quantity of green shaded area (Figure 4).

Finally, it is conductive to the road layout of macroscopic. Click the road drag commands, roads and architectural complex in accordance with urban fabric would be quickly generated, which make planners intuitive to compare the different road schemes to determine urban road skeleton. Meanwhile, the model also visually shows the spatial relationship between transit facilities and areas with residents' frequent activity, like urban community, public centers, etc. Therefore, it would reduce the traffic negative impact on the residents.

Figure 4. Quantitative analysis of greening.

3.2 Applications in meso and micro planning

This is reflected in control and design of urban space in the urban detailed planning and urban design. The emphasis of Cityengine modeling is not on detail and authenticity, while is the expression of the 3D shape of the outline. Therefore, it ideally presents the relationship between physics entities in the urban environment, and meets the need of the detailed planning and city design.

Cityenging presents the urban space at all angle and distance and simulates walking motion process, allowing the designers to experience the design intent and relationship with the real scene. Hence, it improves accuracy and objectivity of designing and planning. More importantly, Cityengine can be edited and adjusted the size, location, layout, materials and other property of buildings, open spaces and green plants, and generated real-time models. Therefore, it assists to design the city skyline, building layout,

street towards and other aspects of the urban detailed planning.

As Cityengine can quickly realize a simulated 3D environment, optimize design ideas and make visualization, it plays an important role in the initial phase of urban planning and urban design.

3.3 Analysis of urban physical environment

Models on Cityengine can import into other software for further analysis, such as analysis of the geographical environment and the physical environment, assisting planners in testing planning effect. In the past, the urban geography environment is mainly analyzed using GIS, lacking of conjoint analysis of urban construction and the terrain. For the physical environment analyzed, models need to be separately generated and imported into the corresponding software. Such models are rough and low utilization, and simply shows the volume of construction and objects.

Urban geography can be analyzed by the combination of Cityengine and GIS. It can realize the functions of storage, query, statistics, analysis, simulation, decision and prediction of terrain data by importing the models into a GIS database and using the geographical processing functions provided by 3D analysis extension module of ArcGIS (Dahal & Chow 2014, Grêt-Regamey 2013). Models made on Cityengine can not only present the spatial relationship of urban construction, but also be imported into other software and analyzed the city physics environment. For example, you can import the models into Ecotect to simulate sunshine and wind environment. What is more, it is expected to be imported into CFD software for wind and thermal environment simulation.

4 SIGNIFICANCE FOR URBAN PLANNING

4.1 To realize the quantitative analysis

The greatest significance of Cityengine is quantitative analysis, like acquisition of object information and simulation of the urban physical environment. The current urban planning is mainly qualitative planning. Although there will be some quantitative indicators in the planning design stage, it is unable to measure the effect of planning. For example, the ratio of green space only reflects the 2D plane. However, in the actual operation process of urban forestation how and what plants be grown would directly affect the urban greening effect. Model made in Cityengine shows vivid green space planning, and extracts the data information of green space (Neuenschwander 2014); it also reactives its effect on the thermal and wind environment through the physical environment simulation. It can be said that Cityengine is a catalyst for urban planning from qualitative to quantitative.

4.2 To achieve combination of planning and practice

Cityengine modeling is not static; it can be updated at any time. In dealing with the urban renewal problems, the conventional urban modeling is usually manual remodeling, which is high cost and time-consuming. Rule based model made in Cityengine can be saved and modified, greatly improving modeling speed and timely giving feedback information of urban construction. Therefore, it promotes the combination of planning and practice.

4.3 To promote public participation in urban planning

Models made in Cityengine can realize a linking editor, scheme comparison and resource sharing, ect. Therefore, it promotes the public participation. With CityEngine2012 release Web scene features, models can be opened through the local Web Scene Viewer and published in ArcGIS Online. Model made on Cityengine can realize the linking editor, program comparison, sharing model, etc. So it can promote the public participation. Through the Web Scene, buildings or rooms can be searched online by automated location system; sunshine can be simulated for hourly change; planning schemes can be compared. Therefore, citizens using the computer can understand the intention of urban planning and participate in the activities of urban planning.

5 CONCLUSIONS

Cityengine makes up the shortcomings of graphic thinking of the conventional urban planning and would be an important technical support for urban planning. This paper teases out the principle, modeling steps and advantages of 3D model on Cityengine and described two applications of Cityengine in urban planning: first, it is a support tool of urban spatial analysis; second, it realizes the simulation of geography and an analysis of physics environmental. These two applications are from the shallow to the deep. This study aimed to give readers a quick understanding of Cityengine, while providing a new way of thinking about how to make full use of it in urban planning.

ACKNOWLEDGMENT

This work was financially supported by the Guangxi Natural Science Foundation under Contract No.2012GXNSFDA053025.

REFERENCES

Yang J. Zhang Y.Q. & Pang B.M. 2003, Three-dimensional modeling techniques for urban planning, Modern Surveying and Mapping (1):122–131.
Barroso S. Besuievsky G. & Patow G. 2013, Visual copy & paste for procedurally modeled buildings by ruleset rewriting. Computers & Graphics 37:238–246.
Zhao X. & Coors V. 2012, Combining system dynamics model, GIS and 3D visualization in sustainability assessment of urban residential development. Building and Environment 47:272–287.
Lv Y. L & Li X.L. 2013, CityEngine modeling based method of high speed railway, Mapping 36(1):19–22.
Laycock S.D. Brown P.G. Laycock R.G. & Day A.M. 2011, Aligning archive maps and extracting footprints for analysis of historic urban environments. Computers & Graphics 35:242–249.
Gao S. & Chen S. 2013, Urban 3 d modeling technology and standard, Bulletin of Surveying and Mapping (3):95–115.
Zhang H. & Wang X.R. 2001, Three-dimensional ecological features of urban green space and ecological functions, China Environmental Science 21(2):101–104.
Dahal K.R. & Chow T.E. 2014. A GIS toolset for automated partitioning of urban lands, Environmental Modelling &Software 55:222–234.
Grêt-Regamey A. Celio E. Klein T.M. & Hayek U.W 2013. Understanding ecosystem services trade-offs with interactive procedural modeling for sustainable urban planning, Landscape and Urban Planning 109: 107–116.
Neuenschwander N. Wissen U.H. & Grêt-Regamey A. 2014. Integrating an urban green space typology into procedural 3D visualization for collaborative planning, Computers, Environment and Urban Systems 48: 99–100.

Energy, Environment and Green Building Materials – Sheng (ed.)
© *2015 Taylor & Francis Group, London, ISBN 978-1-138-02718-3*

On the relationship between walk space form of college campus and summer thermal environment

L. Qin, J. He, Y.W. Luo, Y.Y. Yao, B. Chen, Y.J. Meng & Y.H. Wu
Department of Architecture and Urban Planning, College of Civil Engineering and Architecture, Guangxi University, Guangxi, China

ABSTRACT: In this paper, typical walk space of Guangxi University is selected from the angle of spatial form for thermal environmental test. Spatial form is quantified via Sky View Factor. Through data analysis of the relationship between spatial form and thermal environment, spatial form's impact on thermal environment is summarized and a type of spatial form is proposed for analysis of designer, providing accordance for design of college walk space.

1 INTRODUCTION

Thermal environment is an important weighing standard of comfort of space. At present, the college thermal environment is mainly measured by the instrument, which takes long time and makes it impossible for designers to make quick judgments about the space thermal environment. Solar radiation is a main reason for rise of space temperature;

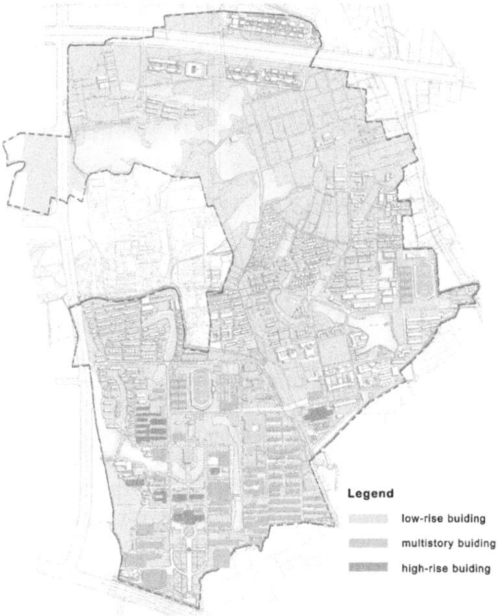

Figure 1. Height distribution of buildings in West Campus of Guangxi University.

therefore, planning of green land and arrangement of buildings is the main influencing factors of college outdoor space thermal environment. Those influencing factors can be presented through Sky View Factor (SVF). However, nowadays, a study of the college thermal environment is hardly started from SVF (Peng 2006, Chen et al. 2012). In this paper, representational space is selected from Guangxi University to measure its WBGT value, temperature, and humidity, which are compared with SVF for analysis, finding out the relationship between the college space thermal environment and SVF and an easier method for judging college thermal environment.

2 INTRODUCTION OF COLLEGE AND CONCEPTS

2.1 *Introduction of Guangxi University*

Guangxi University is located in Nanning, south of the central Guangxi province. The climate of Nanning is subtropical monsoon climate, featured by high temperature and humidity in summer. The extreme maximum temperature in summer reaches as high as 40.4℃. Since there is a high requirement of thermal environment of the summer in the south region, time, section of summer with the highest temperature is selected in this paper for thermal environment test, which is soundly representative (Huang & Huang 2006).

In this paper, west campus of Guangxi University is selected for study. It is of a relatively complete system of teaching area, dormitory area, commercial area and green landscape. It is shown in Figure 1 the distribution of building height in west campus. Buildings are mainly multistory ones (4 to

6 stories), which takes about 80% of all buildings. The rest are low-high-rise buildings and low-rise buildings. From the arrangement of buildings, it is known that the main roads in the college are generally 9m wide and there are relatively plenty of walk space between buildings. The walk space is the place where students frequently carry out daily activities. Since the shading area of buildings is limited, the space receives solar radiation for a long time, which requires more green land. In this paper, reasonableness of combination of space and green land is a main analysis object selected for studying relationship between SVF and thermal environment (Le 2012; Wu et al. 2013).

2.2 *Introduction of concepts*

SVF is a parameter can be used for analyzing shading the condition while studying the thermal environment (Wang &Duan &2013). The heat received by an outdoor space is mainly from direct solar radiation. Shading of trees and buildings is main factors influencing direct solar radiation. The smaller the shading area is, the higher SVF is. In this paper, SVF value comes from fisheye photos of the sky taken from the measuring point. For processing the fisheye photo, magic wand tool in Photoshop, which is of function of selecting specific area, is used for selecting pixels of the sky. Calculate the ratio taken by the sky, i.e. SVF (shown in Figure 2).

Figure 2.   Diagram of obtaining SVF.

Direction of solar radiation, as well as college space, shall be considered for shading of buildings. SVF only presents a visible area of sky at one point, making it unable to analyze shading of space rigorously. In this paper, direction of SVF is summarized as well.

Beside temperature, objective indexes influencing people's thermal sensation in the outdoor thermal environment include humidity, wind speed and thermal radiation etc. (Wang et al. 2008, Lin 2008, Chen 2010) Therefore, in this paper, tests focus on humidity related to solar radiation and comprehensive thermal environmental indices-Wet Bulb Globe Temperature (WBGT) (Yan et al. 2013, Zhang& Meng 2011). Recommended by ISO 7243 to find out the relationship through comparison with SVF.

3 INTRODUCTION OF THERMAL ENVIRONMENT TEST

The method of fixed moving measurement is adopted for test. The finish is measuring and recording data of one measuring point within 10 minutes (including the time of moving). Finish one round of data measurement and recording of 5 measuring points within 1 hour. Take a fisheye photo at every measuring point (fisheye photo is taken at a height of 1.5m from the ground). The test is carried out in 2 days, 7 hours every day from 10:00 to 17:00. There are totally 30 measuring points. Analysis of college thermal environment will be carried out through general condition of all measuring points and eight typical measuring points (shown in Figure 3).

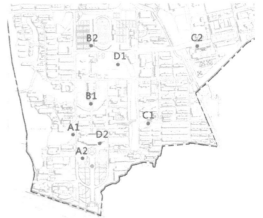

Figure 3.   Distribution of measuring points.

4 PHOTOGRAPHS AND FIGURES

4.1 *General result of environment of measuring points*

Measuring points of walk spaces are selected from road space and the rest space, 15 points each. Distribution of SVF data of measuring points are collected (shown in table 1). From the general distribution of SVF, it is known that the SVF distribution of most rest space and road space is not much different, mostly between 0 to 20%. Compared with that of rest space, SVF of more measuring points of rest space reaches 40%.

4.2 *Comparison and analysis of WBGT and SVF*

From the figure of daily average value of WBGT and SVF distribution at 30 measuring points (Figure 4), it is known that WBGT is roughly directly proportional

Table 1. Different SVF measuring points of rest space and road.

SVF	0-20				20-40	40-60
	0-5	5-10	10-15	15-20		
rest space	2	6	4	2	1	0
road space	2	6	4	1	1	1

to SVF. When SVF is between 5% and 20%, a rise of SVF has great impact on WBGT; when SVF is higher than 20%, the impact of SVF on WBGT is reducing.

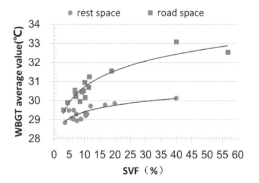

Figure 4. Relationship between WBGT and SVF.

1 Two points A1 and A2, of which the surrounding spaces are similar but SVFise different, are selected for analyzing WBGT (shown in Figure 5). Climate change at noon and solar radiation reduces slightly. It is known for Figure 1 that the temperature of A1 rises slowly when the instantaneous value of solar radiation is high in the morning and afternoon, and the temperature of A1 is lower than that A2, of which the SVF is higher. When solar radiation reduces at noon, temperature of A1 point declines slower than that of A2 point. At 13:00, the temperature of A2 point is lower than that of A1 point. It is thus clear that low SVF is featured by strong heat isolation and weak heat radiation.

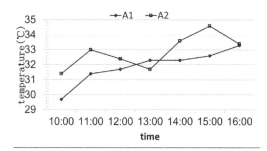

Figure 5. Diagram of temperature change at A1 and A2.

2 Two point B1 and B2, of which the shading directions are different, are selected for analyzing WBGT (Figure 6). B1 is under sunshine in the morning, while B2 in the afternoon. The maximum WBGT value of B1 in the morning is 1.2℃ than that of B2 point at the same time. The maximum WBGT value of B2 in the afternoon is 1.6℃ than that of B1 point. It is thus clear that solar direct radiation has greater impact on thermal environment in the afternoon than in the morning.

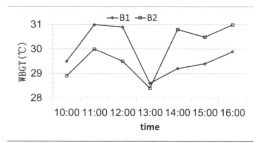

Figure 6. Diagram of WBGT Change at B1 and B2.

3 Two points C1 and C2, which are under solar radiation the whole day and SVF are higher than 40%, are selected for analyzing WBGT (Figure 7). For C1, SVF is lower and the temperature of underlying surface rises faster; for C1, SVF is higher and underlying surface rises slower. It is clear from the figure that for more time, temperature of C2 is lower than that of C1. However, when heat accumulates high in the afternoon, WBGT of C1 is obviously higher than that of C2. It is thus clear that with high SVF, underlying surface absorbs more heat; fast temperature rise is an important influencing factor of space thermal environment.

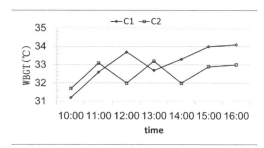

Figure 7. Change of WBGT at C1 and C2.

4.3 *Absolute humidity and SVF*

Two points D1 and D2 of similar SVF are selected for analyzing absolute humidity (Figure 8). Since the

SVF is relatively low, influence of reflection of underlying surface is not considered. D1 is in the trees, and D2 is located at the relatively open area. It is known from the data in the figure that besides SVF, ventilation of surrounding environment influences absolute humidity of space as well.

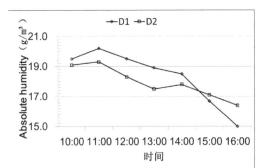

Figure 8.   Change of absolute humidity of D1 and D2.

5   RELATIONSHIPS BETWEEN SVF AND THERMAL ENVIRONMENT

1   When SVF is lower than 20%, quality of thermal environment reduces with an increment of SVF. It is found when SVF is below 20%, WBGT is directly proportional to SVF. For colleges in the south area, where winter is warm and summer heat, over-high SVF influences higher on college campuses. The designer may selection shading trees by considering the value of SVF.
2   Space requires lower west SVF than east SVF. Due to heat accumulation, college space in the afternoon requires more shading and heat radiation. Without directional solar radiation, higher SVF is easier to radiate heat from ground to sky and reduce the temperature. While designing daily college space shading, different shading devices shall be selected for different shading directions. Plant trees with a bushy crown on the west side of the walk way for shading when thermal environment is bad, and plant trees with a sparse crown at east side of the walk way, thus easier to radiate heat from the space to the sky (Figure 9).
3   Reflectivity of underlying surface, water body and SVF together influence the thermal environment. For space with SVF higher than 20%, reflect of the underlying surface and heat radiation brought by high SVF increase heat received from the space. Materials with low reflectivity shall be selected for underlying surface. It shall not use smooth ground with high reflectivity, such as cement or ceramic tiles. Water body increases reflection as well. More

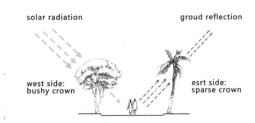

Figure 9.   Case scheme of plant arrangement in college at summer afternoon.

shading shall be arranged around water body to reduce water reflection, which will enhance the temperature reducing function of water.

6   CONCLUSION

Though traditional instrument measurement is able to obtain accurate data while judging thermal environment of college walk space, it takes a long time and the process is complicated. Analysis of value and direction of SVF quantifies college space form in an easier way without longtime real measurement. It provides data for judging thermal environment and accordance for designing college walk space.

ACKNOWLEDGMENTS

This paper is supported by 2014-2016 China National College Students Innovation and Entrepreneurship Training Program No.201410593087.

REFERENCES

Xiaoyun PENG. On Micro Thermal Environment of Organic College Campus [J]. Engineering Construction, 2006, 36 (2):43.
Ailian CHEN & Ranhao SUN & Liding CHEN. Development of Study on Urban Heat Island Based on Landscape Pattern [J]. Acta Ecologica Sinica, 2012, 32(14):4560–4561.
Meili HUANG & Meisong HUANG. Analysis of Climate Condition for Planning and Construction of Nanning [A]. China Association of Science and Technology. Improving Scientific Quality of All People, Constructing Innovation-oriented Country— Proceeding of China Association of Science and Technology 2006 (Vol.2) [C]. China. Association of Science and Technology, 2006:6.
Di LE. Study on the Influence of High-rise Buildings Layout on Urban Regional Thermal Environment [D]. Hunan University, 2012.

Wenzhen WU & Jian TENG & Chunlin PANG. Analysis and Appraisal of the Plant Landscaping of Guangxi University [J]. Anhui Agricultural Science, 2013, 10:4445–4447+4457.

Lin WANG & Yong'an LI & Peilei LIU. Research of the Effect Factors on Campus Thermal Environment [J]. Refrigeration and Air Conditioning (Sichuan), 2008, 04:110–114.

Yansong WANG & Yapeng DUAN, On Assessment of Ecological Suitability of Urban Old Street—Taking Lichuan Old Street of Jiangxi as An Example [J]. Architecture and Culture, 2013, 04:87–88.

Borong LIN. Study on Influence of Greening on Outdoor Thermal Environment [D]. Tsinghua University, 2004.

Zhuolun CHEN. Study on Influence of Greening System on Outdoor Thermal Environment of Building Cluster at Hot and Humid Region [D]. South China University of Technology, 2010.

Zhang YAN, He JIANG, Zhaomen ZENG, Yanwen LUO, Zhao JING. Advanced Materials Research Vols. 671–674 (2013) pp 2414–2419.

Lei ZHANG & Qinglin MENG. Test and Analysis of Summer Thermal Environment of College Campus at Hot and Humid Region [J]. Science of Architecture, 2011, 27(2):48–51.

Energy, Environment and Green Building Materials – Sheng (ed.)
© *2015 Taylor & Francis Group, London, ISBN 978-1-138-02718-3*

Analysis of outdoor thermal environments of universities in summer in hot and humid areas

Y.Y. Yao, J. He, Y.W. Luo, L. Qin, Y.J. Meng, B. Chen & Y.H. Wu
Department of Architecture and Urban Planning, College of Civil Engineering and Architecture, Guangxi University, Guangxi, China

ABSTRACT: In this paper, taking Guangxi University as the object of study, the thermal environment measurement is conducted for the rest space and daily used routes on the campus. On this basis, the summer thermal environment characteristics of outdoor space in a hot and humid area are explored in combination with relevant research results and experimental data. Finally, the conclusions about the comfort of rest space and daily used routes in Guangxi University are drawn, laying foundation for planning and design for green campus.

KEYWORDS: Hot and humid area, thermal environment, rest space, daily used routes

1   INTRODUCTION

The outdoor thermal environment in the summer in a hot and humid area is closely related to the comfort in daily life. However, nowadays, there are few researches on the university campus landscape and comfort concerning the thermal environment, and in "Evaluation Standard for Green Campus" promulgated in 2013, no requirements are raised for thermal environment on the campus. One of the important aspects in improving the quality of summer campus life in a hot and humid area is to create a good outdoor thermal environment through cooling and dehumidification. In Guangxi University, the campus greening rate is higher, and there are many artificial lakes and lotus pond. As its weather conditions in summer are more typical in universities of Guangxi, in this paper, 15 measuring points at rest spaces and three routes frequently used in daily life by teachers and students are selected for study. On hot days of summer, they are measured from these aspects: air temperature, humidity, wind speed, surface temperature, and WBGT value. Through data analysis and comparison, the changing rules and characteristics of the outdoor thermal environment in a hot and humid area are obtained. The status of outdoor thermal environment in Guangxi University is also analyzed and the proposals for transformation of green campus are put forward.

2   OVERVIEW OF GUANGXI UNIVERSITY

Located in south-central Guangxi, Nanning City belongs to the subtropical monsoon climate with ample sunshine, abundant rainfall, long summer and short winters. In summer, it enjoys high temperature and high humidity, and a large amount of solar radiation. In the hottest July and August, the average temperature is 28.2℃, and the extreme temperature can reach up to 40.4℃, and the annual average total amount of radiation is 4412MJ • m-2, and the annual average relative humidity is 79%. (Li 1989, Huang & Huang 2006)

Guangxi University is located in Xixiangtang District of Nanning City, with a total area of 307 hectares. From Fig. 1, it is a flat land with the higher greening rate, abundant water, good environment, and beautiful landscape. Compared to other areas in Nanning, it is cooler and more comfortable, playing a role of "green lung" for surrounding areas. (Jiang et al. 2014; Wu et al. 2013)

3   TEST INTRODUCTIONS

3.1   *Test methods*

The test on thermal environment is conducted in two parts: one is the rest space; the second is the routes frequently used in daily life by students. The rest space is the commonly used areas by teachers and students for recreation, which has considerably higher use frequency especially in hot and humid summer. The quality of thermal environment directly affects people's comfort in use and its own functionality. As the routes daily used by students are longer, and they have higher requirements on greening, sun-shading and ventilation within a specific time, the direction and continuity of sun-shading are mainly studied.

The test on the outdoor thermal environment was conducted from 10:00 – 17:00 on one hot day in hot

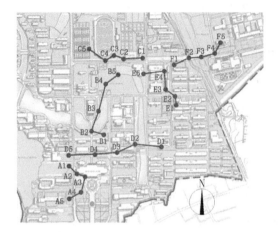

Figure 1. Distribution of measuring points.

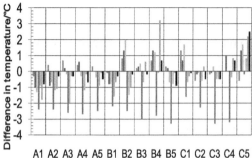

A1 A2 A3 A4 A5 B1 B2 B3 B4 B5 C1 C2 C3 C4 C5

■ 10:00 ■ 11:00 ■ 12:00 ■ 13:00 ■ 14:00 ■ 15:00 ■ 16:00

Figure 2. Difference in temperature between rest space and weather.

Measuring point	Average WBGT value(℃)	Measuring point	Average WBGT value(℃)	Measuring point	Average WBGT value(℃)
A1	28.86	B1	29.04	C1	29.73
A2	28.96	B2	29.79	C2	29.31
A3	29.09	B3	29.51	C3	29.23
A4	29.50	B4	30.14	C4	29.31
A5	29.04	B5	29.37	C5	29.87
Average value	29.09	Average value	29.57	Average value	29.49

Figure 3. Average WBGT values of rest space.

summer- along three routes in three groups each day. For any measurement point, the test and data recording were required to complete in 10 minutes; and also one round of test on five measurement points was required to complete within one hour. In the test, the WBGT index detector was used to test the air temperature, air humidity, black ball temperature and ground temperature and WBGT values of 15 measurement points. (Peng 2006, Zhang&Meng 2011, Zhang et al. 2013)

3.2 Introduction of measurement points

Measuring points are distributed as shown in Figure 1. Route A, B and C are the measuring routes for rest space. For measuring points, these aspects are involved: the presence or absence of water, greening and underlying surfaces. Route D, E and F are the routes in daily life for students to attend classes, relax themselves and study on their own respectively. These three routes are the routes teachers and students use more frequently, and the thermal environment has a greater impact on the comfort the teachers and students feel.

4 TEST RESULTS AND ANALYSIS

4.1 Analysis of rest space test results

For some measuring points, the test environments are not ideal. It can be seen from Figure 2 and 3 that, for most of measuring points at the rest space, their outdoor temperatures are lower compared to the data from the weather station and that WBGT values are between 27℃ ~ 31℃. The rest space on the campus is relatively cool and comfortable. However, due to difference in spatial environments, the temperatures of measuring points vary greatly. For some measuring points, their WBGT values are higher, and their temperatures are higher than the data provided by the weather station.

For the summer heat conditions, Measurement Points A1, A5, B3 and C4 were selected for comparative analysis. Their underlying surfaces are respectively soil, grass-planting bricks, tiles and plastic floor, and their greening is roughly similar to other surrounding environments. (Li et al. 2005) It can be seen from Figure 4 that Measuring Point C4 has a high surface temperature as it is largely influenced by solar radiation because of its plastic floor as the underlying surface and it has a low surface reflectivity. After 13:00, with the increase of accumulated absorbed heat, the heat dissipation becomes slow, and the surface temperature rapidly increases. The surface temperatures at other three measurement points are closer. For Measuring Point A1, as its surface is soil, it has strong water-holding capacity. As the water evaporation can lower the temperature, the surface temperature is low and stable. Obviously, due to difference in reflectivity and water-holding capacity, the underlying surface has a significant effect on surface temperature. Viewed from the perspective of human comfort in the summer in a hot and humid area, the impacts of different underlying surfaces in the thermal environment can be arranged in order (from good to bad): soil, grass-planting bricks, tiles and plastics.

For summer humidity conditions, Measurement Points A4 and B3 were selected for comparison. The environments surrounding Measuring Points A4 and B3 were relatively similar: they were shaded by higher trees and their underlying surfaces were granite floor. There was no water around Measurement Point A4, but there is an artificial lake around Measuring Points B3. It can be seen from Figure5 that, in terms of absolute humidity, Measuring Point B3 is significantly higher than Measurement Point A4. Due to the shade of trees and similarities in other factors, the air temperatures at two measuring points are closer; but the temperature is relatively stable around the artificial lake (Wang et al. 2008).

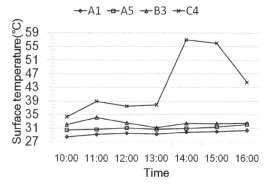

Figure 4. Surface temperatures at typical measuring points of the rest.

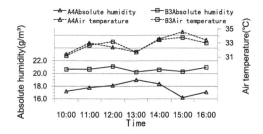

Figure 5. Absolute humidity and air temperature at typical measuring points of the rest space.

4.2 Analysis of test results of daily used routes

Figure 6 and 7 show that the thermal environments of routes daily used on campus are more severe. For more than half of measuring points, the outdoor temperatures are higher than the data from the weather station. The average WBGT value of three daily used routes is close to 31℃; in this case, people will feel uncomfortable. Due to the lack of greenery and strong

radiation from the underlying surface, the highest point is even more than 6℃. In addition, because of the breakage of green belt, the WBGT values of Routes D and F have greater volatility and there exist extremely bad points for thermal environments. At these measuring points, as the greening is lacking and the thermal environments are poor, the degree of comfort in using the whole route is reduced, affecting the overall environment.

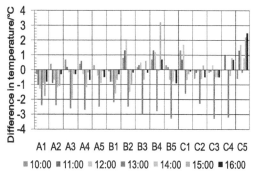

Figure 6. Difference in temperature between daily used routes and weather.

Measuring point	Average WBGT value(℃)	Measuring point	Average WBGT value(℃)	Measuring point	Average WBGT value(℃)
D1	30.21	E1	30.46	F1	29.96
D2	30.40	E2	30.70	F2	29.51
D3	33.09	E3	31.26	F3	29.91
D4	30.53	E4	31.56	F4	32.56
D5	30.56	E5	30.97	F5	30.17
Average value	30.96	Average value	30.99	Average value	30.42

Figure 7. Average WBGT values of daily used routes.

As for thermal environment, these measuring points fall into three kinds: poor, moderate and good. As is shown in Figures 7 and 8, there is almost no shade around Measuring Point D3 and its underlying surface is the concrete floor, so the heat island effect is obvious and the WBGT values are above 31℃. In the afternoon, the WBGT value is even close to 35℃. As a result, there is no continuous good thermal environment along the whole D route, and the degree of comfort is lower for the students using this route for daily study. Along Route E for daily leisure, there are more tall trees, so the shades for measuring points occupy a larger area. In addition, the route is spacious and the wind speed is high. However, as the roads are paved with tiles or bricks, the black ball has higher temperature, indicating that there is a large amount of solar radiation. But the thermal environment along the whole route is more balanced, and the WBGT value is about 31℃. The

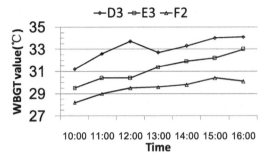

Figure 8. Comparison of WBGT values of typical measuring points in daily.

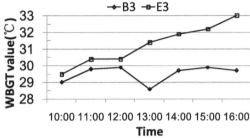

Figure 9. Comparison of measuring points in the rest space and daily used.

thermal environmental conditions along Route F for self-study vary greatly. Measuring Point F2 under Chongzuo Bridge is completely shaded and the average WBGT value is only 29.5℃, so this place is relatively cool and comfortable. But for Measuring Point F4 in front of the library, there is no shade, and the underlying surface is the granite floor with a larger amount of reflection, and the average WBGT value is close to 33℃. It is the worst case in this route, so it brings a strong sense of discomfort to human beings in the summer.

4.3 Comparison of measuring points in the rest space and daily used routes

in order to find the advantages and disadvantages of the rest space and daily used routes, the measuring points in the rest space and daily used routes are compared, exploring whether it is possible to mutually improve their thermal environments. The test results show that the overall thermal environment of rest space is superior to daily used routes. The weather at Measuring Point B3 suddenly becomes cool at 13:00 in the process of measurement, which is shown in Figure 9. For Measuring Point B3 in the rest space, the WBGT value is obviously lower than Measuring Point E3 in daily used routes.

There are shades around Measuring Point B3, and the underlying surface is the tile floor, and B3 is near the artificial lake. Obviously, it is more typical rest space on the campus of Guangxi University. As the shaded area is larger and the microenvironment surrounding the water is cooler, the overall thermal environment of rest space is better. In the summer with a hot climate, it provides the cool and comfortable environment for teachers and students to relax. Measuring Point E3 is often exposed under the blazing sun in the afternoon, and the underlying surface is the tile floor. It is a typical measuring point in daily used routes. Due to the lack of greening or its uneven distribution, for daily used routes, they cannot bring

comfortable thermal environments with overall continuity. In addition, the underlying surface is mostly the material with a large amount of reflection. So the overall thermal environment is in poor condition. Moreover, different areas vary greatly in this aspect. As a result, the comfort in the use of these routes is reduced dramatically.

5 PROPOSALS FOR IMPROVEMENT

In order to improve hot and humid conditions in the summer and create a better thermal environment on campus, Guangxi University can consider the transformation from the following aspects in combination with research results and data analysis:

1 Transform the water loving form of the lotus pond and artificial lake in the rest space, create a water loving space which is more suitable for hot and humid areas in combination with greening and avoid excessive temperature conditions, creating a more beautiful and comfortable open space.

2 Attach importance to uniformity and continuity in greening, improve the role of greening in influencing the environments, and reduce the adverse effects of solar radiation on the thermal environment on campus. Combine the cool and comfortable micro-environment around the water with daily used routes, create an ecological water-loving transportation space through using the revetments of lotus pond and artificial lake, and improve the comfort in daily used routes. Try to use the underlying surface with a small amount of reflection, strong water-holding capacity, and weaker solar radiation, for example, grass-planting brick, avoiding rapid increase in surface temperature caused by direct sunlight in the summer and decrease in comfort of the thermal environment.

6 CONCLUSION

The overall thermal environment on campus is constituted in the form of points, lines, and planes. The measuring points exist in the form of points. Their thermal environments are mutually independent, but interactive, and they are connected in the form of lines. Their interrelation has influence on other surrounding space environments, and they determine the overall thermal environment on campus in the form of the plane. Further, the overall thermal environment on campus is influenced in form of points, lines, and planes.

The thermal environment in the outdoor space on the campus in a hot and humid area has always influenced the daily life of students and teachers. In this study, taking Guangxi University as an example, the thermal environment characteristics of outdoor space on the campus in the summer in a hot and humid area are explored and obtained, providing reference for improvement of thermal environmental conditions and planning and design of green campus in the future.

ACKNOWLEDGMENTS

This paper is supported by 2014-2016 China National College Students Innovation and Entrepreneurship Training Program No.201410593087.

REFERENCES

[1] Rizhong LI. Climate and Greening in Nanning City [J]. Guangxi Meteorology, 1989, 03:57–58.
[2] Meili HUANG & Xuesong HUANG. Analysis of Climate Conditions in Nanning Urban Planning and Construction [A]. China Association for Science and Technology. Improve the Scientific Quality of the Nation, Build an Innovative Country-The Proceedings of Annual Conference for China Association for Science and Technology in 2006 (Volume Two) [C]. China Association for Science and Technology: 2006: 6.
[3] Luoying JIANG, Xianfeng HUANG, Panxi WU, Xiawen MO, Yuyu YANG, Junxin LAN. On Evaluation Method of Thermal Comfort in Gray Space on Campus of Universities — Taking Guangxi University as an Example [J]. Zhonghua Minju, 2014. 08:213–215.
[4] Wenzhen WU & Jian TENG & Chunlin PANG. Analysis and Appraisal of the Plant Landscaping of Guangxi University [J]. Journal of Anhui Agricultural Sciences, 2013, 10:4445–4447 + 4457.
[5] Xiaoyun PENG. Discussion on Thermal Microenvironment of Ecological Campus of College and University [J]. Industrial Construction, 2006, 02: 34–36 + 41.
[6] Lei ZHANG & Qinglin MENG. Measurement and Analysis of Thermal Environment of University Campus in Summer in Hot and Humid Areas [J]. Building Science, 2011, 02:48–51.
[7] Yan ZHANG, Jiang HE, Zhaomen ZENG, Yanwen LUO, Jing ZHAO. Advanced Materials Research. Vols. 671–674 (2013) pp 2414–2419.
[8] Xiong LI, Huiqing DONG, Jiahua HUANG, Junhua LIAO. Characteristics and Forecasting of Temperature on Different Underlying Surfaces [J]. Meteorological Science and Technology, 2005, 06:487–491.
[9] Lin WANG & Yongan LI & Peilei LIU. Research of the Effect Factors on Campus Thermal Environment [J]. Refrigeration &Air Condition (Sichuan), 2008, 04: 110–114.

Energy, Environment and Green Building Materials – Sheng (ed.)
© 2015 Taylor & Francis Group, London, ISBN 978-1-138-02718-3

Study on biogas production of co-digestion of vegetable waste with other materials

R. Feng, J.P. Li, J.Y. Yang & D.D. Zhou
Western China Energy & Environment Research Center, Lanzhou University of Technology, Lanzhou China

ABSTRACT: Anaerobic digestion is a kind of efficient methods of innocent treatment of vegetable wastes, in order to deeply comprehend the effects of co-digestion of vegetable waste (VW) with other materials, the co-digestion tests of mixed vegetable wastes with pig manure, sheep manure, wheat-straw and haulm had been done respectively, under the total solid concentration of 8%, an inoculums concentration of 30%, 37°C fermentation temperature conditions, with different mass ratio. The results showed that the favorable mix ratio of VW with pig manure is 9:1 and 7:3, with sheep manure is 9:1, with haulm is 9:1 and 7:3, VW and wheat-straw was not an ideal choice for co-digestion.

KEYWORDS: Vegetable Waste; Co-digestion; Mixed Ratio; Methane Production Rate

1 INTRODUCTION

With the development of rural economy and continuous improvement of living standards of farmers, the energy demand in rural areas. Biogas occupied an important position in China's rural energy market due to its advantages like convenient and clean (Christiaensen L& Heltberg R 2012). But in recent years, because of the farmer's lifestyle changed and rural industrial structure adjustment, the number of backyard poultry farmers in rural reduced, the source of raw materials for fermentation had been affected (Zhu 2014). However, the plant areas of vegetables increased, lots of vegetable waste (VW) produced. According to estimates, only Lanzhou produces all kinds of vegetable wastes more than 860 thousand tons each year, these wastes, always accumulated in the fields, and then rot, causing serious environmental pollution. Therefore, in combination with the characteristics of high water and organic matter content of VW (Lin 2011), using biogas production systems in rural areas to deal with VW by adopting anaerobic digestion can not only reduce the pollution, but also provide an effective solution to the problems of shortage of raw materials of biogas production systems.

Liu et al(2008) conducted a batch of anaerobic digestion experiments for leaves of cabbage wastes, the feasibility and the effect of different concentration inoculums on anaerobic digestion were studied, the results showed that anaerobic digestion can fit for the characteristics of anaerobic fermentation of VW. They also found that mesophilic fermentation condition was considered suitable for biogas production from cabbage leaves (Liu et al 2009]. But there was a challenging task to VW for fermentation because their high simple sugar content often promotes fast acidification of the biomass with a resulting inhibition of methanogenic bacteria activity (Bouallagui H et al 2005 & Scano E A et al 2014). The technology of co-digestion with VW and other materials could balance the nutrition of slurry. Liu et al (2013) used cauliflower, cabbage, Chinese cabbage, tomato abandoned stem leaves and cow dung as the raw materials of anaerobic fermentation. Beatriz Molinuevo-Salces et al (2013) evaluated in terms of methane yield, Volatile solid removal and lignocellulose material degradation of poultry litter and vegetable processing wastes mixtures, the effect of co-digestion was obviously superior single digestion. Long Wang et al (2014) invested the anaerobic digestion performance of kitchen waste (KW) and fruit/vegetable waste (FVW), the lab-scale experimental results showed that the ratio of FVW to KW at 5:8 present higher methane productivity (0.725 L CH_4/g VS). Yiqing Yao et al (2014)studied the effects of vegetable processing wastes (VPW) and inoculum proportion on water free anaerobic co-digestion of VPW with cattle slurry (CS) for methane production, the results indicated VPW can be co-digested with CS without water addition.

In order to deeply comprehend the effects of co-digestion of vegetable waste (VW) with other materials, the co-digestion tests of mixed vegetable wastes with pig manure (PM), sheep manure (SM), wheat-straw (WS) and haulm (Ha) had been done respectively, under the total solid concentration of 8%, an inoculums concentration of 30%, 37°C fermentation temperature conditions, with different mass ratio. The results could be used for guidance biogas production systems for using and processing vegetable wastes.

2 MATERIALS AND METHODS

The PM, SM and WS were taken from peasant household in Yanjiaping of Lanzhou city, Ha was taken from peasant household near Jingtai, Baiyin City, the VW were taken from the vegetable market in Lanzhou city. The WS and Ha were all smashed less than 2cm. The total solid (TS), volatile solid (VS) of materials is shown in Table 1. Table 2 shows the mass ratio of materials, the VWs mixed with PM, SM, WS and Ha respectively, name group A~D. In each group, the mass ratio was 9:1, 7:3, 5:5, 3:7 and 1:9, denoted by 1~5, then adjusted TS to 8% by changing the water share. The study was conducted in batch mode at lab-scale; the reactor of each digester was 1.5 L. The inoculum was 180mL and the total feed slurry was 600mL. TS were determined by measuring weight loss after heating at 105°C for 24h, VS was determined by measuring weight variation of TS after burning at 550°C for 24h, biogas production was measured by using the method of water displacement, and the methane content was measured by Biogas Check.

Table 1. Physical and chemical properties of substrates.

Materials	TS / %	VS / %
PM	16.39	5.06
SM	91.41	31.26
WS	93.93	25.66
Ha	94.19	25.67
VW	8.15	1.62
Inoculums	2.53	2.06

Table 2. Mass ratio.

Materials	Ratio				
	1 (9:1)	2 (7:3)	3 (5:5)	4 (3:7)	5 (1:9)
A (PM:VW)	A1	A2	A3	A4	A5
B (SM:VW)	B1	B2	B3	B4	B5
C (WS:VW)	C1	C2	C3	C4	C5
D (Ha:VW)	D1	D2	D3	D4	D5

3 RESULTS AND DISCUSSION

Figure 1 shows the daily biogas yield of different materials with different mass ratios. In group A, the accumulated biogas production of 1~5 was 12731 mL,

12896 mL, 5469 mL, and 4547 mL, A1 and A2 were superior than others and good choice for PW and VW co-digestion. In group B, the accumulated biogas production of 1~5 were 11981 mL, 7717 mL, 8308 mL, 8482 mL and 3855 mL, B1 was the best choice. The fermentation period of group C was longer than group A and B, and during the middle time, the daily biogas yields were very little, the accumulated biogas production of 1~5 were 15390 mL, 15476 mL, 13067 mL, 12661 mL and 649 mL, more than group A and B. The fermentation period of group D was similar to group C, and daily biogas yield fluctuate greatly, the accumulated bio gas production of 1~5 were 14603 mL, 13181 mL, 11531 mL, 11535 mL and 8804 mL, D1 and D2 were good for Ha and VW co-digestion.

Figure 2 shows the daily methane content of different materials with different mass ratios. The accumulated methane production of group A1~A5 were 7071 mL, 6962 mL, 3622 mL, 2076 mL and1101 mL, A1 and A2 had the higher methane content of 55.5% and 54.0%. The accumulated methane pro- duction of group B1~B5 were 6578 mL, 3314 mL, 3686 mL, 3786 mL and 1161 mL, B1 had the highest methane content of 54.9%. The accumulated methane production of group

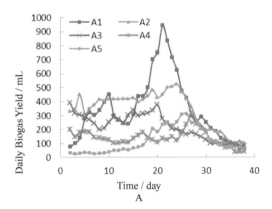

A

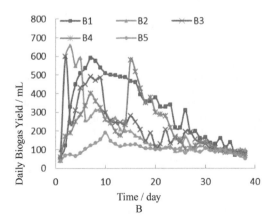

B

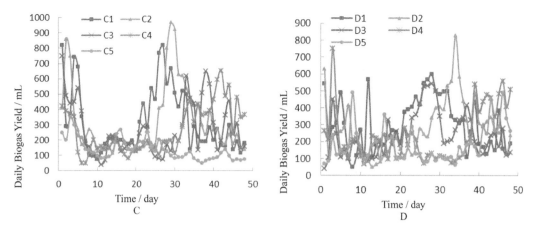

Figure 1.   Daily biogas yield of different materials with different mass ratios.

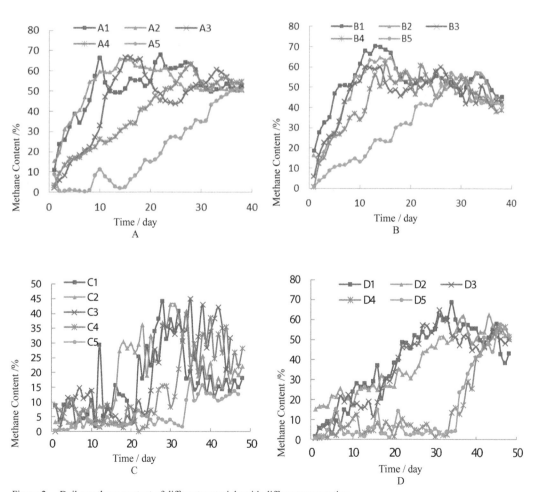

Figure 2.   Daily methane content of different materials with different mass ratios.

C1~C5 were 3154 mL, 3559 mL, 2561 mL, 2478 mL and 2987 mL, these were obviously lower than group A and B, VW and WS didn't suit for co-digestion. The accumulated methane production of group D1~D5 were 5988 mL, 5466 mL, 4786 mL, 2614 mL and 1853 mL, D1 and D2 had the fine methane content of 54.9% in group D. Table 3 shows the average methane production rate of each test, A1, A2, B1, D1 and D2 were favorable ratio for producing methane.

Table 3. The average methane production rate of different materials with different mass ratios (L CH_4 / kg VS).

Materials	Mass ratio				
	1 (9:1)	2 (7:3)	3 (5:5)	4 (3:7)	5 (1:9)
A	457.85	470.05	258.23	159.35	93.98
B	423.67	215.19	242.50	256.09	84.76
C	258.62	292.25	211.37	206.36	255.36
D	490.87	448.73	394.90	217.63	158.38

4 CONCLUSIONS

The co-digestion tests of mixed vegetable wastes with pig manure, sheep manure, wheat-straw and haulm had been done respectively, under the total solid concentration of 8%, an inoculums concentration of 30%, 37°C fermentation temperature conditions, with different mass ratio. The results showed that the favorable mix ratio of VW with PM is 9:1 and 7:3, with SM is 9:1, with Ha is 9:1 and 7:3, VW with WS was not an ideal choice for co-digestion.

ACKNOWLEDGMENTS

This work was funded by the National High Technology Research and Development Program of China (2014AA052801), Funds for Distinguished Young Scientists of Gansu Province (2012GS05601) and "Hongliu Outstanding Talents" Project of Lanzhou University of Technology (Q201101).

REFERENCES

Beatriz Molinuevo-Salces, Xiomar Gómez, Antonio Morán, et al. Anaerobic co-digestion of livestock and vegetable processing wastes: Fibre degradation and digestate stability[J]. Waste Management, 2013, 33:1332–1338.

Bouallagui H, Touhami Y, RBen Cheikh. 2005. Bioreactors performance used in anaerobic digestion of fruit and vegetable wastes: review. Process Biochem, 40:989–995.

Christiaensen L& Heltberg R. 2012. Greening China's Rural Energy: New Insights on the Potential of Smallholder Biogas.

Liu F, Qiu L, Li Z.L, et al. 2013. Gas Characteristics Generated from Vegetables Wastes by Anaerobic Fermentation. Acta Agriculture Boureali-occidentalis Sinica 22(10):162–170. (In Chinese)

Lin J, Zuo J, Gan L.L, et al. 2011. Effects of mixture ratio on anaerobic co-digestion with fruit and vegetable waste and food waste of China. Journal of Environment Sciences, 23 (8):1403–1408.

Liu R.H, Wang Y.Y, Sun, et al. 2008. Experimental study on biogas production from vegetable waste by anaerobic fermentation. Transactions of the Chinese Society of Agricultural Engineering 24(4):209–213. (In Chinese)

Liu R.H, Wang Y.Y, Sun. 2009. Effects of Temperature on Anaerobic Fermentation for Biogas Production from Cabbage Leaves.Transactions of the Chinese Society for Agricultural Machinery 40(9):116–121. (In Chinese)

Scano E.A, Asquer C, Pistis A, et al. 2014. Biogas from anaerobic digestion of fruit and vegetable wastes: Experimental results on pilot-scale and preliminary performance evaluation of a full-scale power plant. Energy Conversion and Management, 77:22–30.

Wang L, Sgeb F, Yuan H.R, et al. 2014. Anaerobic co-digestion of kitchen waste and fruit/vegetable waste: Lab-scale and pilot-scale studie Waste Management, http://dx.doi.org/10.1016/j. wasman.2014.08.005.

Yao Y.Q, Luo Y, YangY.X, et al. 2014. Water free anaerobic co-digestion of vegetable processing waste with cattle slurry for methane production at high total solid content. Energy 74:309–313.

Zhu M. L. 2014. Gas production efficiency of different fermenting materials for household biogas, China Biogas 32(4):62-64. (In Chinese)

Energy, Environment and Green Building Materials – Sheng (ed.)
© *2015 Taylor & Francis Group, London, ISBN 978-1-138-02718-3*

Microstructures and photocatalytic properties of metal ions doped nanocrystalline TiO₂ films

H.Y. Wang & J.M. Yu

School of Chemistry and Chemical Engineering, University of Jinan, Jinan, China

ABSTRACT: To enhance the photocatalytic activity of TiO₂, Fe, Ni, Zn and Cr ions doped TiO₂ film was synthesized by the sol–gel method, respectively. The as-prepared specimens were characterized using X-Ray Diffraction (XRD), high-resolution Field Emission Scanning Electron Microscopy (FE-SEM), Photolumines-cence Spectrum (PL) and UV–vis diffuse reflectance spectroscopy. The photocatalytic activities of the films were evaluated by degradation of Methyl orange solution. The experimental results indicated that the ions doped TiO₂ films were all composed of nanoparticles with the average diameter of ca. 15 nm. Compared with un-doped TiO₂ film, metal ions doped TiO₂ films exhibited more excellent photocatalytic activities under both UV light and visible light. The improvement mechanism by metal ions doped was also discussed.

KEYWORDS: TiO₂ films; metal ions doped; sol-gel; photocatalytic activity

1 INTRODUCTION

Since the water splitting under sunlight irradiation on TiO₂-coated electrodes was reported [Fujishima &Honda 1972], semiconductor-based photocatalysts have attracted considerable attentions for several decades [Li et al. 2011; Hoffmann et al.1995]. Among these semiconductors, TiO₂ is the most widely used photo-catalytic material due to its strong oxidizing power, photo-stability, non-toxicity, chemical and biological inertness, as well as its low-cost [Eshaghi et al. 2011; Bettinelli et al. 2007]. However, the practical application of TiO₂ has been restricted by two main problems. One is of low photocatalytic efficiency due to recombination of photogenerated electrons and holes. The other is of low visible light utilization efficiency due to the wide band gap of TiO₂ (3.2–3.4 eV). To overcome these drawbacks, many approaches were proposed to improve photocatalytic activity of TiO₂ such as various preparations, different carriers and iron doping, etc [Jaiswal et al. 2012; Chang et al. 2011]. Among these methods, metal ions doping is proved to be simple and effective. It has been reported that doping of titania with metal ions can increase photocatalytic performance [Qu et al. 2011; Wang et al. 2008 ; Liu et al. 2013].

In this paper, Fe, Zn, Ni and Cr doped TiO₂ films were prepared by using a simple sol–gel process. The photocatalytic activity was evaluated by photodeg-radation of organic dyes in solution. Compared with the pure TiO₂ film, the degradation rate of Methyl orange under UV irradiation of Fe, Ni, Zn and Cr doped TiO₂ films were increased by 3.07, 2.95, 2.79 and 2.26 times respectively. So the photocatalytic activity of TiO₂ films has improved significantly by doping metal iron. And the mechanism of photoactiv-ity enhancement was also discussed.

2 INTRODUCTION

2.1 *Materials and methods*

Colloidal TiO₂ was prepared by the simple sol–gel method. A total volume of 10 mL TiO(C₄H₉O)₄ was dissolved in 15 mL ethanol absolute under vigorous stirring at room temperature. After 5min, 20 mL diluted nitric acid (0.4 M) was added to the above solution at the speed of 1drop/6 s and kept on stirring till the colloidal suspension could be obtained. Meanwhile 3 mL of acetylacetone was added into the solution. After 16 h aging, the colloidal TiO₂ was obtained with a compo-sition of (C₄H₉O)₄Ti:C₂H₅OH:C₅H₈O₂=1:1.76:0.33 in volume ratio. Colloidal TiO2 films were coated to the surface of glass substrate(25mm×25mm×1mm) by a controllable dip-coating device in an ambient atmosphere with a dipping speed of 2.5 mm s⁻¹. After a repeated dip-coating for different times, the TiO₂ sol-or gel-film with required thickness was obtained. Metal ion doped TiO₂ films were prepared by adding a cer-tain concentration of Ferric nitrate (Fe(NO₃)₃) or other metal ions into the mixture of TiO(C₄H₉O)₄ and ethanol absolute. Finally, the doped TiO₂ gel-films were cal-cined at 450°c in air for 2 h. In our case, the amount of substance of metal ions doping is 0.5 mol and the thickness of TiO₂ films are seven layers.

2.2 *Characteristic test*

The crystalline phase and the crystallite size were measured by Bruker D8 Advance X-ray diffractometer using Cu Kα radiation at 40 kV and 25 mA. The average crystallite sizes were calculated according to the Scherrer equation. Surface morphology of the TiO$_2$ films was detected using high-resolution field emission scanning electron microscopy (FE-SEM) (S-3400N). The PL emission spectra were recorded at room temperature by a FLS 920 spectrometer with a 300 nm line of 450 W Xenon lamp as excitation source. UV–Vis diffuse reflectance spectra of the films were recorded on a UV–Vis spectrophotometer (TU-1901) using blank glass plate as a reference.

2.3 Catalyst test

The photocatalytic activity was evaluated by the degradation of Methyl orange(MO) in aqueous solution under UV-lamp irradiation for 60 min. metal ions doped or undoped TiO$_2$ films were settled in 5 mL MO solution with a concentration of 1.0×10^{-4}g/L. A tungsten halogen lamp equipped with UV cut-off filters(λ>400 nm) was used as avisible light source whose average light intensity was 40mW cm^{-2}, UV–vis Spectrometer (TU-1901) was adopted to assess the photodegradation activity of the film photocatalysts. The degradation rate of photocatalysis can be calculated by formula $D=(A_0-A)/A_0 \times 100\%$, as previously reported [Ma et al.2010].

3 RESULT AND DISCUSSION

3.1 Catalyst characterization

Crystal structure

Figure 1 shows the X-ray diffraction (XRD) patterns of the undoped and ions doped TiO$_2$ samples. The samples were constituted of anatase (25.2°, 37.8°, 48.0°, 54.0°, 55.0°). Note that anatase was the dominant crystal phase in the as-prepared TiO$_2$ samples. According to the Scherrer equation $D = 0.89 \lambda/\beta \cos\theta$, where β is the half-height width

of the diffraction peak of anatase, θ is the diffraction angle, and λ is the X-ray wavelength corresponding to the Cu Kα radiation, the average crystallite sizes of the undoped TiO$_2$, Fe, Cr, Ni and Zn doped TiO$_2$ samples are 18.9, 15.0, 15.6, 16.3 and 14.6 nm, respectively, indicating that metal ions doping hinder the increase of the crystallite size of TiO$_2$. It is well known that the reduction of particles size is favorable to the photocatalytic degradation [Liu et al. 2009].

3.1.1 Surface morphology
Surface morphology of pure TiO$_2$ film and metal ion-doped TiO$_2$ films calcined at 450°c in air for 2 h are shown in Fig. 2. It is clear that the films are composed of round-like nano-particles or aggregates with a size of less than 20nm, and indicating that metal ions doping can effectively prohibit the aggregation of the TiO$_2$ nano-particles, make the surface of the ion-doped film uniform and smooth. A good dispersion or reduced aggregation among particles was expected to increase the active site–reactant contact area, and in turn enhance photocatalytic degradation of organic dyes.

3.1.2 Photoluminescence spectrum (PL)
Photoluminescence emission has been widely used to investigate the efficiency of charge carrier trapping,

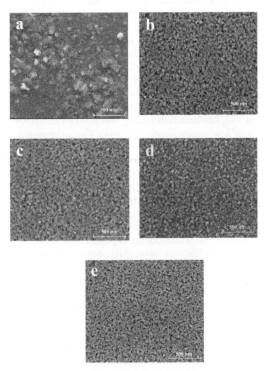

Figure 2. FE-SEM images of (a) pure TiO$_2$ film and (b) Fe doped TiO2 film (c) Zn doped TiO$_2$ film (d) Ni doped TiO$_2$ film (e) Cr doped TiO$_2$ film.

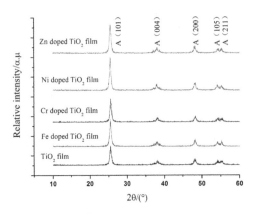

Figure 1. XRD patterns of undoped TiO$_2$ and ion-doped TiO$_2$ samples.

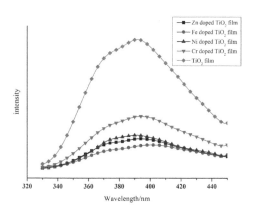

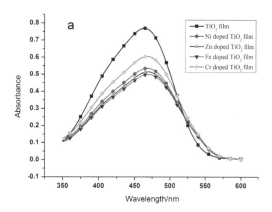

Figure 3. Effect of ions doping methods on PL spectra.

immigration and transfer, and to understand the fate electron-hole pairs in semiconductor particles. The PL emission spectra of the samples were examined in the wavelength range of 330–450 nm in our study, Fig. 3. As can been seen that similar emission peaks at about 390 nm were observed for all of the films, which are assigned to emission of the band gap transition according to the literature reported previously [Zhang et al. 2004]. However, PL intensities is different from each other for the different films. The lowest PL intensity was exhibited from Fe surface doped TiO$_2$ films, indicating that the combination between electron and hole was effectively prohibited compared with pure TiO$_2$ or other doped TiO$_2$.

3.1.3 Photoactivity measurement

Fig. 4 shows UV–vis absorption spectra of the all films and corresponding degradation rate of MO solutions using the Fe, Zn, Ni and Cr doped and undoped TiO$_2$ films on glass substrates, respectively, under UV-lamp irradiation with wavelength of 365 nm for 60 min. As revealed, the degradation rate of MO under UV irradiation with wavelength of 365 nm for 60 min using pure TiO$_2$ film is 14.5%, while that of Fe, Zn, Ni and Cr doped TiO$_2$ films is 44.6%, 42.9%, 40.5% and 32.8%, respectively, under the same conditions. As a result, metal ions doping further improved the photocatalytic activity, and the Fe doped TiO$_2$ film has the best photocatalytic activity under the same conditions. UV–Vis diffuse reflection spectra of pure TiO$_2$ and metal ions doped TiO$_2$ films are shown in Fig. 5. Compared with that of pure TiO$_2$ film at about 370 nm, the absorbance onset of Fe doped TiO$_2$ film red-shifted obviously to about 390 nm, while no apparent red-shift was observed for other metal ions doped films. Consequently, more photo generated carriers can join in the degradation reaction, which may lead to enhanced photocatalytic activity of the metal ions doped TiO$_2$ film under both UV light and visible light irradiations.

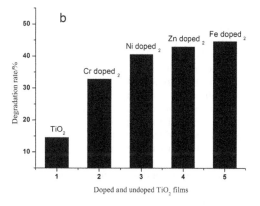

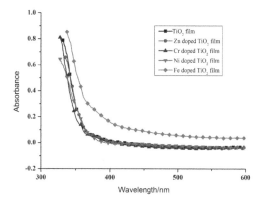

Figure 4. UV–Vis absorption spectra (a) and degradation rate (b) of MO solutions using an appropriate amount of Fe, Zn, Cr Ni doped and undoped TiO$_2$ films on glass substrates after irradiated for 60 min.

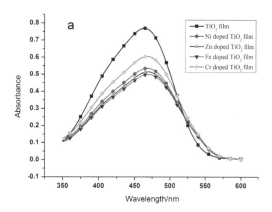

Figure 5. Diffuse reflection UV–Vis spectra of pure TiO$_2$ film and metal ions doped TiO$_2$ film.

Based on the above analyses, Fe doped TiO$_2$ film shows higher photocatalytic activity compared with other metal ions doped TiO$_2$ films. This not only depends on the microstructure of the film such as surface morphology, grain size and distribution, surface area, etc. but also relates to the special electronic structure of dopant Fe^{3+} ($3d^5$). Fe^{3+} is a transition metal ion which d orbits are half full of electronic configuration. When captured electron, stable structure will be broken. Therefore, the captured electrons would easily be released to form a shallow potential capture trap, and thus the life of electron-hole pairs was prolonged. Consequently, the quantum efficiency of TiO_2 should be increased, which further improved photocatalytic activity of the films. In addition, the Fe^{3+} doped into TiO_2 instead of the lattice positions of Ti^{4+}, so TiO_2 lattice will lack of electrons. In order to balance the electrovalence, the oxygen vacancy will be formed inevitably. At the same time, Ti^{4+} was reduced to Ti^{3+}, and the formation of oxygen vacancies makes the symmetry of crystal lattice structure reduced. And with the formation of each oxygen vacancy, TiO_6 octahedrons will reduce four O-O edges shared by adjacent octahedrons, which should be negative contribute to form the co-edged rutile structure with higher symmetry, but positive contribute to the stability of anatase titanium dioxide. Due to the different of the ionic radius, local lattice distortion will be produced, and the stress field formed by this distortion will prevent strongly the mobility of the grain boundary. For compensating this lattice stress, oxygen atoms of the TiO_2 lattice surface are easy to escape the lattice, which can inhibit the transformation of crystalline phase and the growth of the crystalline grain of TiO_2. Both the reduction of particles size and the inhibit the phase transformation of TiO_2 in the solid are achieved, which would be favorable to the photocatalytic degradation[Han et al. 2009; Chen et al. 2010; Liu.2004; Liu.2010].

4 CONCLUSION

Fe, Zn, Ni and Cr doped TiO_2 films were prepared by using a simple sol–gel process. The photocatalytic activity was evaluated by photodegradation of organic dyes in solution. Compared with the pure TiO_2 film, the degradation rate of Methyl orange under UV irradiation of Fe, Ni, Zn and Cr doped TiO_2 films were increased by 3.07, 2.95, 2.79 and 2.26 times respectively.among them Fe doped TiO_2 film shows higher photocatalytic activity compared with other metal ions doped TiO_2 films. The present results would be helpful for the design and development of efficient photocatalytic reactors.

ACKNOWLEDGMENT

This work was financially supported partially by Natural Science Foundation of University of Jinan (XKY1024).

REFERENCES

Bettinelli, M. & Dallacasa,V. et al. 2007. Photocatalytic activity of TiO_2 doped with boron and vanadium [J]. *Journal of Hazardous Materials,* 146:529–534.

Chang, S.M. & Liu, W.S. 2011.Surface doping is more beneficial than bulk doping to the photocatalytic activity of vanadium-doped TiO_2[J].*Applied Catalysis B* 101:333–342.

Chen, C.C., Ma, W.H., Zhao, J.C. 2010. Semiconductor mediated photo-degradation of pollutants under visible–light irradiation[J]. *Chemical Society Reviews* 39:4206–4219.

Eshaghi, A., Mozaffarinia, R., Pakshir, M. & Eshaghi, A. 2011. Photocatalytic properties of TiO_2 sol–gel modified nanocomposite films[J]. *Ceramics International,* 37:327–331.

Fujishima, A. & Honda, K. 1972. Electrochemical photolysis of water at a semiconductor electrode [J]. *Nature,* 238(5338):37–38.

Hoffmann, M.R. & Martin, S.T. 1995. Environmental of semiconductor photocatalysis[J]. *Chemical Reviews,* 95:735–758.

Han, F., Kambala, V.S. R. et al. 2009. Tailored titanium dioxide photocatalysts for the degradation of organic dyes in wastewater treatment[J]. *Applied Catalysis A,* 359: 25–40.

Jaiswal, R. Patel, N., Kothari, D.C. & Miotello, A. 2012. Improved visible light photocatalytic activity of TiO_2 co-doped with vanadium and nitrogen[J]. *Applied Catalysis B* 126:47–54.

Li, P. & Wei, Z. 2011. Au-ZnO hybrid nanopyramids and their photocatalytic properties[J]. *Journal of the American Chemical Society,* 133(15):5660–5663.

Liu, C.C. 2010. Preparation, Modification of TiO_2 films and theirs photocatalytic properties[D].*College of Materials Science and Engineering Zhejiang University,* 3:40.

Liu, Q.H. 2004. Research of Self-cleaning Titanium Dioxide Films by Sol-gel[D].College of Materials Science and Chemical Engineering Zhejiang University, 3:72.

Liu, G., Wang, X. et al. (2009) The role of crystal phase in determining photocatalytic activity of nitrogen doped TiO_2[J]. *Journal of Colloid and Interface Science,* 15: 331–338.

Liu, Z.H. et al. 2013.Band Structure of Metal Ions Doped Modified Nano-TiO_2 and its Photocatalytic Performance[J]. *Journal of the Chinese ceramic society,* 41:402–408.

Ma, X., Wang, X.M. & Deng, Y.X. 2010. Preparation of TiO_2/Diatomite Composite and Its Photocatalytic Activity[J]. *Non-Metallic Mines,* 33:72–77.

Qu, Y.Z. & Yao, M.M. et al.2011. Microstructures and photo-catalytic properties of Fe^{3+}/Ce^{3+} codoped nanocrystal line TiO_2 films[J]. *Water,Air,and Soil Pollution* 221:13–21.

Wang, J., Y. Liu, Z.H. & Cai, R.X. 2008. A new role for Fe^{3+} in TiO_2 hydrosol: accelerated photodegradation of dyes under visible light[J]. *Environmental Science & Technology,* 42:5759–5764.

Zhang, Z.Y., Si, M.S., Wang, Y.H. et al. 2014. Indirect-direct band gap transition through electric tuning in bilayer MoS_2 [J] *J Chem Phys.*17:174707.

Energy, Environment and Green Building Materials – Sheng (ed.)
© 2015 Taylor & Francis Group, London, ISBN 978-1-138-02718-3

Numerical simulation of flow-field coupling with the six degree of freedom topology-changeable motion

Feng–Bo Yang, Da–Wei Ma & Qian–Qian Xia
School of Mechanical Engineering, Nanjing University of Science and Technology, Nanjing, China

ABSTRACT: Based on the ALE (Arbitrary Lagrangian-Eulerian) method, 2nd-order AUSM scheme, and dynamic mesh technology of combining spring-based smoothing method and local remeshing-based zone-moving the method, the simultaneous solving method for hydrodynamic equations and six degrees of freedom motion equations is realized to solve the loosely coupled problem of dynamic flow field coupling with six degrees of freedom relative motion of multi-body. The complicated supersonic jet with one kind of coming stream condition is simulated numerically using the second order AUSM scheme, agreement between the wave structures and experimental schlieren photograph are fairly satisfactory which indicates the fluid dynamics numerical method is reliable. The numerical model of the dynamic separation for a penetrator is established under the consideration of gravity, aerodynamic force, overturning moment, and variable topology movement. The numerical experiments are performed.

KEYWORDS: computational fluid dynamics; six degree of freedom topology-changeable; dynamic mesh; numerical simulation.

1 INTRODUCTION

With the rapid development of high performance computing and numerical algorithms, CFD is becoming a highly effective alternative to acquire more accurate solutions for the complex flow field problems [1–3]. It has been applied to industrial automation, aeronautics, astronautics, and weapon scopes, While the 6DOF problem is international hotspot issue. The AUSM scheme [4–6] is an effective method to solve the complex wave structure such as shear layer, expansion waves, incident shock, mach shock disk and so on. In this paper, the gas jet problem [7] is firstly computed to verificated the accuracy of AUSM scheme in the capturing incident shock, mach shock and so on. Then we give a method to solve the 6DOF problem. Based on the 2nd AUSM scheme, dynamic mesh and mesh reconstruction technology, the Armour-piercing fin-stabilized discarding sabot [8–9] dynamic operating model is established. The numerical simulation of flow field coupling with the six degrees of freedom topology-changeable motion is conducted, and the numerical results are in good agreement with Schlieren picture.

2 NUMERICAL METHODS

2.1 Computational fluid dynamics equation

The governing equations for the compressible flow are continuity, momentum, and energy equations.

The Arbitrary Lagrangian-Eulerian (ALE) from considering control cell moving can be expressed as

$$\frac{\partial}{\partial t} \iiint_{\Omega} Q dV + \oiint_{\partial \Omega} F_c\left(Q, \dot{\mathbf{x}}\right) \cdot \vec{n} dS = 0 \qquad (1)$$

where, Ω is control body, $\partial \Omega$ is control border,

$$F_c\big|_x = \left[\rho U, \rho u U + p, \rho v U, \rho w U, (e+p)U + x_t p\right]$$

$$F_c\big|_y = \left[\rho V, \rho u V, \rho v V + p, \rho w V, (e+p)V + y_t p\right]$$

$$F_c\big|_z = \left[\rho W, \rho u W, \rho v W, \rho w W + p, (e+p)W + z_t p\right]$$

$$Q = \left[\rho, \rho u, \rho v, \rho w\right], \ U = u - x_t, \ V = v - y_t,$$

$W = w - z_t$, x_t, y_t, and z_t are the velocities of mesh at different direction, γ is the specific heat capacity,

$$e = \frac{p}{\gamma - 1} + \rho\left(u^2 + v^2 + w^2\right).$$

All the above formulae constitute the closed system of governing equations. FLUENT employs a cell centered finite volume method to solve them in integral form. The cell centered finite volume is based on the linear reconstruction scheme, which allows the use of computational elements with arbitrary polyhedral topology. A point implicit (block Gauss-Seidel) linear equation solver is used in conjunction with an algebraic multigrid (AMG) method to solve all dependent variables in each cell. A second order implicit AUSM scheme is employed for the convection term. The detailed formulations and methods are available in Ref. [10].

2.2 6DOF rigid body dynamics equations

In the inertial coordinate, according to Newton's second law, the motion equation of rigid body centroid can be expressed as

$$m\frac{dv_G}{dt} = F_G \qquad (2)$$

$$x_G = v_G t + \frac{1}{2}\frac{dv_G}{dt}t^2 \qquad (3)$$

where m is rigid quality, v_G and x_G are the centroid velocity vector and spatial location vector respectively, t is the time variable.

In the non-inertial coordinate, according to Newton's second law, the motion equation of rigid body centroid can be expressed as

$$\frac{\delta H_B}{\delta t} + \omega_B \times H = M_B \qquad (4)$$

where ω_B is the angular velocity of rigid body, H_B and M_B denote the moment of momentum and torque respectively in the non-inertial coordinate.

The relationship of moment of momentum, moment of inertia and angular velocity can be expressed as

$$H_B = I.\omega_B \qquad (5)$$

In the non-inertial coordinate, the kinetic equation of rigid body rotating around the centroid can be expressed as

$$\dot{\omega}_B = -I^{-1}\cdot\omega_B\times(I\cdot\omega_B)+I^{-1}\cdot M_B \qquad (6)$$

$$\begin{cases} M_B = LM_G \\ L \equiv \begin{bmatrix} C_\beta C_\gamma & C_\beta S_\gamma & -S_\beta \\ S_\alpha S_\beta C_\gamma - C_\alpha S_\gamma & S_\alpha S_\beta S_\gamma & S_\alpha C_\gamma \\ C_\alpha S_\beta C_\gamma + S_\alpha S_\gamma & C_\alpha S_\beta S_\gamma - S_\alpha C_\gamma & C_\alpha C_\beta \end{bmatrix} \end{cases}$$

where, $C_x = \cos(x)$: $S_x = \sin(x)$: α, β, γ are the Euler angles that represent rotation about x-axis, y-axis and z-axis, respectively.

In order to obtain the trajectory and attitude of rigid body, the Euler motion equation (7) should be solved

$$\begin{bmatrix} \dot\alpha \\ \dot\beta \\ \dot\gamma \end{bmatrix} = \begin{bmatrix} 0 & \sin\gamma & \cos\gamma \\ 0 & \cos\gamma/\cos\alpha & -\sin\gamma/\cos\alpha \\ 1 & -\tan\alpha\cos\gamma & \tan\alpha\sin\gamma \end{bmatrix}\begin{bmatrix} \omega_x \\ \omega_y \\ \omega_z \end{bmatrix} \qquad (7)$$

Once the angular velocities are computed, the angle change rates are determined using a fourth-order multi-point Adams-Moulton formulation

$$\xi^{t+1} = \xi^t + \frac{\Delta t}{24}\left(9\dot\xi^{t+1} + 19\dot\xi^t - 5\dot\xi^{t-1} + \dot\xi^{t-2}\right) \qquad (8)$$

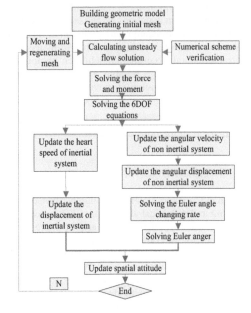

Figure 1. The 6DOF simulating procedures.

3 CASE VERIFICATION FOR AUSM SCHEME

In order to verify the accuracy of numerical format, the numerical experiment is done compared with the reference [7] in the same calculation condition.

The density contours (Fig. 2a), pressure contours (Fig. 2b), the experimental picture under the same flow condition (Fig. 2c) of supersonic gas jet field is shown in Fig2.

As is shown in Fig. 2, the supersonic external flow compresses the nozzle jet in the corner, two oblique shocks, jet shocks and contact discontinuities between the jet shock and oblique shocks occurs. Behind the first oblique shock, there follows the expansion wave while under the second one follows the jet shock. The jet expansion area of the nozzle is compressed by the external flow and reflection occurs when the area

480

reaches the central axis, the reflection shock intersects with the contact discontinuity, the mach disk in the jet structure disappears.

The characteristic of the flow field matches well with the experimental picture in the same flow field condition, so the numerical results obtained by the 2nd AUSM scheme method are credible.

4 6DOF RIGID BODY SEPARATION CASE

With the 2nd AUSM scheme method which could be accurate enough to capture shock, the 6DOF rigid body separation process is simulated. Armor-piercing fin-stabilized discarding sabot consists of bullet and

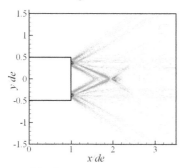

(a) Contours of density

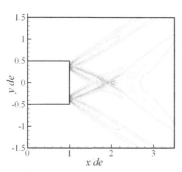

(b) Contours of pressure

(c) Schlieren photograph [7]

Figure 2. Calculation results and Schlieren photograph of supersonic jet.

three discarding sabots, the three discarding sabots separating process is a 6DOF problem. The geometric model is reviewed in Ref. [11–13].

Fig. 3 shows the geometric configuration and the global coordinate system. The centerline of the projectile lies along the x axis with the positive direction toward the projectile tip, y axis points upward along the negative direction of the gravity and z axis is determined by the right hand rule. The origin of the coordinate system is located at the center of the rear circle surface of the projectile.

Before the rigid body separating, the choked flow is the dominant feature since the gaps among the sabots and projectile are not large enough to accommodate the incoming mass flow rate captured by the sabot front scoops. The interaction is characterized by a single detached normal shock (bow shock) formed

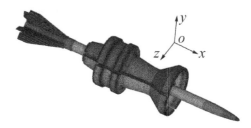

Figure 3. Surface mesh of projectile and sabot.

Pressure/atm

0.50 1.22 1.94 2.66 3.38 4.10 4.82 5.54 6.26 6.98 7.71 8.43 9.15 9.87 10,5911.31

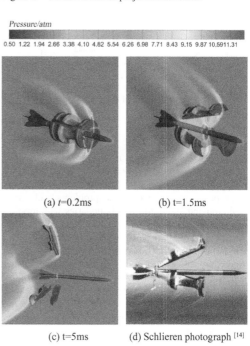

(a) t=0.2ms   (b) t=1.5ms

(c) t=5ms   (d) Schlieren photograph [14]

Figure 4. Pressure distribution compared to Schlieren photograph.

just upstream of the sabot petals and essentially a stagnant volume of air within an annular plenum chamber bounded by the front scoops of the sabot petals, as shown in Fig. 4a.

After the sabots begin to separate, the flow structure is characterized by multiple bow shocks and oblique shocks. The reflected shocks induced by the impinging of the oblique shocks on the projectile can be observed (Fig. 4b).

As mentioned in Phase 2, when the area among the sabots and projectile becomes sufficiently large, the flow structure for each sabot is characterized by its bow shock (Fig. 4c). The sabots rotate and translate independently under the action of aerodynamic forces. Fig. 4d shows an experimental color Schlieren picture of Ref. [14]. It can be clearly seen that our flow structures around the sabots and projectile agree well with it, the Fig. 5 also illustrates the separation process of three sabots, we can see that the dynamic separation process match the laws of physics.

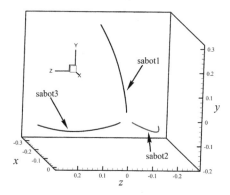

Figure 5.   Separation track of three sabots.

5   CONCLUSION

In this paper, we provide a method to solve the topology-changeable 6DOF problem. The topology-changeable dynamic adaptive grid generation technique is proposed for simulating the problems with large-scale moving and deforming boundaries with topological changes in geometry, in connection with Spring Analogy Method and local grid regeneration method, which has been confirmed to be robust and efficient. According to the numerical simulation of complicated supersonic jet with coming stream, the accuracy of 2nd AUSM scheme which is adopted in this paper is reliable to capture incident shock, λ shock, reflected shock and so on. Based on topology-changeable dynamic adaptive grid technique and the 2nd AUSM scheme, the 6DOF separation procession of Armour-piercing fin-stabilized discarding sabot is simulated. The dynamic analysis shows that the simulation separation process is consistent with the laws of physics.

ACKNOWLEDGMENT

The authors appreciate the support from the defense basic research project foundation of China (NO. B2620110005).

REFERENCES

[1]   H. Zhang, Z. Chen, X. Jiang and H. Li, Investigations on the exterior flow field and the efficiency of the muzzle brake, *Journal of Mechanical Science and Technology*, 27(1) (2013) 95–101.

[2]   R. Gopalapillai, H. D. Kim, T. Setoguchhi and S. Matsuo, On the near-field aerodynamics of a projectile launched from a ballistic range, *Journal of Mechanical Science and Technology*, 21(7) (2007) 1129–1138.

[3]   H. Rehman, S.H. Hwang, B. Fajar, H. Chung, and H. Jeong, Analysis and attenuation of impulsive sound pressure in large caliber weapon during muzzle blast. *Journal of Mechanical Science and Technology*, 25(10) (2011) 2601–2606.

[4]   Meng-Sing Liou and Christopher J. Steffen. JR "A new Flux Splitting Scheme", Joumal of Computational Physics, 1993, 107: 23–39.

[5]   Liou M S. Progress Towards an Improved CFD Methods: AUSM+[R]. AIAA-95-1701-CP, 1995.

[6]   Kim K. H., Rho O H. An important of AUSM Scheme by Introducing the Pressure-Based weight Functions[J] Computers Fluids,1998,27 (3):38–80.

[7]   Agrell J, White.R A. An experiment investigation of supersonic axisy-mmetric flow over boa- ttails containing a centered propulsive jet[R]. 8FFA-TN-AU-913, 1974.

[8]   N. P. Bhange, A. Sen and A. K. Ghosh, Technique to improve precision of kinetic energy projectiles through motion Study, *AIAA Atmospheric Flight Mechanics Conference*. Chicago, Illinois, USA (2009) 1–33.

[9]   E. M. Schmidt, Wind tunnel measurements of sabot discard aerodynamics, *Journal of Spacecraft and Rockets*, 18 (3) (1981) 235–240.

[10]   S. E. Kim, S. R. Mathur, J. Y. Murthy and D. Choudhury, A Reynolds-Averaged Navier-Stokes solver using unstructured mesh-based finite-volume scheme.*36th Aerospace Sciences Meeting and Exhibit*, Reno, NV, USA(1998) 1–8.

[11]   M. J. Nusca, Numerical simulation of sabot discard aerodynamics, *AIAA 9th Applied Aerodynamics Conference*, Baltimore, Maryland, USA, (1991) 423–432.

[12]   K. R. Heavey, J. Despirito and J. Sahu, Computational fluid dynamics flow field solutions for a kinetic energy (KE) projectile with sabot, ARL-MR-572, Army Research Laboratory, Aberdeen Proving Ground, MD, USA (2003).

[13]   M. J. Guillot, R. Subramanian and W. G. Reinecke, A numerical and experimental investigation of sabot separation dynamics, *34th Aerospace Sciences Meeting and Exhibit*, Reno, NV, USA (1995) 1–8.

[14]   Bbs.tiexue.net: Armour Piercing [EB/OL], http://bbs .tiexue.net/post_7005833_1.html.

Energy, Environment and Green Building Materials – Sheng (ed.)
© *2015 Taylor & Francis Group, London, ISBN 978-1-138-02718-3*

Natural conditions suitability analysis of renewable energy building application: Case study of hot-humid climate nanning

Da-Yao Li
College of Civil Engineering and Architecture, Guangxi University, China

Jiang He
Key Laboratory of Disaster Prevention and Structural Safety of Ministry of Education, Guangxi University, China
College of Civil Engineering and Architecture, Guangxi University, China

Xi Xu
College of Civil Engineering and Architecture, Guangxi University, China

Yan-Qing Li
Qinzhou College, Qinzhou, Guangxi, China

ABSTRACT: Factors affecting the renewable energy, specifically local climate conditions and natural resource conditions can be called the most important one. If we want to achieve energy savings in each area under the complex natural conditions and different geographical conditions, we must be adapted to local conditions and adopt different ways of renewable energy application technology in buildings. In this paper, take Hot-Humid Climate Nanning city, for example, doing natural conditions, suitability analysis of renewable energy's building application in Nanning city. Achieving the technology strategy of solar energy resources, surface water resources, land, shallow rock geothermal resources, refer to other similar areas.

KEYWORDS: Natural Resource Conditions; Renewable Energy; Building; Hot-Humid Climate

1 INTRODUCTION

With the rapid development of National urbanization and new industrialization, building energy consumption shows rapid growth, and leads to enormous pressure to the energy supply. Especially the outbreak of the 1970s oil crisis, let the world realize the immediacy of seeking alternative energy, making for the development of renewable energy to become the world's development trend, driving the world to research the new energy.

The use of renewable energy in buildings, mainly the use of solar energy and geothermal energy at present, Specific application should be combined with local climate conditions and natural resource conditions.

2 CLASSIFICATION OF DEVELOPING RENEWABLE ENERGY BUILDING APPROPRIATE TECHNOLOGY NATURAL CONDITIONS IN HOT-HUMID CLIMATE

China has a vast geographical environment so climatic conditions vary widely. According to the Chinese building climate division (Fig. 1), the main area is divided into a very cold area, cold area, hot

Figure 1. China's building climate zoning standards.

summer and cold winter area, hot summer and warm winter area, moderate climate area. Renewable energy technology is not only related to the local climate conditions, but also the natural resource conditions as well. If we want to achieve energy savings in each area under the complex natural conditions and different geographical conditions, we must be adapted to local conditions and adopt different ways of renewable energy application technology in buildings.

Take solar energy resources, for example, China's solar resource distribution has great regional differences (Fig. 2). Solar resource distribution is restricted by the local climate conditions and the natural resource conditions, according to the total amount of year solar radiation broadly divided into four areas. However, the use of solar energy resources not only related to the amount of solar radiation, but also has a close relationship with the local use itself. Reasonable measures can make full use of the local natural resources.

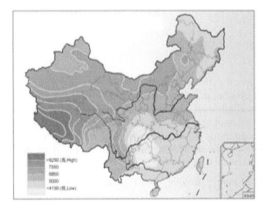

Figure 2.   China's solar energy resources division map.

Developed country cities such as Moscow, London and UK, which solar energy resources amount similar to III and IV area in our country, provides a good example for us.

Table 1.   China's solar energy resources division.

Area level	Annual sun shine hours (h)	Total annual radiation (MJ/m²)	Similar to the foreign area	Note
I	2800–3300	≥6700	India, In northern Pakistan	very rich area
II	3000–3200	5400–6700	Washington. DC	rich area
III	2200–3000	4200–5400	Milan	general area
IV	1400–2200	<4200	Paris, Moscow	poor area

3   NANNING DEVELOPMENT OF RENEWABLE ENERGY APPLICATION TECHNOLOGY ARCHITECTURE ANALYSIS AND INVESTIGATION OF NATURAL CONDITIONS

Nanning is a typical hot and humid weather, climate city with its year-round humidity, summer is long, hot and rainy; winter is short, warm and dry. The annual average sunshine time is 1640.1h; the annual average temperature is 21.6°C; the average relative humidity of 79%; rainfall abundant, with an average annual rainfall of 1173.00mm.

According to the total annual solar radiation, most of Nanning belongs to an IV area (resource-poor area, the average annual total radiation under 4200 MJ/ m²), so Nanning belongs to the solar energy available area. The city built around the river, its rivers, lakes and reservoirs in larger quantities, with a wealth of surface water and groundwater resources. Also urban sewage pipe network construction has also put in place.

According to the natural conditions, Nanning has the best conditions of the application of renewable energy technologies (such as: ground source heat pump technology, ground source heat pump technology, ground source heat pump technology, water source heat pump technology, solar thermal technology, solar photovoltaic technology, etc.). But the wind, biomass and ocean energy, etc. are not suitable in Nanning.

The available renewable energy sources in Nanning are mainly solar, geothermal. Among them, the use of solar energy is mainly in solar thermal systems and solar photovoltaic systems; shallow geothermal energy use in ground source heat pump system, surface water (river, lake) water source heat pump system, and water source heat pump system is based. The following analysis focuses on solar energy resources in Nanning and shallow geothermal resources.

3.1   Solar energy resources

Looking at the total amount of solar radiation in the past years, Nanning belongs of solar energy resources available area only; however, the solar resource of Nanning has its unique application mode. From the table below, the total amount of solar radiation distribution of Nanning is the most in summer, followed by spring and autumn, the least in winter.

From the advantages, first of all, according to the monthly amount of solar radiation in Nanning, the largest proportion of the amount of radiation is summer. Also, the summer is long in Nanning, more concentrated solar energy resources too, so summer is the best time for solar applications.

Second, it's also a good fit with the hot-humid climate. In summer, a large use of the air-conditioning, makes the use of electricity relatively heavy. Solar-thermal systems and its application in the optoelectronic system in hot-humid climate can effectively relieve the electricity shortages. The practice of renewable energy practice in Nanning proved this point as well.

From the disadvantages,

Firstly, according to the amount of radiation in the winter, more cloudiness, but less sunshine throughout the winter, so its radiation of solar energy resources is generally low in winter.

Secondly, the solar energy resource in Nanning winter is very unstable, especially in January and February, cloudy weather appeared multiple times for a week or so.

Solar energy resources in Nanning are intermittent solar energy resources, because its winter solar radiation is less and the solar energy resources is instability, so the city generally needs to add an auxiliary heat source for getting enough heat.

But the design of solar collectors are generally not combined with the local climate, it usually takes the average annual value or experience value to design at present, resulting in improper selection. The proper design of the solar collector should be combined with the local monthly amount of radiation, taking into account each month. And the auxiliary heat source should be chosen in connection with project's size, power consumption, the number of building space volume of the auxiliary heat source account, pollution, risk, price and the like.

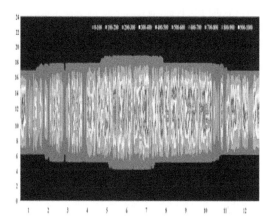

Figure 3.   Nanning's solar energy resources.

3.2   Land shallow rock geothermal resources

3.2.1   Surface water resources

From the natural and geographical conditions, Nanning city is very rich in water resources.

First, Nanning is located in the subtropical regions; it has abundant rainfall, with an average annual rainfall of 1173.00mm, 1327mm, and average annual rainfall of 1327mm for many years.

Second, Nanning has many rivers; there are 39 domain catchment areas over 200 square kilometers. Yongjiang, the largest river in Nanning, full length of 116.4Km, and its river section of Nanning City bed width of 485m, depths of 21m, average surface width of 307m, dry season water depth 8m–9m.

Furthermore, there is an abundant groundwater resource in Nanning, according to the investigation and analysis of groundwater, the average groundwater modulus for many years in Nanning City area is 111,000 cubic meters per square kilometer, the average of shallow groundwater resources recharge for years is 2.5 billion cubic meters. According to the related statistical result, in 2004, water availability of the mainstream in Yongjiang River is only 37%, indicating that there is a large space for development of water resource utilization in Nanning City.

However, water resources utilization status of Nanning city also exists in time and space are not distributed uniformly, which difficult for using water resources. Moreover, the population and urban water consumption are also rising. We should think about that how to improve the utilization of water resources in the wet period, and how to use of water resources reasonably and appropriately in the dry season, which can achieve sustainable development of water resources.

3.2.2   Land shallow rock geothermal resources

In China, the ground source heat pump mainly uses in the hot summer and cold winter area. However, these areas generally have high humidity. Air-cooled evaporator, air source heat pump can easily frost in low temperatures and high humidity weather conditions which causing the effect of heating system is not good in winter. Also, because the ground source heat pump does not heat the air to avoid frost phenomenon, the southern area has high air humidity, the soil thermal parameters will be relatively stable. So the shallow soil temperatures are more stable, and therefore, shallow ground source heat pump is particularly suitable for humid southern regions.

On the temperature, the Nanning has its unique conditions for the development of land in shallow rocks in the shallow hot rock geothermal resources of land. First, Nanning warmer weather, winter is short and high temperatures (average annual temperature is higher than 15°C). Second, Nanning shallow soil temperature about 20°C (close to the annual average ambient temperature). Furthermore, about half the number of days in a year, more than the ambient temperature is higher than 20°C, such as the use of low-temperature heat source heat pump air source provides good conditions. Finally, more than half the number of days the ambient temperature is higher than 12°C in Nanning winter.

Ground source heat pump system development and utilization is closely related to the engineering, geological conditions of the location, the thermal conductivity of rock and soil directly determines the GSHP conditions for development in the region. On the thermal conductivity of the rock layers of view, based on an analysis of Geological Prospecting Bureau of Nanning geological structure of the main city, Nanning shallow surface geological structure mainly miscellaneous fill soil and clay layers, the lower layer of gravel and mainly in the form of mudstone. Miscellaneous fill soil is poor thermal properties, but the thickness is generally not more than eight meters. Average thermal conductivity mostly in the lower part is 2.0 ~ 2.55 w / m°C which is good thermal properties.

4 THE TECHNOLOGY STRATEGY OF RENEWABLE ENERGY TO BE PROMOTING THE APPLICATION IN NANNING

Nanning promotes the use of renewable energy should be taken the following aspects into consideration.

Make full use of solar energy resources in summer and make it a good fit with the hot-humid climate.

Winter application of solar energy resources should be combined with auxiliary heat source or other renewable energy sources. The auxiliary heat source should be chosen in connection with project's size, power consumption, the number of building space volume of the auxiliary heat source account, pollution, risk, price and the like.

Improve the utilization of water resources in the wet season, and use water resources reasonably and appropriately in the dry season, achieve sustainable development of water resources.

Land shallow geothermal hot rock resources should comprehensive analysis evaluated from the land area of geotechnical aspects of each layer, energy efficiency economic and environmental benefits.

5 PROSPECTS

Renewable energy technology is a technology system combines a multi-disciplinary knowledge, many factors affecting the performance of renewable energy technologies. Reliable application of these key technologies designers first need to master knowledge and technology for renewable energy applications, the overall co-ordination, in-depth analysis, reasonable design, in order to lay a solid foundation for the subsequent construction and operation.

Factors affecting the renewable energy, specifically local climate conditions and natural resource conditions can be called the most important one. Therefore, it requires a combination of specific climatic conditions and natural resource conditions around the country to carry out research and analysis based on renewable energy technologies in the practical application of the country, summed up in the practical application of the applicability and advantages and disadvantages of various types of renewable energy systems, which made for in the local development of various renewable technology system and complete technical design application route energy systems. Natural conditions' suitability analysis of renewable energy's building application refers to other similar areas.

ACKNOWLEDGEMENTS

Supported by the Systematic Project of Guangxi Key Laboratory of Disaster Prevention and Structural Safety (2104ZDX05).

REFERENCES

[1] Chedid RB. Policy development for solar water heaters: the case of Lebanon. Energy Conversion and Management 2002; 43:77–86.
[2] Sakkal F, Ghaddar N, Diab J. Solar collectors for Beirut climate. Applied Energy 1993; 45:313–25.
[3] ESCWA. Regional approach for disseminating renewable enrgy technologies. Part I: The regional renewable energy profile. United Nations publications; 2001 [E/ESCA/ENR/2001/10/(Part I)].
[4] Biyuan Li , Sustainable Water Resources Development and Utilization of Nanning. Renminzhujiang, 2008(01): 10–11, In Chinese.
[5] Irfan A G, Mukhtar H S. Review of Modelling Tools for Integrated Renewable Hydrogen Systems[Z]. Singapore: 20115.
[6] Campoccia A, Dusonchet L. Comparative analysis of different supporting measures for the production of electrical energy by solar PV and Wind systems: Four representative European cases [Z]. 2009: 83, 287–297.
[7] Association E W E. The European offshore wind industry – key trends and statistics 2013[Z]. 2013.
[8] Pesnell W D. The Solar Dynamics Observatory (SDO) [Z]. 2012: 275, 3–15.

Energy, Environment and Green Building Materials – Sheng (ed.)
© 2015 Taylor & Francis Group, London, ISBN 978-1-138-02718-3

Leisure space of hospital building design in hot and humid areas—based on the analysis of environment-behavior

Da-Yao Li
College of Civil Engineering and Architecture, Guangxi University, China

Jiang He
Key Laboratory of Disaster Prevention and Structural Safety of Ministry of Education, Guangxi University, China
College of Civil Engineering and Architecture, Guangxi University, China

Xi Xu
College of Civil Engineering and Architecture, Guangxi University, China

Yan-Qing Li
Qinzhou College, Qinzhou, Guangxi, China

ABSTRACT: Along with the development of medical technology and people's living standards improve, "human nature" and the people-oriented design concept attracted more and more attention. Leisure is a comprehensive and complicated process; it exists in all forms of free time and human being, with the purpose of restoring people's physical strength and energy. The open space in the hospital building is an important place for communication, it is the most important part of the public space environment in hospital buildings, Its quality directly affects the use of user comfort, treatment and rehabilitation of patients. The paper, choose leisure space in the Guangxi hospital for research, by using the theory of environment-behavior, summarize the leisure space hospital environment in the hot-humid area. Through the preliminary summary and analysis, reach a conclusion of problems in the current hospital building environment and measures for improvement, discusses the hospital building main method, steps and measures in hot and humid areas, guide the leisure space design of hospital buildings in the future.

KEYWORDS: Hot-Humid Climate; Hospital Buildings; Leisure Spaces; Environment-Behavior

1 THE ORIGIN OF ENVIRONMENTAL BEHAVIOR

With the city modernization and industrialization of high speed propulsion, urban population increases constantly, technology development bring benefits to people. But at the same time, it also brings negative consequences slowly that people doesn't perceive. In the face of the negative effects brought by the development of this, people began to realize that they can't go this way anymore. People realize that there is too little understanding of people for the environment, how to influence their behavior in the past, so environmental-behavior emerges as the times require.

As a branch of psychology, environment-behavior emerged in the 1960s, and then gradually developed in the world. The separation of people and the environment from the understanding of past, environmental-behavior science takes the human behavior and the environment as a whole to research, It believed that human and environment interacted with each other. Environment-behavior science development in the 1970s in the world to form the climax, relates to many subjects as psychology, architecture, city planning, human body engineering etc.

In the field of architecture, environment-behavior science also appeared gradually in the process of architectural design. We should consider of users' demand, not only the pursuit of space to build, but also create a space which can truly meet the demand of consumer psychology and behavior demand.

2 THE CONNECTION BETWEEN LEISURE SPACE OF HOSPITAL BUILDING DESIGN IN HOT-HUMID AREA AND ENVIRONMENTAL-BEHAVIOR

2.1 *The design of hospital buildings*

Hospital is a place where human maintenance of healthy and restore strength. It is also an important place for human survival and fight disease. China is experiencing a huge hospital building changes at present, "medical type" hospital is gradually changing to the complex hospital. Hospital environment becomes the key part of the evaluation of hospital building quality now.

The design of hospital buildings should not only meet the requirements of medical technology, avoid cross-contamination, save energy and protect the environment, but also create a people-oriented medical environment. At the same time, the buildings should reflect the characteristics of the hospital. Besides performing this specification, it should be consistent with the relevant compulsory standards of the state, the norms and other relevant standards, specifications.

Combined with its specialist expertise and management mode and strive to achieve a goal of convenient, practical and beautiful, quiet and comfortable, the interior architecture of flexible space. At last realize sustainable development.

2.2 The design of hospital buildings in hot-humid area

China is vast, geographical environment, climate conditions differs in a thousand ways, if we want to live in such a complicated natural environment comfortable and achieve the goal of saving energy, we must think of a variety of design methods.

Building summer heat-resistant design with hot-humid area should meet the requirement of avoiding cross-contamination. At the same time it needs to pay more attention to ventilation because of health reasons. So hospital building design often uses leisure space to improve the thermal environment in hot-humid area. It can enhance buildings' appearance; produce the psychological effect of pleasing and affectivity for patients, family members and staffs; produce the psychological effect of the emotions; also mitigate pain.

2.3 The connection between leisure space of hospital building design in hot-humid area and environmental-behavior

From the point of patients and their families' need, we shouldn't in pursuit of space construction only, but also create a space truly meet the psychological and behavior needs of patients and their families. Of course, it must be conformed to the related requirements of the hospital architecture design in hot-humid area, satisfied the ventilation requirement of hospital building leisure space in hot-humid area. Then we can really put the "humanity" design

idea of environment-behavior very well in building design.

3 INVESTIGATION OF HOSPITAL BUILDING LEISURE SPACE DESIGN IN HOT-HUMID AREA CURRENT SITUATION

3.1 Definition

Rest is the meaning of stopping (stop things). The "leisure" in oracle words like a tree to rest. A Chinese ancient classics 《Shuo wen》 says, "rest is stopped, for example, people cling to the tree". In a narrow meaning, "leisure" is usually referred to the state of away from the hectic, but in the relaxed, comfortable, comfortable environment, more emphasis on, a quiet and peaceful state of mind. Leisure is a comprehensive and complicated process. It exists in all forms of free time and human being, with the purpose of restoring people's physical strength and energy.

Figure 1.   Hospital building leisure space.

Leisure space is a space where people communicating randomly with a variety of activities in the amateur life by the independent spontaneous way. The leisure space of hospital building is an important place for patient intravenous fluids and communication (Figure 1). Leisure space in hospital building plays the most important part of the hospital building environment. Its quality affects directly the users' comfort level, treatment and rehabilitation of patients.

In order to meet the different psychological needs of patients, leisure space of the hospital is divided into indoor leisure space and outdoor leisure space. Indoor leisure space general as a transfusion room, both have the function of leisure and injection (figure 2). But outdoor leisure space organizations are different.

Figure 2. Indoor leisure space of the First Affiliated Hospital of Guangxi Medical University.

3.2 The fieldwork of hospital building leisure space in hot-humid area

We had a fieldwork of the hospital building in Nanning, Qinzhou city, etc. And had investigated and researched the leisure space of national three levels of first-class hospital in Nanning and Qinzhou.

According to the fieldwork, we can see that the hospital building design is developing very fast in current, but the hospital building design of leisure space has developed slowly. From the early years of the new nation to now, although the design of hospital buildings' leisure space developed from nothing. But after the research, we also learned that there are some problems of the hospital buildings' leisure space design currently. Such as the design methods are generally too single, indoor leisure space is less vitality, the distance between indoor leisure space and outdoor leisure space is too far from each other, and considering the physical environment of leisure space not enough, so it is not implemented in the true sense of "human nature "design.

3.2.1 The design methods are too single
Rich landscape level of leisure space can bring a relaxed mood for the patients and their families. But in the survey, we found that the design methods of leisure space in hospital buildings are too single in current, didn't make consideration of the psychological needs no in-depth. It mainly shows some

aspects like, parts of the texture making people feel "cold", landscape level is quite monotonous and more static scene is given priority to, using less natural elements, artistic thoughtless etc.

3.2.2 The landscape design is too single
In the leisure space perspective, although the leisure space of hospital building is different from other types of buildings, it provides the patients a place for necessary medical activities, the patients want to leave when completed the task, but the landscape design is still playing an important role in the patient pain relief. In the survey, we found that the landscape design of indoor and outdoor leisure space is too monotonous in many hospitals, mainly manifested in the landscape level of a single, static scene is given priority to, generally didn't have a very good landscape view.

3.2.3 Indoor leisure space and outdoor leisure space are too far from each other
If the distance between outdoor leisure space and indoor leisure is too far from each other, it may make the utilization of space insufficiently, make the user feel uncomfortable, and then they don't like to go into the space. What's more, in the course of time, it will destroy the normal aesthetic standards of people, and even affect the physical and mental health. At the same time, a long distance will make the patients produce a sense of lack of psychological safety. Below is an example of a hospital in Qinzhou city. The outdoor leisure space is too far away from the indoor leisure space in this hospital, which makes the patients produce the sense of alienation. There is nobody paying attention to this piece of green space exist, when indoor leisure space was crowded, outdoor leisure space was empty. Space was crowded, outdoor leisure space was empty.

3.2.4 Poor consideration of the physical environment in leisure space
Physical environment plays a key role on the influence of the human body and mind. But from the research we found that, the hospital building design is thoughtless on the acoustic environment, light environment and thermal environment. Take the light environment, for example, the hospital building is a special type of architecture. Bring in the natural light to the indoor leisure space can satisfy the desire of people to nature and sunshine, bringing vigor and vitality to the space, relieving patients' pain. But in the survey, we can find that, many hospital buildings' indoor leisure space is not only the intensity of illumination is insufficient, but also the lack of referencing natural light.

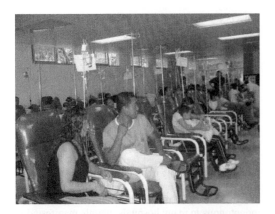

Figure 3. Indoor and outdoor leisure space in a hospital of Qinzhou City.

4 THE MAIN DESIGN METHOD OF HOSPITAL BUILDING LEISURE SPACE IN HOT-HUMID AREA

Although, stander of leisure space environment is not a decisive factor to the quality of hospital at present. But with the development of society and the economy, the quality of leisure space in the hospital will show more and more important role in improving hospital building competitiveness.

If we want the human-nature design to truly reflect on the leisure space of hospital buildings in hot-humid areas, we can mainly consider several aspects, such as the design methods, the functional design, landscape design, rest facilities design, indoor and outdoor leisure space ratio and its contact with our designer.

4.1 Design methods

On the design methods, we should make the "human nature" and the people-oriented concept into the design of every detail.

We should strive to design and think in style, color and material to solve the patient psychological needs.

For example, we can select natural or near natural materials in the range of the user, and use green materials in the room and select diverse plants and flowers.

4.2 The functionality of the leisure space

On the functionality of the leisure space, we should create many kinds of functionalities to meet the diverse needs of people psychology and behavior.

With the transformation of medical mode and function development, arrangements for the non-medical function facilities in the leisure space of the hospital appear. These arrangements can ease the patients' pressure, restlessness and boring. For example, moderate entertainment facilities and ample leisure facilities, all these can relieve their pressure in the leisure time.

4.3 Landscape design in leisure space

The landscape level of leisure space in the hospital building should be created in rich and combination of dynamic and static.

The patient is a hospital service group, so a good landscape effect, wide field of vision is the effective way to alleviate the pressure of waiting. When the outdoor leisure space is a shortage, some design in the interior of the transitional space such as a window seat, also can bring good results. Using the trees and flowers and other natural elements and move scene like fountains, increase the diversity of landscape level, increase the art design of recreational space in the hospital, which is an effective design method to improve the quality of hospital leisure space.

4.4 Privacy design of the leisure space

Privacy design of the leisure space should uniform the privacy and communication.

With the development of the medical model and hospital function, we should strengthen communication between people in the open space and choose some right ways of communication, which can not only guarantee the communication but also protect the patient and family's privacy. It is the trend of the development of space human-nature design.

4.5 The design of the rest facilities

On the design of the rest facilities, firstly, we should ensure adequacy, security, and "human nature", and then we can explore the artistic quality of rest facilities.

We should reference to the hospital scale to set the rest facilities quantity in open space, avoiding the phenomenon of lack of rest facilities. With the improving of life quality and the change of medical mode, design personnel should pay more attention to the design of multi functional of leisure space in the hospital and the convenience in a full range of the patients' need. A design which meets the people's behavior is more important than its fancy space.

4.6 *Indoor and outdoor leisure space ratio and its contact*

On the indoor and outdoor open space, people prefer to the outdoor leisure space, so we should increase the proportion of it. Sunshine, water, flowers and trees and other natural landscape factors are important reasons why people prefer to the outdoor leisure space. At this point, we can lead the natural landscape elements into the design of indoor leisure space when the outdoor leisure space area under the condition of insufficient. In addition, distance between indoor and outdoor leisure space should be referred to the appropriate nursing distance. Strengthen the integrity and unity, also make the patient and accompanying relatives feel secure, too.

5 EPILOGUE

A good leisure space environment should be created by the architect in all kinds of space forms which can meet various kinds of psychological behavior of users on demand. Leisure space environment of hospital buildings should be so. Facing of the patient group who needs care more, we should start from the patient perspective more to understand what kind of space the patient needs, that we can use the environmental behavior of the better in the design.

ACKNOWLEDGMENT

Supported by the Systematic Project of Guangxi Key Laboratory of Disaster Prevention and Structural Safety (2104ZDX05).

REFERENCE

[1] Beehtel RB, Maran, Miehelson WM, et al. Methodsin Environmental & Behavioral Research. NewYork: VanNostrandReinhold，1987.

Energy, Environment and Green Building Materials – Sheng (ed.)
© 2015 Taylor & Francis Group, London, ISBN 978-1-138-02718-3

Applicability of natural ventilation technology to public buildings in South China

H.J. Liu

Department of Architecture and Urban Planning, College of Civil Engineering and Architecture, Guangxi University, Guangxi, China

J. He

National Key Laboratory of Subtropical Building Science, South China University of Technology, Guangdong, China
Department of Architecture and Urban Planning, College of Civil Engineering and Architecture, Guangxi University, Guangxi, China

Y.W. Luo

Department of Architecture and Urban Planning, College of Civil Engineering and Architecture, Guangxi University, Guangxi, China

ABSTRACT: Based on the analysis of climate characteristics in Southe China, and combined with previous scholars research results and the green building demonstration project accepted by Housing and Urban-rural Development in December 2011 – the City Power Union Building in Foshan Guangdong province, the application of natural ventilation technology of public buildings in this area has been summarized, including the commonly used natural ventilation technology and its research status, development trend, inadequate research etc.. The practical application of natural ventilation technology should not only consider the climatic characteristics and local area micro climate, but also combined with spatial layout of building. At the same time, we should consider its influence to air-conditioning, in conjunction with the application of sunshade technology and how to combine with renewable energy. While the lack of natural ventilation research and its application will become the main direction of future research on natural ventilation. The research provides a reference for the application of natural ventilation technology of public buildings in South China.

KEYWORDS: South China; Natural ventilation technology; Application; Usage of renewable energy

1 INTRODUCTION

China has a vast territory, and differs in climate from north to south. The building's green technology used in cold northern is not suitable for moderate South area. Therefore, we should get building energy-saving measures that adapt to climate characteristics of the different climate zones correspondingly. This is also reflected in China's current various countries or local energy-saving design standards, thermal design specifications and other documents. Many scholars and researchers has carried the theory and practice of in-depth study of green technology in various regions with different regional climate environment , and achieved fruitful results (Yang 2003). In this paper, combined with previous scholars research results of public buildings in south China in hot summer and warm winter climate, the application of natural ventilation technology has been summarized. Make a summary of natural ventilation technology suitable

for this area and its existing problems and the future development trend of research and analysis, etc.

2 THE CLIMATE CHARACTERISTICS IN SOUTH CHINA

The South China is located in the most southern of our country, including Guangdong, Guangxi, Hainan, Taiwan, the central and southern of Fujian, Hong Kong and Macao special administrative region. On the thermal partitions, this area belongs to the hot summer and warm winter area, is a subtropical humid monsoon climate, characterized by a long hot summer and hot gentle winter; For high temperature and high humidity, the annual range and daily range of temperature is small; Strong solar radiation and rainfall. The region's building energy consumption is mainly used for summer cooling, refrigeration energy consumption, thus building energy efficiency design gives priority to with the summer heat insulation, basically does not consider the winter heat preservation (Guan 2011).

3 THE RESEARCH STATUS OF THE SUITABLE GREEN TECHNOLOGY IN SOUTH CHINA

Many domestic scholars have studied the appropriate green technology in South China in many aspects and achieved certain results. Research methods are diverse, mainly include climate analysis, comparative study of green building standards, software analysis, etc. Yang Liu has analyzed the country 18 typical city's main meteorological parameters and then established their climate analysis charts. He analyzed on Nanning and Guangzhou as the representative of the climate in South China, and got the strategy of architecture design in this area: shading + natural ventilation plus air conditioning (Yang 2003). Long Enshen got the transition season use of natural ventilation (or mechanical ventilation) from the design is very important to energy efficiency of public building to ensure the strength of the internal heat source, large, strong tightness of greater inner heat and strong enclosed, through the simulation analysis of ventilation and air-conditioning energy consumption of a building in Guangzhou (Long 2008). Li Yi (2012) of the South China University compared the subtropical region of three provinces (Guangdong Province, Guangxi Province, and Shenzhen) Green Building Rating System and the national "green building standards" evaluation, then found that three local standards are proposed the emphasis on outdoor shade, outside the window shade, natural lighting and ventilation. Qin, Dan and Zheng Aijun (2012) also analyzed the local standards of four provinces in southern China (Guangdong, Shenzhen, Fujian, Guangxi) and green building evaluation standard, found in south China the natural ventilation and shading were the great importance of the green building, and put forward some practical design methods. Wu Xiaobo (2012) used Ecotect 5.0 to analysis application time of passive building energy-saving technology in the typical city of each climate zone. And the results show that most of the time in April, May, September and October and, part time in June, July and August it can make indoor comfort under the internal natural ventilation in the hot summer and warm winter area represented by Guangzhou. Liu Yiwei and Feng Wei (2011) demonstrated passive cooling technology such as natural ventilation, shading have good effect for energy saving through the study of a multilayer public building energy saving renovation design in Southern China area, considering the combination of architectural technology and space design, and using the CFD and Ecotect software to have a simulation calculation.

In practice, the green building demonstration project accepted by Housing and Urban-rural Development in December 2011 ----the City Power Union Building in Foshan, Guangdong province, has achieved a good energy saving effect through combining with the characteristics of climate and geographical region in southern China and a good natural ventilation technology application combined with architectural design.

Comprehensive the above research results can be summarized that natural ventilation, building shading is mainly and commonly used green technology in southern China, due to the limited length of this paper, natural ventilation technology were mainly summarized.

4 NATURAL VENTILATION TECHNOLOGY RESEARCH

4.1 The effect of natural ventilation

Has the advantages of energy saving, improve thermal comfort and improve the indoor air quality, indoor natural ventilation is the original means of human history rely on long-term regulation of indoor environment. According to the climate characteristic of Southern China, strengthening the natural ventilation is one of the main means to improve indoor comfort and save the building energy in this area. The use of natural ventilation has significance in two aspects: one is the realization of passive cooling. Natural ventilation can reduce the indoor temperature, take away the damp dirty air, improve the indoor thermal environment in non consumption of energy situation. The other is to provide fresh, clean natural air, which is beneficial to the human's physical and mental health (Zhong & Zeng 2004).

4.2 The research status of natural ventilation

The present research on natural ventilation application at home and abroad, mainly focuses on two related points: one is the use of natural ventilation to control indoor air quality; the other is the use of natural ventilation in summer or in transition season to solve the thermal comfort problems, or replace air conditioning partly. The main research on natural ventilation has three aspects: 1. Research on natural ventilation mode; 2. Comparative study of natural ventilation and mechanical ventilation; 3 effective use of natural ventilation (Zhang & Li 2005). The main research methods adopted include the experimental method, the numerical simulation method, and the specific research methods and theoretical characteristics are shown in Table 1 (Duan & Zhang & Peng 2004).

The use and study of the natural ventilation system to promote the emergence of multiple ventilation system. Multiple ventilation system is a system using natural ventilation and mechanical ventilation in different characteristics with the best way to combine the advantages of natural ventilation and mechanical ventilation in a different time and different seasons. Its basic principle is switching between mechanical ventilation and natural ventilation to maintain good indoor environment, and avoid building annual operation of the air conditioning system cost, the excessive consumption of energy and environmental problems. Li Jin (2011) mainly studied the collaboration of gymnasium, shape factors and wind pressure ventilation technology: based on the basic technology and measurement technology on simulation method in computer, explore the stadium (general layout, environmental factors) and morphology (shape, space, interface, scale) and wind pressure ventilation technology to form more collaborative and effective, and based on further exploration of the traditional sports venue, form new design strategy. In consideration of the building envelope, material, natural ventilation, ventilation control and service life under the premise of development of special urban climate building integrated within the natural ventilation technology which need to explore more in the long run, but it will be the developing direction of natural ventilation.

Many space combined natural ventilation technologies are used in the City Power Union Building of South China. Many space combined natural ventilation technologies are mainly from the analysis and research of Southern China area natural conditions and climate characteristics of the building, according to the surrounding environment, the leading direction, the architectural layout and the space combination of different forms of organization and the induced natural ventilation, to achieve the effect of natural ventilation by air pressure, wind pressure and hot pressing and hot pressing combination. Through the simulation of quantitative analysis of the effects of spatial patterns and composition of indoor ventilation with CFD computer software in this engineering practice, the results show that: the 4 different building space form and combination of simple, straight through space, dislocation of the straight through space, space and courtyard space in the lobby can enhance the summer indoor natural ventilation in the South China. The specific design and application combined with the windows can be opened and closed set (Open casement window to the prevailing wind direction in summer, The aisle room height window etc.), make interior space not only making full use of natural ventilation to reduce building energy consumption in the summer and transition season, but also providing a very good solution to solve the

Table 1. Research methods of natural ventilation (Duan, Zhang & Peng 2004).

Research method	Theory	Feature
Theoretical analysis	Analysis of natural ventilation by using fluid mechanics principle of the jet, Bernoulli equation and energy conservation equation	Simple, fast predicting indoor speed, temperature distribution
Experimental investigations (The wind tunnel model, thermal buoyancy experimental model and tracer gas method)	The wind tunnel model simulates the pressure field and velocity field of building surfaces and around with similarity theory, and determining the coefficient of wind pressure; The thermal buoyancy experimental model is mainly used for the simulation of the physical process of natural ventilation driven by hot pressing; The tracer gas method was studied according to the mass conservation equation of tracer gas.	The accurate prediction of the indoor air distribution and understand of the indoor air flow situation.
Network method (multi zone model method and regional model method)	Using the energy balance concept to form a heat network and ventilation network simulation method;	The method is simple, easy to master, the calculated amount is relatively small, but the calculation is rough and the accuracy is low.
The computational fluid dynamics method (CFD method)	Simulate experiments on the computer; combined with the actual boundary conditions, the numerical solution of algebraic equations governing partial differential equation of discrete income	Predict of air flow and temperature distribution of each area, and analysis of thermal comfort in rooms with a different position
The combination of network method and CFD method	A reasonable combination of the network method and CFD method, firstly calculate the flow boundary condition of the monomer area with the network model, and then use the CFD method to do further analysis and calculation.	Synthesizes the advantages of network method and CFD method

problem of winter wind. The small wind generator and a battery are arranged in the 45m high pull wind towers the building, through the automatic intelligent control switch, providing power to night lighting. The example suggests that the natural ventilation research in South China is not only according to the functional layout of the building but also considering the use of renewable energy.

4.3 Deficiency of natural ventilation research

Natural ventilation has its certain deficiencies, such as difficult to control, unstable, easily affected by the surrounding environment, especially by the outdoor environment has a high concentration of particles and gaseous pollutants, outdoor noise influence, rain effect etc. And Long Enshen (2008) has studied the relationship between cooling consumption of ventilation and air-conditioning of south building found that ventilation is possible with energy-saving effect in cooling dominated the south, and may make the air conditioning energy consumption increases, so in the opening of air conditioning equipment to control ventilation, in addition to still can make outdoor damp air into the indoor.

The main existing problems in the study and application of natural ventilation are: 1. Most of the natural ventilation research are empirical, qualitative or semi quantitative, having yet to form a system, quantitative theory and application of achievements; 2. The effective use of natural ventilation, such as in the building adopts the structure of the atrium, solar chimney, sun shading and the lack of complete theoretical analysis, a lot of research based on the experimental basis of scattered, the lack of practical guiding significance; 3. A study on the influence of meteorological conditions on the strength of relevant natural ventilation is not enough; 4. Research and application of control air flow entrance are less mature.

The lack of the study on natural ventilation itself and the application is becoming the main direction of future research on natural ventilation.

5 CONCLUSIONS

Having made a summary of natural ventilation technology suitable for this area and combined with engineering practice of the City Power Union Building in Foshan Guangdong province, the following conclusions are obtained:

1 The practical application of natural ventilation technology should not only consider the climatic characteristics and local area micro climate, but also combined with the spatial layout of the building. In addition, the region should be fully considered, so that the green technology can fully reflect the cultural and historical characteristics of the area in the practical application.

2 We should fully consider the impact on other aspects, such as the influence between natural ventilation and air conditioning energy consumption, and carry out the comprehensive utilization research, to achieve the objective of a number of technologies. An environment optimization system of a number of technologies can achieve better energy-saving effect.

3 Combined with the development trend of modern building energy saving, natural ventilation, passive energy saving measures should be fully considered in combination with the use of renewable energy, such as combined with wind power, building shading, and solar photovoltaic, and should be integrated design according to the actual situation of the building.

4 The natural ventilation technology suitable for this area are not necessarily universal, we should not only fully analyze the project locations, weather conditions, but also conduct a comprehensive analysis with the construction surrounding environment (including the existing environment and forecast the future) to determine the feasibility and effectiveness of green technology.

5 Computer simulations can guarantee the validity of the technology and improve the design efficiency in a certain extent, so it is necessary to further strengthen the study and development of the use of simulation software on the computer. It is convenient for building designers to use better in consideration of green building design stage of technology.

This paper, has summarized the application of natural ventilation technology combined with previous scholars research results of public buildings in south China in hot summer and warm winter climate and made a summary of natural ventilation technology suitable for this area and its existing problems and the future development trend of research and analysis, etc. This research provides a reference for the application of natural ventilation technology of public buildings in South China.

ACKNOWLEDGMENT

The authors acknowledge the support of Funds from the National Key Laboratory of Subtropical Building Science Open Subject (Project number: 2014KA03).

REFERENCES

Yang Liu. 2003, Strategy analysis and design research on climate construction, Xi'an University Of Architecture And Technology.

Guan Xuanhui. 2011, Research and application of green building technology in Southern China area, China Engineering Consulting (09):35–38.

Long Enshen. 2008, The air-conditioning energy saving potential of south building ventilation, Heating Ventilating & Air Conditioning (05):33–37.

Li Yi. 2012, Analysis of green energy saving technologies of public buildings in hot summer and warm winter zone, South China University of Technology.

Qin Dan, Zhan Aijun. 2012, Analysis and Discussion on the embodiment of green building evaluation criteria related to residential buildings in Southern China area, The 8th International Green Building and Building Energy Conference.

Wu Xiaobo, Liu Meng, Yang Qiaoxia. 2012, Analysis applicability of passive building energy-saving technology of each climatic region, Civil and Environmental Engineering (S2):51–53.

Liu Yiwei, Feng Wei. 2011, The design application and simulation of passive cooling strategies in the transformation of old buildings in Southern China area, Huazhong Architecture (12):27–29.

Energy, Environment and Green Building Materials – Sheng (ed.)
© 2015 Taylor & Francis Group, London, ISBN 978-1-138-02718-3

Architectural design strategies for bio-climatic design in hot-summer and cold-winter region

H.J. Liu
Department of Architecture and Urban Planning, College of Civil Engineering and Architecture, Guangxi University, Guangxi, China

J. He
National Key Laboratory of Subtropical Building Science, South China University of Technology, Guangdong, China
Department of Architecture and Urban Planning, College of Civil Engineering and Architecture, Guangxi University, Guangxi, China

Y.W. Luo
Department of Architecture and Urban Planning, College of Civil Engineering and Architecture, Guangxi University, Guangxi, China

ABSTRACT: The hot-summer and cold-winter region in the transition zone of China's climate division was selected as the study area, and its climate characteristics of this area were summarized. The changing characteristics of climatic factors (the air temperature and humidity, solar radiation, wind etc.) with a greater impact on the architecture were analysed on the basis of the typical meteorological year data of the main cities and previous research. According to the climate analysis results, the strategies for an architecture, climate design that widely applicable to this area were presented: thermal insulation + sunshade + ventilation, and the commonly used measures of each climate design strategy were analyzed. The research contents in order to provide reference for architecture, climate design in hot-summer and cold-winter region, and contribute to the development of building energy saving in this area.

KEYWORDS: Hot-summer and Cold-winter Region; Climatic analysis techniques; Climatic building design; Energy-saving

1 INTRODUCTION

The changes of temperature, humidity, solar radiation, wind, light and other climate factors and interaction form a different climate, the climate is not only closely related to the people's production and life, but also affect the form of architecture layout, material and the way of using (Chen 2009). Our country is vast in territory, from north to south, from east to west, the climate difference is great, has a cold climate in the north, also has a warm climate, humid and rainy south subtropical area. Architectural forms adapt to the various climatic zones have been grown up gradually in the long history of the evolution process. Such as the northern courtyard, cave and the south of the Ganlan style architecture, arcade and so on. These forms of traditional building are well adapted to the region of the construction climate, can provide a comfortable building environment with low energy consumption for people.

The building adapting to climate is the functional a birthright demand of building, but in the middle of the 20th Century the rapid development of small heating and air conditioning equipment broke this traditional rule. More and more buildings use artificial devices to create a comfortable indoor environment, and bring a lot of adverse effects at the same time, the most important is the large consumption of non renewable energy, and the health problems brought by artificial environment. The outbreak of the world energy crisis in 1973 made people begin to pay attention to building energy consumption, and carried out much environmental protection by reducing building energy consumption, building energy conservation technology research, the formation of the current known as the energy conservation and environmental protection as the core to design the green building, energy-efficient building, ecological building, the construction of climate.

The hot-summer and cold-winter region is located in the Middle East China, between cold area and hot-summer and warm-winter area in the building climate demarcation, and its climate is also integrated at both ends of the characteristics with high temperature and hot in summer, cold winter. Thermal protection in the summer and keeping warm in winter of this region must be considered in building thermal design. Many scholars have done research on the building energy saving technology in this area: Fu Xiangzhao's (2002) Building energy saving technology in hot-summer and cold-winter region, Li Baofeng (2004) studied building skin adapt to the climate of hot summer and cold winter area; Li Jie (2013) et al. studied the passive building energy conservation technology based on the analysis of climate characteristics in hot summer and cold winter area; Feng Ya(2012), Li Junge(2006), Yang Liu(2008) et al. researched the indoor environment and comfort in hot summer and cold winter area. And the construction of energy-efficient development in hot summer and cold winter area in our country is in a leading position, especially in Jiangsu province, carry out a comprehensive building energy-saving, is promoting energy-saving pace in this area. Therefore, combining with the existing work, literature, green architectural examples, the climatic features of this region were analyzed and the climate adaptability technology of building were summarized, providing reference to choose convenient climatic building design strategy in this area.

2 ANALYSIS OF CLIMATE IN HOT-SUMMER AND COLD-WINTER REGION

2.1 Regional division of hot-summer and cold-winter region

According to China's "Civil Building Thermal Design Code" GB 50176-93, the classification index of hot-summer and cold-winter region is: the lowest average monthly temperature is 0~10°C, the hottest month average temperature is 25~30°C, and the requirements of architectural design in this area must meet the summer heat requirements, and balance winter insulation properly. The hot-summer and cold-winter region is mainly included the middle and lower reaches of the Yangtze River, that is to the north of Nanling region and the south of the Yellow River. Including Shanghai, Chongqing , Hubei, Hunan, Jiangxi, Anhui, Zhejiang Province, the eastern half of Sichuan and Guizhou Province, the southern half of Jiangsu, Henan Province, the northern half of Fujian Province, the southern end of Shaanxi, Gansu Province, and the northern end of Guangdong, Guangxi Province. The area is about 1,800,000 square kilometers, with a population of about 550,000,000, accounted for about 48% of the country's GDP, is a densely populated, economically developed area. Because the area is located in the area between a cold region of northern China and the south hot area, so it's known as the "transitional area".

2.2 Climatic characteristics of hot-summer and cold-winter region

The climate characteristics of the area are: stuffy and hot in summer, Wet and cold in winter, sunshine slants

small, wet and rainy, and the rainfall changes greatly; hot summer and cold winter with four distinctive seasons. The hottest month average temperature is 25~30°C, and the average relative humidity is around 80%. By the impact of the pacific subtropical in the summer, the highest temperature can reach 40°C, the daily minimum temperature is over 28°C, a day without a cool moment. In the daytime, the sunlight is strong with high temperature and wind speed, wind heat (fire) is rampant as a stove, warming the air making air and the surface of object. The region's coldest month average temperature is 0~10°C, the average relative humidity is around 80%, and its temperature in winter while is higher than the north, but the sunshine rate is far below the north. The northern winter sunshine rate is more than 60%, while this area's is only 1/6 ~ 1/2 of the northern's. And it with large internal differences, winter sunshine rate decreases rapidly from east to west, the west's is only 1/4 ~ 1/2 of the east's. It is highest in the east, only about 40%; that is 30% of the central, 20% of the west, and only 13% of Chongqing (as shown in Fig. 1).

The weather in winter is always snow and rain, cloudy, almost do not see the sun, the basic characteristics of climate is cold and wet. The annual average relative humidity is high, around 70% ~ 80%, the four seasons' are similar, the annual number of rainy days is about 150D, much more than 200D, the annual precipitation is 1000 ~ 1800mm. The area of the two seasons of winter and summer is wet, with the relative humidity at around 80%. Attention is required that the results of wet in winter and summer are not the same: summer is because the air contents too much water vapor; winter is because the air temperature is low, and a serious shortage of sunshine.

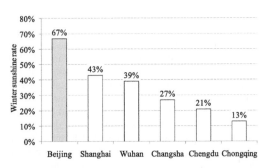

Figure 1. The sunshine rate comparison in winter between Northern city (Beijing) and the cities in hot-summer and cold-winter region.

2.3 Analysis of temperature and humidity

According to the typical annual hourly parameters in "China building thermal environment analysis of meteorological data"(2005) ,the summer and winter weather parameters of 10 cities (include Shanghai, Hangzhou, Nanjing, Hefei, Nanchang, Wuhan, Changsha, Guilin,Chongqing, Chengdu)were sorted out. The average temperature in the hottest month is 25~30°C, and mostly at 28~30°C. In addition to Guilin, Chongqing, Chengdu, the average minimum temperature of the hottest month reaches 25°C in other 7 cities, Nanchang, Wuhan reach more than 26°C. The hot weather higher than 35°C is for half a month to 1 months in the majority of the local. The hot-summer and cold-winter region is also a water network area, is very damp, and the relative humidity is often as high as 80%, the minimum average humidity in the hottest month is 58% ~ 70%; Lasts weather of overcast and rainy season of Huangmei often appears in the middle and lower reaches of the Yangtze River in summer. Although the temperature is not high, but because of the low air pressure and high humidity, the body feels very hot.

The average temperature of the coldest month is 2 ~ 8°C, mostly in 4~6°C, the minimum average temperature of the lowest month in these cities is 3°C except for Guilin and Chongqing. The days with average daily temperature lower than 5°C above one and a half months or even nearly 2 and a half months in the middle and lower reaches of Yangtze River and north coastal area (Shanghai, Hangzhou, Nanjing, Wuhan, Hefei). The relative humidity in winter is still very high, up to 74% ~ 85%, the minimum average relative humidity of the coldest month is 53% ~ 76%, thus the body feels cold.

2.4 Analysis of solar radiation

Figure 2 shows the monthly total radiation variation of 3 cities Shanghai, Changsha, Chongqing in hot-summer and cold-winter region. The solar radiation of this region is strong in summer and weak in winter on the whole. Contrasting these three cities, in addition to the most hot months July and August, it is Shanghai > Changsha > Chongqing in other months.

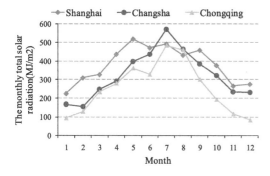

Figure 2. Changes of annual, monthly total radiation in 3 big cities in hot-summer and cold-winter region.

2.5 Analysis of wind environment

The wind speed in this area is low in hot summer, that is not conducive to dissipate heat and pollutant; and it is high in winter, exacerbating the outdoor cold and the cold intrusion load. Therefore, we should pay attention to the layout of the building design and the windows open design, to get good natural ventilation in summer. The unfavorable plane layout, entrance design will lead to serious cold air intrusion and the sharp decline in air conditioning heating effect.

As mentioned above, climate characteristics in hot summer and cold winter areas were summarized. And based on the analysis of the typical meteorological year data of main cities in this area and previous studies on the changes of four main climatic factors temperature and humidity, solar radiation and wind speed, the following conclusions are obtained : 1. In hot-summer and cold-winter region, the summer temperature is higher at 25 ~ 30°C, winter changes at 2 ~ 8°C, the relative humidity is larger both in winter and summer, at around 80%, the resulting heat and wet summer, cold and wet winter; 2. The solar radiation resource distribution in general, including Chongqing, Sichuan area is relatively poor, and the sun radiation intensity gradually reduces from the east to the west in winter; 3. The area of wind speed in summer is too small, resulting in stuffy and hot, slants big in winter, increasing the apathetic and causing the cold air into the building easily.

3 THE STRATEGY OF CLIMATIC BUILDING DESIGN

Combining with the analysis of the climate characteristics in hot summer and cold winter area on the front part, the architecture, climate design points in this area is heat proof in the summer, heat preservation in winter and dehumidification in each period. The main design point is heat insulation + sunshade + ventilation, several aspects of this climate design strategy are summed up as follows.

3.1 Sunshade and ventilation in summer

With a high temperature, large humidity and strong solar radiation in this area, it is necessary to give full consideration to the construction of thermal cooling ventilation, shading and ventilation. The climate design principle is to control the solar radiation, ventilation and dehumidification. It is essential to pay attention to building envelope (walls, roof, windows and doors) design, but does not do too much summary of this part in the paper.

Sunshade: In order to block out the sun radiation in summer, does not affect the sunshine into the room in winter, shading device should be designed into movable shading. In outdoor landscape design, deciduous trees are chosen in favor of shading in summer and let the sunlight into the interior in winter.

Ventilation: The location of doors and windows, apartment layout design should make full use of the natural convection to adjust indoor temperature, keep the indoor air circulation, improve the comfort of living environment and reduce the energy consumption of air conditioning. Setting a patio or using the stairs to ventilate in the public space; Effective cooling reasonable application of night ventilation.

Besides, gray space should be set in the graphic layout of east, west and south, outdoor activity space layout in the north, and choose light colored material as building facade material.

3.2 Thermal insulation and windproof in winter

Winter temperatures in the region are low, require heating, for determining the heating period base on the comprehensive consideration of sunshine and temperature, while the winter sunshine rate is small and has large differences In different cities, it needs comprehensive consideration of sunshine and outdoor temperature to determine the heating period.

Table1. Winter sunshine rate and the heating period in hot-summer and cold-winter region.

Region	Winter sunshine rate	The daily average temperature of the outdoors
The western region	Less than 20%	Below 10°C into the heating period
The central region	Between 20%~30%	Below 9°C into the heating period
	Between 30%~40%	Below 8°C into the heating period
The eastern region	Between 40%~50%	Below 7°C into the heating period

Climate design principle is mainly thermal insulation, wind proof, that can be carried out through the following two aspects

In plane layout, the main function of the space layout of the south wing for solar heating; To the arrangements north of the small scale division room, the large space south room will be more conducive to maintain the indoor heat stability. Rational layout of buffer space, on the north side (the stairwell, storage etc.), south (sunshine room).

In the wind proof design: the space layout of the wind on the dominant wind assisted in winter; Using the sunshine room or sun porch as buffer space, the openings are arranged under the direction of the wind; using large roof slope.

4 CONCLUSIONS

Based on the analysis and summary of change characteristics of main climatic factors (including air temperature, humidity, solar radiation intensity, wind) in hot-summer and cold-winter region, put forward the strategy of architecture, climate design is widely applicable to the area, and provide a reference for the climatic building design in this region. However, there are some differences in climate between different cities and regions, and architecture also will be affected by the micro climate, process should be combined with the local climatic characteristics for more detailed analysis of the climate in the actual design and to determine a more suitable building climate strategy.

ACKNOWLEDGMENT

The authors acknowledge the support of Funds from the National Key Laboratory of Subtropical Building Science Open Subject (Project number: 2014KA03).

REFERENCES

Chen Fei. 2009, Building wind environment—study on wind environment and building energy conservation design in hot summer and cold winter climate zone, Chinese Building Industry Press.
Fu Xiangzhao. 2002, Building energy saving technology in hot-summer and cold-winter region, Chinese Building Industry Press.
Li Baofeng. 2004, The research on climatic-active design strategy of building skin in hot-summer and cold-winter zone, Tsinghua University.
Li Jie, Wu Changyou, Qu Wanying. 2013, Passive design strategies based on climate characteristics in hot-summer and cold-winter zone: taking Wuhan city for example, Building Energy Saving(07):54–56.
Feng Ya, Nan Yanli, Zhong Huizhi. 2012, Energy efficiency and indoor thermal environment design of rural buildings in hot summer and cold winter zone, Building Science(12):21–24.
Li Junge. 2006, An adaptive thermal comfort model for hot summer and cold winter context, Xi'an University Of Architecture And Technology.
Li Junge, Yang Liu, Liu Jiaping. 2008, Adaptive thermal comfort model for hot-summer and cold-winter area, Heating Ventilating & Air Conditioning(07):20–24.
Yang Liu. 2010, Building climatology, Building Industry Press.
Yang Liu. 2003, Climate Analysis and Architectural Design Strategies for Bio-climatic Design, Xi'an University Of Architecture And Technology.
Fu Xiangzhao. 2000, Building energy saving technology in Chinese hot-summer and cold-winter area, New building materials(06):13–17.

Energy, Environment and Green Building Materials – Sheng (ed.)
© *2015 Taylor & Francis Group, London, ISBN 978-1-138-02718-3*

On strategies of developing eco-tourism around Poyang Lake eco-tourism circle

J.B. He & Z. Wang
Jiangxi Science & Technology Normal University, Nanchang, Jiangxi, China

ABSTRACT: The construction of ecological tourism in the tourism industry has become the current development direction of China's tourism industry. Its development exerts great significance in the construction of Poyang Lake ecological tourism circle. At present, during the process of developing ecological tourism of Poyang Lake, problems such as the imperfect infrastructure, weak environmental awareness and misuse of resources are surfacing. Based on the present situation of ecological resources in Poyang Lake eco-economic circle, this article stresses on analyzing the measures and strategies of development of Poyang Lake ecological tourism to promote the sustainable development of Poyang Lake ecological tourism circle.

KEYWORDS: Poyang Lake Eco-tourism Circle; Environmental Protection; Resource; Sustainable Development

1 INTRODUCTION

With the rapid development of the national economy, people's quality of life and income increase, the traditional way of tourism has not been satisfied with human requirements. Thus, a variety of needs related to tourism appear, and ecological tourism is among them. In 1983, The International Eco-tourism Association defines it as a travel activity with two responsibilities of protecting the natural environment and defending the local people's life. Under the rapid development of both spiritual and material civilization, international resource crisis such as environmental and resource issue heighten the vigilance. At the same time, the world blew up a whirl of wild wind of "green activities" and "green consumptions".

Ecological tourism is a newly tour activity, which appeared and promoted by the side effect on the ecological environment, causing by tourism. It is based on the ecological tourism resources, under the premise of not destroying the integrality of the ecosystem, making the tourists know, appreciate and enjoy the tour with the original natural ecological system and worthy local culture, putting emphasis on the mutualism interaction between human beings and nature.

As the construction of ecological tourism around Poyang Lake is raised to be a national strategy, Poyang Lake also meets with its golden development chances. A great prospect is sure waiting for it. Rational development of Poyang lake tourism will be better facilitated the construction around there. Whether the development a success or failure will directly exert influence on the local development change of tourism, even the economic development. Therefore, we have the responsibility to observe the objective law, pay attention to the unification of economic, social and ecological benefits.

2 THE PRESENT SITUATION OF ECOLOGICAL RESOURCES IN POYANG LAKE ECO-ECONOMIC CIRCLE

Poyang Lake is the largest fresh water lake around the world, which is situated in the south of the middle and lower reaches of Yangtze River, north of Jiangxi province. Among the four fresh water lake, Poyang lake is the only none eutrophication lake as well as the wetland which exerts high importance over the world. The drainage area of Poyang Lake is 162,000 square kilometers, of which 10% within the borders of Nanchang, capital of Jiangxi province. Lake area occupies 5100 square kilometers, which can store water around 34 billion square kilometers. It is the largest wetland of Asia, the biggest migratory protection reservation among the world, the important regulating lake in the stem of Yangtze River. The amount of water run into the Yangtze River normally reaches to 146 billion square kilometers, occupies 15.5% of the Yangtze River runoff, reputed as the "mainland's kidney of China. Moreover, it is one of the ten ecological protection reserves. And the world wildlife fund delimited it as one of the most important ecological areas in the world with affluent resources.

The land area of ecological development of ecological tourism circle around Poyang Lake is more than 53,000 square kilometers, concluding Nanchang, Jiujiang, Shangrao, Yingtan, Fuzhou and Jingdezhen which lie in Jiangxi province. And it accounts for over 30% of Jiangxi's total land area. The prior places are the 15 lakeside counties around Poyang Lake.

2.1 *Good ecological environment*

Good ecological environment is the basic condition of ecological tourism. Thanks to its wide area and it links to Yangtze River, Poyang lake has a good self purification capacity and its water is still of good quality. And its average clarity is 0.65 meters, average water quality is the second level standard. Wucheng County, the core reservation of migratory birds, has a population of 15,000. It is based on agriculture and fishing. In addition, the traffic there is inconvenient and affected by the human influence of social and economic activities, the dry season shallow depressions provide safe habitat and abundant food for the rare migratory birds.

2.2 *Abundant bird resources*

The most familiar compositions known by people in the reservations are the birds. The already known birds in Poyang Lake reached to 310 kinds, of which the first-degree national protection birds is 10, the second-degree one is 44. Poyang Lake is an important habitat and wintering sites for birds, cranes and other rare waterfall land forest. Cranes are the first-degree national animal, about 3,200 live in wild, and 90% of them winter on Poyang Lake. This place has the largest number (more than 60,000) of swan geese and small swans (more than 70,000) winter here. Furthermore, it is also the migratory corridor and rest place for a great deal of rare and precious birds. It becomes a spectacular scene here in winter as so many birds migrating. Thus, Poyang Lake is reputed as "The word of crane" and "the country of precious bird". The bird resources in Poyang Lake have a high value both in appreciating and developing.

2.3 *Various ecological scenes of wetlands*

A great number of wildlife breed, live and grow here because of the unique geographical advantage and weather condition, some endangered and precious waterfowls such as white-flag dolphin, cowfish and Chinese sturgeon in particular. It lake-centered when the water level is high, marsh—centered when low,

presenting a scene that alternates of water or land. Thus did the natural wetlands stay in an annually regular fluctuation which is rich the ecological resources. Poyang lake national wetland park is one of the six largest wetland parks. It is a national wetland park, which has the largest area, the most abundant wetlands, the most spectacular scenery, and the stately wetland culture.

3 THE SUSTAINABLE DEVELOPMENT STRATEGY ON THE ECOLOGICAL DEVELOPMENT OF POYANG LAKE ECO-TOURISM CIRCLE

Since the approval of the Poyang Lake Economic Zone by the state council of the People's Republic of China from now, the development of Poyang lake has taken 5 years. At present, ecological resource development in this area has made tremendous progress, stepping out of a sustainable development of rapid economy, efficient utilization of resources, and environmental protection. But at the same time, there are problems such as environmental pollution, ecological tourism resources, imperfect tourism facilities construction, and weak awareness of environmental protection still exists in its development and construction. How to develop and use the ecological tourism resources around the District of Poyang Lake rationally to realize the ecological resource protection and sustainable development of the tourism industry has become the urgent issue to be solved.

3.1 *The combination of strict protection and appropriate development*

As a natural tourism relying on nature, eco-tourism emphasizes on the nature and the original nature of tourism object, putting the theory of "harmony between human beings and nature" and "sustainable utilization of resources" in the first place. All localities must walk the combined road of protection, development and stability. During The development, we should firmly establish the "beautiful scenery is golden hill" concept. Protecting the green mountains and rivers is the basis of ecological tourism, and also the key measure of controlling soil erosion and improving agriculture as well as industrial ecological environment. At the same time, we must be strict to protect migratory birds and wetland ecological environment where migratory birds survive. Thoroughly implement the guideline of "comprehensive protection, ecological priority, greater focus, rational use, sustainable development".

3.2 Improve legal system and increase government supervision

Attach great importance to control tourism. Ecological environmental protection of the Poyang Lake nature reserve, resource protection of migratory birds' area and development of ecological tourism management, all these depend on the perfection of the legal system. Our government should formulate regulations of suitable wetland protection, utilization and management, and strengthen the supervision and its inspection. They should take the initiative to host and participate in the overall planning of ecological development, preparation of the ecological tourism project, and the work of demonstration and management. Moreover, we should also use market operation in the development and construction of the project.

3.3 Attach importance to environmental protection and realize the sustainable development of ecological resources

Manage the tourism resource in the way of sustainable development, and we must strictly implement the environmental impact assessment system and the "three simultaneous" system for the tourism project construction. Regional tourism treatment facilities of sewage, dust, waste must be synchronized with the development of tourism, we need to make close coordination between design, construction and operation, organically combined the development of ecological tourism with ecology, environment and protection of historical and cultural resources. Tourism activity is controlled in the range of resources and environmental capacity, establish and improve the ecological compensation mechanism, realize the sustainable development of ecological tourism industry and adapt to resource, environment. Additionally, we should control tourism activity in the range of resources and environmental capacity, establish and improve the ecological compensation mechanism, realize the sustainable development of ecological tourism industry as well as the environment.

3.4 Rationally plan and create ecological tourism products

The lake is called Crane Park and the kingdom of rare birds, monopoly both in domestic and foreign. Ecological environment, there is good with the beautiful landscape of lakes and mountains, together with village folk with profound cultural accumulation, we can carry a variety of tourism activities.

According to the conditions and development goals of the Poyang Lake Nature Reserve, scenic spot can be integrated into the following four tourist areas:

Firstly, migrant bird viewing area. Mainly concentrated on lakes and wetland birds stayed, this district is the main habitat and a viewing area for migratory birds. Bird watching activities must adhere to moderate, scientific, orderly principle. There can be no bird watching which would affect the perched phenomenon of migratory birds.

Secondly, grass island recreational area. Focus on seasonal grassland, and the main activity are entertainment, sports as well as athletics. In addition, it also includes the distribution in the vicinity of the sand hills and the development of grassland.

Thirdly, custom area for folk water is mainly composed of the Ganjiang River and Xiushui interchange water village. The major projects of Poyang Lake are the production (fishing, farming) life (pastoral people), etiquette and taboo, clothing and diet and festival activities etc. There is a great attraction to the tourists, and at the same time, the tourism products can be integrated together.

Fourthly, tourist reception center district. As representative of the reception town, Wuchen, located in the protected area, is the largest protected area of town and administrative center. It also can be individually as a scenic spot. The main content there is shopping.

3.5 Increase the promotion and publicity

Character is the soul of tourism, which makes tourism an attractive and competitive ability. We must enrich tourism propaganda to create Poyang Lake district tourism brand. The development of ecological tourism in the Poyang Lake district needs a distinctive tourism image, highlight the characteristics of its biodiversity, migratory bird observation and science education. Aiming at the two the biggest bright spot -protection of migratory birds and wetland ecology, in order to build a distinctive feature of the famous International Birding base and ecological tourism, the tourism image positioning is as the: "birds paradise", "white crane kingdom" and "world of wetland ecological tourism". In addition, publication must be diversified. We must exploit the cultural heritage of Poyang Lake, combined the Poyang Lake main tourism resources with tourism culture which includes the local economic development, social customs, historical events, legends and landscape characteristics. Establish tourism province and domestic as well as foreign oriented marketing network system. Give comprehensive, in-depth publicity and introduce the regional tourism around Poyang Lake and tourism services.

4 CONCLUSION

Ecological resources are the foundation and prerequisite of the development of ecological tourism, once destroyed, it will lose the interdependent condition, and thus it will be difficult to realize sustainable development of resources. We should focus on the characteristics of the ecological tourism resources area around Poyang Lake, use it rationally, make protection and development of the Poyang Lake District coexist, push to walk the road of sustainable development of ecological tourism.

Under the good tendency of development of ecological tourism, Poyang Lake should make full use of government resources, location advantages and good unique support conditions. Make efforts to overcome the shortcomings in their development to pull Poyang Lake district tourism to a new path of ecological tourism. What's more, we should try to make the tourism industry become a pillar industry in Jiangxi national economy, and fatherly promote the construction of Jiangxi, becoming an ecological tourism province with prosperous tourism industry.

ACKNOWLEDGMENT

This work was financially supported by 2012 Scientific Research Program of Jiangxi Science & Technology Normal University (KY2012SY10), 2011 Jiangxi Provincial Art Science Program (YG2011079).

REFERENCES

David A. fennel. Ecotourism: an Introduction, London: Routlege Press, 1999.
David Weaver. Ecotourism. John Wiley & Sons, 2002.
Gunn C A. Tourism Planning (3rd edition). New York: Taylor & Francis, 1994.
Stephen Wearing and John Neil. Ecotourism: Impacts, Potentials and Possibilities. Butterworth Heinemann, Oxford, 1999.
Jones B, Tear T. Australian national ecotourism strategy. UNEP Industry and Environment, 1996, 18(1):55.
Freeman R E. Strategic Management: A Stakeholder Approach. Boston: Pitman, 1984.
Chris Cooper, Stephen Wanhill. Tourism Development: Environmental and Community Issuers. John Wiley & Sons, 1997.
World Tourism Organization (WTO). Voluntary Initiatives for Sustainable Tourism. Madrid: World Tourism Organization, 2002.
Priskin J. Assessment of natural resources for nature based tourism: The case of the Central Coast Region of Western Australia. Tourism Management, 2001, 22(6):637–646.

Energy, Environment and Green Building Materials – Sheng (ed.)
© 2015 Taylor & Francis Group, London, ISBN 978-1-138-02718-3

The analysis of properties of oxide scale on the surface of typical superheater material

Min-Cong Zheng
State Grid Anhui Electric Power Corporation Research Institute, Anhui, China

Jian-Feng Xiao, Sen Liu & Zhi-Ping Zhu
Changsha University of Science & Technology, Changsha, China

ABSTRACT: Superheater withstands the highest temperature in the heat exchanger of boiler, so the material's performance directly affects the safe operation of the boiler. In this paper, it's necessary to make a brief introduction about the selection of superheater. The structure of the oxide scale and its impact were analyzed. Summarize primarily the several typical superheater material, such as T23, T91, TP347H, and Analyze these three performance characteristics of each material, also its characteristics of oxide scale were analyzed.

KEYWORDS: superheater; material; oxide scale

1 INTRODUCTION

The main way to increase efficiency is to increase the steam power plant parameters, in another way, It increases the vapor pressure and temperature. It's very clear that developing ultra-supercritical thermal power units and improving steam parameters can improve the efficiency of thermal power plants [1]. With the improvement of the steam temperature and pressure, the efficiency of power plants is increased, and the coal consumption of electricity supply is declined significantly. While the main technical problems that we meet in improving steam parameters are metal materials, high temperature and high pressure are also problem parameters.

In the power plant boiler accident, heating tube's surface (water wall tubes, superheater tubes, reheater tubes, economizer tubes, also known as the "four tubes") is the most serious accident in burst damage and leak common. These accidents account for 71.7% of boiler accident, 40% of unplanned downtime and more than 50% in less generating capacity, are also the main factors in affecting the safe operation of the generator units. As the part of high temperature and pressure, superheater, reheater that due to its harsh environment service cause severe oxidative corrosion of pipe [2]. In particular, the presence of oxide scale on the inner wall results in the reduction of the effective wall thickness of the tube wall, and the tube wall is also correspondingly increased with the stress. At the same time, oxide scale caused thermal deterioration of the tube wall. The average operating temperature of tube wall increases, which keep the tube in a long period of over-temperature service status. When developed to a certain extent, it will eventually lead to boiler tube burst and leakage accident [3]. So, that we research the surface oxide scale properties of the superheater, and reheater materials have a great significance in the safe operation of boilers and energy saving..

2 SUPERHEATER METAL AND OXIDE SCALE PROFILES

2.1 *Superheater metal selection*

Flue gas out superheater tube has high temperature in the ultra (ultra) supercritical boiler, and the flue gas is very corrosive. Metallic materials must meet the high rupture strength, creep strength, while meeting anti-corrosion from the flue pipe wall, erosion properties of fly ash, anti-oxidant properties from steam of pipe inner wall, and also has better cold working process and welding properties. Superheater tubes of metal wall temperature have more 50 °C than the steam temperatures. In the steam parameters 605 °C/ 603 °C, the final superheater and furnace platen superheater generally use HR3C that has better resistance to corrosion and oxidation resistance, because of its outer ring in the high temperature corrosion zone. The tube rows in the low-temperature region use the new ferritic heat resistant steel, for example T23, T91, T92, T122, TP347H(use more), etc.

2.2 Structure and properties of the oxide scale

In terms of metallic high temperature oxidation, the nature that metal oxide scale form on the surface (corrosion product film) is essential for the use of metal at a high temperature. On the one hand, the amount and the rate of formation of corrosion products is a sign of high-temperature corrosion, on the other hand, the nature of the product decided that course conducted corrosion can prevent the corrosion of the metal to continue or not. From a macro point of view, the corrosion product scale has three categories: compact integral scale; scale has cracks or voids; liquid or gaseous product.

The stability of the oxide scale is that it not melted, decompoed, evaporated, cracked, peeled and so on under the oxidation conditions, which is associated with environmental factors and also depending on the oxide film itself, the physical- mechanical and chemical properties.

In severe corrosive environments, heat-resistant corrosion of metallic materials can mainly realized by adding Cr, Al, and Si and other metal elements. The affinity of these metal elements and oxygen is very strong, which can form a stable oxide scale, for example, Cr_2O_3, Al_2O_3, SiO_2, etc. As we all know, the size of the thermodynamic stability and the diffusion coefficient, Cr2O3, Al2O3 and SiO2 are excellent protective oxide scale. SiO_2 scale can protect strongly metal from H_2-H_2S atmosphere. Al_2O_3 scale is an ideal protective scale, because Al content has an important role in the formation of Al_2O_3 scale and single Al_2O_3 layer need enough Al content. The diffusion coefficients of these metal ions in the oxide scale are small, so that the oxide scale formed on the surface can cover the surface of the metal. It is a very excellent protective oxide scale.

3 SEVERAL TYPICAL SUPERHEATER MATERIAL AND OXIDE SCALE CHARACTERISTICS

The following describes two typical supercritical units superheater material: T23, T91. Their component analysis is shown in Table 1.

3.1 T23 steel

T23 steel is a new boiler steel developed by Japan's Mitsubishi Heavy Industries and Sumitomo Metal together. On the base of T22 steel, we can get bainitic heat-resistant steel with low carbon multiple composite, high strength, high toughness by reducing the C content, adding W, V, Nb, B content, micro-alloying and diffusion precipitation strengthening.

Table 1. The mass fraction of three common steels of supercritical boiler heat-resistant (%).

Steel and components	T23	T91
C	0.04	0.08–0.12
Si	≤0.50	0.20–0.50
Mn	0.10–0.60	0.30–0.60
S≤	0.01	0.01
P≥	0.03	0.02
Cr	1.90–2.60	8.00
Mo	0.05–0.30	9.50
W	1.45–1.75	–
V	0.20–0.30	0.18–0.25
Nb	0.08–0.20	0.60–0.10
Ni	–	≤0.40
N	≤0.030	0.03–0.07
Al	≤0.030	≤0.04
B	0.0005	–

In recent years, T23 steel has been applied, on the some extent, in our ultra (ultra) supercritical unit. A boiler factory first designs the 600MW supercritical boiler that its final superheater and high temperature reheater tubes use a T23 material. After the practice of 6 000h to 10 000h shows that T23 steel final superheater tube inner wall generates oxide easily to peel. And the blockage at the elbow causes overheating pipe explosion incident [4].

According to the analysis of pipe sample analysis of a power plant, T23 steel pipe sample of final superheater occurred oxide scale off the inner wall [5]. The native oxide of T23 steel pipe sample is a double layer structure with a outer layer of thick columnar crystals of Fe_3O_4 and an inner layer of isometric and fine grained spinel containing W and Cr. Fe easily spread to the outer layer forming Fe_3O_4 or Fe_2O_3. W atomic has a big radius, which is difficult to diffusion-deposition in the inner and forming W-rich region. Inner layer of native oxide scale exists one or more hole-chain arranging in the circumferential direction. The oxide easily separated along the hole-chain, which results oxide off.

3.2 T91 steel

T91 (Europe grades X10CrMoVNb9-1, American grades ASME SA-213 T91 / P91) is successfully developed by American Oak Ridge National Laboratory in 1980s. It, as 10Cr9Mo1VNb grades, was included China Standard GB5310-1995 high-pressure boiler tube. It also became indispensable steels for developing supercritical generating units in our country.

Taking pipe sample a power plant superheater and analysing oxide scale, the results as follows:

1 The oxide scale of T91 steel steam side of superheater tube inner wall after running 85 000h is the

two-layer structure with about 167.6μm thickness. Outer oxide, the thickness is 79. 1μm, is rich in Fe element and poor in Cr elements. The main components are FexOy. The oxide thickness is 88.5μm, by(Fe, Cr)$_3$O$_4$ composition. The interface of two layers is as the original metal surface.

2 The transition region with a layer of 10 ~ 15μm present between the oxide film and the metal substrate. And in this region, Cr is oxidated by O and Fe, because of spreading to the outer, is diffused and lost. So the region of the metal substrate flows in the steam side direction, the Cr, O concentration is increased and the Fe concentration is reduced. The (Fe,Cr)$_3$O$_4$ formed is rich of Cr2O3. Since the T91 steel metal Cr content is insufficient to form a protective Cr$_2$O$_3$ oxide scale, metallic matrix in the transition region occurs an intragranular oxidation phenomenon [6].

4 CONCLUSION

It is inevitable that oxide scale on the high-temperature heating surface lied the steam side. The spelling of oxide scale has become a worldwide problem. With the rapid development of ultra (ultra) supercritical power generation technology, oxide scale and the related issues will become more significant. But the replacement of material is slow. Focus on the comparison and summary multidimensional structural growth law and falling characteristics of oxide scale formed on the T23,T91 and new austenitic steels tube that lay in the running boiler. Study how various operating conditions, such as temperature, dissolved oxygen, cooling rate, operation time and so on, affect the thickness of oxide scale, structural morphology and the time of falling. Reveals the common law from a large number of oxide scale phenomena and incidents of many running power plant's units. That all above is a realistic way to predict and prevent oxide scale hazards from the ultra (ultra) supercritical unit operation mode, maintenance strategies and control many concepts and many other aspects.

REFERENCES

[1] A. V. Dub,V. N. Skorobogatykh,et al.New heat-resistant Chromium Steels for a promising objects of power engineering[J]. Thermal Engineering, 2008, 55(7):594–601.

[2] Minami Yusuke, Toovama Akira, Seki Mikihito, let at Steam-oxidation resistance of shot blasted stainless steel tubing after lO-year service [J]. NKK Technical Review, 1996, 1(75): 10.

[3] H Nickel.T, Wouters.M, Thiele W, Quadakkers. The effect of water vapor on the oxidation behavior of 9% Cr steels combustion gases [J]. Fresenius J Anal Chem (1998)361: 540–544.

[4] Zhao Huichuan, Jia Jianmin, Chen Jigang, et a1. Failure analysis of superheater tube used in supercritical boiler[C] Supercritical(ultra) Supercritical Boiler Steel and Welding Technology Collaboration Proceedings Conference of the Third Forum. Tianjin: Power Industry Boiler and Pressure Vessel Safety Supervision and Management Committee, 2009: 571–577.

[5] Huang Xingde, Zhou Xinya,You Zhe, etal, Supercritical(ultra) supercritical steam boiler heating surface temperature oxide growth and spalling characteristics[J]. Journal of Power Engineering, 2009, 29(6): 602–608.(in Chinese)

[6] P J Ennis, W J Quadakkers. Mechanisms of steam oxidation in high artensitic steels[J]. International Journal of Pressure Vessels and Piping, 2007, 84(1–2):75–81.

Energy, Environment and Green Building Materials – Sheng (ed.)
© 2015 Taylor & Francis Group, London, ISBN 978-1-138-02718-3

Adsorption of Cr(VI) from aqueous solution by attapulgite

J. Ren, F.J. Zhang, Y.N. Li, L. Tao & Y.Z. Zhang
School of Environmental and Municipal Engineering. Lanzhou Jiaotong University, Lanzhou, P. R. China

ABSTRACT: The potential of attapulgite was assessed for sorption of chromium from aqueous solution. Adsorption of Cr^{6+} onto attapulgite was investigated with respect to temperature, initial concentration and contact time. The adsorption processes for attapulgite can be well described by the pseudo-second-order kinetic model. The temperature has not much effect on adsorption performance. The equilibrium adsorption isotherm for natural attapulgite was closely fitted with the Langmuir model. The adsorption data for natural attapulgite were described using Freundlich isotherm equation and the Correlation is close to 1. The sorption of Cr^{6+} on attapulgite were spontaneous process, its value in the range of 0 to 1, and thesorption was endothermic.

1 INTRODUCTION

Chromium is a toxic metal ion present in wastewater and most of chromium is used in steel factories, electroplating and ceramic industry, battery and accumulator manufacturing. Sorption of Cr^{6+} on clay minerals, which are ubiquitous in the environment, has been widely studied in the last decades. Compared with conventional technologies for the removal of heavy metals in waste water, such as ion exchange treatment, coagulation and precipitation, adsorption and coprecipitation/adsorption (Dabrowski 2004). Recently, much attention has been paid to inorganic materials, such as montmorillonite (Lee et al., 2004), attapulgite (Lee et al., 2005), kaolin (Wu et al., 2003).

Attapulgite is a hydrated octahedral layered magnesium aluminum silicate mineral with large surface area, excellent chemical stability and strong adsorption. Although, ATP possesses high adsorption capabilities, the modification of its structure can successfully improve its capabilities. In the present study, attapulgite original soil was applied as an adsorbent for the removal of heavy metal Cr^{6+} from aqueous solutions. The adsorption properties such as equilibrium, kinetics and thermodynamics were demonstrated by batch mode adsorption experiments.

2 MATERIAL AND METHODS

Attapulgite was obtained from the Gansu Shuangtai Attauplgite Co. Ltd. The Cr^{6+} ion solutions were prepared by dissolving $K_2Cr_2O_7$ in distilled water. Attapulgite underwent extrusion, drying, crushing.

The attapulgite was then put through a screen to obtain particles of 200 mesh (0.075 mm).

The attapulgite was treated in 120°C drying for2 hours of solid potassium dichromate. 2.828 g to 250 ml beaker, adding suitable amount of distilled water, to dissolve and transferred to 1000 mL volumetric flask inside. Solution Cr^{6+} quality concentration is 1000 mg/L.

20, 40, 60, 100, 150, 200 mg/L Cr^{6+} were used. The simulation Cr^{6+} wastewater pH value is 5, with electronic analytical balance of 1 g the optimum craft treated adsorbent, join simulation Cr^{6+} waste water, quickly into the multifunction bath thermostat oscillator to a certain speed respectively in temperature is 25°C, 30°C, 40°C, 50°C, 60°C, respectively, under the condition of oscillation 5, 10, 20, 30, 40, 50, 60, 90, 120, 150 min rapidly after treatment with different temperature (placed in temperature controlling magnetic stirrer in a magnetic stirring speed mixing time), transfer to 10 ml centrifuge tube and in 3000 r/min conditions centrifugal for 5 min. The upper clarified filtrate were measured.

The initial and the final concentration of Cr^{6+} in the aqueous solution were measured with ICP-AES. The adsorption capacity of adsorbent was calculated through the following equation:

$$q_e = \frac{\left(c_0 - c_e\right)v}{m} \tag{1}$$

Where q_e is the adsorption capacity of Cr^{6+} on adsorbent (mg/g), C_0 is the initial concentration of Cr^{6+} (mg/L), C_e is the equilibrium Cr^{6+} concentration in solution (mg/L), m is the mass of adsorbent used (g) and V is the volume of Cr^{6+} solution (L).

3 RESULTS AND DISCUSSION

3.1 *Adsorption kinetics*

To examine the controlling mechanism of adsorption processes such as mass transfer and chemical reaction, pseudo-first-order and pseudo-second-order kinetic equations were used to test the experimental data. The pseudo-first-order kinetic model was suggested by Lagergren for the adsorption of solid/liquid systems and its linear form can be formulated as (Chen & Wang, 2009):

$$\log(q_e - q_t) = \log q_e - \frac{k_1 t}{2.303} \tag{2}$$

where q_t is the adsorption capacity at time t (mg/g) and k_1 (min^{-1}) is the rate constant of the pseudo-first adsorption, was applied to the present study of Cr^{6+} adsorption. The k_1 and correlation coefficient were calculated from the linear plot of $\log(q_e-q_t)$ versus t.

Table 1. The pseudo-first-order kinetic model parameters for Cr^{6+} ions adsorbed onto adsorbents of attapulgite.

T(K)	C$_0$(mg/L)	q$_e$(exp) mg/g	k$_1$ ×10^{-2}min^{-1}	q$_{e1}$(cal) mg/g	r^2
298	20	0.48	1.67	0.11	0.0075
	40	0.80	2.35	1.03	0.3021
	60	1.21	—	1.00	—
	100	2.10	2.09	0.81	0.1789
	150	3.36	2.45	0.38	0.0362
	200	5.01	2.65	0.23	0.1192
303	20	0.48	0.02	0.18	0.0272
	40	0.96	1.55	0.28	0.0061
	60	1.15	2.87	0.49	0.0765
	100	2.03	2.31	0.35	0.0114
	150	3.30	3.32	0.50	0.1454
	200	4.48	2.46	0.26	0.1005
313	20	0.34	0.35	0.51	0.0054
	40	0.65	−567.24	0.58	0.0672
	60	1.03	2.89	0.47	0.0862
	100	1.91	—	1.00	—
	150	2.91	6.78	0.59	0.0047
	200	4.98	14.98	0.66	0.1190
323	20	0.52	3.55	0.63	0.0802
	40	1.02	1.93	0.18	0.108
	60	1.51	1.77	0.11	0.0675
	100	2.42	5.30	0.23	0.1416
	150	3.77	17.78	0.41	0.1777
	200	5.18	—	—	—
333	20	0.38	1.57	0.63	0.0033
	40	0.75	1.61	0.18	0.134
	60	1.10	2.89	0.11	0.0029
	100	1.83	2.69	0.23	0.0127
	150	2.98	1.80	0.40	0.3480
	200	4.08	—	—	0.6535

The kinetic data were further analyzed using Ho's pseudo-second-order kinetics model. It can be expressed as (Liu et al., 2010):

$$\frac{t}{q_t} = \frac{1}{k_2 q_e^2} + \frac{t}{q_e} \tag{3}$$

Where k_2(g/mg·min) is the rate constant of the pseudo-second-order adsorption. The k_2, the calculated q_e value and the corresponding linearregression correlation coefficient r^2 are given in follows.

Table 2. The pseudo-second-order model parameters for Cr^{6+} ions adsorbed onto adsorbents of attapulgite.

T(K)	C$_0$(mg/L)	q$_e$(exp) mg/g	k$_2$ ×10^{-2}g/ mg·min	q$_{e2}$(cal) mg/g	r^2
298	20	0.48	1195.37	0.48	0.9999
	40	0.80	−392.85	0.80	1.0000
	60	1.21	−177.87	1.21	1.0000
	100	2.10	−83.13	2.08	0.9999
	150	3.36	−31.80	3.31	0.9997
	200	5.01	−510.77	5.01	0.9999
303	20	0.48	−294.11	0.46	0.9992
	40	0.96	−4823.00	0.96	0.9996
	60	1.15	−162.14	1.15	0.9998
	100	2.03	−91.73	2.01	0.9999
	150	3.30	−212.34	3.26	0.9998
	200	4.48	−123.48	4.44	0.9999
313	20	0.34	−151.04	0.34	0.9979
	40	0.65	−67.68	0.65	0.9980
	60	1.03	−89.50	1.03	0.9989
	100	1.91	185.57	1.92	0.9999
	150	2.91	−45.68	2.93	0.9991
	200	4.98	−121.08	4.98	0.9999
323	20	0.52	−1027.19	0.52	0.9999
	40	1.02	487.90	1.01	0.9999
	60	1.51	2131.84	1.50	0.9999
	100	2.42	−107.07	2.42	0.9998
	150	3.77	94.59	3.77	0.9997
	200	5.18	−412.30	5.16	1.0000
333	20	0.38	−447.43	0.38	0.9999
	40	0.75	−57.21	0.75	0.9914
	60	1.10	−128.87	1.08	0.9994
	100	1.83	−82.80	1.82	0.9998
	150	2.98	−180.67	2.98	1.0000
	200	4.08	−62.84	4.03	0.9999

The experiment data were analyzed by the three kinetic models and the parameters calculated from above equations and correlation coefficients were listed. The low correlation coefficient values obtained for the pseudo-first-order model indicate that sorption is not occurring exclusively onto one site per ion. It was found that the correlation coefficient for the intra-particle diffusion kinetics model is lower than those of the pseudo-first-order and the pseudo-second-order models. This indicates that adsorption of Cr^{6+} onto the adsorbent does not follow the intra-particle diffusion kinetics. However, a clearly difference of q_e between the experiment and calculation was observed, indicating a poor pseudo first-order and the intra-particle diffusion kinetics model fit to the experimental data (Mazzotti, 2006). The results suggested that the pseudo-second-order adsorption mechanism dominated the adsorption process, and adsorption rates of Cr^{6+} onto the adsorbents were probably controlled by the chemical process (Ho & Mckay, 1999). Moreover, the calculated q_e value also agrees with the experimental data in the case of pseudo-second-order kinetics (Tables 1 and 2).

3.2 Adsorption isotherms

It is important to establish the most suitable correlations for the adsorption equilibrium data using isotherm equations as it plays a crucial role in designing adsorption system and optimizing experiment process. Adsorption isotherms are also efficacious to understand adsorption mechanism. Many isotherm models have been proposed to explain adsorption equilibrium, and the most commonly used isotherm models for liquid-solid adsorption are Langmuir, Freundlich and Dubinin-Radusckevich(D-R) isotherms. The equilibrium data obtained were tested with respect to the three isotherm models.

The Langmuir isotherm assumes that the adsorbent surface homogeneous and the adsorption sites are energetically identical indicating that the adsorbed molecules do not react with each other. The linear form of Langmuir equation can be depicted as (Liu et al., 2010):

$$\frac{c_e}{q_e} = \frac{c_e}{q_{m1}} + \frac{1}{b \times q_{m1}} \quad (4)$$

Where q_{m1} represents the maximal adsorption capacity to form a monolayer (mg/g) in the system, q_e represents the amount of solute adsorbed per unit weight of adsorbent (mg/g) at equilibrium, C_e is the equilibrium solute concentration (mg/L) in solution and b is the Langmuir constant (L/mg). The plot of C_e/q_e against C_e would be a straight line, and then q_{m1} and b can be obtained from the slope and intercept of the plot.

The essential feature of the Langmuir isotherm is R_L, which is a dimensionless separation factor or equilibrium parameter, defined as (Mazzotti, 2006):

$$R_L = \frac{1}{1 + bc_0} \quad (5)$$

Where band C_0 are the same as defined above. The value of R_L is indicator of the shape of adsorption isotherm to be favorable or unfavorable. The value of R_L between 0 and 1 indicates favorable adsorption. While $R_L > 1$ suggests unfavorable adsorption and the adsorption process is linear adsorption if $R_L = 1$. while $R_L = 0$ represents irreversible adsorption.

The Freundlich isotherm is applied for multilayer adsorption on heterogeneous adsorbent, and it assumes that the adsorption sites increase exponentially with respect to the heat of adsorption and the Freundlich equation is an empirical equation. The linear form of the Freundlich equation is expressed as (Ho & Mckay, 1999):

$$\ln q_e = \ln K_F + \frac{1}{n} \ln c_e \quad (6)$$

Where K_F ((mg/g)(L/mg) $^{1/n}$) and l/n are Freundlich constants, related to adsorption capacity and adsorption intensity, respectively, and the other symbols are the same as defined above. If value of l/n is in the range of 0-1, the adsorption process is feasible and favorable. From the slope and intercept of the plot of $\ln q_e$ versus $\ln C_e$, l/n and K_F can be obtained, respectively. The D-R isotherm can be applied for depicting adsorption on both homogeneous and heterogeneous surfaces (Allen 2004). A linear form of D-R equation is given by:

$$\ln q_e = \ln q_{m2} - \beta \varepsilon^2 \quad (7)$$

Where q_{m2} is the D-R monolayer capacity (mg/g). β is a constant correlated to sorption energy $((mol/J)^2)$, and ε is the Polanyi potential related to the equilibrium concentration (J/mol), illustrated as follows:

$$\varepsilon = RT \ln(1 + \frac{1}{c_e}) \quad (8)$$

Where R is the universal gas constant (8.314J/(mol·K)) and T is the absolute temperature (K). The mean free energy E (J/mol), which is defined as adsorption per molecule of the adsorbate when it is transferred to the surface of the solid from infinity in the solution,can be calculated from constant β,and the relationship is presented as:

$$E = (2\beta)^{-0.5} \quad (9)$$

The constant parameters and correlation coefficients were calculated from the Langmuir, Freundlich

Table 3. The constant parameters and correlation coefficients of Langmuir model for Cr^{6+} ions adsorbed onto adsorbents of attapulgite under different temperature.

T/K	Linear fitting regression equation	R^2	qm (mg/g)	b (L/mg)
25°C	1/ q$_e$ =0.12+20.88/C$_e$	0.950	7.73	0.006
30°C	1/ q$_e$ =0.11+21.36/C$_e$	0.980	9.52	0.005
40°C	1/ q$_e$ =-0.20+46.46/C$_e$	0.994	5.01	0.004
50°C	1/ q$_e$ =0.03+18.25/C$_e$	0.999	34.60	0.002
60°C	1/ q$_e$ =0.03+31.23/C$_e$	0.997	44.24	0.001

Table 4. The constant parameters and correlation coefficients of Freundlich model for Cr^{6+} ions adsorbed onto adsorbents of attapulgite under different temperature.

T/K	Linear fitting regression equation	R^2	k$_f$	1/n
25°C	Log q$_e$ = 0.92logC$_e$ +1.41	0.958	25.51	0.92
30°C	Log q$_e$ = 1.05logC$_e$+1.40	0.973	25.20	1.05
40°C	Log q$_e$ = 0.78logC$_e$+1.55	0.967	35.85	0.78
50°C	Log q$_e$ = 1.04logC$_e$+1.28	0.998	19.04	1.04
60°C	Log q$_e$ = 0.96logC$_e$+1.51	0.993	32.17	0.96

Table 5. The constant parameters and correlation coefficients of D-R model for Cr^{6+} ions adsorbed onto adsorbents of attapulgite under different temperature.

T/K	qm^2 mg/g	β ×10-6(mol/J)^2	E J/mol	r^2
298	2.42	30.00	0.0077	0.5855
303	2.41	30.00	0.0077	0.6912
313	2.24	60.00	0.0109	0.6634
323	2.89	30.00	0.0077	0.7405
333	2.21	40.00	0.0089	0.7066

and D-R equations mentioned above, as summarized. It can be seen that the values of correlation coefficients of Langmuir equation were higher than the other two isotherm values, which indicated the Langmuir isotherm correctly fitted the equilibrium data confirming the monolayer coverage of Cr^{6+} onto the modified attapulgite. And the D-R isotherm values were lowest over the entire concentrate on range, which indicated that the D-R equation represented the poorest fit for the equilibrium data than the other

isotherm equations. It was obvious that the values of R$_L$ and 1/n in the range of 0–1confirmed the favorable uptake of the adsorption process (Tables 3–5).

4 CONLUSIONS

Kinetics experiments show that modified attapulgite offered fast kinetics for adsorption of Cr^{6+}, and the adsorption processes follow pseudo-second order type adsorption kinetics. Equilibrium data werewell described by Langmuir model, indicating the adsorption was monolayer adsorption on a surface containing a finite number of identical sites. Moreover, thethermodynamic analysis presented the endothermic, spontaneous and entropy gained nature of the process.

REFERENCES

Chen, H., Wang A.Q. (2009). Adsorption characteristics of Cu (II) from aqueous solution onto poly (acrylamide)-attapulgite composite. Journal of Hazardous Materials. 165, 223–231.

Dabrowski, A., Hubicki, Z., Podkoscielny, P., Robns, E. (2004). Selective removal of the heavy metal ions from waters and industrial wastewaters by ion-exchangemethod. Chemosphere. 56, 91–106.

Ho, Y.S., McKay, G. (1999). Pseudo-second order model for sorption processes. Process Biochemistry. 34, 451–465.

Lee, W.F., Chen, Y.C. (2005). Effect of intercalated reactive mica on water absorbency for poly(sodium acrylate) composite superabsorbents. European Polymer Journal. 41, 1605–1612.

Lee, W.F., Yang, L.G. (2004). Superabsorbent polymeric materials. XII. Effect of montmorillonite on water absorbency for poly(sodium acrylate) and montmorillonite nanocomposite superabsorbents. Applied Polymer Science. 92, 3422–3429.

Liu, Y., Wang, W.B., Wang, A.Q. (2010). Adsorption of lead ions from aqueous solution by using carboxymethyl cellulose-g-poly (acrylic acid)/attapulgite hydrogel composites. Desalination, 259(1), 258–264.

Mazzotti, M. (2006) Equilibrium theory based design of simulated moving bed processes for a generalized Langmuir isotherm. Journal of Chromatography. 1126, 311–322.

Liu, Y., Wang, W.B., Wang, A.Q. (2010). Adsorption of lead ions from aqueous solution by using carboxymethyl cellulose-g-poly (acrylic acid)/attapulgite hydrogel composites. Desalination. 259, 258–264.

Wu, J.H., Wei, Y.L., Lin, J.M., Lin, S.B. (2003). Study on starch-graft-acrylamide/mineral powder superabsorbent composite. Polymer. 44, 6513–6520.

Energy, Environment and Green Building Materials – Sheng (ed.)
© 2015 Taylor & Francis Group, London, ISBN 978-1-138-02718-3

Numerical simulation of airflow distribution in air-conditioned room of office building based on ANSYS

R.H. Ma

School of Civil Engineering and Architecture, Panzhihua College, China

ABSTRACT: Based on the computational fluid dynamics and numerical heat transfer theory, the physical and mathematical models of the air-conditioned room of an office building are set up with ANSYS software, and the indoor air distribution is simulated by k-eequation turbulence model. Air temperature field and air velocity field be simulated by ANSYS software are presented, and the thermal comfort of persons in different areas are achieved which are used to instruct the improvement of air-conditioning and ventilation in the office.

KEYWORDS: Numerical simulation; Temperature field; Velocity field.

1 INTRODUCTION

With the development of society and economy, people's living standards continue to improve, so the requirements to the environment of the building are increasingly high. Air-conditioning as an important means to improve the construction environment is widely used in the gymnasium, cinema, shopping malls, office buildings and other public buildings and residential buildings (Doha Z. Jinhui Z & Lijun Z 2013). Comfort and energy saving has become a basic subject in today's construction, and equipment design, airflow in the room directly affects the indoor air-conditioning effect, which must be considered highly in engineering design. The traditional method of airflow organization revealing the distribution law is obtained through model experiments, empirical or semi-empirical formula, and then the airflow is organized based on these equations, but it takes much manpower, material resources and long, difficult period, furthermore the conditions are limited by experimental model, sometimes it is difficult to simulate all the air flow feature of the complex space. CFD technology can overcome these weaknesses of experimental method, on one hand, it can be researched the influence of the indoor air velocity field and temperature field under different air flow conditions by the CFD technology, and the numerical results of the velocity field and temperature field are presented, so this technology is not only visual, but also it saves a lot of manpower and material force, on the other hand, it is good for the plans' comparison. In the end the design cycle is greatly shortened with the CFD technology, and the technology improves the indoor air quality, which is very important for the realization of optimization design (Pin L. Shu, Yongbao X & ke D 2005).

2 COMPUTATIONAL MODEL

A representative office building room in Chengdu city is selected. The size of the room is 4.6m*3.6m*3m, and the air-conditioner whose size is 0.5m*0.2m*0.2m is 2m away from the ground, 0.8m away from the wall. There is one table in the room which is 2m away from the back wall, and the table whose size is 1.4m*0.8m*0.8m is in the center in the X direction.

Assuming that in the room four walls whose temperature is 312.15K are in contact with the external environment. The upper and lower walls contact other rooms whose temperature is 301.15K. The refrigeration temperature of the air-conditioner is 293.15K, and the inlet velocity is 0.1m/s.

The following assumptions to simplify the room (Fujun W 2004. Hongguang Z & Guang C 2009. Wu Z 2011).

a. The indoor air is continuous, incompressible medium.
b. The indoor flow field is unsteady.
c. The room is no gap leakage.
d. Not considering the radiation of solar radiation and the radiation heat transfer between the internal wall of the room.

The control equations of the model are shown as follows (Xin x & Haifeng c 2013).

a. Continuity equation

$$\frac{\partial u_i}{\partial x_i} = 0$$

In the equation: u_i refers to time averaged velocity in the x_i direction.

b. Momentum equation

$$\frac{\partial}{\partial}(\rho u_i u_j)=\frac{\partial p}{\partial x_i}+\frac{\partial}{\partial x_j}\left[\mu\left(\frac{\partial u_i}{\partial x_j}+\frac{\partial u_j}{\partial x_i}\right)-\frac{3}{2}\mu\frac{\partial u_i}{\partial x_i}\delta_{ij}\right]+\rho g_i+F_i$$

In the equation: ρ refers to air density; P refers to air pressure; ρg_i refers to volume force in the i direction; F_i refers to heat source.

c. Energy conservation equation

$$\frac{\partial(\rho u_i h)}{\partial x_i}=\frac{\partial}{\partial x_i}\left[\left(\frac{u}{P_r}+\frac{\mu_t}{\sigma_t}\right)\frac{\partial h}{\partial x_i}\right]+S_h$$

In the equation: μ_t refers to transmission viscosity; S_h refers to volume source.

d. Turbulent kinetic energy K equation

$$\frac{\partial(\rho k)}{\partial t}+\frac{\partial(\rho k u_i)}{\partial x_i}=\frac{\partial}{\partial x_j}\left[\left(\mu+\frac{\mu_t}{\sigma_k}\right)\frac{\partial k}{\partial x_j}\right]+\mu_t\frac{\partial u_i}{\partial x_j}\left(\frac{\partial u_i}{\partial x_j}+\frac{\partial u_j}{\partial x_i}\right)-\rho\varepsilon$$

e. Turbulence kinetic energy dissipation rate ε equation

$$\frac{\partial(\rho\varepsilon)}{\partial t}+\frac{\partial(\rho\varepsilon u_i)}{\partial x_i}=\frac{\partial}{\partial x_j}\left[\left(\mu+\frac{\mu_t}{\sigma_e}\right)\frac{\partial\varepsilon}{\partial x_j}\right]+\frac{C_{1\varepsilon}}{k}\frac{\partial u_i}{\partial x_j}\left(\frac{\partial u_i}{\partial x_j}+\frac{\partial u_j}{\partial x_i}\right)-C_{2\varepsilon}\rho\frac{\varepsilon^2}{k}$$

The simplified physical model of the room is shown in Fig. 1.

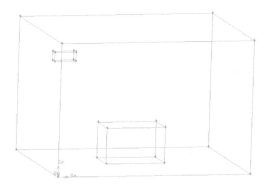

Figure. 1.   Physical model.

3   ANALYSIS OF THE RESULTS OF NUMERICAL SIMULATION

The planes of x=1m, x=3m, z=0.9m, z=1.8m is obtained for analysis the temperature distribution of the room, and the temperature distribution is shown in Fig 2 to 5. The plane of x=1m, x=2m, y=1m, y=2m,

z=0.9m, z=1.8m is obtained for analysis the velocity distribution of the room, and the velocity distribution is shown in Figs 6 to 11.

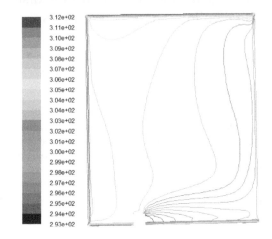

Figure 2.   Air temperature contour of X=1m.

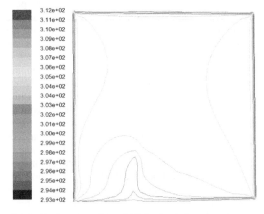

Figure 3.   Air temperature contour of X=3m.

Figure 4.   Air temperature contour of Z=0.9m.

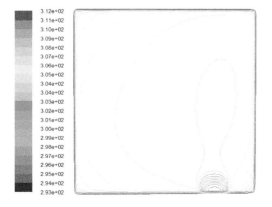

Figure 5.   Air temperature contour of z=1.8m.

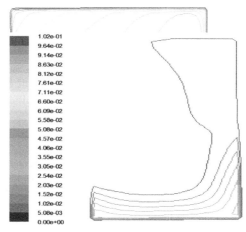

Figure 8.   Air velocity contour of Y=1m.

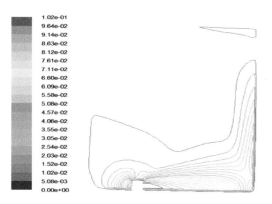

Figure 6.   Air velocity contour of x=1m.

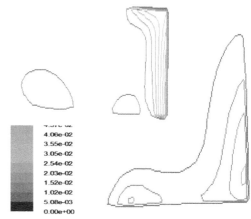

Figure 9.   Air velocity contour of Y=2m.

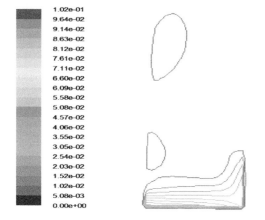

Figure 7.   Air velocity contour of x=2m.

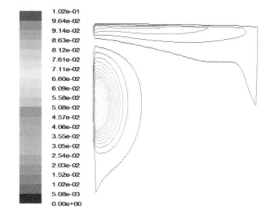

Figure 10.   Air velocity contour of z=0.9m.

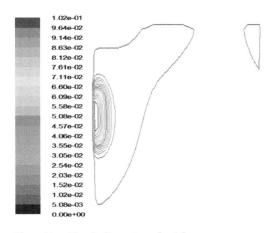

Figure 11.　Air velocity contour of z=1.8m.

From the temperature contours at different positions of the figures can be seen that the air temperature below the air-conditioner and the space close to the ground is lower than the rest space of the room, and the temperature is greatly different. The temperature contours of the space in the horizontal direction from the air-conditioner about 1.5m are density, and the temperature gradient is larger. The temperature of the space upper the air-conditioner and range from the air-conditioner exceed 3m in the horizontal direction is higher, so the thermal comfort is poor.

From the velocity figures can be seen that when the air at a certain speed from the inlet at low temperature gets into the room, the airflow in the air-conditioning room can easily flow down because of its low temperature and high density. Meanwhile the airflow appears semi-affix jet shape near the inlet. The airflow in the area near the room wall appears affix jet shape obviously. The airflow below the affix jet area has a downward velocity because of the induction effect, as a result the jet area expands unceasingly. But with the attenuation of the jet velocity, the range of action is becoming smaller and smaller, and the jet core velocity is becoming smaller and smaller too.

At the same time, due to in the middle of the room there is one table, the room exists a no wind area in the middle of the room. Airflow from the air—conditioner jetting out, due to the hindrance by the walls and table, forms several recirculation zones and vortexes which are easily produce dead end, meanwhile it is not conductive to the discharge of the pollutants.

4　CONCLUSIONS

The airflow distribution of the air-conditioned room of an office building is numerically simulated, and the temperature field and velocity field are presented, after analyzing the following conclusions can be obtained:

a. The temperature field distribution in different position of the room is different, also the thermal comfort is different. The temperature at the bottom of the outlets of the air-conditioner and range near the ground is lower, and the farther away from the air-conditioner in the horizontal direction, the higher the temperature is, so the thermal comfort is much worse.

b. The scope of the low temperature, airflow generated by the air-conditioner will expand in the process of moving, but the velocity attenuation is fast. Due to the hindrance by the walls and table, there are several recirculation zones and vortexes in the room, as a result the comfort of people will be affected in the air-conditioning room.

REFERENCES

Dahai Z. Jinhui Z & Lijun Z. 2013. Numerical simulation of airflow distribution in air- conditioned room. energy conservation technology 31(5):420–425.
Pin L. Shu, Yongbao X & ke D. 2005.The research on numerical simulations of velocity field and temperature field of airflow in air-conditioned room and application. Journal of Anhui University of science and technology (natural science) 25(4):22–26.
Fujun W. 2004. Computational fluid dynamics analysis: Theory and application of CFD software.Beijin. Tsinghua University press:97–121.
Hongguang Z & Guang C. 2009. Numerical simulation of air distribution in air-conditioned room. Journal of Anhui Vocational college of metallurgy and technology 19(1):41–45.
Wu Z. 2011. The numerical simulation of air flow distribution in air-conditioning office in summer. Refrigeration and air conditioning 25(3):304–308.
Xin x & haifeng c. 2013. Indoor airflow simulation of a performance building. Building energy & environment 32(1):92–94.

Energy, Environment and Green Building Materials – Sheng (ed.)
© 2015 Taylor & Francis Group, London, ISBN 978-1-138-02718-3

A dual interfaces and high-speed solid state recorder design

S. Li & Q. Song
National Space Science Center, Chinese Academy of Sciences, Beijing, China
University of Chinese Academy of Sciences, Beijing, China

Y. Zhu & J.S. An
National Space Science Center, Chinese Academy of Sciences, Beijing, China

ABSTRACT: With the development of the space, Solid State Recorder (SSR) has become more and more important. More space explore tasks means more data, so the capacity and throughput of SSR should be larger and higher. We designed a high-speed SSR structure which is flexible and configurable in both capacity and throughput. The SSR used a domestic CPU, LONGSON, and FPGA as control unit. The interface between the two chips is PCI and HPI alternatively. HPI, which is a low speed bus, is used for ordinary tasks. PCI, which is a high speeding bus, is used for high throughput application. Also, with the PCI bus the capacity and throughput of the system can be extended. The two interfaces design enhances the flexibility of the system, and also realizes the flexibility of equipment.

KEYWORDS: LONGSON, HPI, PCI, large capacity storage

1 INTRODUCTION

1.1 Solid State Recorder

Data storage devices are one of the key devices for spacecraft, which can store data for a variety of space experiments and space exploration and also is an important part of data processing platform in-orbit. With the development of aerospace electronics technologies, solid state record (SSR) has become the mainstream scheme of the spacecraft data storage system. NAND FLASH is the basic storage medium for the solid state record (SSR), because it has the advantages of small size, low cost, low power consumption, long life, anti-vibration, and wide temperature range.

1.2 PCI

PCI stands for Peripheral Component Interconnect, which is rather commonly used in industry, for the high speed and steady structure. At a PCI bus clock frequency of 33MHz, a transfer rate of 132Mbytes/second may be achieved. A transfer rate of 264Mbytes/second may be achieved in a 64-bit implementation when performing 64-bit transfers during each data phase. A 66MHz PCI bus implementation can achieve 264 or 528Mbytes/second transfer rates using 32-bit or 64-bit transfers [1].

1.3 Traditional scheme

The current mainstream schemes for SSR are the three combinations as: CPU + FPGA + memory unit [2], DSP + FPGA + memory unit and FPGA + FPGA + memory unit. The first scheme is more suitable for a system which needs to realize a variety of control algorithms because of CPU. The second scheme is weaker in system control or management, while its capacity in the signal processing algorithm is better because of the DSP. The third scheme is weaker in both system control or management and signal processing algorithm, but better in the interface expansion, which could connect more NAND FLASH and got a very large capacity. This article aims to design a flexible, high-performance and high-capacity storage platform which has the configurable interface to adapt more application situation, so we develop the platform based on the third scheme, CPU + FPGA + memory unit.

2 STRUCTURE OF SSR WITH DUAL INTERFACES

2.1 Introduction to the system

The structure of the system is CPU + FPGA + memory unit. Among all the components, CPU

2.2 Characteristics of CPU—LONGSON

A CPU is the control core of the SSR, which controls all the operations of the memory unit. While the embargo of high-performance CPU, the CPUs we could buy from overseas are all low-end products. In order to make great effort in the development of the space industry, we should develop our own CPU, LONGSON, which is developed by Institute of Computing Technology, Chinese Academy of Science and have got some intellectual property rights. They develop a series of LONGSON for different usage. There is one series designed towards the space application with forward design.

It contains interrupt controllers, timers, RS232 serial port controller, floating-point processor, PCI and some memory interface supporting SDRAM and Flash ROM. The internal frequency is 50MHz, while the external frequency is 100MHz. Fixed point 300MIPS and floating-point 50MIPS. To make it competent using in the radiation harsh environment, it has been fastened in modules. It can stand the anti-radiation dose not less than 168Krad, and the single particle lockout threshold of it is not less than 74Mev.cm2/mg [3].

2.3 Startup circuit of CPU

The system of CPU is composed of program storage unit, the buffer unit and the interface unit. All the unit has an allocated space in the CPU, which must be followed when designing. The space table is listed in Table 1.

Table 1. Space allocation of LONGSON.

Starting address	Space	description
0x0000_0000	256M	SDRAM
0x1000_0000	128M	PCI MEM space
0x1800_0000	32M	NOR Flash space
0x1e00_0000	16M	HPI space
0x1f00_0000	128	EMI configuration register (excluded 0x1f00_0040)
0x1f00_0040	4	NAND Flash data I/O
0x1f00_2000	4K	PCI configuration register
0x1f00_3200	256	HSB_MISC configuration register
0x1f00_4000	4	SPI configuration register
0x1f00_4080	16	UART #0 configuration register
0x1f00_4090	16	UART #1 configuration register
0x1f00_40D0	8	I2C configuration register
0x1f00_4100	256	LPB_MISC configuration register
0x1fc0_0000	1M	Boot space
0x1fd0_0000	1M	PCI IO space
0x1fe0_0000	1M	PCI configuration read and write space

The program store unit is important for the system. The EEPROM is used to store the startup program and some application program. While the program may be changed a lot when testing, a two-way alternative program store unit is employed. A NOR flash chip with a chip holder is used for program storage when testing. The nor flash chip is cheap and can be rewritten millions of times. Only one of EEPROM chip or nor flash chip can be selected when starting up. A 3-8 decoder decodes one-of-eight lines based upon the conditions at the three binary select inputs and the three enable inputs [4]. So we adopt a 3-8 decoder to form an alternative selection circuit, which is showed in figure 1. The J14 determines which chip will be selected when CPU starts up. When testing, the 1-2 of J14 is connected and J15 is dis-connected, so the nor flash chip will be selected. When programming fixed program into EEPROM chip, the 1-2 of J14 is connected and J15 is also connected, so all of EEPROM chip and nor flash chips are selected. When starting up, the nor flash would be read at once. After the system built, it will recognize the EEPROM chips as a normal buffer chip and then programming fixed program into the EEPROM chip from the nor flash chip. Next time, J15 could be left unconnected and the 2-3 of J14 could be connected, so the EEPROM chip is the first chip selected when starting up. The alternative start up circuit design makes the whole design flexible.

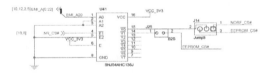

Figure 1. Alternative circuit for system starts up.

2.4 Clock structure of the CPU

Usually, a CPU needs an SDRAM chip to as a buffer for the application program to run on the platform. LONGSON's SDRAM interface support 256MB space. While, to save pins, LONGSON's SDRAM interface output data width is no longer shielded signal DQM0, DQM1, DQM2, and DQM3. So those pin of SDRAM should directly connect to ground. LONGSON could support the EDAC of SDRAM because in the radiation harsh environment the SDRAM chip is more prone to get SEU effect. There is a pin, A24, responsible for ECC SDRAM data shielding, as shown in the specific implementation. To synchronize the operation with SDRAM, CPU must share the same clock with it. Usually, to get more quick processing ability, there is a multiplier inside CPU and CPU's core is running on a clock from the multiplier. However, the interface with SDRAM is running under the primary clock. To simplify the design of PCB and to minimize the occupied pins of the CPU, the clock of both CPU and SDRAM are from the crystal. A dual matching resistor design is employed to reduce the interference between the signals. The specific implementation is shown in Figure 2.

Figure 3. PCI connection between CPU and FPGA with an additional connector.

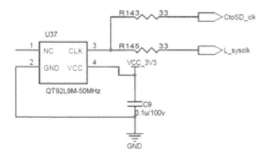

Figure 2. Alternative circuit for system starts up.

2.5 Dual interface of CPU

LONGSON supports 8-bits and 16-bits of the HPI bus operation. The specific operation pattern can be set by EMI_A17 pin. When starting up, CPU will read the value of EMI_A17 and set the HPI operation mode. If the value of EMI_A17 is '1', this time the HPI is 8-bits operation. If the value of EMI_A17 is '0', this time the HPI is 16-bits operation. In our design the EMI_A17 is controlled by a double-pole switch, so the operation mode could be changed easily. We connected all the signals of HPI to FPGA, which include HPI_RDY, HPI_CS#, HPI_RST#, NR_CS#, 16-bits address and 16-bits data.

The HPI interface of LONGSON has 25 address lines and 32 data lines in all, while the number of address lines and data lines used can be different for different usage. Usually, the number of address lines will affect the number of register for communicating with FPGA. In our design we use 16 address lines and 16 data lines.

The PCI bus of the CPU is also connected to the FPGA. Considering the compatibility of the PCI bus and CPU, the 32-bits data with 33MHz PCI clock is chosen. A PCI IP core [] is placed inside FPGA to control all the operation of PCI instead of PCI Protocol Chip. The PCI bus is connected to a PCI connector on board, too. In electric structure, PCI signal could not be bifurcate, which would impact the signal transmission characteristics, may lead to miscarriage of justice, transfer wrong data or other issues. So we use the FPGA as a signal buffer, which makes full use of the FPGA. The connection is shown detailed in Figure 3.

The PCI bus with CPU will transfer the data and instructions from CPU to FPGA for the controlling process of data storage and replay. The PCI with PCI connector can be used to connect this board to other boards or systems. The capacity of one board is limited by the size of the board and chips. So we can use the PCI bus to connect some other SSR board to form a bigger SSR with larger capacity and higher throughput. Moreover, the PCI bus could connect a test system. The PCI bus could transform data from SSR to the test system. Usually, the capacity of SSR is so large, so a high speed interface is important for the SSR test. The data would be received and written into NAND FLASH, then replay the data inside. Use the test system to compare the data read back from PCI helps the SSR test process a lot.

The instruction and data can be transferred by both HPI and PCI. The two interfaces design got two interfaces physically, but when operating the memory unit only one interface can be used for the limited resource. So the two interfaces provide the user two choices, one is HPI and another is PCI, for them to attain the best performance of SSR at the lowest cost. HPI, which is a low speed bus, is used for ordinary tasks. PCI, which is a high speeding bus, is used for high throughput application.

One system with more function modules is more easily to be re-developed, so this design can deduce the developing time of the SSR while enhancing the reliability of it.

2.6 Structure of the memory unit

SSR includes memory unit, a computer unit, a data interface unit and cache units. We evaluate the function and principle of every part and optimize the design of any vital part to ensure the SSR has high throughput, large capacity and flexibility interface.

The memory unit is formed by NAND FLASH chips. The NAND FLASH chip cannot be rewritten in the same place. It must be erased before rewriting. The write operation of NAND FLASH contains two steps. The first step is transforming the data from outside into the buffer in a chip by the FPGA. The second step is transforming the data from buffer inside into the memory cell by itself. Assume the working frequency is 32M and the time for the first step is 138us, while the time for the second step which is also called programming time is 200us-700us. So the time for writig operation is 338us-838us, when the chip will not respond to any other operation expect reset, which is the bottle neck for the throughput of NAND FLASH. Some measures must be taken to break it. A parallel extensions and pipeline operation are employed to increase the throughput evidently.

Considering the operation in CPU is 32-bit, the NAND FLASH area uses 32-bit parallel extensions. The pipeline operation would decrease the gap between the write operations for one chip, for it transforms data into buffer in different chips in sequence which is making full use of the no-operation time. So considering the normal programming time for the NAND FLASH is 300 us- 400us. The four pipeline operation is used. Under this control scheme the throughput could increase 16 times, which is over 1 Gbps.

Figure 4. The principle of pipe-line operation for the NAND FLASH.

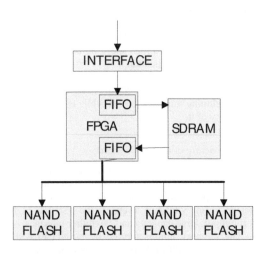

Figure 5. The structure of memory unit.

While the control of NAND FLASH is so complex that the memory unit controller need some buffer. The capacity of the buffer is proportional to the throughput of the whole system. However, the more buffer used inside the FPGA, the harder layout will be. To solve the problem of buffer, an SDRAM is employed

as a data buffer in the memory unit. The structure is shown in figure 5. The black arrowhead stands for the data flow. The data goes into the FIFO inside the FPGA from some interface chips. Then the data will be written into SDRAM. When data accumulated to some amount, FPGA transfers it into NAND FLASH through another FIFO. The SDRAM can be working in higher clock frequency than NAND FLASH, so the peak throughput can be higher.

In this structure, the throughput of the whole system is determined by NAND FLASH. When adapting the structure into a PCI stack structure, one board could be treated as a module. With the help of PCI, all modules could be powered up and used for recording data. The throughput and capacity would be the addition of all modules.

3 CONCLUSION

In our design, the CPU has two interfaces with FPGA, which are an HPI bus and a PCI bus. The system determines to use, which on by the throughput of this board. There also a PCI connector connects with FPGA, which could be used to extend the single board to an SSR system with bigger capacity and higher throughput.

SSR is playing a significant role in space exploration, so a flexible and high-speed architecture is very important for SSR. Moreover, some crucial chips are monopolized by foreign countries, which strongly restrict the development of China's space industry. We design a flexible, high-speed and configurable SSR with domestic CPU, which could be a landmark in the SSR developing history of China. Tough our design may have some shortcomings. We hope this design can be enlightening for others, so the home-made SSR will develop fast, even catch up with those of developed countries.

ACKNOWLEDGMENT

This paper is supported by the "Strategic Priority Research Program" of the Chinese Academy of Sciences under Grant No.XDA04060300.

REFERENCES

[1] MindShare, Inc, Shanley T, etal. PCI System Architecture (4th Edition) [M]. Addison-Wesley Professional, 1999.

[2] Xu Yujie. design and implementation of high-speed and huge-capacity storage system[D].Xian: Xidian University, 2014.

[3] LONGSON processor user manual. LONGSON Technology Co., Ltd. [S], 2011.07.

[4] Texas Instruments.74LS138 datasheet [R].USA: Texas Instruments, 1988.

Energy, Environment and Green Building Materials – Sheng (ed.)
© 2015 Taylor & Francis Group, London, ISBN 978-1-138-02718-3

Research on urban residential pattern oriented to the aging society

Q. Liu
Northwest University for Nationalities, Lanzhou, China
University of Chinese Academy of Science, Beijing, China
Institute of Geographic Sciences and Natural Resources Research, Chinese Academy of Science, Beijing, China

L.C. Fang
Institute of Geographic Sciences and Natural Resources Research, Chinese Academy of Science, Beijing, China

ABSTRACT: Firstly, the current issue of residential design in aging society is discussed, and the living characteristic of the elderly and their special requirements on residential environment is revealed. Secondly, the design patterns and principles are put forward by the analysis of the physiological and psychological residential environment. Finally, the experiences of the residential design for the aged are described in the conclusion part.

KEYWORDS: Aging society, Residential design, Residential pattern

1 INTRODUCTION

With the acceleration of the urbanization process in China, aging population is rapidly growing in cities. Population aging has become a social problem which cannot be avoided in the urban development process. With the trend of the growing number of aging population, it is essential to Discuss on residential design how to adapt to the aging society. And to establish a reasonable residential system for the elderly becomes an urgent problem in the current construction industry. In this context, the study of residential patterns of the aging society plays a key role to improve the quality of life for hundreds older people.

2 THE ISSUES OF LIVING ENVIRONMENT FOR THE ELDERLY

The demand for living space is changing with people's age and lifestyle. Today, the demand for China's urban elderly living environment is different from the past, but also different foreign elderly. Even the two similar size communities, there are also many different demands because of different ages. This difference is generated by its own special characteristics and behavior of the elderly live. These special requirements are mainly in the following aspects.

2.1 *Requirements for the allocation change of living time*

After leaving to work, great changes have taken place at the time allocated lives of the elder. That means time using for work and study is greatly reduced, but accounted for the major part of their leisure time. According to statistics, the elderly leisure activity time is 2.6 times as active employees. With the increase in free time, it is reasonable arrangements require to enrich their own old age, improving the life quality of the older people. In order to arrange leisure time, it must provide the necessary space environment. For example, in order to meet the interests of older persons to carry out activities, it would be arranged special space for the related to activities at home or in the community. In order to allow the elderly to exercise, it is necessary in the vicinity of residential areas with sports venues and equipment. These are the special needs of the elderly because of the time structure changing for the elderly living style.

2.2 *Requirements for the structural change of the living space*

Retirement living of elders not only in terms of time allocation changes, but also in the structure of living space has changed. It means the range of the main space and living space is getting narrower than before, which is around the family as the core. According to statistics, in-service personnel spent an average of 14 hours at home, and about 10 hours of outings time. While the elderly daily time spent at home is about 20 hours, outings only about 4 hours approximately. As the family became the main living space for the elderly, the housing situation will become an important factor affecting the quality of life of older persons.

With the increase in the elderly age, their enthusiasm the downward trend in activities, shows a continuous downward trend, and the scope of activity space also presents the tendency of narrowing. Therefore, more and more elderly people confined space within the range of the community. For this reason, to create a good environment for the elderly community extraordinarily important. Meanwhile, the living conditions of the elderly are not only related to the quality of life of individuals, but also to the communication and social interaction between the older and their children. Only providing sufficient quantity and better quality housing for the elderly, it can ensure the comfort of well-being in later life for them.

2.3 Requirements for the physiological and psychological change of elderly individual

The limbs, sensory or intelligence, of the elderly will show different degrees of decline. They began to appear in all kinds of chronic diseases and aging due to physiological function caused by a certain body organ disorder, such as walking inconvenience, bradykinesia, visual acuity decreasing, identification ability reducing, hearing losing, unresponsive, poor memory and so on. More seriously, even is hemiplegia, dementia and other diseases in the elderly. These diseases which enable the elderly to respond to the environment and the ability to adapt to gradually reduce, emotionally fragile, asks people to give more consideration and care.

At the same time, old people, due to exit the occupation life or action inconvenience reduces their chance to take part in social communication significantly. And there also for children have married and separated, there is a sense of loss and abandonment in Psychology, or produces depression and panic feeling because of its ageing. These negative psychological and in turn influence human body health, forming a vicious spiral. In order to adapt to changes in the physiological and psychological, they need the extensive contacts, exchanges and help from their relatives, friends and neighbors. This requires a safe and comfortable living environment, both for the convenience of the elderly independent activities, but also get the timely care of children in need. Accordingly, it should require the facilities equipped with completely in the community environment, which not only benefits to the elderly, social communication, keeps a good state of mind, and also can get all kinds of necessary social services in a timely manner. For this means, only with the perfect community service and medical facilities to support the home environment, is really an old people's home.

3 CAUSE ANALYSIS OF HOME-BASED CARE

The Home-based care for aged is the main endowment pattern in China, and corresponding with the endowment pattern, there is a type of special residential

building for the aged. The elderly residence consists two kinds of buildings, one is only the elderly living alone, the other is the elderly living together with their children. Home-based care for the aged is a mode developed out of the society's demand of making good use of social resources to meet the demands of the aging population and the changing family structure and therefore many countries attach importance to this mode. The elderly in our country will be in a long period of time to Home-based care as the core. This is mainly decided by the following factors:

3.1 The restrictions on the economy and the living conditions

The pay and the savings of old people in our country is generally low, the welfare housing distribution system is the primary way for most of them to solve the housing problem. However, most of the distribution of housing is now aging, and is not conducive to the two level housing market. At the same time, the market of housing makes real estate prices rise sharply, most children cannot afford to buy a new house and had to live with their parents. So many old people choose Home-based care is also helpless initiatives.

3.2 The influence of the traditional culture and morality

Providing for the aged, advocating social virtues of the Chinese nation, is the fine tradition of the Chinese nation.It is our traditional virtue to respect and honor the aged, and filial piety spirit is an important heritage of our traditional culture. "Raise children to provide against old age" concept has won support among the people, the happiness of a family union enjoy old age is as unalterable principles. Therefore, "old-age home" is not only the choice of the aged, but also is the hope of children filial piety requirement.

4 RESIDENTIAL TYPES FOR THE ELDERLY

Depending on the model of family pension and family demographic, the residential layout can be varied selection. Based on the degree of separation of the elderly live with their children can be divided into three types as follows.

4.1 Shared type

The elderly and children live in a convenient communication and care in the dwelling. According to the degree of separation of the dedicated section of the elderly living in residential dwelling, it can be divided into three categories combination (Fig. 1).
- Separate living room, including two forms of shared toilet and elderly special toilet.

- Separate living room and kitchen, including two forms of shared bathroom and elderly special bathroom.
- Separate main living space, mainly living in a duplex unit.

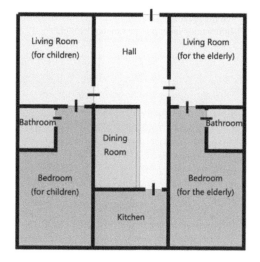

Figure 1. Shared type.

4.2 *Neighbor type*

This type allows two generations live independent, but living space in close proximity. It ensures the privacy of their activities, but also conducive to the exchange of two generations of life and emotional care of each other (Fig. 2).

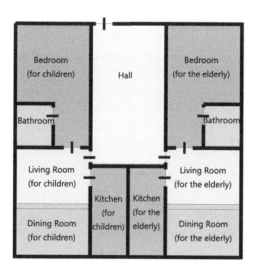

Figure 2. Neighbor type.

- Two residential condominiums in the same building, as un-adjacent neighbors.
- Two residential dwelling not in the same building, but at the same block.
- Two residential condominiums in different closely blocks, convenient transport links.

4.3 *Divisible type*

Elderly families and young families completely independent in different neighborhoods, but at the same residential area. This pattern is in line with contemporary living an ideal life style, which live closely, share open, keep in touch and is easy to take care of each other. This type are also very suitable for those large numbers of Empty-nest-family currently (Fig.3).

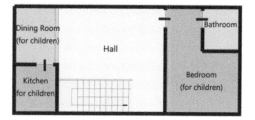

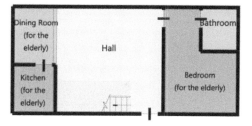

Figure 3. Divisible type.

5 SUMMARY

With the high speed urbanization, the family miniaturization is becoming more and more serious, which result in the traditional family structure and idea changing gradually. The difficulties of family pension will inevitably lead to the demand of the elderly for increasing social welfare pension. The weakening of the family and society forced pension mode to diversified development. Based on this special situation, the residential design for the elderly must be placed emphasis on the design concept of the coexistence of a variety of patterns.. Architecture and environment should be consistent with the characteristics of the elderly physiological and psychological behavior, which to create a supportive environment for them. This kind of architectural environment can be made old people in safely activities, but also can be pleased and motivated their senses of place.

ACKNOWLEDGMENTS

This work was financially supported by Construction Department of Gansu Province (JK2009-15), and by National Natural Science (51208243).

REFERENCES

[1] Cai Qin: The city elderly residential toilet environment research (Tsinghua University Press, Beijing 2004).

[2] Renlu Hu, and Ma light: The living environment design (Southeast University Press, Nanjing 2004).

[3] Harbin Architecture University: Design standard architecture for the elderly (China Building Industry Press, Beijing 2005).

[4] Kai Yan: submitted to Journal of Housing Science and technology (2010).

[5] Yan Chunlin: submitted to Journal of New buildings (2011).

[6] Zhang Jianmin: submitted to Journal of New buildings (2012).

[7] Information on http://www.mohurd.gov.cn/

Author index

An, J.S. 519

Bai, Y.T. 361
Bian, Z. 115

Cai, C.F. 247, 251
Cai, Z.Y. 311
Cao, C.C. 1
Chen, B. 13, 259, 373, 459, 465
Chen, C. 25
Chen, C.J. 161
Chen, C.Y. 389
Chen, F.F. 129
Chen, G.P. 217
Chen, H.R. 447, 451
Chen, J. 311
Chen, L. 227, 335
Chen, M. 427
Chen, Q. 139
Chen, S.F. 451
Chen, X.M. 43, 203
Chen, Y. 293
Cheng, D.S. 183
Cheng, G.S. 199, 239
Cheng, J. 1
Cheng, P. 183
Cheng, Y.K. 39

Da, W.F. 59
Dai, B.Q. 273
Dai, G.H. 35
Dang, X.Z. 53
Deng, G.L. 53
Di, H.Y. 417
Di, Y.Q. 417
Dong, C.Q. 239
Dong, S.Y. 399
Du, X.J. 385
Du, X.Z. 79
Duan, J. 43, 135, 203
Duan, J.F. 119
Duan, J.K. 243
Duan, Y.B. 39

Fan, X.Y. 19
Fan, Z.C. 443
Fang, L.C. 523

Feng, H.T. 145
Feng, J.G. 119
Feng, R. 471
Feng, T. 331
Fu, C. 339
Fu, S.J. 5

Gao, Z.J. 119
Ge, J.X. 273
Gen, J.J. 385
Gong, Y.F. 35
Gu, J.C. 315
Gu, Q. 349
Guo, T.Y. 193

Han, L.J. 223
Han, R.R. 207
Han, Z. 9
Hao, G.F. 255
Hao, J. 79
Hao, Y.H. 5
He, B.Q. 89
He, J. 455, 459, 465, 483, 487,
    493, 499
He, J.B. 263, 503
He, X.F. 311
He, Z.G. 29
Hong, W. 9
Hu, D.H. 101
Hu, P.L. 35
Hu, W.J. 93
Hu, Y. 39
Huang, H.M. 301
Huang, Y.Y. 207
Huang, Z.Q. 63

Ji, C.M. 399
Jiang, B.H. 379
Jiang, L. 243, 247, 251
Jiang, R.G. 19
Jiang, W. 123
Jing Q.R. 223

Le, G.G. 285
Li, B. 179, 349
Li, D.Y. 223, 433, 483, 487
Li, G.B. 79

Li, G.Q. 167, 173, 255
Li, H.Y. 289
Li, J.P. 471
Li, J.Y. 63
Li, L. 217
Li, L.L. 281
Li, L.Y. 69
Li, M. 411, 427
Li, M.T. 19
Li, N. 207
Li, P. 367
Li, S. 519
Li, S.J. 223
Li, W.C. 255
Li, W.H. 145
Li, Y. 9, 243
Li, Y.G. 43, 135, 203, 311
Li, Y.N. 511
Li, Y.P. 9
Li, Y.Q. 433, 487
Li, Y.R. 19
Li, Z.R. 105
Liang, B. 385
Liao, W.Z. 411
Liu, G.F. 451
Liu, H.C. 379
Liu, H.J. 455, 493, 499
Liu, K. 277
Liu, L. 443
Liu, M.J. 149
Liu, Q. 183, 523
Liu, S. 507
Liu, X.D. 223
Liu, X.Y. 273
Liu, Y. 145, 281
Liu, Y.N. 123
Liu, Y.S. 403
Liu, Z.C. 367
Lu, G. 59
Lu, L. 331
Lu, Y. 129
Luo, C.Y. 199
Luo, Y.W. 455, 459, 465, 493, 499
Luo, Z.G. 233
Lv, J.F. 179
Lv, L. 293
Lv, X.H. 327

Ma, D.W. 479
Ma, J.N. 335
Ma, L. 13, 259, 373
Ma, R.H. 515
Mao, Y.H. 289
Meng, X.H. 59
Meng, Y.J. 459, 465

Nguyen, N.L. 105
Nian, F.Z. 155
Nian, T.F. 367

Pan, W.C. 149
Peng, Y.Z. 139
Peng, Z.X. 29

Qi, F.Z. 247, 251
Qi, H. 43, 203
Qi, J.C. 207
Qian, W.B. 1
Qin, L. 459, 465
Qiu, Q.L. 73
Qiu, X.G. 13, 259, 373, 443
Qiu, Z.Y. 223

Ren, J. 511

Sang, H.T. 233
Shao, C.H. 139
Shao, D.W. 155
Shao, S.X. 243
Shen, S.J. 411
Shi, B. 269
Shi, K.Q. 243
Shi, M.J. 119
Shi, S.H. 123
Shi, S.S. 403
Shu, S. 349
Song, Q. 519
Song, Z.C. 259
Su, Y. 183
Sun, B. 105
Sun, J.H. 277
Sun, J.Y. 203
Sun, X.Q. 393
Sun, Y.L. 233

Tang, C. 193
Tang, L.S. 233
Tang, W.H. 79
Tang, Z. 427
Tao, L. 511

Tian, Q. 385
Tian, Z.S. 301

Wang, C.S. 379
Wang, C.X. 139, 189
Wang, C.Y. 167, 173
Wang, F.M. 53
Wang, H.B. 79
Wang, H.J. 255, 411
Wang, H.Y. 475
Wang, J.F. 97
Wang, J.P. 385
Wang, L. 281, 361
Wang, M. 97
Wang, M.H. 399, 403
Wang, N.Z. 173
Wang, R. 315
Wang, S.L. 281
Wang, T. 25
Wang, W. 139
Wang, X.N. 161
Wang, X.Y. 361
Wang, X.Z. 355
Wang, Y. 239
Wang, Y.J. 243
Wang, Y.L. 101
Wang, Z. 263, 503
Wang, Z.B. 9
Wang, Z.M. 89
Wang, Z.Q. 189
Wei, J. 101, 129
Wei, W.B. 269
Wei, Z. 269
Wu, C.F. 251
Wu, Y. 73
Wu, Y.H. 459, 465
Wu, Y.N. 335

Xia, Q. 39
Xia, Q.Q. 479
Xia, Z.Y. 93
Xiang, Y. 239
Xiao, J.F. 507
Xiao, X. 19
Xiao, Y.X. 315
Xie, F.X. 227
Xie, M.M. 393
Xie, X. 439
Xie, X.F. 385
Xing, J. 49
Xiong, R.X. 97
Xu, J. 79
Xue, R. 385

Xu, J.P. 247, 361
Xu, X. 483, 487

Yan, J.P. 355
Yan, P. 319, 323
Yan, X.D. 97
Yang, B. 289
Yang, F.B. 479
Yang, J. 101, 129
Yang, J.Y. 471
Yang, L. 367
Yang, Q.R. 319, 323
Yang, X.L. 427
Yang, X.M. 85
Yang, X.T. 399, 403
Yang, Y. 167
Yao, Y.Y. 459, 465
Yao, Z.F. 269
Ye, C.F. 73
You, Z.J. 335
Yu, J.C. 433
Yu, J.M. 475
Yuan, G.X. 63
Yuan, J.J. 273
Yuan, L. 423
Yuan, M. 129
Yuan, W.C. 53
Yuan, Z.W. 301

Zeng, Q.J. 433
Zeng, X. 39
Zhan, P.H. 105
Zhang, Ch.W. 145
Zhang, F.J. 511
Zhang, G.F. 319, 323
Zhang, H.L. 239
Zhang, L. 25
Zhang, L.B. 13, 373
Zhang, L.L. 93
Zhang, M. 161
Zhang, M.Y. 145
Zhang, P. 13, 373, 443
Zhang, Q.L. 307
Zhang, W. 217
Zhang, X. 139
Zhang, X.J. 93
Zhang, Y. 239
Zhang, Y.L. 367
Zhang, Y.M. 63
Zhang, Y.W. 115
Zhang, Y.Z. 511
Zhang, Z.L. 379
Zhao, H.G. 427

Zhao, H.L. 379
Zhao, H.Y. 79
Zhao, J. 109
Zhao, J.Y. 179
Zhao, S.Y. 5
Zhao, W. 417
Zhao, X.Y. 443
Zhao, Y. 199, 239
Zheng, M.C. 507

Zheng, Z.M. 239
Zhong, J.L. 285
Zhong, Z.G. 407
Zhou, A. 227
Zhou, B.H. 207
Zhou, D.D. 471
Zhou, M. 301
Zhou, T. 277
Zhou, X. 259, 349, 373

Zhou, Z.Q. 35
Zhu, G.Q. 217
Zhu, H.C. 247
Zhu, Q.W. 361
Zhu, X.M. 105
Zhu, Y. 519
Zhu, Z.P. 507
Zuo, J. 73

Printed and bound by CPI Group (UK) Ltd, Croydon, CR0 4YY

18/10/2024

01776219-0013